TEACHER'S EDITION

PHYSICAL SCIENCES FOR NGSS

Contents

CLASSROOM GUIDE

STUDENT EDITION

APPENDIX

ISBN 978-1-927309-82-7

First Edition 2020

Copyright © 2020 Richard Allan

Published by BIOZONE International Ltd

Printed by THOMSON PRESS (INDIA) LTD
using environmentally friendly paper
www.thomsonpress.com

Purchases of this book may be made direct from the publisher:

BIOZONE Corporation
USA and Canada
FREE phone: 1-855-246-4555
FREE fax: 1-855-935-3555
Email: sales@thebiozone.com
Web: www.thebiozone.com

FAQs ABOUT PHYSICAL SCIENCES FOR NGSS

Next Generation Science Standards (NGSS) is a registered trademark of Achieve. Neither Achieve nor the lead states and partners that developed the Next Generation Science Standards were involved in the production of this product and do not endorse it.

The Contents: A Planning Tool

The contents pages are not merely a list of the activities in the student edition. Encourage your students to use them as a planning tool for their program of work. Students can identify the activities they are to complete and then tick them off when completed. The teacher can also see at a glance how quickly the student is progressing through the assigned material.

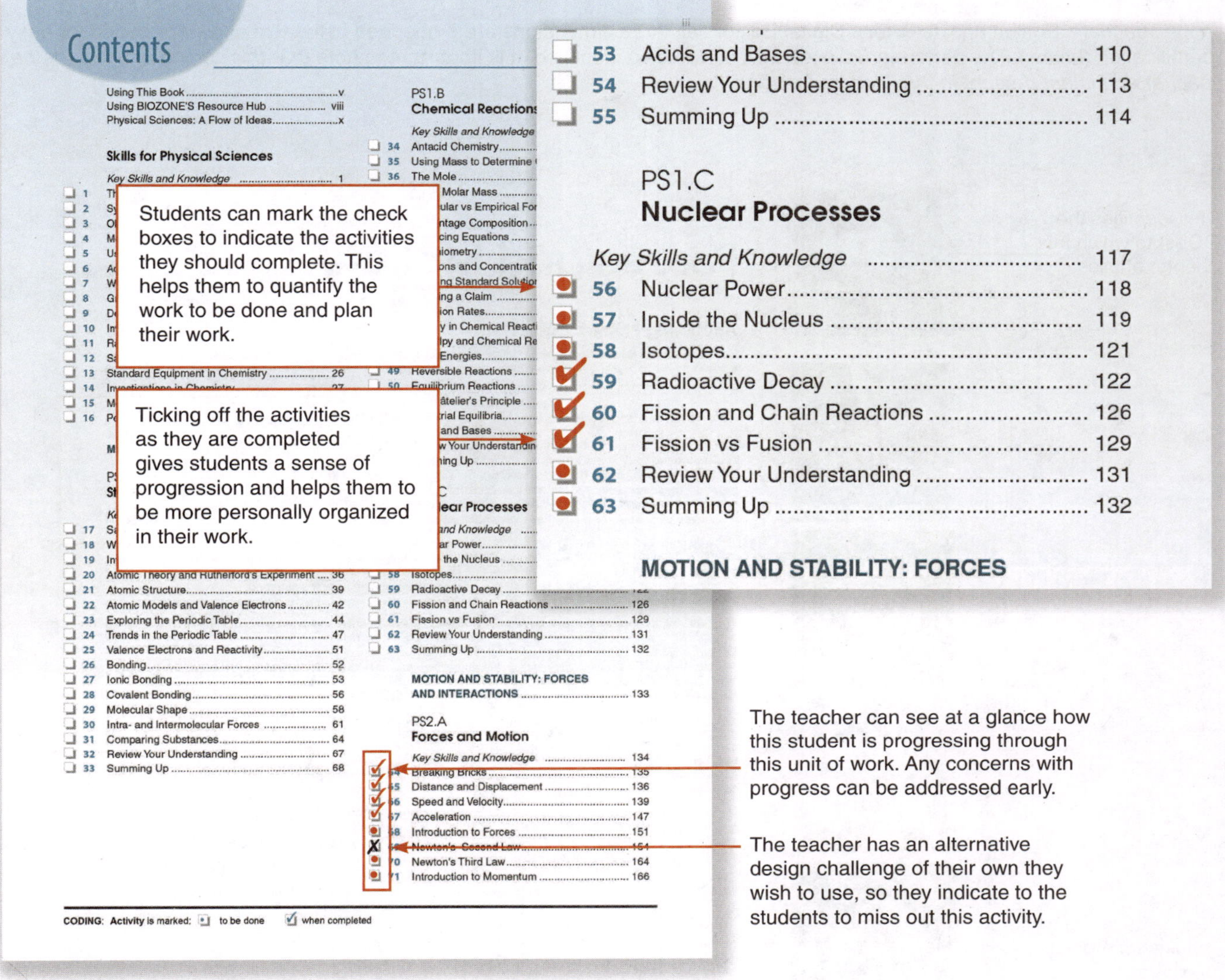

What about a pacing guide?

The 9-12 NGSS framework is fluid in terms of the grade in which each program is offered, so in many respects defies a rigid pacing guide. Within grade, other variables contribute to changes in pacing:

- There are opportunities for students to spend longer on some activities (e.g. in improving or refining their design solutions or in exploring simulations beyond the minimum). These elaborations will demand more time.
- The time allocated for investigations will depend on (1) how you choose to organize the class (which may be determined by available resources) and (2) how far students take the investigation. Adjust your lesson plan to incorporate more or less material as needed. You may have investigations you already like to use, so you could choose to leave out equivalent investigations in the book. To help you, activities including a practical investigation are identified with a green dot (●) in the contents of the student book.
- For computer modeling activities, completed models are available on **BIOZONE's Resource Hub** and the Teacher's Digital Edition, so students can save time by exploring the model, but not building it themselves.
- The pace may quicken as students complete more of the book. Later chapters draw on knowledge and understanding of previous chapters, as well as exploring new concepts. Students gain increasing levels of competence and learn valuable skills that enable them to arrive at solutions more quickly. That aside, teachers may appropriately choose to do the physics block of chapters before chemistry.

©2020 **BIOZONE** International

Identifying Learning Intentions and Goals

In developing *Physical Sciences for NGSS*, we have embraced the three dimensions of the NGSS framework, emphasizing the application of ideas and skills to new scenarios. The activities in *Physical Sciences for NGSS* have been specifically designed to address the **disciplinary core ideas (DCIs), science and engineering practices, and crosscutting concepts** in a way that helps students to meet specific performance expectations.

In the Teacher's Edition and Teacher's Digital Edition, all three dimensions are embedded in the text and color coded for easy identification (below). The performance expectations are also identified. It is important to note that *this coding is a tool for the teacher and is not present in the Student Edition.*

©2020 BIOZONE International

Scaffolded Learning with the 5 Es

In developing *Physical Sciences for NGSS* we have utilized the 5Es instructional model as a basis for developing materials to address all three dimensions of the NGSS framework: **disciplinary core ideas** (DCIs), **science and engineering practices**, and **crosscutting concepts**. By successfully completing the activities, students can demonstrate competence in all three dimensions. This is central to meeting the performance expectations for *Physical Sciences for NGSS* with confidence.

BIOZONE encourages the development of the NGSS learner profile using the 5Es model

The Five Es

Engage:	make connections between past and present learning experiences.
Explore:	become actively involved in the activity.
Explain:	communicate the learning experience.
Elaborate:	expand on the concepts learned.
Evaluate:	assess understanding of the concepts.

BIOZONE's series for NGSS is **phenomenon-based**. Students engage with phenomena through their own investigations and observations, through modeling and data analysis, and through collaborative work and discussion.

Using phenomena to drive inquiry promotes discussion and the sharing of ideas. The iterative approach presents opportunities to look at phenomena from several different perspectives. This allows students of all abilities to expand their thinking and understanding, increasing understanding each time the phenomenon is revisited.

Each chapter begins with an **anchoring phenomenon** (right). In each instance, we have chosen a phenomenon that the student is probably familiar with, but which they cannot explain (or cannot explain fully). Teachers can use this activity to find out what the students already know (or think they know) before delving into the content more fully.

The subsequent activities in a chapter take the students, step by step, through phenomena that explore the ideas inherent in the anchoring phenomenon. By the time students revisit the anchoring phenomenon, they should be able to fully explain it.

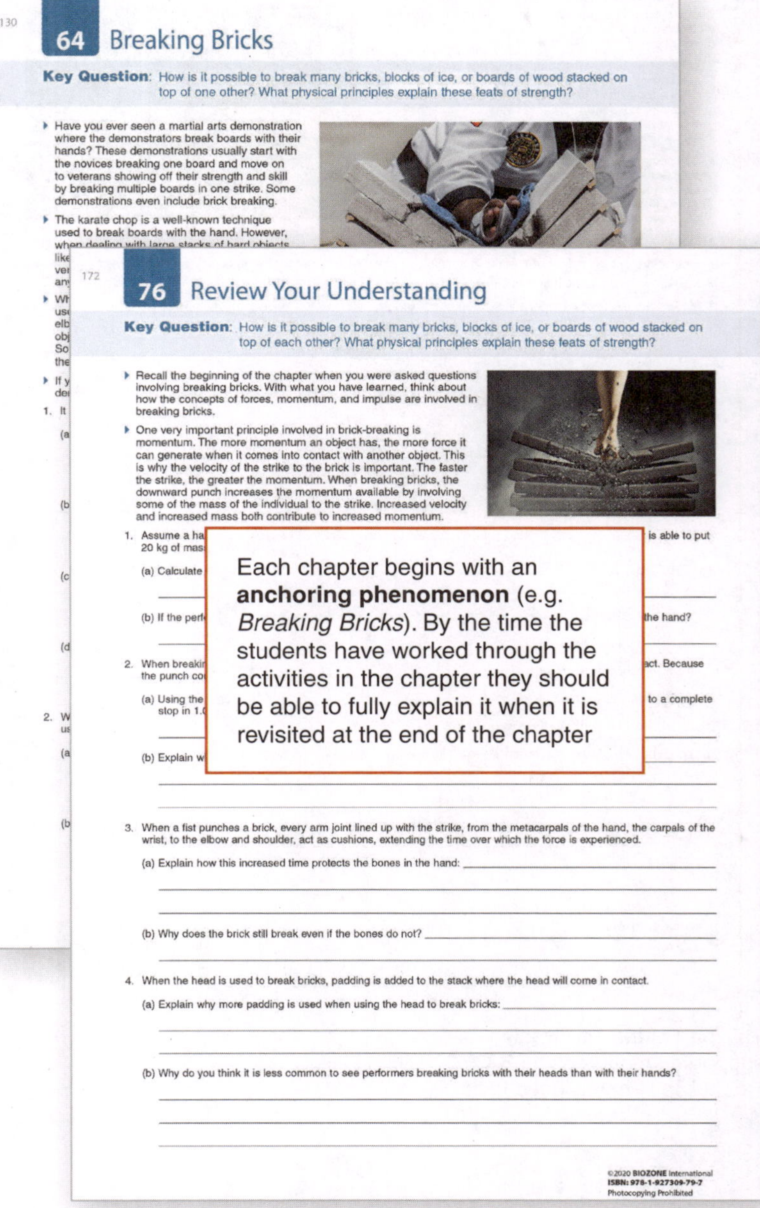

130

64 Breaking Bricks

Key Question: How is it possible to break many bricks, blocks of ice, or boards of wood stacked on top of one other? What physical principles explain these feats of strength?

- Have you ever seen a martial arts demonstration where the demonstrators break boards with their hands? These demonstrations usually start with the novices breaking one board and move on to veterans showing off their strength and skill by breaking multiple boards in one strike. Some demonstrations even include brick breaking.
- The karate chop is a well-known technique used to break boards with the hand. However, when dealing with large stacks of hard objects ...

172

76 Review Your Understanding

Key Question: How is it possible to break many bricks, blocks of ice, or boards of wood stacked on top of each other? What physical principles explain these feats of strength?

- Recall the beginning of the chapter when you were asked questions involving breaking bricks. With what you have learned, think about how the concepts of forces, momentum, and impulse are involved in breaking bricks.
- One very important principle involved in brick-breaking is momentum. The more momentum an object has, the more force it can generate when it comes into contact with another object. This is why the velocity of the strike to the brick is important. The faster the strike, the greater the momentum. When breaking bricks, the downward punch increases the momentum available by involving some of the mass of the individual to the strike. Increased velocity and increased mass both contribute to increased momentum.

1. Assume a ha... is able to put 20 kg of mas...
   (a) Calculate ...
   (b) If the perf... the hand?
2. When breaki... act. Because the punch co...
   (a) Using the ... to a complete stop in 1.0...
   (b) Explain w...
3. When a fist punches a brick, every arm joint lined up with the strike, from the metacarpals of the hand, the carpals of the wrist, to the elbow and shoulder, act as cushions, extending the time over which the force is experienced.
   (a) Explain how this increased time protects the bones in the hand:
   (b) Why does the brick still break even if the bones do not?
4. When the head is used to break bricks, padding is added to the stack where the head will come in contact.
   (a) Explain why more padding is used when using the head to break bricks:
   (b) Why do you think it is less common to see performers breaking bricks with their heads than with their hands?

©2020 BIOZONE International
ISBN: 978-1-927309-79-7
Photocopying Prohibited

Each chapter begins with an **anchoring phenomenon** (e.g. *Breaking Bricks*). By the time the students have worked through the activities in the chapter they should be able to fully explain it when it is revisited at the end of the chapter

©2020 **BIOZONE** International

The content of the *Physical Sciences for NGSS* is organized into 11 chapters based on the DCIs of the High School Physical Sciences framework. Chapter 1 addresses basic skills for students in physical sciences. Chapters 2 - 11 each begin with an introduction outlining learning goals, which is immediately followed by the anchoring phenomenon. Activities make up the bulk of each chapter, with each one focusing on the student investigating and developing understanding of a phenomenon, applying that understanding to new scenarios, and developing (or practicing) a skill or essential science practice, such as graphing, data analysis, modeling, or evidence-based explanation.

Annotated diagrams and photographs are a major part of most activities and the student's understanding of the information is evaluated through questions and/or tasks involving data handling and interpretation. Tabs at the bottom of the page identify crosscutting concepts, science and engineering practices, and disciplinary core ideas as appropriate. Tabs in the margin also indicate if the activity is supported via **BIOZONE's Resource Hub**, which provides online teacher and student support for specific aspects of the activity.

Concepts are presented as a logical sequence, which may be divided among several consecutive activities. Understanding of phenomena is developed progressively through exploration and explanation.

ENGAGE with phenomena
Activities normally begin with a brief task, observation, or example to engage student thinking, allowing them to review their current understanding of a phenomenon, or providing an interesting (if not yet fully explained) piece of information that relates to the concept about to be studied. This is a chance for teachers to assess prior knowledge or engage students by posing challenging questions about seemingly simple phenomena.

161

71 Introduction to Momentum

Key Question: How do we describe the quantity of motion in an object and how do we use it to explain why some objects are more difficult to stop than others?

Momentum

Ever played pool or gone bowling? These games operate on the same principle: using the movement of one ball to move or knock over other objects.

- In the game of pool (right) the white ball is used to maneuver the others around the table. Sometimes when it hits a target ball, the target will move away while the white ball stops moving. Other times, the white ball continues to move but more slowly after hitting the target ball, which also moves but more slowly.
- A important part of pool or bowling, is the concept of **momentum**. In pool, momentum for the white ball is transferred to the target ball to make it move. How this is done will determine the way the target ball and the white ball move after colliding.
- Momentum is a physics term referring to the quantity of motion an object has.
- Momentum can be defined as "mass in motion". All objects have mass, so if an object is moving, it has momentum.
- Momentum is dependent on two variables: mass and velocity.
- Momentum (p) is equal to the mass of an object (m) multiplied by the velocity of the object (v):

$$p = mv$$

p / m | v

- Momentum has units of kilogram meters per second (kg m/s).
- Because momentum has a velocity component, it is a vector with the same direction as the velocity.

1. Determine the momentum of:

(a) a 60 kg running back running east at 10 m/s: ____

(b) a 36,000 kg semi traveling north on interstate 95 at 30 m/s: ____

(c) a 5.0 gram snail moving south at 2.0 meters per hour: ____

2. A car has 30,000 units of momentum. What would the car's new momentum be if:

(a) its velocity was halved: ____

(b) its velocity was doubled: ____

(c) its mass was doubled: ____

(d) its velocity was doubled and its mass halved: ____

3. Which has the greater momentum: a 25,000 kg truck moving at 5 m/s, or a 1200 kg car moving at 21 m/s?

4. A 990 kg car reduces its velocity from 22 m/s to 13 m/s. Calculate the change in the car's momentum:

©2020 BIOZONE International
ISBN: 978-1-927309-79-7
Photocopying Prohibited

SSM PS2.A

Important formulas that students should be able to understand and use are given in a dark blue box. Fact family triangles make use of these formulae simpler.

Students are given enough information to complete the activity's tasks. To progress through the activity they may need to apply knowledge and information developed earlier in the activity. Answers to questions are not always directly available on the page. Students may need to analyze data or information and draw conclusions to answer the questions and progress to the next part of the activity. Students are sometimes asked to do further research or carry out their own research or investigation.

This part of the activity also has supporting resources on **BIOZONE's Resource Hub** assigned to it.

Relevant SEPs, DCIs, and CCCs are identified through the tab system.

Students can **EXPLORE** the phenomenon via practical activities, creating their own models, analyzing or using second hand data, or interpreting diagrams. In this example, students use marbles of different mass and size to explore the concept of momentum. They also explore momentum using second-hand data of carts on an air track.

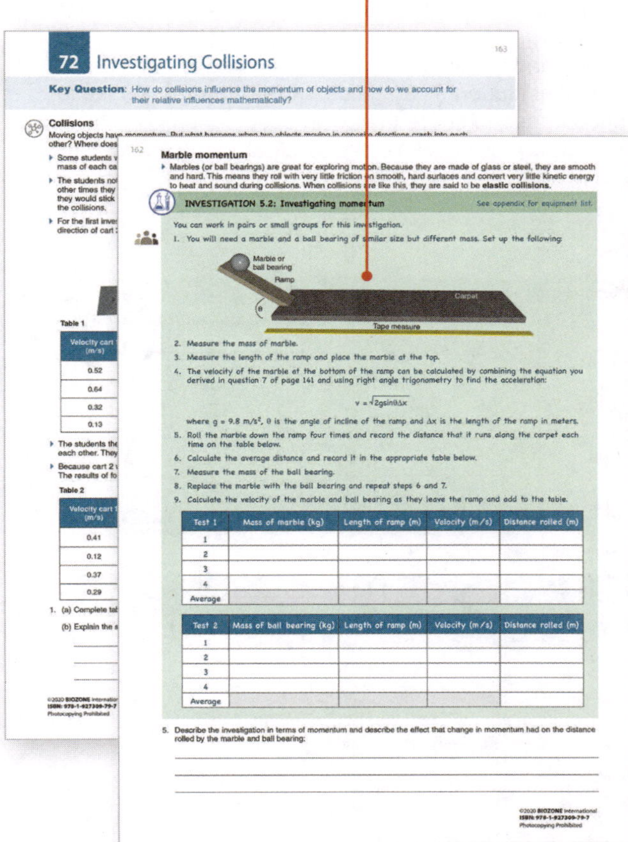

Students **EXPLAIN** phenomena by building on what they discovered through exploration. They are encouraged to use scientific principles and logical reasoning to construct explanations and devise solutions to the problems presented to them. Here students analyze data of initial and final velocities to find and explain the underlying relationship between initial and final momentum.

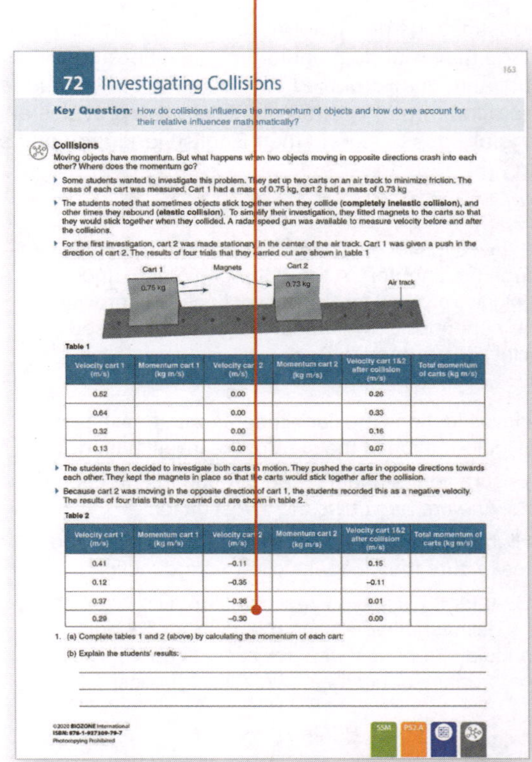

After sound explanations of phenomena are developed, students have opportunity to **ELABORATE**, applying their understanding to new phenomena or using their experience to develop or refine engineering solutions to relevant problems. Here students apply the concepts of momentum and impulse to design a landing device that will protect an egg from a fall.

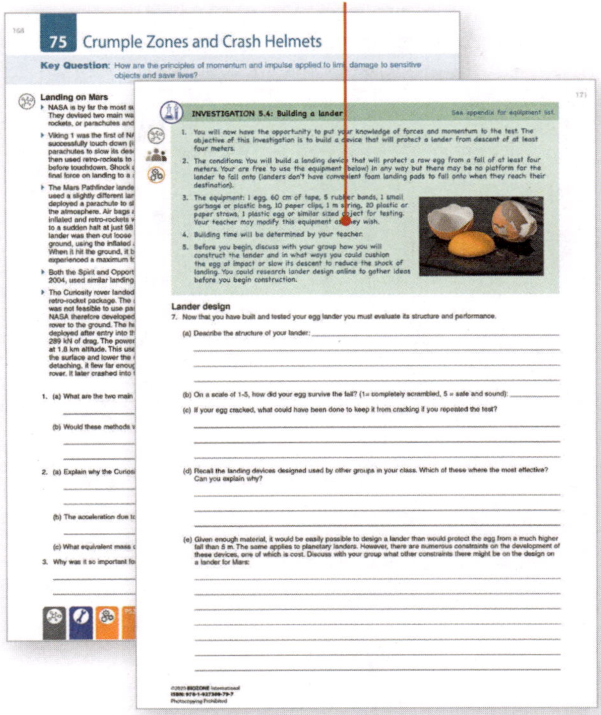

Students **EVALUATE** their investigations. This can provide opportunities for **formative assessment** (if you wish). In this example, students evaluate their lander design and decide if there are any improvements that could be made. There is opportunity here for peer assessment.

©2020 **BIOZONE** International

Practical Investigations

Throughout *Physical Sciences for NGSS*, students are given opportunities to explore phenomena through experimentation. These **investigative phenomena** are opportunities for students to develop competency in laboratory procedures, to practice and refine skills in observation and analysis, and to manipulate data. Some investigations act as stimulus material while others require students to take what they have already learned and apply their knowledge to a more complex scenario.

The investigations provide an excellent opportunity for collaborative work and will stimulate discussion and the sharing of ideas. You may wish to pair students of different abilities together. Confident students can guide and encourage less able students and, in this relaxed environment, striving students will be encouraged to share their own observations and thoughts. Collaboration through paired practical work provides an excellent opportunity for English language learners to interact in meaningful ways to extend their English language and scientific vocabulary.

Each investigation is clearly numbered (sequentially through the chapter). A list of the equipment and chemicals required for each investigation is provided in the appendix. Only standard equipment is used (no special kits are required).

The investigations provide students with health and safety information at the start of the investigation.

- Ensure your students read through the procedure fully *before beginning* the investigation.
- Highlight any hazardous or important steps, and make sure the students follow your directions.
- Chapter 1 (Science Practices) contains basic safety guidelines for working in a laboratory. Students should complete these activities as appropriate before they begin their own investigations.

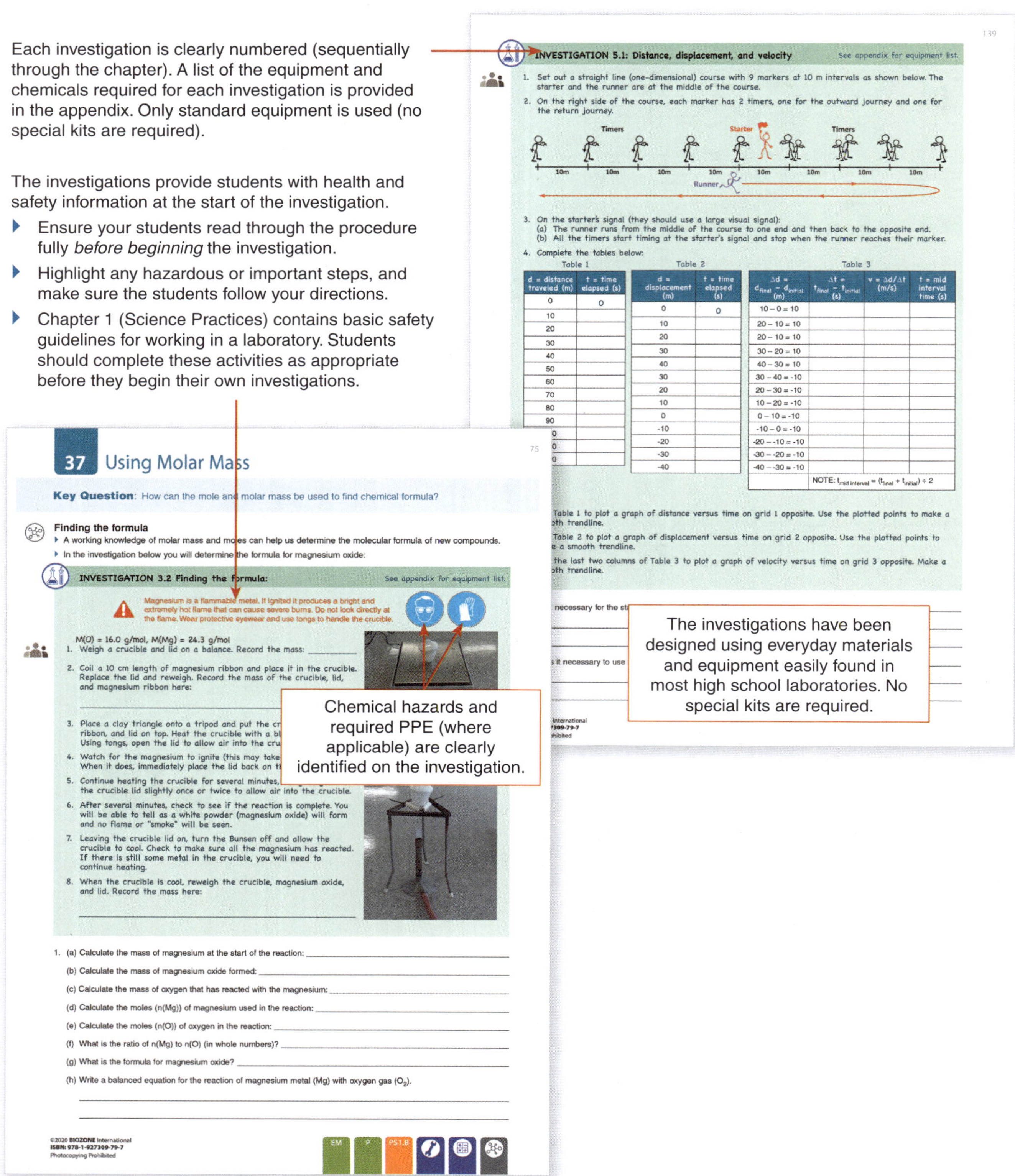

©2020 **BIOZONE** International

Teaching Strategies for Classroom Use

Achieving effective differential instruction in classes is a teaching challenge. Students naturally have mixed abilities, varying backgrounds in the subject, and different language skills. Used effectively, BIOZONE's student books and supporting resources can make teaching a mixed ability class easier. Here, we suggest some approaches for differential instruction.

MAKING A START

Regardless of which activity you might be attempting in class, a short introduction to the task by the teacher is a useful orientation for all students. For collaborative work, the teacher can then divide the class into appropriate groups, each with a balance of able and less able students. Depending on the activity, the class may regroup at the end of the lesson for discussion.

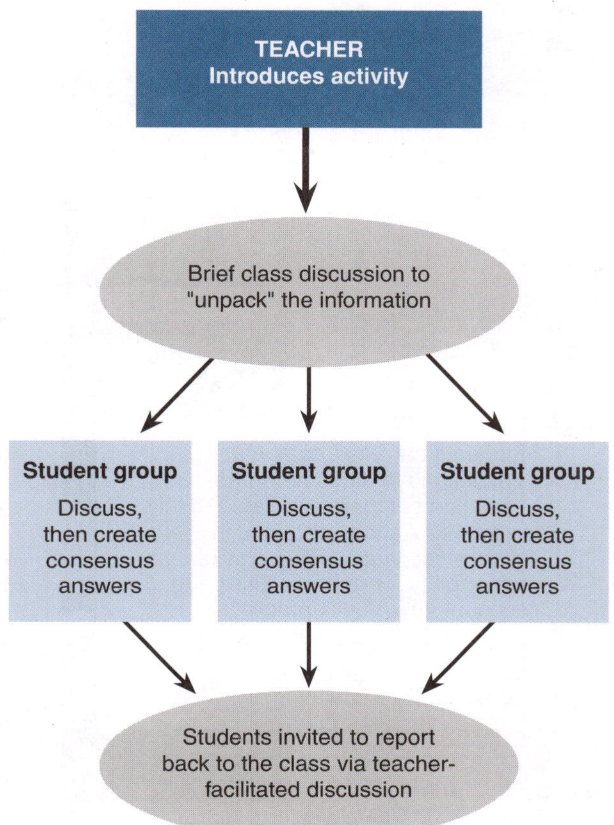

Using collaboration to maximize learning outcomes

- The structure of *Physical Sciences for NGSS* allows for a flexible approach to unpacking the content with your students.
- The content can be delivered in a way to support collaboration, where students work in small groups to share ideas and information to answer and gain a better understanding of a topic, or design a solution to a problem.
- By working together to ask questions and evaluate each other's ideas, students maximize their own and each other's learning opportunities. They are exposed to ideas and perspectives they may not have come up with on their own.
- Collaboration, listening to others, and voicing their own ideas is valuable for supporting English language learners and developing their English and scientific vocabularies.
- Use a short, informal collaborative learning session to get students to exchange ideas about the answer to a question. Alternatively, collaboration may take a more formal role that lasts for a longer period of time (e.g. assign groups to work together for a practical activity, to research an extension question, or design a solution to a problem).

The teacher introduces the topic. They provide structure to the session by providing background information and setting up discussion points and clear objectives. Collaboration is emphasized to encourage participation from the entire group. If necessary, students in a group can be assigned specific tasks.

Students work in small groups so everyone's contribution is heard. They collaborate, share ideas, and engage in discourse. The emphasis is on discussing questions and formulating a consensus answer, not just sharing ideas.

At the end of the session, students report back on their findings. Each student should have enough knowledge to report back on the group's findings. Reporting consists primarily of providing answers to questions, but may involve presenting a report, model, or slide show, or contributing to a debate.

©2020 **BIOZONE** International

Peer to peer support

- **Peer-to-peer learning** is emphasized throughout the book, and is particularly valuable for more challenging activities in which the content is more complex or the questions require students to draw on several areas of their knowledge to solve a problem.
- Stronger students can assist their peers and both groups benefit from verbalizing their ideas. Students for whom English is a second language can ask their classmates to explain unfamiliar terms and this benefits the understanding of both parties.
- **Practical investigations** are an ideal vehicle for peer-to-peer learning. Students can work together to review their results, ask and answer questions, and describe phenomena. There are also opportunities for students to collaborate using online simulations (e.g. **Energy2D** shown below).

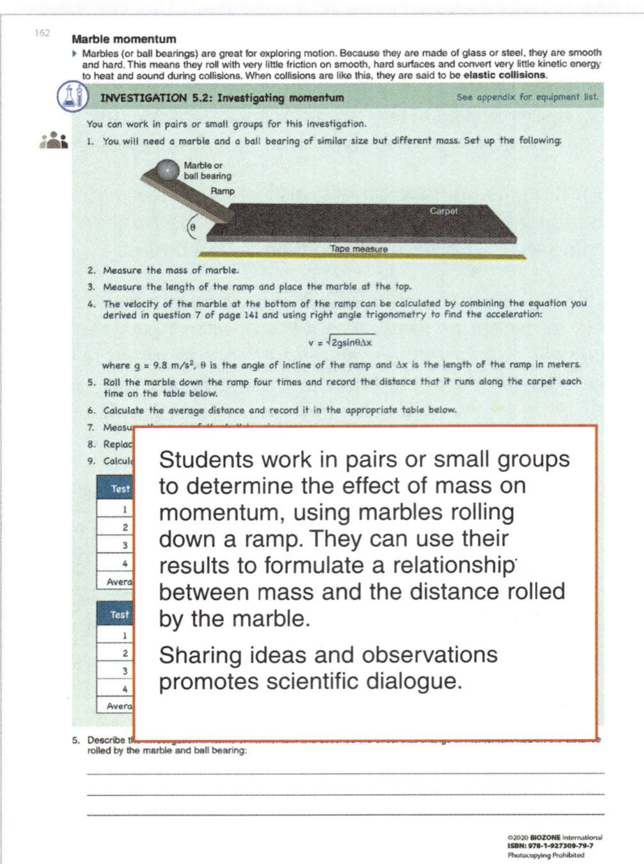

162

Marble momentum

- Marbles (or ball bearings) are great for exploring motion. Because they are made of glass or steel, they are smooth and hard. This means they roll with very little friction on smooth, hard surfaces and convert very little kinetic energy to heat and sound during collisions. When collisions are like this, they are said to be **elastic collisions**.

INVESTIGATION 5.2: Investigating momentum — See appendix for equipment list.

You can work in pairs or small groups for this investigation.

1. You will need a marble and a ball bearing of similar size but different mass. Set up the following:

2. Measure the mass of marble.
3. Measure the length of the ramp and place the marble at the top.
4. The velocity of the marble at the bottom of the ramp can be calculated by combining the equation you derived in question 7 of page 141 and using right angle trigonometry to find the acceleration:

$$v = \sqrt{2g\sin\theta\Delta x}$$

where g = 9.8 m/s², θ is the angle of incline of the ramp and Δx is the length of the ramp in meters.

5. Roll the marble down the ramp four times and record the distance that it runs along the carpet each time on the table below.
6. Calculate the average distance and record it in the appropriate table below.

Students work in pairs or small groups to determine the effect of mass on momentum, using marbles rolling down a ramp. They can use their results to formulate a relationship between mass and the distance rolled by the marble.

Sharing ideas and observations promotes scientific dialogue.

5. Describe ... rolled by the marble and ball bearing:

©2020 BIOZONE International
ISBN: 978-1-927309-79-7
Photocopying Prohibited

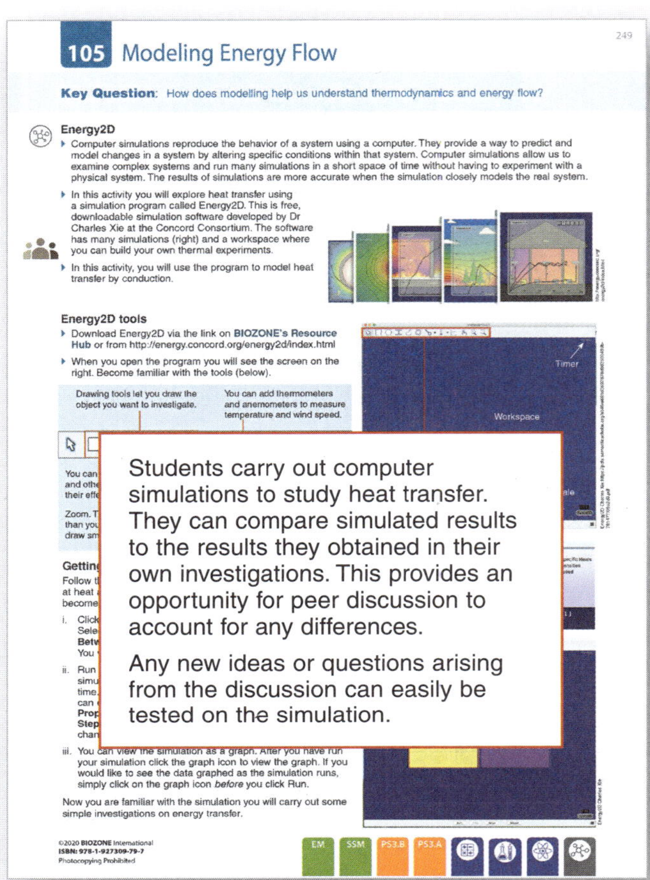

249

105 Modeling Energy Flow

Key Question: How does modelling help us understand thermodynamics and energy flow?

Energy2D

- Computer simulations reproduce the behavior of a system using a computer. They provide a way to predict and model changes in a system by altering specific conditions within that system. Computer simulations allow us to examine complex systems and run many simulations in a short space of time without having to experiment with a physical system. The results of simulations are more accurate when the simulation closely models the real system.
- In this activity you will explore heat transfer using a simulation program called Energy2D. This is free, downloadable simulation software developed by Dr Charles Xie at the Concord Consortium. The software has many simulations (right) and a workspace where you can build your own thermal experiments.
- In this activity, you will use the program to model heat transfer by conduction.

Energy2D tools

- Download Energy2D via the link on **BIOZONE's Resource Hub** or from http://energy.concord.org/energy2d/index.html
- When you open the program you will see the screen on the right. Become familiar with the tools (below).

Drawing tools let you draw the object you want to investigate.

You can add thermometers and anemometers to measure temperature and wind speed.

Timer

Workspace

Students carry out computer simulations to study heat transfer. They can compare simulated results to the results they obtained in their own investigations. This provides an opportunity for peer discussion to account for any differences.

Any new ideas or questions arising from the discussion can easily be tested on the simulation.

iii. You can view the simulation as a graph. After you have run your simulation click the graph icon to view the graph. If you would like to see the data graphed as the simulation runs, simply click on the graph icon *before* you click Run.

Now you are familiar with the simulation you will carry out some simple investigations on energy transfer.

©2020 BIOZONE International
ISBN: 978-1-927309-79-7
Photocopying Prohibited

EM SSM PS3.B PS3.A

Interactive revision of tasks in class

- The **Teacher's Digital Edition** provides a digital rights managed (DRM) version of the student book as PDF files. It features useful HIDE/SHOW answers, which can be used to review activities in class using a data projector or interactive whiteboard (left).
- Students benefit from the feedback in class, where questions can be addressed, and teachers benefit by having students self-mark their work and receive helpful feedback on their responses.
- This approach is particularly suited to activities with questions requiring a discussion, as students will be able to clarify some aspects of their responses. Stronger students can benefit by contributing to the explanatory feedback and class discussion.

©2020 **BIOZONE** International

NGSS as collaboration and discovery

- BIOZONE's *Physical Sciences for NGSS* provides multiple chances for collaboration and discovery. By working together and sharing ideas, students are exposed to different perspectives and levels of knowledge about phenomena.
- NGSS requires deeper student engagement with less emphasis on facts and more on understanding. By exploring principles and concepts within a context students are more easily able to apply these principles to new phenomena.
- BIOZONE's *Physical Sciences for NGSS* uses the NGSS framework to develop student understanding by providing a range of activities. These include getting students to think about and share what they already know and then build on this knowledge by exploring and explaining phenomena.

Student A is capable. He helps to lead the discussion and records the discussion in a structured way.

Students B and C are also capable but less willing to lead discussion they will add ideas to the discussion but need a little direction from A to do so.

Student D is less able but gains ideas and understanding from the discussion of students A, B, and C. She may add to the discussion as she gains confidence in the material being studied.

Student challenges

Do you ever need that little extra fun challenge for your more able students? The **Teacher's Edition** includes an extra page immediately preceding or following the **Teacher's Notes** at the beginning of each chapter (placement depends on pagination constraints). This page provides a challenge question for gifted and talented students in particular (or any students keen to have go!). It can be downloaded from **BIOZONE's Resource Hub**, where it is the first link for each chapter.

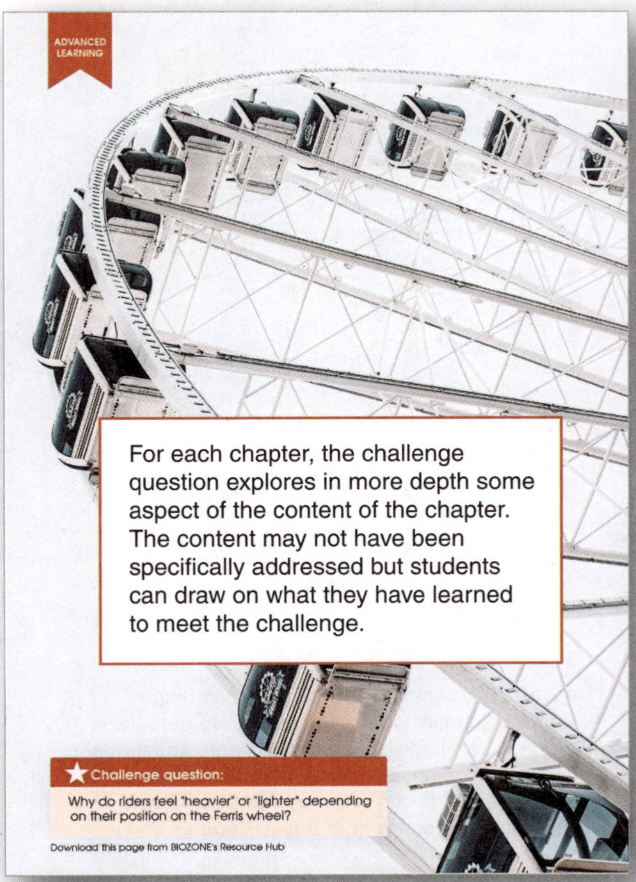

©2020 **BIOZONE** International

Differentiated Learning

The structure of *Physical Sciences for NGSS* promotes differentiated instruction, and has been designed to cater for students of all abilities. BIOZONE's collaborative approach to science inquiry encourages students of all abilities to share their ideas and knowledge with their peers while at the same time broadening their own understanding of phenomena.

There are several ways you can use *Physical Sciences for NGSS* to implement differential instruction in your classroom:

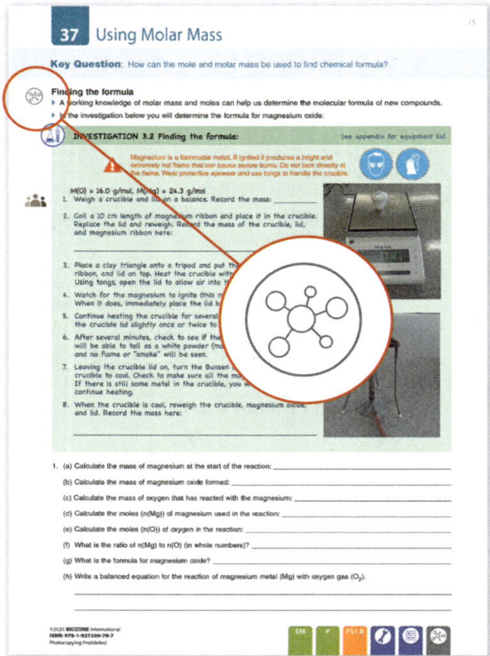

Animations and videos: Use the videos and animations on BIOZONE's Resource Hub to help striving learners with their English language skills and understanding of content. The Resource Hub also provides material tagged for gifted and talented students (also see p. CG11).

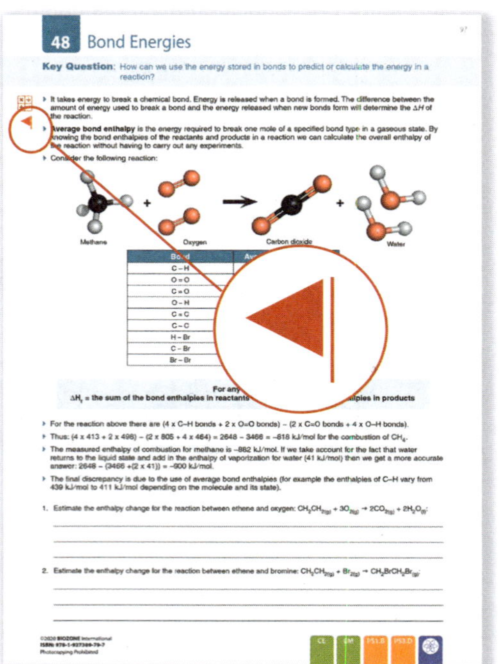

Red flag codes beside a section or question (on the Teacher's Edition or Teacher's Digital Edition) indicate that students may need extra guidance from the teacher. These questions are also suitable as challenges for more able students to tackle on their own. For able students, also see our **challenge question** pages (see p. CG11)

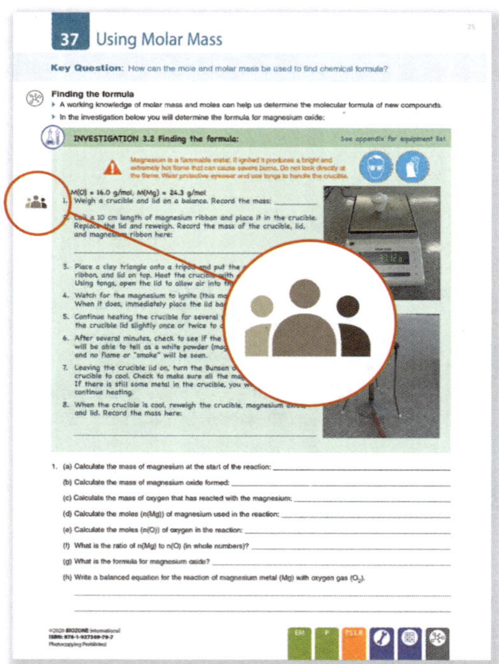

A group symbol indicates where students can work together. Group work provides opportunities for student collaboration and peer-to-peer support to explore phenomena. Working in groups, students can experiences the benefits of collaboration in the scientific process of discovery. By speaking and listening to each other, English language skills and scientific vocabulary are extended.

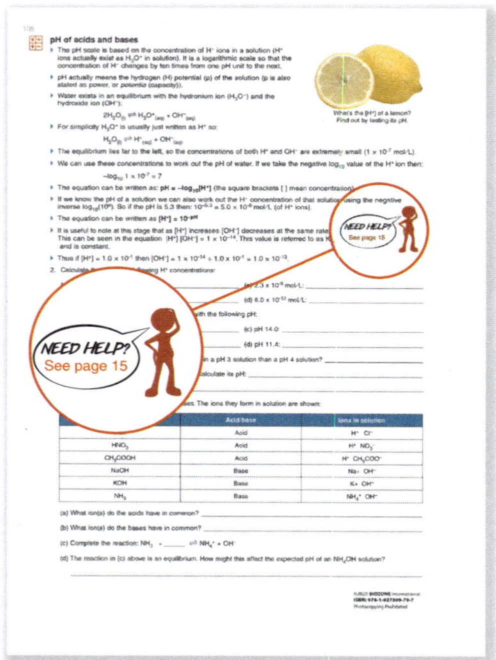

A red figure with a NEED HELP? icon helps students identify where they can go to get help with a specific skill. Skills and tips for computation, data analysis, plotting, statistical analysis, and aspects of experimental design are provided in the Basic Skills chapter at the start of the book. Students can visit this chapter regularly, or you can assign activities as homework before they attempt a specific task in class.

©2020 **BIOZONE** International

The Concept Maps

The concept maps in Physical Sciences for NGSS have two broad purposes: to provide a map of ideas covered in the program and to provide a vehicle for students to make their own connections between those ideas. They are particular useful as graphic organizers for striving students and visual learners. The introductory map provides an overview of the structure of the *NGSS Physical Sciences* program. Section concept maps divide the book into four parts, each providing a visual summary of one of four broad areas within the program, corresponding to PS1-PS4. Students can make their own connections between ideas on the concept maps as they work through the topics.

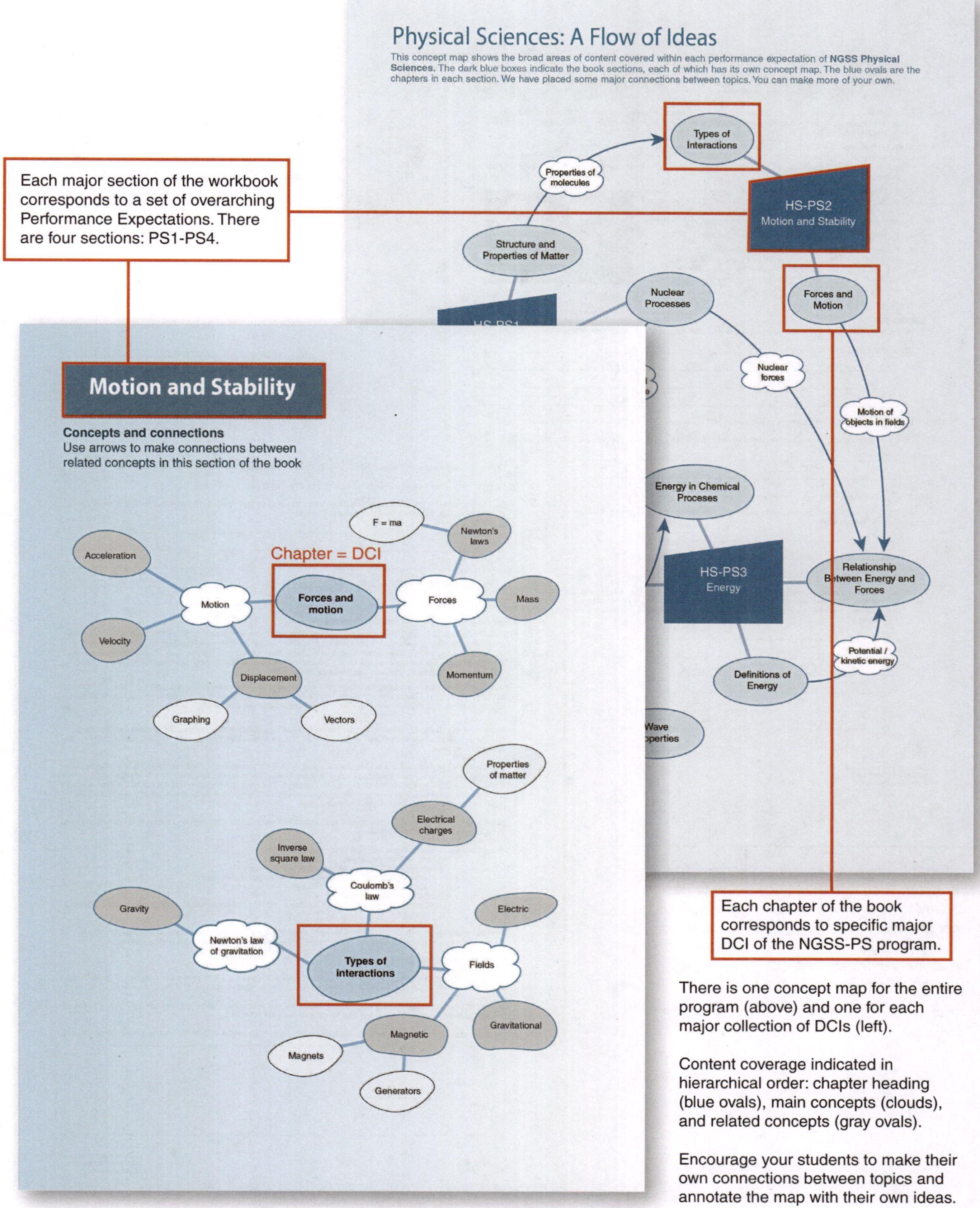

There is one concept map for the entire program (above) and one for each major collection of DCIs (left).

Content coverage indicated in hierarchical order: chapter heading (blue ovals), main concepts (clouds), and related concepts (gray ovals).

Encourage your students to make their own connections between topics and annotate the map with their own ideas.

©2020 **BIOZONE** International

Engineering Design Solutions

ETS SEPs, DCIs, and PEs as indicated in the NGSS framework are met through appropriately integrated engineering and design challenges. Typically tasks include analyzing problems, developing solutions using engineering, evaluating a design solution based on costs and benefits, or modeling a design solution.

The three dimensions of the NGSS framework appropriate to each design challenge are indicated in the chapter introduction of the Teacher's Edition (and Teacher's Digital Edition) and indicated through the tab system on the activity itself. Such tasks are usually examples of 'ELABORATE' as they involve the students applying what they have learned to solve a problem. As such, they also make good tasks for formative or summative assessment.

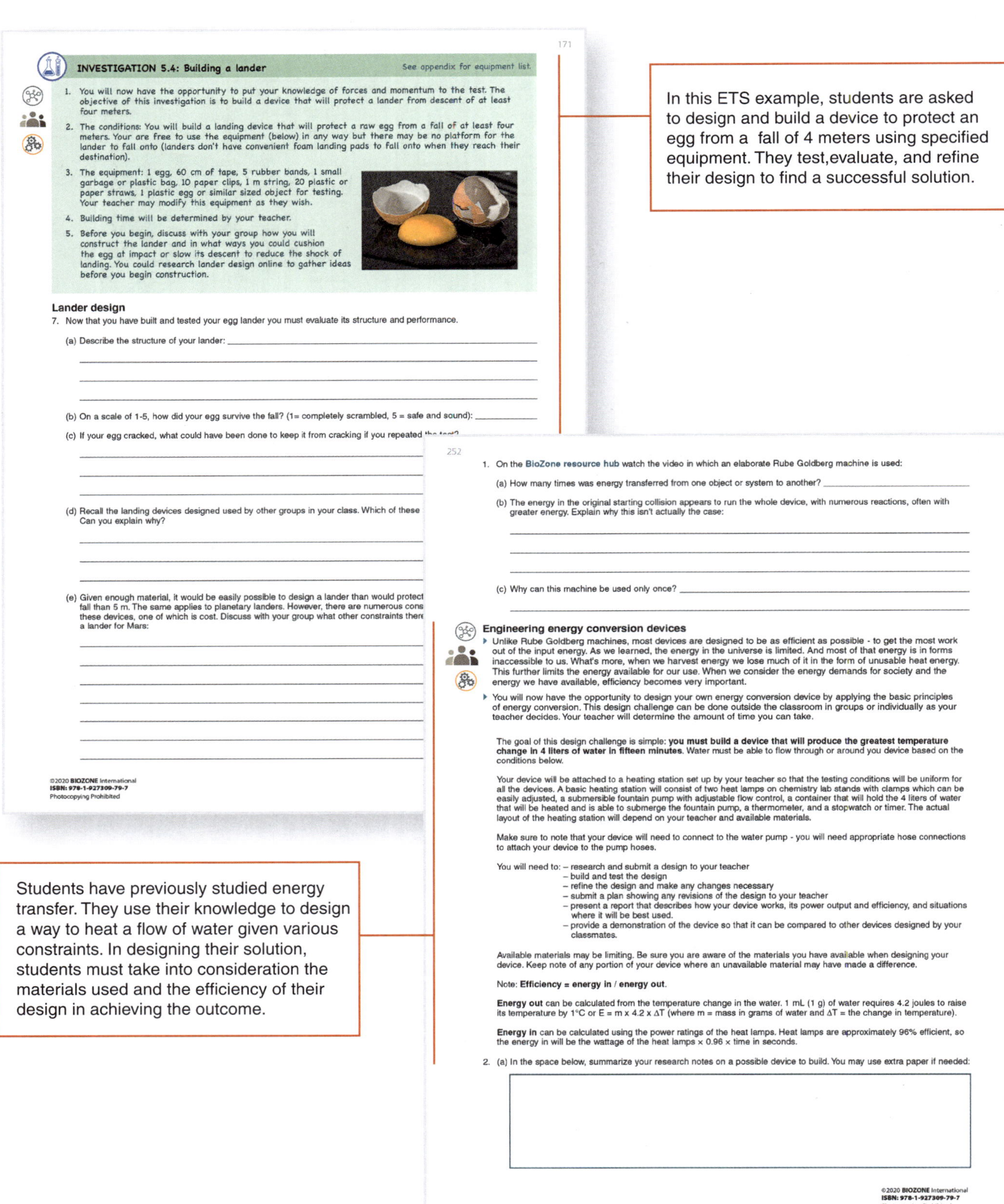

171

INVESTIGATION 5.4: Building a lander

See appendix for equipment list.

1. You will now have the opportunity to put your knowledge of forces and momentum to the test. The objective of this investigation is to build a device that will protect a lander from descent of at least four meters.
2. The conditions: You will build a landing device that will protect a raw egg from a fall of at least four meters. Your are free to use the equipment (below) in any way but there may be no platform for the lander to fall onto (landers don't have convenient foam landing pads to fall onto when they reach their destination).
3. The equipment: 1 egg, 60 cm of tape, 5 rubber bands, 1 small garbage or plastic bag, 10 paper clips, 1 m string, 20 plastic or paper straws, 1 plastic egg or similar sized object for testing. Your teacher may modify this equipment as they wish.
4. Building time will be determined by your teacher.
5. Before you begin, discuss with your group how you will construct the lander and in what ways you could cushion the egg at impact or slow its descent to reduce the shock of landing. You could research lander design online to gather ideas before you begin construction.

Lander design

7. Now that you have built and tested your egg lander you must evaluate its structure and performance.

(a) Describe the structure of your lander:

(b) On a scale of 1-5, how did your egg survive the fall? (1= completely scrambled, 5 = safe and sound):

(c) If your egg cracked, what could have been done to keep it from cracking if you repeated ...

(d) Recall the landing devices designed used by other groups in your class. Which of these ... Can you explain why?

(e) Given enough material, it would be easily possible to design a lander than would protect ... fall than 5 m. The same applies to planetary landers. However, there are numerous cons... these devices, one of which is cost. Discuss with your group what other constraints there ... a lander for Mars:

©2020 BIOZONE International
ISBN: 978-1-927309-79-7
Photocopying Prohibited

In this ETS example, students are asked to design and build a device to protect an egg from a fall of 4 meters using specified equipment. They test,evaluate, and refine their design to find a successful solution.

252

1. On the BioZone resource hub watch the video in which an elaborate Rube Goldberg machine is used:

(a) How many times was energy transferred from one object or system to another?

(b) The energy in the original starting collision appears to run the whole device, with numerous reactions, often with greater energy. Explain why this isn't actually the case:

(c) Why can this machine be used only once?

Engineering energy conversion devices

- Unlike Rube Goldberg machines, most devices are designed to be as efficient as possible - to get the most work out of the input energy. As we learned, the energy in the universe is limited. And most of that energy is in forms inaccessible to us. What's more, when we harvest energy we lose much of it in the form of unusable heat energy. This further limits the energy available for our use. When we consider the energy demands for society and the energy we have available, efficiency becomes very important.
- You will now have the opportunity to design your own energy conversion device by applying the basic principles of energy conversion. This design challenge can be done outside the classroom in groups or individually as your teacher decides. Your teacher will determine the amount of time you can take.

The goal of this design challenge is simple: **you must build a device that will produce the greatest temperature change in 4 liters of water in fifteen minutes.** Water must be able to flow through or around you device based on the conditions below.

Your device will be attached to a heating station set up by your teacher so that the testing conditions will be uniform for all the devices. A basic heating station will consist of two heat lamps on chemistry lab stands with clamps which can be easily adjusted, a submersible fountain pump with adjustable flow control, a container that will hold the 4 liters of water that will be heated and is able to submerge the fountain pump, a thermometer, and a stopwatch or timer. The actual layout of the heating station will depend on your teacher and available materials.

Make sure to note that your device will need to connect to the water pump - you will need appropriate hose connections to attach your device to the pump hoses.

You will need to:
- research and submit a design to your teacher
- build and test the design
- refine the design and make any changes necessary
- submit a plan showing any revisions of the design to your teacher
- present a report that describes how your device works, its power output and efficiency, and situations where it will be best used.
- provide a demonstration of the device so that it can be compared to other devices designed by your classmates.

Available materials may be limiting. Be sure you are aware of the materials you have available when designing your device. Keep note of any portion of your device where an unavailable material may have made a difference.

Note: **Efficiency = energy in / energy out.**

Energy out can be calculated from the temperature change in the water. 1 mL (1 g) of water requires 4.2 joules to raise its temperature by 1°C or $E = m \times 4.2 \times \Delta T$ (where m = mass in grams of water and ΔT = the change in temperature).

Energy in can be calculated using the power ratings of the heat lamps. Heat lamps are approximately 96% efficient, so the energy in will be the wattage of the heat lamps × 0.96 × time in seconds.

2. (a) In the space below, summarize your research notes on a possible device to build. You may use extra paper if needed:

©2020 BIOZONE International
ISBN: 978-1-927309-79-7
Photocopying Prohibited

Students have previously studied energy transfer. They use their knowledge to design a way to heat a flow of water given various constraints. In designing their solution, students must take into consideration the materials used and the efficiency of their design in achieving the outcome.

The Nature of Science

The Nature of Science combines established information with new knowledge to constantly refine what we know about the natural world. Eight Nature of Science understandings are presented in the NGSS document. Four are associated most closely with Science and Engineering Practices, and four with the Crosscutting Concepts. The Nature of Science understandings have been incorporated into most activities in *Physical Sciences for NGSS*, so we have not identified them specifically on the activity pages. Some examples of activities relating to the eight Nature of Science understandings are illustrated below. The subheading to which they relate is also given.

Nature of science understandings most closely associated with science and engineering practices

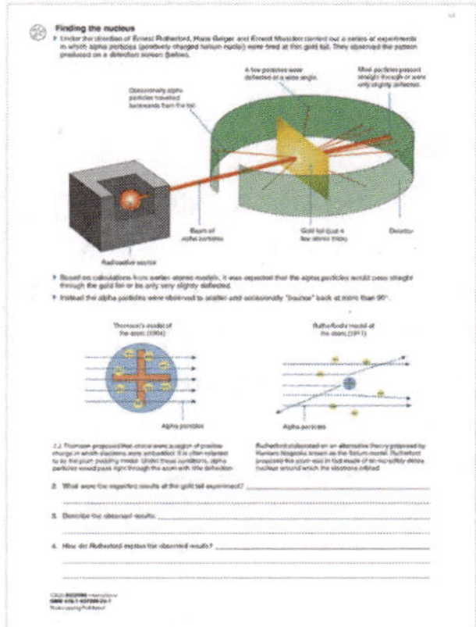
Finding the nucleus

Scientific investigations use a variety of methods.

- Scientific investigations use a variety of methods, tools, and techniques to revise and produce new knowledge

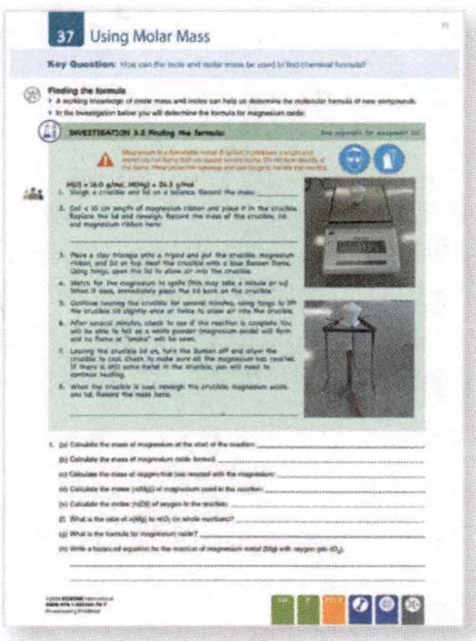
37 Using Molar Mass

Scientific knowledge is based on empirical evidence.

- Scientific knowledge is based on empirical evidence.

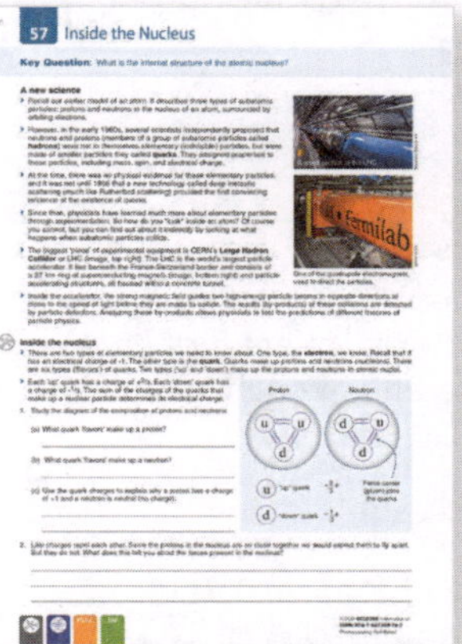
57 Inside the Nucleus

Scientific knowledge is open to revision in light of new evidence.

- Most scientific knowledge is quite durable but is subject to change based on new evidence and/or reinterpretations of existing evidence.

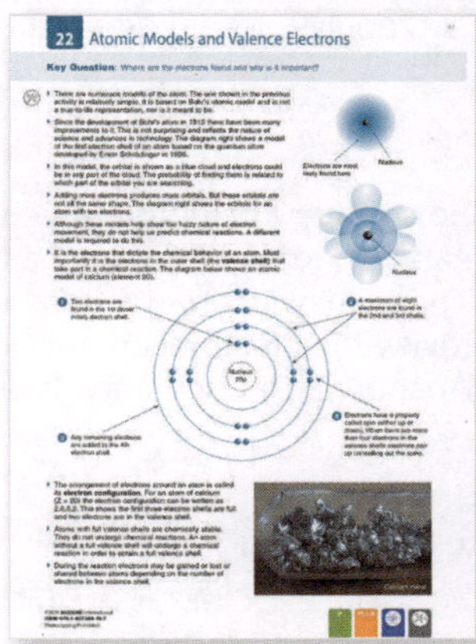
22 Atomic Models and Valence Electrons

Science models, laws, mechanisms, and theories explain natural phenomena.

- Models, mechanisms, and explanations collectively serve as tools in the development of a scientific theory.

Nature of science understandings most closely associated with crosscutting concepts

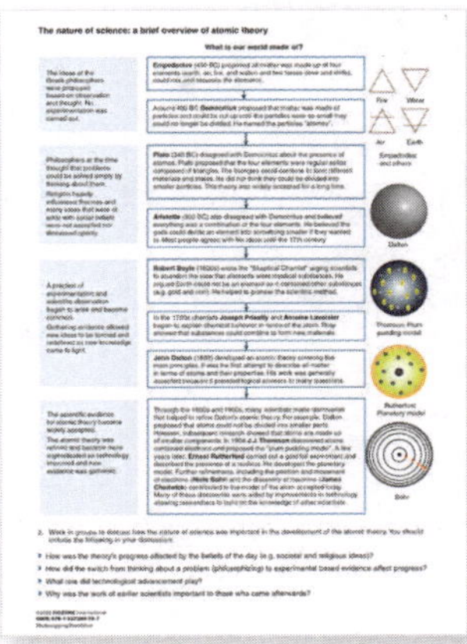
The nature of science: a brief overview of atomic theory

Science is a way of knowing.

- Scientific knowledge has a history that includes refinement of, and changes to, theories, ideas, and beliefs over time.

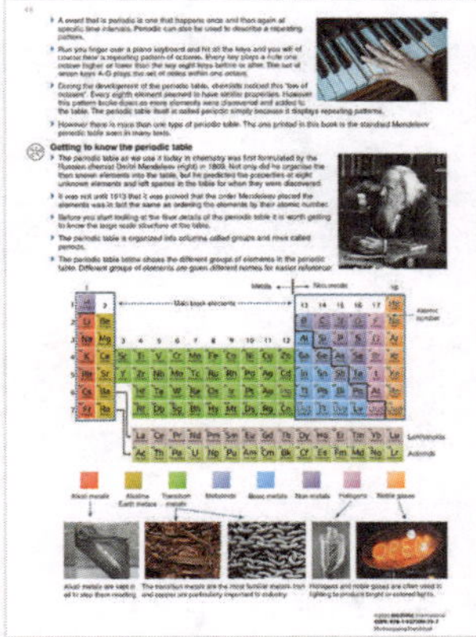
Getting to know the periodic table

Scientific knowledge assumes an order and consistency in natural systems.

- Scientific knowledge is based on the assumption that natural laws operate today as they did in the past and will continue to do so in the future.

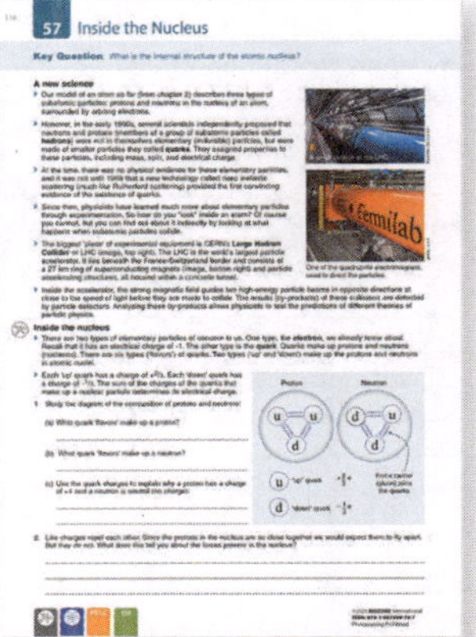
57 Inside the Nucleus

Science is a human endeavor.

- Technological advances have influenced the progress of science and science has influenced advances in technology.

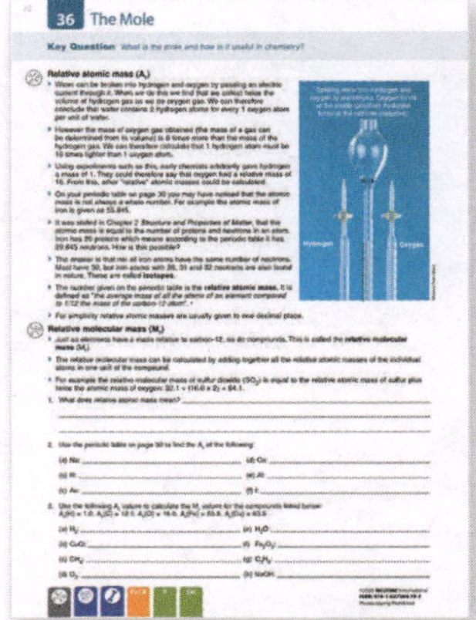
36 The Mole

Science addresses questions about the natural and material world.

- Scientific knowledge indicates what can happen in natural systems - not what should happen. The latter involves ethics, values, and human decisions about the use of knowledge.

©2020 BIOZONE International

Evaluating Student Performance

Physical Sciences for NGSS provides ample opportunity for students to demonstrate their understanding and proficiency in all three dimensions of the standards. Opportunities for both formative and summative and assessment are provided.

All activities (including assessments) have been designed to be three-dimensional in their approach, with the goal to enabling achievement of specific performance expectations. Performance expectations (PE) are not always met through completion of one activity or assessment, but through completion of a connected suite of tasks (as intended by the framework).

Assessments involve a variety of tasks as appropriate to a 3D approach, e.g., constructing models, analyzing and interpreting data, explaining, and communicating understanding through short and long answers, drawings, calculations, group work, design, and problem solving. The structure of the tasks is such that students use specific science and engineering practices and apply relevant crosscutting concepts to demonstrate their understanding of disciplinary core ideas.

Formative assessments can be chosen by the teacher to determine how a student's knowledge is progressing within a selected topic. We suggest that 'ELABORATE' and 'EVALUATE' sections of activities be used for formative assessment. These may incorporate some aspect of a performance expectation with the goal being to build confidence. Teachers can revise their instruction, revisit material, or set further tasks if a student is having difficulty with the material. Revisiting the Anchoring Phenomenon near the end of each instructional segment also provides a way to evaluate student understanding.

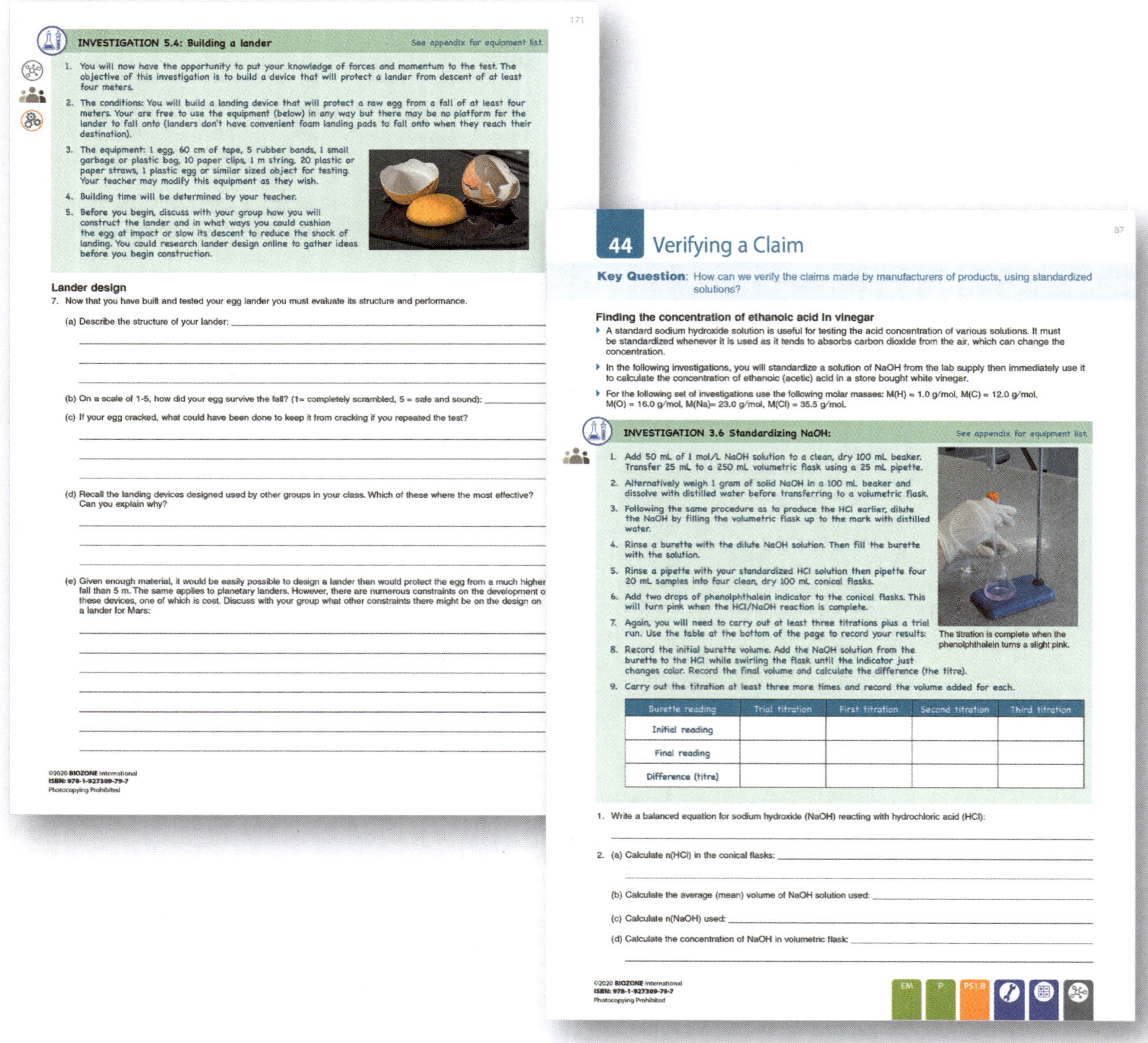

171

INVESTIGATION 5.4: Building a lander — See appendix for equipment list.

1. You will now have the opportunity to put your knowledge of forces and momentum to the test. The objective of this investigation is to build a device that will protect a lander from descent of at least four meters.
2. The conditions: You will build a landing device that will protect a raw egg from a fall of at least four meters. Your are free to use the equipment (below) in any way but there may be no platform for the lander to fall onto (landers don't have convenient foam landing pads to fall onto when they reach their destination).
3. The equipment: 1 egg, 60 cm of tape, 5 rubber bands, 1 small garbage or plastic bag, 10 paper clips, 1 m string, 20 plastic or paper straws, 1 plastic egg or similar sized object for testing. Your teacher may modify this equipment as they wish.
4. Building time will be determined by your teacher.
5. Before you begin, discuss with your group how you will construct the lander and in what ways you could cushion the egg at impact or slow its descent to reduce the shock of landing. You could research lander design online to gather ideas before you begin construction.

Lander design

7. Now that you have built and tested your egg lander you must evaluate its structure and performance.

(a) Describe the structure of your lander:

(b) On a scale of 1-5, how did your egg survive the fall? (1= completely scrambled, 5 = safe and sound):

(c) If your egg cracked, what could have been done to keep it from cracking if you repeated the test?

(d) Recall the landing devices designed used by other groups in your class. Which of these where the most effective? Can you explain why?

(e) Given enough material, it would be easily possible to design a lander than would protect the egg from a much higher fall than 5 m. The same applies to planetary landers. However, there are numerous constraints on the development o these devices, one of which is cost. Discuss with your group what other constraints there might be on the design on a lander for Mars:

©2020 **BIOZONE** International
ISBN: 978-1-927309-79-7
Photocopying Prohibited

87

44 Verifying a Claim

Key Question: How can we verify the claims made by manufacturers of products, using standardized solutions?

Finding the concentration of ethanoic acid in vinegar

- A standard sodium hydroxide solution is useful for testing the acid concentration of various solutions. It must be standardized whenever it is used as it tends to absorbs carbon dioxide from the air, which can change the concentration.
- In the following investigations, you will standardize a solution of NaOH from the lab supply then immediately use it to calculate the concentration of ethanoic (acetic) acid in a store bought white vinegar.
- For the following set of investigations use the following molar masses: M(H) = 1.0 g/mol, M(C) = 12.0 g/mol, M(O) = 16.0 g/mol, M(Na)= 23.0 g/mol, M(Cl) = 35.5 g/mol.

INVESTIGATION 3.6 Standardizing NaOH: — See appendix for equipment list.

1. Add 50 mL of 1 mol/L NaOH solution to a clean, dry 100 mL beaker. Transfer 25 mL to a 250 mL volumetric flask using a 25 mL pipette.
2. Alternatively weigh 1 gram of solid NaOH in a 100 mL beaker and dissolve with distilled water before transferring to a volumetric flask.
3. Following the same procedure as to produce the HCl earlier, dilute the NaOH by filling the volumetric flask up to the mark with distilled water.
4. Rinse a burette with the dilute NaOH solution. Then fill the burette with the solution.
5. Rinse a pipette with your standardized HCl solution then pipette four 20 mL samples into four clean, dry 100 mL conical flasks.
6. Add two drops of phenolphthalein indicator to the conical flasks. This will turn pink when the HCl/NaOH reaction is complete.
7. Again, you will need to carry out at least three titrations plus a trial run. Use the table at the bottom of the page to record your results:
8. Record the initial burette volume. Add the NaOH solution from the burette to the HCl while swirling the flask until the indicator just changes color. Record the final volume and calculate the difference (the titre).
9. Carry out the titration at least three more times and record the volume added for each.

The titration is complete when the phenolphthalein turns a slight pink.

Burette reading	Trial titration	First titration	Second titration	Third titration
Initial reading				
Final reading				
Difference (titre)				

1. Write a balanced equation for sodium hydroxide (NaOH) reacting with hydrochloric acid (HCl):

2. (a) Calculate n(HCl) in the conical flasks:

(b) Calculate the average (mean) volume of NaOH solution used:

(c) Calculate n(NaOH) used:

(d) Calculate the concentration of NaOH in volumetric flask:

©2020 **BIOZONE** International
ISBN: 978-1-927309-79-7
Photocopying Prohibited

EM P PS1.B

Summing up tasks at the close of each instructional segment can be used as a formal summative testing moment to evaluate student skills, understanding, and application of knowledge. These tasks are designed to meet part or all of one or more performance expectations. Material to address specific performance expectations is identified with a red tab in the margin throughout the Teacher's Edition. Performance expectations are also identified in the chapter introduction for the instructional segment, and in the tables summarizing BIOZONE's 3D approach by chapter earlier in this guide.

Note: All coding associated with assessment is hidden from the student and is available only in teacher's materials.

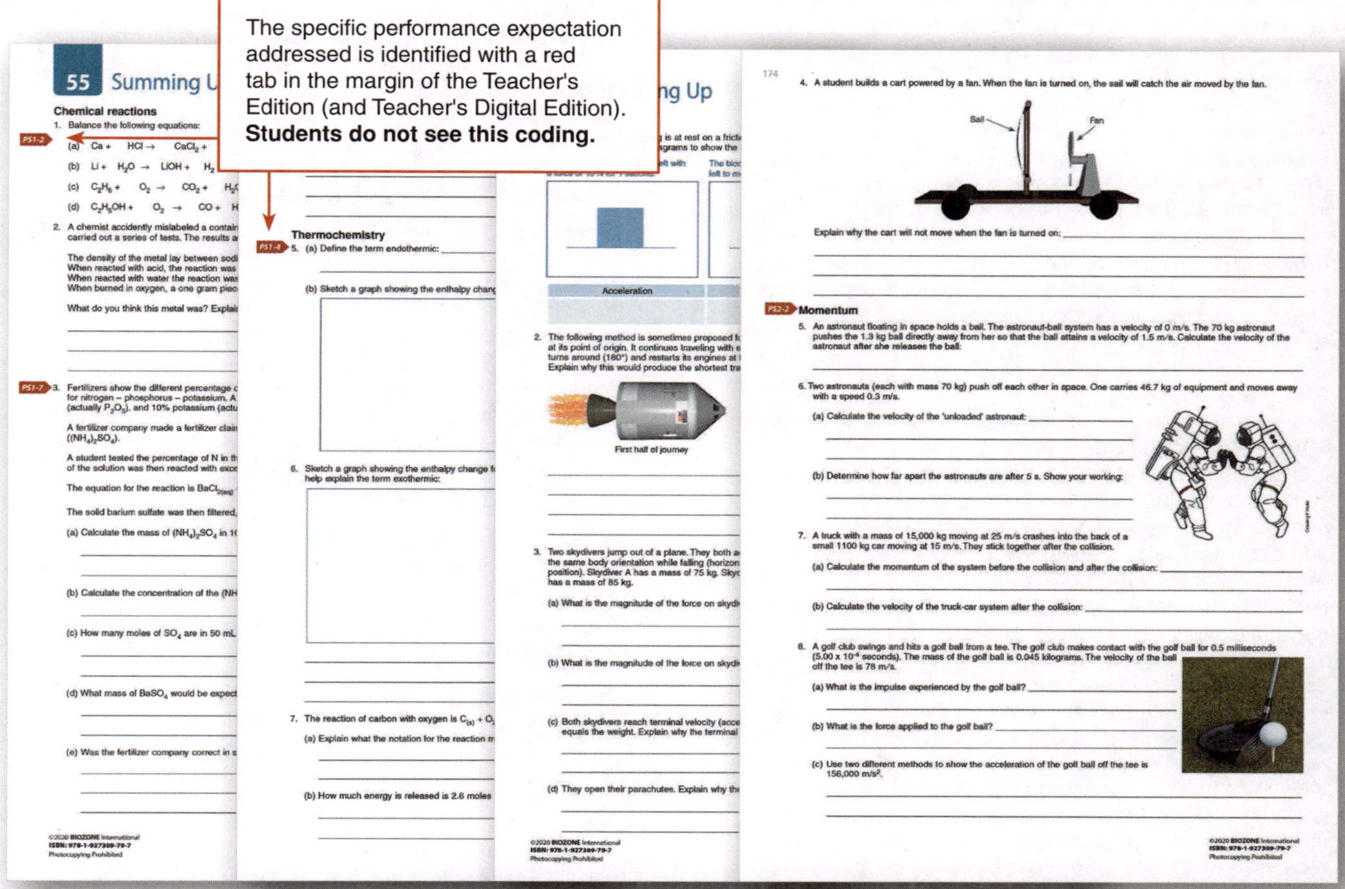

Summative assessments are three dimensional assessments of student understanding, including but not restricted to:

• Short answer questions • Long answer questions • Graphing • Data analysis and interpretation • Modeling

Does BIOZONE provide test banks?

- We are currently developing test banks to test the DCI content within each chapter.
- These will test content knowledge, and take the form of:
  - Multiple choice
  - True/False
  - Modified True/False
  - Multiple response
  - Matching
  - Short answer
  - Yes/No
  - Numeric response
- Test bank questions will be formatted for ingestion into test generator software such as ExamView.
- Standard and credit recovery options will be available.

Credit recovery questions

Science Practices

True/False

1. Investigations require you to make assumptions about the system you are working with.

ANS: T

2. You should always record and report the units of measurement during investigations.

ANS: T

Modified True/False

3. Base units are independently expressed and cannot be used on their own.

ANS: F
Modified answer: Base units are independently expressed and **can** be used on their own.

4. The color of a Bunsen flame has no relationship to its temperature.

ANS: F
Modified answer: The color of a Bunsen flame **gives an indication of its temperature.**

5. The density of a substance is the relationship between the mass of a substance and its volume.

ANS: T

Multiple choice

6. The inverse square law applies to:
a. Gravity
b. Electric fields
c. Light intensity
d. Sound intensity
e. All of the above

ANS: e

7. Light intensity at 3x the distance from the source is equal to:
a. 1/4 the brightness of the source
b. 1/8 the brightness of the source
c. 1/9 the brightness of the source
d. 1/2 the brightness of the source

ANS: c

©2020 **BIOZONE** International

Identifying CCCs and SEPs by Number

CROSSCUTTING CONCEPTS (CCCs)

CCCs are unifying ideas that apply across all disciplines of science. A CCC connects topics where the same unifying concept underpins the content. A statement for each numbered CCC is provided below. CCCs are identified by number in the tables following and in the embedded coding in the chapter introductions (Teacher's Edition). Statements are paraphrased.

1: Patterns
In grades 9-12, students observe patterns in systems at different scales and cite patterns as evidence for causality in supporting explanations of phenomena. They recognize that classifications or explanations at one scale may need revision using a different scale, thus requiring improved investigations and experiments. They identify and analyze patterns, and use analysis to re engineer and improve designed systems.

2: Cause and effect
In grades 9-12, students understand that empirical evidence is required to differentiate between cause and correlations and to make claims about cause and effect. They suggest cause and effect relationships to explain and predict behaviors in natural and designed systems. They also propose causal relationships by examining what is known about smaller-scale mechanisms within the system. They recognize changes in systems may have various causes that may not have equal effects.

3: Systems and system models
In grades 9-12, students investigate or analyze a system by defining its boundaries and initial conditions, as well as its inputs and outputs. They use models to simulate the flow of energy, matter, and interactions within and between systems at different scales. They also use models and simulations to predict the behavior of a system and recognize why these predictions have limited precision and reliability. They also design systems to do specific tasks.

4: Structure and function
In grades 9-12, students investigate systems by examining the properties of different materials, the structures of different components, and their interconnections to reveal a system's function and/or solve a problem. They infer the functions and properties of natural and designed objects and systems from their structure, the way their components are shaped and used, and the molecular substructures of their various materials.

5: Scale, proportion, and quantity
In grades 9-12, students understand that the significance of a phenomenon depends on the scale, proportion, and quantity at which it occurs. They recognize that patterns observable at one scale many not be observable or exist at other scales and that some systems can only be studied indirectly. Students use orders of magnitude to understand how a model at one scale relates to a model at another scale. They use algebraic thinking to examine scientific data and predict the effect of a change in one variable on another.

6: Energy and matter
In grades 9-12, students learn that the total amount of energy and matter in closed systems is conserved. They can describe changes of energy and matter in a system in terms of energy and matter flows into, out of, and within that system. They also learn that energy cannot be created or destroyed, only transferred and transformed. In nuclear processes, atoms are not conserved, but the total number of protons plus neutrons is conserved.

7: Stability and change
In grades 9-12, students understand that much of science deals with constructing explanations of how things change and how they remain stable. They quantify and model changes in systems over short or very periods of time. They see that some changes are irreversible and that negative feedback can stabilize a system, while positive feedback can destabilize it. They recognize that systems can be designed for more or less stability.

SCIENCE & ENGINEERING PRACTICES (SEPs)

SEPs for NGSS are overlapping and interconnected practices that students should know and understand. A statement for each numbered SEP is provided below. SEPs are identified by number in the tables following and in the embedded coding in the chapter introductions (Teacher's Edition).

1: Asking questions and defining problems
"Asking questions and defining problems in 9-12 builds on K-8 experiences and progresses to formulating, refining, and evaluating empirically testable questions and design problems using models and simulations."

2: Developing and using models
"Modeling in 9-12 builds on K-8 experiences and progresses to using, synthesizing, and developing models to predict and show relationships among variables between systems and their components in the natural and designed world(s)."

3: Planning and carrying out investigations
"Planning and carrying out investigations in 9-12 builds on K-8 experiences and progresses to include investigations that provide evidence for and test conceptual mathematical, physical, and empirical models".

4: Analzying and interpreting data
"Analyzing data in 9-12 builds on K-8 experiences and progresses to introducing more detailed statistical analysis, the comparison of data sets for consistency, and the use of models to generate and analyze data."

5: Using mathematics and computational thinking
"Mathematical and computational thinking in 9-12 builds on K-8 experiences and progresses to using algebraic thinking and analysis, a range of linear and non-linear functions, including trigonometric functions, exponentials and logarithms, and computational tools for statistical analysis to analyze, represent, and model data. Simulations are created and used based on mathematical models of basic assumptions."

6: Constructing explanations and designing solutions
"Constructing explanations and designing solutions in 9-12 builds on K-8 experiences and progresses to explanations and designs that are supported by multiple and independent student-generated sources of evidence consistent with scientific ideas, principles, and theories".

7: Engaging in argument from evidence
"Engaging in argument from evidence in 9-12 builds on K-8 experiences and progresses to using appropriate and sufficient evidence and scientific reasoning to defend and critique claims and explanations about the natural and designed world(s). Arguments may also come from current scientific or historical episodes in science."

8: Obtaining, evaluating, and communicating information
"Obtaining, evaluating, and communicating information in 9-12 builds on K-8 experiences and progresses to evaluating the validity and reliability of the claims, methods, and designs."

Summary of BIOZONE's 3D Approach By Chapter

Science and Engineering Practices (SEPs), Crosscutting Concepts (CCCs), Disciplinary Core Ideas (DCIs), and Performance Expectations (PEs) for each chapter of *Physical Sciences for NGSS* are listed in the tables following. An introductory "Science Practices" chapter is also included. Performance Expectations are met within the chapter and/or the *Summing Up* activity.

1: SCIENCE PRACTICES

Activity	SEP	DCI	CCC	PE
1	1, 6, 7, 8	NA	4	NA
2	2	NA	4	NA
3	1	NA		NA
4	3,5	NA		NA
5	5,8	NA		NA
6	3, 4, 5	NA		NA
7	4, 5	NA		NA
8	3,4,5,8	NA	1, 2	NA
9	3,4,5	NA	1	NA
10	3	NA		NA
11	5	NA		NA
12	3	NA		NA
13	2, 3, 6	NA		NA
14	3	NA		NA
15	2, 3, 6	NA		NA
16		NA	3	NA

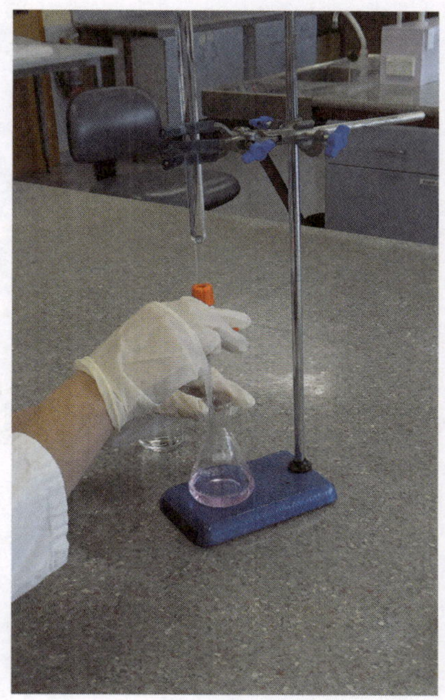

2: STRUCTURE AND PROPERTIES OF MATTER

Activity	SEP	DCI	CCC	PE
18-21	2	PS1.A	1	
22	2	PS1.A	1	HS-PS1-1
23	2, 4	PS1.A	1	
24-25	2	PS1.A	1	HS-PS1-1
26-28	2	PS1.A	1	
29, 30	3, 6	PS1.A	1	
31	2, 3	PS1.A	1	HS-PS1-3
33	2	PS1.A		HS-PS1-1

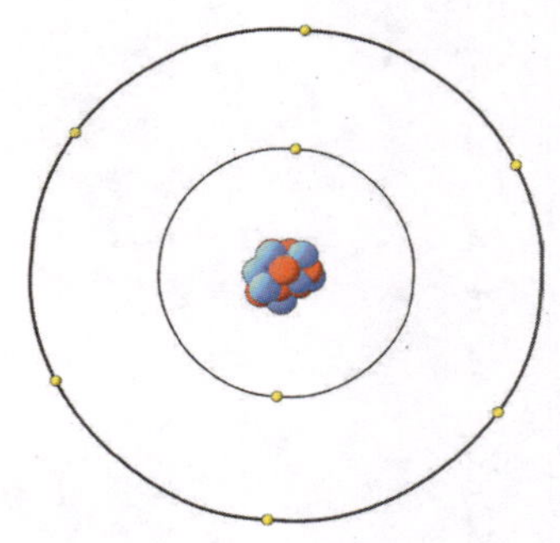

3: CHEMICAL REACTIONS

Activity	SEP	DCI	CCC	PE
35-36	5, 6	PS1.B	5	HS-PS1-7
37	3, 5, 6	PS1.B	1, 5	HS-PS1-7
38-40	5, 6	PS1.B	1, 5	HS-PS1-7
41	5, 6	PS1.B	1, 3, 5	HS-PS1-2, HS-PS1-7
42	5, 6	PS1.B	1, 5	HS-PS1-2, HS-PS1-7
43	2, 3, 5, 6	PS1.B	1, 5	HS-PS1-2, HS-PS1-7
44	3, 5, 6	PS1.B	1, 5	HS-PS1-2, HS-PS1-7
45	3, 6	PS1.B	1	HS-PS1-5
46	2, 3	PS1.B, PS3.D	5	HS-PS1-4
47	2, 3, 5	PS1.B, PS3.D	5	HS-PS1-4
48	2	PS1.B, PS3.D	5	HS-PS1-4
49	3, 6	PS1.B	4, 7	HS-PS1-5
50-51	6	PS1.B	4, 7	HS-PS1-5
52	6	PS1.B	7	HS-PS1-6
53	3, 5, 6	PS1.B	7	
55	2, 5, 6	PS1.A, PS1.B	1, 5	HS-PS1-2, HS-PS1-4, HS-PS1-5, HS-PS1-6 HS-PS1-7

4: NUCLEAR PROCESSES

Activity	SEP	DCI	CCC	PE
57	2, 3	PS1.C	5	HS-PS1-8
58	2	PS1.C	5	HS-PS1-8
59	2, 3, 8	PS1.C	5	HS-PS1-8
60	2	PS1.C	5, 7	HS-PS1-8
61	2	PS1.C	5, 7	
63	2	PS1.C	5	HS-PS1-8

5: FORCES AND MOTION

Activity	SEP	DCI	CCC	PE
65-66	3, 4	PS2.A	2	
67-68	4	PS2.A	2	HS-PS2-1
69-70	2, 5, 4	PS2.A	2	
71	3, 5	PS2.A	4	
72-73	5	PS2.A	4	HS-PS2-2
74	3, 5	PS2.A	4	HS-PS2-2
75	3, 5, 6	PS2.A, ETS1.A, ETS1.C	2	HS-PS2-2, HS-PS2-3, HS-EST-1-2, HS-EST1-3
77	4, 5	PS2.A	2, 4	HS-PS2-1, HS-PS2-2

6: TYPES OF INTERACTIONS

Activity	SEP	DCI	CCC	PE
79	5	PS2.B	1	
80	2, 5, 6	PS2.B	1	HS-PS2-4
81	3, 5	PS2.B	1	
82	2, 5	PS2.B	1, 4	HS-PS2-4
83	2	PS3.C	2	HS-PS3-5
84	2, 3	PS2.B, PS3.C	1, 2	
85	2, 3	PS2.B, PS3.C	2	
86	3	PS2.B	2	HS-PS3-5
87	2	PS3.C	2	HS-PS3-5
88	3, 6	PS2.B, PS3.A	2, 5	HS-PS2-5, HS-PS3-5
89	3, 6, 8	PS2.B	6	HS-PS2-6
91	2, 5, 8	PS2.B, PS3.C	1, 2, 6	HS-PS2-4, HS-PS2-6, HS-PS3-5

7: DEFINITIONS OF ENERGY

Activity	SEP	DCI	CCC	PE
93	2, 6	PS3.A	4, 5	
94	2, 5	PS3.A	4, 5	
95	2, 5	PS3.A	4, 5	HS-PS3-1
96	2, 5	PS3.A	4, 5	HS-PS3-2
97	2, 3, 5	PS3.A	4, 5	HS-PS3-2
98	2, 5	PS3.A	2, 4, 5	HS-PS3-2
99	2, 5	PS3.A	4, 5	
101	2	PS3.A	4	HS-PS3-1

8: CONSERVATION OF ENERGY AND ENERGY TRANSFER

Activity	SEP	DCI	CCC	PE
103	2, 5, 6	PS3.B, PS3.D	4, 5	HS-PS3-1, HS-PS3-2 HS-PS3-3, HS-PS3-4
104	3, 6	PS3.B, PS3.D	4, 5	HS-PS3-1, HS-PS3-3, HS-PS3-4
105	2, 3, 5	PS3.A, PS3.B	4, 5	HS-PS3-1
106	3, 5	PS3.B	4	
107	5	PS3.B	4	
108	2, 3, 6	PS3.B, PS3.D, ETS1.1	4, 5	HS-PS3-1, HS-PS3-3, HS-PS3-4, HS-ETS1-1
110	2	PS3.A, PS3.B	4, 5	HS-PS3-1, HS-PS3-2

Chocolateoak CC 3.0

9: WAVE PROPERTIES

Activity	SEP	DCI	CCC	PE
112-113	3, 5	PS4.A	2	HS-PS4-1
114	5	PS4.A	2	HS-PS4-1
115	5	PS4.A	2	
116	3, 5	PS4.A	2	HS-PS4-1
117	3, 7	PS4.A	2, 4	
118	1, 4, 8	PS4.A	2, 7	HS-PS4-2, HS-PS4-5
120	5	PS4.A	2	HS-PS4-1

10: ELECTROMAGNETIC RADIATION

Activity	SEP	DCI	CCC	PE
122	7	PS4.B	4, 5	
123	3, 7	PS4.B	4, 5	HS-PS4-3
124	8	PS4.B	2	HS-PS4-4
125	8	PS4.B	2	HS-PS4-4
127		PS4		HS-PS4-3

11: INFORMATION TECHNOLOGY AND INSTRUMENTATION

Activity	SEP	DCI	CCC	PE
128	8	PS4.C	2	
129-130	8	PS4.C	2	
132	8	PS4.C	2	HS-PS4-5

Identifying CCSS Connections

The activities in *Physical Sciences for NGSS* provide many opportunities to address the Common Core State Standards (CCSS) for numeracy, and literacy, and English language development (ELD). The incorporation of these standards allows students to practice and develop these key skills while exploring science.

Activities incorporating the CCSS Math Connections, ELA/ literacy, and ELD Connections specified in the NGSS Science Framework are identified by codes (right) in the **Teacher's Edition** and **Teacher's Digital Edition**.
Note that this coding is a tool for the teacher and is not present in the Student Edition.

- A red calculator indicates a math connection.
- A blue pencil indicates an ELA/literacy or ELD connection.

A list of the specific Math Connections, ELA/ Literacy Connections and ELD Standards addressed in the NGSS framework can be found in the tables at the bottom of this page and on the following pages.

BIOZONE recognizes that ELD Standards are not to be used in isolation, and are intended to be implemented in conjunction with ELA/Literacy and other academic content standards. This is why you will see them appearing along with the relevant ELA/literacy connection in the following tables.

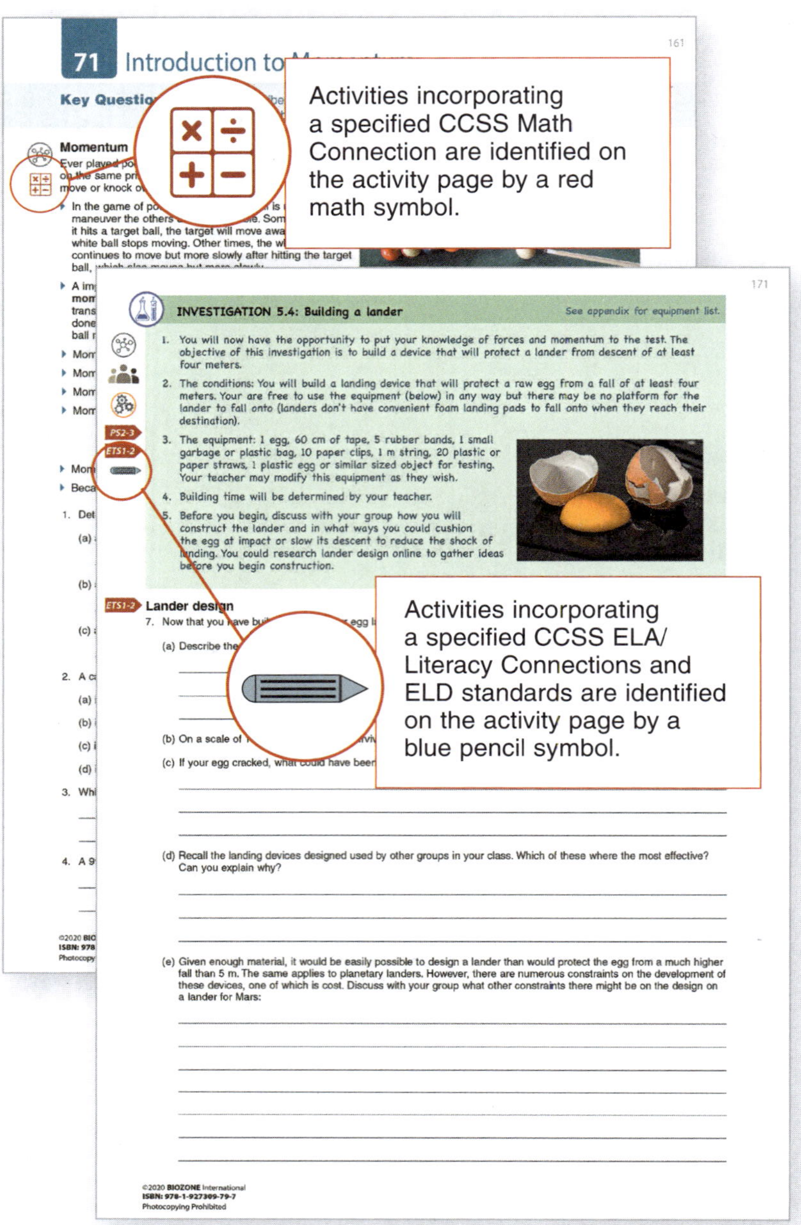

1: SCIENCE PRACTICES

Activity number	Activity	CCSS Math connection	CCSS ELA/Literacy & ELD connection
1	The Nature of Science		WHST.9-12.2, SL.9-12.1, ELD.P1.11-12.1, 5
6	Accuracy and Precision	MP.6, HSN-Q.A., HSS-ID.A.2	
7	Working With Numbers	MP.4, HSA-CED.A.4	
8	Graphical Analysis	MP.4, HSS-ID.A.1, HSS.ID.C.7	
9	Describing the Data	MP.4, HSS.ID.A.2, HSS.ID.A.2	
10	Investigations in Physics	MP.4, HSS-ID.A.1	
15	Measurement and Quantitative Analysis		SL.9-12.1, ELD.P1.11-12.1, 5

2: STRUCTURE AND PROPERTIES OF MATTER

Activity number	Activity	CCSS Math connection	CCSS ELA/Literacy & ELD connection
20	Atomic Theory and Rutherford's Experiment		WHST.9-12.8
24	Trends in the Periodic Table	MP4	

3: CHEMICAL REACTIONS

Activity number	Activity	CCSS Math connection	CCSS ELA/Literacy & ELD connection
36	The Mole	MP.2, HSN-Q.A.1	
37	Using Molar Mass	MP.2, HSN-Q.A.2	
38	Empirical Formulae	MP.2, HSN-Q.A.2	
39	Percentage Composition	MP.2, HSN-Q.A.2	
40	Balancing Equations	MP.1, MP.2, HSN-Q.A.1	
41	Stoichiometry	MP.2, HSN-Q.A.1, HSN-Q.A.3	
43	Creating Standard Solutions	MP.2, HSN-Q.A.1 , HSN-Q.A.3	
44	Verifying a Claim	MP.2, HSN-Q.A.1 , HSN-Q.A.3	
47	Enthalpy and Chemical Reactions	MP.2, HSN-Q.A.1, HSN-Q.A.3	
48	Bond Energies	MP.4	
49	Reversible Reactions	MP.4, HSN-Q.A.1	
52	Industrial Equilibria	MP.4, HSN-Q.A.1	
53	Acids and Bases	MP.4, HSN-Q.A.1	
55	Summing Up	MP.4, HSN-Q.A.1	

4: NUCLEAR PROCESSES

Activity number	Activity	CCSS Math connection	CCSS ELA/Literacy & ELD connection
59	Radioactive Decay	MP.4	
61	Fission vs Fusion	MP.4	

5: FORCES AND MOTION

Activity number	Activity	CCSS Math connection	CCSS ELA/Literacy & ELD connection
65	Distance and Displacement	MP.2, MP.4, HSN-Q.A.1, HSN-Q.A.2	RST.11-12.7
66	Speed and Velocity	MP.2, MP.4, HSN-Q.A.1, HSN-Q.A.2, HSA-SSE.A.1, HSA-CED.A.4, HSS-ID.A.1	
67	Acceleration	MP.2, MP.4, HSN-Q.A.1, HSN-Q.A.2, HSA-SSE.A.1, HSA-CED.A.1, HSA-CED.A.2 HSA-CED.A.4, HSA-SSE.B.3	
69	Newton's Second Law	MP.2, MP.4, HSN-Q.A.1, HSN-Q.A.2, HSA-SSE.A.1, HSA-CED.A.1, HSA-CED.A.2 HSA-CED.A.4	
71	Introduction to Momentum	MP.2, MP.4, HSN-Q.A.1	
72	Investigating Collisions	MP.2, MP.4,	
73	Law of Conservation of Momentum	MP.2, MP.4, HSA-CED.A.1	
74	Impulse	MP.2, MP.4, HSN-Q.A.1	
75	Crumple Zones and Crash Helmets	MP.2, MP.4, HSN-Q.A.1	
76	Review Your Understanding	MP.2, MP.4, HSN-Q.A.1	
77	Summing Up	MP.2, MP.4, HSN-Q.A.1	

6: TYPES OF INTERACTIONS

Activity number	Activity	CCSS Math connection	CCSS ELA/Literacy & ELD connection
79	Gravity	MP.4	
80	Newton's Law of Gravitation	MP.2, MP.4, HSN-Q.A.1, HSA-SSE.A.1, HSA-CED.A.1, HSA-CED.A.2 HSA-CED.A.4, HSA-SSE.B.3	
82	Coulomb's Law	MP.2, MP.4, HSN-Q.A.1, HSA-CED.A.1, HSA-CED.A.2 HSA-CED.A.4,	
83	Electric Fields	MP.2, MP.4, HSA-SSE.B.3, HSA-CED.A.4	
87	Magnetic Fields	MP.2, MP.4, HSA-SSE.B.3, HSA-CED.A.1	
89	Forces in Materials		RST.11-12.7, WHST.11-12.2, 7
91	Summing Up	MP.2, MP.4, HSA-SSE.A.1, HSA-SSE.B.3	

7: DEFINITIONS OF ENERGY

Activity number	Activity	CCSS Math connection	CCSS ELA/Literacy & ELD connection
94	The Physics of Work	MP.2, MP.4, HSN-Q.A.1	
95	The Work-Kinetic Energy Theorem	MP.2, MP.4, HSN-Q.A.1	
96	Stored Energy	MP.2, MP.4, HSN-Q.A.1	
97	Conservation of Energy	MP.2, MP.4, HSN-Q.A.1	
98	Pendulums	MP.2, MP.4	
99	Efficiency	MP.2, MP.4	
100	Review Your Understanding	MP.2, MP.4, HSN-Q.A.1, HSN-Q.A.2	
101	Summing Up	MP.2, MP.4, HSN-Q.A.1, HSN-Q.A.2, HSN-Q.A.3	

8: CONSERVATION OF ENERGY AND ENERGY TRANSFER

Activity number	Activity	CCSS Math connection	CCSS ELA/Literacy & ELD connection
104	The Second Law of Thermodynamics	MP.2, MP.4, HSS-IS.A.1	
105	Modeling Energy Flow	MP.2, MP.4	
106	Work and Power	MP.2, MP.4	
107	Energy and Power Plants	MP.2, MP.4	
108	Energy Conversion Devices		WHST.11-12.2
110	Summing Up	MP.2, MP.4, HSN-Q.A.1	

9: WAVE PROPERTIES

Activity number	Activity	CCSS Math connection	CCSS ELA/Literacy & ELD connection
112	Properties of Waves	MP.4	
113	The Speed of Sound	MP.2, MP.4, HAS-SSE.A.1, HAS-SSE.B.E, HSA.CED.A.4	
118	Digitizing Waves	MP.2, MP.4	

10: ELECTROMAGNETIC RADIATION

Activity number	Activity	CCSS Math connection	CCSS ELA/Literacy & ELD connection
123	Evaluate Claim		RST.9-10.8, RST.11-12.7

11: INFORMATION TECHNOLOGY AND INSTRUMENTATION

Activity number	Activity	CCSS Math connection	CCSS ELA/Literacy & ELD connection
132	Summing Up		RST.11-12.7, WHST.9-12.2

Teacher's Notes

Extended teacher's notes are found at the front of each chapter in the Teacher's Edition and Teacher's Digital Edition of *Physical Sciences for NGSS*. These notes appear immediately after the chapter introduction and provide context for the material and additional detail for the learning points (matched point for point). Where appropriate, opportunities to incorporate group work, practical activities, or design challenges are explained. Suggestions for differentiated instruction are also provided, including ways to support striving learners, e.g. through peer-to-peer support. For gifted and talented students we have included a "challenge question" and an associated image on the page immediately preceding or following the teacher's notes (see page CG11). Most activities are supported by material on **BIOZONE's Resource Hub**. The hub provides access to a large collection of free resources to supplement your teaching. Where they are provided, they are identified with a hub icon in the margin of both the Student Edition and Teacher's Edition. Where the resource is integral to the delivery of the activity (e.g. online data sets, computer simulations, or spreadsheets) we have indicated this in the teacher's notes.

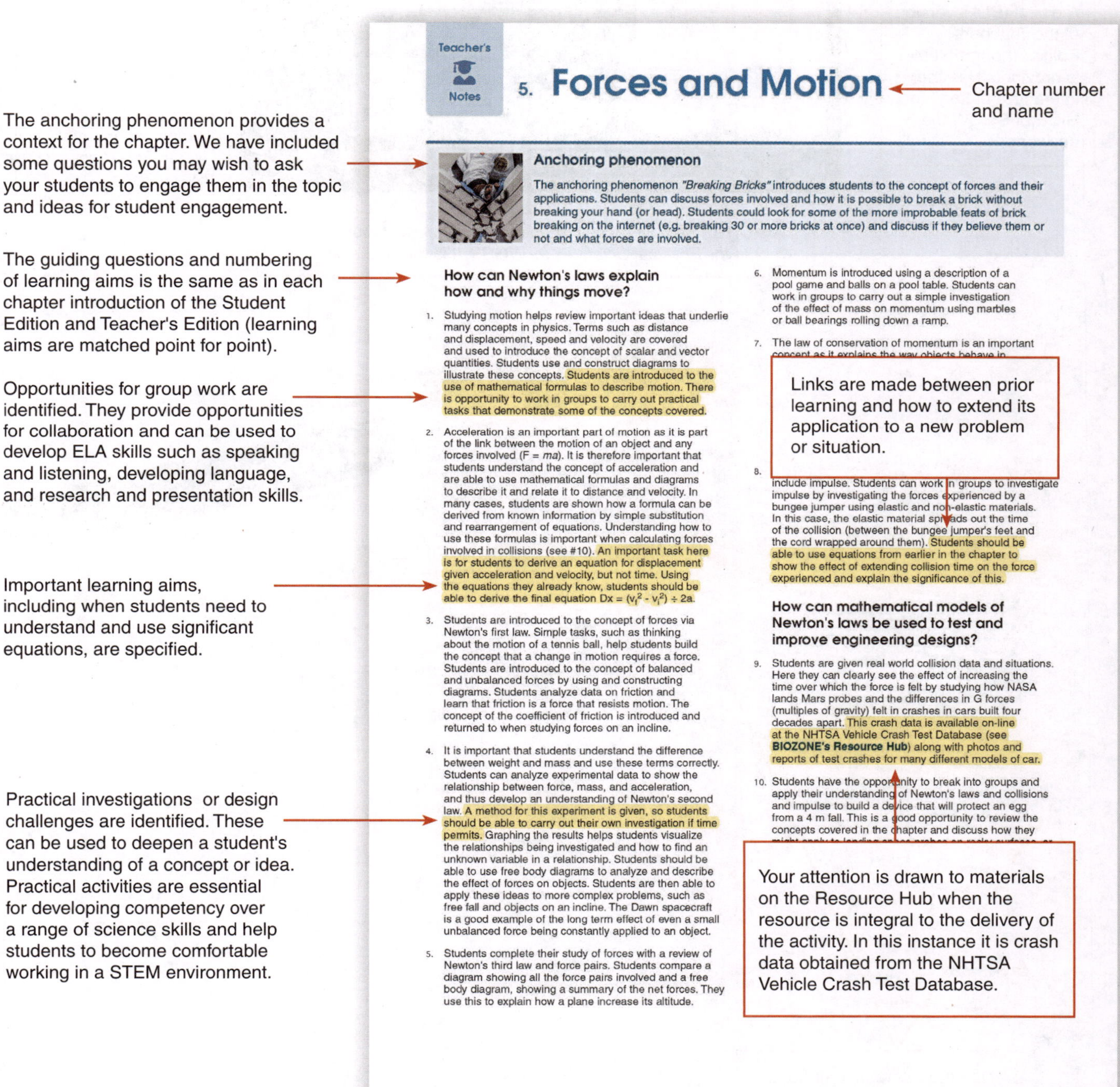

Teacher's Notes

5. Forces and Motion

Anchoring phenomenon

The anchoring phenomenon *"Breaking Bricks"* introduces students to the concept of forces and their applications. Students can discuss forces involved and how it is possible to break a brick without breaking your hand (or head). Students could look for some of the more improbable feats of brick breaking on the internet (e.g. breaking 30 or more bricks at once) and discuss if they believe them or not and what forces are involved.

How can Newton's laws explain how and why things move?

1. Studying motion helps review important ideas that underlie many concepts in physics. Terms such as distance and displacement, speed and velocity are covered and used to introduce the concept of scalar and vector quantities. Students use and construct diagrams to illustrate these concepts. Students are introduced to the use of mathematical formulas to describe motion. There is opportunity to work in groups to carry out practical tasks that demonstrate some of the concepts covered.
2. Acceleration is an important part of motion as it is part of the link between the motion of an object and any forces involved ($F = ma$). It is therefore important that students understand the concept of acceleration and are able to use mathematical formulas and diagrams to describe it and relate it to distance and velocity. In many cases, students are shown how a formula can be derived from known information by simple substitution and rearrangement of equations. Understanding how to use these formulas is important when calculating forces involved in collisions (see #10). An important task here is for students to derive an equation for displacement given acceleration and velocity, but not time. Using the equations they already know, students should be able to derive the final equation $Dx = (v_f^2 - v_i^2) \div 2a$.
3. Students are introduced to the concept of forces via Newton's first law. Simple tasks, such as thinking about the motion of a tennis ball, help students build the concept that a change in motion requires a force. Students are introduced to the concept of balanced and unbalanced forces by using and constructing diagrams. Students analyze data on friction and learn that friction is a force that resists motion. The concept of the coefficient of friction is introduced and returned to when studying forces on an incline.
4. It is important that students understand the difference between weight and mass and use these terms correctly. Students can analyze experimental data to show the relationship between force, mass, and acceleration, and thus develop an understanding of Newton's second law. A method for this experiment is given, so students should be able to carry out their own investigation if time permits. Graphing the results helps students visualize the relationships being investigated and how to find an unknown variable in a relationship. Students should be able to use free body diagrams to analyze and describe the effect of forces on objects. Students are then able to apply these ideas to more complex problems, such as free fall and objects on an incline. The Dawn spacecraft is a good example of the long term effect of even a small unbalanced force being constantly applied to an object.
5. Students complete their study of forces with a review of Newton's third law and force pairs. Students compare a diagram showing all the force pairs involved and a free body diagram, showing a summary of the net forces. They use this to explain how a plane increase its altitude.
6. Momentum is introduced using a description of a pool game and balls on a pool table. Students can work in groups to carry out a simple investigation of the effect of mass on momentum using marbles or ball bearings rolling down a ramp.
7. The law of conservation of momentum is an important concept as it explains the way objects behave in ...
8. ... include impulse. Students can work in groups to investigate impulse by investigating the forces experienced by a bungee jumper using elastic and non-elastic materials. In this case, the elastic material spreads out the time of the collision (between the bungee jumper's feet and the cord wrapped around them). Students should be able to use equations from earlier in the chapter to show the effect of extending collision time on the force experienced and explain the significance of this.

How can mathematical models of Newton's laws be used to test and improve engineering designs?

9. Students are given real world collision data and situations. Here they can clearly see the effect of increasing the time over which the force is felt by studying how NASA lands Mars probes and the differences in G forces (multiples of gravity) felt in crashes in cars built four decades apart. This crash data is available on-line at the NHTSA Vehicle Crash Test Database (see **BIOZONE's Resource Hub**) along with photos and reports of test crashes for many different models of car.
10. Students have the opportunity to break into groups and apply their understanding of Newton's laws and collisions and impulse to build a device that will protect an egg from a 4 m fall. This is a good opportunity to review the concepts covered in the chapter and discuss how they ...

The Teacher's Digital Edition

The *Teacher's Digital Edition* is a DRM product, sold separately, and aimed primarily at extending the pedagogical tools at a teacher's disposal. Many of the features of this resource have been developed in response to requests from teachers themselves.

The **Classroom Guide** is provided as a printable PDF.

A digital (PDF) version of the Teacher's Edition (non-printable) is provided. Use the interactive buttons to HIDE or SHOW the answers.

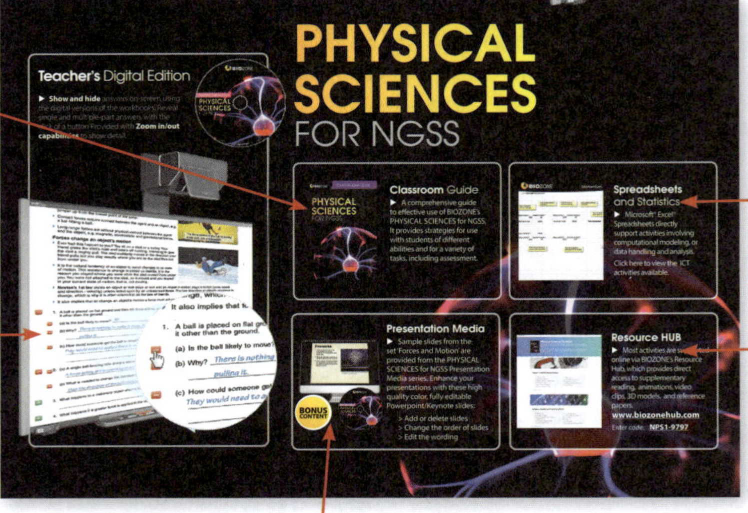

Link to ***Excel***® **spreadsheets** for selected activities with a data analysis or computer modeling component.

Access **BIOZONE's Resource Hub** directly from this link for a range of resources to support the activities.

A **BONUS** sample copy from the planned presentation media for *Physical Sciences for NGSS*. It is fully editable.

Use the interactive buttons to reveal the answers as you work through the activity on-screen.

Activities that manipulate data using formulae are supported by spreadsheets. These include all data and comments on graphical analysis.

©2020 **BIOZONE** International

Student Edition

With answers in place

PHYSICAL SCIENCES FOR NGSS

PHYSICAL SCIENCES

FOR NGSS

Meet the Writing Team

Kent Pryor
I have a BSc from Massey University majoring in zoology and ecology and taught secondary school biology and chemistry for 9 years before joining BIOZONE as an author in 2009.

Benjamin J. Westleigh
I have a BSc from the University of Miami and a Masters in Teaching and Learning from Waikato University. I have taught high school science and mathematics since 2009 and was based in Maine, in the USA, before moving to New Zealand.

Tracey Greenwood
I have been writing resources for students since 1993. I have a Ph.D in biology, specialising in lake ecology and I have taught both graduate and undergraduate biology.

Lissa Bainbridge-Smith
I worked in industry in a research and development capacity for 8 years before joining BIOZONE in 2006. I have a MSc from Waikato University.

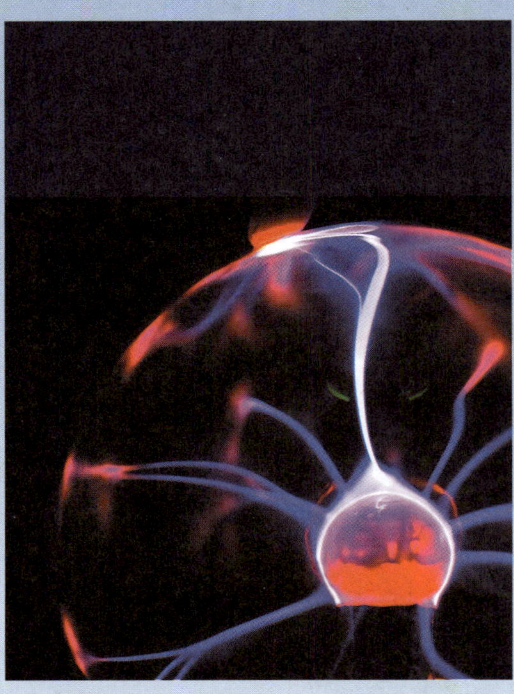

Cover photograph

Plasma globe

Plasma globes produce streams of colored plasma that radiate from the inner to the outer globe. High voltage from the AC Tesla coil inside the inner globe excites electrons, producing a high frequency radiowave. The radiowave ionizes the noble gases inside the outer globe producing a plasma (separated ions and electrons). A difference in charge between the inner globe and outer globe results in a flow of excited electrons from the inner to outer globe. As the electrons become less excited they lose energy and emit a photon of light, producing colored streams.

PHOTO: tarasylo https://stock.adobe.com/nz/images/plasma-globe/89439371

Thanks to:

The staff at BIOZONE, including Clare Mansfield for design and graphics support, Paolo Curray for IT support, Anu Chauhan for logistics, Felix Hicks for illustration, Allan Young for office handling, and the BIOZONE sales team.

Next Generation Science Standards (NGSS) is a registered trademark of Achieve. Neither Achieve nor the lead states and partners that developed the Next Generation Science Standards were involved in the production of this product and do not endorse it.

All rights reserved. No part of this publication may be reproduced, stored in a retrieval system, or transmitted in any form or by any means, electrical, mechanical, **photocopying**, recording or otherwise, without the permission of BIOZONE International Ltd. This book may not be **re-sold**. The conditions of sale specifically prohibit the photocopying of exercises, worksheets, and diagrams from this book for any reason.

Contents

CODING: **Activity** is marked: ☐ to be done ☑ when completed ● Activity includes practical

Contents

CODING: **Activity** is marked: ☐ to be done ☑ when completed ● Activity includes practical

Using This Book

Activities make up most of this book. These are presented as integrated instructional sequences, which may extend over multiple pages, allowing you to build a deeper understanding of phenomena as you progress through each chapter. Each chapter begins with a broad **anchoring phenomenon**. This is something you may have seen, heard about, or experienced (e.g. breaking bricks by hand) but may not necessarily be able to explain. You will come to understanding of this phenomenon through an exploration of related everyday and investigative phenomena in the activities that follow and then test your own understanding in the review activity at the end of the chapter.

Structure of a chapter

Chapter introduction

Identifies the activities relating to the guiding questions.

Introductory activity

The first activity acts as an anchoring phenomenon. It introduces a phenomenon that can be explained by the rest of the activities in the chapter.

Introductory activity revisited

Once you have completed the activities in the chapter, you should be able to explain the anchoring phenomenon more fully.

Summing Up

Find out what you know about the ideas, connections, and skills you have explored in the chapter.

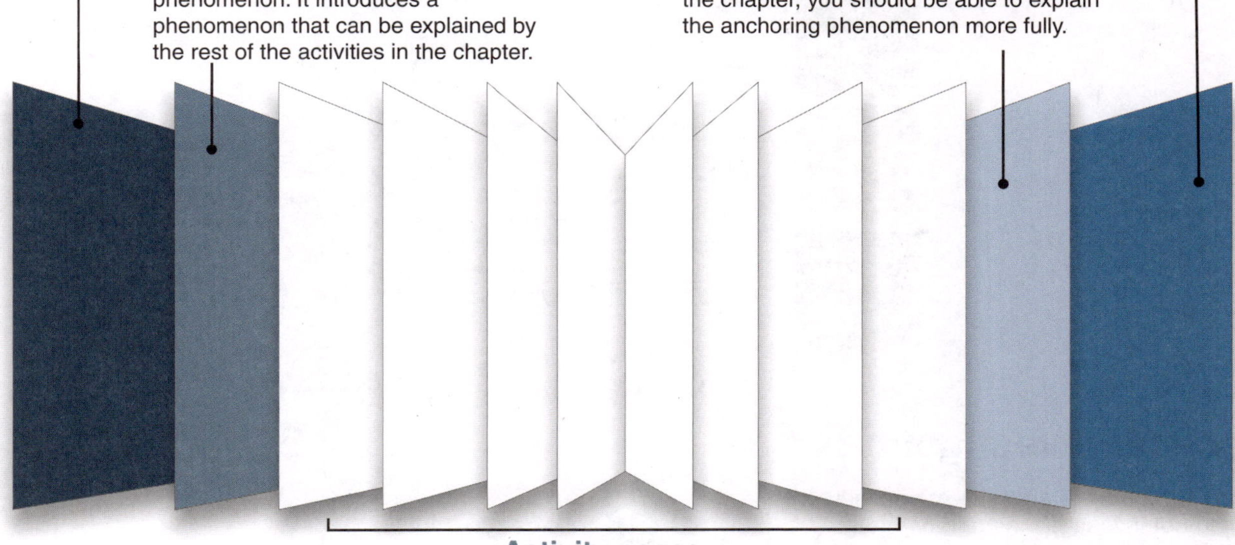

Activity pages

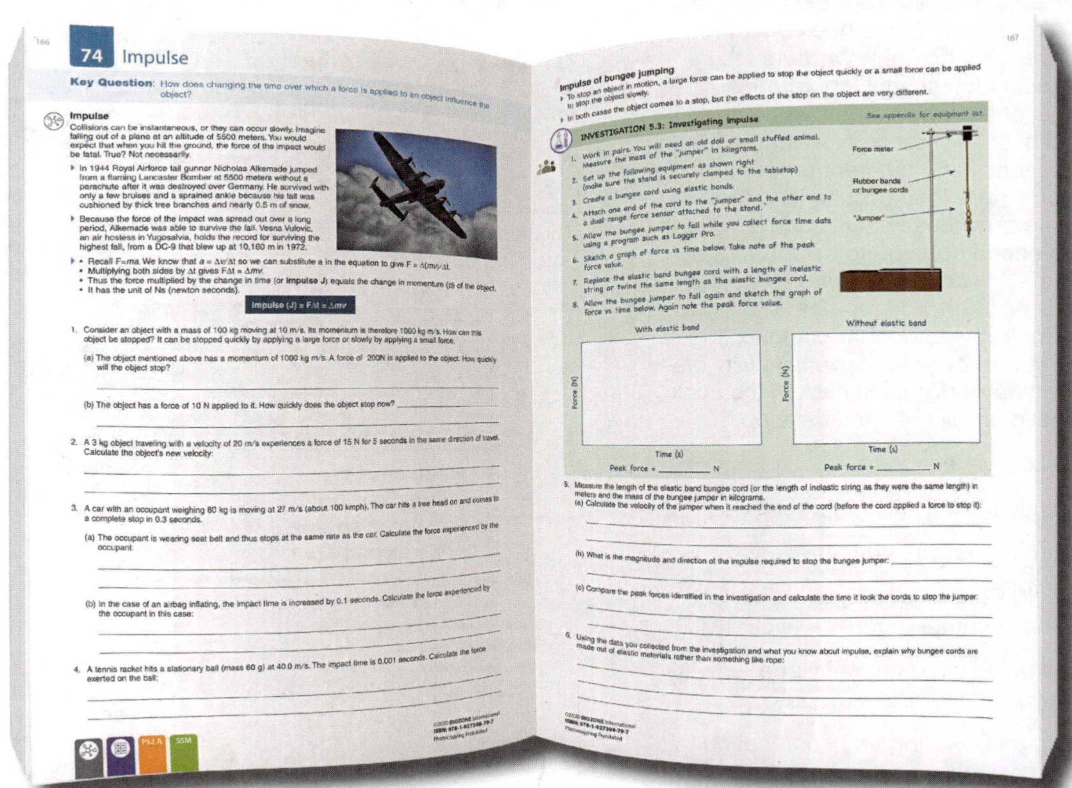

©2020 **BIOZONE** International
Photocopying Prohibited

Chapter Introductions

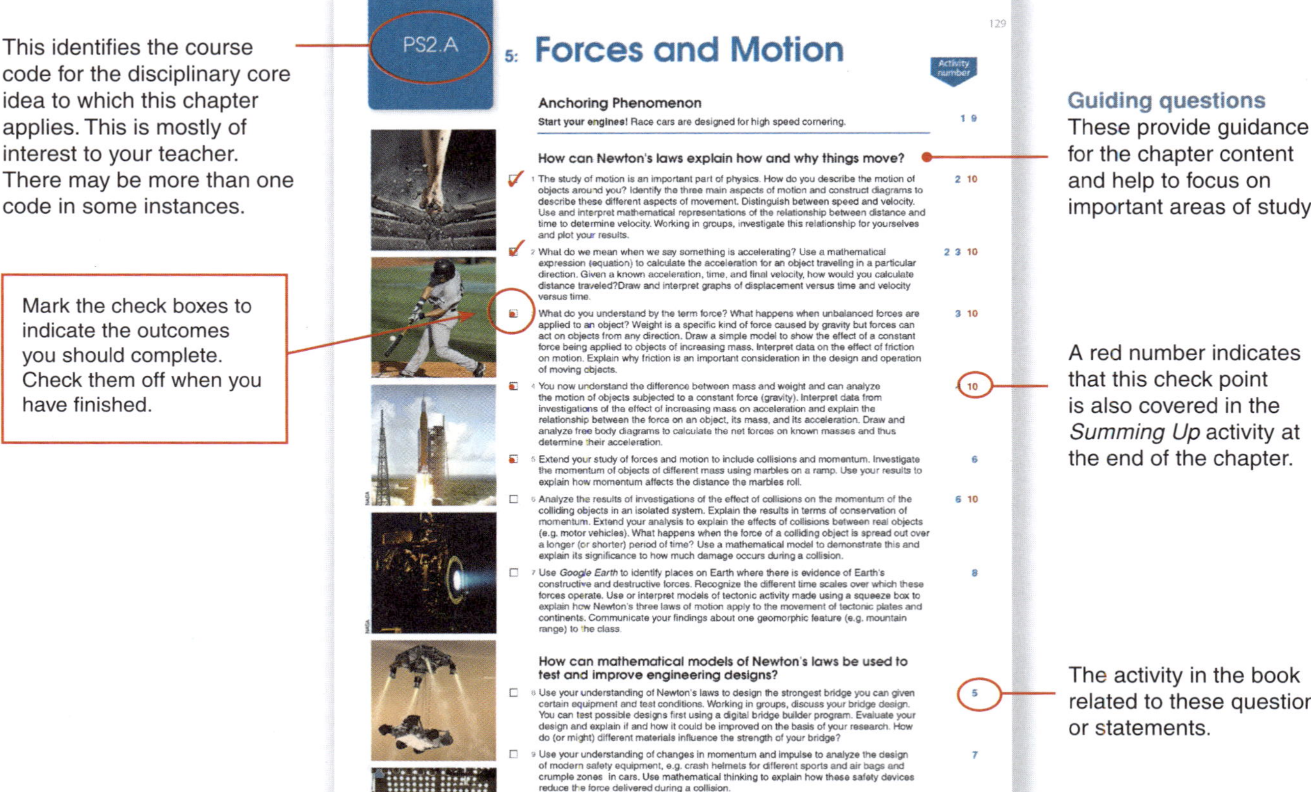

PS2.A

129

5: Forces and Motion

Activity number

Anchoring Phenomenon

Start your engines! Race cars are designed for high speed cornering. — 1 9

How can Newton's laws explain how and why things move?

1. The study of motion is an important part of physics. How do you describe the motion of objects around you? Identify the three main aspects of motion and construct diagrams to describe these different aspects of movement. Distinguish between speed and velocity. Use and interpret mathematical representations of the relationship between distance and time to determine velocity. Working in groups, investigate this relationship for yourselves and plot your results. — 2 10
2. What do we mean when we say something is accelerating? Use a mathematical expression (equation) to calculate the acceleration for an object traveling in a particular direction. Given a known acceleration, time, and final velocity, how would you calculate distance traveled?Draw and interpret graphs of displacement versus time and velocity versus time. — 2 3 10
3. What do you understand by the term force? What happens when unbalanced forces are applied to an object? Weight is a specific kind of force caused by gravity but forces can act on objects from any direction. Draw a simple model to show the effect of a constant force being applied to objects of increasing mass. Interpret data on the effect of friction on motion. Explain why friction is an important consideration in the design and operation of moving objects. — 3 10
4. You now understand the difference between mass and weight and can analyze the motion of objects subjected to a constant force (gravity). Interpret data from investigations of the effect of increasing mass on acceleration and explain the relationship between the force on an object, its mass, and its acceleration. Draw and analyze free body diagrams to calculate the net forces on known masses and thus determine their acceleration. — 4 10
5. Extend your study of forces and motion to include collisions and momentum. Investigate the momentum of objects of different mass using marbles on a ramp. Use your results to explain how momentum affects the distance the marbles roll. — 6
6. Analyze the results of investigations of the effect of collisions on the momentum of the colliding objects in an isolated system. Explain the results in terms of conservation of momentum. Extend your analysis to explain the effects of collisions between real objects (e.g. motor vehicles). What happens when the force of a colliding object is spread out over a longer (or shorter) period of time? Use a mathematical model to demonstrate this and explain its significance to how much damage occurs during a collision. — 6 10
7. Use *Google Earth* to identify places on Earth where there is evidence of Earth's constructive and destructive forces. Recognize the different time scales over which these forces operate. Use or interpret models of tectonic activity made using a squeeze box to explain how Newton's three laws of motion apply to the movement of tectonic plates and continents. Communicate your findings about one geomorphic feature (e.g. mountain range) to the class. — 8

How can mathematical models of Newton's laws be used to test and improve engineering designs?

8. Use your understanding of Newton's laws to design the strongest bridge you can given certain equipment and test conditions. Working in groups, discuss your bridge design. You can test possible designs first using a digital bridge builder program. Evaluate your design and explain if and how it could be improved on the basis of your research. How do (or might) different materials influence the strength of your bridge? — 5
9. Use your understanding of changes in momentum and impulse to analyze the design of modern safety equipment, e.g. crash helmets for different sports and air bags and crumple zones in cars. Use mathematical thinking to explain how these safety devices reduce the force delivered during a collision. — 7
10. Demonstrate your understanding of forces and momentum by designing a landing device to protect a raw egg from a fall of at least 5 meters. Working in groups, discuss how you will make the lander and the different designs you could use to cushion the egg at impact and/or slow its descent. Justify your design choice in terms of Newton's laws of motion. — 7

Practical Investigations

An important part of physical science involves carrying out investigations and carefully observing and recording what occurs during them. Throughout the book you will notice green investigation panels (like the one shown right). Each investigation has been designed using simple equipment found around the home or in most high school laboratories. The investigations provide opportunities for you to investigate phenomena for yourself. The investigations have different purposes depending on where they occur within the chapter. Some provide stimulus material or ask questions to encourage you to think about a particular phenomenon before you study it in detail. Others build on work you have already carried out and provide a more complex scenario for you to explain. Equipment lists are provided as an appendix at the back of the book. The investigations will help you develop:

- Skills in observation
- Skills in critical analysis and problem solving
- Skills in mathematics and numeracy
- Skills in collecting and analyzing data and maintaining accurate records
- Skills in working independently and collaboratively as part of a group
- Skills in communicating and contributing to group discussions

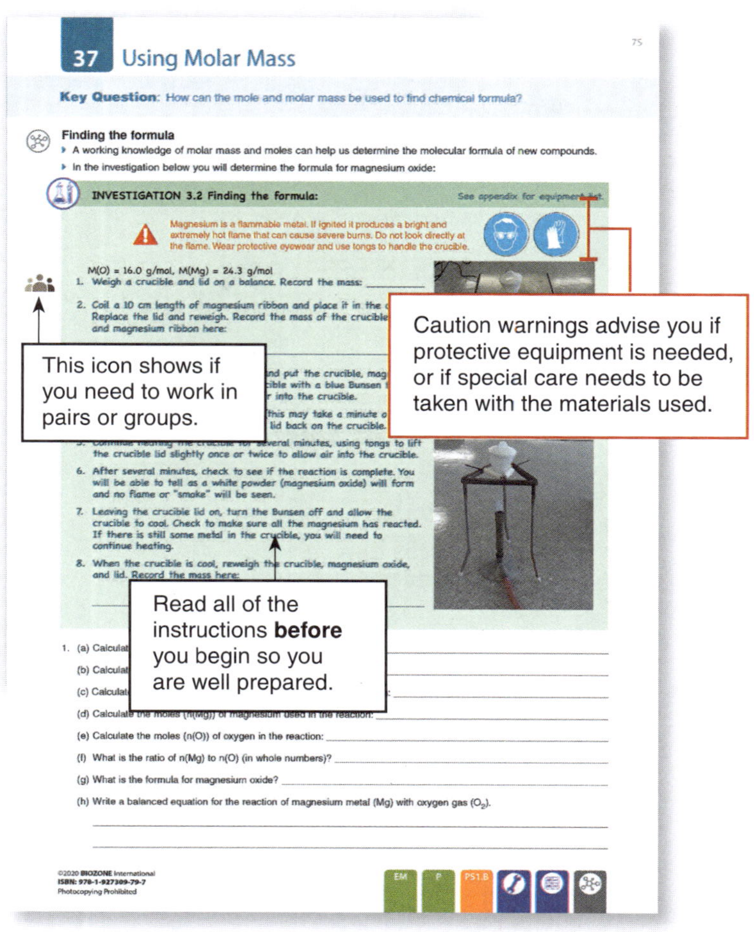

75

37 Using Molar Mass

Key Question: How can the mole and molar mass be used to find chemical formula?

Finding the formula

- A working knowledge of molar mass and moles can help us determine the molecular formula of new compounds.
- In the investigation below you will determine the formula for magnesium oxide:

INVESTIGATION 3.2 Finding the formula: See appendix for equipment list.

Magnesium is a flammable metal. If ignited it produces a bright and extremely hot flame that can cause severe burns. Do not look directly at the flame. Wear protective eyewear and use tongs to handle the crucible.

M(O) = 16.0 g/mol, M(Mg) = 24.3 g/mol

1. Weigh a crucible and lid on a balance. Record the mass: ______
2. Coil a 10 cm length of magnesium ribbon and place it in the c... Replace the lid and reweigh. Record the mass of the crucible and magnesium ribbon here:

... nd put the crucible, mag... ible with a blue Bunsen ... r into the crucible.

... this may take a minute o... lid back on the crucible.

5. ... everal minutes, using tongs to lift the crucible lid slightly once or twice to allow air into the crucible.
6. After several minutes, check to see if the reaction is complete. You will be able to tell as a white powder (magnesium oxide) will form and no flame or "smoke" will be seen.
7. Leaving the crucible lid on, turn the Bunsen off and allow the crucible to cool. Check to make sure all the magnesium has reacted. If there is still some metal in the crucible, you will need to continue heating.
8. When the crucible is cool, reweigh the crucible, magnesium oxide, and lid. Record the mass here:

1. (a) Calculat...
   (b) Calculat...
   (c) Calculat...
   (d) Calculate ... of magnesium used in the reaction: ______
   (e) Calculate the moles (n(O)) of oxygen in the reaction: ______
   (f) What is the ratio of n(Mg) to n(O) (in whole numbers)? ______
   (g) What is the formula for magnesium oxide? ______
   (h) Write a balanced equation for the reaction of magnesium metal (Mg) with oxygen gas (O_2).

©2020 BIOZONE International
ISBN: 978-1-927309-79-7
Photocopying Prohibited

EM P PS1.B

©2020 **BIOZONE** International
Photocopying Prohibited

Using the Tab System

The tab system is a useful way to quickly identify the crosscutting concepts, and science and engineering practices embedded within each activity. The tabs also indicate whether or not the activity is supported online.

The **orange** disciplinary core idea (DCI) tabs indicate the core ideas that are covered in the activity. These are covered in the introduction to each chapter, under the guiding questions. The code itself is just a guide for your teacher.

The gray hub tab indicates that the activity is supported online at the **BIOZONE RESOURCE HUB**. Online support may include videos, animations, articles, 3D models, and computer models.

The ETS icon indicates an engineering design DCI is included in the activity.

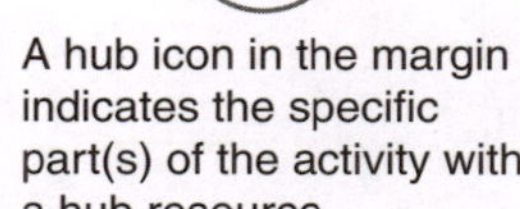

A hub icon in the margin indicates the specific part(s) of the activity with a hub resource.

An ETS icon in the margin indicates that a specific part of the activity includes some aspect of engineering design.

The **green** crosscutting concepts tabs indicate activities that share the same crosscutting concepts. You will become familiar with the concepts that connect all areas of science.

The **blue** science and engineering practices tabs use picture codes to identify the science and engineering practices (SEPs) relevant to the activity. You will use science and engineering practices in the course of completing the activities.

Science and Engineering Practices

Asking questions (for science) and defining problems (for engineering)
Asking scientific questions about observations or content in texts helps to define problems and draw valid conclusions.

Developing and using models
Models can be used to represent a system or a part of a system. Using models can help to visualize a structure, process, or design and understand how it works. Models can also be used to improve a design.

Planning and carrying out investigations
Planning and carrying out investigations is an important part of independent research. Investigations allow ideas and models to be tested and refined.

Analyzing and interpreting data
Once data is collected it must be analyzed to reveal any patterns or relationships. Tables and graphs are just two of the many ways to display and analyze data for trends.

Using mathematics and computational thinking
Mathematics is a tool for understanding scientific data. Converting or transforming data helps to see relationships more easily while statistical analysis can help determine the significance of the results.

Constructing explanations (for science) and designing solutions (for engineering)
Constructing explanations for observations and phenomena is a dynamic process and may involve drawing on existing knowledge as well as generating new ideas.

Engaging in argument from evidence
Scientific argument based on evidence is how new ideas gain acceptance in science. Logical reasoning based on evidence is required when considering the merit of new claims or explanations of phenomena.

Obtaining, evaluating, and communicating information
Evaluating information for scientific accuracy or bias is important in determining its validity and reliability. Communicating information includes reports, graphics, oral presentation, and models.

Cross Cutting Concepts

Patterns
We see patterns everywhere in science. These guide how we organize and classify events and organisms and prompt us to ask questions about the factors that create and influence them.

Cause and effect
A major part of science is investigating and explaining causal relationships. The mechanisms by which they occur can be tested in one context and used to explain and predict events in new contexts.

Scale, proportion, and quantity
Different things are relevant at different scales. Changes in scale, proportion, or quantity affect the structure or performance of a system.

Systems and system models
Making a model of a system (e.g. physical, mathematical) provides a way to understand and test ideas.

Energy and matter
Energy flows and matter cycles. Tracking these fluxes helps us understand how systems function.

Structure and function
The structure of a substance or object determines many of its properties and functions.

Stability and change
Science often deals with constructing explanations of how things change or how they remain stable.

©2020 **BIOZONE** International
Photocopying Prohibited

Using BIOZONE's Resource Hub

- BIOZONE's Resource Hub provides links to online content that supports the activities in the book. From this page, you can also check for any errata or clarifications to the book since printing.
- The external websites are, for the most part, narrowly focused animations and video clips directly relevant to some aspect of the activity on which they are cited. They provide great support to help your understanding.

Then enter the code in the text field **NPS1-9797**

Search for an activity here.

BIOZONE

Search activity number, title, keyword...

Physical Sciences for NGSS

BIOZONE's Resource Hub provides links to online content that supports the activities in the book. From this page, you can also check for any errata or clarifications to the book or model answers since printing.

The external websites are, for the most part, narrowly focused animations and video clips directly relevant to some aspect of the activity on which they are cited. They provide great support to help your understanding.

Chapter title based on the Disciplinary Core Ideas.

Chapter 1 - Skills for Physical Sciences

- 1 The Nature of Science
- 2 Systems and System Models
- 3 Observations, Hypotheses, and Assumptions
- 4 Measurement and Units
- 5 Useful Concepts in Physical Science
- 11 Ratio and Proportion in Chemistry
- 12 Safety in the Lab
- 13 Standard Equipment in Chemistry
- 14 Investigations in Chemistry
- 15 Measurement and Quantitative Analysis
- 16 The Periodic Table

Click on an activity title to go directly to the resources available for that activity.

View resources →

©2020 **BIOZONE** International
Photocopying Prohibited

Search for an activity here.

Activity

BIOZONE

Search activity number, title, keyword...

Physical Sciences for NGSS

Activity you are viewing

Physical Sciences for NGSS / Chapter 5 / Forces and Motion

69 Newton's Second Law

Weight is not your mass

Mass vs. weight and the normal force

Force, mass and acceleration

Force, mass and acceleration

Force, mass, acceleration and Newton's Second Law

STEMonstrations: Newton's Second Law of Motion

Simulation: Forces and motion basics

How to calculate force

Resources available for this activity. Hyperlink to an external website. Resources are tagged if they are teacher-only or specifically for gifted students.

Free body diagrams

Khan Academy: Free body diagrams

Terminal velocity

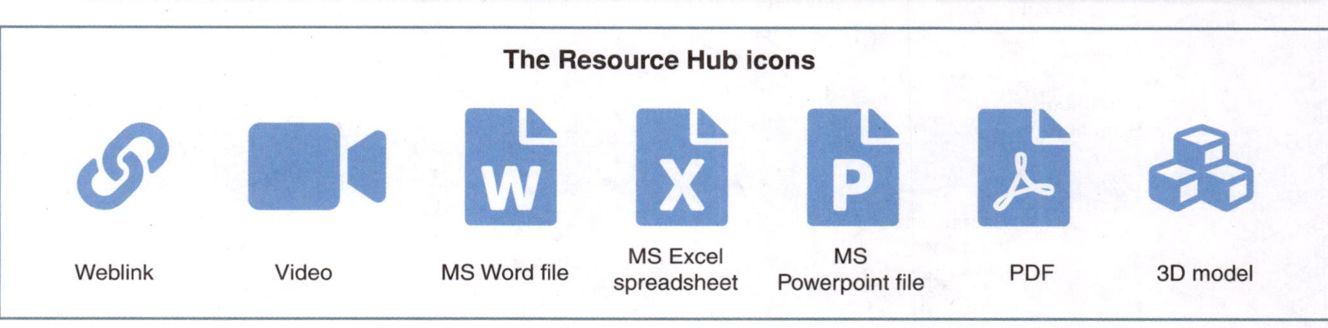

Explore videos

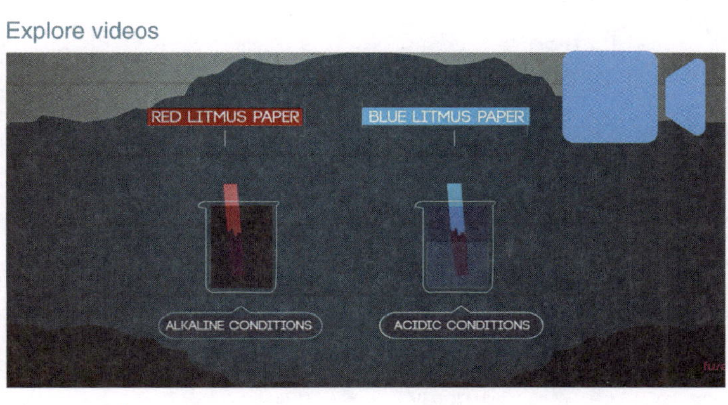

Explore spreadsheet modeling

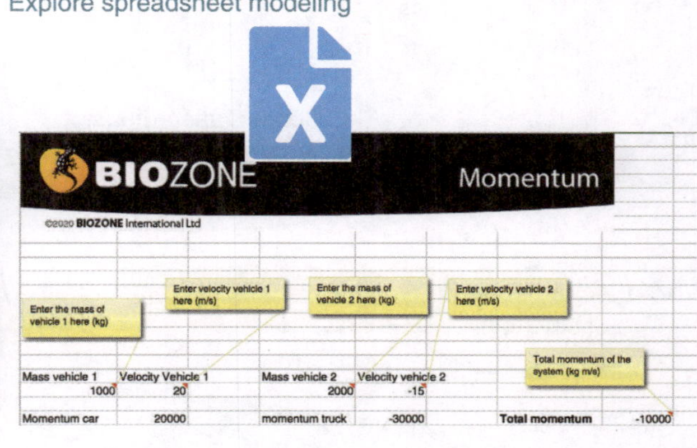

Explore web based resources

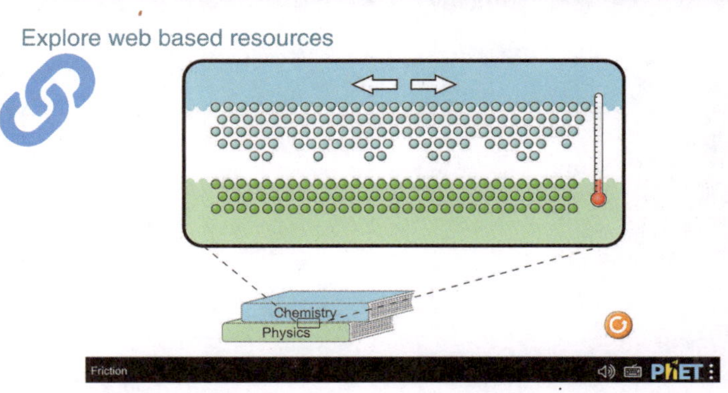

Explore 3D models

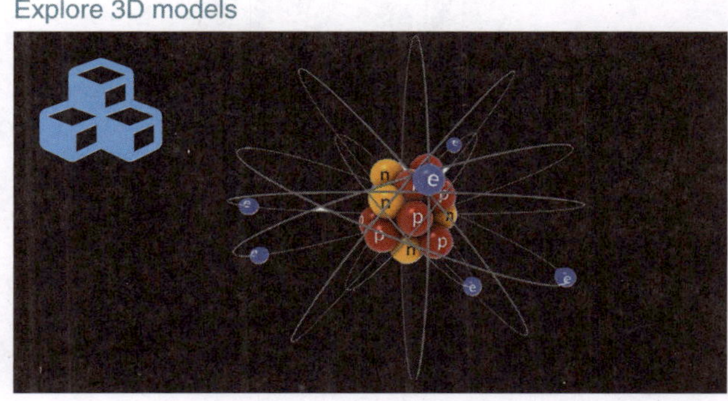

©2020 **BIOZONE** International
Photocopying Prohibited

Physical Sciences: A Flow of Ideas

This concept map shows the broad areas of content covered within each performance expectation of **Physical Sciences for NGSS.** The dark blue boxes indicate the book sections, each of which has its own concept map. The blue ovals are the chapters in each section. We have placed some major connections between topics, but you can make more of your own.

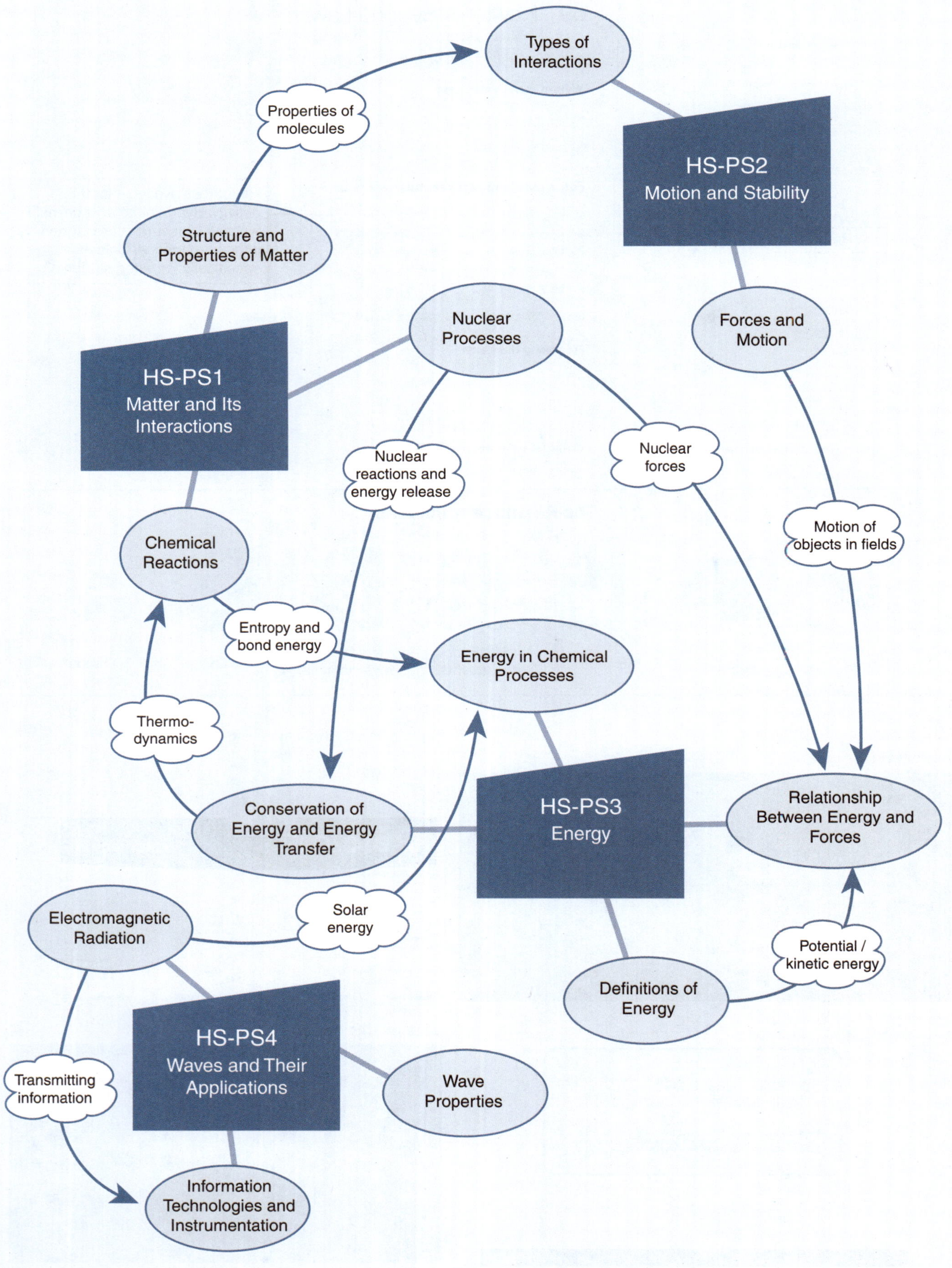

SEP support

1: Science Practices

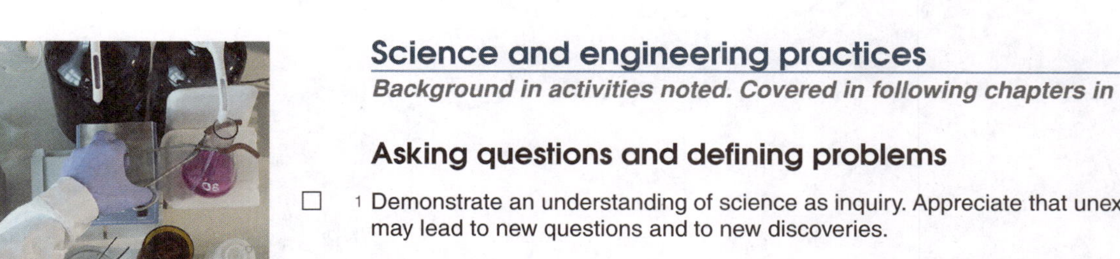

Activity number

Science and engineering practices

Background in activities noted. Covered in following chapters in context.

Asking questions and defining problems

- ☐ 1 Demonstrate an understanding of science as inquiry. Appreciate that unexpected results may lead to new questions and to new discoveries. — **1**
- ☐ 2 Ask and evaluate questions that you are able to investigate with the resources you have available. Ask questions that arise from observation or examining models or theories, or to find out more information, determine relationships, or refine a model. — **3 5**

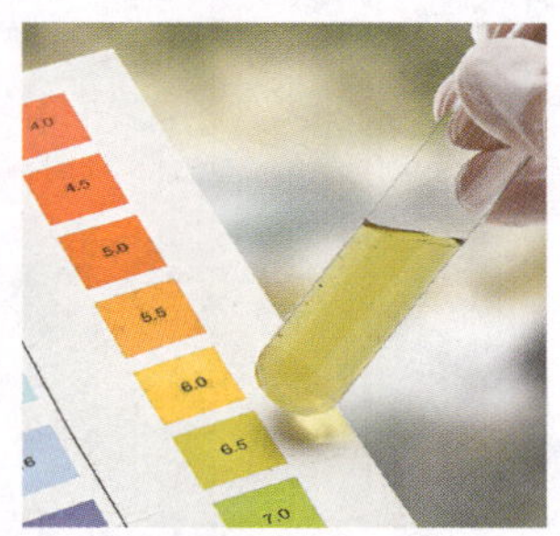

Developing and using models

- ☐ 3 Develop and use models to describe systems and how they work, to explain or make predictions about phenomena, or to generate data to support explanations or solve problems. — **2**
- ☐ 4 Develop a model that allows you to manipulate and test a system or process. — **2**

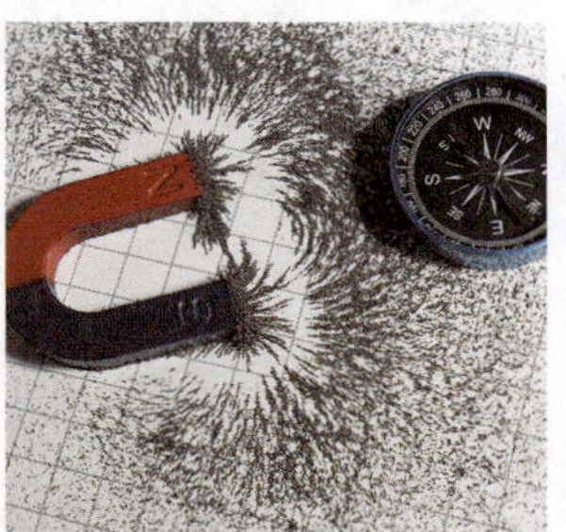

Planning and carrying out investigations

- ☐ 5 Plan and carry out investigations to provide data to test a hypothesis, support an explanation, or test a solution to a problem. Identify and evaluate the importance of any assumptions in the design of your investigation. — **3 10 12 14**
- ☐ 6 Use appropriate tools to collect, record, analyze, and evaluate data. — **4 13 15**
- ☐ 7 Make and test hypotheses about the effect on a dependent variable when an independent variable is manipulated. Understand and use blanks, controls, and trial runs appropriately. — **3**
- ☐ 8 Consider and evaluate the accuracy and precision of the data that you collect. — **6**

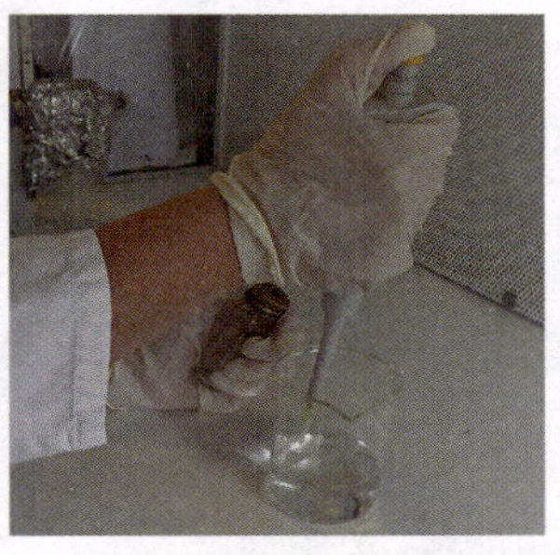

Analyzing and interpreting data

- ☐ 9 Analyze data in order to make valid and reliable scientific claims. Consider limitations of data analysis (e.g. measurement error, bias) when analyzing and interpreting data. — **7-9 15**
- ☐ 10 Apply concepts of statistics and probability to answer questions and solve problems. — **9**

Use mathematics and computational thinking

- ☐ 11 Demonstrate an ability to use mathematics and computational tools to analyze, represent, and model data. Recognize and use appropriate units in calculations and demonstrate an ability to apply unit conversions. — **7**
- ☐ 12 Create and use simple computational simulations based on mathematical models. Understand the use of logarithms, e.g. as in pH scale. — **2 7**
- ☐ 13 Demonstrate an ability to apply ratios and proportional reasoning when mixing solutions with specific molarities, making dilutions, and analyzing molecular structure. Calculate rates and use algebra and graphs to analyze reaction rates (*chemistry*). Apply ratios, percentages, and unit conversions in the case of complex measurement problems involving quantities with derived or compound units, e.g. kg/m^3 (*physics*). — **7 8 11**

Construct explanations and design solutions

- ☐ 14 Apply scientific evidence, ideas, and principles to explain phenomena and solve problems. — **1**

Engage in argument from evidence

- ☐ 15 Use evidence to defend and evaluate claims and explanations about science. — **1**
- ☐ 16 Provide and/or receive critiques on scientific arguments by using scientific methodology. — **1**

Obtain, evaluate, and communicate information

- ☐ 17 Evaluate the validity and reliability of designs, methods, claims, and/or evidence. Communicate scientific and/or technical information in multiple formats. — **5 6 8**
- ☐ 18 Demonstrate an ability to read critically and compare, integrate, and evaluate sources of information in different media and formats. — **1**

Download this page from BIOZONE's Resource Hub

1. Science Practices

Supporting skills in science and engineering practices

This introductory chapter has a dual purpose for students. It covers the basic computational, analytical, and practical skills they will need in physical sciences and so provides valuable information for students lacking sufficient background or competency in some areas. It also provides a reference for them throughout their program of work. If students have questions, they can refer to this chapter at any time. A red "Need Help" figure appears in the text to direct them to specific help.

Ask questions and define problems

1. Students should be able to understand science as a discipline involving rigor but also creativity and serendipity. They should understand the role of observation; it is through observation that scientists collect data to test hypotheses. Students should also understand that observation often involves the use of tools (e.g. taking measurements). They should apply a critical, open minded approach to their own investigations and recognize that unexpected results during their own investigations do not represent failure, but an opportunity to ask new questions.

2. Hypotheses put forward to explain observations about a system should be based on sound prior knowledge. For students, this usually involves drawing on their knowledge from earlier grades, provided information, and background reading. Students should formulate questions and hypotheses that are appropriate to the system they are investigating and that they can feasibly investigate with the time and resources available. While recognizing its limitations, a sound approach at this level could be: 1) to ask a question about a phenomenon, 2) to make observations about that phenomenon, 3) to construct a hypothesis to explain the phenomenon, 4) to test, 5) to collect and interpret data, 6) to draw conclusions and communicate findings, 7) to discuss (peer review).

Develop and use models

3. The model is central to science. A model is any representation, simplification, or substitute for what you are actually studying or trying to predict. A good model must be able to explain as many characteristics of the observed system as possible, but also be as simple as possible. Students should recognize the range of things that act as models of a system or its components, e.g. physical models, abstracts, analogies, drawings, and simulations. Students should be aware of the limitations of models and understand that models, like the laws and explanations they represent, are constantly being revised in science.

4. Throughout *Physical Sciences for NGSS* students will have opportunities to develop models that will describe or explain a system or its components.

Plan and carry out investigations

5. Students should plan investigations with consideration to making a fair test where possible or accounting for factors that are beyond their ability to control. They should be aware of assumptions they are making about the system and evaluate how reasonable these are.

6. Students should distinguish quantitative data (numbers) from qualitative data (descriptions), and recognize that some data are categorical by nature (e.g. color). They should be able to explain why it is desirable to collect quantitative data and can think about how they could convert qualitative to quantitative data. For example, qualitative data, such as color, can instead be measured as a wavelength (quantitative).

7. Students should distinguish between and explain the range and the purpose of independent, dependent, and controlled variables in a controlled experiment. Students should understand that the purpose of the experimental control is to establish the effect of one variable of interest on a system. Thus the control (or control groups) is not exposed to the changes in the independent variable of interest.

8. Students should understand that accuracy refers to how close a value is to the true value. With widespread use of dataloggers and measuring instrumentation, accuracy is often a feature of instrument calibration. Precision is an indication of how close measurements are to each other. To improve precision, students should always assign the same person to make measurements in an experiment.

Analyze and interpret data

9. Students should be able to present data (e.g. in a table) in a way that makes trends and patterns evident to the reader. They should be able to explain any apparent trends in graphs.

10. Students should understand how statistics and probability can be used to evaluate the reliability of findings and so increase confidence in statements of cause and effect.

Use mathematics & computational thinking

11. Students should be able to recognize and use appropriate units in calculations and convert between units and between decimal and standard form. They should be comfortable using computational tools such as spreadsheets to analyze, represent and model data. The Teacher's Digital Edition provides sample spreadsheets that can be used for this purpose.

12. Students should be able to construct simple mathematical models (e.g. equations) to simulate and predict outcomes of various events (e.g. effect of pH or change of speed).

13. Simple data manipulations, including percentages, ratios, rates, frequencies, means, and logarithms are a way to summarize data and enable samples to be meaningfully compared. Students should be comfortable with these basic calculations.

Construct explanations & design solutions

14. Many of the activities in the book require students to explain results based on real evidence presented and using their understanding of basic principles previously covered.

Engage in argument from evidence

15. Many of the activities present real world second hand data to support the concepts presented. Students are often asked to evaluate explanations based on the evidence and are encouraged to propose explanations for any inconsistencies in the data (as real world data is seldom a perfect story).

16. Students will construct their own methods to carry out investigations. They will receive feedback from teachers and use this information to improve their method.

Obtain, evaluate, & communicate information

17. Throughout the book, students are encouraged to evaluate the validity and reliability of designs, methods, claims, and evidence. These evaluations are integral to most of the data handling and interpretation activities presented.

18. Students will need to research information to complete some tasks, They will need to critically evaluate the information they use to determine its appropriateness.

1 The Nature of Science

Key Question: What is the "scientific method' and how does it help us identify, explore, and understand natural phenomena?

- Science is a way of understanding the universe we live in: where it came from, the rules it obeys, and how it changes over time. Science distinguishes itself from other ways of understanding the universe by using empirical standards, logical arguments, and skeptical review. What we understand about the universe changes over time as more information is gathered.
- Science is a human endeavor and requires creativity and imagination. New research and ways of thinking can be based on the well argued idea of a single person. It could be said that the scientific method is '*the art of embracing failure*', because apparent failures can lead to new discoveries and ways of thinking.
- Science influences and is influenced by society and technology, both of which are constantly changing. Scientists build on the ideas and work of their contemporaries and those that went before them. **"If I have seen further it is by standing on the shoulders of Giants"**.....*Isaac Newton*
- Science can never answer questions about the universe with absolute certainty. It can be confident of certain outcomes, but only within the limits of the data. Science might help us predict with 99.9% certainty a system will behave a certain way, but that still means there's one chance in a thousand it won't. For example Newton's Law of Universal Gravitation is limited to macroscopic objects up to the mass of large stars. For very small or distant objects (e.g. protons) it provides answers which are unverifiable. For very massive objects (e.g. black holes) gravity is better described by general relativity.

1. Science is not a linear process, it is dynamic and progressive. Results may answer some questions but it may also raise new questions that require investigation. New discoveries can be made by accident or because unexpected results occur. Nor is science an isolated process. Throughout history, the work of many has been has been important to explaining or developing ideas. Collaborators bring new findings, ways of thinking, and new directions to research.

   Using the circles below, construct a model or mind map to show how the nature of science is dynamic and progressive.

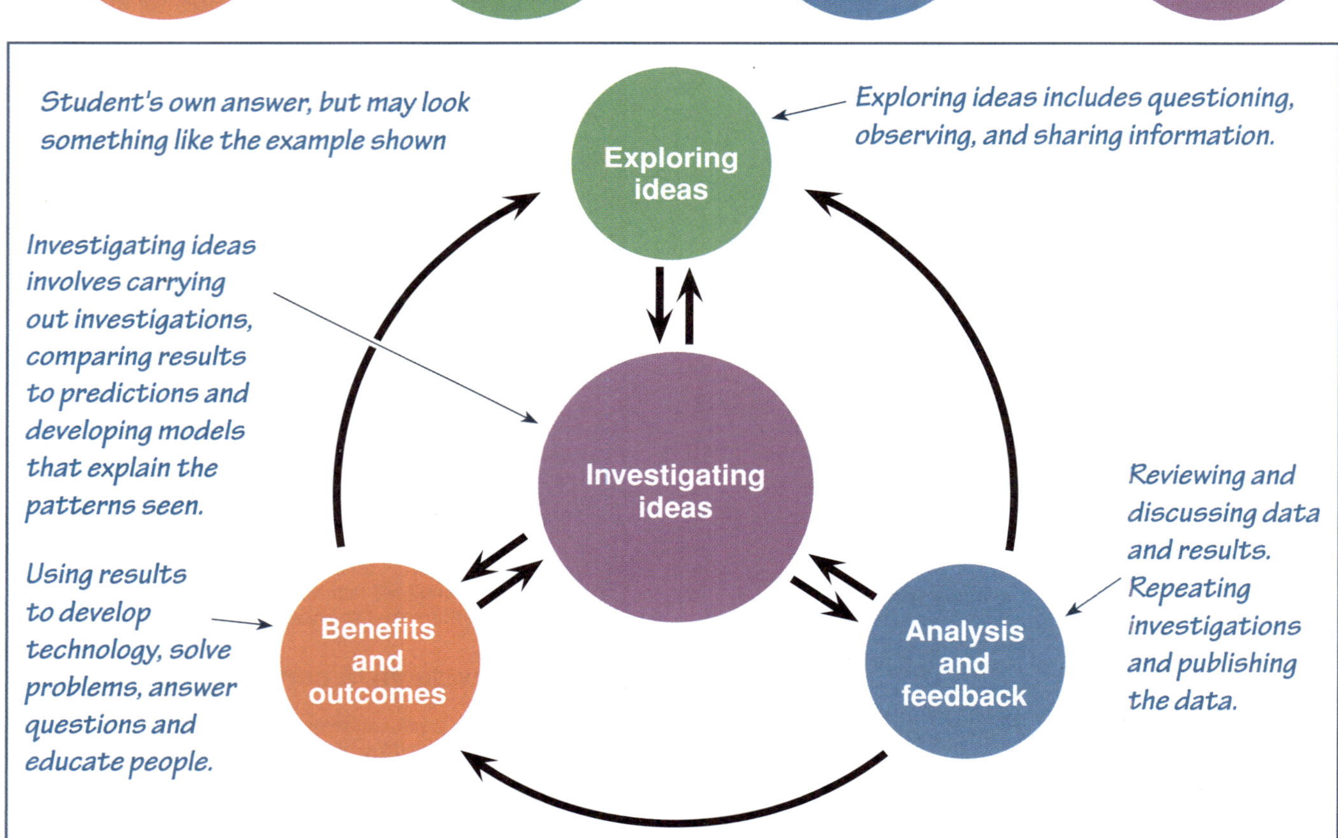

SSM

©2020 **BIOZONE** International
ISBN: 978-1-927309-82-7
Photocopying Prohibited

The nature of science: a brief overview of atomic theory

What is our world made of?

The ideas of the Greek philosophers were proposed based on observation and thought. No experimentation was carried out.

Empedocles (450 BC) proposed all matter was made up of four elements (earth, air, fire, and water) and two forces (love and strife), could mix and separate the elements.

Around 400 BC **Democritus** proposed that matter was made of particles and could be cut up until the particles were so small they could no longer be divided. He named the particles "atomos".

Fire Water Air Earth

Empedocles and others

Philosophers at the time thought that problems could be solved simply by thinking about them.

Religion heavily influenced theories and many ideas that were at odds with social beliefs were not accepted nor discussed openly.

Plato (340 BC) disagreed with Democritus about the presence of atomos. Plato proposed that the four elements were regular solids composed of triangles. The triangles could combine to form different materials and states. He did not think they could be divided into smaller particles. This theory was widely accepted for a long time.

Aristotle (300 BC) also disagreed with Democritus and believed everything was a combination of the four elements. He believed the gods could divide an element into something smaller if they wanted to. Most people agreed with his ideas until the 17th century.

Dalton

A practice of experimentation and scientific observation began to arise and become common.

Gathering evidence allowed new ideas to be formed and redefined as new knowledge came to light.

Robert Boyle (1600s) wrote the "Skeptical Chemist" urging scientists to abandon the view that elements were mystical substances. He argued Earth could not be an element as it contained other substances (e.g. gold and iron). He helped to pioneer the scientific method.

In the 1700s chemists **Joseph Priestly** and **Antoine Lavoisier** began to explain chemical behavior in terms of the atom. They showed that substances could combine to form new materials.

John Dalton (1803) developed an atomic theory covering five main principles. It was the first attempt to describe all matter in terms of atoms and their properties. His work was generally accepted because it provided logical answers to many questions.

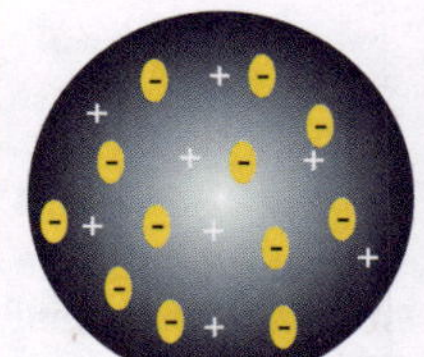

Thomson: Plum pudding model

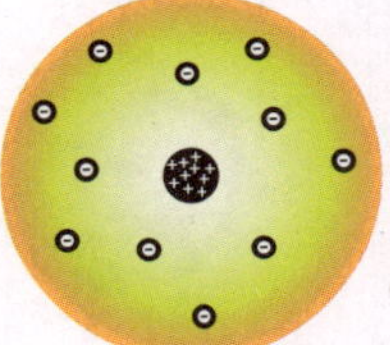

Rutherford: Planetary model

The scientific evidence for atomic theory became widely accepted.

The atomic theory was refined and became more sophisticated as technology improved and new evidence was gathered.

Through the 1800s and 1900s, many scientists made discoveries that helped to refine Dalton's atomic theory. For example, Dalton proposed that atoms could not be divided into smaller parts. However, subsequent research showed that atoms are made up of smaller components. In 1904 **J.J. Thomson** discovered atoms contained electrons and proposed the "plum pudding model". A few years later, **Ernest Rutherford** carried out a gold foil experiment and described the presence of a nucleus. He developed the planetary model. Further refinements, including the position and movement of electrons (**Niels Bohr**) and the discovery of neutrons (**James Chadwick**) contributed to the model of the atom accepted today. Many of these discoveries were aided by improvements in technology allowing researchers to build on the knowledge of other scientists.

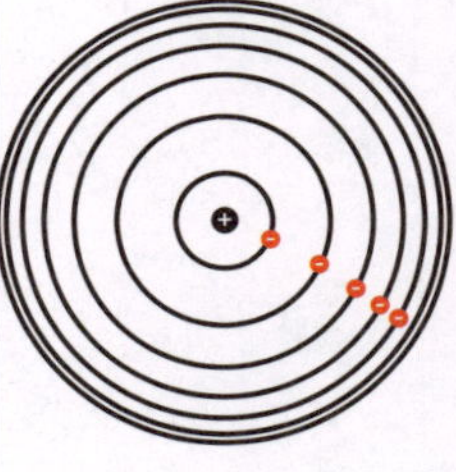

Bohr

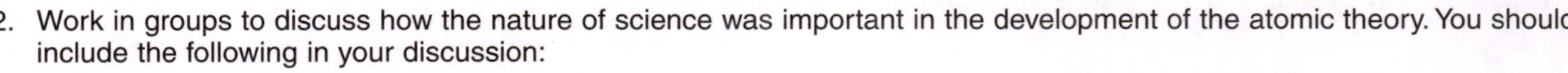

2. Work in groups to discuss how the nature of science was important in the development of the atomic theory. You should include the following in your discussion:

- How was the theory's progress affected by the beliefs of the day (e.g. societal and religious ideas)?
- How did the switch from thinking about a problem (philosophizing) to experimental based evidence affect progress?
- What role did technological advancement play?
- Why was the work of earlier scientists important to those who came afterwards? *See long answer section*

©2020 **BIOZONE** International
ISBN: 978-1-927309-82-7
Photocopying Prohibited

2 Systems and System Models

Key Question: Why do we model systems, and how do we do it?

- **Systems** are assemblages of components working or moving together in some related way. Energy put into the system will be changed in some way to produce an output. An example of the system could be the simple electrical circuit shown on the right. Energy from a battery is transferred through wires to the light bulb where it is converted to light and heat energy.
- Systems may be open (able to exchange matter, energy and information with their surroundings), closed (exchange energy with their surroundings, but not matter) or isolated. Isolated systems exchange no energy, information or matter with their surroundings. No such systems are known to exist (except possibly the entire universe).
- Scientists often used models to study how a system works. A **model** is a representation of a system and is useful for breaking a complex system down into smaller parts that can be studied more easily. Often only part of a system is modeled. As scientists gather more information about a system, more data can be put into the model so that eventually it represents the real system more closely.

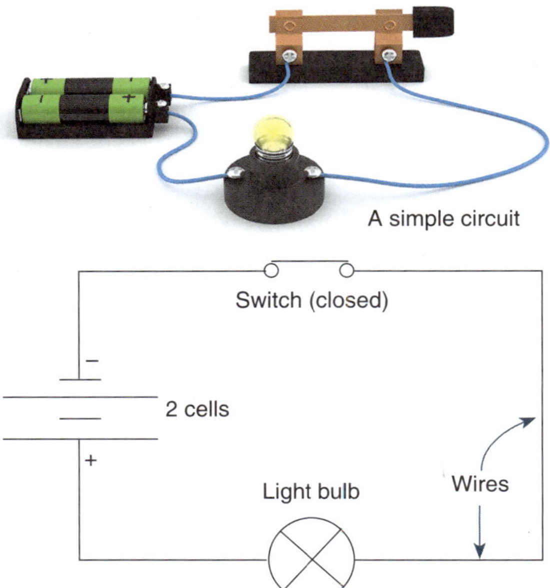

A graphical model of the circuit showing the four labeled components of the system.

Types of models

Visual models can include drawings, such as these plant cells on the right. Three dimensional models can be made out of materials such as modeling clay and ice-cream sticks, like this model of a water molecule (below) or they could be simple drawings.

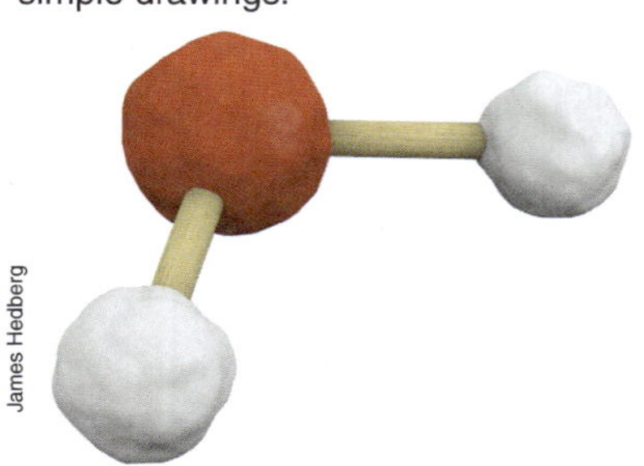

Displaying data in a **graph** or as a **mathematical equation**, as shown below for the distance time graph of a moving object, often helps us to see relationships within a system.

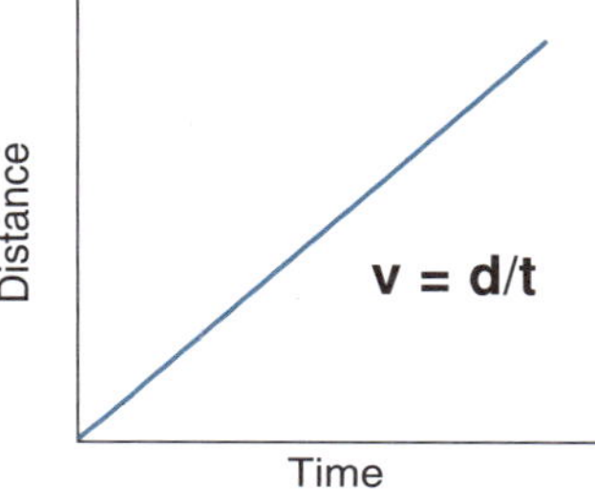

An **analogy** is a comparison between two things. Sometimes comparing physical systems can help us to understand them better. For example, electricity flowing through wires in a circuit is sometimes compared to water flowing through pipes (below).

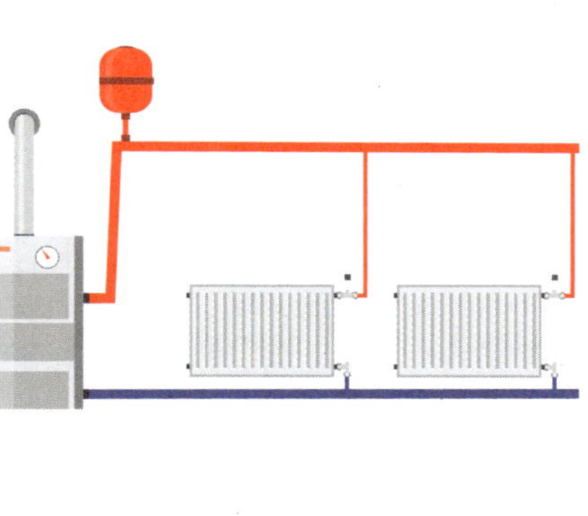

1. Do you think the simple circuit shown at the top of the page is an open, closed, or isolated system. Give a reason for your answer:

   It is a closed system as it is exchanging only energy (as light waves) with the environment.

2. Why do scientists often model just a single part of a system at a time rather than the whole system at once?

   Breaking a model down into single parts allows us to understand the role of each component within the whole system.

©2020 **BIOZONE** International
ISBN: 978-1-927309-82-7
Photocopying Prohibited

3 Observations, Hypotheses, and Assumptions

Key Question: What is the importance of making observations, producing hypotheses, and recognizing assumptions?

Observations and hypotheses

- An observation is watching or recording what is happening. Observation is the basis for forming hypotheses and making predictions. An observation may generate a number of hypotheses (tentative explanations for what we see). Each hypothesis will lead to one or more predictions, which can be tested by investigation.
- A hypothesis is often written as a statement to include the prediction: "If X is true, then if I do Y (the experiment), I expect Z (the prediction)". Hypotheses are accepted, changed, or rejected on the basis of investigations. A hypothesis should have a sound theoretical basis and should be testable.

Observation 1: Objects with the same mass fall at the same rate.

Observation 2: Two sheets paper, flat and crumpled, have the same mass.

Assumptions

- Any investigation requires you to make assumptions about the system you are working with. Assumptions are features of the system you are studying that you assume to be true but that you do not (or cannot) test.
- Assumptions are made all the time, but we should be careful of what assumptions are being made when we carry out scientific investigations.

1. Based on only the two observations above, write a hypothesis or a prediction about how the flat and crumpled paper will behave when falling:

   Based on the two observations: Crumpled paper will fall at the same rate as flattened paper.

2. What assumptions are being made in your hypothesis?

   - The shape of the paper does not affect how the paper falls.

   - Air resistance will have no effect.

3. If we were to drop the flat piece of paper and the crumpled piece of paper from the same height in a closed room, what result would you actually expect? What does this tell us about assumptions?

   From experience in the real world ,we would expect that the crumpled piece of paper would fall faster and hit the ground first. We must be careful to take assumptions into account and know when they apply. The hypothesis is only true if the two assumptions are correct. This only applies in a vacuum.

4. What would you need to do to prove your hypothesis correct (without fiddling with the data!).

   - The test would need to take place in a vacuum so air resistance does not affect the result.

©2020 **BIOZONE** International
ISBN: 978-1-927309-82-7
Photocopying Prohibited

4 Measurement and Units

Key Question: What is the difference between base units and derived units? How do units help us standardize our measurements?

Measurement and units

- You will take many measurements during your study of physical science. One of the most important things to remember is to always record and report the units of measurement. Without units, the measurements are meaningless (below).

16

- A student measured the length of wood (above) and recorded the length as 16. 16 what? Without the units we know nothing about the physical quantity. We have no idea of the missing unit. As soon as you add the unit (e.g. 16 cm) we can immediately quantify the measurement. A few measures (e.g. temperature in Kelvin) have no units because they are already calibrated against another scale (°C).

DID YOU KNOW?

Non-metric units can be confusing! A US ton is 2000 lb and different to the British ton (2240 lb). A metric ton or tonne (1000 kg) is ~2204 lb. With a metric ton, there is no confusion.

SI units

Different units are used to measure quantities in different countries. For example, in the US, miles are commonly used to measure distance but most other countries use kilometers. To standardize measurement, scientists use SI units (International System of Units) to remove these differences.

Base units

Base units are the building blocks of the SI system. These are units that can be used on their own. For example: Kilogram (kg) is a base unit because it is independently expressed (other units are not required for it to make sense).
The seven base units are:

Quantity name	Unit name	Unit symbol
length	meter	m
mass	kilogram	kg
time	second	s
electric current	ampere	A
thermodynamic temperature	kelvin	K
amount of substance	mole	mol
luminous intensity	candela	cd

Derived units

Derived units are expressed using combinations of base units. They are formed by powers, products or quotients of the base units and cannot be expressed in the absence of basic units. In SI units, inverse notation is used as a rule, although we have replaced it here with the solidus (/).
Some commonly used derived units in physics are:

Quantity name	Unit name	Expression	Unit symbol
speed, velocity	meter per second	m/s	*v*
acceleration	meter per second squared	m/s^2	*a*
force, weight	newton	$kg.m/s^2$	N
pressure, stress	pascal	N/m^2	Pa
energy, work, heat	joule	Nm	J
power	watt	J/s	W
electrical charge or quantity of electricity	coulomb	A.s	C
frequency	hertz	s^{-1} (per second)	Hz

1. Explain why using standardized units is important in science:

Standardized units mean that a measurement has the same value anywhere in the world. For example, using a metric ton removes any confusion about the measurement. This means that results are standardized and can easily be compared.

©2020 BIOZONE International
ISBN: 978-1-927309-82-7
Photocopying Prohibited

Distance

The units used to measure distance depend greatly on the distance being measured. Distances in the classroom may be measured in meters. In the lab distance is commonly measured in meters or centimeters. Outside over longer distances it might be measured in kilometers.

Tape measures and rulers are usually accurate to 1 mm.

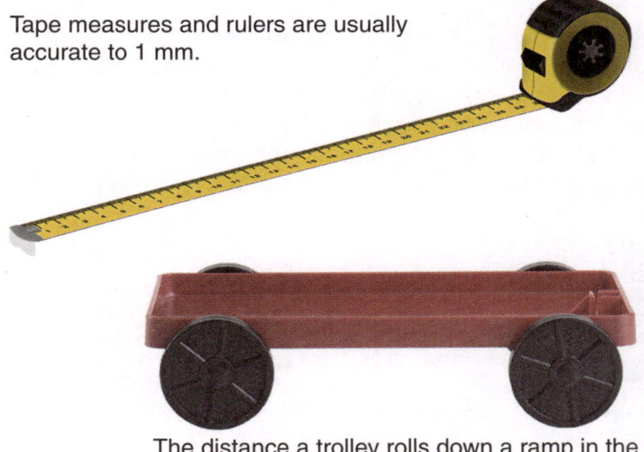

The distance a trolley rolls down a ramp in the physics lab might be measured in meters to two decimal places.

The distance a car travels is commonly measured in kilometers and their speed in km/h. Neither of these are SI units. Long distances are normally written in meters in standard form (see Activity 7) and the speed of a car is usually recalculated to m/s.

Force

A force is a push or a pull, and the push or pull needs to be exerted on an object. The unit of force is called a newton and is represented by the symbol N.

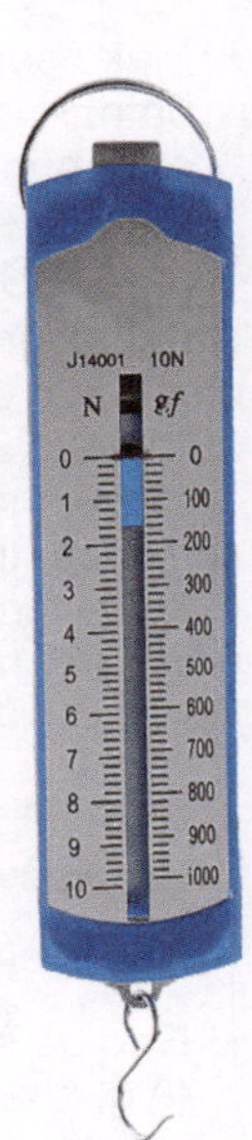

Force can be measured by force meters (sometimes called force gauges). They can be very simple and consist of a spring in a marked cylinder (right) or they can be more complicated digital devices. As force is applied, the spring is stretched and the force is read off the scale. A spring-based force meter is a reliable measure of force because the extension or compression of a spring is proportional to the applied force.

A force diagram can be drawn to show all the forces acting on an object, and indicate the force's direction and magnitude. It is usually shown with the object represented by a dot, and the vectors are labeled by the type of force, the object exerting the force, and the object receiving that force. In a force diagram, the longer the arrow, the bigger the force. Several types of force are typically used:

- Gravity: F_{grav}, direction downward.
- Applied: F_{app}, applied to an object.
- Support (normal): F_{norm}, perpendicular to surface.
- Tension: F_{tens}, along a string/rope/chain.
- Friction: F_{fric}, direction opposing relative motion.

2. Name the basic SI units in the following derived SI units:

   (a) speed, velocity (v): Meter and second

   (b) force, weight (N): Kilogram, meter and second

3. Measuring meters to two decimal places (2 d.p.) is the same as measuring in what other common measurement?

   Centimeters (e.g. 1.22 m = 122 cm).

4. Recalculate 100 km/h to m/s: 100 km = 100,000 m. 1 hour = 3600 s. 100,000 ÷ 3600 = 27.8 m/s

5. A trolley roles 156 cm. How many meters is this? 1.56 m.

6. A very simplified force diagram is shown below. Describe what it shows:

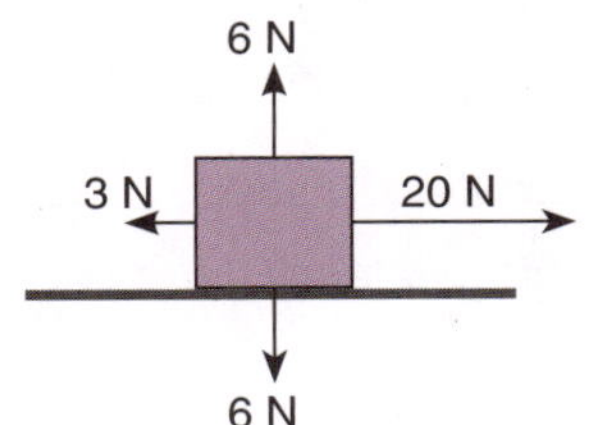

The 6 N forces opposite each other cancel each other out, there is no acceleration in the vertical direction. The 20 N force acting to the right is 17 N greater than the 3 N force acting on the left. The object is accelerating to the right.

©2020 BIOZONE International
ISBN: 978-1-927309-82-7
Photocopying Prohibited

Recording your results accurately is very important in any type of scientific investigation. If you have recorded your results accurately and in an organized way, it makes analyzing and understanding your data easier. Log books and dataloggers are two methods by which data can be recorded.

Log books

- A log book records your ideas and results throughout your scientific investigation. It also provides proof that you have carried out the work.
- A lined exercise book is a good choice for a log book. Write ideas, record results and paste in photos or extra material (such as printouts).
- Each entry must have the date recorded.
- Your log book is a full record of your work. Include any mishaps, failed experiments, or changes in methodology in your logbook. Where possible, explain the reasons for the failure or change. Sometimes failed experiments can be just as valuable as successful experiments in understanding a result.
- Make sure that you can read what you write at a later date. A log book entry is meaningless if it is incomplete or cannot be read.

Dataloggers

Graphtecinst CC4.0

Dataloggers (right) are electronic device that automatically records data over time. In physical sciences, they can be used to measure motion, pressure, heat, sound, light, or radioactivity among others. Some advantages of a datalogger are:

- Recordings have a high degree of precision and accuracy (which is known and factory-set).
- Can be left without needing to be monitored.
- Can be set to take readings over a long period of time (e.g. hourly readings every day) or many readings in a short period of time.
- Can be used when there is a safety risk involved (e.g. radiation exposure or extreme heat).
- Data collected can be downloaded to a computer so that the data can be accessed and analyzed.

Tables

A table (below) is a good way to record and present results. Patterns, trends, or anomalies can be easier to see. You may want to change your experimental conditions as a result of emerging trends. For example, increase the frequency of data collection if changes are occurring quickly.

09 / 21 / 2019

Saturation density of water vapor in the air

Temperature (°C)	Saturation density (kg/m³)
0	0.0049
5	0.0068
10	0.0094
15	0.00128
20	0.0173
25	0.0228
30	0.0304
35	0.0396

- Title, row and column headings must state clearly and accurately what the table is about.
- Tables allow you to systematically record and condense a large amount of information. They provide an accurate record of your data.
- Columns can be added for calculated values such as density, rate, and summary statistics (e.g. mean).
- Summary statistics make it easier to identify trends and compare treatments. Rates are useful in comparing multiple data sets, e.g. if recordings were made over different time periods.

7. Why is it important to keep a detailed logbook during a scientific investigation?

Logbooks are a record of an investigation. They are important in that they allow an experimenter to record results and observations for later analysis. Recording results is important so that information is not lost of forgotten. A logbook also provides a record of the work for scrutiny or reference by peers.

8. (a) Describe some advantages of using a datalogger over a person manually recording the data: *Dataloggers can measure more accurately and precisely than humans, they do not need constant monitoring, and can be placed in hazardous environments. They can be programmed to take many readings at set times and over different periods. Data can be downloaded for analysis.*

(b) What do you think might affect the accuracy of readings made with a datalogger? *Datalogger sensors need to be calibrated. If the calibration was poor the data collected may not be accurate.*

©2020 BIOZONE International
ISBN: 978-1-927309-82-7
Photocopying Prohibited

5 Useful Concepts in Physical Science

Key Question: What concepts are useful to know in physical sciences?

Temperature

- Temperature is a measure of the energy of an object. Molecules constantly vibrate (kinetic energy) and temperature is a measure of these vibrations. The faster and larger the vibrations the higher the object's temperature. Temperature rises as heat is added to a system and lowers when heat is removed.
- There are many temperature scales used throughout the world, but the most common are the Celsius, Fahrenheit, and Kelvin scales. The Kelvin scale is commonly used in science, although because it uses the same magnitude of scale as the Celsius scale, Celsius is also commonly used.

Scale	Basis	Freezing/boiling point of water
Celsius	Boiling and freezing points of pure water.	Freezing 0°C Boiling 100°C
Fahrenheit	Zeroed on a mixture of ice and salt. Original scale set freezing point of water at 30°F and body temperature at 90°F.	Freezing 32°F Boiling 212°F
Kelvin	Absolute zero; the point at which all vibrations stop. Same scale as Celsius. An absolute scale so no degree symbol is used.	Freezing 273 K Boiling 373 K

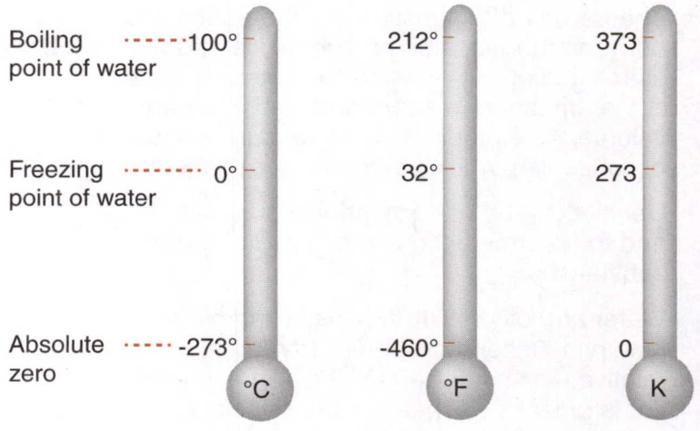

- Temperature can be inferred from color. A heated piece of iron will initially glow red, then orange-yellow as it gets hotter, then white.
- In chemistry, the color of a Bunsen flame gives an indication of temperature. A yellow (or safety) flame is the coolest at 300°C. With increasing oxygen supply, a hotter blue flame is produced, with a temperature of ~1500°C (right). A blue-colored flame indicates that there is no soot (uncombusted material) in the flame.

Jan Fjaldowski CC3.0

1. What is temperature measuring? *Temperature is measuring the amount of kinetic energy (vibrations of molecules) contained within an object. The more kinetic energy (greater vibrations), the higher the temperature.*

2. How can flame color be used to predict temperature? *Objects change color as they are heated, and emit a particular color depending on their temperature. The greater energy electrons have they higher the frequency of light they emit as they return to lower energy levels. High frequencies of light produce blue light, lower frequencies produce red. Thus energy (and so temperature) can be linked to color.*

3. For each of the temperature scales above, write down the temperature at which the following occurs:

   (a) The boiling point of water:

   (°C) *100°C* (°F) *212°F* (K) *373 K*

   (b) The freezing point of water:

   (°C) *0°C* (°F) *32°F* (K) *273 K*

©2020 BIOZONE International
ISBN: 978-1-927309-82-7
Photocopying Prohibited

Pressure

- Pressure is an important concept in the physical sciences. Pressure is a measure of the force being applied per unit surface area. The pressure of the air pressing on the surface of the Earth is 101.3 Pa (pascals).
- Pressure is one of the most measurable properties of a gas. The pressure (P), volume (V), number of moles, and temperature (T) of an ideal gas are related by a simple formula called the ideal gas law. This equation allows us to predict the behavior of a gas given certain conditions. For example, if you keep the volume constant and heat a gas, the pressure increases is a linear way. $P \propto T$ for a given mass and volume of gas. $\propto$ means proportional to.

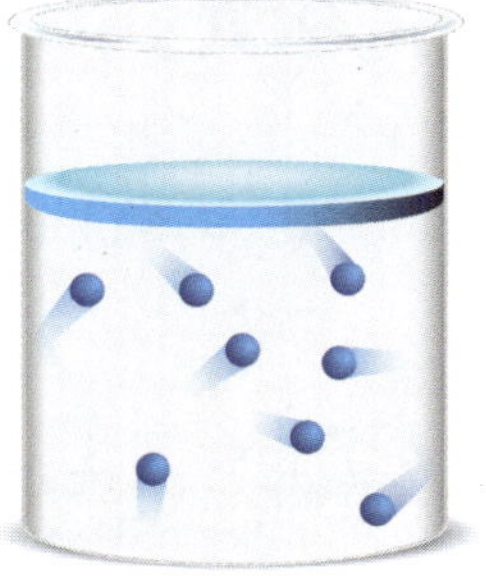

Low temperature - slow movement of gas particles

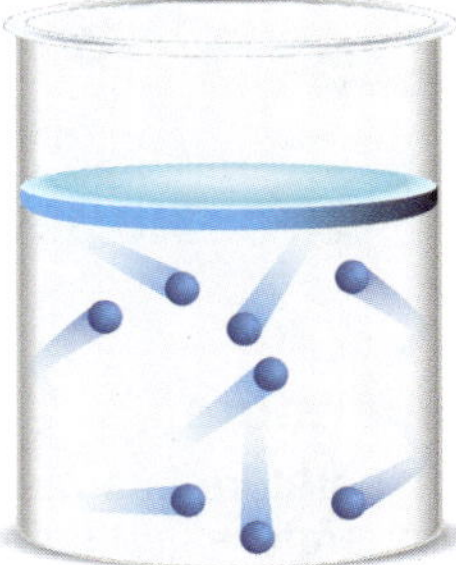

High temperature - fast movement of gas particles

Density

- The density of a substance is the relationship between the mass of the substance and how much space it takes up (volume). In other words, density is the amount of matter contained in a particular volume. An element such as lead is very dense because its particles are very close together.
- Density can be used to identify pure substances and to determine the composition of mixed substances.
- Water and oil provide an easy example of observing differences in density (right). An object's relative density to water will determine whether it floats or sinks. Oil has a lower density than water so it floats above the water. An object with a higher density will sink. In glass A to the right, the density of the water is greater than that of the cork, but less than that of the metal.

4. What is pressure? *Pressure is a measure of the force being applied per unit surface area.*

5. (a) Explain why the pressure in a sealed container of gas increases when heat is added to the system:
Pressure is proportional to temperature, so as temperature increases so does the pressure as the gas has nowhere to escape. Gas molecules collide with the container with more force as the temperature increases.

(b) Explain why the pressure in a sealed container of gas decreases when the system is cooled:
Pressure is proportional to temperature, so as temperature decreases so does the pressure within a system. The gas molecules collide with the container with less force as the temperature drops.

6. Fresh water has a density of 1000 kg/m^3. Will the following objects sink or float if placed in it?

(a) A cork (density = 240 kg/m^3): *Float*

(b) A piece of lead (density = 11,340 kg/m^3): *Sink*

(c) Air (density = 1.2 kg/m^3): *Float*

7. In glass B (above right) what can you say about the densities of the three substances present?
The fluid at the bottom is more dense than either of the other two fluids. The fluid on the top is less dense than either of the other fluids. The density of the middle fluid is the between the densities of the others.

©2020 **BIOZONE** International
ISBN: 978-1-927309-82-7
Photocopying Prohibited

Inverse square law

The inverse square law describes how the intensity of an effect varies with distance. Specifically, the intensity of an effect (e.g. light) changes in inverse proportion to the square of the distance from the source. To put it simply, the further the distance between two objects, the less intense the effect (see diagram below).

A number of physical properties reduce in magnitude as they become more distant in a way that can be represented by an inverse square law. These include:

- Gravity
- Electric field
- Light intensity
- Radiation
- Sound intensity

The inverse-square law can be written as:

Intensity is proportional to: $\frac{1}{\text{distance squared } (d^2)}$

If the intensity at one distance is known the intensity at a second distance can be calculated using the following equation:

$$\text{intensity}_1 \times \text{distance}_1^2 = \text{intensity}_2 \times \text{distance}_2^2$$

Where:

I_1 = Intensity 1 at D_1

I_2 = Intensity 2 at D_2

D_1 = Distance 1 from source

D_2 = Distance 2 from source

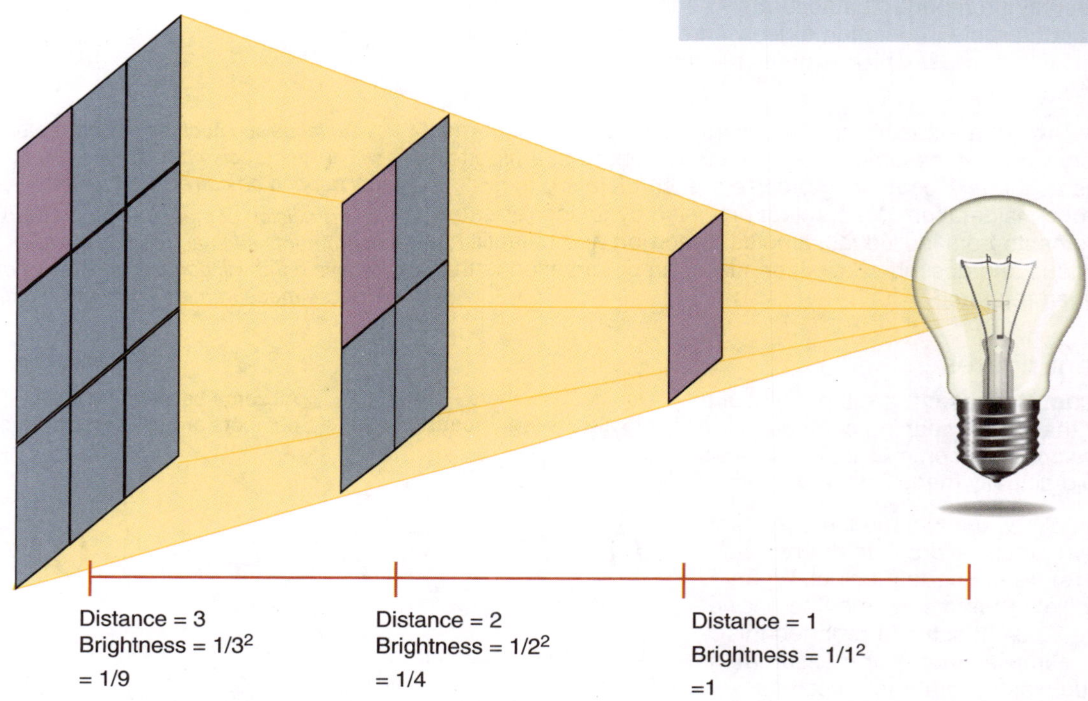

Distance = 3
Brightness = $1/3^2$
= 1/9

Distance = 2
Brightness = $1/2^2$
= 1/4

Distance = 1
Brightness = $1/1^2$
=1

8. If a flashlight has a light intensity of 15.0 candela (cd) at a distance of 1 m from the lens what is the intensity of the light at 100 m from the lens? Show your working in the space below.

$15\text{ cd} \times 1\text{ m}^2 = I_2 \times 100\text{ m}^2$

$I_2 = (15\text{ cd} \times 1\text{ m}^2) \div 100\text{ m}^2$

$I_2 = 0.0015\text{ cd}$

9. The intensity of an iridium 192 source was 62 milliroentgen/hour at 100 m. What is the intensity at 1 m? Show your working in the space below.

$62\text{ mR/hr} \times 100\text{ m}^2 = I_2 \times 1\text{ m}^2$

$I_2 = (62\text{ mR/hr} \times 100\text{ m}^2) \div 1\text{ m}^2$

$I_2 = 620{,}000\text{ mR/hr}$

©2020 BIOZONE International
ISBN: 978-1-927309-82-7
Photocopying Prohibited

6 Accuracy and Precision

Key Question: What do accuracy and precision mean, how are they different, and why are they important when taking measurements?

- **Accuracy** refers to how close a measured or derived value is to its true value. Simply put, it is the correctness of the measurement. The accuracy of a measurement can be increased by increasing the number of measurements taken. For example, the accuracy of determining how long a ball takes to roll down a ramp can be increased by increasing the number of times the experiment is carried out or the accuracy of determining the concentration of an analyte in a solution can be increased by increasing the number of titrations carried out.
- **Precision** refers to how close repeated measurements are to each other, i.e. the ability to be exact. A balance with a fault in it could give very precise (repeatable) but inaccurate (untrue) results. Data can only be reported as accurately as the measurement of the apparatus allows. It is often expressed as significant figures (the digits in a number which express meaning to a degree of accuracy).
- The precision of a measurement relates to its repeatability. In most laboratory work, we assume a piece of equipment (e.g. a pipette) performs accurately, so making precise measures is the most important consideration. We can test precision by taking repeated measurements from individual samples. Precision and reliability are synonymous and describe how dependably an observation is the same when repeated.

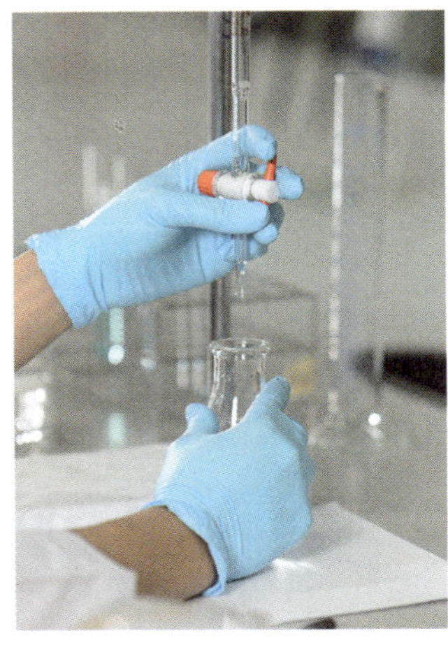

A digital device such as the pH meter (above left) will deliver precise measurements, but its accuracy will depend on correct calibration.

The precision of measurements taken with instruments such as reading the level in a burette (above right) will depend on the skill of the operator. Precise measurements give reliable data.

Significant figures

- **Significant figures** (sf) are the digits of a number that carry meaning contributing to its precision. They communicate how well you could actually measure the data.
- For example, you might measure the time of 20 toy parachute drops to the nearest second (s) from a certain height. When you calculate their mean time, the answer might be 25.3412 s. If you reported this number, it implies that your measurement technique was accurate to 4 decimal places. You would have to round the result to the number of significant figures you had accurately measured. In this instance the answer is 25 s.

Non-zero numbers (1-9) are always **significant**.

All zeros between non-zero numbers are always **significant**.

0.005704510

Zeros to the left of the first non-zero digit after a decimal point are **not significant**.

Zeros at the end of number where there is a decimal place are **significant** (e.g. 4600.0 has five sf).

BUT

Zeros at the end of a number where there is no decimal point are **not significant** (e.g. 4600 has two sf).

Visualizing accuracy and precision

The analogy of golfers trying to get their golf balls in the cup is a good one for explaining accuracy and precision. Imagine four golfers each hit five golf balls. The results from each golfer are shown below.

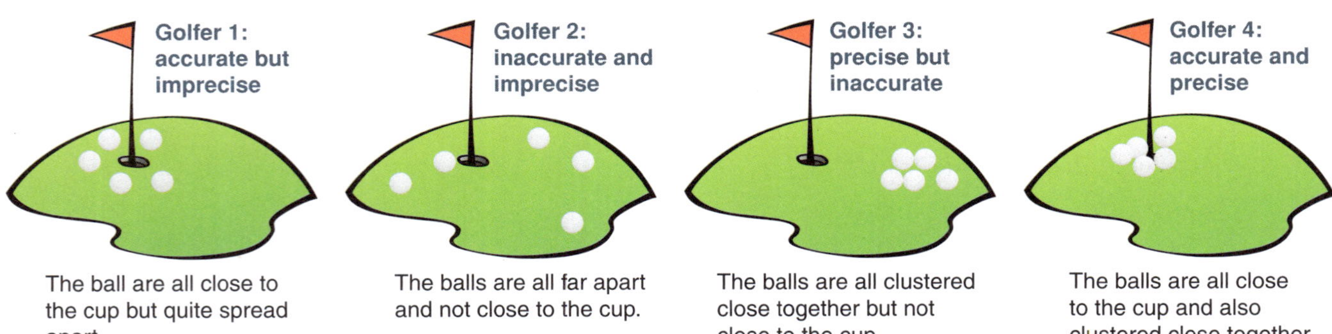

The ball are all close to the cup but quite spread apart.

The balls are all far apart and not close to the cup.

The balls are all clustered close together but not close to the cup.

The balls are all close to the cup and also clustered close together.

©2020 **BIOZONE** International
ISBN: 978-1-927309-82-7
Photocopying Prohibited

Reducing error

Sometimes reducing error requires taking more measurements over a longer period of time. For example, waves breaking on a shore do so with a relatively regular frequency, e.g. 1 per 5 s. Recording the time between one wave breaking and the next (the period) may be difficult to determine precisely for an individual wave and the waves may be breaking too quickly to enable accurate recordings.

To increase the accuracy of measuring the period between each wave, it is best to record the time for a larger number of waves to break (e.g. 10) and divide by that number to obtain the period between each wave. This allows for slight variations in the period and reduces the total error in the measurement.

Example: Actual wave period: 5.0 seconds.
Accuracy of timer (i.e. reaction speed) 0.3 seconds

Measurements of individual periods (in seconds):
5.4, 5.7, 5.7, 5.8, 4.5, 4.6, 5.7, 5.8, 5.1, 5.3
Mean: 5.4

In each measurement above, the error is about 0.3 s producing an error of up to 6.7% (0.3 ÷ 4.5 x 100) of the recorded value of a wave period.

If the time recorded for 10 waves to break was 51.1 s, then the time for one wave to break is 5.1 s. The error is spread over the whole 51.1 s (0.3 ÷ 51.1) and thus is much smaller at just 0.6% of the wave period.

1. The period of a pendulum is based on the length of the pendulum and the mass at its end. Two students measure the time it takes for a pendulum to swing back and forth (its period). Student A measures three individual swings and calculates a mean (average) value. Student B measures three sets of ten swings and calculates a mean. Each student measures the accuracy of the timer as 0.2 seconds. The results are shown below:

Student A	
	Time for swing (s)
1	2.7
2	2.1
3	2.5
Mean (1 swing)	2.4 s

Student B	
Set	Time for 10 swings (s)
1	20.3
2	20.1
3	19.8
Mean (10 swings)	20.1 s
1 swing	2.0 s

(a) Calculate the mean for each student's results and the time for one swing for student B.

(b) Why are student B's results more accurate than student A's:

Student B allowed 10 swings for the pendulum. They are run continuously so the only error is in starting and stopping the timer. This results in a lower error. Student A recorded error with every swing thus has more error in the final mean.

2. Distinguish between accuracy and precision: Accuracy refers to the correctness of the measurement, or how close a measurement is to the true value. Precision refers to the repeatability of the measurements, i.e. how close the measured values are to each other.

3. Describe why it is important to take measurements that are both accurate and precise:

Measurements that are both accurate and precise give data that will provide statistics closer to the true parameters of the population being sampled, e.g. mean volume delivered by a pipette.

4. A researcher is trying to determine the melting point of a substance. Their temperature probe is incorrectly calibrated. How would this affect the accuracy and precision of the data collected?

The researcher may still obtain precise data but the data will not be accurate. Because the temperature probe is incorrectly calibrated, it will give measurements that are either higher or lower than the actual melting point.

5. In the example above of the time between waves on a beach the numerical mean 5.36 s. Why is 5.4 s reported as the mean instead?

Measurements were only taken to 1 d.p. To report the mean as 2 d.p. would be beyond the limit of accuracy for the data.

©2020 BIOZONE International
ISBN: 978-1-927309-82-7
Photocopying Prohibited

7 Working With Numbers

Key Question: How can converting and manipulating numbers make them easier to understand?

- Using correct mathematical notation and being able to carry out simple calculations and conversions are fundamental skills in science. Mathematics is used to analyze, interpret, and compare data. It is important that you are familiar with mathematical notation (the language of mathematics) and can confidently apply some basic mathematical principles and calculations to your data.
- Much of our understanding of the physical sciences is based on our ability to use mathematics to interpret the patterns seen in data and express laws of the universe in simple notation.

Commonly used mathematical symbols

In mathematics, universal symbols are used to represent mathematical concepts. They save time and space when writing. Some commonly used symbols are shown below.

- $=$ Equal to
- $<$ The value on the left is **less than** the value on the right
- $>$ The value on the left is **greater than** the value on the right
- $\propto$ Proportional to. $A \propto B$ means that $A = (\text{a constant}) \times B$
- $\sim$ Approximately equal to
- ∞ Infinity
- $\sqrt{b}$ The square root of b
- b^2 b squared ($b \times b$)
- b^n b to the power of n (b × b... n times)
- Δ The change in. For example $\Delta T / \Delta d$ = the change in T ÷ the change in d (see rates below right).

Length

Kilometer (km)	1000 m
Meter (m)	1000 mm

Volume

Liter (L)	1000 mL
Milliliter (mL)	= 1 mm^3

Area

Square kilometer	1,000,000 m^2
Hectare	10,000 m^2
Square meter	1,000,000 mm^2

Mass

kilogram (kg)	1000 g
1 tonne (t) (also called a metric ton)	1000 kg

Decimal and standard form

Decimal form (also called ordinary form) is the longhand way of writing a number (e.g. 15,000,000). Very large or very small numbers can take up too much space if written in decimal form and are often expressed in a condensed **standard form**. For example, 15,000,000 is written as 1.5×10^7 in standard form.

In standard form a number is always written as $A \times 10^n$, where A is a number between 1 and 10, and n (the exponent) indicates how many places to move the decimal point. n can be positive or negative.

For the example above, A = 1.5 and n = 7 because the decimal point moved seven places (see below).

$$15\,000\,000 = 1.5 \times 10^7$$

Small numbers can also be written in standard form. The exponent (n) will be negative. For example, 0.00101 is written as 1.01×10^{-3}.

$$0.00101 = 1.01 \times 10^{-3}$$

Adding numbers in standard form

Numbers in standard form can be added together as long as they are both raised to the same power of ten.
Example: $1 \times 10^4 + 2 \times 10^3 = 1 \times 10^4 + 0.2 \times 10^4 = 1.2 \times 10^4$

Rates

Rates are expressed as a measure per unit of time and show how a variable changes over time. Rates are used to provide meaningful comparisons of data that may have been recorded over different time periods.

Often rates are expressed as a mean rate over the duration of the measurement period, but it is also useful to calculate the rate at various times to understand how rate changes over time. The table below shows the distance traveled by a rolling ball. A worked example for the rate at 2 seconds is provided below.

Time (s)	Distance traveled (m)	Rate of movement (speed) (m/s)
0	0	0
2	34	17*
4	42	4
6	48	3
8	50	1
10	50	0

* meters moved between 0 - 2 seconds: 34 m − 0 m = 34 m
Rate of movement (speed) between 0 - 2 seconds
34 m ÷ 2 seconds = 17 m/s

1. Use the information above to complete the following calculations:

(a) $\sqrt{9}$: 3

(b) 4^3: 64

(c) Write 6,340,000 in standard form: 6.34×10^6

(d) Write 0.00103 in standard form: 1.03×10^{-3}

(e) Convert 10 cm to millimeters: 100 mm

(f) Convert 4 liters to milliliters: 4000 mL

(g) Write 7.82×10^7 as a number: 78,200,000

(h) $4.5 \times 10^4 + 6.45 \times 10^5$: 45,000 + 645,000 = 690,000

©2020 BIOZONE International
ISBN: 978-1-927309-82-7
Photocopying Prohibited

Using Logarithms

- Physical sciences often deal with very large numbers or scales. Numerical data indicating scale can often increase or decrease exponentially. Large scale changes in numerical data can be made more manageable by transforming the data using logarithms.

Exponential function

- Exponential growth or decay occurs at an increasingly rapid rate in proportion to the increasing or decreasing total number or size.
- In an exponential function, the base number is fixed (constant) and the exponent is variable.
- The equation for an exponential function is $y = c^x$.
- An example of exponential decay is radioactive decay. Any radioactive element has a half-life, the amount of time required for its radioactivity to fall to half its original value.

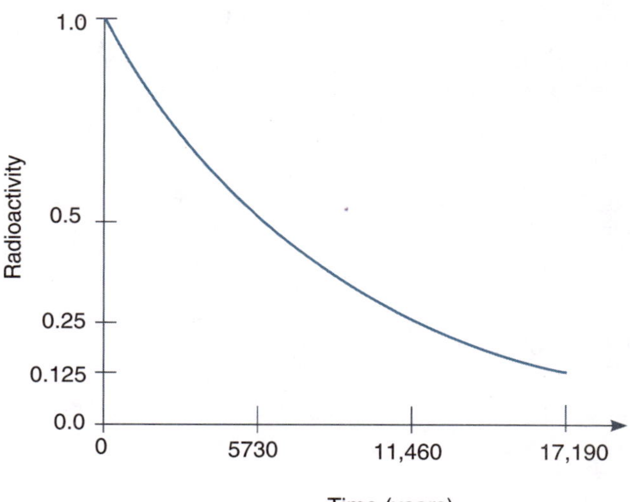

Example above: Carbon-14 (^{14}C) has a half life of 5730 years. If a sample with a mass of 10 g was left for 5730 years half the sample will have decayed, leaving 5 g of radioactive material. After another 5730 years, 2.5 g of radioactive carbon will be left.

Log transformations

- A log transformation can make very large numbers easier to work with.
- The log of a number is the exponent to which a fixed value (the base) is raised to get that number. So $\log_{10}(1000) = 3$ because $10^3 = 1000$.
- Both $\log_{10}$ and $\log_e$ (natural logs or *ln*) are commonly used.
- Log transformations are useful for data where there is an exponential increase or decrease in numbers. In this case, the transformation will produce a straight line plot.
- To find the $\log_{10}$ of a number, e.g. 32, using a calculator, key in log 32 =. The answer should be 1.51.
- It is also important to be able to return an number from its log back to the original number. This is done using the inverse log.
- For example using the number 12.34. The log of 12.34 is 1.09. To return 1.09 back to the original 12.34 on a calculator key in 10^x 1.09 = .
- Logs are important in chemistry as they used to calculate pH and H^+ concentrations in acids.

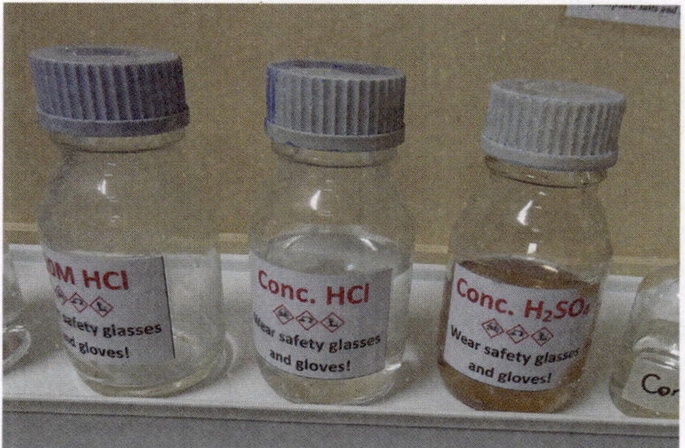

2. (a) What is the $\log_{10}$ of 567? *2.75*

   (b) What is 10^x (inverse $\log_{10}$) of 0.067? *1.17*

3. The pH scale measures the acidity of a substance. It is a negative logarithmic scale. A pH of 3 has a hydrogen ion (H^+) concentration (which is responsible for acidity) ten times greater than a pH of 4.

   (a) How many times greater is the hydrogen ion concentration of a pH 2 solution than a pH 6 solution?

   The hydrogen ion concentration is 10,000 times more in a pH 2 solution than a pH 6 solution.

   (b) The pH of a solution is simply the negative of the $\log_{10}$ of the H^+ concentration. If the H^+ concentration of a solution was 1.6×10^{-3} (ignoring units) what would the pH of the solution be?

   Log of $1.6 \times 10^{-3} = -2.8$. The negative of a negative is positive : pH = 2.8

4. Carbon-14 (^{14}C) is found in living organisms. It has a half life of 5730 years. When an organism dies it stops taking in ^{14}C and this results in a change in the ratio of ^{14}C to ^{12}C. Using these pieces of information explain how we can calculate how long ago an organism died:

   ^{14}C decays at a known rate so the ratio of ^{14}C to ^{12}C changes. By measuring the ratio in the fossil or sample and comparing it to a living organism we can calculate the age of the fossil or sample. The older the sample the less ^{14}C it will have.

©2020 **BIOZONE** International
ISBN: 978-1-927309-82-7
Photocopying Prohibited

Rearranging equations

- Sometimes you will need to rearrange an equation to find the answer for an unknown variable. For example if $x \times y = z$ what is x if we already know z and y?
- Here we can say that if $x \times y = z$ then $z \div y = x$. We can also write the equation in a fact family triangle:

Fact family triangles can make working out unknown variables in formulas simpler. For the example here, to work out the value of x cover x in the triangle with your finger. The remaining figures show you that you need to divide z by y to find x. Similarly to find z, covering z in the triangle shows you need to multiply x by y.

z = xy

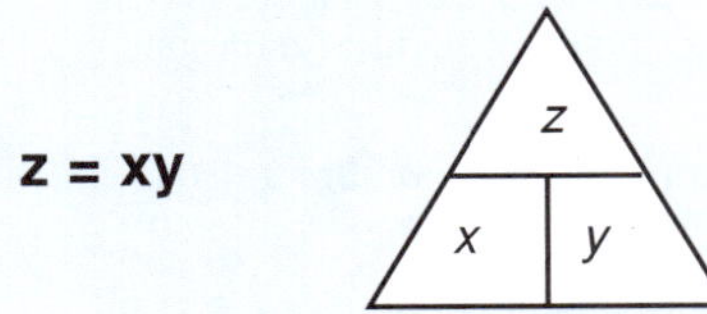

- Fact family triangles are very useful when working with simple (and sometimes even complex) equations.
- Although the first equation here is shown as $x \times y = z$ you often find formulae with the single variable written first: $z = xy$ (as shown above right). Similar formulas you will come across are $v = d/t$, $E_k = ½mv^2$, and $F = ma$.
- The rules of mathematics sometimes need to be exploited to rearrange equations. For example take the equation $1/x \times y = z$. How can we rearrange the equation to find x? There are various ways of doing this. One is shown below:
- First we can use the **commutative law of multiplication.** This states that if $x \times y = z$ then $y \times x = z$. The order of the multipliers does not matter (this is not the case for division!). So if $1/x \times y = z$ then $y \times 1/x = z$.
- Now we can also use the **reciprocal rule of division** that states that dividing by a number is the same as multiplying by the number's reciprocal, thus $y \times 1/x = y \div x$. We can now write the equation $1/x \times y = z$ as $y \div \; x = z$. Rearranging to find x we get $\mathbf{x = y \div z}$ (similarly $y \div 1/x = y \times x$).

5. An equation can be written as: $W = ½DP$:

   (a) Rearrange the equation to solve for P: *P = W ÷ 0.5D. (Using a triangle D and P would be on the bottom and W on the top. Covering P leaves us with W and D).*

   (b) Rearrange the equation to solve for D: *W ÷ P = 0.5D. D = 2(W÷P)*

6. (a) Use an example to show that $x \times y$ is the same as $y \times x$: *If X = 3 and Y = 4 then: 3 × 4 = 12 , 4 × 3 = 12*

   (b) Use an example to show that x/y is not the same as y/x: *If x = 10 and y = 5 then: 10 ÷ 5 = 2, 5 ÷ 10 = 0.5*

7. In the space below produce a simple fact family triangle for $E_k = ½mv^2$:

8. An equation can be written as: $K = 1/V \times T^2$:

   (a) Rearrange the equation to solve for V (show your working):

   $K = 1/V \times T^2$ $\quad K = T^2 \times 1/V$ $\quad K = T^2/V$ $\quad V = T^2/K$

   (b) Rearrange the equation to solve for T (show your working):

   $V = T^2/K$ $\quad T^2 = K \times V$ $\quad T = \sqrt{(K \times V)}$

   (c) If V = 3 and T = 5, calculate K: *$K = 1/3 \times 5^2 = 8.33$*

   (d) If K = 6 and T = 3.5, calculate V: *$V = 3.5^2 \div 6 = 2.04$*

   (e) If K = 4 and V = 2.4 calculate T: *$T = \sqrt{(4 \times 2.4)} = 3.10$*

©2020 **BIOZONE** International
ISBN: 978-1-927309-82-7
Photocopying Prohibited

8 Graphical Analysis

Key Question: Why do we use different styles of graphs and what information do they show?

- Graphs are a good way of visually showing trends, patterns, and relationships without taking up too much space. Complex data sets tend to be presented as a graph rather than a table.
- You should plot your data as soon as possible, even during your experiment, as this will help you to evaluate your results as you proceed and make adjustments as necessary (e.g. to the sampling interval).
- Give your graph a title and appropriately labeled axes so the information displayed is clearly communicated.

Line graph	Scatter plot
The data must be continuous for both variables (i.e. not counts and not categories).	The data must be continuous for both variables (i.e. not counts and not categories).
The response variable is dependent on the independent variable. The independent variable is often time or the experimental treatment.	There is no manipulated (independent) variable but the variables are often correlated, i.e. they vary together in a predictable way.
The points are connected point to point.	The points on the graph should not be connected but a **line of best fit** can be drawn through the points. A line of best fit should follow the trend of the data with roughly half the data points above the line and half below.
Used to illustrate the response to a manipulated variable, e.g. distance versus time.	Used to show the relationship between two correlated variables, e.g. atomic number vs atomic radius.
Distance versus time graph for a freely falling body Line connecting points Distance (meters): 0, 5, 10, 15, 20, 25, 30, 35, 40 Time (seconds): 1, 2, 3, 4, 5, 6, 7	Relationship between recession velocity and distance Line of best fit Outlier: a data value that lies outside the main spread of data Recession velocity (km/s): 0, 500, 1000 Distance (Mpc): 0, 1, 2

The graph on the right illustrates three types of gradients for a line graph.

- **Positive gradients** (blue line): the line slopes upward to the right (y is increasing as x increases).
- **Negative gradients** (red line): the line slopes downward to the right (y is decreasing as x increases).
- **Zero gradients** (green line): the line is horizontal (y does not change as x increases).

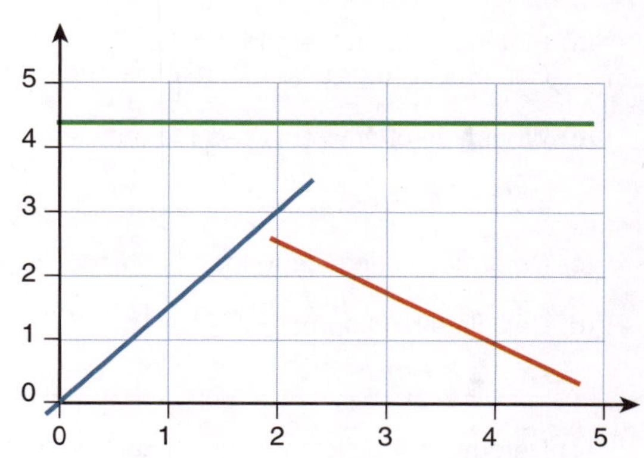

©2020 BIOZONE International
ISBN: 978-1-927309-82-7
Photocopying Prohibited

Measuring gradients and intercepts

- Data plotted as a linear (straight) line can give us information about the system we are observing.
- A linear line can be described by the equation: **y = mx + c**.
- The equation can be used to calculate the gradient (slope) of a straight line and tells us about the relationship between x and y (how fast y is changing relative to x). For a straight line, the rate of change of y relative to x is always constant.

The equation for a straight line is written as:

y = mx + c

Where :

y = the y-axis value

m = the slope (or gradient)

x = the x-axis value

c = the y intercept (where the line crosses the y-axis).

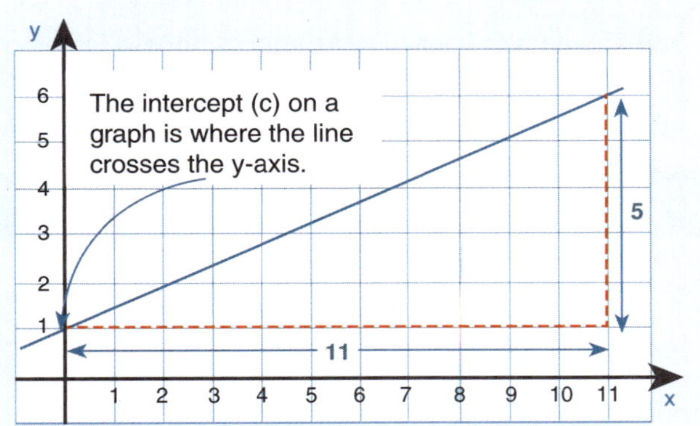

Determining "m" and "c"

To find "c" just find where the line crosses the y-axis.

To find m:

1. Choose any two points on the line.
2. Draw a right-angled triangle between the two points on the line.
3. Use the scale on each axis to find the triangle's vertical length and horizontal length.
4. Calculate the gradient of the line using the following equation:

$$\frac{\text{change in y (rise)}}{\text{change in x (run)}}$$

For the example above:

c = 1

m = 0.45 (5 ÷11)

Once c and m have been determined you can choose any value for x and find the corresponding value for y.

For example, when x = 9, the equation would be:

y = 9 x 0.45 + 1

y = 5.05

1. A student measured the how far their classmate ran when she ran at a steady pace for five seconds. Distance was measured every second and their results are shown in the table below.

Time (seconds)	Distance (meters)
0	0
1	7
2	14
3	21
4	28
5	35

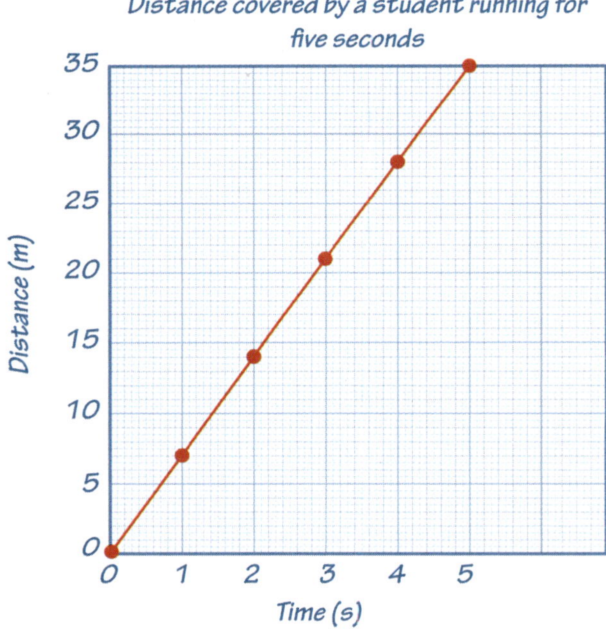

(a) Plot the data on the grid (right). Remember to give your plot a title and axes labels:

(b) What type of gradient does the data show?

Positive

(c) Determine c (intercept) for this graph: *0*

(d) Calculate m (slope): *m = 7. For example between 1 and 3 seconds: rise = 21 m/ 7 m = 14 m. run = 3 s - 1 s = 2 s. So 14 ÷ 2 = 7.*

(e) Determine the distance the student ran in 3.5 seconds: *y= 7 × 3.5 + 0, so y = 24.5 meters*

©2020 BIOZONE International
ISBN: 978-1-927309-82-7
Photocopying Prohibited

9 Describing the Data

Key Question: How do we use simple statistics to describe or summarize patterns in data?

- Most data shows variability. Descriptive statistics (e.g. mean and standard deviation) are used to summarize important features of a data set such as central tendency (the mid-point of the data's distribution) and how the data values are distributed around this value.

- **Mean**: The mean is a single value representing the central position in a set of data with a normal distribution (see plot at the bottom of the page). In biology, the mean is often used to describe a variable in a population (e.g. mean height). Data sets are often large. In physics, the mean is most often applied to smaller data sets (e.g. the mean lap time of a race car over 10 laps).

How is the mean calculated?

The sample mean ($\bar{x}$) is calculated by summing all the data values (x) and dividing them by the total number of data points (n). Outliers (very extreme values) are usually excluded from calculations of the mean.

For very skewed data sets, it is better to use the median as a measure of central tendency. This is the middle value when the data values are placed in rank order. If two values share the central position, the sum of the two values is divided by two.

DID YOU KNOW?

It is not always appropriate to calculate a mean. **Do not calculate a mean if**:

- The values are already means (averages) themselves.
- The data are ratios, such as percentages.
- The measurement scale is not linear (e.g. pH units).

1. Write a mathematical expression for how to calculate a mean:

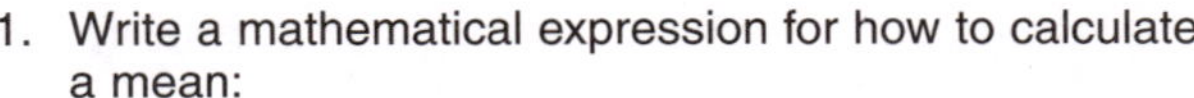

$\bar{x} = (\sum x \div n)$

2. A handball player wanted to know if she could throw the ball faster using an above-head release or a side-release throw. She threw the ball five times using each method. The ball release speed is recorded in the table, right.

(a) Complete the table by calculating the sum and the mean for each data set:

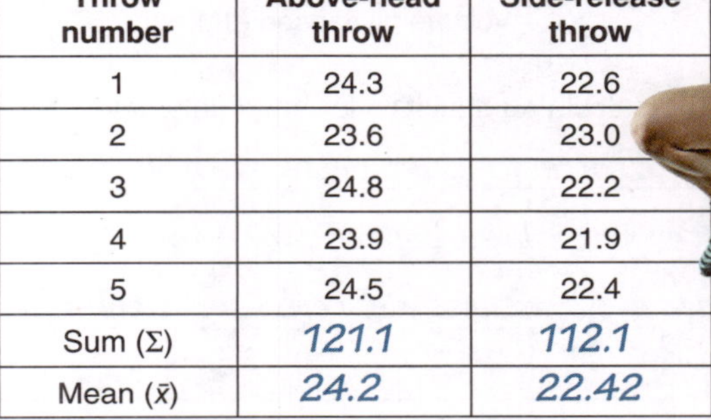

Throw number	Ball release speed (m/s)	
	Above-head throw	Side-release throw
1	24.3	22.6
2	23.6	23.0
3	24.8	22.2
4	23.9	21.9
5	24.5	22.4
Sum (Σ)	121.1	112.1
Mean ($\bar{x}$)	24.2	22.42

(b) Which type of throw produced the fastest release speed? Above head throw.

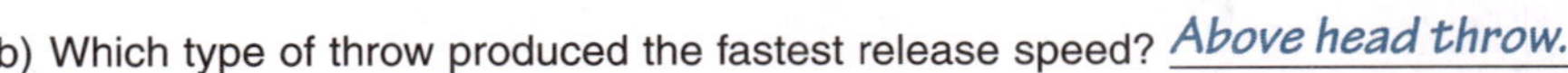

3. During throw 3 of the above-head throws, the ball slipped from the player's hand and recorded a speed of 10.2 m/s. The player decided not to use this value and rethrew the ball.

(a) Recalculate the mean using 10.2 instead of 24.8 m/s: $106.5 \div 5 = 21.3$

(b) Was retesting throw 3 the correct choice? Explain your reasoning: Yes. If they hadn't discounted throw 3 the mean would have been artificially low and it would seem that the side release throw was the fastest.

Standard deviation

While it is important to know the mean of a data set, it is also important to know how much spread there is within the data. For normally distributed data (right) this is measured using standard deviation. It provides a way to evaluate the reliability of estimates of the true mean.

- Sample standard deviation (s) is usually presented as $\bar{x} \pm s$. In normally distributed data, 68% of all data values will lie within one standard deviation ($1s$) of the mean and 95% of all values will lie within two standard deviations ($2s$) of the mean (right).
- The lower the standard deviation, the more closely the data values cluster around the mean.

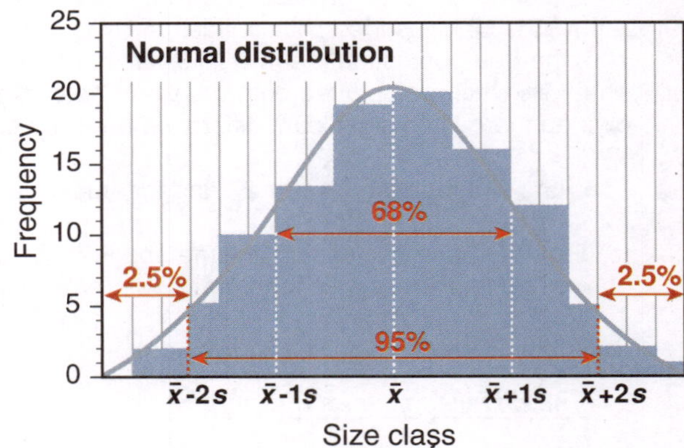

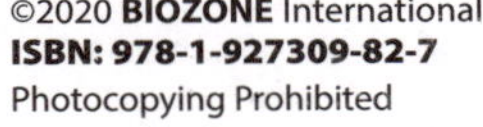
©2020 BIOZONE International
ISBN: 978-1-927309-82-7
Photocopying Prohibited

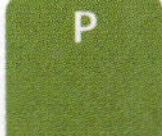

Sample standard deviation is very easily calculated using the equation on the right. It is often calculated using a spreadsheet by entering the data into columns and typing the formula using standard spreadsheet formula rules (see the **Resource Hub**).

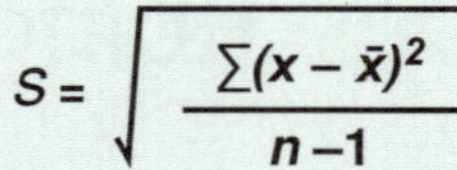

$$s = \sqrt{\frac{\sum(x - \bar{x})^2}{n-1}}$$

$\sum(x - \bar{x})^2$ = the sum of squared deviations from the mean

n = sample size (number of data values).

n -1 provides a unbiased s for small sample sizes (large samples can use n).

Two students tested to see how pipette accuracy is affected by pipette size. Both students dispensed 200 µL 50 times and weighed the dispensed volume each time. Student A used a 200 µL pipette, student B used a 1000 µL pipette. Their results are shown below.

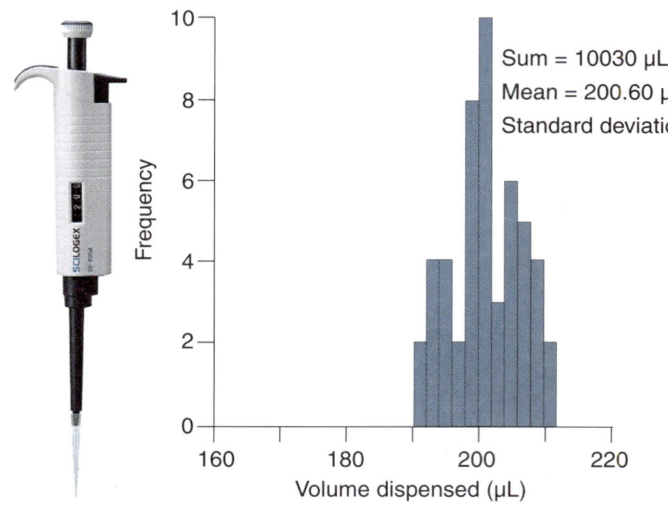

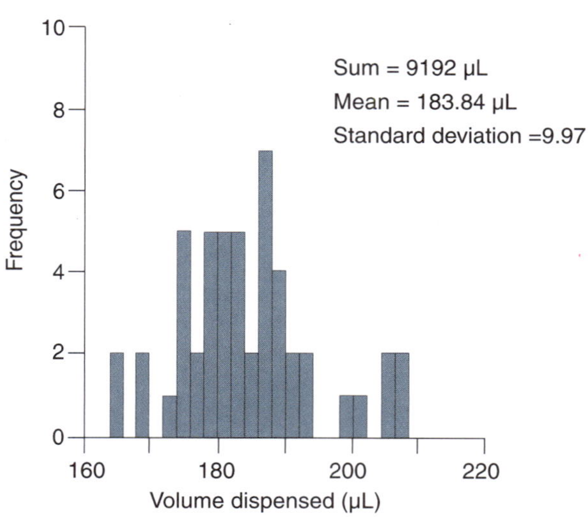

4. (a) Use the calculated statistics for each data set to comment on how pipette size affected the dispensing accuracy:

Using the correct sized pipette has a big effect on the accuracy of the volume being dispensed. Larger pipettes have a large capacity for error. In the example above, the 200 µL pipette has a much more closely grouped set of results clustered about the actual value wanted than the 1000 µL pipette.

(b) Set A has a lower standard deviation than set B. What does this tell us about the reliability of the data?

Set A data is more reliable than set B.

(c) How could you use this information to select the appropriate equipment for your experiments?

This data would tell us that the size of the pipette and the volume needed should be matched. That is, if 200 µL or less is needed then a 200 µL pipette should be used rather than a larger pipette.

5. Calculating a 95% confidence interval (95% CI) provides a good estimate of how close the sample mean is to the true population mean. The population here is the entire pool of values from which the sample is drawn. The mean ± the 95% CI gives the 95% confidence limits. On average, 95 times out of 100, the true population mean will lie within the confidence limits. The 95% CI can be plotted onto graphs to determine significance. If they overlap there is no significant difference between the data.

The 95% CI for student A is 1.27 and for student B is 2.37 at p=0.05.

(a) Plot the mean dispensing volume and 95% CI for both sets of data on the grid right). Plot as a column graph with error bars.

(b) Is the difference significant? Why or why not?

The difference is significant because the bars showing the 95% CIs do not overlap.

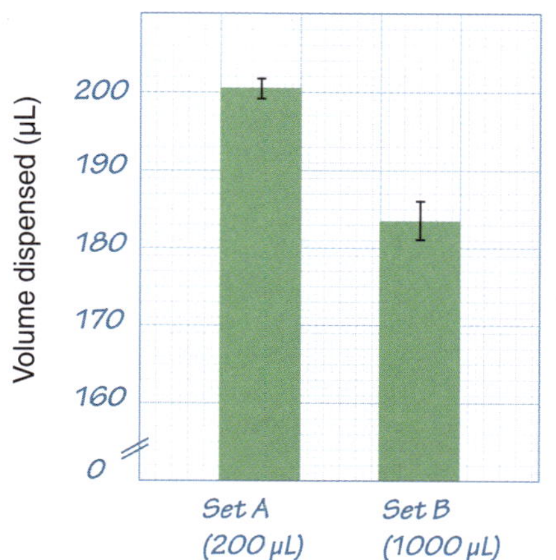

©2020 **BIOZONE** International
ISBN: 978-1-927309-82-7
Photocopying Prohibited

10 Investigations in Physics

Key Question: What things should we think about when carrying out an investigation in physics?

- Physics experiments are often used to find out some fundamental rule or relationship between different variables within a system.
- To do this all variables must be controlled. One variable (the independent variable) is then changed systematically and the effect on the system is observed.

Variables and controls

- The **independent variable** is set by the experimenter. It is the variable that is deliberately altered and is recorded on the x axis of a graph. Where a response is recorded over time, the independent variable is time.
- The **dependent variable** is the response variable, measured during the course of the experiment. It is recorded on the y axis of a graph.
- **Controlled** (or constant) **variables** are variables that you keep the same during your experiment.
- To be a fair test, an experiment should alter only one variable at a time and there should be a **control**, which acts as a standard or reference treatment. A control is identical to the original experiment except it lacks the altered variable, so it provides confidence that the analytical method is working correctly.

Focus on one variable at a time

- Finding a relationship between variables means focusing your measurement and recording on just those variables which are important in the relationship.
- The data below shows the result of an investigation into the relationship between how the curvature of a lens affects the focal length of the lens.

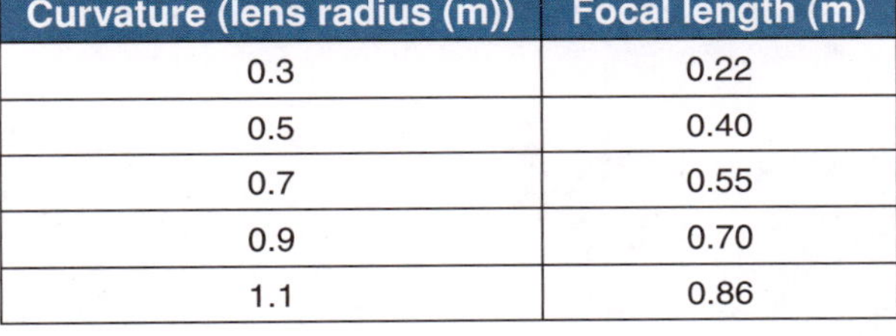

Curvature (lens radius (m))	Focal length (m)
0.3	0.22
0.5	0.40
0.7	0.55
0.9	0.70
1.1	0.86

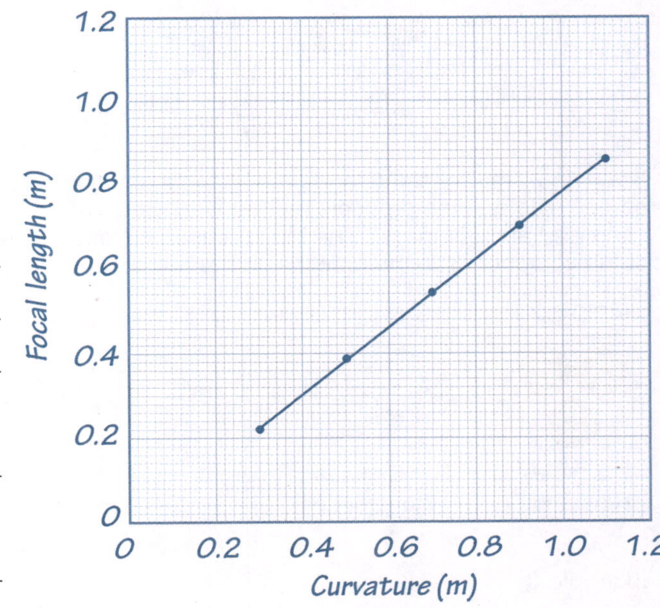

1. For the experiment above list three things that will need to be controlled:

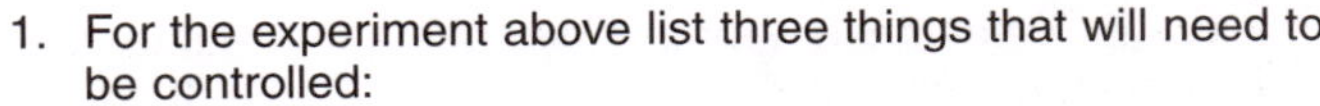

The diameter of the lens. The material the lens is made of. The distance the source of light is from the lens. The color of the light used.

2. (a) What was the independent variable for the data above?

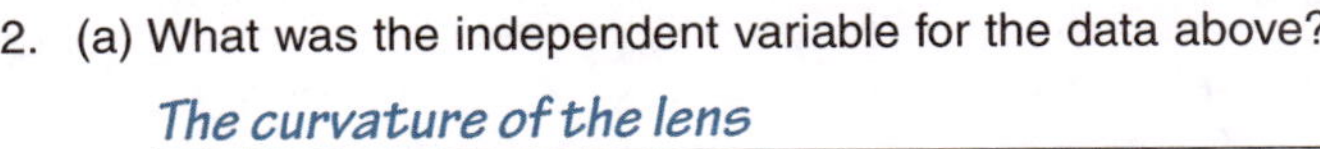

The curvature of the lens

(b) What was the dependent variable for the data above?

The focal length

3. Graph the data on the grid provided:

4. Describe the relationship between lens curvature and focal length:

As the radius of the curve of the lens increases (the lens becomes less curved) the focal length also increases.

5. Try producing a simple equation that describes this relationship: *Gradient = rise / run. G = FL / C = 0.77*

Focal length = 0.77 × Curvature

©2020 BIOZONE International
ISBN: 978-1-927309-82-7
Photocopying Prohibited

11 Ratio and Proportion in Chemistry

Key Question: What is the importance of ratio and proportion in chemistry and how can we use them to work out required masses in experiments?

- Ratio and proportion are important concepts in chemistry. A ratio compares quantities of two different categories, e.g. in a bread recipe, we use 2 cups of flour for every 1 cup of water. The ratio of flour to water is 2 : 1. We use ratios a lot in chemistry where atoms and reagents are found in fixed proportions (they are proportional).
- Proportion is a mathematical concept, which states **the equality of two ratios or fractions**. You use proportions every time you double or halve a recipe because the ingredients go up or down in proportion. You do the same thing in chemistry!

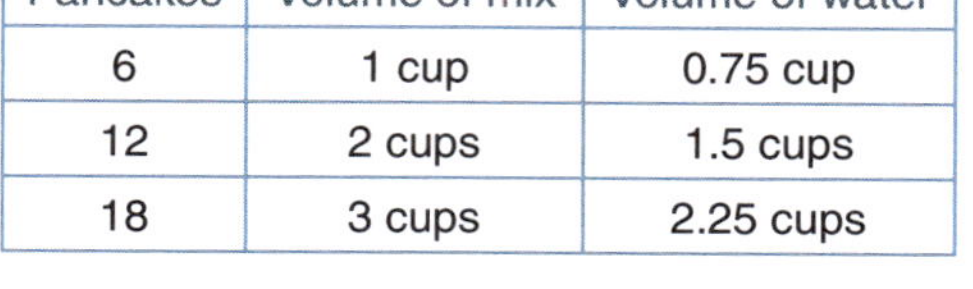

Pancakes	Volume of mix	Volume of water
6	1 cup	0.75 cup
12	2 cups	1.5 cups
18	3 cups	2.25 cups

Try it out!

A box of pancake mix gives instructions for mixing the pancakes (right):

1. What is the ratio of mix to water in each case? *1: 0.75 (4:3)*

2. How can we work out how much mix we need to make 30 pancakes? The answer is we cross multiply using the ratio of mix to pancakes (1 : 6):

$$\frac{1}{6} = \frac{x}{30} \qquad 30 = 6x \quad 5 = x \qquad \text{Answer} = 5 \text{ cups}$$

Now use the ratio of mix to water to calculate the amount of water needed:

Mix / water = 1.33 = 5/water. 1.33 x water = 5

Water = 5/1.33 = 3.76 cups water

Now let's apply these basic principles to some important aspects of chemistry.

Ratios in stoichiometry

Example 1

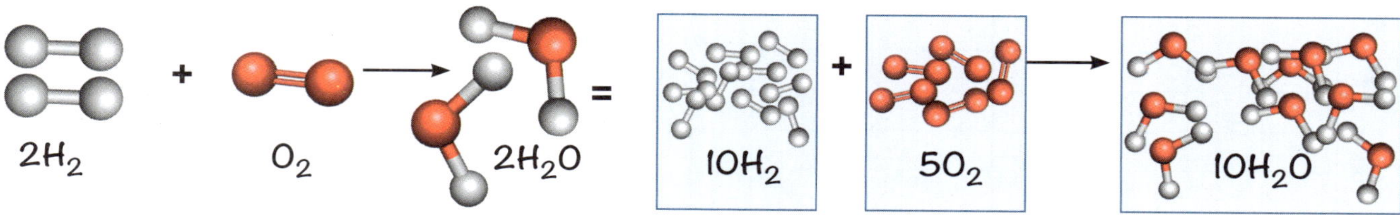

If you know the reactants and products in a reaction, you can write an unbalanced equation by putting the reactants on one side and the products on the other. Both sides of the equation must have the same number of atoms (the law of conservation of mass). The multiplication factors required to balance the equation tells you the mole ratios of each of the compounds in the reaction. In the reaction above, 2 moles H_2 and 1 mole O_2 give 2 moles of water.

Example 2

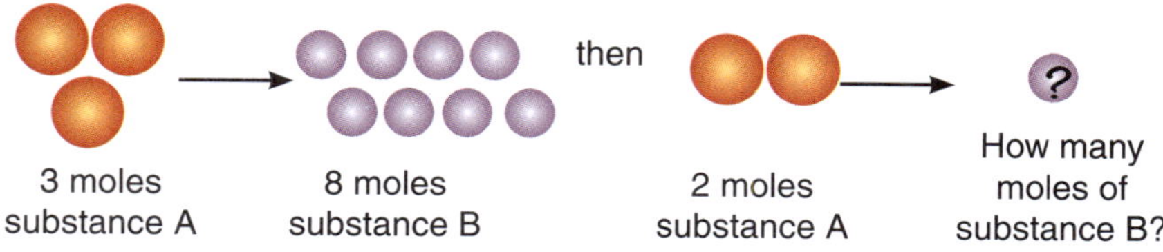

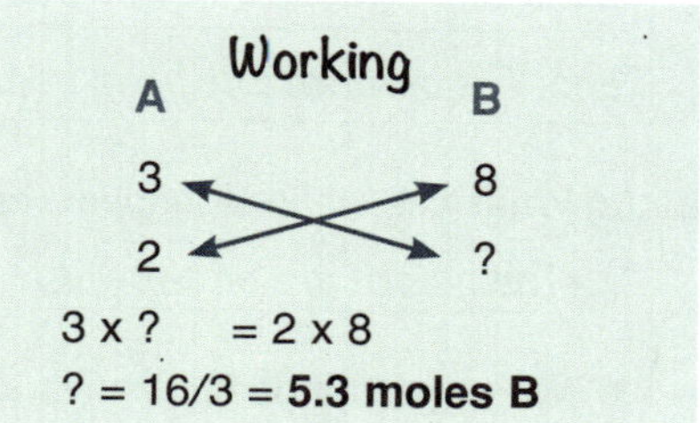

Example 3

- Consider the formula for propanol: C_3H_5OH. The ratio of carbon to hydrogen atoms is 3:6. The mass ratios will be in proportion.
- First we use molar mass (M) in mol/g and the equation **n=m/M** to convert the mass (m) to number of moles (n). We can then cross multiply to calculate that if there is 8 g C in a propanol sample, there will be 1.33 g of H.
- Note that the units mol and mol/g cancel, leaving the final unit (grams).

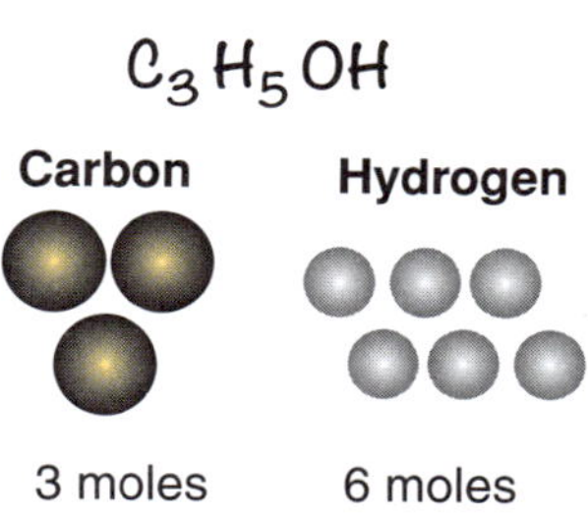

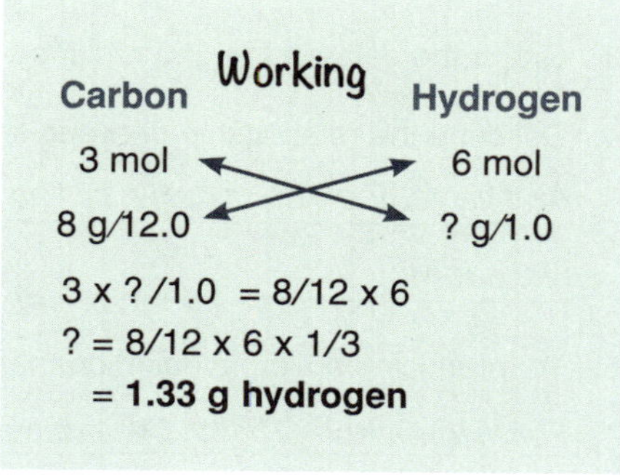

©2020 BIOZONE International
ISBN: 978-1-927309-82-7
Photocopying Prohibited

Using ratios to concentrations and pressures

▸ Now you can apply your understanding of ratios and proportions to some other common situations.

3. **Concentration**: If a solution of the desired concentration contains 5 g of a salt in 30 mL of water, how many grams of salt would you dissolve in 75 mL to have the same concentration? If you want, draw arrows on the diagram to help you.

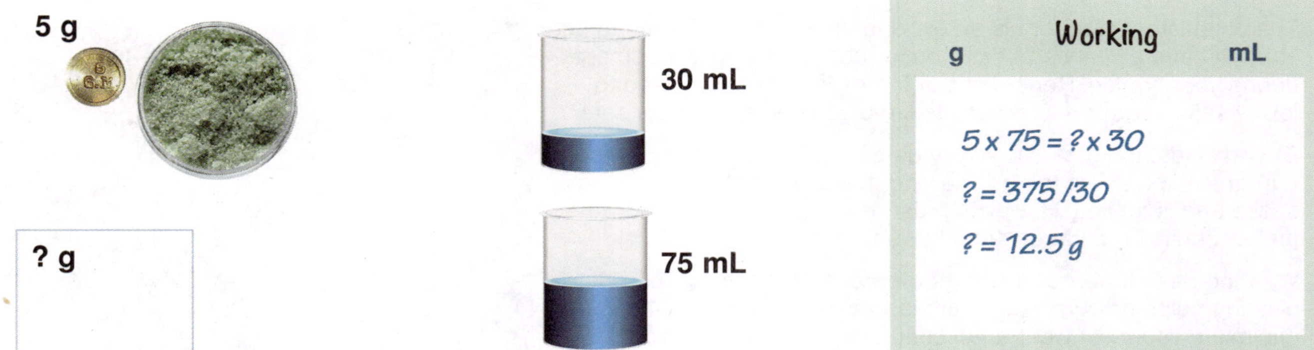

4. **Pressures**: A pressure of 759 mm Hg gives 101.2 kPa. If a balloon is inflated to 60 kPa, what is the pressure in mm Hg? If you want, draw arrows on the diagram to help you.

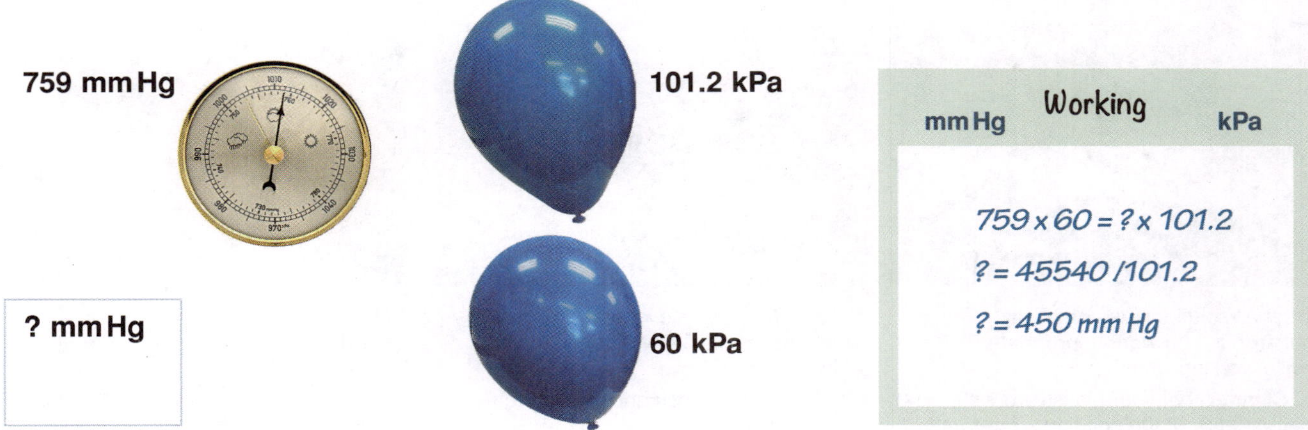

5. Ammonia (NH_3) is a compound with 1 : 3 ratio of nitrogen to hydrogen atoms.

   (a) If a sample of ammonia contains 1500 nitrogen atoms, determine the number of hydrogen atoms present?

   $1500 \times 3 = 4500$

   (b) If a sample of ammonia contains 1410 hydrogen atoms, determine the number of nitrogen atoms present?

   $1410 \div 3 = 470$

6. Methane (CH_4) is a gas.

   (a) What is the ratio of carbon to hydrogen atoms? 1 : 4

   (b) If a methane sample contains 1575 atoms, how many carbon and hydrogen atoms are present? Show your working:

   C + H = 1575. C/H = 1/4. 4C = 1H. Substitute 4 C for 1 H in first equation: C + 4C = 1575

   C = 1575/5 = 315. H = 1575 – 315 = 1260

   (c) Can a sample of methane contain 1577 atoms? Explain your answer:

   No. This number is not a simple 4 : 1 ratio. There could only be 1575 or 1579 atoms.

7. The mass to charge ratio of an isotope was found to be 2.0×10^{-7} kilograms per coulomb. If the charge on the isotope is 1.6×10^{-19} coulombs, what is the mass of the isotope? HINT: Determine the ratio and cross multiply as usual.

   Mass/charge = 2.0×10^{-7}. Charge = 1.6×10^{-19}. Cross multiply mass/1.6×10^{-19} = 2.0×10^{-7}

   mass = $1.6 \times 10^{-19} \times 2.0 \times 10^{-7}$. Mass = 3.2×10^{-26}

©2020 **BIOZONE** International
ISBN: 978-1-927309-82-7
Photocopying Prohibited

12 Safety in The Lab

Key Question: What do we need to be aware of when in the chemistry lab?

- The chemistry laboratory is an exciting but hazardous place to be. You will have a chance to carry out or observe many experiments during your course. Regardless of the experiment, you need to follow safe laboratory practices to keep yourself and others safe.
- Strong acids and alkalis are very damaging to the skin. A Bunsen burner is explosive if the gas flowing to it builds up in a confined space and is ignited. However, most risks can be eliminated if proper safety procedures are followed.
- Wearing **personal protection equipment** (PPE) and appropriate clothing helps prevent injury from laboratory chemicals. At a minimum, you must wear a lab coat, gloves, and safety goggles. You must also tie back long hair and wear shoes that fully enclose your feet. Depending on the experiment, you may have to wear additional PPE. For example, if you are working with hot objects you should wear heat resistant gloves.

Measuring

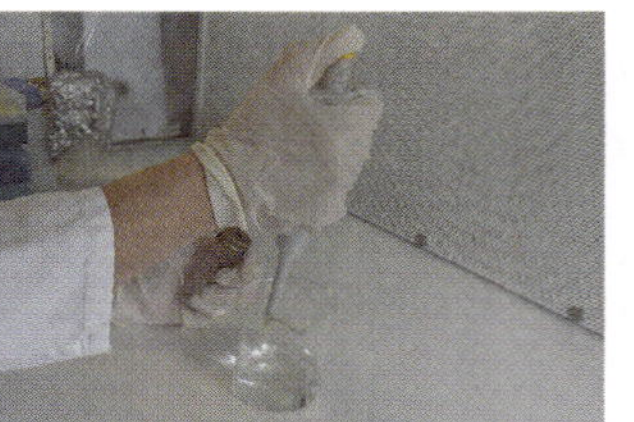

Treat all solutions as potentially hazardous and wear appropriate PPE while working with them.

- Clearly label your glassware with the name of the solution. Water and concentrated acid look just the same, you don't want to confuse them!
- Keep your work space free of clutter. This decreases the chances of accidently knocking things over while you work.
- Choose appropriately sized glassware and pipettes to measure the volumes you are working with. This decreases measurement errors.
- Measure especially corrosive or noxious chemicals in a fume hood.
- Some chemicals are very toxic and must be weighed out in a fume hood.

Heating

Wear PPE, use safe laboratory practices and common sense whenever you heat a substance.

- Tie long hair back and do not wear loose clothing.
- Never reach across a flame.
- Keep the area clear, especially of papers and flammable substances.
- Never leave a lit Bunsen burner unattended.
- Make sure you never point the open end of a heating vessel at anyone. Never smell or look directly into a heating tube.
- Heated materials can stay hot for a long time. Handle them using tongs or thermal gloves. Place them on a hot mat, not directly on to your desk.

Weighing

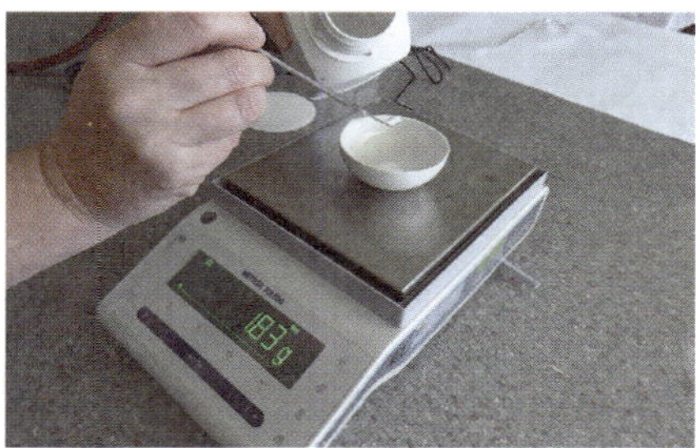

Treat all chemicals as potentially hazardous and wear appropriate PPE.

- Choose a balance and weighing vessel suitable for the type and amount of chemical you are weighing out.
- Place your balance on an even, stable surface where it will not be bumped and will not move.
- Keep the electrical cord safely out of the way.
- Never over-fill the weighing vessel.
- Clean up any spills immediately.
- Make sure your sample is secure when you move from the balance to your work station. Cover it if necessary to stop it falling out of the container.

In the event of an emergency......... do NOT panic!

Good laboratory practices prevent most accidents from happening. But occasionally an accident will occur. Knowing how to deal with it can protect you and your classmates.

- Tell your teacher straight away if an accident has occurred.
- Listen to their instructions and follow them.
- Keep calm. Panicking, yelling, and running can make the situation worse.
- Be prepared. Know where the safety stations (eyewash and shower) and fire extinguishers are. Know how to use them.
- Know where the exits and fire alarms are located.

©2020 **BIOZONE** International
ISBN: 978-1-927309-82-7
Photocopying Prohibited

1. The cartoon below shows students working in a laboratory. Identify at least 8 safety hazards in the cartoon above and provide solutions for how you could reduce risk:

- Test tube being heated but pointed at face (x 2). Test tubes should have the opening pointed away from people in case of explosion.
- Eating in the lab. Food should never be consumed in the lab. It is unsafe and leads to lab contamination.
- Unsafe wires. Wiring should be curled up and tied or put away in a cupboard.
- Bare feet. Always wear covered footwear in the lab.
- Silly behavior. Labs are not places for be silly or playing tricks, people can get seriously hurt.
- Over flowing sink. Equipment should be monitored carefully, including taps and sinks.
- Mouth pipetting. Never put anything in your mouth in the lab. A pipette filler should be used.
- Inattention to Bunsen flame, loose paper beside Bunsen, hair not tied back near Bunsen. Bunsen flames must be carefully attended.
- Loose clothes, papers, or hair can catch alight. Hair should be tied back and papers kept well away from the flame.

©2020 **BIOZONE** International
ISBN: 978-1-927309-82-7
Photocopying Prohibited

13 Standard Equipment in Chemistry

Key Question: What kind of equipment is used in the chemistry lab?

- Most high school chemistry experiments can be carried out using just a few basic pieces of laboratory equipment. In addition to the laboratory supplies, it is important that personal protective equipment (PPE) is worn at all times.
- Some basic laboratory equipment is shown below. It is important that you know how to use it correctly and also how to maintain and clean it so that it works properly. Clearly label any containers so you know what is in them and at the end of the experiment dispose of the waste correctly. Clean your equipment thoroughly after use. This prolongs the life of the equipment and prevents contamination when the equipment is used next.

Laboratories have an assortment of glassware in a variety of sizes. This includes beakers, conical flasks, volumetric flasks and measuring cylinders (L→R above). They all measure liquids, but their accuracy varies.

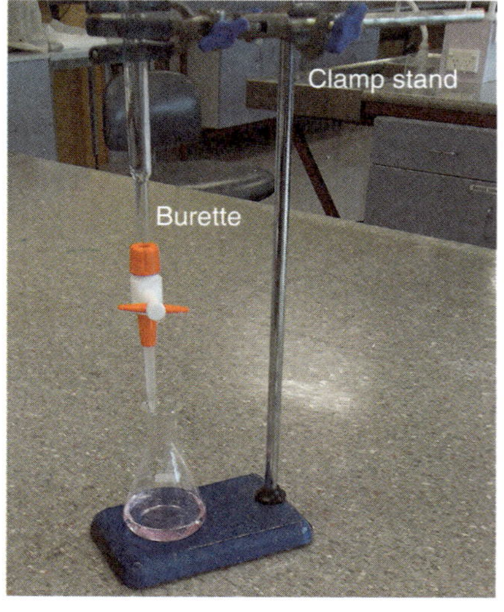

Titrations are commonly performed in the laboratory. They require several pieces of equipment including a burette, clamp stand, conical flask, funnel, and chemical indicator. Phenolphthalein is a commonly used indicator. It changes color when the end point is reached (above).

Balances, pH meters (or indicator paper) and thermometers are used in nearly every chemistry experiment. Remember if you are using pH meters you must calibrate them with standards first. This will tell you if the machine is working correctly. There are many different types of balances, make sure you choose the correct one for the mass of sample you want!

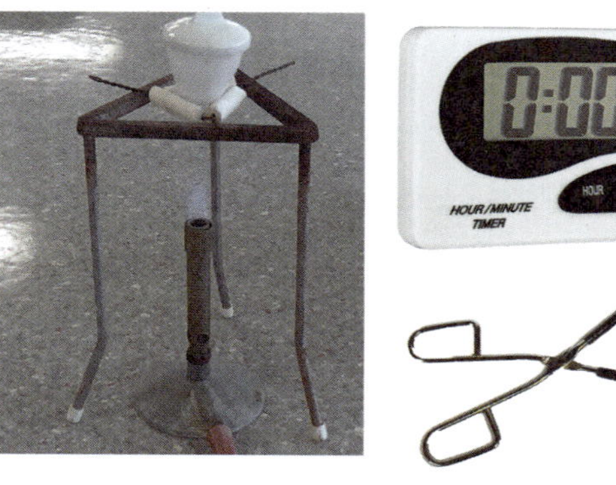

Make sure you know how to use each piece of equipment properly *before you begin your experiment.* If you are trying to figure out how to use equipment while carrying out your experiment the chances of mistakes (and collecting unreliable data) are high. Correct use also keeps you and your classmates safe.

1. Work in pairs or small groups to identify the chemistry equipment you use most frequently. Take some time to ask yourself if you know how to use it correctly? Are there any special instructions you need to know before you use it? Write your ideas here (you many want to use extra paper and staple it to this page):

Students will likely use beakers, Bunsen burners, and tripods most often. Test tubes are also commonly used. Burettes and conical flasks are commonly used for specific measuring and testing (volumetric). Balances are used when weighing masses (gravimetric).

©2020 BIOZONE International
ISBN: 978-1-927309-82-7
Photocopying Prohibited

14 Investigations in Chemistry

Key Question: What things should we think about when carrying out an investigation in chemistry?

- Chemistry is an experimental science and all chemical statements are based on experiment. In your study of chemistry, you will investigate the properties of substances and the interactions between different types of matter. Chemistry helps you to understand how things work.
- You can investigate in two broad ways: by using experimentation to explore the properties of a substance or substances and by conducting controlled experiments. In a controlled experiment you alter just one variable at a time in order to answer a question you have asked about a phenomenon.
- Remember a variable is a factor that can be changed during an experiment (e.g. temperature, mass, pressure). Investigations often look at how changing one variable affects another.

Cooking is chemistry!

A chemistry experiment is not unlike a recipe. Ingredients are added in specific proportions and, given certain conditions, there is a result.

- In bread-making, dried baker's yeast is activated by adding it to a mix of warm water and sugar, and then mixed with flour to form a dough.
- The dough is worked (kneaded) for 5-10 minutes and then incubated at around 25-27°C. The growing yeast produces CO_2, which causes the dough to rise.
- Imagine you wanted to find out if the amount of yeast added affected the doubling time of the dough (the time taken for the volume of dough to double in size).

New dough, beginning to rise

The dough, 40 minutes later

Photos: ElinorD cc 3.0

1. In the experiment to investigate the effect of added yeast mass on the doubling time of dough (above right):

   (a) Identify the independent variable: *The mass of the yeast*

   (b) Identify the dependent (response) variable: *The time taken for the volume of dough to double*

   (c) What would the controlled (constant) variables be?

   Temperature, mass of flour and other ingredients used (water, sugar), amount of kneading.

   (d) What would the control be? *Having no yeast.*

2. In the space right, draw and label the axes on which you could plot your results.

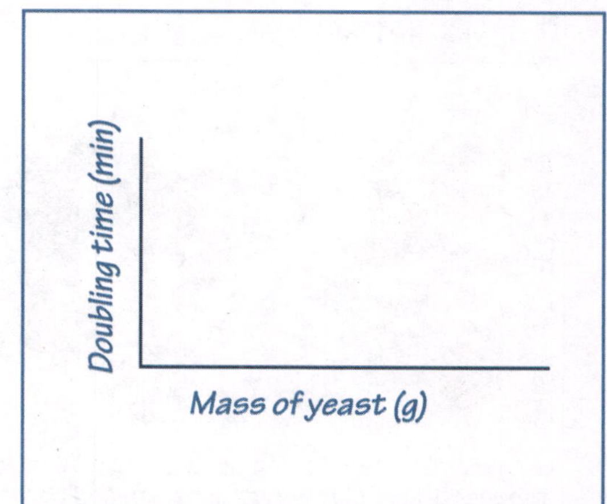

3. What difficulties or sources of error are most likely to interfere with obtaining a reliable result in this cooking experiment?

   Providing the exact same conditions for yeast growth. Yeast, being a living organism, can behave unpredictably and this can affect results.

4. How would you need to change the experimental method if you wanted to measure the effect of changing temperature on doubling time instead of the amount of yeast?

   The mass of yeast used would now need to be kept the same. Instead the temperature at which the dough was incubated at would need to be varied.

©2020 **BIOZONE** International
ISBN: 978-1-927309-82-7
Photocopying Prohibited

15 Measurement and Quantitative Analysis

Key Question: Why do we need to ensure we use the correct equipment and techniques during a chemistry experiment?

- In experimental work, we use the term 'experimental error' to describe the uncertainty created by variations in the data. These are often systematic errors (consistent over- or underestimation), which arise because of instrument errors or errors in the technique of the experiment.
- Systematic errors can be reduced by choosing apparatus that is appropriate for the experiment or analysis and calibrating and using it correctly. Calibration ensures that the equipment is delivering the correct response. This is done by measuring the actual response against a standard that you know to be accurate (true).
- Inaccurate results are produced if equipment is not calibrated. In a high school laboratory, inaccurate results are annoying but in industry they could have serious effects. For example, a faulty pH meter in a food processing plant may result in not enough acid being added to a food, which could result in spoilage and costly product recalls.

Selecting the correct equipment

It is important that you choose equipment that is appropriate for the type of measurement you want to take. For example, which of the glassware would you use if you wanted to measure 225 mL?

The 500 mL graduated cylinder has graduations every 5 mL whereas the 500 mL beaker has graduations every 50 mL. It would be more accurate to measure 225 mL in a graduated cylinder.

Different types of **graduated glassware** have different accuracies. A beaker is less accurate than a measuring cylinder and a measuring cylinder is less accurate than a pipette. Volumetric glassware is the most accurate. Can you think why?

What about pipettes?

Pipettes are a common tool for volumetric work in chemistry and you are likely to come across several different types, from simple, single unit glass pipettes (right) to adjustable micropipettes for delivering very small volumes (up to 1 mL).

In titrations, you will frequently use a volumetric pipette of 10-25 mL volume.

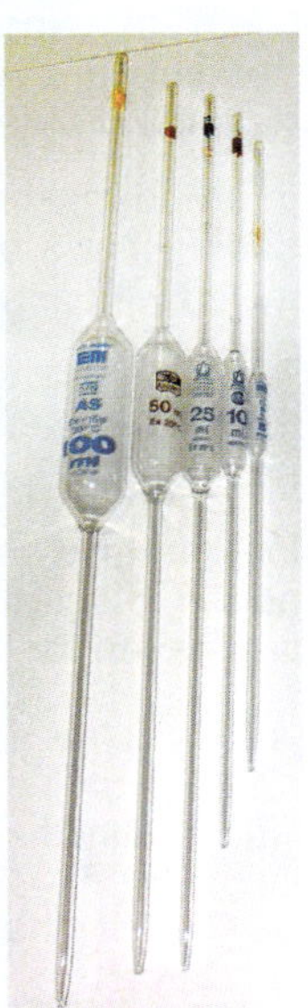

Volumetric pipettes have a large bulb and a long narrow part above that has one graduation mark. They are calibrated to deliver accurately a fixed volume of liquid (just like a volumetric flask).

Different devices, such as bulbs, are used to draw the liquid into the pipette.

Percentage errors

Percentage error allows you to express how far your result is from the ideal (expected) result. The equation for calculating percentage error is:

$$\frac{\text{experimental value} - \text{ideal value}}{\text{ideal value}} \times 100$$

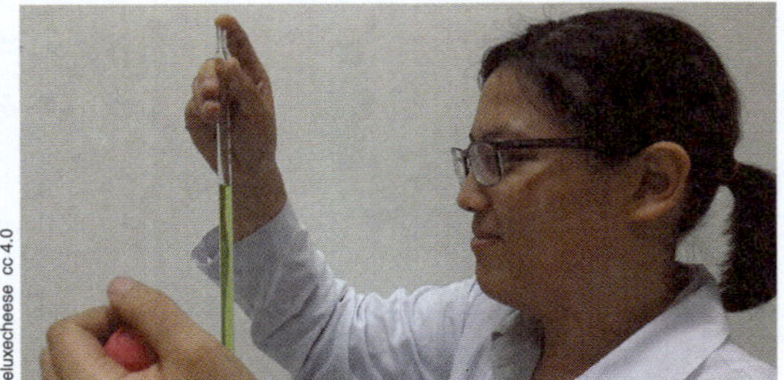

Example: To determine the accuracy of a 10 mL pipette, dispense 10 mL of water from the pipette and weigh the dispensed volume. The mass (g) = volume (mL). Imagine the volume is 10.02 mL.

$$\frac{10.02 - 10.0}{10.0} \times 100$$

The percentage error = 0.2% (the pipette is dispensing **more** than it should).

1. Assume that you have the following measuring devices available: 50 mL beaker, 50 mL cylinder, 50 mL volumetric pipette, 25 mL cylinder, 10 mL pipette, 10 mL beaker, 25 mL volumetric pipette. What would you use to most accurately measure:

(a) 21 mL: 25 mL cylinder

(b) 48 mL: 50 mL cylinder

(c) 9 mL: 10 mL pipette

(d) 25 mL: 25 mL volumetric pipette

(e) 50 mL: 50 mL volumetric pipette

2. Calculate the percentage error for the following situations (show your working):

(a) A 1 mL pipette delivers a measured volume of 0.98 mL:

$((0.98-1)/1) \times 100 = -2\%$

(b) A 5 mL pipette delivers a measured volume of 4.98 mL:

$((4.98-5)/5) \times 100 = -0.4\%$

(c) A 50 mL volumetric pipette delivers a measured volume of 50.025 mL:

$((50.025-50)/50) \times 100 = 0.05\%$

©2020 BIOZONE International
ISBN: 978-1-927309-82-7
Photocopying Prohibited

Using titration in chemistry

- Chemistry often involves titration. As this involves using solutions, it is also called volumetric analysis. During a titration, a solution of known concentration (called a **standard solution**) is used to determine the concentration of a solution with an unknown concentration.
- A standard solution contains a known and accurate amount of a substance or element and is made by dissolving a **primary standard** in a suitable solvent (often distilled water). A primary standard is typically a soluble solid compound that is very pure, with a consistent formula and a relatively high molar mass.
- The solution being added from the burette is called the titrant. The solution in the receiving conical flask contains 1-2 drops of **indicator** that changes color at a known point (e.g. pH). The concentration of the unknown can be calculated from how much titrant was added to reach an end point (usually neutralization in an acid-base titration).
- The end color depends on the indicator. Methyl orange and phenolphthalein are common indicators for acid-base titrations. The titration is repeated several times to obtain values within 0.5 mL and then a mean value is calculated.

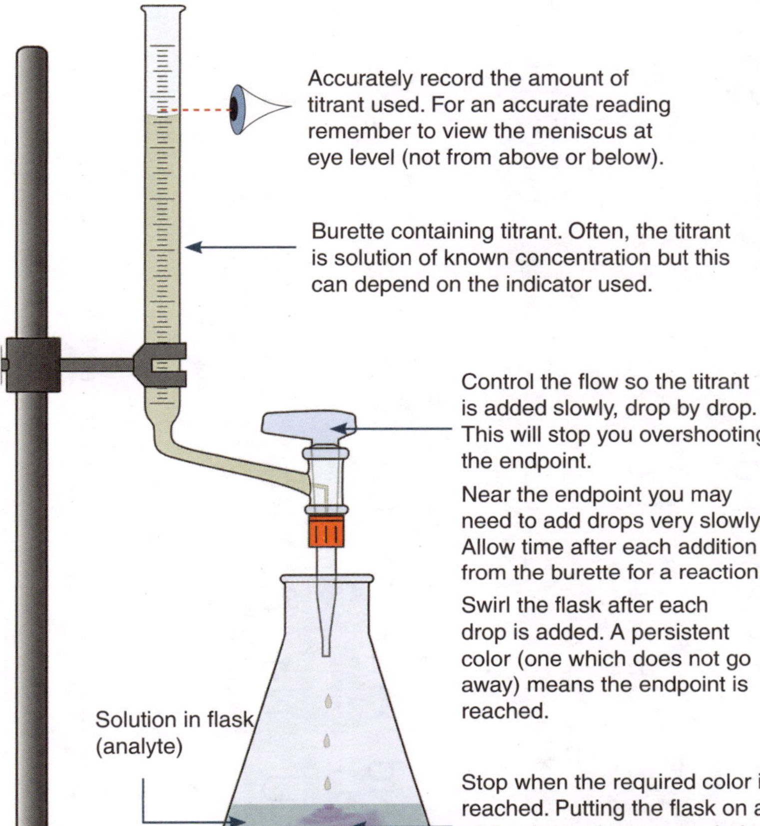

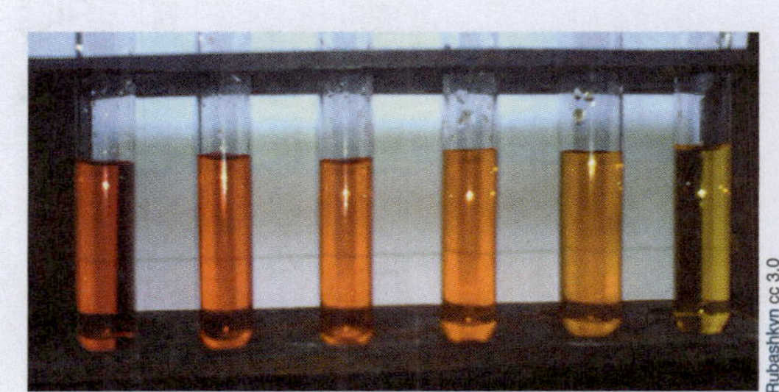

Methyl orange shows a red color below pH 3.1 and a yellow color above pH 4.4.

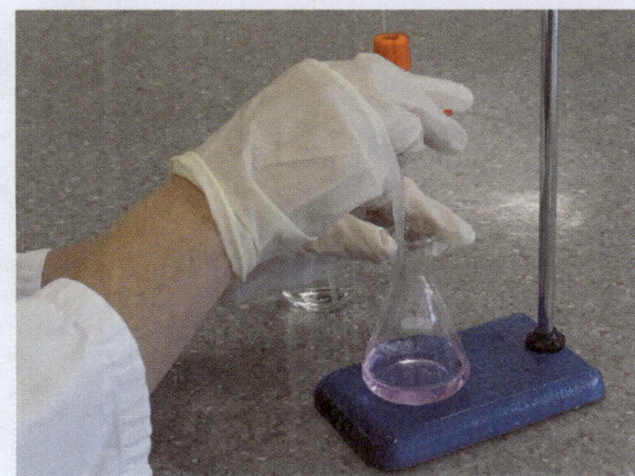

The phenolphthalein molecule is colorless but the ion is pink. When a base is added to a solution containing phenolphthalein, more of it becomes ionized and the solution turns pink. Titrate just to the end point (color change), no further!

3. Indicators themselves are weak acids or weak bases. Why do you think then that only a 1-2 drops of indicator is used in an acid-base titration?

   Adding too much may affect the overall acidity of the solution and affect the results.

4. Why do you think a primary standard typically has a high molar mass?

   High molar masses reduce measuring errors when weighing out the solid.

5. Discuss potential sources of error during a volumetric analysis with your lab partners (from making the standard solution to performing the titration). Name them and describe how you would minimize them. Use more paper if you need to:

   Potential errors come from weighing the mass, pipetting of the sample, the use of the burette for measuring the reactant used, the time when the color change is said to have occurred. These sources of error can be reduced by being careful and accurate when measuring and in some cases by repeating the measurement and using means (averages) where appropriate.

©2020 **BIOZONE** International
ISBN: 978-1-927309-82-7
Photocopying Prohibited

16 Periodic Table

Atomic number → 26 | 1.83 ← Electronegativity
Chemical symbol → Fe
Name → Iron
Relative atomic mass → 55.845

* = unknown

STATE OF MATTER
SOLID GAS LIQUID UNKNOWN

	1	2	3	4	5	6	7	8	9	10	11	12	13	14	15	16	17	18
1	1 2.20 H Hydrogen 1.00794																	2 * He Helium 4.002602
2	3 0.98 Li Lithium 6.941	4 1.57 Be Beryllium 9.012182											5 2.04 B Boron 10.811	6 2.55 C Carbon 12.0107	7 3.04 N Nitrogen 14.0067	8 3.44 O Oxygen 15.9994	9 3.98 F Fluorine 18.998403	10 * Ne Neon 20.1797
3	11 0.93 Na Sodium 22.98976	12 1.31 Mg Magnesium 24 3050											13 1.61 Al Aluminium 26.98153	14 1.90 Si Silicon 28.0855	15 2.19 P Phosphorus 30.97696	16 2.58 S Sulfur 32.065	17 3.16 Cl Chlorine 35.453	18 * Ar Argon 39.948
4	19 0.82 K Potassium 39.0983	20 1.0 Ca Calcium 40.078	21 1.36 Sc Scandium 44.95591	22 1.54 Ti Titanium 47.867	23 1.63 V Vanadium 50.9415	24 1.66 Cr Chromium 51.9962	25 1.55 Mn Manganese 54.93804	26 1.83 Fe Iron 55.845	27 1.88 Co Cobalt 58.93319	28 1.91 Ni Nickel 58.6934	29 1.90 Cu Copper 63.546	30 1.65 Zn Zinc 65.38	31 1.81 Ga Gallium 69.723	32 2.01 Ge Germanium 72.64	33 2.18 As Arsenic 74.92160	34 2.55 Se Selenium 78.96	35 2.96 Br Bromine 79.904	36 2.0 Kr Krypton 83.798
5	37 0.82 Rb Rubidium 85.4678	38 0.95 Sr Strontium 87.62	39 1.22 Y Yttrium 88.90585	40 1.33 Zr Zirconium 91.224	41 1.6 Nb Niobium 92.90638	42 2.16 Mo Molybdenum 95.96	43 1.9 Tc Technetium 98.907	44 2.2 Ru Ruthenium 101.07	45 2.28 Rh Rhodium 102.9055	46 2.20 Pd Palladium 106.42	47 1.93 Ag Silver 107.8682	48 1.69 Cd Cadmium 112.441	49 1.78 In Indium 114.818	50 1.96 Sn Tin 118.710	51 2.05 Sb Antimony 121.760	52 2.1 Te Tellurium 127.60	53 2.66 I Iodine 126.9044	54 2.6 Xe Xenon 131.293
6	55 0.79 Cs Caesium 132.9054	56 0.89 Ba Barium 137.327	57 - 71	72 1.3 Hf Hafnium 178.49	73 1.5 Ta Tantalum 180.9478	74 2.36 W Tungsten 183.84	75 1.9 Re Rhenium 186.207	76 2.2 Os Osmium 190.23	77 2.20 Ir Iridium 192.217	78 2.28 Pt Platinum 195.084	79 2.54 Au Gold 196.9665	80 2.0 Hg Mercury 200.59	81 1.62 Tl Thallium 204.3833	82 2.33 Pb Lead 207.2	83 2.02 Bi Bismuth 208.9804	84 2.0 Po Polonium 208.982	85 2.2 At Astatine 209.987	86 * Rn Radon 222.018
7	87 0.7 Fr Francium (223)	88 0.9 Ra Radium (226)	89 - 103	104 * Rf Rutherfordium (261)	105 * Db Dubnium (262)	106 * Sg Seaborgium (266)	107 * Bh Bohrium (264)	108 * Hs Hassium (277)	109 * Mt Meitnerium (268)	110 * Ds Darmstadium (271)	111 * Rg Roentgenium (272)	112 * Cn Copernicium (285)	113 * Nh Nihonium (284)	114 * Fl Flerovium (289)	115 * Mc Moscovium (288)	116 * Lv Livermorium (293)	117 * Ts Tennessine (294)	118 * Og Oganesson (294)

lanthanoids	57 1.10 La Lanthanum 138.9054	58 1.12 Ce Cerium 140.116	59 1.13 Pr Praseodymium 140.9076	60 1.14 Nd Neodymium 144.242	61 * Pm Promethium (145)	62 1.17 Sm Samarium 150.36	63 * Eu Europium 151.964	64 1.2 Gd Gadolinium 157.25	65 * Tb Terbium 158.9253	6 1.22 Dy Dysprosium 162.500	67 1.23 Ho Holmium 164.9303	68 1.24 Er Erbium 167.259	69 1.25 Tm Thulium 168.9342	70 * Yb Ytterbium 173.054	71 1.27 Lu Lutetium 174.9668
actinoids	89 1.1 Ac Actinium (227)	90 1.3 Th Thorium 232.0380	91 1.5 Pa Protactinium 231.0358	92 1.38 U Uranium 238.0289	93 1.36 Np Neptunium (237)	94 1.28 Pu Plutonium (244)	95 1.3 Am Americium (243)	96 1.3 Cm Curium (247)	97 1.3 Bk Berkelium (247)	98 1.3 Cf Californium (251)	99 1.3 Es Einsteinium (252)	100 1.3 Fm Fermium (257)	101 1.3 Md Mendelevium (258)	102 1.3 No Nobelium (259)	103 * Lr Lawrencium (262)

nonmetals | alkaline earth metals | metalloids | halogens | lanthanoids
alkali metals | transition metals | post-transition metals | noble gases | actinoids

P

©2020 BIOZONE International
ISBN: 978-1-927309-82-7
Photocopying Prohibited

Matter and its Interactions

Concepts and connections

Use arrows to make connections between related concepts in this section of the book

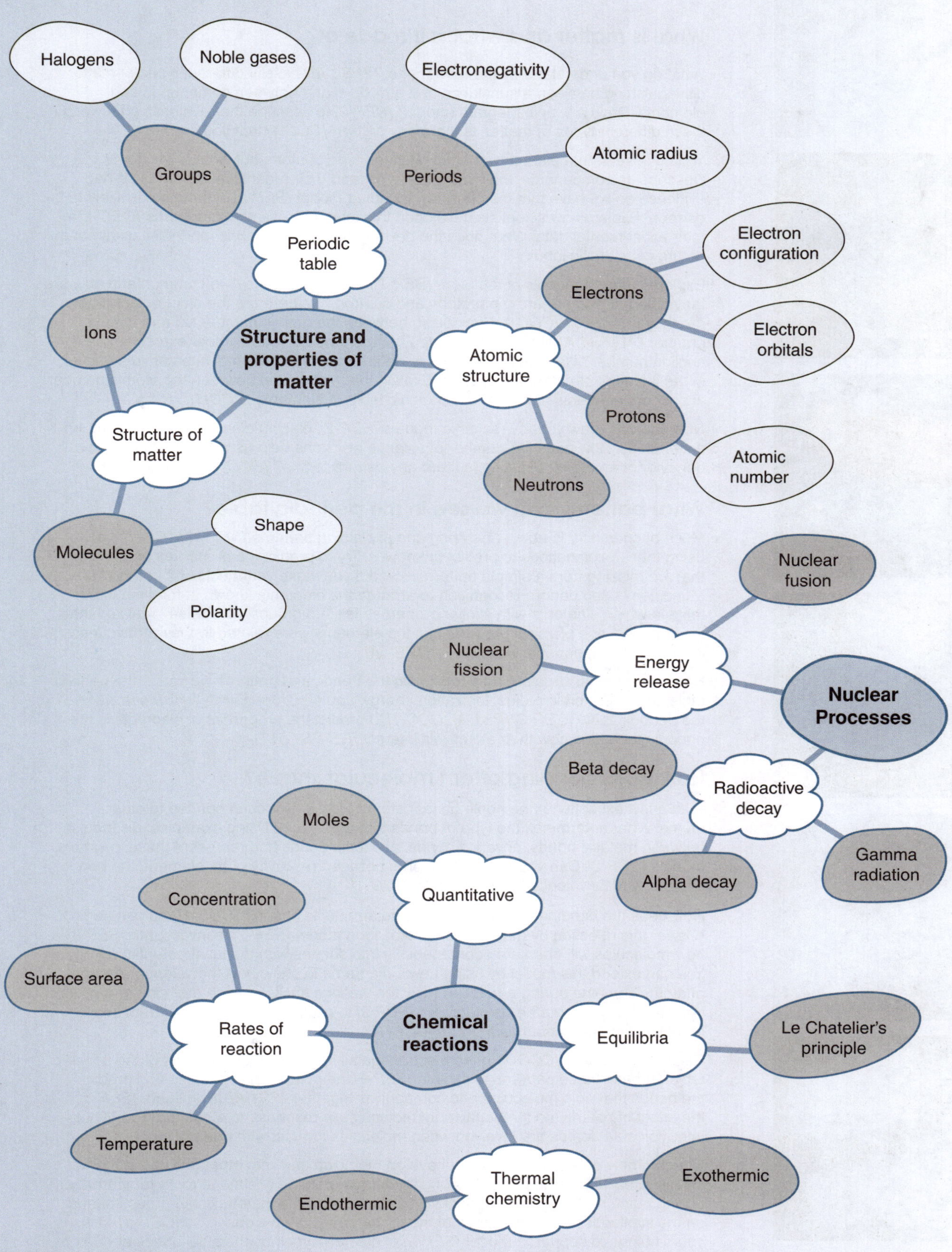

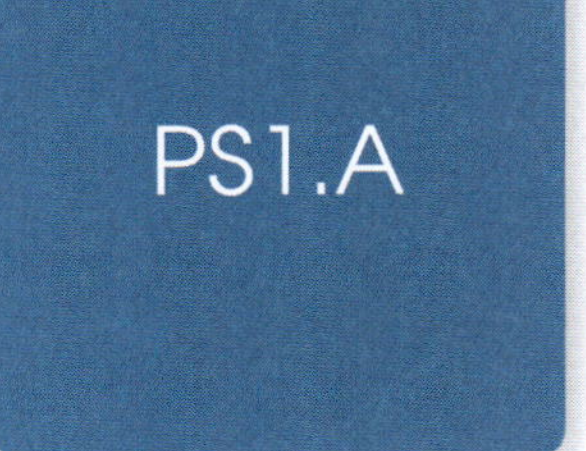

2: Structure and Properties of Matter

Anchoring Phenomenon

Salt on those? What are the underlying causes of the different properties of compounds? — 17 32

What is matter and what is it made of?

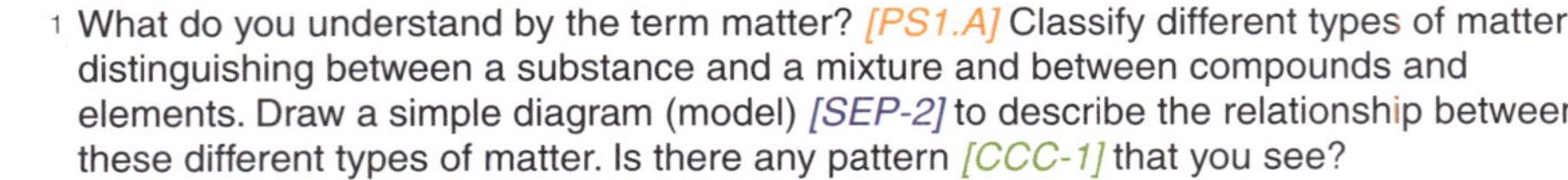

☐ 1 What do you understand by the term matter? *[PS1.A]* Classify different types of matter distinguishing between a substance and a mixture and between compounds and elements. Draw a simple diagram (model) *[SEP-2]* to describe the relationship between these different types of matter. Is there any pattern *[CCC-1]* that you see? — 18

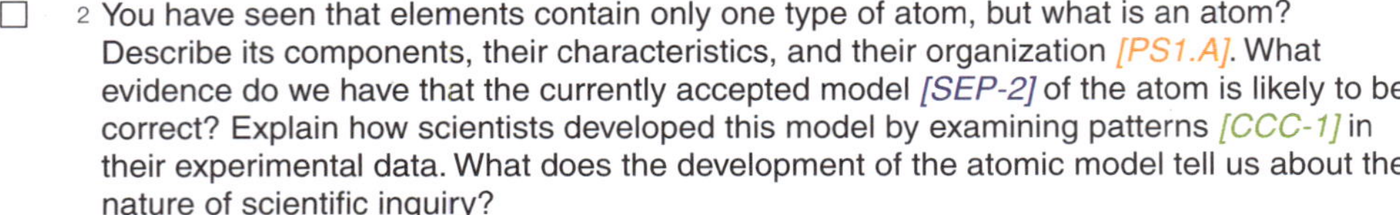

☐ 2 You have seen that elements contain only one type of atom, but what is an atom? Describe its components, their characteristics, and their organization *[PS1.A]*. What evidence do we have that the currently accepted model *[SEP-2]* of the atom is likely to be correct? Explain how scientists developed this model by examining patterns *[CCC-1]* in their experimental data. What does the development of the atomic model tell us about the nature of scientific inquiry? — 19 20

Bionerd CC 3.0

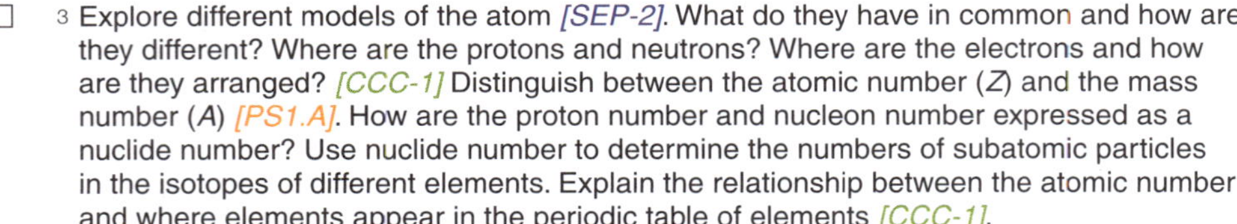

☐ 3 Explore different models of the atom *[SEP-2]*. What do they have in common and how are they different? Where are the protons and neutrons? Where are the electrons and how are they arranged? *[CCC-1]* Distinguish between the atomic number (Z) and the mass number (A) *[PS1.A]*. How are the proton number and nucleon number expressed as a nuclide number? Use nuclide number to determine the numbers of subatomic particles in the isotopes of different elements. Explain the relationship between the atomic number and where elements appear in the periodic table of elements *[CCC-1]*. — 21

☐ 4 Now you can draw your own atomic models *[SEP-2]*, placing electrons in shells around the nucleus. How does the number of valence electrons help us to predict the chemical behavior of an atom? *[PS1.A]* Is there any pattern to this? *[CCC-1]* *[HS-PS1-1]* — 22 25 33

What patterns can we see in the periodic table?

☐ 5 Much of chemistry is about analyzing and explaining patterns *[SEP-4]* *[CCC-1]* and using these explanations to predict unknowns. Try to predict the properties of objects that are missing from a simple pattern. How did you make your decision? In the 1800s, Mendeleev used pattern recognition to arrange the elements known at the time into a table in which elements with similar properties fell into groups. Mendeleev realized that the chemical and physical properties of the elements were related to their atomic mass in a predictable, periodic way *[SEP-2]* *[PS1.A]*. — 23

☐ 6 How does atomic structure help to explain the trends and patterns we see in the periodic table, such as atomic radius, ionization energy, and electronegativity? How can we use the periodic table *[SEP-2]* *[PS1.A]* *[CCC-1]* to predict the properties of elements in the periodic table and how their atoms will interact *[HS-PS1-1]*. — 24 25 33

How does bonding affect molecular shape?

☐ 7 With a few exceptions (the noble gases) atoms are always found bonded to other atoms. What determines the type of bonds that are made? Using examples, distinguish between metallic bonds, covalent bonds, and ionic bonds and represent these in simple models *[SEP-2]*. Can you see any patterns between reactivity of an element, the bonds it forms, and its valence state? *[PS1.A]* *[CCC-1]* — 26-28

☐ 8 How does the bonding within a molecule determine its shape? *[CCC-1]* You can begin to answer this question by making a simple balloon model *[SEP-2]* to investigate *[SEP-3]* how molecules will orientate spontaneously into shapes with the least potential energy. You can extend this model by using Lewis diagrams to show how the valence electrons of atoms combine during a reaction to fill the valence shells of each atom in the reaction. You can then use your Lewis diagrams *[SEP-2]* to explain *[SEP-6]* the different shapes of molecules or compounds *[PS1.A]* *[CCC-1]*. — 29

☐ 9 Recall the patterns *[CCC-1]* in the electronegativities of the first 20 atoms of the periodic table and use these patterns to explain *[SEP-6]* the polarity of molecules and predict the bonds that hold molecules and compounds together *[PS1.A]*. Investigate *[SEP-3]* the effect of polarity on the behavior of molecules in response to a charged rod. What intermolecular forces are involved when molecules interact with one another? — 30

☐ 10 Observe how different materials behave on their own and with other substances. Investigate *[SEP-3]* the properties of different materials (e.g. state at room temperature, boiling point, melting point, conductivity) and organize the materials into groups based on the similarities and differences of the properties you have observed *[CCC-1]*. Use your findings to support a model *[SEP-2]* of different types of chemical bonds and attractions *[PS1.A]* *[HS-PS1-3]*. — 31

2. Structure and Properties of Matter

Anchoring phenomenon

The anchoring phenomenon *"Salt on Those?"* introduces students to the idea that different compounds have different crystalline shapes. The compounds shown are chosen to contain over-lapping ions or groups of ions (e.g. copper sulfate, calcium sulfate, and calcium chloride) so that students can see that changing just one group can have a significant effect on the properties of a compound. It is also an opportunity to remind students of the difference between everyday use of language and scientific use of language.

What is matter and what is it made of?

1. Students review the differences between substances, compounds and elements, identifying the patterns involved. They develop a model of these patterns.

2. The atom is the functional building block of matter. The story of atomic theory illustrates the nature of science well (science is based on empirical evidence). Scientists use new technologies to test current models and then revise models (or develop new ones) when new evidence adds to or contradicts existing models. Rutherford's gold foil experiment is a particularly good example of asking "What if?" in which he asked what if we looked in front of the gold foil, not just behind it?

3. Students need to understand the difference between protons, neutrons, and electrons. They will need to distinguish between the atomic number and the atomic mass of an atom and realize the atomic number reflects the position of the atom on the periodic table.

4. The octet model of electron configuration works well for predicting the valence electrons of atoms up to atomic number 20. Beyond this it becomes less useful. The octet model can be used to show that some atoms have the same number of valence electrons. The significance of this is that these atoms will have similar behaviors during chemical reactions and form bonds with other atoms in similar ways.

What patterns can we see in the periodic table?

5. The periodic table is something students of chemistry need to be familiar with. It is not necessary for them to memorize the table, but they should understand its structure and how this structure reflects the properties of the elements. It is useful for students to realize that the standard periodic table used in this book is again a type of model, which shows some aspects well but is limited in other ways. An internet search of alternative periodic tables should show students that there are other ways of presenting the table based on different end uses and aims.

6. Plotting atomic number against various properties of elements reflects the periodic structure of the periodic table. Students initially plot and describe these trends (e.g. atomic number vs atomic radius). Students will later explain these trends by relating them to the structure of the periodic table and electron configuration. Students can use these explanations to predict behavior of other elements.

How does bonding affect molecular shape?

7. Students are introduced to bonding using the example of sodium metal reacting with chlorine gas to form a crystalline solid, sodium chloride. Students must use their knowledge of the periodic table to explain what has happened during the reaction. Ionic and covalent bonds are covered in consecutive activities. These types of bonds, as well as metallic bonds, are reviewed in chapter 6. There are a number of non-main block elements that students are introduced to. These form ions in a way that is difficult to explain using the octet rule, but they are commonly used in school laboratories and so are useful for students to know. Covalent bonding introduces Lewis dot diagrams. These are useful for predicting the number of covalent bonds and the shape of the molecule produced.

8. Understanding molecular shape is important for later discussions of interactions between molecules. Valence shell electron pair repulsion theory (VSEPR) explains why molecules take up certain shapes. Balloons can be used to model molecular shape. Using a different colored balloon to represent non bonding electrons extends this model. Students will build a tetrahedral model (representing CH_4 for example) but adding more balloons is possible for extension (such as modeling SF_6).

9. Bonding polarity is important in explaining intermolecular interactions. Students should realize that differences in electronegativities lead to unequal sharing of electrons in covalent bonds. In symmetrical molecules these differences cancel out, but in non-symmetrical molecules they do not, producing polar molecules. Small polar molecules (e.g. water) experience relatively strong electrostatic forces and are drawn together. They are usually liquids at room temperature. Small non-polar molecules (e.g. methane) experience much weaker forces and are usually gases at room temperature. Students should be able to explain these observations.

10. Students compare the physical properties of substances. They analyze the data they gather and group the substances based on their observations. Students should be able to identify ionic, covalent and metallic substances and explain why they have placed the substances in those groups.

Challenge question:

Why do plasma globes contain noble gases?

Download this page from BIOZONE's Resource Hub

17 Salt On Those?

Key Question: What are the underlying causes of the different properties of compounds?

- As with many commonly used words, the common use of the word is different to the scientific use. Examples include alien, mole, belt, and family. Salt is another example.
- Talk to someone in your classroom about salt and you are most likely talking about table salt or sodium chloride. In everyday language this is what salt means. In chemistry, however, a salt is defined as an ionic compound that can be formed from the neutralization of an acid by a base. Sodium chloride is just one of hundreds of different salts.
- The images below show a selection a crystallized salts. Note that many of these are hydrated salts, meaning they have water bound inside them. A cleavage plane describes how a crystal naturally splits along a flat surface.

Sodium chloride: colorless, cubic crystal. Cleaves in three planes (at right angles).

Hans-Joachim Engelhardt CC.4.0

Calcium chloride: colorless, rhombohedral (diamond-shaped) crystal. Cleaves in three planes (but not at right angles).

Géry Parent CC.3.0

Calcium fluoride. Colorless octahedral crystal. Cleaves in four planes.

Eurico Zimbres CC 2.0

Hydrated calcium sulfate dihydrate: Colorless elongated prisms. Cleaves in three planes (but not at right angles).

Hydrated copper sulfate: blue rhombohedral crystal. Cleaves in three planes (but not at right angles).

Hydrated copper chloride: green elongated prisms. Cleaves in three planes.

1. Have you seen any of these types of salts before? List the examples above that you know about:

   Student's answer

2. List the elements in the salts you listed:

   Student's answer based on their answer above. e.g. sodium chloride: sodium and chlorine

3. What do you think causes the different shapes and planes of cleavage of the crystals?

   Student's answer. Some students may realize that the shape of the crystal is related to the arrangement of the particles in the crystal.

4. What do you think causes the color of the crystals?

   Student's answer. Some may realize that the color is related to the types of the particles in the crystal.

©2020 BIOZONE International
ISBN: 978-1-927309-82-7
Photocopying Prohibited

18 What is Matter?

Key Question: How do we define matter? What are the different types of matter and how can we classify them?

You may be familiar with the concept of matter. It includes everything you can touch, and a lot of things you can't. However, it is worth reviewing your understanding as an introduction to the structure and properties of matter.

In physical sciences two words describe all the physical objects around you: **matter** and **substance**.

Matter

- Matter is defined as anything that has mass and takes up space. Anything you can touch is matter, e.g. rocks, water, this book. Some things you can't touch, but can probably still feel, are also matter, such as exhaust gases from a rocket engine and plasma.

Substances

- A **substance** is a pure form of matter. Water is an example of a pure form of matter (a substance). So is iron. Steel is not. Neither is air. Air and steel are **mixtures**; they contain more than one type of substance.

Iron is hardly ever used in its pure form. Steel, a mixture of iron and carbon, is much more common due to its greater strength.

Classifying matter

- Compare the two substances mentioned above for a moment: water and iron. What is water made of? What is iron made of? You may be aware that water has the chemical formula H_2O and that iron has the formula Fe. But what do these formulas actually mean? In simple terms they mean that water is composed of two different **elements** (H and O) and that iron is composed on only one type of element (Fe). An element is a substance composed of just one type of atom.
- This means we can separate substances into two more groups, those that have just one kind of element and those that have more than one kind of element. Substances with more than one kind of element are called **compounds**.

1. Use the bold words in the text to produce a flow chart in the space below to classify the different groups of matter. You will need to add a short definition to each category:

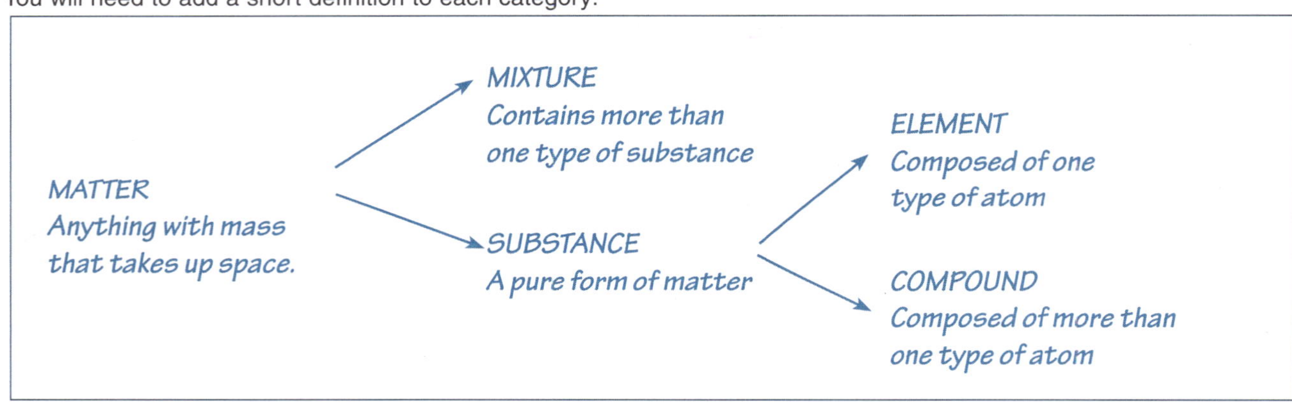

2. Classify the following using your chart:

(a) 9 ct gold: *Mixture*

(b) Sugar: *Compound*

(c) A soda drink: *Mixture*

(d) Sodium chloride (table salt): *Compound*

(e) Oxygen gas: *Element*

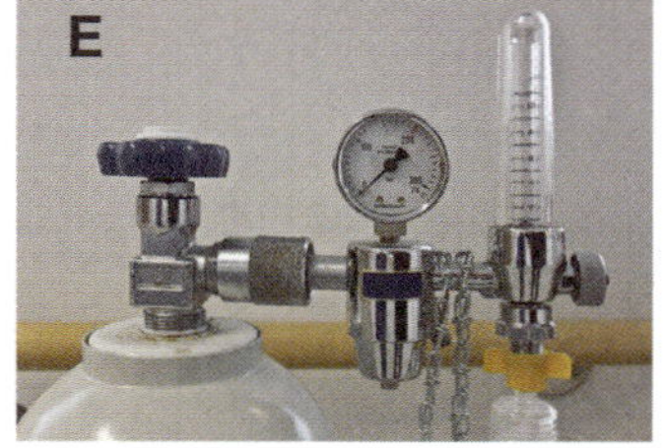

©2020 **BIOZONE** International
ISBN: 978-1-927309-82-7
Photocopying Prohibited

19 Introduction to the Atom

Key Question: What is an atom and what is it made of?

The image below is of the rough cuts of the Cullinan diamond, found in South Africa in 1905 and now part of the Crown Jewels of the United Kingdom.

- Diamond is a substance made of the element carbon. The diamond that these smaller diamonds were cut from was originally 3106.75 carats, or 621.35 grams. Now imagine cutting the larger diamond shown here (which weighs 106.08 grams) in half, then in half again. How long could we keep doing this before we reached something that couldn't be cut? Is that even possible?
- This is same problem the ancient Greek philosophers Leucippus and Democritus thought of around 400 BCE. They believed that eventually you would get to a point where you could not divide something any further. They called this *atomos*, meaning uncuttable, from which we get the word atom.
- The atom of the ancient Greeks was a theoretical construction. Today we know that there is indeed a point where you can no longer cut or divide a substance or element into something smaller.
- Thus the atom is defined as the smallest unit of an element (e.g. carbon) that can be identified in terms of its chemical properties. If the atom is divided it loses the chemical properties associated with that element, (i.e. it is no longer a carbon atom).

The atom

- Atoms are the basic unit of an element that can carry out a chemical reaction. Each element (e.g. gold, iron, hydrogen) is made up of only one type of atom. The image on the right is from a scanning tunneling microscope. Each small circular structure is a gold atom with a radius of about 135 picometers (about 135 trillionths of a meter).
- Atoms are made up of three components and have a precise structure. Change it in any way and the behavior or even the type of atom changes.
- The components of atoms are protons, neutrons, and electrons. The table below summarizes their properties:

Particle	Symbol	Charge	Mass (g)
Proton	p	+1	1.673×10^{-24}
Neutron	n	0	1.675×10^{-24}
Electron	e-	-1	9.109×10^{-28}

Gold atoms

1. The mass of a carbon atom is about 1.99×10^{-23} grams. How many times would you need to cut the largest diamond in the picture, top, in half to reach this mass? (Hint, try using a spreadsheet for a quick answer).

   About 86 times

2. (a) What is the smallest (least massive) component of an atom? The electron

   (b) What is the largest (most massive) component of an atom? The neutron

   (c) What is the charge on a proton? +1

3. Compare and contrast an atom and an element: An atom is the basic unit of the element. It is the smallest unit with the characteristics of the element. An element is a substance with all the same atoms.

©2020 BIOZONE International
ISBN: 978-1-927309-82-7
Photocopying Prohibited

P PS1.A

20 Atomic Theory and Rutherford's Experiment

Key Question: How was the current model of the atom developed?

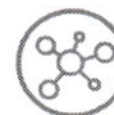

The history of the atom

- The idea of the atom has been around for many centuries and thought of in many cultures. Greek and Indian philosophers are known to have proposed the idea many centuries BCE.
- The idea of the atom reached Europe in the late 14th century and was popularized in France in the late 15th century. During the 17th century the idea of atoms began to gain acceptance in scientific communities as research into the make-up of chemicals became more rigorous.
- In the 19th century, John Dalton used the concept of atoms to explain why certain elements always reacted together in the same ratios. Dalton's atomic theory was verified by Jean Perrin in the early 20th century using Einstein's explanation of Brownian motion.
- The theory of indivisible atoms was overturned in 1904 by J.J. Thomson with the discovery of the electron. In 1911, experiments led by Ernest Rutherford (notably the gold foil experiment) discovered that atoms have an extremely dense nucleus around which electrons "orbit".
- In 1913, Niels Bohr adapted Rutherford's model of the atom to produce the Bohr model in which electrons are found around the nucleus at specific orbits.
- In 1932, James Chadwick discovered the neutron, adding the last subatomic particle to the atomic model.

Ernest Rutherford carried out some of the most important experiments into atomic structure. He was awarded the Nobel prize in 1908 and is sometimes called the father of nuclear physics.

1. In the space below produce a time line of the development of the atomic model. Research and add any other relevant developments not mentioned in the text above:

400 BCE: Democritus theorizes atom make up all matter. They are indivisible.

1803: John Dalton proposes that elements consisted of atoms that were identical and had the same mass and that compounds were atoms from different elements combined together.

1869 Dmitri Mendeleev creates the periodic table.

1897 J.J. Thomson determines the charge to mass ratio of electrons..

1898 Rutherford discovers alpha, beta, and gamma rays in radiation.

1903 Hantaro Nagaoka proposes an atomic model called the Saturnian Model to describe the structure of an atom.

1906 Hans Geiger invents a device that could detect alpha particles.

1911 Rutherford's gold foil experiment. He proposes an atomic structure of a central positive nucleus orbited by negative electrons.

1913 Niels Bohr modifies Rutherford's model using quantum theory, proposing electrons are found in shells.

1926 Erwin Schrödinger modifies Bohr's shells into electron clouds, areas of probability of where electrons will be.

1932 James Chadwick discovers neutron in the atom's nucleus.

©2020 BIOZONE International
ISBN: 978-1-927309-82-7
Photocopying Prohibited

Finding the nucleus

- Under the direction of Ernest Rutherford, Hans Geiger and Ernest Marsden carried out a series of experiments in which alpha particles (positively charged helium nuclei) were fired at thin gold foil. They observed the pattern produced on a detection screen (below).

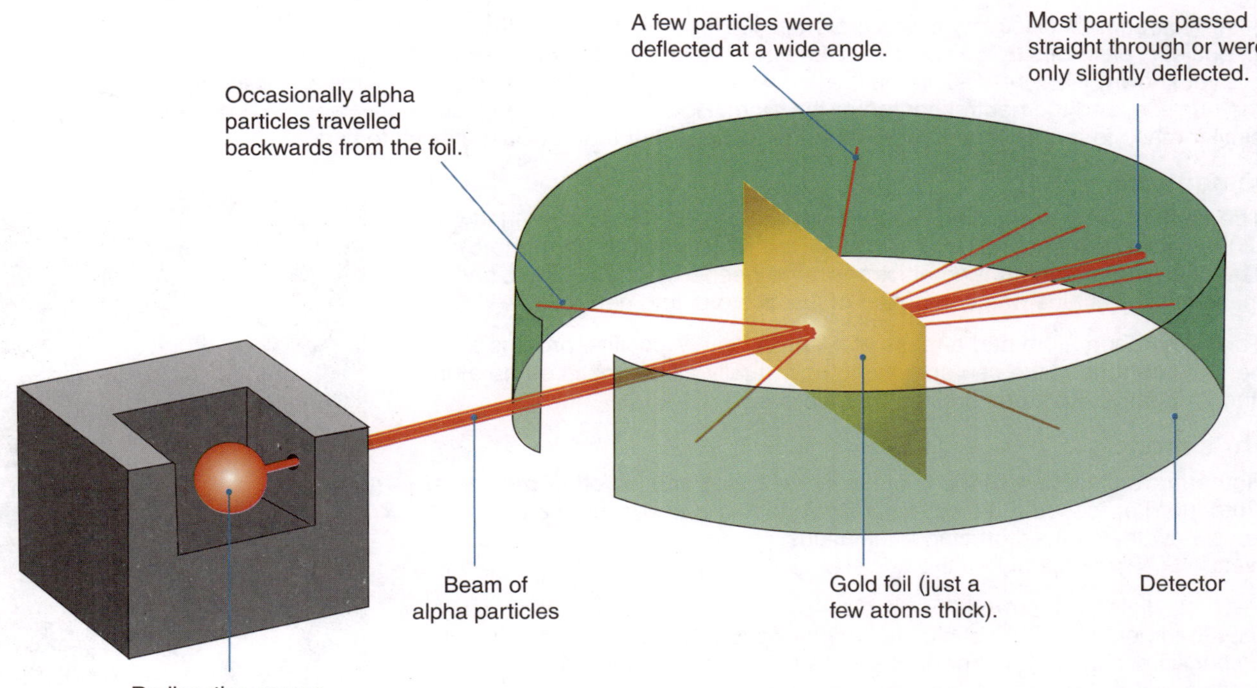

- Based on calculations from earlier atomic models, it was expected that the alpha particles would pass straight through the gold foil or be only very slightly deflected.
- Instead the alpha particles were observed to scatter and occasionally "bounce" back at more than 90°.
- Rutherford later famously described the observation as like "firing a 15 inch shell at a piece of tissue paper and having it come back and hit you".

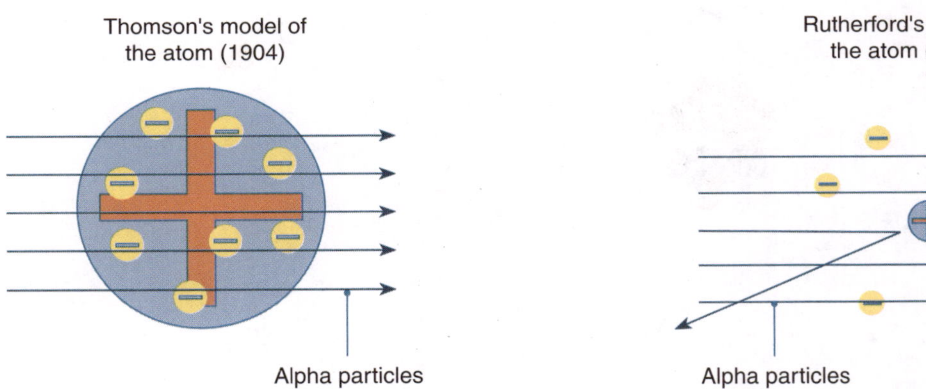

J.J. Thomson proposed that atoms were a region of positive charge in which electrons were embedded. It is often referred to as the plum pudding model. Under these conditions, alpha particles would pass right through the atom with little deflection.

Rutherford elaborated on an alternative theory proposed by Hantaro Nagaoka known as the Saturn model. Rutherford proposed the atom was in fact made of an incredibly dense nucleus around which the electrons orbited.

2. What were the expected results of the gold foil experiment? *It was expected that the alpha particles would pass straight through the atoms without interacting in them.*

3. Describe the observed results: *The alpha particles deviated from a straight path, some by a very wide angle. A few particles bounced back in the way they had come.*

4. How did Rutherford explain the observed results? *He placed all the positive charge of the atom at its center, explaining that the positive alpha particles were repelled or deflected by this dense positive charge.*

©2020 BIOZONE International
ISBN: 978-1-927309-82-7
Photocopying Prohibited

Problems with Rutherford's model

- Although Rutherford managed to show the separate nucleus of an atom, he knew there were problems with his model. For example, it was known that the mass of a nitrogen atom was 14 units, but the charge in the nucleus was +7. It was also known that electrons were negative and tiny compared to the mass of the nucleus.
- After Rutherford's discovery of the proton in 1920 it was theorized that in a neutral nitrogen atom there must be 7 "nuclear" electrons in the nucleus to balance the charge of (the presumed)14 positive protons (neutrons were not yet known) (-7 + 14 = +7 in total) and 7 electrons orbiting the nucleus. This preserved the mass and charge of the nucleus, and the neutral charge of the atom. However, even then Rutherford believed a neutral particle with a similar mass to the proton might exist in the nucleus, but that it would difficult to find.

The neutron

- The neutron, as it was called, was discovered in 1932 by James Chadwick at the Cavendish Laboratory (then under the directorship of Rutherford). The discovery meant that no electrons were needed in the nucleus, and the mass of the nucleus was composed of just protons and neutrons.
- Thus a nitrogen atom had a mass of 14 units, with 7 positive protons and 7 neutral neutrons in the center, making up virtually all the mass of the atom, and 7 negative electrons orbiting the nucleus.

James Chadwick

Los Alamos National Laboratory

Bohr's model

- Rutherford's discovery of the nucleus in 1911 also caused other problems. Rutherford proposed the electrons must orbit the nucleus, but he didn't know how they stayed orbiting the nucleus. Theory at the time suggested electrons should spiral into the nucleus.
- In 1913, Niels Bohr solved this problem by proposing that the electrons must be in orbitals with specific energy levels. Each orbital could only hold a specific maximum number of electrons and electrons could jump between orbitals, but never be found in between them. His model not only solved Rutherford's problem but helped explain how elements react with each other.

5. Use the information above to produce a simple model of a nitrogen atom in the space below:

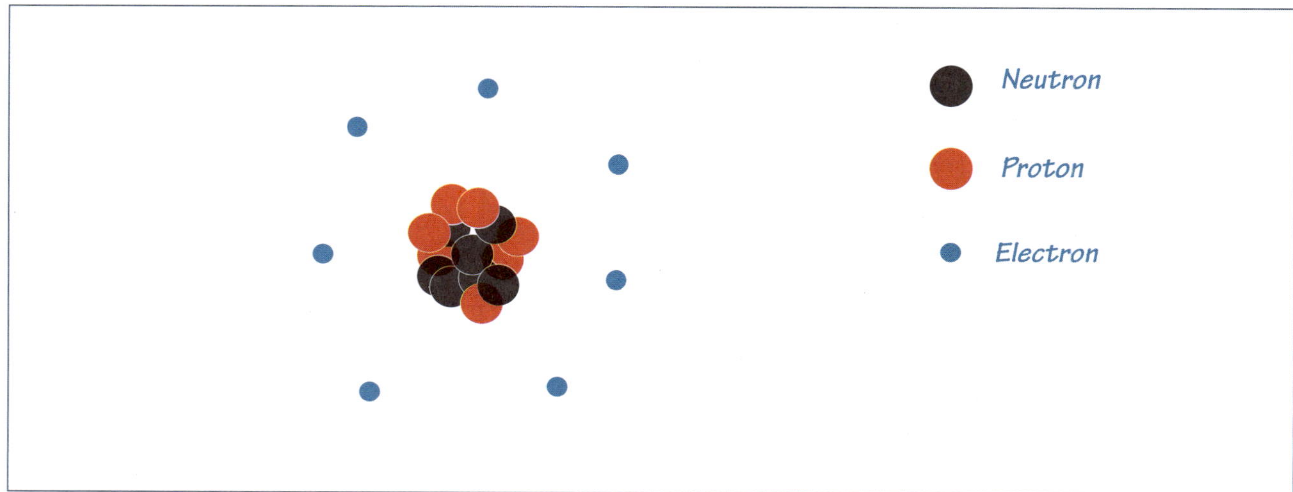

6. Describe how the discovery of the neutron changed Rutherford's model:

The neutron removed the idea that electrons might be found in the nucleus of an atom. It showed that the mass of an atom was caused by the number of neutrons and protons in the nucleus and that electrons were found only orbiting the nucleus.

7. Use the information above to illustrate the scientific process: *The scientific process can be seen in the development of the atom from JJ Thompson's plum pudding model to the model produced after Chadwick's discovery of the neutrons. Each model helped show how the atom might look or behave but at each time allowed scientists to ask questions about that model that were then tested, answered, and added to (or removed from) the model.*

©2020 **BIOZONE** International
ISBN: 978-1-927309-82-7
Photocopying Prohibited

21 Atomic Structure

Key Question: What kinds of model of the atom are there and how are they useful?

- Models of the atom are important for being able to explain the behavior of nuclear and chemical reactions. Different models suit different purposes. Some are useful for explaining the placement of charges and matter in the atom, but not for explaining chemical reactions. The models we will use in this book are relatively simple, but are appropriate for explaining the reactions and behavior of atoms.
- The diagram below shows a stylized representation of the structure of a carbon-12 atom. Note that at the atomic scale atoms do not look like this, but the model is useful in understanding their behavior.

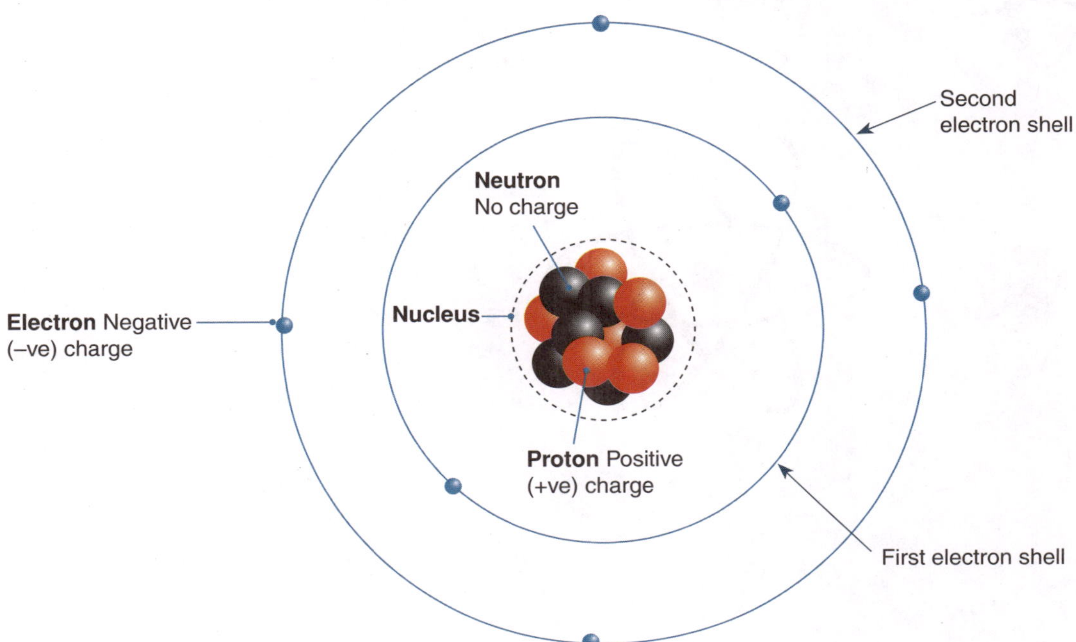

- **Neutral atoms** contain the same number of protons as electrons.
- In the neutral carbon-12 atom shown above there are 6 protons and 6 electrons. Electrons orbit the nucleus in shells. Two electrons are found close to the nucleus in the first electron shell. The other four are found orbiting further away from the nucleus in the second electron shell.
- Importantly it is the electrons in this outer shell that determine an atom's behavior during a chemical reaction.
- The number of protons in the atom determines the element. For example, six protons is carbon, while seven protons is nitrogen. A carbon-12 atom has 6 protons and 6 neutrons.
- The nucleus contains protons and neutrons. The number of protons in the nucleus is called the **atomic number** (symbol ***Z***). The number of protons and neutrons together is called the **mass number** (symbol ***A***).

Protons and their importance

- The number of protons is an important property of an atom. The images below show four different elements and the number of protons in each of the element's atoms.

Copper: 29 protons

Uranium: 92 protons

Gold: 79 protons

USAF
Oxygen: 8 protons

- If the number of protons changes, the element changes. The elements shown above are those elements because of the number of protons they have.

©2020 **BIOZONE** International
ISBN: 978-1-927309-82-7
Photocopying Prohibited

1. The atom is depicted in many ways in books or on the internet. In the space below, choose six models or depictions of the atom you can find in books or on the internet. Draw them or print and paste them into the space below. Describe similarities and differences between them:

All show a central nucleus with electrons around it. Some models show the nucleus as being made of neutrons and protons but others do not. In some models the orbitals are shown and some show electrons in the orbitals. In other models the electrons are just floating around the nucleus.

Nuclide notation

What is the nuclide number? The proton number (or **atomic number**) (**Z**) is the same as the number of electrons in orbit around a neutral atom. The nucleon number (or **mass number**) (**A**) is the total number of protons and neutrons (**nucleons**) in the nucleus. This information is shown by the notation shown left, where X is the chemical symbol of the element. Isotopes of the same element have the same number of protons but different numbers of neutrons.

2. Complete the table below (you may need to use the periodic table on page 30 or 45):

	Element	Atomic number (Z)	Mass number (A)	Number of neutrons	Number of electrons	Nuclide notation
(a)	Hydrogen	*1*	1	*0*	1	$^{1}_{1}H$
(b)	*Aluminium*	13	*27*	14	*13*	$^{27}_{13}Al$
(c)	Potassium	*19*	*39*	*20*	*19*	$^{39}_{19}K$
(d)	Argon	*18*	40	*22*	18	$^{40}_{18}Ar$
(e)	*Calcium*	20	*40*	*20*	*20*	$^{40}_{20}Ca$
(f)	Helium	2	*4*	*2*	*2*	$^{4}_{2}He$

©2020 **BIOZONE** International
ISBN: 978-1-927309-82-7
Photocopying Prohibited

3. (a) What is meant by the term neutral in relation to atoms? A neutral atom has the same number of protons as electrons. The positive charges on the protons cancel the negative charges on the electrons, giving no or neutral charge over all.

   (b) If a neutral atom has 18 protons how many electrons would it have? 18

   (c) If a neutral atom has 10 protons how many electrons would it have? 10

4. Use the periodic table on page 30 or 45 to answer the following questions:

   (a) If a carbon atom spontaneously lost a proton what element would be formed? Boron

   (b) If a sulfur atom gained 3 protons what atom would be formed? Potassium

   (c) Name the element that has atoms with 4 protons in the nucleus: Beryllium

   (d) Name the element that has atoms with 20 protons in the nucleus: Calcium

   (e) Name the element with an atomic number of 12: Magnesium

   (f) Name the element with an atomic number of 1: Hydrogen

5. (a) How many neutrons are there in an atom with an atomic number of 14 and a mass number of 29? 15

   (b) How many neutrons are there in an atom with an atomic number of 8 and a mass number of 16? 8

   (c) Does the number of neutrons in the nucleus affect the type of element? No

6. (a) If an atom had 3 protons but only 2 electrons how would this atom be affected?
   There would be a +1 charge on the atom (3 – 2 = 1). It is now a positive ion.

   (b) If an atom had 17 protons but 18 electrons how would this atom be affected?
   There would be a –1 charge on the atom (17 – 18 = –1). It is now a negative ion.

7. In the space below draw a model of the following atoms:

   (a) Hydrogen (Z = 1, A = 1) (b) Lithium (Z= 3, A = 7) (c) Fluorine (Z = 9, A = 19)

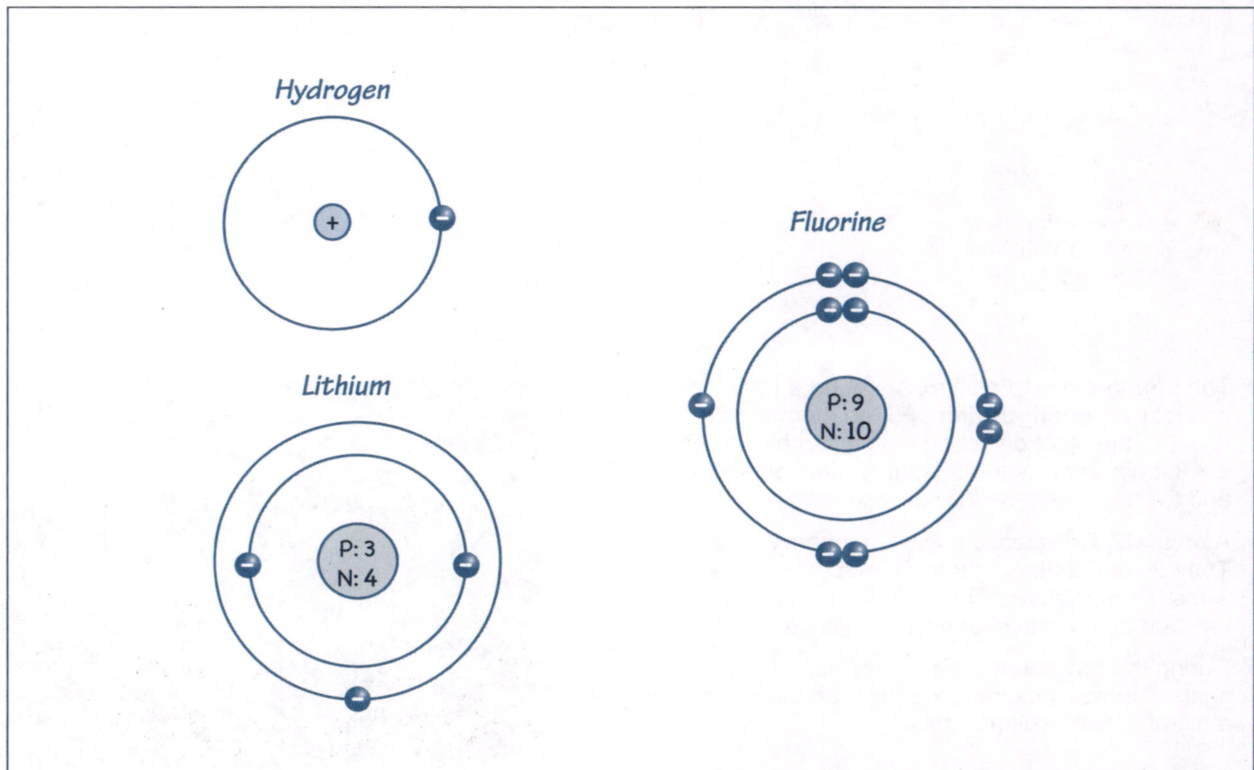

©2020 **BIOZONE** International
ISBN: 978-1-927309-82-7
Photocopying Prohibited

22 Atomic Models and Valence Electrons

Key Question: Where are the electrons found and why is their location important?

- There are numerous models of the atom. The one shown in the previous activity is relatively simple. It is based on Bohr's atomic model and is not a true-to-life representation, nor is it meant to be.
- Since the development of Bohr's atom in 1913 there have been many improvements to it. This is not surprising and reflects the nature of science and advances in technology. The diagram right shows a model of the first electron shell of an atom based on the quantum atom developed by Erwin Schrödinger in 1926.
- In this model, the orbital is shown as a blue cloud and electrons could be in any part of the cloud. The probability of finding them is related to which part of the orbital you are searching.
- Adding more electrons produces more orbitals. But these orbitals are not all the same shape. The diagram right shows the orbitals for an atom with ten electrons.
- Although these models help show the fuzzy nature of electron movement, they do not help us predict chemical reactions. A different model is required to do this.
- It is the electrons that dictate the chemical behavior of an atom. Most importantly it is the electrons in the outer shell (the **valence shell**) that take part in a chemical reaction. The diagram below shows an atomic model of calcium (element 20).

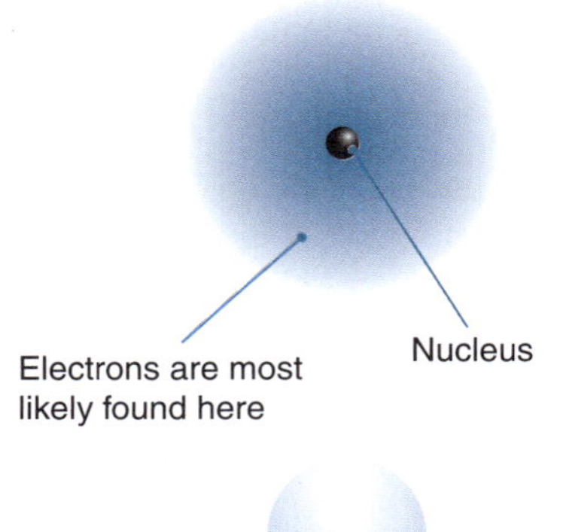

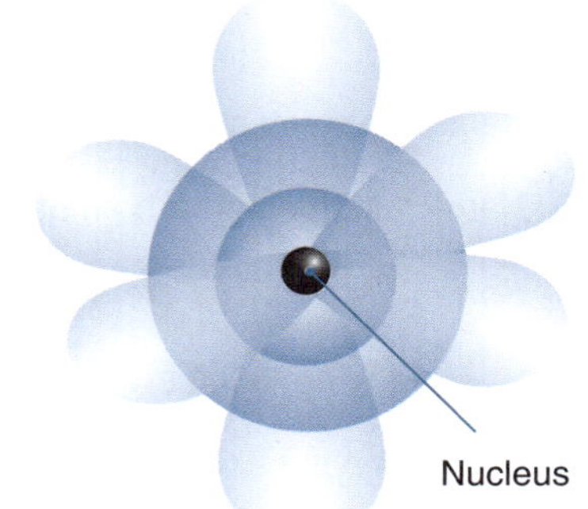

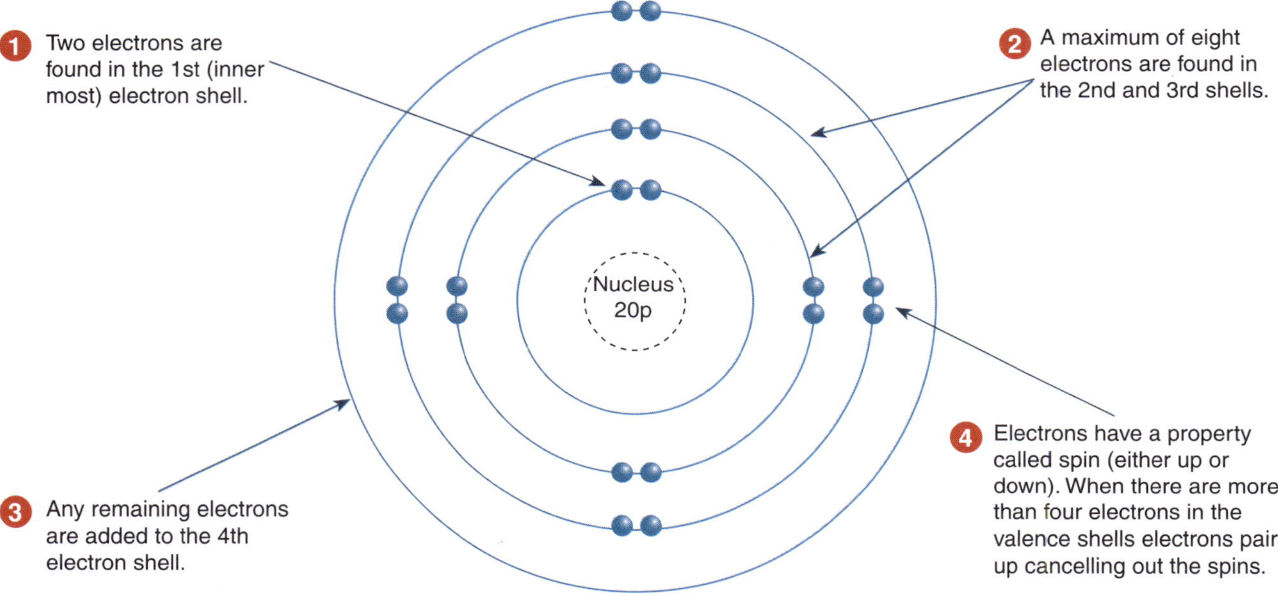

- The arrangement of electrons around an atom is called its **electron configuration**. For an atom of calcium (Z = 20) the electron configuration can be written as 2,8,8,2. This shows the first three electron shells are full and two electrons are in the valence shell.
- Atoms with full valence shells are chemically stable. They do not undergo chemical reactions. An atom without a full valence shell will undergo a chemical reaction in order to obtain a full valence shell.
- During the reaction electrons may be gained or lost or shared between atoms depending on the number of electrons in the valence shell.

Calcium metal

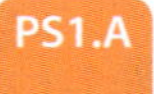

©2020 BIOZONE International
ISBN: 978-1-927309-82-7
Photocopying Prohibited

1. What are the electron configurations for:

   (a) Oxygen (Z = 8): 2,6

   (b) Magnesium (Z = 12) 2,8,2

2. (a) Draw a diagram of the electrons around an oxygen atom:

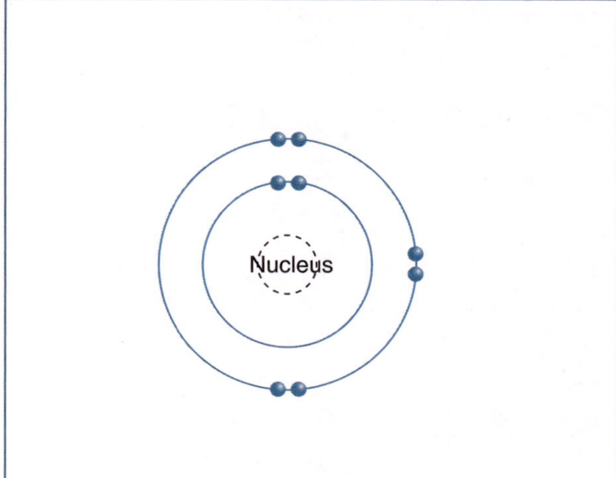

   (b) Draw a diagram of the electrons around a magnesium atom:

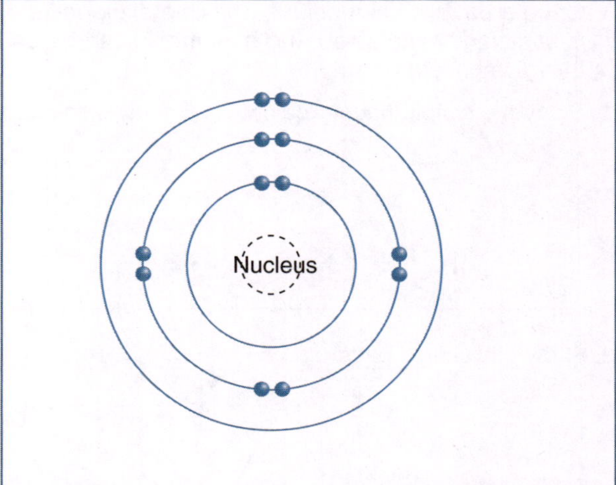

3. Write down the electron configuration of the following atoms:

   (a) Lithium (Z = 3): 2,1

   (b) Sodium (Z = 11): 2,8,1

   (c) Potassium (Z = 19): 2,8,8,1

   (d) What do you notice about the valence electrons of these atoms?

   They all have the same number of valence electrons.

   (e) What might the significance of this be in terms of the way these elements react?

   Because it is the valence electrons that are involved in chemical reactions, the fact that all these atoms have the same number of electrons means they will all have similar chemical properties.

4. Write down the electron configuration of the following atoms:

   (a) Helium (Z = 2): 2

   (b) Neon (Z = 10): 2,8

   (c) Argon (Z = 18): 2,8,8

   (d) What do you notice about the valence electrons of these atoms? The valence shell is full

   (e) How might this change they way these elements react compared to the elements in question 3?

   Because the valence shell is full, the atoms do not "need" any extra electrons or have any spare to exchange during a reaction. Thus they will be unreactive.

5. Predict the following:

   (a) Will lithium gain or lose electrons in a reaction? Lose

   (b) Will chlorine lose or gain electrons in a reaction? Gain

   (c) How reactive would you expect neon to be? Unreactive (it doesn't need to gain or lose electrons)

   (d) Which is more reactive: sodium or magnesium? Sodium

©2020 **BIOZONE** International
ISBN: 978-1-927309-82-7
Photocopying Prohibited

23 Exploring the Periodic Table

Key Question: What is the periodic table and what is the significance of its structure?

- Many aspects of science rely on making observations and looking for patterns within a system or its parts.
- Once a pattern is identified, possible explanations for it (hypotheses) can be proposed and tested. If the pattern is sufficiently explained, the hypothesis can be used to predict the properties of other parts of the system whether they are known or unknown.

1. Study the simple pattern below. How many sides does the missing object have?

Five

2. Study this simple pattern:

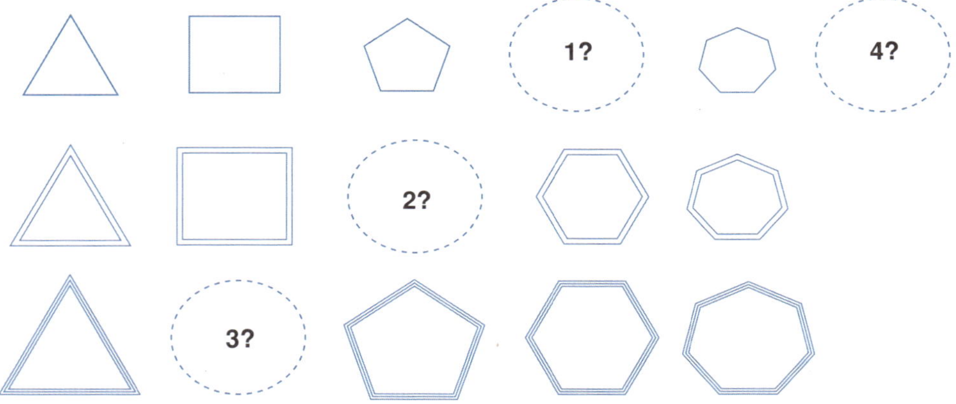

(a) Describe where would you find the largest object with the least sides? *In the bottom left corner*

(b) What would you expect to find at 3? *A large four sided shape with an outline of three lines*

(c) Is 1? larger or smaller than 2? *Smaller*

(d) Patternologists think there may be a shape yet to be discovered at 4. Predict the properties of this shape:

The shape would have eight sides. It would be smaller than any of the other shapes in the table. It would have an outline with one line.

3. Shops often use patterns when displaying items on the shelf. Study the shelves right. Write down the rules by which the display has been made:

Student answers may vary. Some rules are:

- Price generally increases left to right

- Blue is beside red.

- Price increases bottom to top.

- Lower rows have more colors.

©2020 BIOZONE International
ISBN: 978-1-927309-82-7
Photocopying Prohibited

- An event that is periodic happens once and then again at specific time intervals. Periodic can also be used to describe a repeating pattern.
- Run you finger over a piano keyboard and hit all the keys and you will of course hear a repeating pattern of octaves. Every key plays a note one octave higher or lower than the key eight keys before or after. The set of seven keys A-G plays the set of notes within one octave.
- During the development of the periodic table, chemists noticed this "law of octaves". Every eighth element seemed to have similar properties. However this pattern broke down as more elements were discovered and added to the table. The table itself is called periodic because it displays repeating patterns.
- However there is more than one type of periodic table. The one printed in this book is the standard Mendeleev periodic table seen in many texts.

Getting to know the periodic table

- The periodic table as we use it today in chemistry was first formulated by the Russian chemist Dmitri Mendeleev (right) in 1869. Not only did he organize the then known elements into the table, but he predicted the properties of eight unknown elements and left spaces in the table for when they were discovered.
- It was not until 1913 that it was proved that the order Mendeleev placed the elements was in fact the same as ordering the elements by their atomic number.
- Before you start looking at the finer details of the periodic table it is worth getting to know the large scale structure of the table.
- The periodic table is organized into **columns called groups** and **rows called periods**. The periodic table below shows the different groups of elements. Different groups of elements are given different names for easier reference:

Metals ← | → Non metals

← Main block elements →

Atomic number →

Period	1	2	3	4	5	6	7	8	9	10	11	12	13	14	15	16	17	18
1	1 H Hydrogen																	2 He Helium
2	3 Li Lithium	4 Be Beryllium											5 B Boron	6 C Carbon	7 N Nitrogen	8 O Oxygen	9 F Fluorine	10 Ne Neon
3	11 Na Sodium	12 Mg Magnesium											13 Al Aluminium	14 Si Silicon	15 P Phosphorus	16 S Sulfer	17 Cl Chlorine	18 Ar Argon
4	19 K Potassium	20 Ca Calcium	21 Sc Scandium	22 Ti Titanium	23 V Vanadium	24 Cr Chromium	25 Mn Manganese	26 Fe Iron	27 Co Cobalt	28 Ni Nickel	29 Cu Copper	30 Zn Zinc	31 Ga Gallium	32 Ge Germanium	33 As Arsenic	34 Se Selenium	35 Br Bromine	36 Kr Krypton
5	37 Rb Rubidium	38 Sr Strontium	39 Y Yttrium	40 Zr Zirconium	41 Nb Niobium	42 Mo Molybdenum	43 Tc Technetium	44 Ru Ruthenium	45 Rh Rhodium	46 Pd Palladium	47 Ag Silver	48 Cd Cadmium	49 In Indium	50 Sn Tin	51 Sb Antimony	52 Te Tellurium	53 I Iodine	54 Xe Xenon
6	55 Cs Caesium	56 Ba Barium		72 Hf Hafnium	73 Ta Tantalum	74 W Tungsten	75 Re Rhenium	76 Os Osmium	77 Ir Iridium	78 Pt Platinum	79 Au Gold	80 Hg Mercury	81 Tl Thallium	82 Pb Lead	83 Bi Bismuth	84 Po Polonium	85 At Astatine	86 Rn Radon
7	87 Fr Francium	88 Ra Radium		104 Rf Rutherfordium	105 Db Dubnium	106 Sg Seaborgium	107 Bh Bohrium	108 Hs Hassium	109 Mt Meitnerium	110 Ds Darmstadium	111 Rg Roentgenium	112 Cn Copernicium	113 Nh Nihonium	114 Fl Flerovium	115 Mc Moscovium	116 Lv Livermorium	117 Ts Tennessine	118 Og Oganesson

Lanthanoids	57 La Lanthanum	58 Ce Cerium	59 Pr Praseodymium	60 Nd Neodymium	61 Pm Promethium	62 Sm Samarium	63 Eu Europium	64 Gd Gadolinium	65 Tb Terbium	66 Dy Dysprosium	67 Ho Holmium	68 Er Erbium	69 Tm Thulium	70 Yb Ytterbium	71 Lu Lutetium
Actinoids	89 Ac Actinium	90 Th Thorium	91 Pa Protactinium	92 U Uranium	93 Np Neptunium	94 Pu Plutonium	95 Am Americium	96 Cm Curium	97 Bk Berkelium	98 Cf Californium	99 Es Einsteinium	100 Fm Fermium	101 Md Mendelevium	102 No Nobelium	103 Lr Lawrencium

Alkali metals | Alkaline Earth metals | Transition metals | Metalloids | Basic metals | Non metals | Halogens | Noble gases

Alkali metals are kept in oil to stop them reacting.

The transition metals are the most familiar metals. Iron and copper are particularly important to industry.

Halogens and noble gases are often used in lighting to produce bright or colored lights.

©2020 BIOZONE International
ISBN: 978-1-927309-82-7
Photocopying Prohibited

4. Which side of the periodic table are the metals on? *The left hand side*

5. Which group number are the noble gases in? *18*

6. Which group number are the halogens in? *17*

7. What is the name given to the metals that includes copper and gold? *Transition metals*

8. How does the atomic number (number of protons) change along a period? *The number increases left to right*

9. In your own words describe the patterns and features you can see in the periodic table:
Student's answer. Students might describe the positions of different types of atoms, and the structure in terms of groups and periods.

10. There are many versions of the periodic table. Try typing "alternative versions of the periodic table" into an internet search engine and explore that numerous different versions. Print some out and paste them into the space below:

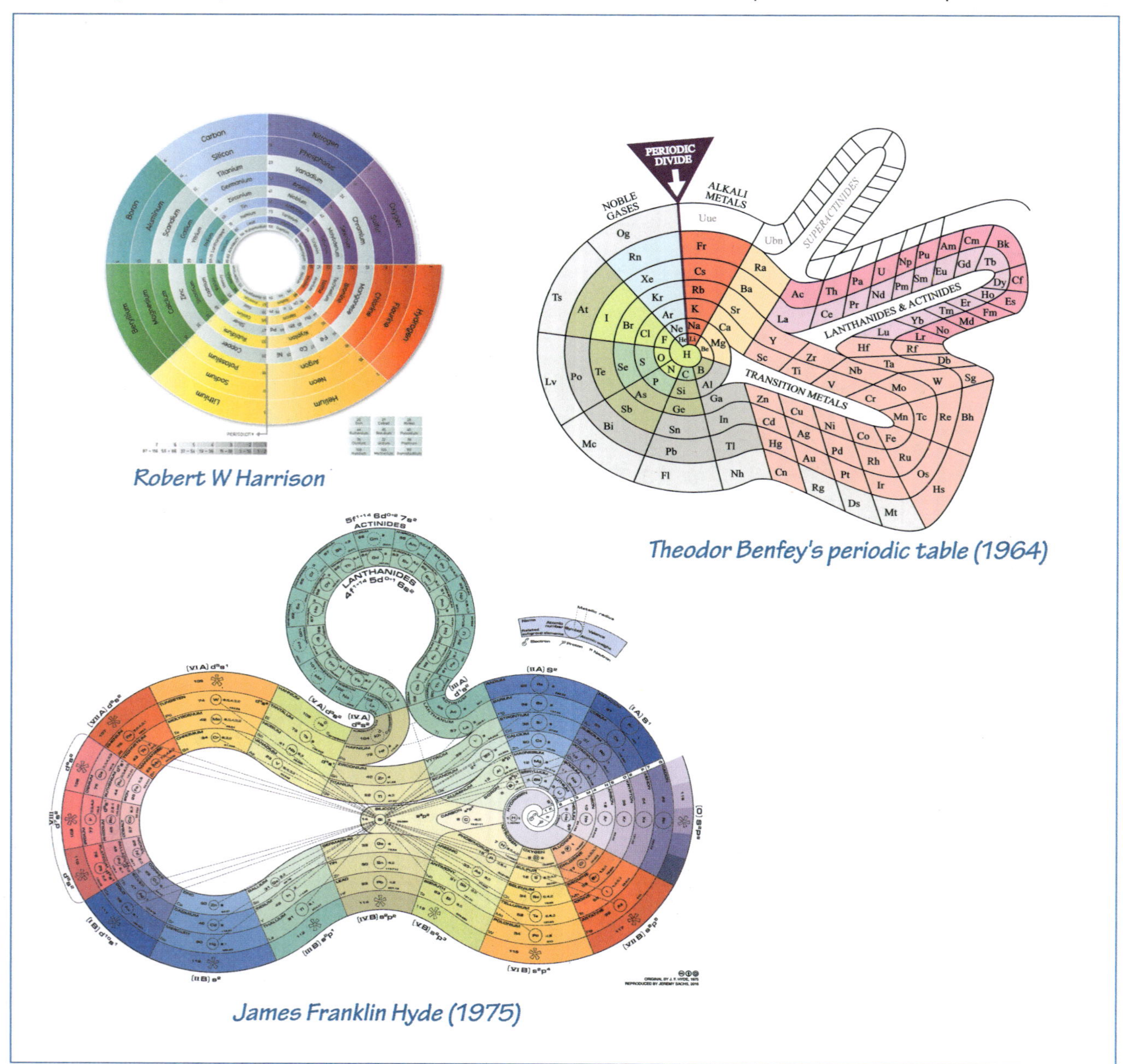

Robert W Harrison

Theodor Benfey's periodic table (1964)

James Franklin Hyde (1975)

©2020 **BIOZONE** International
ISBN: 978-1-927309-82-7
Photocopying Prohibited

24 Trends in the Periodic Table

Key Question: What trends can be seen in the periodic table and what do they mean?

Atomic radius

- The atomic radius at its simplest definition is the distance from the nucleus to the edge of the electron cloud. Since the electron cloud has no fixed edge, a more definitive measure of the atomic radius is half the distance between two identical atoms' nuclei in a covalent bond (the covalent radius) (right).

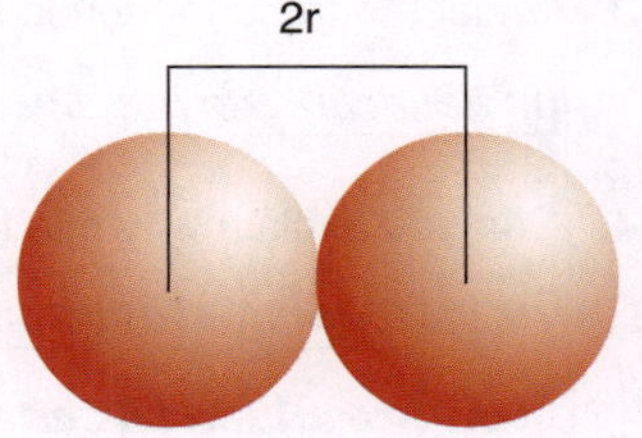

Ionization energy

- The first ionization energy is the energy required to remove one mole* of the most loosely held electrons from one mole of neutral atoms in a gas state to produce one mole of ions each with a charge of +1. Similarly there can be second and third ionization energies for atoms.

*A mole is the unit of measurement for amount of substance in SI units. We will explore the mole later.

1. Plot the data on the grid provided:

Atomic number	Atomic radius (pm)
1	53
2	31
3	167
4	112
5	87
6	67
7	56
8	48
9	42
10	38
11	190
12	145
13	118
14	111
15	98
16	88
17	79
18	71
19	243
20	194
21	184
22	176
23	171
24	166
25	161
26	156
27	152
28	149
29	145
30	142
31	136
32	125
33	114
34	103
35	94
36	88

Atomic radius (pm)

Atomic number

©2020 BIOZONE International
ISBN: 978-1-927309-82-7
Photocopying Prohibited

2. Describe any trends or pattens you can see in the data: *Atomic radius decreases within the elements of a period. It increases from periods higher on the table to those lower on the table (low number periods to high number periods).*

3. Predict the trend in atomic radius across row five of the periodic table: *Atomic radius will be larger than row 4 (in general). Across the period atomic radius will decrease in a similar way to row 4.*

4. Can you explain why this trend in occurring? (Hint think about the effect of the increasing positive charge in the nucleus). *The increasing atomic charge pulls the electrons closer to the nucleus, decreasing the atomic radius. As shells are added to the atoms (each new period) the radius increases again.*

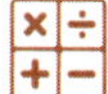

5. Plot the data on the grid provided:

Atomic number	1st ionization energy (kJ/mol)
1	1312
2	2372
3	520
4	899
5	801
6	1086
7	1402
8	1314
9	1681
10	2081
11	496
12	738
13	578
14	786
15	1012
16	1000
17	1251
18	1521
19	419
20	590

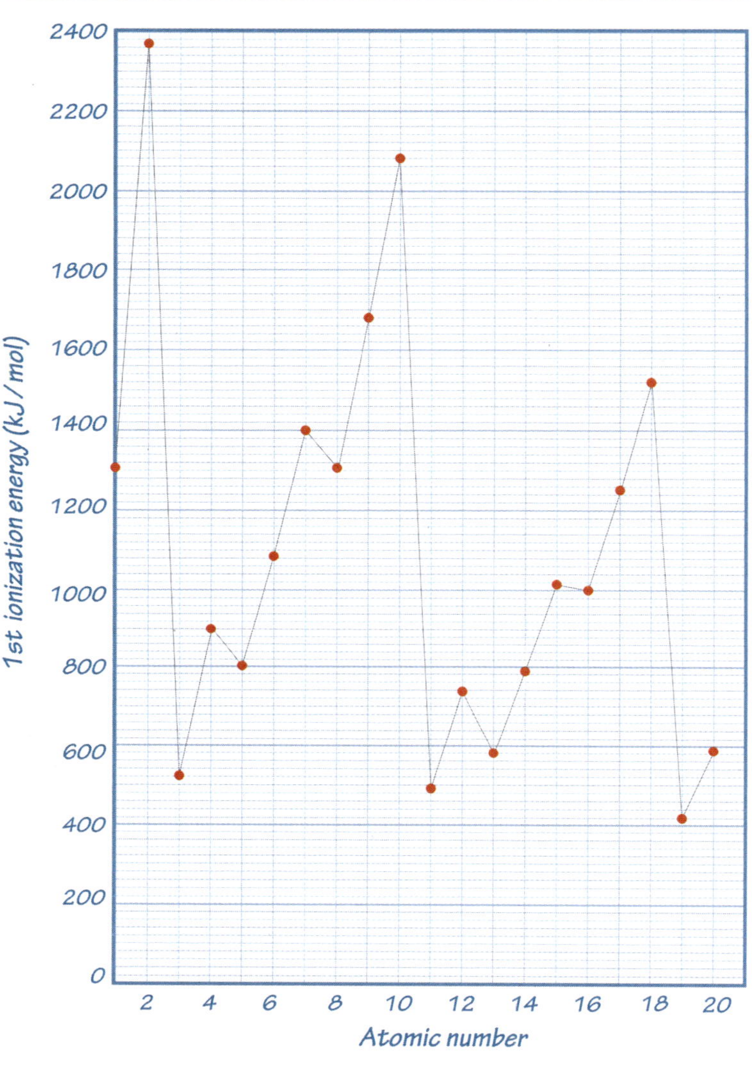

6. What is the trend in ionization energy across the first twenty atoms? *Ionization energy increases across a row. It decreases down a group. Note the zigzag nature of the line is associated with filled suborbitals.*

7. How does the atomic number and the electron configuration help to explain the pattern in 1st ionization energy? *The ionization energy can be explain by the number of electrons in the valence shell and the number of protons in the nucleus (atomic number). If only one electron is in the valence shell it is simple to remove. However a greater number of electrons in the valence shell becomes harder to remove because of the increased attractive force of the larger nucleus.*

©2020 **BIOZONE** International
ISBN: 978-1-927309-82-7
Photocopying Prohibited

Electronegativity

- There are many other trends and patterns that can be seen in the periodic table. These include electronegativity.
- Electronegativity is the ability to attract a pair of bonding electrons in a chemical bond. The electronegativity of atoms affects how they share electrons when forming a chemical bond and this in turn affects the properties of the product produced in a reaction.
- The graph right shows how the electronegativity changes for the first twenty atoms.

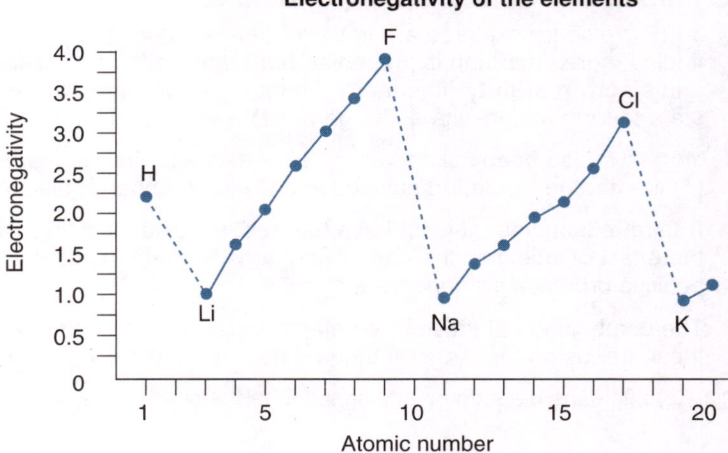

Trends down a group

- While it is informative to plot trends right across a number of periods, it is also informative to plot trends down a group. The graphs below show the boiling point and density of the halogens (group 17).

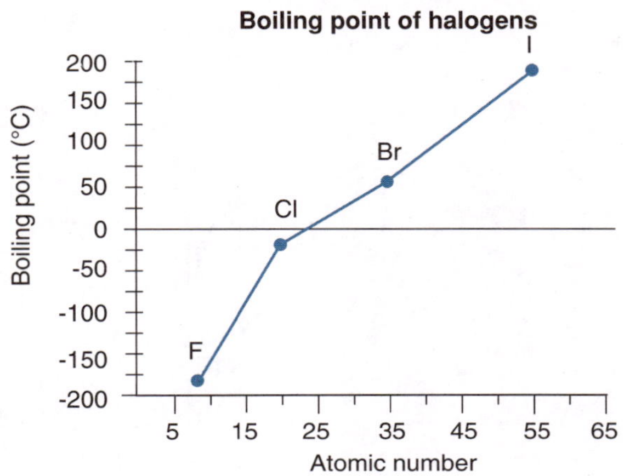

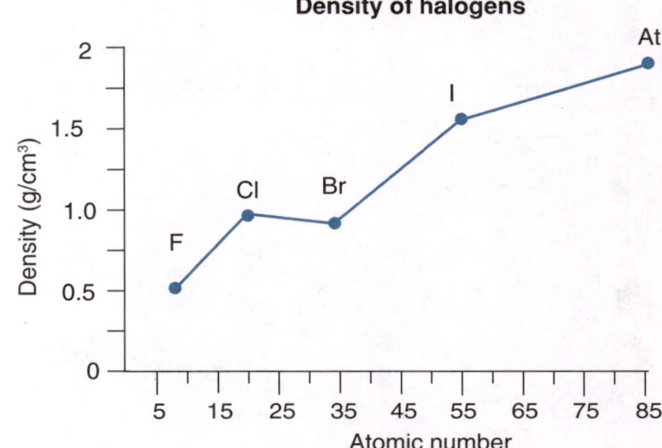

8. (a) If a reaction occurs between an element with a high electronegativity and one with a low electronegativity (e.g. lithium and fluorine) what do you think will happen in terms of the sharing of electrons in the product (e.g. lithium fluoride)?

An electron(s) will be removed from lithium (low ability to hold its electron) and transferred to fluorine. The number of electrons transferred depends on the number needed to obtain a full valence shell.

(b) If a reaction occurs between two elements with similar electronegativities what do you think will happen in terms of the sharing of electrons in the product?

The electrons will be evenly shared as neither atom has the ability to attract or remove electrons from the other atom.

9. Suggest why the electronegativity of the noble gases can not be calculated: *The noble gases have full valence shells. They do not attract electrons from other atoms and thus electronegativity has no meaning for them.*

10. How does electronegativity change down a group and across a period (left to right)?

Electronegativity increases across a period and decreases down a group.

11. What can be said about the trends in boiling point and density of the halogens?

Boiling point increases down the group as does the density. This may be because the atoms increase in size and mass (iodine is heavier than fluorine and takes more energy to make it move).

©2020 **BIOZONE** International
ISBN: 978-1-927309-82-7
Photocopying Prohibited

Summarizing trends in the periodic table

- You have so far explored a number of trends across the periodic table and attempted to explain some of them. By finding more data than is presented here many other trends in the table can be seen. These include **ionic radius** and **electron affinity** (the energy change that occurs when an electron is accepted by an atom in the gaseous state to form an ion with a charge of –1).
- You have also been asked to suggest reasons for the trends. You may have realized that the trends are a result of (1) the charge inside the nucleus and (2) the number of electrons in the valence shell.
- A third reason you may not have realized is called shielding. As the number of electron shells increases, they have the effect of shielding the attractive charge from the nucleus and reducing the electrostatic attraction between the positive protons and negative electrons.
- The combination of these three effects can account for the trends we see in the periodic table. We can summarize these trends on the diagram below. The arrows show increasing strength of the property.

12. Complete the diagram by adding the labels *Atomic radius, 1st ionization energy, Electonegativity:*

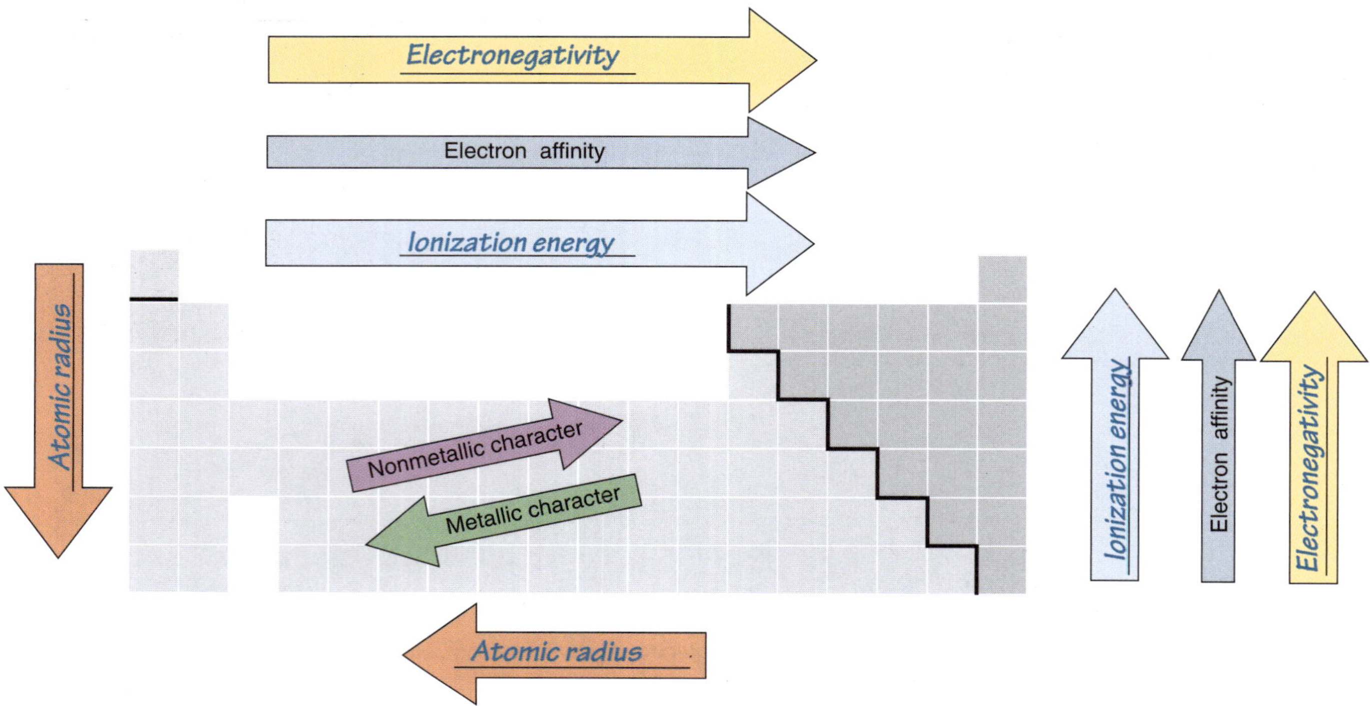

13. Using the information in this and the previous activity answer the following questions:

Element	Electron configuration	Electronegativity	Metal / nonmetal
A	1	2.20	Nonmetal (but has metallic properties)
B	2,8,2	1.31	Metal
C	2,8,3	1.63	Metal
D	2,4	2.55	Nonmetal
E	2,8,6	2.58	Nonmetal

14. Use the periodic table on page 30 or 45 to help you answer the following questions:

(a) Which atom has the largest radius, potassium or bromine? Potassium

(b) Which atom would react most strongly with fluorine? Cesium (or a group 1 element)

(c) Which has a smaller radius, an oxygen atom or a fluorine atom? Fluorine

(d) Which atom is more electronegative, iodine or chlorine? Chlorine

©2020 **BIOZONE** International
ISBN: 978-1-927309-82-7
Photocopying Prohibited

25 Valence Electrons and Reactivity

Key Question: Why are some metals more reactive than others?

- The photographs below show the metals sodium, potassium, and cesium reacting with water. From left to right the size of the piece of metal placed in the water is getting smaller (can you guess why?).
- The primary reaction that occurs is the metal reacting with water to produce hydrogen gas and a hydroxide (which dissolves in water). The hydrogen gas ignites due to the heat of the reaction and reacts with oxygen in the air, producing more heat which can cause the metal to melt, burn (also react with oxygen), and often explode.
- In photograph 1 the size of the piece of sodium is about half a cubic centimeter. The piece of cesium used in photograph 3 is about a quarter of that size. It explodes on contact with the water.

1: Sodium with water

Tavoromann CC 3.0

2: Potassium with water

Cyberchemist CC 2.0

3: Cesium with water

1. Complete the table below to show the electron configuration of the group 1 elements:

Element	Simplified electron configuration
Lithium	2, 1
Sodium	2, 8, 1
Potassium	2, 8, 8, 1
Rubidium	2, 8, 8, 18, 1
Cesium	2, 8, 8, 18, 18, 1
Francium	2, 8, 8, 18, 18, 32, 1

2 (a) Which of these elements appears to have the largest atom. Explain your answer: Francium is the largest atom. It has 87 electrons and therefore 87 protons (plus a larger number of neutrons). It therefore not only has a large nucleus but many layers of electrons, making its radius very large.

(b) Compare the size of the sodium and cesium atom. Which would have electrons furthest from its nucleus?

Cesium has more electron orbitals than sodium and its valence electrons are therefore further away.

(c) How might this affect the attractive force of the protons acting on the valence (outermost) electron?

The attractive force between the cesium nucleus and its valence electrons is much less than that of the sodium nucleus and its valence electrons.

(d) How does this explain the change in the reactive nature between sodium and cesium?

Cesium valence electrons will be lost much more quickly than sodium's. It will therefore react much more quickly, producing an explosion, rather than a fast burn.

(e) Would you expect lithium to be more or less reactive than sodium? Less reactive

©2020 BIOZONE International
ISBN: 978-1-927309-82-7
Photocopying Prohibited

26 Bonding

Key Question: What are some of the ways atoms bind together?

Sticking together

- Apart from a small group of elements called the "noble gases", elements are never found in nature as singular free-floating atoms. Their atoms are always found bonded to other atoms. These can be either the same kind of atom (as in hydrogen gas) or they can be different atoms (as in carbon dioxide).

Wilco Oelen Creative Commons Attribution-ShareAlike 3.0 Unported License.

In its pure form, the element sodium is a silvery metal. Its atoms share their mobile electrons and are held together by **metallic bonds**. It is a very reactive metal.

Wilco Oelen Creative Commons Attribution-ShareAlike 3.0 Unported License.

Chlorine is a gaseous element with a yellow tinge. In its pure form, the atoms are found **covalently** bonded together in pairs. Chlorine is highly toxic and reactive.

Sodium chloride (table salt) is a highly stable crystal made of sodium and chloride ions held together by **ionic** bonds.

- Atoms without full valence shells are reactive because having unpaired electrons and vacant orbitals is energetically unfavorable. Vacant orbitals can be filled by either sharing electrons (e.g. covalent bonding) or by gaining or losing electrons. When an atom gains or loses an electron (or electrons) it becomes an **ion**.
- In the example above of sodium and chlorine, both elements are highly reactive in their pure form. Although their atoms are sharing electrons, it is energetically more favorable for sodium atoms to lose an electron and chlorine atoms to gain an electron and form ions. When sodium and chlorine react, a large amount of thermal energy is released. The resulting product, sodium chloride, is stable and unreactive.

Sodium reacting with chlorine in the presence of water (which "kick starts" the reaction).

Wilco Oelen Creative Commons Attribution-ShareAlike 3.0 Unported License.

- The diagram below shows the changes in bonding that occurs during the reaction between sodium and chlorine.

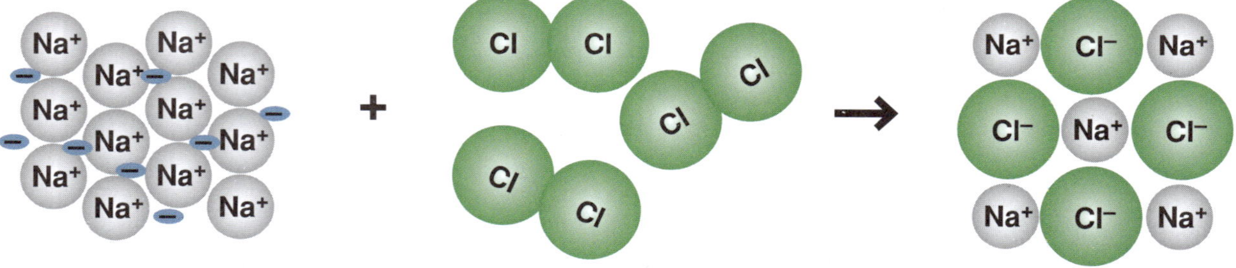

Metallic bonds in sodium

Covalent bonds in chlorine

Ionic bonds in sodium chloride

1. (a) What has happened to the charge on the chlorine after it became a chloride ion?
   The charge has gone from 0 on the Cl atom to −1 on the Cl ion.

   (b) How has this happened? *An electron has been exchanged from a sodium atom to a chlorine atom.*

   (c) Where did this charge come from? *The exchange of an electron, which carries a negative charge.*

2. Compare the positions of sodium and chlorine on the periodic table. What does this say about their electronegativity and the reaction between them?
   Sodium has a low electronegativity and chlorine has a high electronegativity. Chlorine will remove the valence electron from sodium and form a negative ion. Sodium will form a positive ion.

PS1.A P

©2020 BIOZONE International
ISBN: 978-1-927309-82-7
Photocopying Prohibited

27 Ionic Bonding

Key Question: How do ions form and how do they bind together to form ionic compounds?

- In the previous activity sodium and chlorine were shown reacting together to form sodium chloride. But what happens during a reaction between sodium and chlorine? It helps to look at the electron configurations of each. In the diagram below, we see that sodium has a single electron in the valence shell and chlorine has seven.
- Earlier, at the end of Activity 22, you were asked to predict if chlorine would lose or gain an electron in a reaction. Here we can ask the question, "*is it easier (does it take less energy) to gain one electron to fill chlorine's valence shell or to lose seven electrons and thus lose a shell entirely?*"
- A similar question must be asked of sodium. Is it easier to lose one electron (and thus the shell) or gain seven electrons to fill the valence shell?
- Of course it is easier for chlorine to gain one electron and sodium to lose one electron. The atoms then form ions. Ions are atoms that have gained or lost electrons and thus have positive or negative charges.

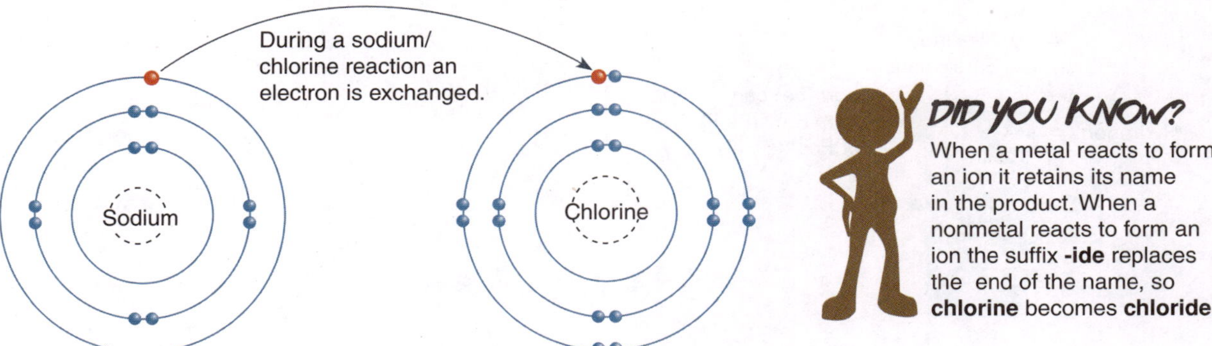

1. (a) Write the electron configuration of the sodium ion: 2,8

   (b) Write the electron configuration of the chloride ion: 2,8,8

2. (a) What charge does the sodium ion have? +1

   (b) What charge does the chloride ion have? –1

3. Explain why sodium loses an electron during a reaction:
   Sodium loses an electron to produce a stable valence shell. By losing the electron it loses the original valence shell. The new valence shell has eight electrons and is stable.

4. Explain why chlorine gains an electron during a reaction Chlorine as seven electrons in its valence shell. By gaining one electron it forms a full valence shell (with eight electrons) which is stable.

5. Explain why sodium chloride is a highly stable substance in comparison to sodium and chlorine:
   Sodium chloride is composed of ions that have full valence shells. The ions are therefore energetically stable. Sodium metal and chlorine gas as elements share electrons but this is not as energetically stable as having complete valence shells. They therefore react to full more stable ions (which do not share electrons).

6. Name the following:

   (a) The ion that forms from the fluorine atom: Fluoride

   (b) The ion that forms from the sulfur atom: Sulfide

   (c) The compound that forms from the reaction between lithium and chlorine: Lithium chloride

   (d) The compound that forms when a potassium atom and a sulfur atom combine: Potassium sulfide

©2020 **BIOZONE** International
ISBN: 978-1-927309-82-7
Photocopying Prohibited

Important ions

- You should realize by now that metals form positive ions and non metals form negative ions. Positive and negative ions are also referred to as cations and anions respectively.

Cations are positive

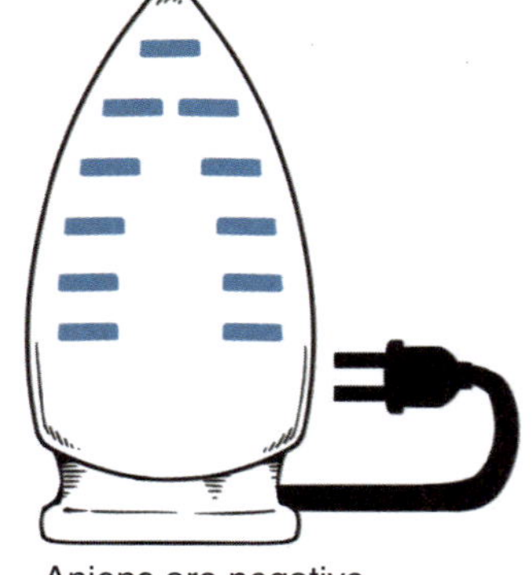
Anions are negative

Images: F. Hicks

- Important monoatomic (singular) ions you should know are shown below:

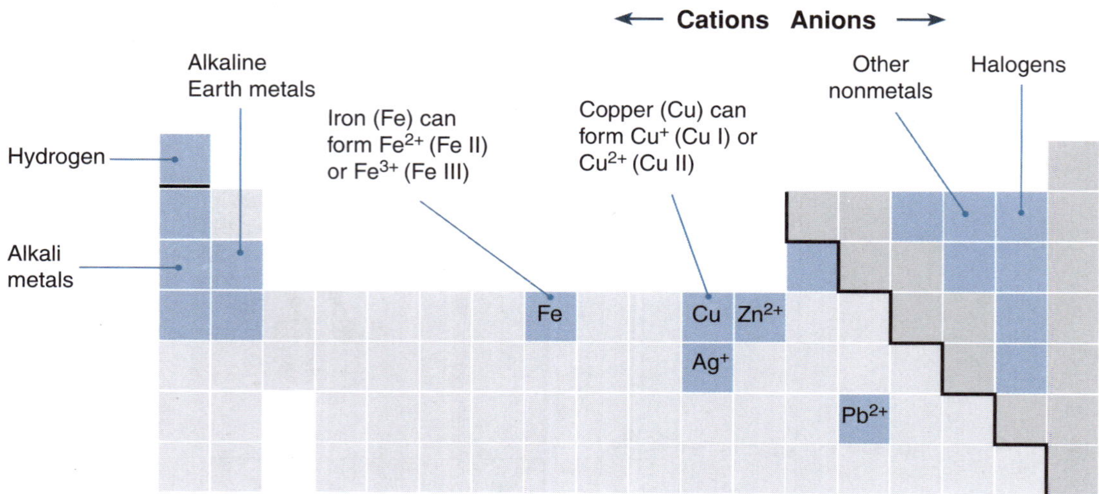

- Sometimes ions can be found as groups. These are called polyatomic ions.
- Polyatomic ions you will commonly come across include the sulfate ion (SO_4^{2-}), the hydrogen carbonate (bicarbonate) ion (HCO_3^-), the carbonate ion (CO_3^{2-}), the hydroxide ion (OH^-), the nitrate ion (NO_3^-), and the ammonium ion (NH_4^+).

7. (a) Draw a diagram to show the electron configuration of a sulfur atom and a sulfide ion:

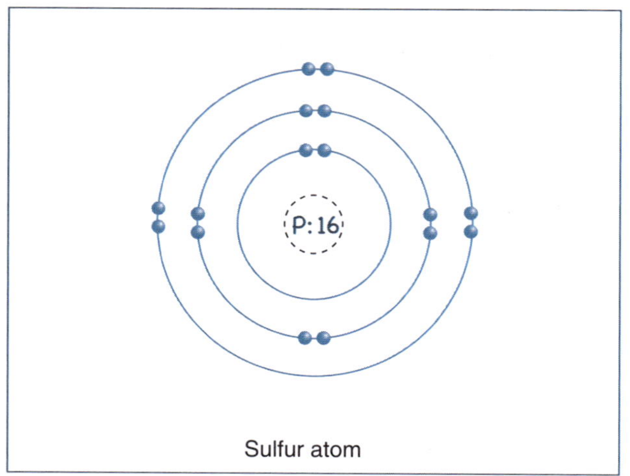

Sulfur atom

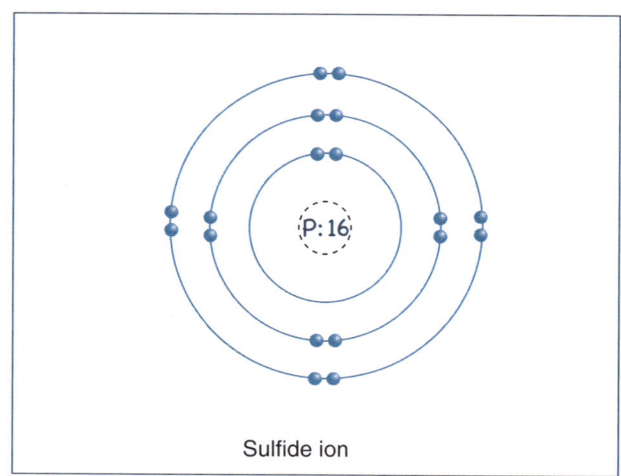

Sulfide ion

(b) What is the change on the sulfur atom? 0

(c) What is the charge on the sulfide ion? -2

(d) How many sodium ions would be needed to react with the sulfide ion in order to produce a neutral compound?
2

©2020 **BIOZONE** International
ISBN: 978-1-927309-82-7
Photocopying Prohibited

8. Explain why group I elements always form an ion with a single positive charge:

The first group elements all have one electron in their valence shell. When they react it requires less energy to lose this one electron than gain seven more. As such they will have one more positive charge (from the protons) than negative charges (from the electrons resulting a +1 charge overall).

9. Would you expect the element carbon to form a cation or an anion? Explain your reasoning:

It will form neither. Because it has four electrons in the valence shell it is energetically no different to lose or gain four electrons. It therefore shares electrons in covalent bonding with other atoms.

10. Explain why nonmetals do not form cations:

Nonmetals all need to gain fewer electrons to fill valence shells than to lose electrons to remove the shell entirely. Gaining electrons forms anions. The electronegativity of nonmetals is also relatively high and this causes them to tend to attract electrons.

11. What can be said about any elements that are in the same group number of the periodic table?

All elements in the same group will have similar properties or properties will change in a regular way (e.g. boiling point). They will all have the same number of electrons in the valence shell. Atoms will gain an electron shell down the group (increase in a period).

12. The chemical equation for sodium reacting with chlorine can be written in two ways:
As a word equation: sodium + chlorine → sodium chloride
And as a chemical equation: Na + Cl → NaCl

(a) Write a word equation for the reaction of lithium and fluorine:

Lithium + fluorine → lithium fluoride

(b) Write a word equation for the reaction of magnesium and oxygen:

Magnesium + oxygen → magnesium oxide

13. Ionic compounds are held together by the opposite electrostatic charges of the ions (positive and negative charges). At a high enough temperature ionic compounds will melt. Consider the ions in sodium chloride and the ions in magnesium sulfide. Which do you think would have the highest melting point. Explain your reasoning:

The sodium and chloride ions have a +1 and -1 charge respectively. The electrons are close to the nucleus. Magnesium and sulfide have a +2 and -2 charge respectively the electrons are slightly further away from the sulfide ion than the magnesium ion. We can theorize that the +1-1 charges will hold the Na^+ and Cl^- tightly together, but the +2-2 charges of Mg^{2+} and S^{2-} much more so. However the electrons are slightly further away so they will be held slightly less tightly. Thus MgS will have a higher melting point.

14. Name the compound that forms between the following:

(a) The potassium and sulfate ions: Potassium sulfate

(b) The ammonia and nitrate ions: Ammonium nitrate

(c) Copper and chlorine: Copper chloride

(d) Sodium and oxygen: Sodium oxide

©2020 BIOZONE International
ISBN: 978-1-927309-82-7
Photocopying Prohibited

28 Covalent Bonding

Key Question: How can atoms complete their valence shell without exchanging electrons?

- Not all atoms form ions when they react with other atoms. In the case of nonmetals the atoms may share electrons in order to obtain a complete valence shell.
- Consider the water molecule (a **molecule** is a discrete group of bound atoms, normally with a neutral charge). It is made up of two hydrogen atoms bound to an oxygen atom.
- Because it is the valence electrons that are involved in reactions we can draw a diagram to show just the valence electrons of the oxygen atom:

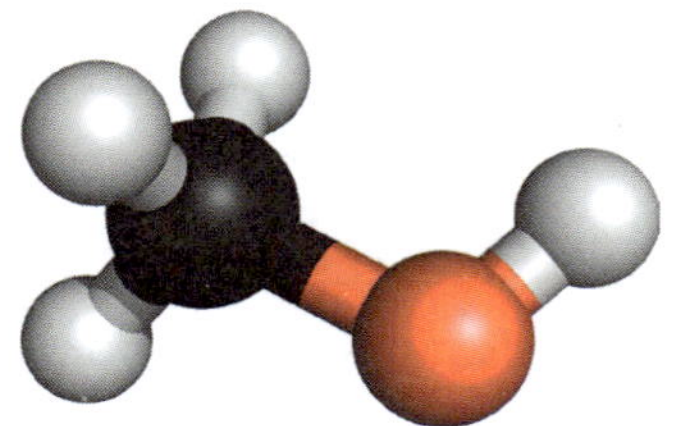

The methanol molecule has a specific number of atoms in the molecule.

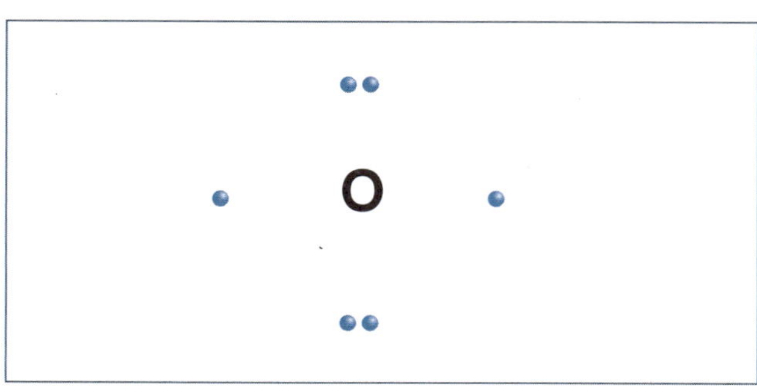

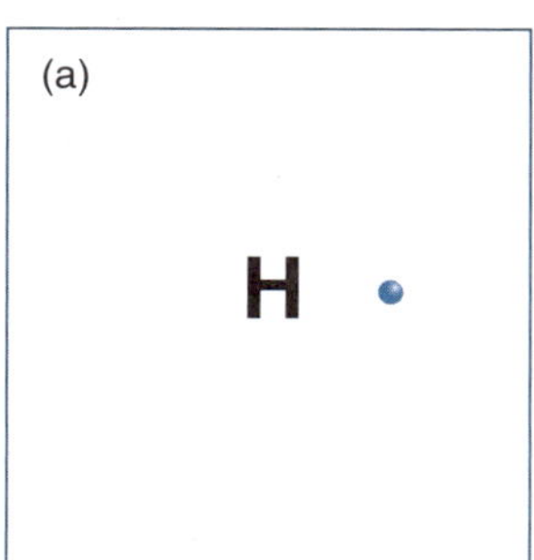

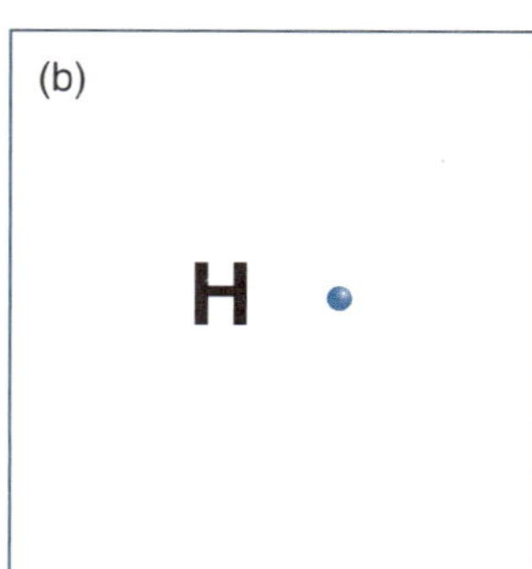

- The same thing can be done with the hydrogen atom. The diagrams can be combined to produce **Lewis structures** (also called Lewis dot diagrams, Lewis dot formulas, or electron dot diagrams) Lewis structures are a useful diagrammatic tool in chemistry. They are used to show how the valence electrons of atoms combine during a reaction to fill the valence shells of each atom in the reaction.

1. Fill boxes (a) and (b) with a diagram of the hydrogen atom:

2. (a) How many valence electrons does a neutral hydrogen atom have? *1*

   (b) How many electrons does a neutral hydrogen atom need to fill its valence shell? *1*

   (c) How many valence electrons does a neutral oxygen atom have? *6*

   (d) How many electrons does a neutral oxygen atom need to fill its valence shell? *2*

   (e) What is the total number of electrons in the valence shells of two hydrogen atoms and one oxygen atom?

   1 + 1 + 6 = 8

   (f) What is the total number of pairs of electrons in the valence shells of two hydrogen atoms and one oxygen atom?

   8 ÷ 2 = 4 pairs

   (g) In the space below hydrogen and oxygen have been placed with oxygen as the central atom. Place the pairs of electrons (use dots to represent electrons) around the hydrogen and oxygen according these rules: Only eight electrons around oxygen. Only two electrons around each hydrogen. The electrons must be in pairs:

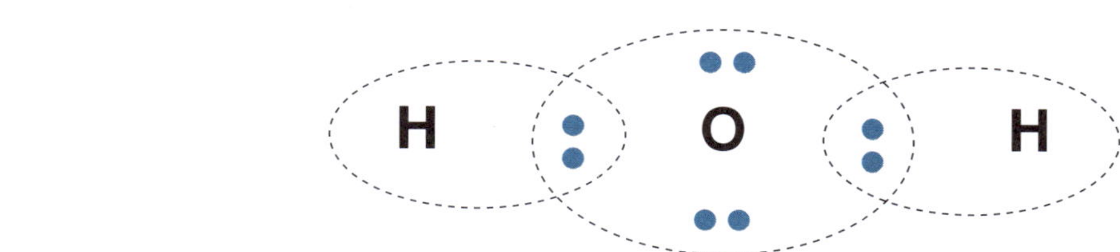

Note: The dotted ovals simply show each atom has a full valence shell.

   (h) Is the water molecule charged or neutral? Explain your answer: *The number of electrons and protons are equal therefore there is no charge on the molecule due to a difference in charges. Some students may realise the unpaired electrons could cause a polarity in the molecule.*

PS1.A P

©2020 BIOZONE International
ISBN: 978-1-927309-82-7
Photocopying Prohibited

- Covalent bonding is the sharing of electrons by atoms. It is restricted to bonding in the nonmetals. The sharing of electrons in a water molecule can be shown in numerous ways. Two models you may come across are shown below:

Space filling model. The space the atoms' electrons take up is represented as spheres. Note how the spheres overlap.

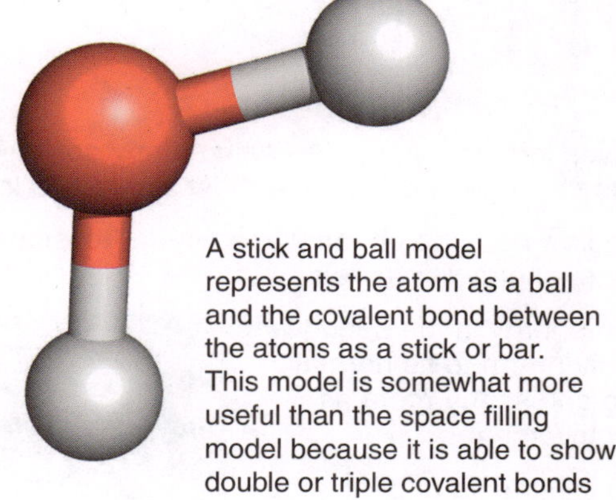

A stick and ball model represents the atom as a ball and the covalent bond between the atoms as a stick or bar. This model is somewhat more useful than the space filling model because it is able to show double or triple covalent bonds

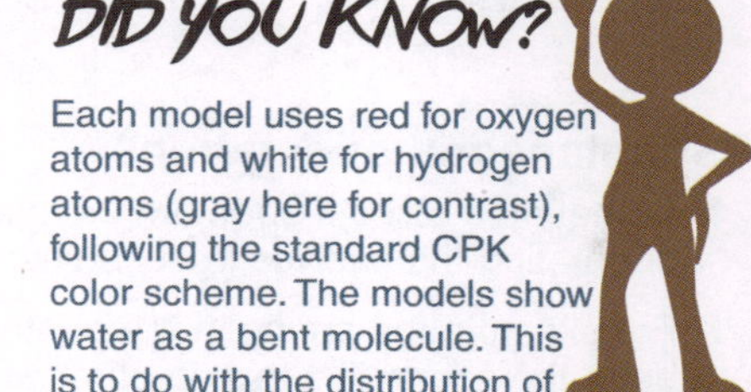

Each model uses red for oxygen atoms and white for hydrogen atoms (gray here for contrast), following the standard CPK color scheme. The models show water as a bent molecule. This is to do with the distribution of the electrons.

- Now consider the structure for carbon dioxide (CO_2).

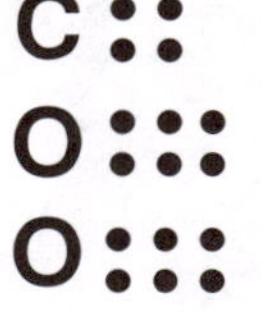

Carbon has an atomic number of 6, so it also has six electrons. Two of those fill the inner electron shell, leaving four in the valence shell. The two oxygen atoms have 6 electrons in their valence shells. This gives a total of 4 + 6 + 6 = 16 electrons or eight pairs to place around the atoms.

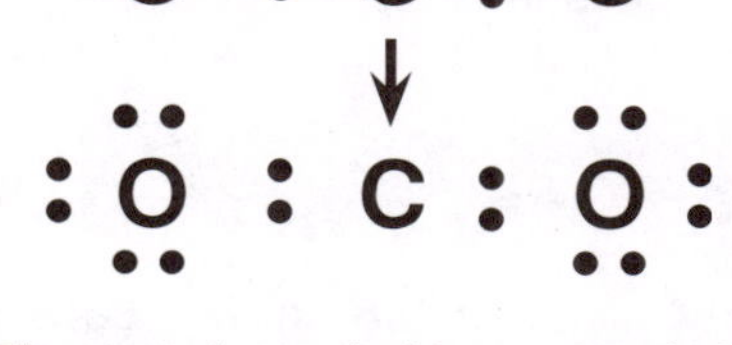

The central atom in this structure is C as this has the lowest electronegativity. Placing the electrons around the atoms is more tricky this time but it is worth starting by adding one pair of electrons to each bond (between the C and the Os), leaving 6 pairs. If we distribute three pairs to each oxygen atom, then both Os have full valence shells but C does not.

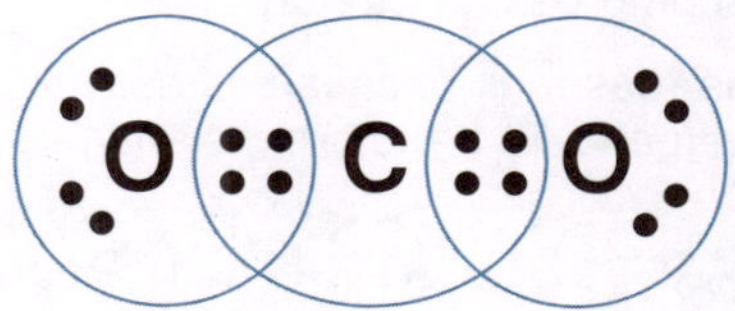

But atoms can share more than one pair of electrons. If we move one of the three pairs from each O to the bonding area between C and O then we find that all three atoms now have four pairs of electrons surrounding them. In this case C and O are bonded by **double** covalent bonds, whereas H_2O had two single covalent bonds.

3. Try drawing Lewis structures for these molecules:

(a) O_2

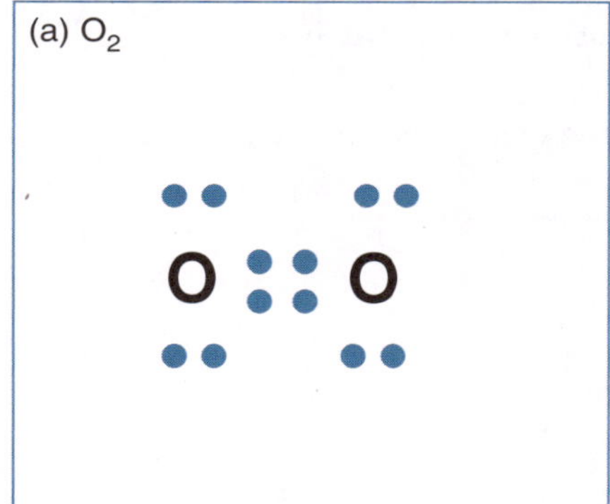

(b) CH_4

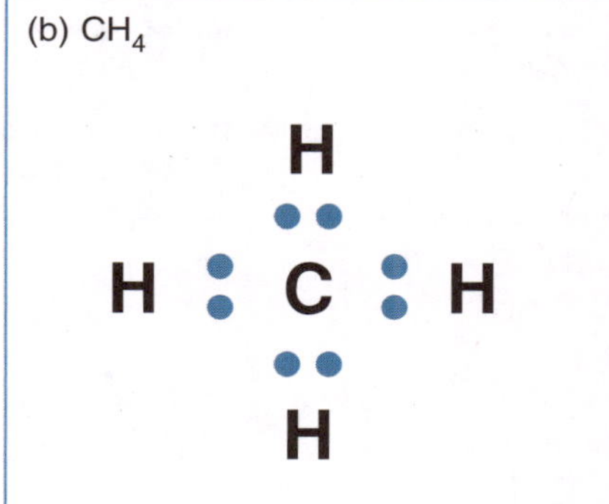

(c) NH_3

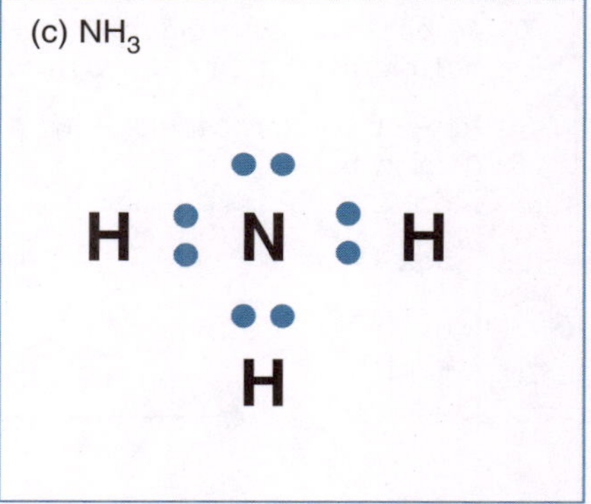

(d) C_2H_6

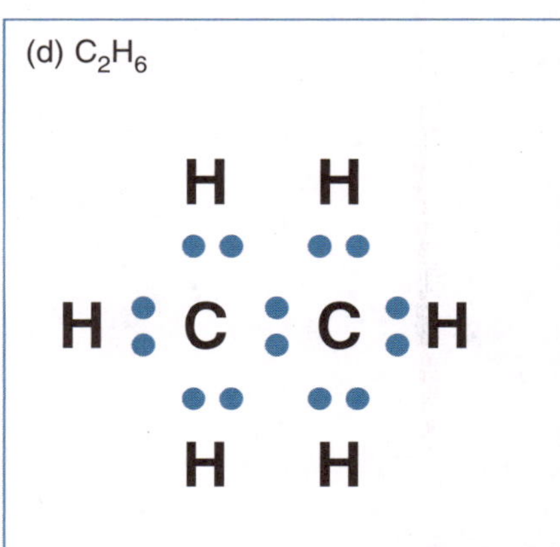

(e) C_2H_4

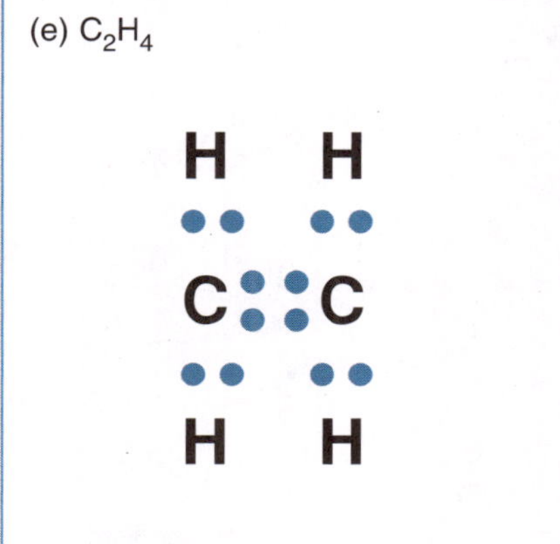

(f) NH_4^+

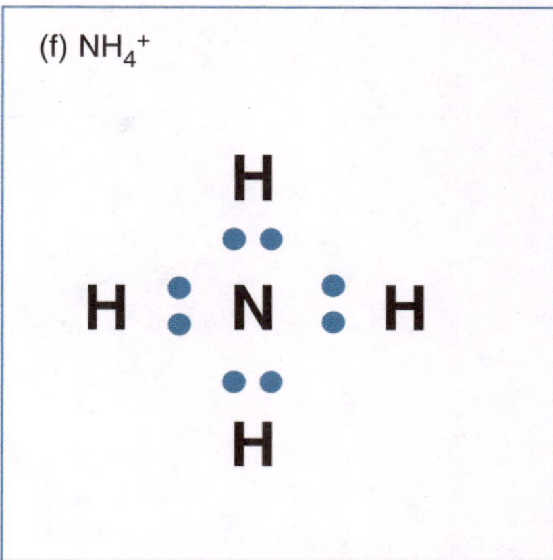

©2020 **BIOZONE** International
ISBN: 978-1-927309-82-7
Photocopying Prohibited

29 Molecular Shape

Key Question: How does the sharing of electrons in molecules affect a molecule's shape?

Spontaneous orientations

- Sometimes objects orientate themselves into certain shapes without any apparent input of energy. In fact by orientating themselves in such shapes the objects are in their most stable and least energetic form.
- For example the spring in the photo below right, has been placed under tension and is bent. Work must be done to bend the spring like this. In other words, energy is needed.
- If the person removes their fingers the energy in the spring will be released and the spring will instantly return to its normally straight shape (and probably go flying across the room as some of the energy it converted into movement).
- But note that the spring will remain in its straight unbent shape as long as no one puts energy into it by bending, stretching, or compressing it. Why doesn't the spring spontaneously bend, or compress, or stretch?
- The reason is because its normal straight shape is the least energetic. To transform its shape energy most be put into it.

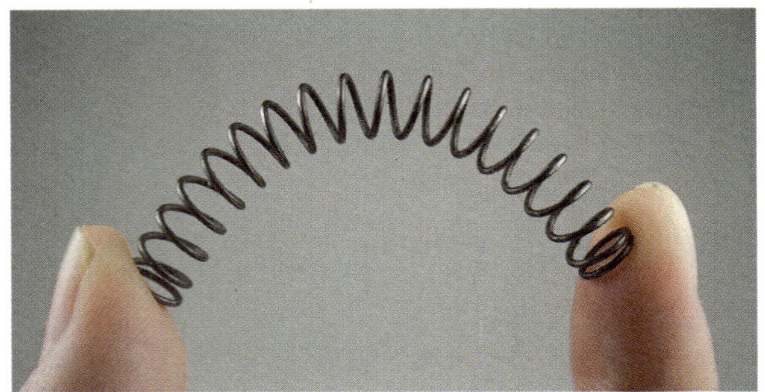

INVESTIGATION 2.1: Repulsion theory

See appendix for equipment list.

1. Inflate a balloon and tie it closed. Draw a dot at the top and bottom of the balloon with a marker.
2. Imagine a line connecting the dots you have drawn. Box 1 below shows a simple drawing of the two dots and the line connecting them.
3. Inflate a second balloon and tie it closed. Draw a dot on the top. Tie the end of the second balloon to the end of the first balloon.
4. In box 2 draw a diagram (similar to box 1) to show how the three dots are connected.
5. Bend the balloons at the point where they are joined. What happens when you let them go?
6. Inflate a third balloon and tie it closed. Again draw a dot on the top and tie it to the join of the first and second balloons.
7. In box 3 draw a diagram to show how the four dots (the top of three balloons and the one at the bottom of the first balloon) are connected.
8. Repeat this procedure with a fourth balloon and draw the diagram of the shape connecting all five dots in box 4.

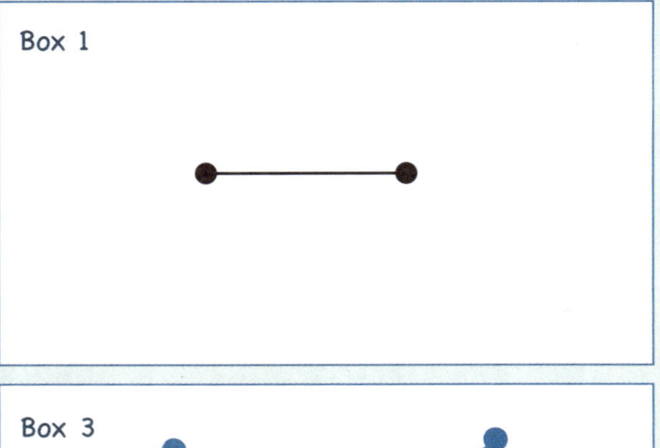

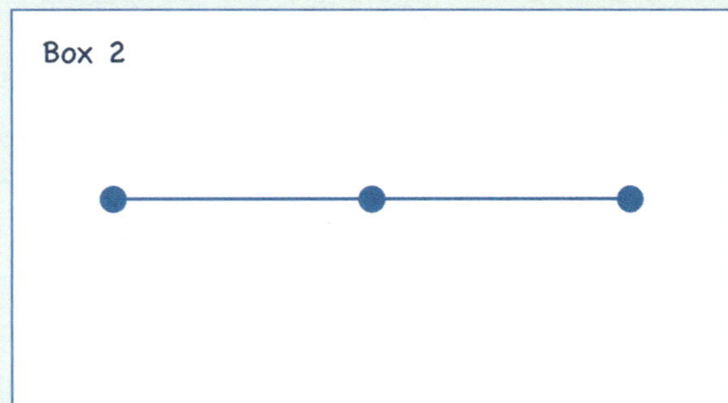

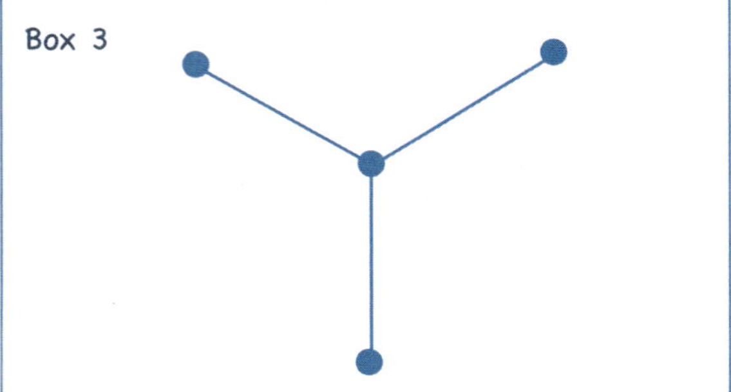

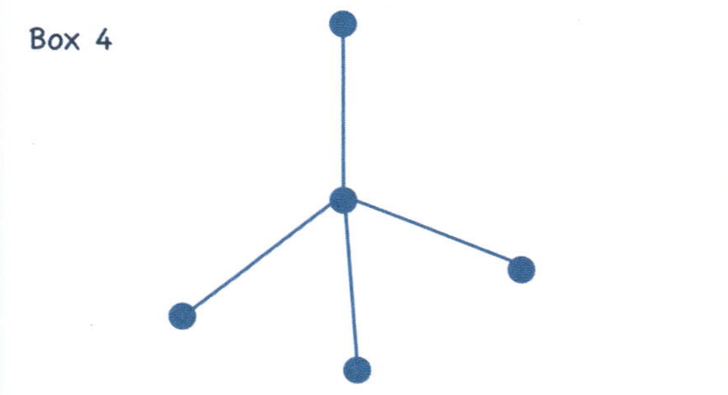

PS1.A P

©2020 BIOZONE International
ISBN: 978-1-927309-82-7
Photocopying Prohibited

Electron repulsion and molecular shape

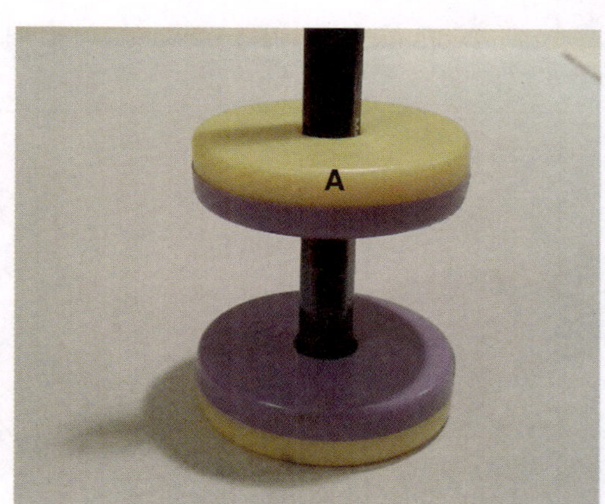

- What exactly causes the balloons to spontaneously produce the shape you drew in box 4?
- The two magnets shown right have reached an equilibrium between the repulsive force of the magnets and the attractive force of gravity. At this point the total potential energy of the system is at its lowest. Work is needed to move the top magnet (A) either up or down.
- Just as similar poles of the magnets repel each other, so the like charges of the electron pairs repel each other. Around a sphere, the four pairs of electrons orientate to positions that are 109.5° apart.
- Knowing that electron pairs repel each other this way (called Valence Shell Electron Pair Repulsion (VSEPR) theory) we can predict the shape of various molecules.

1. Why do the balloons orientate themselves in the shapes you drew above?

   The balloons press against each other. The greater the pressure, the more potential energy in the system. The configurations above reduce the pressure to the minimum and are thus the most stable shapes.

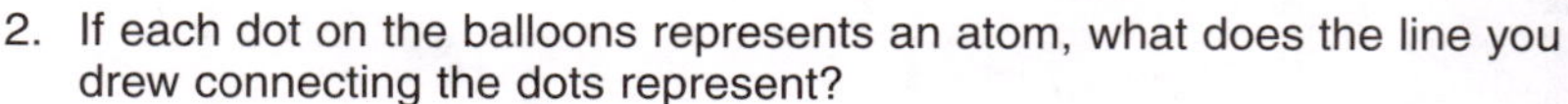

2. If each dot on the balloons represents an atom, what does the line you drew connecting the dots represent?

   The bonds between the atoms.

3. (a) The molecule methane (CH_4) has carbon as its central atom with the hydrogen atoms surrounding it. Which box (1-4) contains the shape of a methane molecule?

   Box 4. C is central the 4 H atoms arrange themselves evenly around the C atom.

   (b) Why would the molecule orientate to that shape?

   The tetrahedral shape keeps the H atoms (and the electrons in the bonds) as far away from each other as possible, reducing stress and potential energy in the molecule.

Bonding and shape

- Recall the Lewis structures you drew in the previous activity. The bonding electrons can also be represented by lines as in the diagram of NH_3 below:

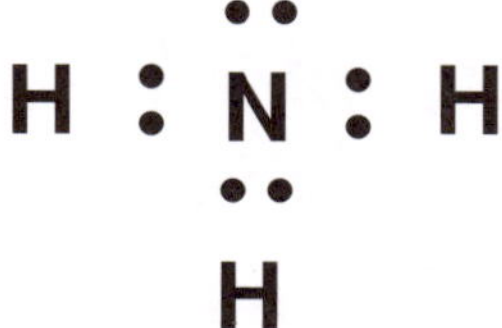

Lewis structure of NH_3

H — N — H
|
H

Bonded electron pairs represented as lines

- Recall the balloon investigation when only three balloons were attached to the central atom (Box 3). This formed a flat shape called a trigonal planar. Is this the shape that NH_3 would form? No. The balloon model is missing something that NH_3 has.
- Look at the structure for NH_3 above. Notice that it has one pair of non-bonded electrons. These act the same way as the other electron pairs and repel the other three bonded pairs.
- NH_3 therefore has a shape similar to the tetrahedral shape of the four balloons, but is missing one of the balloons (shown in red in the image right). This shape is called trigonal pyramidal. The non-bonded pair of electrons are held slightly closer to the N and force the three legs of the pyramid down a little more than in a tetrahedral structure such as CH_4.
- But again NH_3 forms that shape because this is the most stable and least energetic structure.

©2020 **BIOZONE** International
ISBN: 978-1-927309-82-7
Photocopying Prohibited

4. Draw diagrams to predict the shape of the following molecules. Use the following terms to label your diagrams: *linear, bent, tetrahedral.*

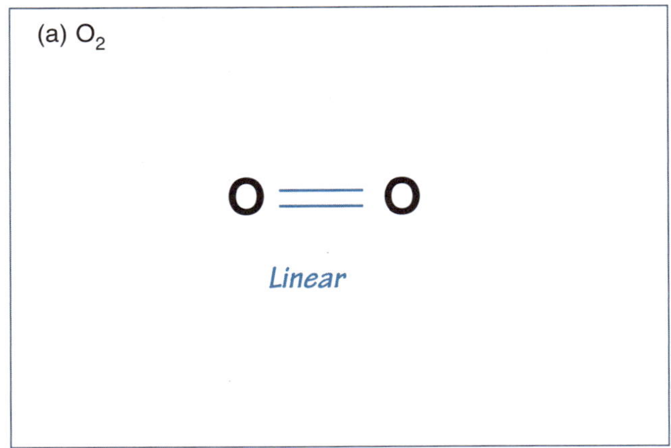

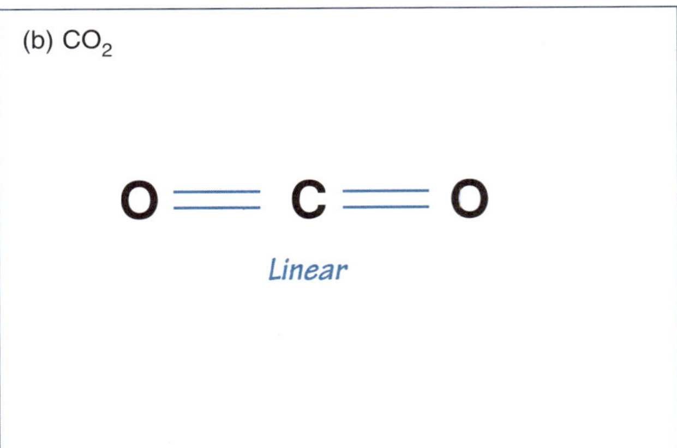

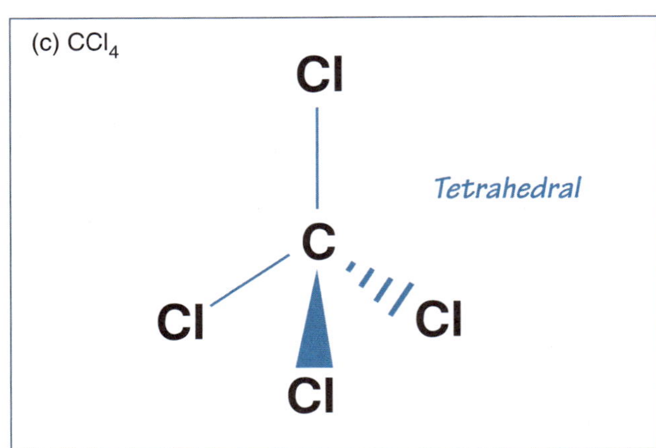

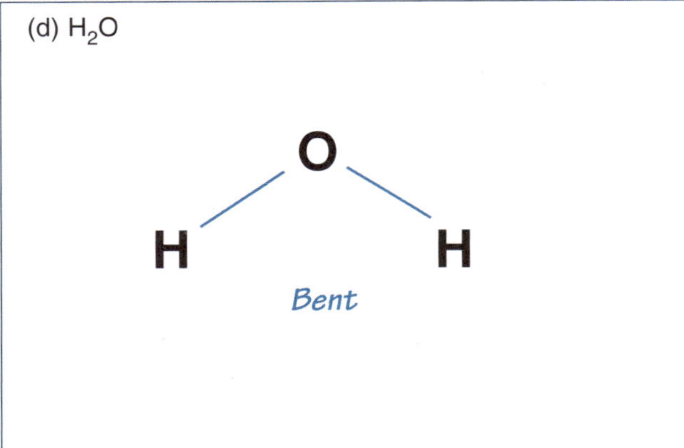

5. Explain why CH_4 has a tetrahedral shape, while H_2O is bent:

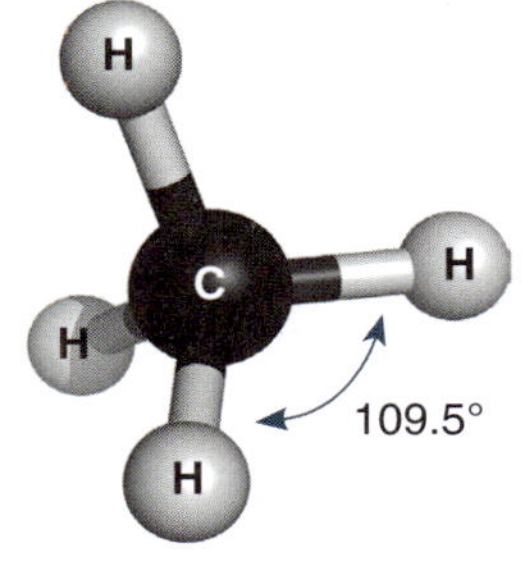

In CCl_4 has four bonded atoms around it. The bonding electrons repel each other and thus the bonds are arranged in such a way that the electron pairs are as far away from each other as possible. This produces a tetrahedral shape. H_2O is bent because it is missing two bonded atoms. It still has four electron pairs evenly spread around it, but two of those pairs have no bonded atoms. Thus a bent shape is what is seen. The non bonding electrons are held slightly closely that the others and this forces the bonded pairs slightly closer together that in the CCl_4.

6. Consider the shapes of the molecules in question 4. Label them as *symmetrical* or *not symmetrical*:

(a) O_2: Symmetrical

(b) CO_2: Symmetrical

(c) CCl_4: Symmetrical

(d) H_2O: Not symmetrical

7. How might the symmetry of a molecule affect the way it interacts with other molecules?

In a symmetrical object all points will behave in the same way. The behavior of non symmetrical objects will depends on their orientation. Electronegativity is important in symmetry. Atoms attract electrons. In symmetrical molecules attractions are even. In non-symmetrical molecules attractions are not even.

©2020 BIOZONE International
ISBN: 978-1-927309-82-7
Photocopying Prohibited

30 Intramolecular and Intermolecular Forces

Key Question: What kinds of forces are there within and between molecules and how does this affect the way they interact?

Electronegativity and bonding

- Recall that the electronegativity of an atom is its ability to attract electrons. If the difference in electronegativity between two reacting atoms is very high, then the electron(s) will be entirely transferred from the atom with the lowest electronegativity to the atom with the highest electronegativity and **ions** will form.
- If the electronegativity is equal the electrons will be shared relatively evenly to form a **covalent bond**.
- What if there is only a slight difference in electronegativity? How will this affect the distribution or sharing of electrons between the atoms? In this case the electrons will be shared, but unevenly, with the electrons being closer to the atom with the highest electronegativity. This results in a **polar covalent bond**.
- For example, HCl is a polar molecule. The polarity can be represented as a diagram:

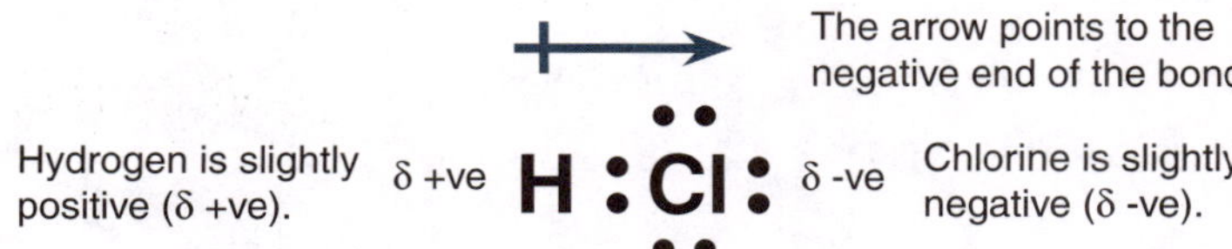

The uneven sharing of electrons produces a permanent **dipole**, with one end of the bond being negative and the other positive.

1. Consider the following substances: NaCl, H_2O, MgO, Na_2O, CO_2, HI, CH_4, Cl_2, O_2, Li_2O.
Use your periodic table to find the electronegativities of each of the atoms for each of the substances above, calculate the difference between the electronegativities, and decide whether the bond is ionic, polar, or covalent.

(a) Na:	0.93	Cl:	3.16	Difference:	2.23	ionic
(b) H:	2.20	O:	3.44	Difference:	1.24	polar covalent
(c) Mg:	1.31	O:	3.44	Difference:	2.13	ionic
(d) Na:	0.93	O:	3.44	Difference:	2.51	ionic
(e) C:	2.55	O:	3.44	Difference:	0.89	covalent
(f) H:	2.20	I:	2.66	Difference:	0.46	covalent
(g) C:	2.55	H:	2.20	Difference:	0.35	covalent
(h) Cl:	3.16	Cl:	3.16	Difference:	0	covalent
(i) O:	3.44	O:	3.44	Difference:	0	covalent
(j) Li:	0.98	O:	3.44	Difference:	2.18	Ionic

2. In the molecule H_2O, around which atom are you most likely to find the electrons at any one moment in time? Explain why the electrons are more likely to be found in the location you have predicted:

The electrons will be found around the oxygen. This is because the oxygen has a higher electronegativity and attracts the electrons more than hydrogen.

3. For each of the molecules below, draw a Lewis structure and draw on the polarity of the bonds (as shown top of page):

(a) CO_2

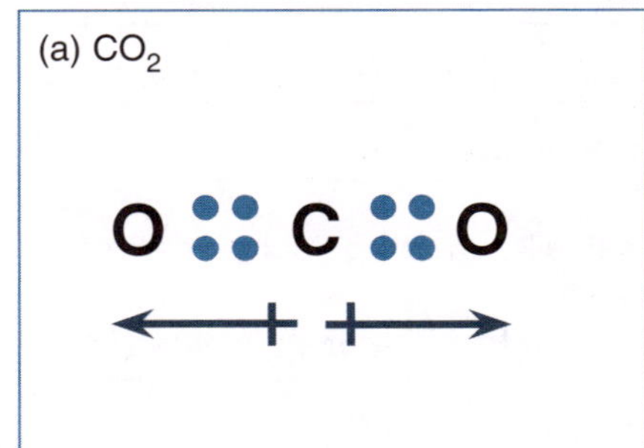

(b) H_2O

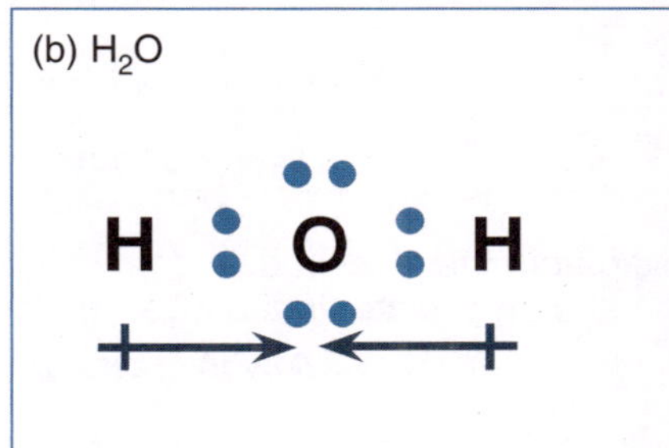

(c) CH_4

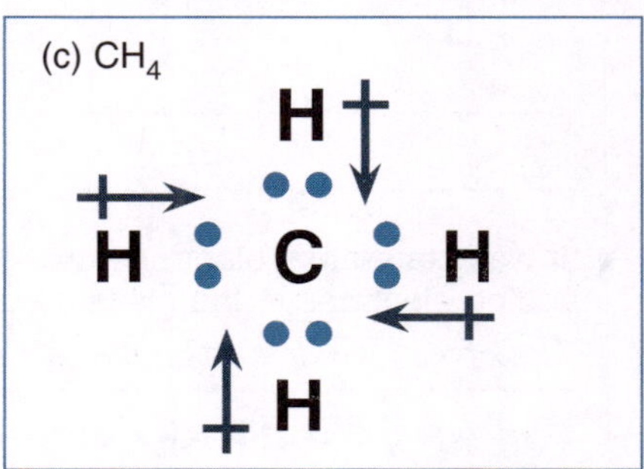

©2020 BIOZONE International
ISBN: 978-1-927309-82-7
Photocopying Prohibited

Shape and polarity

- Look at the shape of the molecules below. You have already encountered these molecules in the previous activity. The polarity of the bonds from question 3 have been added to the molecules.

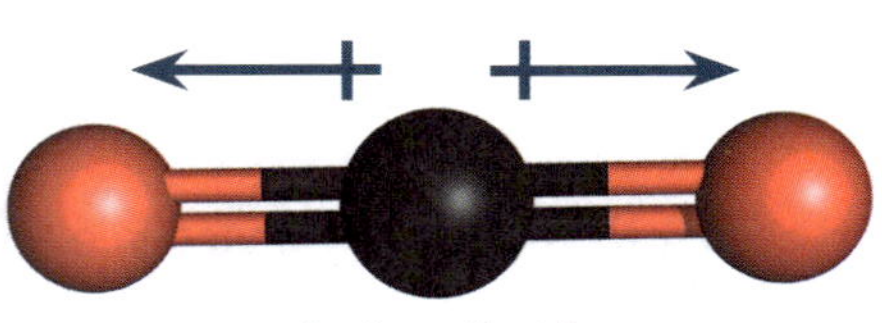
Carbon dioxide

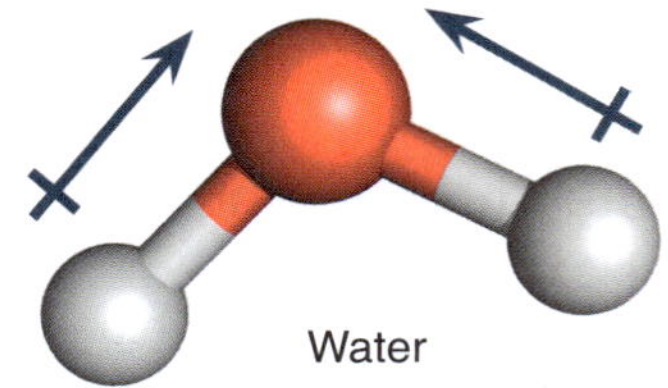
Water

- How does the shape of the molecule affect polarity? We can see that carbon dioxide is symmetrical and so even though the bonds are slightly polar the molecule itself is not. Water is not symmetrical. There are no bonds to balance out the slightly polar O–H bonds. As a result a water molecule is polar over all.

Testing polarity

INVESTIGATION 2.2: Polarity See appendix for equipment list.

1. Fill a 50 mL burette with distilled water. Fill another burette with cyclohexane.
2. Rub a glass rod with a silk or polyester towel. This removes electrons from the glass.
3. Turn on the burette containing water so that it flows into a beaker.
4. Hold the rod close to the burette and observe the effect.
5. Repeat with the burette of cyclohexane and observe the result.

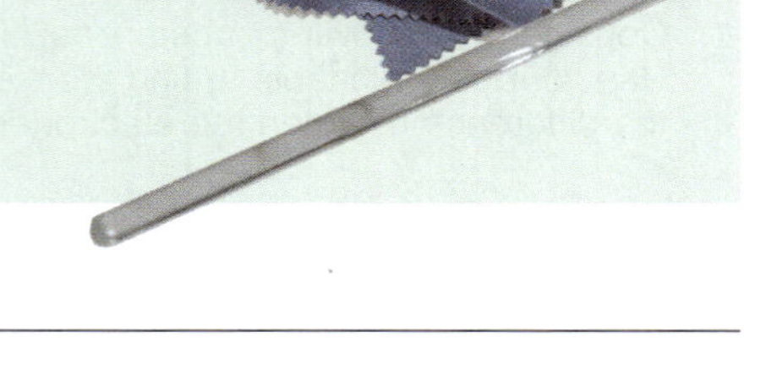

4. What was the effect of holding the rod close to the stream of water?

The stream of water bends towards the rod.

5. What was the effect of holding the rod close to the stream of cyclohexane?

There is no effect. The stream of cyclohexane continues to flow straight down.

6. The cyclohexane molecule is shown right. Explain why water bends towards the charged glass rod and cyclohexane does not. Draw a diagram in the box below to help your explanation:

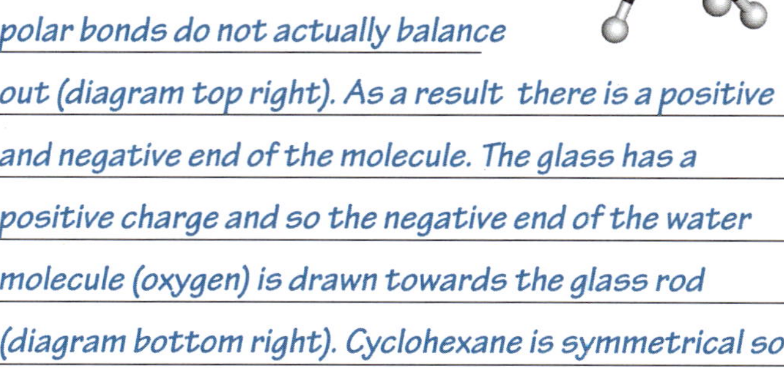

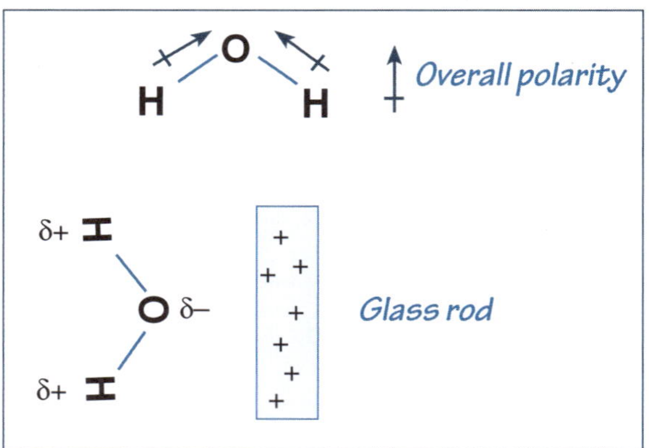

The water molecule is polar. The molecule is bent and so the two polar bonds do not actually balance out (diagram top right). As a result there is a positive and negative end of the molecule. The glass has a positive charge and so the negative end of the water molecule (oxygen) is drawn towards the glass rod (diagram bottom right). Cyclohexane is symmetrical so that there is no positive or negative end.

7. It was stated that electrons are removed from the glass rod when rubbing it with silk. What would the effect of adding electrons to the glass be on the stream of water in the investigation above?

The effect would be the same. Water would still bend towards the rod. However, in this case the positive (hydrogen) end of the water molecule would be attracted to the rod, causing the bending.

©2020 BIOZONE International
ISBN: 978-1-927309-82-7
Photocopying Prohibited

Intermolecular forces

- Molecules interact with each other via intermolecular forces. These are very weak compared to the intramolecular forces that bond the atoms within molecules together.
- Collectively intermolecular forces are called **van der Waals** forces. Two important van der Waals forces are hydrogen bonding and induced dipole-dipole forces.

Hydrogen bonds

- When hydrogen forms a covalent bond with either oxygen, nitrogen, or fluorine, those atoms attract the electrons much more than the hydrogen, producing a polar covalent bond. The dipole that forms allows other similar molecules to be attracted. Hydrogen bonds are the strongest intermolecular force.

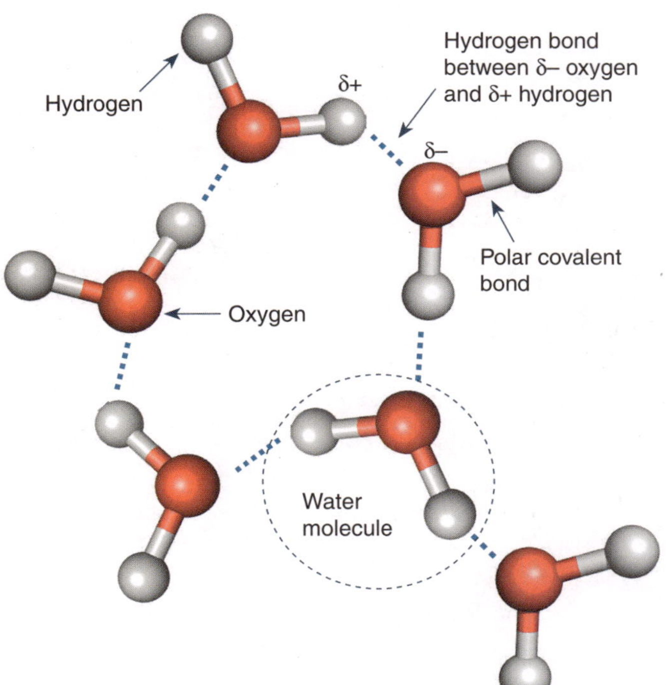

Induced dipole-dipole forces

- Electrons are always in motion around an atom, even in atoms covalently bonded to other atoms. The motion is random and this results in instances where there are more electrons on one side of the molecule than another. This results in a slight polarity from one side of the molecule to the other (below).

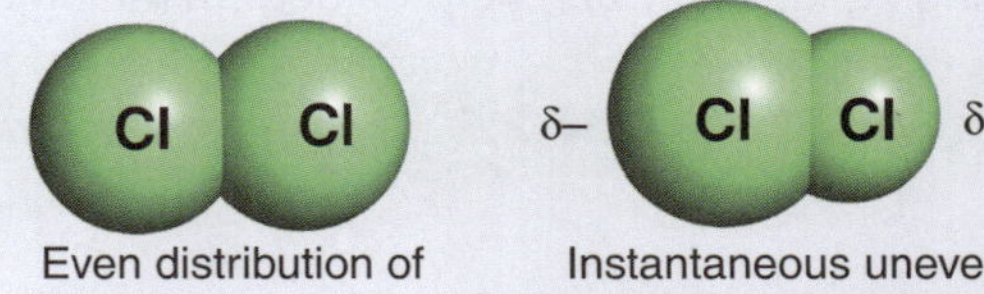

Even distribution of electrons

Instantaneous uneven distribution of electrons

- The instantaneous dipole can affect nearby molecules, so that they also form dipoles, resulting in temporary attractions between the molecules (below):

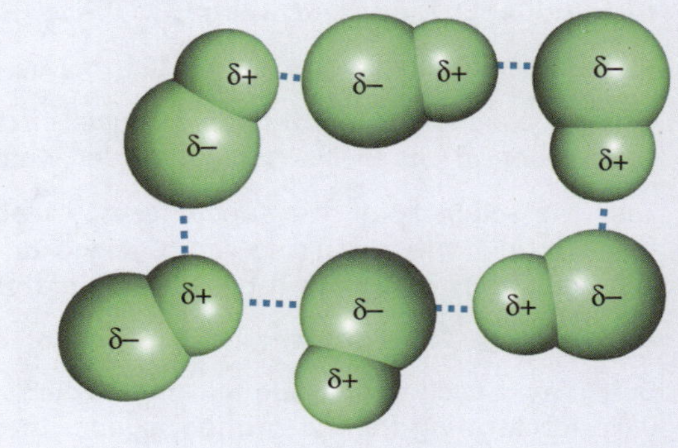

Van der Waals forces play an important role in sticking things together. The ability of geckos (near right) to walk up seemingly smooth surfaces is due, in part, to van der Waals forces. The small effect is multiplied millions of time over by the microscopic hairs on the gecko's feet.

Scientists have been inspired by the gecko's feet to create dry glues that can be used multiple times or sticky pads that can be used to hold objects to a wall. In 2014, scientists at Stanford University created an adhesive device the size of a human hand and based on gecko's feet, that allowed a 70 kg man to climb a vertical glass wall (far right).

Biomimetics and Dexterous Manipulation Lab, Stanford University

8. Explain why water is a liquid at 25°C but oxygen must be cooled to less than –183°C before it forms a liquid:

Oxygen is a non-polar molecule. There are very few intermolecular interactions between these molecules, so oxygen has a very low boiling point. On the other hand, as a polar molecule, the hydrogen bonds in water molecules helps hold the molecules together in a substance. This is why water is held together as a liquid at much higher temperatures than oxygen.

©2020 **BIOZONE** International
ISBN: 978-1-927309-82-7
Photocopying Prohibited

31 Comparing Substances

Key Question: What does the comparison of the properties of substances tell us about their structure?

PS1-3 Properties of matter

- Everything around us is made up of a particular combination of atoms or ions. That combination gives a particular substance certain properties, such as a liquid's viscosity or the hardness of the solid.
- Exploring the properties of various substances can tell us a lot about the atoms or ions in the substances and how they are held together. This helps us to categorize materials. We can then use the properties of the materials in those categories to predict the properties of other substances and how they might be used.
- Testing conductivity can help us determine if there are free charges in the substance (e.g. ions or electrons).

INVESTIGATION 2.3 Properties of matter

See appendix for equipment list.

Magnesium (right) is a flammable metal. It burns with a bright, extremely hot flame, which can cause severe burns. Do not look directly at the flame. Sulfur is flammable. Igniting it produces sulfur dioxide gas, which is dangerous to inhale. Cyclohexane is volatile and flammable. Wear protective eyewear.

1. You will work in groups to investigate various properties of the following substances: sodium chloride (NaCl), magnesium (Mg), magnesium sulfate ($MgSO_4$), sulfur (S), copper, (Cu) copper sulfate ($CuSO_4$), distilled water (H_2O), ice (H_2O), cyclohexane (C_6H_{12}), quartz or glass (SiO_2).
2. Place a small sample (1 gram or 1-2 mL) of each substance on a watch glass or in a beaker.
3. Using a conductivity meter (or a simple circuit) test the conductivity of each of the substances. Record your observations in the table provided below.
4. Test the solubility of the substances by placing 1 gram of each substance into 50 mL of distilled water and stirring with a stirring rod. Cyclohexane should be used in a fume hood. Add 1 mL to 10 mL of distilled water in a test tube. Record in the table whether or not each substance dissolves in water.
5. At the same time test the conductivity of any solutions formed. Record the results in the table.
6. Sulfur and sodium chloride will melt when heated by a Bunsen flame. Place a small amount of sulfur in a deflagrating (def-la-grating) spoon. In a fume hood, heat the sulfur over a low Bunsen flame until the sulfur begins to melt. **DO NOT** let the sulfur ignite and burn (this will produce toxic sulfur dioxide gas). If the sulfur does ignite, remove the spoon from the heat and cover it with a heat resistant pad or glass plate to cut off the air supply or place it into a gas jar to trap the gas given off. Close the fume hood and wait for the gas to clear.
7. While the sulfur is still molten, carefully test its conductivity. Don't touch the spoon with the electrodes or you will get a false reading! Let the sulfur cool and harden before disposing of it.
8. Clean the deflagrating spoon and repeat the procedure with sodium chloride. Record the conductivity in the table.
9. Glass tubing will melt if held over a strong Bunsen flame. Your teacher may demonstrate the heating of glass and test its conductivity when molten.

Substance	Solid conducts ✔ ✘	Liquid conducts ✔ ✘	Solution forms in water ✔ ✘	Solution conducts ✔ ✘
NaCl	✘	✔	✔	✔
Mg	✔	✔	✘	NA
$MgSO_4$	✘	✔	✔	✔
S	✘	✘	✘	NA
Cu	✘	✔	✘	NA
$CuSO_4$	✘	NA	✔	✔
H_2O	✘	✔ weakly	NA	NA
C_6H_{12}	NA	✘	✘	NA
SiO_2	✘	✘	✘	NA

PS1.A P

©2020 BIOZONE International
ISBN: 978-1-927309-82-7
Photocopying Prohibited

- Before you interpret your results, other properties for the substances you tested should be noted. Some you may already know from experience, others you may need to test, normally by simple observation of the macroscopic structure of the substance (i.e. what is looks like, how it feels etc).

1. Fill in the table below. You may need to do some research or testing of the substance. Note: malleable means workable:

Substance	Appearance (color, shine etc)	State at room temperature	Malleable ✓ ×	Brittle ✓ ×	Melting point / boiling point (°C)
NaCl	White crystalline solid	Solid	✘	✔	801 / 1465
Mg	Silver lustrous solid	Solid	✔	✘	650 / 1091
$MgSO_4$	White crystalline solid	Solid	✘	✔	1124 / Decomposes
S	Yellow powder	Solid	NA	NA	113 / 445
Cu	Orange lustrous solid	Solid	✔	✘	1085 / 2927
$CuSO_4$	Blue crystals	Solid	✘	✔	Decomposes
H_2O	Colorless liquid	Liquid	NA	NA	0 / 100
C_6H_{12}	Colorless liquid	Liquid	NA	NA	6.47 / 807
SiO_2	White/clear crystalline solid	Solid	✘	✔	1710/ 2230

2. Now put the substances into groups based on the similarities and differences of the properties you have investigated:

Students may have various answers as the properties are not always indicative of the correct groups. Students should be able to group substances that conduct both as a solid and a liquid (metals), or as a liquid and in solution (ionic) and those that do not conduct (molecular). They may note that all the metals are lustrous (shine) and malleable and many are gray/silver (although copper is not). Ionic substances tend to be solid and brittle.

Sulfur exists a discrete molecule while SiO_2 is a giant lattice. Covalent networks have variable conductivity, e.g diamond does not conduct, but graphite does, and silicon is a semiconductor.

3. Explain the underlying properties of the substances in each of the groups you placed the substances into:

Student's answer based on the groups they have put the substances into. For example, students might explain that the metals are lustrous (shine) and malleable. Other explanations as above:

©2020 **BIOZONE** International
ISBN: 978-1-927309-82-7
Photocopying Prohibited

4. You have investigated the properties of various substances and should now be able to explain the difference between different types of substances. Explain the differences between the following substances:

(a) Explain why water (H_2O) is liquid at room temperature but hydrogen sulfide (H_2S) is a gas: *O is more electronegative than H. In water, this causes the bonds to be polar. The shape of the water molecule it is also polar so the water molecules are strongly attracted to each other. A lot of energy is needed to separate the molecules so water is liquid at room temperature. H and S have similar electronegativities so the bonds between H and S are only weakly polar. The molecule has the same shape as water but is less polar. The molecules only weakly interact and thus vaporize easily and so is a gas at room temperature.*

(b) Surface tension is an effect where the surface of a liquid is strong. It is caused by cohesive forces between the liquid's molecules. Explain why the surface tension of water (H_2O) is much greater than that of cyclohexane (C_6H_{12}): *Water is a polar molecule. Its molecules are strongly attracted to each other. Water molecules on the surface are attracted to ones below them and to the side of them causing the surface to be very strong and difficult to part. Cyclohexane is a non-polar molecule. Its molecules are only weakly attracted to each other and so the surface tension is less and the molecules at the surface are easy to part.*

(c) Why is solid NaCl ($NaCl_{(s)}$) unable to conduct electricity but liquid NaCl ($NaCl_{(l)}$) and NaCl in solution ($NaCl_{(aq)}$) can? *The ions in solid NaCl are unable to move and so cannot conduct a charge from terminal to terminal. In liquid NaCl, the ions can move about and so can conduct a charge. In a solution of NaCl, the ions are also free to move about and conduct a charge.*

5. Fill in the table below with the particles, forces, examples, melting points (very high to very low) and characteristics of the substances listed:

Substance		Particles	Attractive forces involved	Examples	Melting point (high / low)	Characteristics
Ionic		*Ions*	*Ionic bonds*	*NaCl, LiCl, Na_2SO_4*	*High*	*Solid at standard pressure & temperature. Brittle. Non-conductive when solid.*
Metal		*Metal atoms*	*Metallic bonds*	*Silver, iron, copper*	*High*	*Mostly solid, lustrous, malleable, conductive.*
Molecule	Polar	*Molecules*	*Hydrogen bonds*	*Water, HCl*	*Low*	*Often liquids, usually slightly conductive*
	Non-polar	*Molecules*	*Van der Waals*	*Iodine, chlorine, oxygen*	*Very low*	*Often gas, non conductive*

©2020 BIOZONE International
ISBN: 978-1-927309-82-7
Photocopying Prohibited

32 Review Your Understanding

Key Question: What are the underlying causes of the different properties of compounds?

- As the start of this chapter you were shown some examples of salts and asked what might cause the various properties of these salts including the way they cleave and the reason for their shape and color.
- A crystal of sodium chloride is shown right:
- Note that the crystal forms a cube. The cleavage planes are shown in the image, far right. If the crystal is given a hard strike with a razor blade along these planes, it will cleave in straight lines. The cleavage planes are at 90° to each other.

Hans-Joachim Engelhardt cc.4.0

1. You should now be able to explain how these cleavage planes occur and the reason for the shape of the crystal.

   (a) In the box below produce a diagram that shows how the particles in a sodium chloride crystal are arranged. Use circles to represent the particles in the sodium chloride crystal.

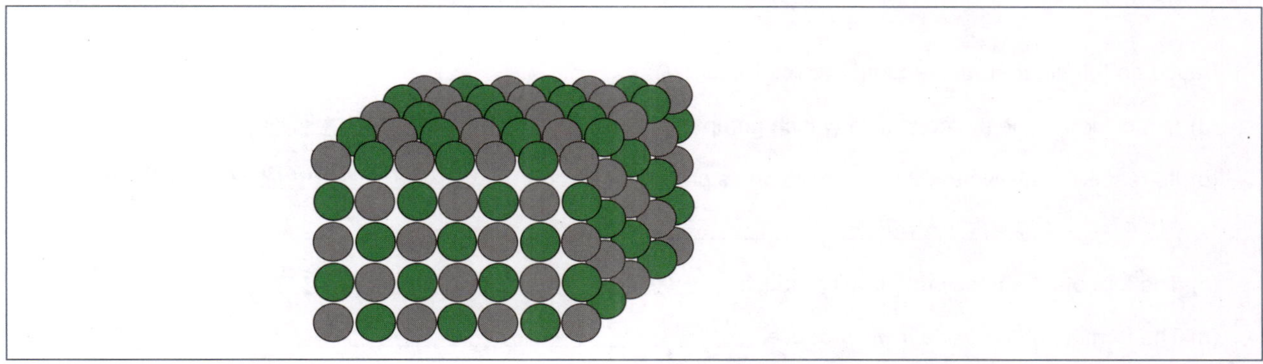

   (b) Use your diagram to explain why the crystal will cleave in flat planes 90° to each other if struck with a razor blade along these planes:

   The diagram is a 3D lattice with the ions aligned at right angles to each other. Thus when struck with a blade the crystal cleaves between the ions, producing planes at 90° to each other.

2. Explain why the crystals shown below have quite different shapes and cleavage planes to the sodium chloride crystal:

   The ions are aligned differently to NaCl and are in different ratios. The different ratios and alignments produces different angles of cleavage planes.

Géry Parent CC3.0

3. What causes the various colors of the different crystals?

   Each ionic crystal has its own arrangement of electrons. The interaction of light with these electrons is what gives each crystal its characteristic color.

©2020 BIOZONE International
ISBN: 978-1-927309-82-7
Photocopying Prohibited

33 Summing Up

PS1-1 Atomic structure and bonding

For the following summative assessment you may refer to your periodic table on page 30 of this book.

1. Study the diagram of the atom labelled X below:

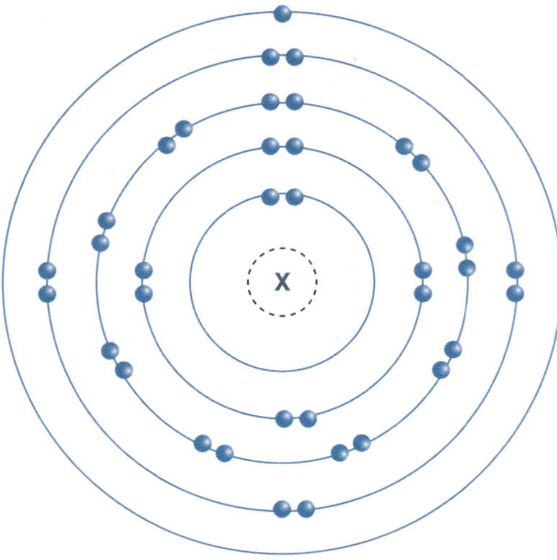

Based on the diagram of the atom predict the following:

(a) Its position in the periodic table (which group and period): *Group 1 Period 5 (1 valence electron, 5 shells)*

(b) Its reactivity relative to other elements in its group and period: *Highly reactive compared to other elements in both its group and its period.*

(c) The formula of the chloride it will produce: *XCl (a ratio of 1:1 as X will form an X^+ ion)*

(d) The formula of the oxide it will produce: *X_2O*

(e) Whether it is a metal or nonmetal: *X is a metal. Specifically it is an alkali metal.*

(f) Its electronegativity relative to other atoms on the periodic table: *Low electronegativity*

(g) Its ionization energy relative to other atoms in its group: *It will be lower than other's in its group*

2. (a) The graph below shows the atomic radii of the alkali metals. Sketch a line on the graph to show where you think **ionic radii** for these metals would go.

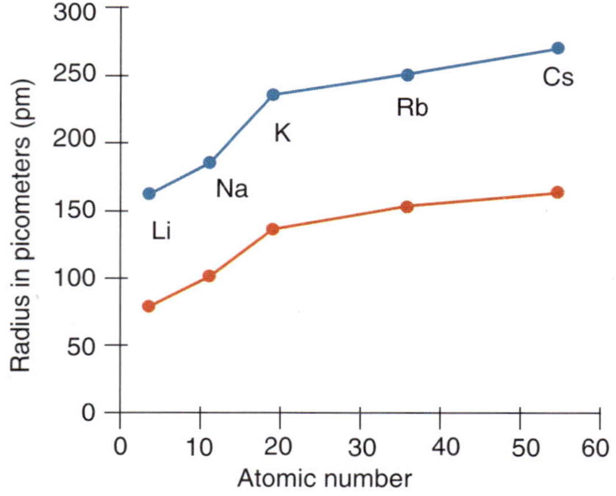

(b) Explain the placement of your sketch: *The graph of ionic radii will follow the same pattern as atomic radius but the values are smaller because the atoms have lost the valence electron. The electrons left in the new valance shell are also held closer to the nucleus due to the increased relative positive charge.*

PS1.A P

©2020 **BIOZONE** International
ISBN: 978-1-927309-82-7
Photocopying Prohibited

3. (a) An atom has 17 electrons. Write down its electron configuration: 2, 8, 7

   (b) Will the atom form a positive or negative ion? Negative

   (c) What would the charge on the ion be? 1–

   (d) Is this atom a metal or nonmetal? Nonmetal

4. An element has atoms with 20 protons, 20 neutrons, and 20 electrons:

   (a) What is the atomic number of this element? 20

   (b) What is the mass number of the element? 40

   (c) In the boxes draw the electron arrangement of the atom and its ion:

Atom

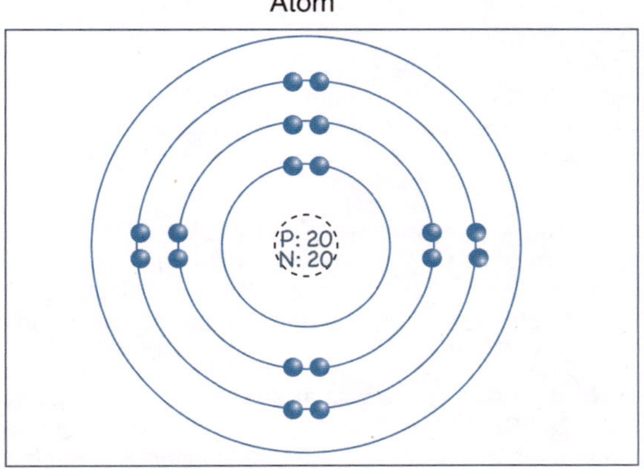

Ion

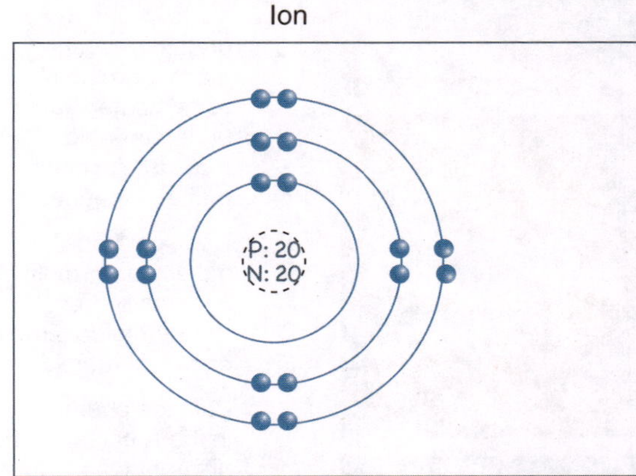

5. Complete the Lewis structure for:
   (a) Fluorine gas (F_2)

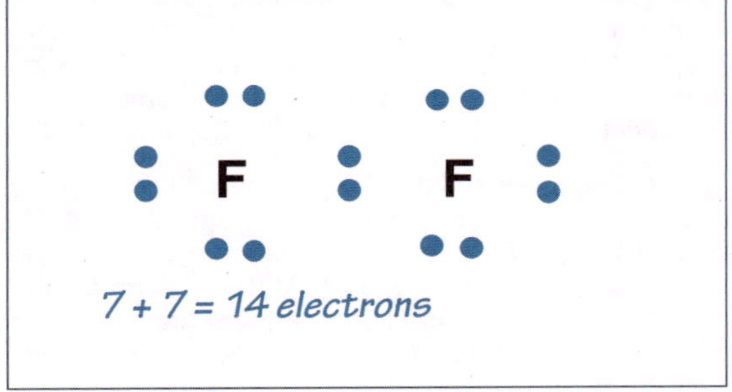

   (b) Hydrogen cyanide (HCN)

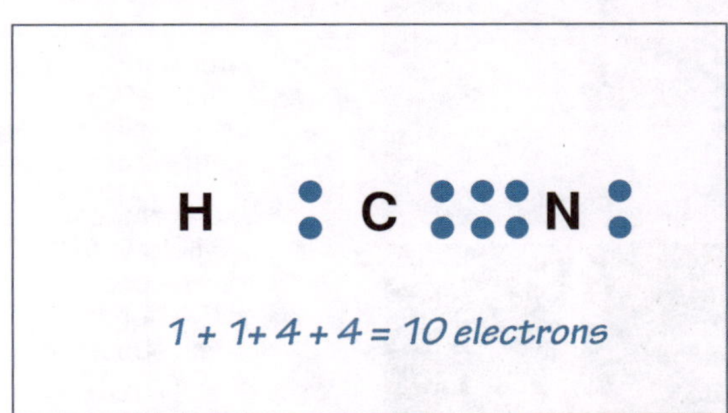

6. Use your knowledge of the periodic table to explain the following reactions:

   (a) One carbon atom reacts with 4 hydrogen atoms to form methane: $C + 4H \rightarrow CH_4$.

   Carbon requires four electrons to obtain a full valence shell. Hydrogen, in the 1st group has one extra electron. Four hydrogens (more importantly their electrons) are then needed to fill the four empty spaces in carbon's valence shell, producing CH_4.

   (b) One magnesium atom reactions with one chlorine gas molecule to form magnesium chloride: $Mg + Cl_2 \rightarrow MgCl_2$.

   Magnesium is in group 2 of the periodic table and therefore has 2 electrons in its valence shell. Chlorine, in group 17, requires 1 electron to fill its valence shell. Chlorine gas contains 2 chlorine atoms. The 2 electrons from magnesium can be transferred to the chlorine atoms, filling their shells, forming $MgCl_2$.

   (c) Any alkali (group I) metal will react with any halogen to form a salt with a 1:1 ratio of metal ions to halogen ions:

   All group 1 metal have 1 electron in their valence shells. All halogens require 1 electron to fill their valence shell. Thus an electron can be exchanged between the metal and the halogen, so that each has a full valence shell and forms a +1 and -1 ion respectively. The ions attract in a 1:1 ratio.

©2020 **BIOZONE** International
ISBN: 978-1-927309-82-7
Photocopying Prohibited

PS1.B
PS3.D

3: Chemical Reactions

Anchoring Phenomenon

Antacid chemistry: There are many antacid formulas on the market. How do they work? 34 54

Toby Hudson CC 3.0

Displacement

What is a mole and why is it important?

☐ 1 The mole (symbol = mol) is a number, specifically 6.02 x 10^{23}. You can easily find the mass of one mole of atoms in any element because it is the relative atomic mass in grams. Describe the relationship between the number of moles of a substance, the mass, and the molar mass mathematically *[SEP-5]*. Use the periodic table *[CCC-1]* to calculate *[SEP-5]* relative molecular masses for different compounds. Use your understanding of the mole to interpret chemical formulas, account for products and reactants in chemical equations, and explain *[SEP-6]* the conservation of energy and matter in chemical reactions *[PS1.B]* *[CCC-5]* *[HS-PS1-7]*. 35 36 55

Decomposition

☐ 2 Use your knowledge of moles and molar mass to determine the empirical and molecular formula of new compounds and investigate *[SEP-3]* the percentage composition of a substance (such as an ore or a hydrated crystal). Use your calculations *[SEP-5]* to construct an explanation *[SEP-6]* for how energy and matter are conserved in chemical reactions *[PS1.B]* *[CCC-5]* *[HS-PS1-7]*. 37-39 55

☐ 3 Balance chemical equations for ionic compounds to show that all the atoms and elements in the reactants are present in the products *[SEP-5]* *[PS1.B]* *[CCC-5]* *[HS-PS1-7]*. Identify different types of reactions by how the reactants interact and the products formed. Explain *[SEP-6]* different types of reactions, using equations to show that energy and matter are conserved *[PS1.B]* *[CCC-5]*. Demonstrate understanding of state (e.g. gas) during reactions by using the appropriate notation in chemical equations. 40

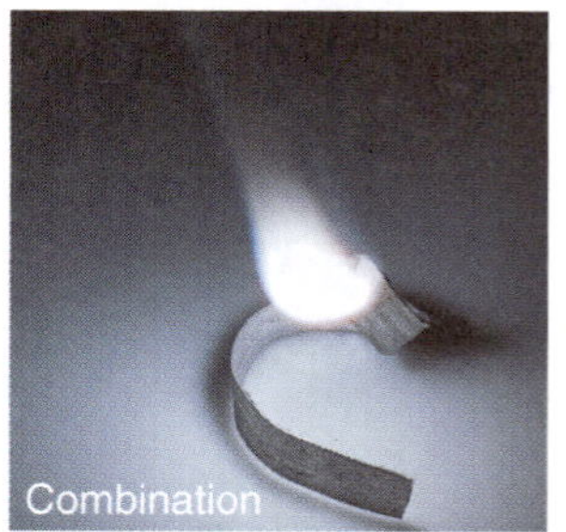
Combination

☐ 4 Stoichiometry is a word for how we use a balanced equation to calculate the amounts of reactants and products in a chemical reaction. It is based on the *law of conservation of mass* *[CCC-5]* and *law of definite proportions* *[CCC-3]*. Use mathematics *[SEP-5]* and apply basic principles of stoichiometry *[CCC-1]* to balance equations and determine quantitative values for reactants and products *[SEP-6]* *[PS1.B]* *[HS-PS1-2]* *[HS-PS1-7]*. 41 55

☐ 5 Use a mathematical model *[SEP-2]* and your understanding of the periodic table *[CCC-1]* to calculate *[SEP-5]* the mass of a substance required to produce a standard solution of a known concentration. Use titration *[SEP-3]* to determine the concentration of an identified analyte for which the concentration is unknown. Show how you would use titration to test claims about the composition of a store-bought product such as vinegar or orange juice *[SEP-6]* *[PS1.B]* *[CCC-5]* *[HS-PS1-2]* *[HS-PS1-7]*. Volume measurements have a key role in titration, so chemical analysis by titration is also called volumetric analysis. 42 43 44

Complete combustion

What happens during a chemical reaction?

☐ 6 Investigate *[SEP-3]* the rate of a chemical reaction under different conditions (e.g. temperature, concentration of reactants, surface area) *[PS1.B]*. Explain *[SEP-6]* any patterns *[CCC-1]* you see *[HS-PS1-5]*. What effect do catalysts have on reaction rates? 45 55

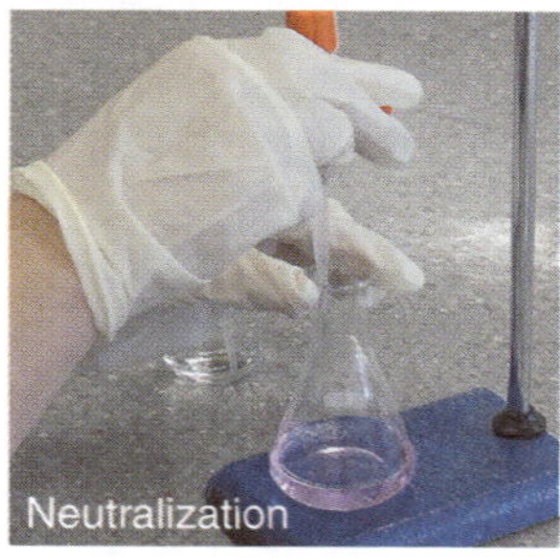
Neutralization

☐ 7 Develop a model *[SEP-2]* of how energy is absorbed and released during chemical reactions *[SEP-5]* *[CCC-5]*. Identify changes of state as either exothermic or endothermic *[PS1.B]* *[PS3.D]* *[CCC-5]* *[HS-PS1-4]*. 46-48 55

☐ 8 Enthalpy change (ΔH) is the name given to the amount of heat released or absorbed in a reaction *[PS1.B]* *[CCC-5]*. Investigate *[SEP-3]* the change in enthalpy for some common reactions, including combustion ($\Delta_c H$) and product formation ($\Delta_f H$) *[HS-PS1-4]*. 47 48

☐ 9 Many reactions in chemistry are reversible. A dynamic equilibrium is an example of stability *[CCC-7]* where reactions in each direction are equal and opposite. Investigate *[SEP-3]* a dynamic equilibrium in a model system *[PS1.B]* *[CCC-4]* to show how changes to reactants, products, or reaction rate affect the equilibrium that is established. Use the equilibrium reaction between Fe^{3+} and SCN^- to explain *[SEP-6]* how changing the conditions in a chemical system can alter the equilibrium in favor of reactants or products *[PS1.B]* *[CCC-7]* *[HS-PS1-5]*. Explain the results using Le Châtlier's principle. 49-51 55

Incomplete combustion

☐ 10 Apply your understanding of Le Châtlier's principle to processes in industrial chemistry to explain how you would increase the amount of a product in a chemical reaction *[SEP-6]* *[PS1.B]* *[CCC-7]* *[HS-PS1-6]*. 52

☐ 11 Investigate *[SEP-3]* the pH of common acids and bases. What does the pH scale represent? To understand it better, calculate *[SEP-5]* the pH of solutions with known H^+ concentration and the (conversely) the H^+ concentration of solutions of known pH. Use your understanding of chemical equilibria to explain *[SEP-6]* why some acids and bases are weak (do not fully dissociate when in water) *[PS1.B]* *[CCC-7]*. 53

3. Chemical Reactions

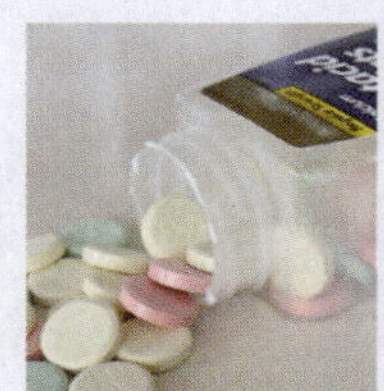

Anchoring phenomenon

The anchoring phenomenon *"Antacid Chemistry"* uses the reactions of various types of antacids to introduce chemical reactions. Students are introduced to different aspects of chemical reactions, including pH and concentration. Students are asked to write and balance equations for different types of antacid reaction, but they should not be expected to get these correct at this stage. They serve as a simple pretest and are revisited at the end of the chapter.

What is a mole and why is it important?

1. The mole is an important concept in chemistry. Students sometimes find it difficult to understand so it is important to point out it is simply a number, the same as saying "a dozen", or "10 dozen". Although the number of atoms in a mole is given as 6.02×10^{23}, this number is in a way irrelevant, as the mole is the unit used in chemical reactions. It is important for students to understand that the mass in grams of 1 mole of an element is equal to the relative atomic mass of that element. They should be able to calculate relative molecular mass from relative atomic mass and then relate the figure to molar mass (g/mol).

2. Students carry out two different investigations, using their knowledge of molar mass to work out firstly, the formula of magnesium oxide, and then the number of water molecules incorporated in the hydrated copper sulfate crystal. This is done using the equation $n = m/M$, which relates mass to number of moles and molar mass. You can remind students to go to page 16 in the introductory skills chapter if they want guidance on how to rearrange equations using fact family triangles. Thus the mass gained when magnesium is heated in air is related to the number of moles of oxygen that react. Similarly, the mass lost by hydrated copper sulfate is related to the moles of water lost. Students also explore empirical formula and percentage composition using their knowledge of molar masses. In all cases, the mathematics are simple cases of calculating an unknown given known quantities. A certain amount of logic is required such as "the mass of X in molecule 1 is the same as the mass of X in molecule 2".

3. There are numerous ways to teach students how to balance equations. Different students respond to different techniques. We have presented a number of ways, both graphically and numerically. Teachers could use the jigsaw graphic on page 81 as a template to design other jigsaw ions for students to manipulate. A similar exercise could be done using the block on page 82. Students should understand that an ionic unit (e.g. MgO) has no overall charge and that this is because the positive and negative charges balance. Students should further understand that all atoms on one side of the equation should be found on the other side. They can only balance equations by changing the number in front of the compound (the number of moles), not be changing subscript numbers within the compound.

4. Students should now be able to use balanced equations to calculate the expected mass of a product given the masses of the reactants. Teachers could help students verify the calculations in activity 41, question 2 by performing a precipitation reaction using known masses or concentrations of reactants. Note that in the chemical cycling investigation in activity 41, it is important to note if the copper carbonate being used is in fact basic copper carbonate ($Cu_2CO_3(OH)_2$) as this will affect the initial calculation of the mass of copper in the sample.

5. Solutions and concentrations extend the concept of molar mass by adding the step of dissolving a mass into water. Students should realize that if 1 mole of substance Y with a mass X is placed into 1 L of water, then the concentration can be both X g/L and 1 mol/L of substance Y. From here, they can investigate the concentration (and ultimately mass) of dissolved substances, using the extra equation $c = n \div V$. Students create standards and use them to verify a claim as to the concentration of acid in vinegar.

What happens during a chemical reaction?

6. Students investigate the effect of changing conditions on reaction rates. Increases in temperature, concentration, and surface area all increase the rate of a reaction. Understanding the effect of these on reaction rates is useful for later discussions of equilibrium.

7. Understanding energy changes in chemical reactions is sometimes difficult for students to visualize. This is especially the case for endothermic reactions. Creating simple column graphs as in activity 46 is a step towards a more complete understanding of the energy changes in a reaction.

8. Students should understand that a negative enthalpy means that energy has been released or lost from the system. The products have less internal energy than the reactants had. The temperature of the external environment will increase. A positive enthalpy means energy was absorbed from the environment. The external temperature will go down. The internal energy of the products is more than that of the reactants. The simplest way to quantify these energy changes is to measure the temperature change of water in a closed system (a calorimeter). This can be done by measuring the temperature change when mixing solutions of known volume, or using the heat from a known mass of burning substance (e.g. alcohol) to heat a known volume of water. Students should also understand that breaking chemical bonds requires energy, and forming chemical bonds releases energy.

9. Not all reactions go to completion. In some cases, as the products are formed, they form back into the reactants. When these reaction rates become equal, an equilibrium is reached. Students are given information on various examples of equilibria. There is an opportunity for teachers to demonstrate the iron thiocyanate equilibrium as well as the potassium dichromate equilibrium as these are relatively simple solutions to produce.

10. Various industrial processes using equilibria are explored. This gives students the opportunity to see how even the apparently simple chemistry they are studying can be applied to commercially important industries. The final question in activity 52 assesses the student's understanding of equilibrium conditions by asking them to manipulate the ethene-ethanol equilibrium in order to produce more ethanol.

11. Equilibrium can be extended to acids and bases and used to explain the differences in pH of various acids and bases. Students can study how acids and bases that completely dissociate have more extreme pHs than acids and bases that do not dissociate as readily.

ADVANCED LEARNING

Nikthestunned cc 3.0

CaesiumFluoride cc 3.0

★ Challenge question:

Write a balanced equation for the thermite reaction of aluminum powder heated with iron (III) oxide (above). Classify the reaction and suggest when it might be used.

34 Antacid Chemistry

Key Question: There are many antacid formulas on the market. How do they work?

- The stomach naturally produces acid to aid digestion. Indigestion or heartburn are symptoms of damage to the lining of the stomach caused by overproduction of HCl (hydrogen chloride or hydrochloric acid).
- The overproduction can be eased by using antacid tablets. These can be easily bought from the local drugstore. There are many types but all work on the principle of neutralizing the acid in the stomach.

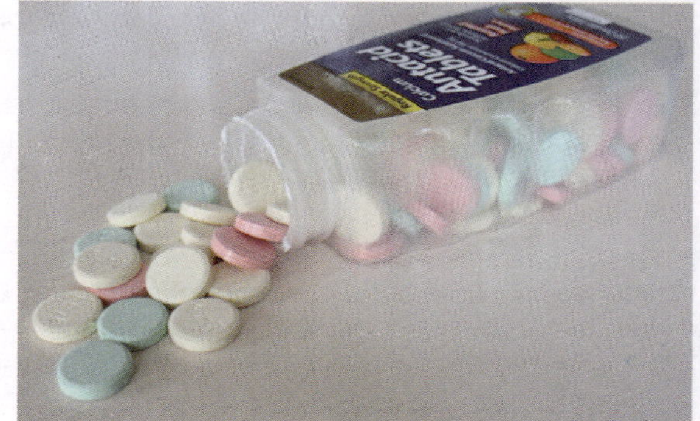

- There are many different ways of doing this and some are more effective than others. Some antacids produce carbon dioxide as a product and can make the user burp. Others produce no gas in the reactions.
- The various types of antacids generally contain one of three main ingredients to produce the neutralizing effect: calcium carbonate, sodium bicarbonate (sodium hydrogen carbonate), or magnesium hydroxide.

1. Do you know any one who uses antacids? *Student's answer*

2. Do they make that person burp? *Student's answer*

3. If you can, find out what the ingredients are in an antacid tablet. Can you figure out why they do or do not make that person burp?

   Ingredients could include: calcium carbonate, sodium bicarbonate, magnesium hydroxide, magnesium carbonate, or aluminum hydroxide. If a reaction in the stomach creates a gas, it might make the person burp.

4. What is the pH of stomach acid? *Stomach acid has a pH of around 2*

5. What is the concentration of stomach acid? *Around 0.1 to 0.01 mol/L of HCl*

- The following questions relate to the subject matter you will be learning in this chapter. It is **not expected** that you will know all the answers now, but by the end of the chapter you should be able to answer them all (and more)! The questions will be revisited as a check on your understanding at the end of the chapter.

6. Complete the word equation for each of the following reactions:

   (a) Hydrochloric acid + calcium carbonate → *water + carbon dioxide + calcium chloride*

   (b) Hydrochloric acid + sodium hydrogen carbonate → *water + carbon dioxide + sodium chloride*

   (c) Hydrochloric acid + magnesium hydroxide → *water + magnesium chloride*

7. Complete and balance the chemical equations for each of the following reactions:

   (a) $2 HCl + CaCO_3 \rightarrow H_2O_{(l)} + CO_{2(g)} + CaCl_{2(aq)}$

   (b) $HCl + NaHCO_3 \rightarrow NaCl_{(aq)} + H_2O_{(l)} + CO_{2(g)}$

   (c) $2 HCl + Mg(OH)_2 \rightarrow MgCl_{2(aq)} + 2H_2O_{(l)}$

8. In groups, discuss how you might work out the amount of the active ingredient calcium carbonate in an antacid tablet. (Hint: calcium carbonate doesn't dissolve in water, but it does dissolve in acid). Summarize your ideas here:

   Dissolve the calcium carbonate in a known excess volume of a known concentration of HCl. Next react the left over acid with a sodium hydroxide solution of known concentration until the solution is neutral. From the known volume and concentration of base used, the amount of acid used up can be calculated and then the amount of calcium carbonate reacted can be calculated.

©2020 BIOZONE International
ISBN: 978-1-927309-82-7
Photocopying Prohibited

35 Using Mass To Determine Quantity

Key Question: How can we use mass to determine quantity and how might this be useful?

Mass is related to the number of objects

- If we know the mass of a particular object then we can work out the number of individual objects in a great pile simply by weighing the pile and applying a little mathematics.
- Have you ever been to a fairground or amusement park and tried to guess the number of jelly beans in a jar? If you know the mass of one jelly bean then you could make a good guess at the number of jelly beans in the jar by weighing the jar (minus the mass of the jar).
- The same applies to chemistry. If we know the mass of a sample of a chemical then we can work out the number of atoms in the sample if we know the mass of the atoms. This will then help us work out the exact amount of product we would get if we were to carry out a reaction with that sample.

Certain numbers have special names

- Sometimes names are given to certain amounts or numbers of objects. We are all familiar with the term "dozen". One dozen is of course equal to twelve. There are many other names used for particular amounts.
- We give these special numbers names because it makes using them simpler in everyday language. We might say "Can I buy a dozen eggs?", or "there's a couple of chairs over there," etc.
- In chemistry one number in particular is given a name because it makes it easier to use in chemical calculations: **the mole**. This is a number that all measurement in chemistry is based on. When we say "one mole of sodium" we mean an exact number of sodium atoms. It is like saying "one dozen sodium atoms" except that a mole is several trillion times larger.

1. Write down all the ways you can think of to express a number. Check with your classmates. What did they come up with?

   Answers could include:

   Couple (2), brace (2), pair (2), dozen (12), baker's dozen (13), score (20), gross (144), grand (1000).

2. You are clearing out the garage and come across a plastic bag of golf balls. You want to find out how many there are but don't want to have to count them all. Instead you weigh one golf ball and then the entire bag of golf balls. One golf ball weighs 46 grams and the bag of balls weighs 2.442 kg. How many golf balls are there in the bag?

   2442 ÷ 46 = 53.1. The mass of the bag will be small so there are likely 53 golf balls in the bag.

3. A friend comes over with three dozen hamburger patties for the night's party.

   (a) What number of hamburger patties were there? *36*

   (b) Your friend states that the patties are "a quarter pound each": What mass of patties are there (in grams)? (1 pound = 453.6 grams)

   453.6 ÷ 4 = 113.4 grams

   (c) Your friend also confidently states that the patties are "really healthy. They only have 2% fat per patty". You decide to test his claim by cooking a patty and collecting and weighing the fat that runs off. You find that there is 5 grams of fat per patty. Is your friend's claim correct?

   5 ÷ 113.4 × 100 = 4%. Your friend's claim is incorrect.

PS1.B EM

©2020 BIOZONE International
ISBN: 978-1-927309-82-7
Photocopying Prohibited

36 The Mole

Key Question: What is the mole and how is it useful in chemistry?

Relative atomic mass (A_r)

- Water can be broken into hydrogen and oxygen by passing an electric current through it. When we do this we find that we collect twice the volume of hydrogen gas as we do oxygen gas. From this we can conclude that water contains 2 hydrogen atoms for every 1 oxygen atom per unit of water.
- However the mass of oxygen gas obtained (the mass of a gas can be determined from its volume) is 8 times more than the mass of the hydrogen gas. We can therefore calculate that 1 hydrogen atom must be 16 times lighter than 1 oxygen atom.
- Using experiments such as this, early chemists arbitrarily gave hydrogen a mass of 1. They could therefore say that oxygen had a relative mass of 16. From this, other "relative" atomic masses could be calculated.
- On your periodic table on page 30 you may have noticed that the atomic mass is not always a whole number. For example the atomic mass of iron is given in the periodic table as 55.845.
- In the previous chapter, we saw that the atomic mass equals the number of protons and neutrons in an atom. Iron has 26 protons, so according to the periodic table it has 29.845 neutrons. How is this possible?

*The answer is that not all iron atoms have the same number of neutrons. Most have 30, but iron atoms with 28, 31 and 32 neutrons are also found in nature. These are called **isotopes**.*

- The number given on the periodic table is the **relative atomic mass** (A_r). It is defined as "*the average mass of all the atoms of an element compared to 1/12 the mass of the carbon-12 atom*".
- For simplicity A_r values are usually given to one decimal place.

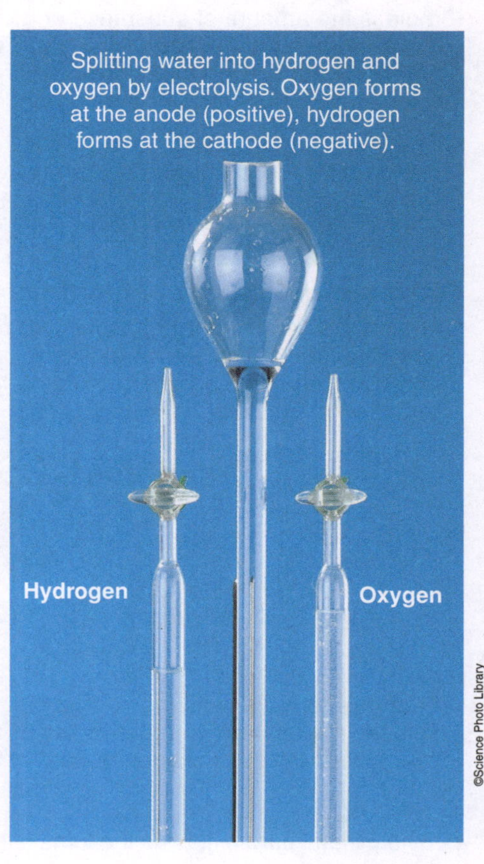

Splitting water into hydrogen and oxygen by electrolysis. Oxygen forms at the anode (positive), hydrogen forms at the cathode (negative).

Relative molecular mass (M_r)

- Just as elements have a mass relative to carbon-12, so do compounds. This is the **relative molecular mass** (M_r).
- The relative molecular mass can be calculated by adding together all the relative atomic masses of the individual atoms in one unit of the compound.
- For example the relative molecular mass of sulfur dioxide (SO_2) is equal to the relative atomic mass of sulfur plus twice the atomic mass of oxygen: 32.1 + (16.0 x 2) = 64.1.

1. What does relative atomic mass mean?

   It is the average mass of an elements atoms relative to the 1/12 the mass of a carbon-12 atom (a carbon-12 atom has exactly 6 protons and 6 neutrons and has a mass of 12 atomic units).

2. Use the periodic table on page 30 to find the A_r of the following:

   (a) Na: 22.99
   (d) Ca: 40.08
   (b) N: 14.01
   (e) Al: 26.98
   (c) Au: 196.97
   (f) I: 126.90

3. Use the following A_r values to calculate the M_r values for the compounds listed below:
   A_r(H) = 1.0, A_r(C) = 12.0, A_r(O) = 16.0, A_r(Fe) = 55.8, A_r(Cu) = 63.5

   (a) H_2: 2.0
   (e) H_2O: 18
   (b) CuO: 79.5
   (f) Fe_2O_3: 159.6
   (c) CH_4: 16.0
   (g) C_3H_8: 44
   (d) O_2: 32
   (h) NaOH: 40

©2020 **BIOZONE** International
ISBN: 978-1-927309-82-7
Photocopying Prohibited

EM P PS1.B

The mole is a number

- Making the relative atomic mass of all elements relative to carbon-12 gives us a way to measure equal numbers of atoms of different elements.
- In 12 grams of carbon-12 there is a certain number of atoms. We call this number of atoms the mole. But how many atoms are there in a mole?
- In 2011, the International Bureau of Weights and Measures stated that "The mole is the amount of substance of a system which contains as many elementary entities as there are atoms in 0.012 kg of carbon-12". This number, 6.02×10^{23} (to 2 significant figures), is also called Avogadro's constant (named after the Italian scientist Amedeo Avogadro).
- The figure of 6.02×10^{23} is directly linked to the mass of carbon-12 (the standard for the measurement of relative atomic masses).

You can easily find the mass of one mole (1 mol) of atoms in any element because it is the relative atomic mass in grams.

The mole and chemical formulas

- One of the most important aspects of chemistry is understanding chemical formulas. Every compound can be written as a chemical formula, based on the number and type of atoms present in the compound. We have already come across some simple formula earlier and is it is now useful to fully explain what they mean. Consider the formula for 2 units of aluminum sulfate:

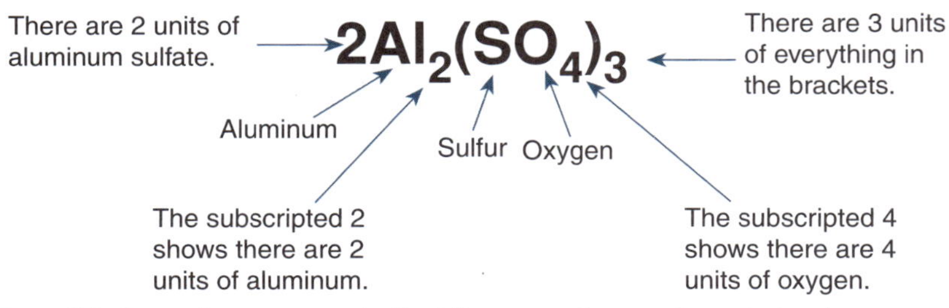

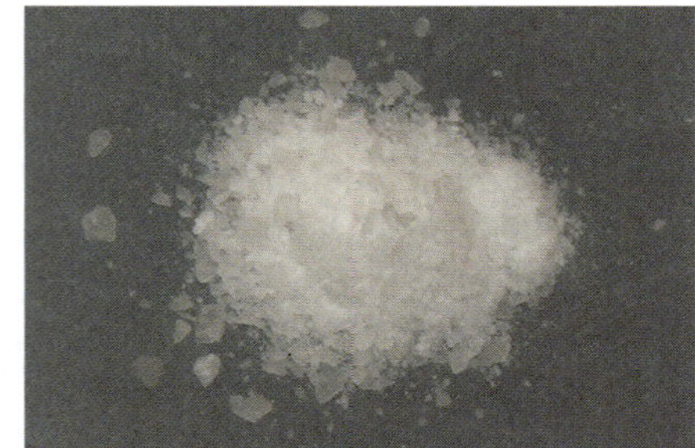

Aluminium sulfate is used as a clumping agent (flocculent) in water purification.

- From this formula we can see that there are two moles of aluminum sulfate. In total there are four moles of aluminum, six moles of sulfur, and twenty four moles of oxygen, a total of thirty four moles of atoms.

4. The mineral dawsonite has the formula $NaAlCO_3(OH)_2$. Explain what this formula means:

   The formula tells us in 1 unit of dawsonite there is 1 Na, 1 Al, 1 C, and 3 O atoms (plus 2 of the OH group which behaves as a single unit) and 2H atoms (also in the OH group).

   We can then say that there are 1 Na, 1 Al, 1 C, and 5 O and 2 H atoms in a single dawsonite unit.

5. Calculate the number of moles of atoms in the following :

   (a) NaOH: *3 (1 + 1 + 1)*

   (b) Al_2O_3: *5 (2 + 3)*

   (c) $2CaCl_2$: *6 (2 × (1 + 2))*

   (d) $2Al(OH)_3$: *14 (2 × (1+ (1 + 1) × 3))*

6. You also need to be able to interpret chemical equations in terms of moles of products and reactants. Consider the equation below. The arrow in the equation means "goes to form". Explain the meaning of the equation in terms of moles of products and reactants:

$$C_2H_5OH + 3O_2 \rightarrow 2CO_2 + 3H_2O$$

   The reactions can be explained as 1 mole of C_2H_5OH reacts with 3 moles of O_2 to form 2 moles of CO_2 and 3 moles of H_2O. At length it could be explained as 2 moles of C, 6 moles of H and 7 moles of O react together reform new bonds (forming new products). We can also say 4 moles of reactants reacts to form 5 moles of product.

7. Write the following as a chemical equation: Two moles of hydrogen chloride plus two moles of sodium hydroxide go to form two moles of water and sodium chloride:

   $2HCl + 2\ NaOH \rightarrow 2H_2O + NaCl$

©2020 **BIOZONE** International
ISBN: 978-1-927309-82-7
Photocopying Prohibited

Molar mass (M) and the mole

- The molar mass links the relative atomic and relative molecular masses to the mole. The molar mass is the mass of one mole of a substance in grams. So if the relative atomic mass of lithium is 6.9, then the molar mass is 6.9 g/mol. Similarly if the relative molecular mass of sulfur dioxide is 64.1 then its molar mass is 64.1 g/mol.
- In this text molar mass is rounded to 1 decimal place for ease of calculation.

8. Calculate the molar mass of the following compounds. Show your working:
   M(H) = 1.0 g/mol, M(C) = 12.0 g/mol, M(N) = 14.0 g/mol, M(O) = 16.0 g/mol,

   (a) CH_3COOH: 60 g/mol

   (b) C_2H_5OH: 46 g/mol

   (c) HNO_3: 63 g/mol

   (d) $(NH_4)_2CO_3$: 96 g/mol

- It is now useful to compare moles of substances in real terms. The investigation below is a simple observation task that may be set up by your teacher or you may weigh out the substances yourself.

INVESTIGATION 3.1: Molar mass

See appendix for equipment list.

Caution is required when handling chemicals in the lab. Avoid contact with skin. Wear protective eyewear and gloves.

M(H) = 1.0 g/mol, M(C) = 12.0 g/mol, M(O) = 16.0 g/mol, M(Na) = 23.0 g/mol, M(S) = 32.1 g/mol, M(Cl) = 35.5 g/mol, M(Fe) = 55.8 g/mol, M(Cu) = 63.5 g/mol.

1. Here, you will compare one mole of the following substances: sodium chloride (NaCl), sulfur (S), hydrated copper II chloride ($CuCl_2.2H_2O$), calcium carbonate ($CaCO_3$), carbon (C), and glucose ($C_6H_{12}O_6$).
2. To begin, calculate the molar masses (M) of the six substances:

   (a) NaCl: 58.5 g/mol

   (b) S: 32.1 g/mol

   (c) $CuCl_2.2H_2O$: 170.5 g/mol

   (d) $CaCO_3$: 71.8 g/mol

   (e) C: 12 g/mol

   (f) $C_6H_{12}O_6$ 180 g/mol

3. Obtain six clean petri dishes or 6 sheets of filter paper.
4. Weigh 1 mole of sodium chloride onto the first petri dish or filter paper.
5. Continue weighing out the other five substances onto separate petri dishes or filter paper.
6. Observe the differences in mass and volume of one mole of the different substances. Remember there is the same number of atoms in each of the samples you have weighed out.

By now you may have noticed a relationship between the number of moles of a substance, the mass, and the molar mass.

If we have a sample of one mole of sodium chloride, its mass is 58.5 g. If we have a sample of two moles of sodium chloride its mass is 2 x 58.5 = 117 g.

The mass of a sample in grams (**m**) equals the number of moles (**n**) multiplied by the molar mass of the substance (**M**).

This is also written as: **n = m ÷ M**

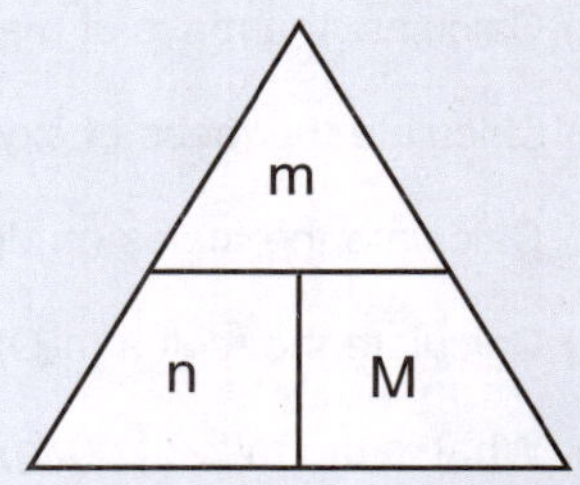

9. Use the formula above and the following molar masses answer the questions below: M(Li) = 6.9 g/mol, M(O) = 16.0 g/mol, M(Zn) = 65.4 g/mol

   (a) The number of moles in 29.8 grams of Li_2O: M Li_2O = (6.9 × 2) + 16 = 29.8 g/mol n = 29.8/29.8 = 1.0 mole

   (b) The mass of 1.5 moles of ZnO: M ZnO = 65.4 + 16 = 81.4 m = 1.5 × 81.4 = 122.1 g

©2020 **BIOZONE** International
ISBN: 978-1-927309-82-7
Photocopying Prohibited

37 Using Molar Mass

Key Question: How can the mole and molar mass be used to find chemical formula?

Finding the formula

- A working knowledge of molar mass and moles can help us determine the molecular formula of new compounds.
- In the investigation below you will determine the formula for magnesium oxide:

INVESTIGATION 3.2: Finding the formula

See appendix for equipment list.

Magnesium is a flammable metal. If ignited it produces a bright and extremely hot flame that can cause severe burns. Do not look directly at the flame. Wear protective eyewear and use tongs to handle the crucible.

M(O) = 16.0 g/mol, M(Mg) = 24.3 g/mol

1. Weigh a crucible and lid on a balance. Record the mass: *e.g. 37.12 g*
2. Coil a 10 cm length of magnesium ribbon and place it in the crucible. Replace the lid and reweigh. Record the mass of the crucible, lid, and magnesium ribbon here:

   e.g 37.68 g
3. Place a clay triangle onto a tripod and put the crucible, magnesium ribbon, and lid on top. Heat the crucible with a blue Bunsen flame. Using tongs, open the lid to allow air into the crucible.
4. Watch for the magnesium to ignite (this may take a minute or so). When it does, immediately place the lid back on the crucible.
5. Continue heating the crucible for several minutes, using tongs to lift the crucible lid slightly once or twice to allow air into the crucible.
6. After several minutes, check to see if the reaction is complete. You will be able to tell as a white powder (magnesium oxide) will form and no flame or "smoke" will be seen.
7. Leaving the crucible lid on, turn the Bunsen off and allow the crucible to cool. Check to make sure all the magnesium has reacted. If there is still some metal in the crucible, you will need to continue heating.
8. When the crucible is cool, reweigh the crucible, magnesium oxide, and lid. Record the mass here:

   e.g 38.02 g

1. (a) Calculate the mass of magnesium at the start of the reaction: *m(Mg) = 37.68 – 37.12 = 0.56 g*

   (b) Calculate the mass of magnesium oxide formed: *m(MgO) = 38.02 – 37.12 = 0.90 g*

   (c) Calculate the mass of oxygen that has reacted with the magnesium: *m(O) = 38.02 – 37.68 = 0.34 g*

   (d) Calculate the moles (n(Mg)) of magnesium used in the reaction: *n(Mg) = 0.56 ÷ 24.3 = 0.023 moles*

   (e) Calculate the moles (n(O)) of oxygen in the reaction: *n(O) = 0.34 ÷ 16 = 0.21*

   (f) What is the ratio of n(Mg) to n(O) (in whole numbers)? *1:1*

   (g) What is the formula for magnesium oxide? *MgO*

   (h) Write a balanced equation for the reaction of magnesium metal (Mg) with oxygen gas (O_2).

   $2Mg + O_2 \rightarrow 2MgO$

PS1.B P EM

©2020 BIOZONE International
ISBN: 978-1-927309-82-7
Photocopying Prohibited

- Let us now compare the formula of magnesium oxide to the formula of sodium oxide. Burning sodium in air is too dangerous in a lab situation because sodium is highly reactive. The reaction does not form pure sodium oxide in any case (some sodium peroxide is formed).
- Sodium oxide can be formed by the thermal decomposition of sodium carbonate (Na_2CO_3). When sodium carbonate (right) is heated to above 800°C under one atmosphere of pressure it decomposes to sodium oxide and carbon dioxide (CO_2).

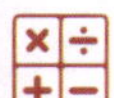

2. 3.00 grams of sodium carbonate was thermally decomposed. The remaining sodium oxide powder was weighed and had a mass of 1.75 grams. M(C) = 12.0 g/mol, M(O) = 16.0 g/mol, M(Na) = 23.0 g/mol.

   (a) Write a word equation for the reaction: *Sodium carbonate +heat → Sodium oxide + carbon dioxide*

   (b) Calculate $M(Na_2CO_3)$: *$M(Na_2CO_3) = 106$ g/mol*

   (c) Calculate $n(Na_2CO_3)$ that were heated: *$n(Na_2CO_3) = 3.00 \div 106 = 0.028$ moles*

   (d) Calculate n(Na) in the sample of Na_2CO_3: *$n(Na) = 0.028 \times 2 = 0.057$ moles*

   (e) How many moles of sodium must therefore be in the sodium oxide? *0.057 moles*

   (f) How many grams of sodium must therefore be in the sodium oxide? *$m(Na) = 0.057 \times 1.30$ g*

   (g) How many grams of oxygen must therefore be in the sodium oxide? *$m(O) = 1.75 - 1.30 = 0.45$ g*

   (h) Calculate n(O) in sodium oxide: *$n(O) = 0.45 \div 16.0 = 0.028$ moles*

   (i) What is the ratio of sodium to oxygen atoms in the sodium oxide (in whole numbers)?
   $0.057 \div 0.028 = 2.03$. Ratio is therefore 2:1

   (j) What is the formula of sodium oxide? *Na_2O*

3. Use your understanding of the periodic table and ion formation to explain why magnesium oxide and sodium oxide have the formula you determined:
   Sodium is a group 1 element. It therefore forms a +1 ion. Oxygen is a group 16 element and therefore forms a 2– ion. When these come together two Na ions are needed to balanced the –2 charge of the O ion. Therefore the formula is Na_2O.

4. 10.0 g of solid coal (which in this example will contain pure carbon) is completely combusted in oxygen to form carbon dioxide. The only product detected was carbon dioxide. The gas was collected in a cylinder and compressed and refrigerated before the solid was weighed. The mass of carbon dioxide was found to be 36 grams. A small amount of gas was lost during the refrigeration process. M(C) = 12.0 g/mol, M(O) = 16.0 g/mol

   (a) Calculate the n(C) in the coal: *$n(C) = 10 \div 12 = 0.83$ moles*

   (b) Calculate the number of moles of carbon dioxide: *(n carbon dioxide) = $36 \div 0.82$ moles*

   (c) Calculate the mass of oxygen that must be in the carbon dioxide: *36 - 10 = 26 grams oxygen*

   (d) Calculate n(O) in the carbon dioxide: *$n(O) = 26 \div 16 = 1.62$ moles*

   (e) What is the whole number ratio of carbon to oxygen? *0.82:1.62 = 1:2*

   (f) Based on the ratio what is the molecular formula of carbon dioxide: *CO_2*

   (g) Based on your knowledge of the periodic table determine if your formula makes sense and justify your reasoning:
   Carbon needs four electrons to fill its valence shell. Oxygen needs two electrons to fill its valence shell. By combining electrons of one carbon and 2 oxygens carbon can fill its valence shell as can both oxygen atoms. Thus the formula CO_2 calculated above makes sense.

©2020 **BIOZONE** International
ISBN: 978-1-927309-82-7
Photocopying Prohibited

Determining the ratio of water to copper sulfate in hydrated copper sulfate

- Many ionic compounds contain water bound into the crystal structure (water of crystallization). This must be taken into account when weighing out the substance. Different compounds contain different ratios of water in their crystal structures. The formula for a hydrated compound is **ionic formula·xH_2O**, where x is the number of water molecules incorporated into the crystal. It is important to realize that the water is not part of the compound, it is simply bound inside the crystal structure.
- In the investigation below you will determine the ratio of copper sulfate to water in hydrated copper sulfate and hence the value of x in $CuSO_4 \cdot xH_2O$. M(H) = 1.0 g/mol, M(O) = 16.0 g/mol, M(S)= 32.1 g/mol,M(Cu) = 63.5 g/mol.

INVESTIGATION 3.3: Hydrated copper sulfate

See appendix for equipment list.

Be careful when handling copper sulfate. It can irritate the skin and ingestion and inhalation of the dust is harmful. The crucible will become extremely hot. Wear eyewear and gloves, and use tongs to lift the crucible lid.

1. Weigh a crucible and lid on a balance. Record the mass here: e.g 21.3 g
2. Add about 6 grams of hydrated copper sulfate to the crucible. Reweigh: e.g 27.4 g
3. Record the color of the hydrated copper sulphate: Blue
4. Place the crucible on a clay triangle. Tilt the lid slightly to allow water to escape. Heat gently for a minute or so over a Bunsen burner then heat strongly for about 5 minutes.
5. Close the lid and allow the crucible to cool. Reweigh the crucible when it is cool enough to touch. Any hotter than this and it could damage the balance. Record the mass of the first heating: e.g 25.3 g

Wilco Oelen Creative Commons Attribution-ShareAlike 3.0 Unported License.

6. Reheat the crucible and copper sulfate for 2 more minutes, tilting the lid to let any water escape. Close the lid, cool, and reweigh. Record the mass after the second heating: e.g 25. 2 g
7. Repeat if the mass is not constant (to within 0.02 g depending on the accuracy of your balance). Final mass: e.g 25.2 g
8. Record the color of the dehydrated copper sulfate: Pale blue, almost white.

5. (a) Calculate $m(CuSO_4 \cdot xH_2O)$ in the crucible before heating: $m(CuSO_4.xH_2O) = 6.1$ g
   (b) Calculate $m(CuSO_4)$ in the crucible after heating: $m(CuSO_4) = 25.2 - 21.3 = 3.9$ g
   (c) Calculate the molar mass of anhydrous (without water) $CuSO_4$: $M(CuSO_4) = 159.6$ g/mol
   (d) Calculate the $n(CuSO_4)$ left: $n(CuSO_4) = 3.9 \div 159.6 = 0.024$ moles
   (e) Calculate the mass of water lost: $m(H_2O) = 6.1 - 3.9 = 2.2$ g
   (f) Calculate $M(H_2O)$: $M(H_2O) = 18$ g/mol
   (g) Calculate the $n(H_2O)$: $n(H_2O) = 2.2 \div 18 = 0.122$ moles
   (h) Determine the ratio of $CuSO_4$ to H_2O: $0.122 \div 0.024 = 5.09 = 5$. Ratio = 1:5
   (i) What is the formula of hydrated copper sulfate? $CuSO_4.5H_2O$
6. Why is it important to include the number of bound waters when weighing out a compound? The water adds to the mass of the compound so must be taken in account or else the incorrect moles are calculated.

©2020 **BIOZONE** International
ISBN: 978-1-927309-82-7
Photocopying Prohibited

38 Molecular vs Empirical Formulas

Key Question: How can an empirical formula be used to determine a molecular formula?

- Finding the formula for chemical compounds is an important part of chemistry. The formulas you have determined for magnesium oxide and sodium oxide are **empirical** formulas. An empirical formula is the simplest whole number ratio of atoms in a compound. For ionic compounds, e.g MgO, the formula is always given in the simplest ratio.
- However molecules can have both an empirical and a **molecular** formula. For example, the molecule hydrogen peroxide has the molecular formula H_2O_2. The empirical formula is the simplest ratio, HO. If a compound's molecular formula cannot be reduced any more, then the empirical formula is the same as the molecular formula.
- The diagram below shows a method for determining the products of combustion of cyclohexane and therefore its composition and empirical formula. Cyclohexane is a hydrocarbon so contains only hydrogen and carbon atoms.
- From the diagram we can see that the only products of the reaction are water (H_2O) and carbon dioxide (CO_2).
- It is possible to calculate the formula of cyclohexane from the mass of the products produced. This is based on that fact that you know the elements in cyclohexane, and that it fully reacts with oxygen. Thus the mass of H in H_2O must be the mass of H in cyclohexane and the mass of C in CO_2 mass also be the mass of C in cyclohexane.

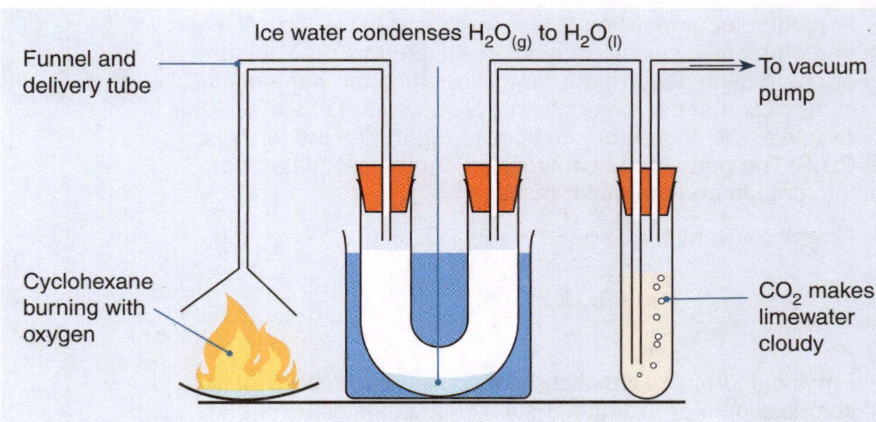

1. 5.6 grams of cyclohexane was completely combusted and the products passed through a condenser. The mass of water obtained was 7.2 grams. M(H) = 1.0, M(C)= 12.0, M(O) = 16.0.

(a) Calculate $n(H_2O)$: $n(H_2O) = 7.2/18 = 0.4$ moles

(b) Calculate n(H) in the cyclohexane that was combusted: $n(H) = 2(0.4/1) = 0.8$ moles

(c) Calculate the mass of hydrogen in cyclohexane: $m(H) = 0.8 \times 1 = 0.8$ g

(d) Calculate the mass of carbon in cyclohexane: $m(C) = 5.6 - 0.8 = 4.8$ g

(e) Calculate n(C) in the cyclohexane that was combusted: $n(C) = 4.8 \div 12 = 0.4$ g

(f) What is the whole number ratio of C to H to cyclohexane? 1 : 2

(g) The molar mass of cyclohexane is known to be 84 g/mol. What is the molecular formula of cyclohexane?

C : H = 1 : 2, $M(CH_2) = 14$, $14 \times 6 = 84$ g/mol therefore C_6H_{12}

2. 10.0 grams of glucose were completely combusted and the products passed through a condenser. Glucose contains the elements C, H, and O. The mass of water obtained was 6.0 grams. From the volume of CO_2 produced it was found the mass of CO_2 was 14.7 g. M(H) = 1.0, M(C)= 12.0, M(O) = 16.0.

(a) Calculate $n(H_2O)$: $n(H_2O) = 6 \div 18 = 0.33$ moles

(b) Calculate n(H) in the glucose: $n(H) = 0.33 \times 2 = 0.66$ moles

(c) Calculate $n(CO_2)$: $n(CO_2) = 14.7 \div 44 = 0.33$ moles

(d) Calculate n(C) in the glucose: C in CO_2 comes from glucose therefore n(C) = 0.33 moles

(e) Calculate the mass of oxygen in the glucose: 0.66 moles H = 0.66 g. 0.33 moles C = 3.96 g. 10 − 4.62 = 5.38 g

(f) Calculate n(O) in the glucose $n(O) = 5.38 \div 16 = 0.34$ moles

(g) Calculate the ratio of C, H , and O in the glucose in whole numbers: 0.33: 0.66: 0.33 = 1:2:1

(h) The molar mass of glucose is 180 g/mol. What is the molecular formula of glucose?

12 + 2 + 16 = 30. 180 ÷ 30 = 6 so $C_6H_{12}O_6$

©2020 **BIOZONE** International
ISBN: 978-1-927309-82-7
Photocopying Prohibited

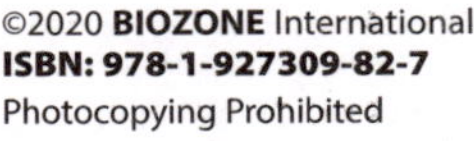

39 Percentage Composition

Key Question: What is percentage composition and how can it be used?

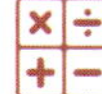

- Percentage composition is commonly used in industry. The label on the right shows the percentage composition of the various components in cat food.
- Percentage composition can be calculated from the mass of a component ÷ the total mass of the substance x 100 or:

$$\% \text{ composition} = \frac{\text{Mass of component}}{\text{Total mass}} \times 100$$

GUARANTEED ANALYSIS:	
Crude Protein (Min)	34.0%
Crude Fat (Min)	14.0%
Crude Fiber (Max)	4.0%
Moisture (Max)	12.0%
Linoleic Acid (Min)	2.3%
Calcium (Ca) (Min)	1.0%
Phosphorus (P) (Min)	0.9%
Zinc (Zn) (Min)	150 mg/kg
Selenium (Se) (Min)	0.35 mg/kg
Vitamin A (Min)	14,000 IU/kg
Vitamin E (Min)	450 IU/kg
Taurine (Min)	0.15%
Omega-6 Fatty Acids* (Min)	2.5%

- Percentage composition is also important in mining. Minerals and ores are a mixture of elements. Determining whether or not to mine somewhere may depend on the percentage composition of the elements in the ores found at the site. For example, the copper ore chalcocite (right) has the formula Cu_2S. The percentage composition can be calculated from the formula and the molar mass:

$$M(Cu_2S) = (63.5 \times 2) + 32 = 159$$

$$\% \text{ Cu} = \frac{(63.5 \times 2)}{159} \times 100 = 79.9\ \%$$

- Empirical formulas can also be calculated from percentage composition. For example a substance contains 20% hydrogen and 80% carbon by mass. By assuming that there is 100 grams of the sample we can turn the percentages into grams, i.e. 20 g of H and 80 g of C. It is then a simple matter of calculating the moles of each component and obtaining the ratio. If the molar mass is known, the molecular formula can also be calculated.

1. Using the percentages of C and H given above:

   (a) Calculate n(C) and n(H): *n(H) = 20/1 = 20 moles*

   n(C) = 80/12 = 6.67 moles

   (b) Determine the ratio of C to H: *1 : 3*

   (c) Write the empirical formula of the compound: *CH_3*

   (d) If the molar mass is of the compound is 30 g/mol what is its molecular formula?

   $M(CH_3)$ = 15 × 2 = 30 g/mol so molecular formal is C_2H_6 (ethane)

2. A compound was found to have the following percentages of elements: 49.5% C, 5.0% H, 28.9% N, and 16.3% O.

   (a) Find the empirical formula of the compound: *Turn % into grams. Therefore: n(C) 49.5 ÷ 12 = 4.13 moles,*

   n(H) = 5 ÷ 1 = 5.00 moles. n(N) = 28.9 ÷ 14 = 2.06, n(O) = 16.3 ÷ 16 = 1.02.

   Determine mole ratios: 4.13 ÷ 1.02 = 4, 5.00 ÷ 1.02 = 5, 2.04 ÷ 1.02 = 2, 1.02 ÷ 1.02 = 1,

   Mole ratios: 4:5:2:1 Empirical formula: $C_4H_5N_2O$

   (b) The molar mass of the substance was found to be 194 g/mol. What is the molecular formula of the compound?

   48 + 5 + 28 + 16 = 97. 194/97 = 2 Therefore formula = $C_8H_{10}N_4O_2$ (caffeine)

PS1.B P EM

©2020 BIOZONE International
ISBN: 978-1-927309-82-7
Photocopying Prohibited

40 Balancing Equations

Key Question: What notation can we use to show that all the atoms and elements in the reactants before a reaction are present in the products after a reaction?

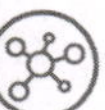

Ionic formula

- So far we have investigated the ratios of ions in ionic compounds (MgO, Na_2O, and $CuSO_4 \cdot 5H_2O$) and we have explained why these ratios occur.

- In a reaction, Mg forms a 2+ ion and oxygen forms a 2– ion. The electrostatic forces of each ion attract each other in a 1:1 ratio to form MgO. In Na_2O, the Na forms a 1+ ion, so two Na^+ ions are needed to balance the charge on O^{2-}.
- It is important to be able to write ionic formula so that the charges on the ions balance. For example, aluminum oxide has the formula Al_2O_3. Al forms a 3+ ion, while O forms a 2– ion. To balance these charges we need to find a common multiple. This can be done by multiplying the charges: 3 × 2 = 6. So there must be enough of each ion to add up to either 6+ or 6–.
- Therefore 2 × 3+ = 6+ and 3 × 2– = 6–. So 2 Al react with 3 O to form Al_2O_3.
- We can look at this as a model of jigsaw pieces:

1. We can line up the pieces and join them. The pieces do not form a regular rectangle.

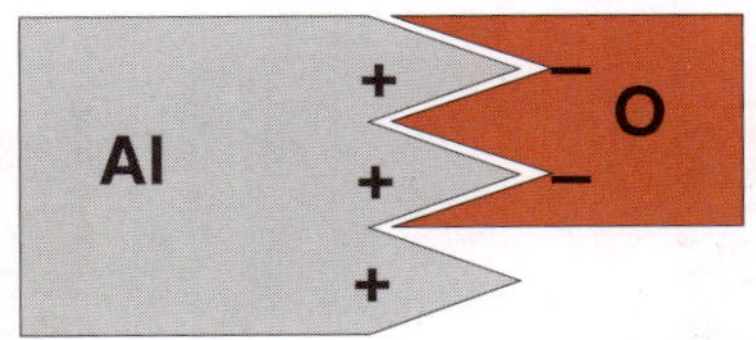

2. Let's add another O to fill up the empty space on Al. The pieces still do not form a regular rectangle:

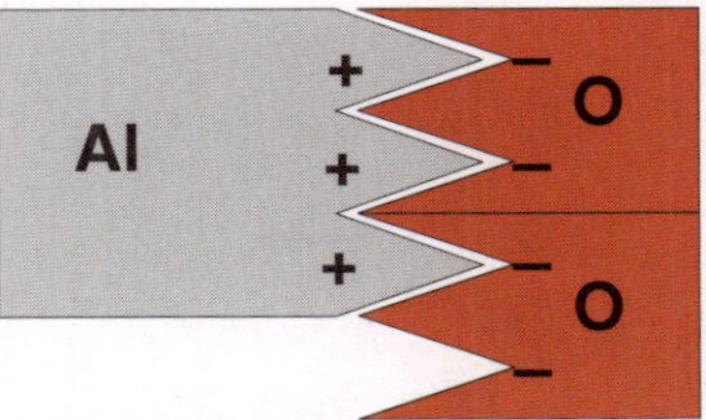

3. Let's add an Al to fill up the empty space on O. Not quite there.

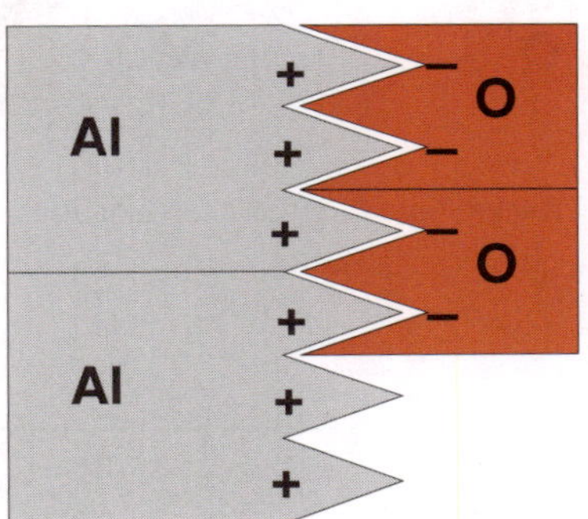

4. Let's add another O. Perfect! To make a rectangle we need 2 Al and 3 O. So we get Al_2O_3.

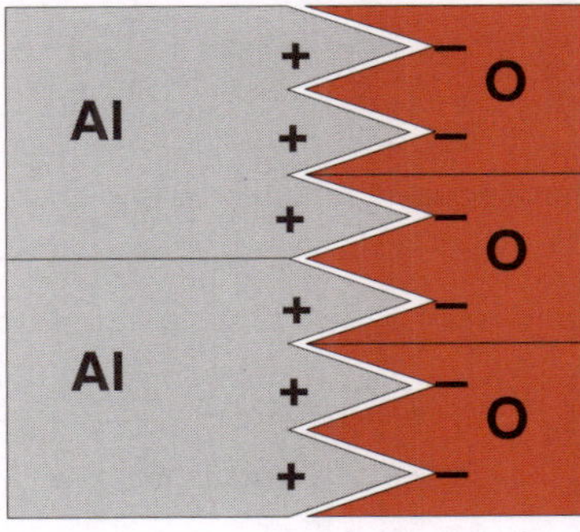

- Another method is called the "swap and drop" as shown below:

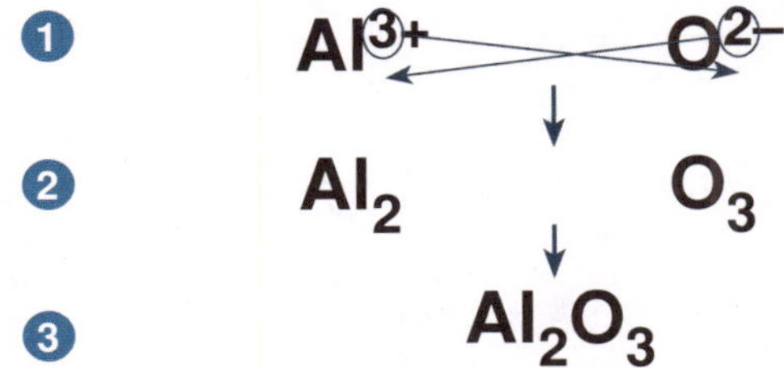

Take the numbers (not the charge) and place it behind and below the opposite ion.

Place the ions together

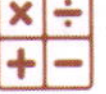

1. Write balanced ionic formula for each of the following:

(a) Calcium hydroxide: $Ca(OH)_2$

(b) Iron III chloride: $FeCl_3$

(c) Sodium sulfide: Na_2S

(d) Potassium sulfate: K_2SO_4

©2020 BIOZONE International
ISBN: 978-1-927309-82-7
Photocopying Prohibited

Balancing equations for ionic compounds

- Hopefully, you will have noticed in the previous example that oxygen was referred to as O (the singular atom) as opposed to O_2 (molecular oxygen as it is found in the gas state). If we were to write an equation for the reaction of aluminum with oxygen to form aluminum oxide we would need to write oxygen as O_2 (as this is the oxygen in the air that is reacting with the aluminum).
- So our equation is $Al + O_2 \rightarrow Al_2O_3$.
- But we can see the number of atoms on each side of the equation are not equal. There is one Al and two O on the left but two Al and three O on the right. We need to write a number in front of the Al and O_2 to increase the number of atoms and balance in the equation.
- Let's set out the units of Al, O_2, and Al_2O_3:

1. We need one Al, 1 O_2, and one Al_2O_3

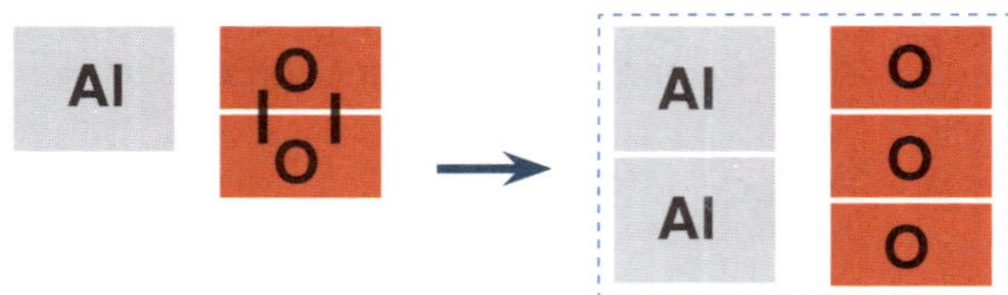

2. We can increase the amount of Al by adding another Al:

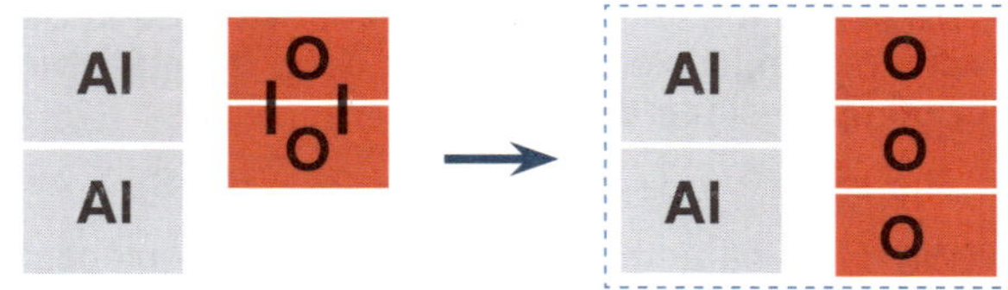

3. This balances the Als but the O_2s are still not balanced. Lets add another O_2 to the left side of the equation. To do that we also need another unit of Al_2O_3 to the right side of the equation:

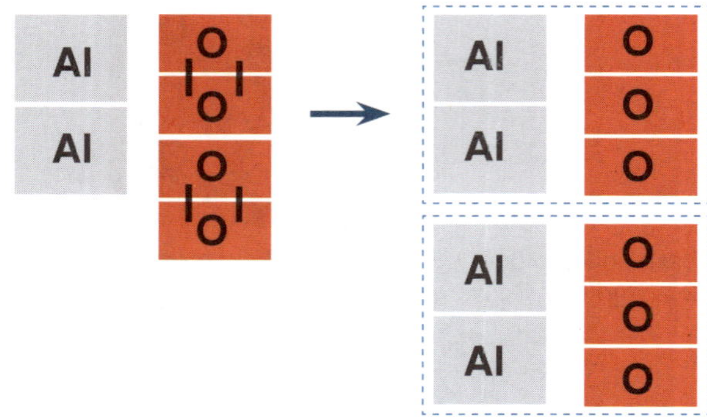

4. There are now 4 Als and 6 Os on the right. We can balance those by adding 2 more Als and 1 O_2 to the left.

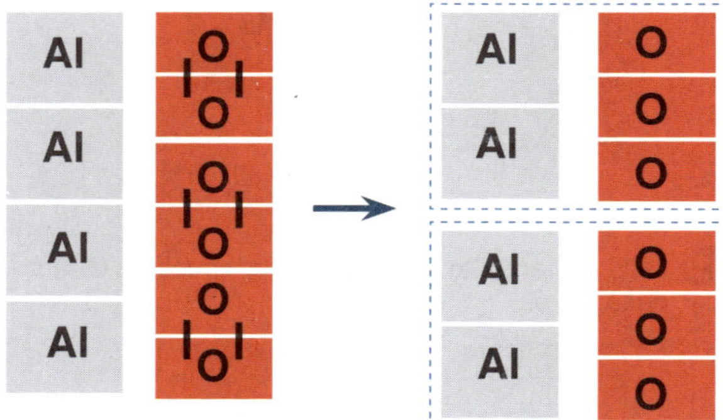

5. We now have equal numbers of Al and O on the left and right sides of the equations. The final equation for the reaction is:

$$\mathbf{4Al + 3O_2 \rightarrow 2Al_2O_3}$$

2. Balance and rewrite the following equations:

(a) $ZnO + HCl \rightarrow ZnCl_2 + H_2O$: $ZnO + 2HCl \rightarrow ZnCl_2 + H_2O$

(b) $Al + Cl_2 \rightarrow AlCl_3$: $2Al + 3Cl_2 \rightarrow 2AlCl_3$

(c) $Na_2O + H_2O \rightarrow NaOH$: $Na_2O + H_2O \rightarrow 2NaOH$

(d) $H_2SO_4 + NaOH \rightarrow NaSO_4 + H_2O$: $H_2SO_4 + 2NaOH \rightarrow Na_2SO_4 + 2H_2O$

(e) $HCl + CaCO_3 \rightarrow CaCl_2 + CO_2 + H_2O$: $2HCl + CaCO_3 \rightarrow CaCl_2 + CO_2 + H_2O$

(f) $Fe_2(SO_4)_3 + BaCl_2 \rightarrow BaSO_4 + FeCl_3$: $Fe_2(SO_4)_3 + 3BaCl_2 \rightarrow 3BaSO_4 + 2FeCl_3$

3. The equations you have balanced above all involve the movement of ions. Not all reactions involve ions, but the same rule that what appears on one side of the equation must appear on the other side still applies. The combustion of simple organic compounds (e.g. methane) usually produces just CO_2 and H_2O as the products. Try balancing the equations below using these steps: **balance the carbons, then the hydrogens on the right and finally the oxygens on the left.**

(a) $CH_4 + O_2 \rightarrow CO_2 + H_2O$: $CH_4 + O_2 \rightarrow CO_2 + 2H_2O$

(b) $C_3H_8 + O_2 \rightarrow CO_2 + H_2O$: $C_3H_8 + 5O_2 \rightarrow 3CO_2 + 4H_2O$

(c) $C_2H_5OH + O_2 \rightarrow CO_2 + H_2O$: $C_2H_5OH + 3O_2 \rightarrow 2CO_2 + 3H_2O$

(d) $C_3H_6 + O_2 \rightarrow CO_2 + H_2O$: $C_3H_6 + 4\frac{1}{2}O_2 \rightarrow 3CO_2 + 3H_2O$

(e) $C_6H_{12} + O_2 \rightarrow CO_2 + H_2O$: $C_6H_{12} + 9O_2 \rightarrow 6CO_2 + 6H_2O$

©2020 **BIOZONE** International
ISBN: 978-1-927309-82-7
Photocopying Prohibited

Types of reactions

- Chemical reactions can be classified according to how the reactants interact and the types of products formed. The reactions below will be covered in this book (some you have already seen, others you will study soon). There are others that are outside the scope of this book. You may come across these other reaction types in further studies.

Combination reactions

- An example of a combination or (synthesis) reaction you have already come across is that of magnesium and oxygen combining to form magnesium oxide (right). Two reactants (magnesium and oxygen) have combined to form one product (magnesium oxide): $2Mg + O_2 \rightarrow 2MgO$.

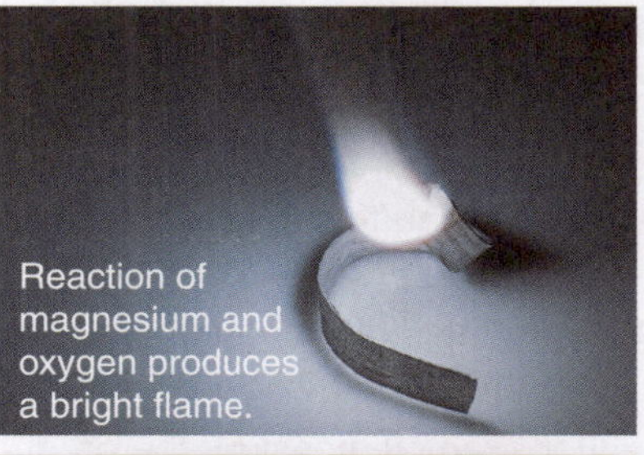

Reaction of magnesium and oxygen produces a bright flame.

Decomposition reaction

- In a decomposition reaction a single product breaks up (decomposes) into two or more simpler products. This often happens when heat is applied to a compound. For example copper carbonate decomposes when heated to form copper oxide and carbon dioxide: $CuCO_3 \rightarrow CuO + CO_2$.

Combustion of paraffin wax

Combustion

- Combustion occurs when a fuel reacts with oxygen. Common forms of combustion involve a carbon-based compound such a methane, coal, oil, or wood. Combustion produces heat (it is exothermic) and can be complete or incomplete. Complete combustion of a carbon compound such as paraffin wax will produce just carbon dioxide and water. Incomplete combustion will produce carbon (soot) or carbon monoxide, and water.

Displacement reactions

- These occur when one reactant replaces or is replaced by another to form a new product. This can happen as a single displacement reaction or as a double displacement reaction.
- An example of a single displacement reaction occurs when an copper wire is placed in a solution of silver nitrate. The copper ions move into solution and the silver ions combine to form solid silver: $Cu + 2AgNO_3 \rightarrow 2Ag + Cu(NO_3)_2$.
- Double displacement reactions often occur when two ionic solutions are mixed and a solid (a precipitate), gas, or water forms out of the solution. For example mixing a solution of potassium iodide with a solution of lead nitrate produces a yellow precipitate of lead iodide and a solution of potassium nitrate: $2KI + Pb(NO_3)_2 \rightarrow PbI_2 + 2KNO_3$.

Copper displaces silver from a solution of silver nitrate.

Neutralization reactions

- A neutralization reaction occurs between an acid (e.g. hydrogen chloride, HCl) and a base (e.g. sodium hydroxide, NaOH). The reaction produces neutral products (they are not acidic or basic): $HCl + NaOH \rightarrow NaCl + H_2O$.

Identifying state in reactions: a note on notation

- So far when we have referred to chemical compounds or equations we have simply used the chemical formula. However we know that various chemicals can be either solid, or liquid, or a gas. Some can dissolve into water to make a solution. This is important as it affects our understanding of the equation.
- In a chemical equation the state of the reactants and products are shown as subscripts after the formula. A solid is shown as solid (s), a liquid (l), a gas (g) or a dissolved compound (aqueous, aq). For example the displacement reaction of potassium iodide and lead nitrate earlier can be shown as: $2KI_{(aq)} + Pb(NO_3)_{2(aq)} \rightarrow PbI_{2(s)} + 2KNO_{3(aq)}$.
- For a combustion reaction the reaction may be: $2CH_3CH_2OH_{(l)} + 6O_{2(g)} \rightarrow 4CO_{2(g)} + 6H_2O_{(g)}$. Note that the water is in the gaseous form. It doesn't condense to a liquid until it is cooled.

4. Balance the following equations (if needed) and identify the type of reaction as noted above:

(a) $CH_3OH_{(l)} + O_{2(g)} \rightarrow CO_{2(g)} + H_2O_{(g)}$: $CH_3OH_{(l)} + 2O_{2(g)} \rightarrow CO_{2(g)} + 2H_2O_{(g)}$ Complete combustion

(b) $H_2SO_{4(aq)} + KOH_{(aq)} \rightarrow K_2SO_{4(aq)} + H_2O_{(l)}$: $H_2SO_{4(aq)} + KOH_{(aq)} \rightarrow K_2SO_{4(aq)} + H_2O_{(l)}$ Neutralization

(c) $C_2H_{6(g)} + O_{2(g)} \rightarrow CO_{(g)} + H_2O_{(g)}$: $C_2H_{6(g)} + 3\frac{1}{2}O_{2(g)} \rightarrow 2CO_{(g)} + 3H_2O_{(g)}$ Incomplete combustion

(d) $Li_{(s)} + F_{2(g)} \rightarrow LiF_{(s)}$: $2Li_{(s)} + F_{2(g)} \rightarrow 2LiF_{(s)}$ Combination

(e) $Fe_{(s)} + CuCl_{2(aq)} \rightarrow FeCl_{2(aq)} + Cu_{(s)}$: $Fe_{(s)} + CuCl_{2(aq)} \rightarrow FeCl_{2(aq)} + Cu_{(s)}$ Displacement reaction

(f) $Na_2CO_{3(s)} + HCl_{(aq)} \rightarrow NaCl_{(aq)} + CO_{2(g)} + H_2O_{(l)}$: $Na_2CO_{3(s)} + 2HCl_{(aq)} \rightarrow 2NaCl_{(aq)} + CO_{2(g)} + H_2O_{(l)}$ Neutralization

(g) $HgO_{(s)} \rightarrow Hg_{(l)} + O_{2(g)}$: $2HgO_{(s)} \rightarrow 2Hg_{(l)} + O_{2(g)}$ Decomposition

©2020 BIOZONE International
ISBN: 978-1-927309-82-7
Photocopying Prohibited

41 Stoichiometry

Key Question: How does balancing a chemical equation help us work out the yield of product that should be present after a reaction?

Cjp24 CC 4.0

- A balanced chemical equation provides the stoichiometry for a reaction.
- Stoichiometry studies the amount of each substance in a reaction. You have already been doing this to some extent by using mole ratios to calculate an empirical or molecular formula.
- From the stoichiometry of a reaction it is possible to calculate the expected **yield** of a product from a set of reactants.
- For example consider the production of ammonium nitrate, an important part of fertilizer. The reaction for its production is $HNO_{3(aq)} + NH_{3(g)} \rightarrow NH_4NO_{3(aq)}$.
- From the equation we can see the stoichiometric ratios or stoichiometry is 1:1→1. That is, 1 mole of nitric acid + 1 mole of ammonia reacts to form 1 mole of ammonium nitrate.
- Suppose we use 1.00 kg (1000 g) of ammonia to make ammonium nitrate. Using the stoichiometry, we can calculate the mass of ammonium nitrate that we can expect to produce (the yield) (assuming it is all used and HNO_3 isn't limited).
- How much product is expected compared to how much is actually produced can be expressed as the **percentage yield** (% yield). This simply the mass of the product produced ÷ the expected mass of the product × 100.

$$\text{Percentage (\%) yield} = \frac{\text{Mass product produced}}{\text{Expected mass of product}} \times 100$$

1. (a) Using the following molar masses calculate the moles of ammonia in 1 kg of ammonia: M(O) = 16.0 g/mol, M(N) = 14.0 g/mol, M(H) = 1.0 g/mol.

   $n(NH_3) = 1000 \div (14 + 3) = 58.8$ mol

   (b) How many moles of ammonium nitrate would be produced? 58.8 mol

   (c) What is the mass of ammonium nitrate that could be expected to be produced?

   $m = n \times M$, $m = 58.8 \times (14 + 4 + 14 + 48) = 58.8 \times 80 = 4704$ g $= 4.7$ kg

   (d) Suppose the reaction only produced 3500 g. What was the percentage yield of ammonium nitrate?

   $3500 \div 4704 \times 100 = 74.4\%$

2. Barium sulfate ($BaSO_4$) is an insoluble ionic compound. It has many uses in industry including oil well drilling fluid, paper brightener, as part of white pigments in paint, and as a radiocontrast agent for X-ray imaging. It can be produced by mixing together solutions of sodium sulfate (Na_2SO_4) and barium chloride ($BaCl_2$). This produces a precipitate (a solid forming out of solution) of solid barium sulfate.

Benutzer:the-viewer CC 3.0

   M(O) = 16.0 g/mol , M(S) = 32.1 g/mol, M(Na) = 23.0 g/mol, M(Cl) = 35.5 g/mol, M(Ba) = 137.3 g/mol,

   (a) Balance the equation: $BaCl_{2(aq)} + Na_2SO_{4\,(aq)} \rightarrow BaSO_{4(s)} + 2NaCl_{(aq)}$

   (b) If there was 2.0 grams of $BaCl_2$ in solution, what yield of barium sulfate (in grams) would you expect?

   $M(BaCl_2) = 208.3$ g/mol. $n(BaCl_2) = 2 \div 208.3 = 9.60 \times 10^{-3}$ moles. Therefore also 9.60×10^{-3} moles of $BaSO_4$ (1:1 ratio). $M(BaSO_4) = 233.4$ g/mol. $m(BaSO_4) = 9.60 \times 10^{-3} \times 233.4 = 2.2$ g

   (c) Determine the percentage yield if the realized yield was only 1.5 grams: $1.5 \div 2.2 \times 100 = 68\%$

   (d) If the solution was filtered to remove the $BaSO_4$, then evaporated to produce NaCl, what mass of NaCl would be expected if all the $BaCl_2$ and Na_2SO_4 reacted?

   $n(BaCl_2) = 9.60 \times 10^{-3}$ moles. Therefore $n(NaCl) = 2 \times 9.60 \times 10^{-3} = 0.0192$ moles (1:2 ratio).

   $M(NaCl) = 58.5$ g/mol. $m(NaCl) = 0.0192 \times 58.5 = 1.1$ g

PS1.B | P | SPQ | EM

©2020 BIOZONE International
ISBN: 978-1-927309-82-7
Photocopying Prohibited

Cycling matter in chemical reactions

- You should have realized by now that matter is not lost in a chemical reaction. All the atoms or ions in the reactants are present in the products.
- In the investigation below you will investigate the movement of copper ions through a series of reactions.

INVESTIGATION 3.4: The cycling of copper ions through a series of reactions

See appendix for equipment list.

Copper carbonate can irritate the skin and ingestion and inhalation of dust is harmful. H_2SO_4 is corrosive. Wear eyewear and gloves.

Buchner funnel

1. Weigh precisely about 2 g of copper II carbonate ($CuCO_3$) on to a piece of filter paper. Record the mass here:
   e.g. 2.00 g
2. What color is the copper carbonate? *Pale green*
3. Weigh a clean and dry test tube and record the mass: *e.g. 5 g*
4. Carefully place all the copper II carbonate into the test tube (don't leave any behind!). Attach the test tube to a clamp stand. Place a rubber cork and delivery tube on the end of the test tube and place the delivery tube in a beaker of limewater.
5. Place a Bunsen burner under the test tube and carefully heat the copper carbonate. Observe the change in the limewater. What gas is being produced?
   Carbon dioxide
6. Note the color change in the copper carbonate as it is heated. What color does the copper carbonate turn? What is this new powder?
   The powder turns black. This is copper II oxide (CuO)
7. When the color change is complete (the powder is all the same color) turn off the Bunsen and let the test tube and powder cool.
8. Reweigh the test tube and powder. Calculate the mass of the powder: *e.g. 1.29 g*
9. Weigh and record the mass of a clean, dry 100 mL beaker and place all the powder from the test tube into the beaker. Reweigh the beaker and powder and record the mass:
   Beaker = 40 g + 1.29 g = 41.29 g.
10. Add 80 mL of 1 mol/L H_2SO_4 to the beaker. Stir the solution with a clean glass stirring rod. Record the color change in the solution. What is this new solution?
    The color turns blue. This is copper sulfate.
11. Use steel wool or sandpaper to clean any coating from a large iron nail. Record the mass of the nail:
    e.g. 2.00 g
12. Add the cleaned nail to the solution and leave the solution overnight.
13. Record the color change of the solution and any other observations of the reaction:
    The solution will be a pale green color. This is iron II sulfate. A coating of red copper covers the nail.
14. Filter the solution to recover the new powder that has formed on the iron nail. Rinse the filtered powder with distilled water. The powder will need to be dried. Place the powder and filter paper onto a Buchner funnel to remove any remaining distilled water. If you have a drying oven, place the powder and filter paper into the oven just long enough to dry the powder completely.
15. Record the mass of the powder. What is this new powder?
    1.03 g. The new powder is copper.

©2020 BIOZONE International
ISBN: 978-1-927309-82-7
Photocopying Prohibited

16. Record the mass of a test tube: *e.g. 5g*

17. Place all the powder into the test tube. Hold the test tube with tongs and heat it strongly over a Bunsen burner. When all the powder has changed color let the test cool and weigh the test tube and power (if possible try to weigh the powder on it own). This new powder is the same as the powder you produced in step 6.

e.g. 1.29 g. The black powder is copper oxide. It should have the same mass as in step 8.

- You must now try to explain the changes in the series of reactions you carried out. If everything has worked correctly and you have been very careful to collect all the powder at the different stages, the powder you weighed in step 8 should have the same mass as the powder in step 17. It is unlikely the masses will be identical, but you should be close.
- For the calculations below use the following molar masses M(Cu) = 63.5 g/mol, M(Fe) = 55.9 g/mol, M(S) = 32.1 g/mol, M(O) = 16.0 g/mol, M(C) = 12.0 g/mol, M(H) = 1.0 g/mol.

3. (a) From the mass of copper carbonate calculate n($CuCO_3$): *n($CuCO_3$) = 2.00 ÷ 123.5 = 0.016 moles*

(b) Heating the $CuCO_3$ produced a **decomposition reaction**. Carbon dioxide (CO_2) was lost from the $CuCO_3$ and black CuO remained. Write an equation for the decomposition of $CuCO_3$.

$CuCO_3$ + heat → CuO + CO_2

(c) What mass of CuO would you have expected to be yielded from the $CuCO_3$?

$CuCO_3$:CuO = 1:1 so 0.016 moles CuO. m = n x M = 0.016 x 79.5 = 1.29 g

(d) What mass of CuO did you actually produce?: *Student answer*

(e) Write an equation for the reaction of the CuO with the H_2SO_4: *CuO + H_2SO_4 → $CuSO_4$ + H_2O*

(f) When the iron nail was placed in the solution a **displacement reaction** occurred. The iron in the nail displaces the copper ions in solution causing them to form red copper metal (although it is powdery). From the original mass of $CuCO_3$ calculate the expected yield of Cu:

If all the original copper is displaced: n(Cu) = 0.016.

m(Cu) = 0.016 x 63.5 = 1.03 g

(g) What was the actual mass of Cu you produced? *Student's answer*

(h) Heating the powdered copper caused it to oxidize (react with oxygen) producing the same black powder as you produced after heating the $CuCO_3$. What was this powder?

Copper oxide

(i) What mass of this powder should you have yielded and what was your actual yield?

m(CuO) = 0.016 x 79.5 = 1.29 g

4. Explain the reactions above and the cycling of copper through the reactions in terms of ions present, stoichiometry of the reactions, and conservation of mass. Include balanced equations for all the reactions.

The reaction scheme is as follows:

Copper carbonate decomposes to copper oxide: $CuCO_3$ + heat → CuO + CO_2

Copper oxide reacts with sulfuric acid: CuO + H_2SO_4 → $CuSO_4$ + H_2O

Iron displaces the copper from solution: Fe + $CuSO_4$ → $FeSO_4$ + Cu

Copper powder oxidizes to copper oxide: 2Cu + O_2 → 2CuO

The moles of copper (and therefore mass) is established in the original mass of copper carbonate. If the reactions are carried out carefully then the should be no loss of copper. Copper ions pass from reaction to reaction but are not lost from the system.

©2020 BIOZONE International
ISBN: 978-1-927309-82-7
Photocopying Prohibited

42 Solutions and Concentration

Key Question: What is concentration and how do we express the concentration of solutions?

Solutions

- Have you ever made a flavored drink by adding powder from a sachet into a larger volume of water, e.g. a liter?
- In doing this you have produced a solution. It will have had a certain concentration, perhaps 100 grams of powder per litre or 100 g/L.
- Sometimes you may have added a concentrate to water to form a diluted solution. This might be adding a cordial (right) or household ammonium solution to water.

_indica Garcinia indica cc1.0

Concentration

- So far in chemistry when we have referred to the amount of a substance we have used its mass in grams. In reality many substances you will come across will not be able to be easily weighed out because they are dissolved in solution. We must refer to their concentration, the amount of substance per liter. This might be as grams per liter (g/L) or, specifically in chemistry, moles per liter (mol/L).

1. Consider the label on the right from a carton of orange juice:

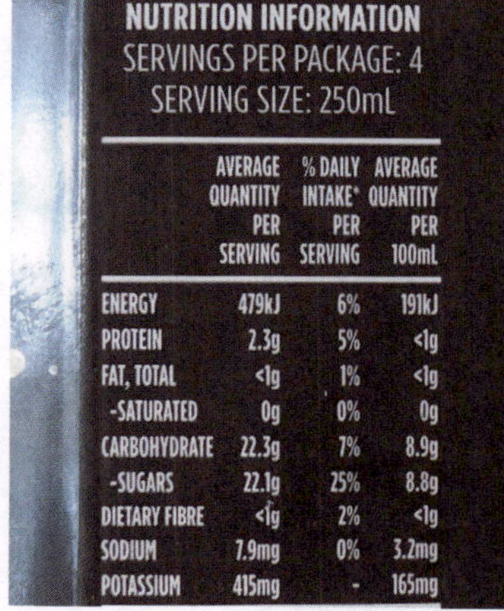

NUTRITION INFORMATION
SERVINGS PER PACKAGE: 4
SERVING SIZE: 250mL

	AVERAGE QUANTITY PER SERVING	% DAILY INTAKE* PER SERVING	AVERAGE QUANTITY PER 100mL
ENERGY	479kJ	6%	191kJ
PROTEIN	2.3g	5%	<1g
FAT, TOTAL	<1g	1%	<1g
-SATURATED	0g	0%	0g
CARBOHYDRATE	22.3g	7%	8.9g
-SUGARS	22.1g	25%	8.8g
DIETARY FIBRE	<1g	2%	<1g
SODIUM	7.9mg	0%	3.2mg
POTASSIUM	415mg	-	165mg

(a) What is the concentration of potassium in g/100 mL (1 g = 1000 mg)?

0.165 g/100mL

(b) What is the concentration of potassium in g/L (1 g = 1000 mg)?

1.65 g/L

(c) What is the concentration of sodium per liter in g/L (1 g = 1000 mg)?

0.0032 g/100mL

(d) What is the concentration of sugars per liter in g/L (1 g = 1000 mg)?

0.032 g/L

Salt from the sea

- Sodium chloride is one of the most common chemicals used by humans. It is used to enhance flavor in food, in medicine (e.g. saline solution), thousands of tonnes are spread on roads every year to prevent ice forming, and it is an important ingredient in many industrial reactions. Approximately 280 million tonnes are produced every year.
- Much of this comes from the sea or salt lakes by evaporating seawater in huge shallow ponds (shown right). Seawater has a concentration of sodium chloride of about 35 g/L. Different seas and oceans have very slightly different salinities due to their position. For example the Mediterranean Sea is mostly enclosed and has a concentration of 38 g/L.
- Some lakes have very high salt concentration. The Great Salt Lake in Utah has a salt concentration of up to 317 g/L (right).

Jeff Kubina CC 2.0

2. (a) Sodium chloride has a molar mass of 58.5 g/mol. How many moles of sodium chloride are in one liter of sea water?

$n = m \div M = 35 \div 58.5 = 0.598$ moles

(b) How many times more concentrated than seawater is the Great Salt Lake? About 9 times more concentrated.

(c) What is the concentration of the Great Salt Lake in moles per liter? $n = m \div M = 317 \div 58.5 = 5.42$ mol/L

- Many metals can be found in seawater. Gold has a concentration of 1×10^{-11} grams per liter of seawater. That's 1 gram per 100 million tonnes of seawater.

3. Gold has a molar mass of 197.0 g/mol. What is the concentration of gold in seawater in moles per liter?

$n(Au) = m \div M = 1 \times 10^{-11} \div 197 = 5.08 \times 10^{-14}$ mol/L

©2020 BIOZONE International
ISBN: 978-1-927309-82-7
Photocopying Prohibited

43 Creating Standard Solutions

Key Question: How can the concentration of a solution be calculated? How does knowing the concentration of one solution help calculate the concentrations of another solution?

Creating standards

- Standards are important in everyday life. We use standards for measuring other things against. Standards can be anything but should be a specific value that can be compared against.
- Examples include standards for distance (e.g. miles, kilometers etc), mass (grams, ounces, kilograms, pounds) or volume (liters, pints, gallons).
- When trying to find out the concentration of a solution we first need a standard to measure the unknown concentration against. This is called a **standard solution**. Standard solutions are made up in volumetric flasks. Placing a known mass of a compound in the flask and then filling up to the mark with distilled water will produce a volume with a known and precise concentration against which other solutions can be measured.
- This can be demonstrated by making a standard solution of anhydrous sodium carbonate (Na_2CO_3) and using it to standardize a solution of hydrochloric acid.

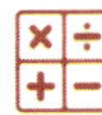

1. (a) The molar mass of Na_2CO_3 is 106 g/mol. How many moles of Na_2CO_3 are needed to make up a 250 mL solution of 0.05 mol/L Na_2CO_3?

   $n(Na_2CO_3) = c \times V$

   $= 0.05 \times 0.25 = 0.0125$ moles

   (b) What mass of Na_2CO_3 is your answer in (a) equal to?

   $m(Na_2CO_3) = n \times M = 0.0125 \times 106 = 1.325$ g

In chemistry, it is the number of atoms or ions involved in a reaction that matters. To know this, we must know the number of moles of reactants. In a solution, the number of moles is expressed as moles per liter (mol/L). When we refer to the concentration of a solution we are referring to mol/L. To calculate a solution's concentration we can use a simple equation:

$$c = n/V$$

Or concentration (in mol/L) = number of moles ÷ volume (in liters).

Learn how to use the equation c = n/V to calculate concentrations. You will be able to solve for any variable in the equation if you know the other two. Cover the variable you want to know and you are left with how to calculate it.

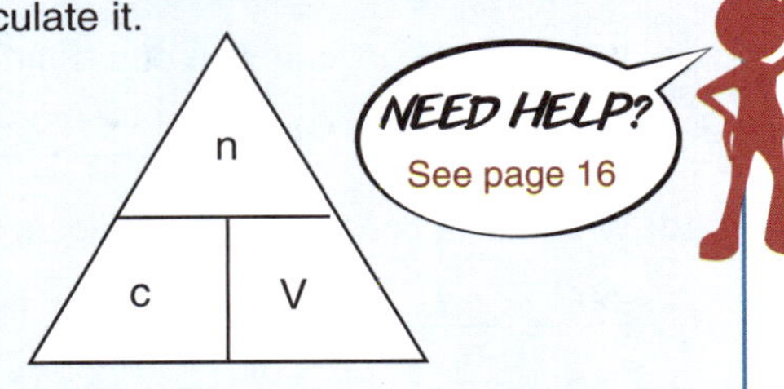

- For the following set of investigations use the following molar masses: M(H) = 1.0 g/mol, M(C) = 12.0 g/mol, M(O) = 16.0 g/mol, M(Na)= 23.0 g/mol, M(Cl) = 35.5 g/mol.

INVESTIGATION 3.5: Making a standard

See appendix for equipment list.

1. Add about 1.3 grams of anhydrous Na_2CO_3 to a clean, dry beaker. Record the precise mass you have weighed:
   e.g. 1.30 g
2. Dissolve the powder in distilled water.
3. Clean and rinse a 250 mL volumetric flask with distilled water. Using a funnel, transfer the Na_2CO_3 solution to the volumetric flask, rinsing the beaker with distilled water to ensure all the Na_2CO_3 is transferred.
4. Add distilled water to the flask until it is about half full. Stopper the flask and invert several times to ensure the Na_2CO_3 solution is thoroughly mixed.
5. Fill the flask almost to the mark. Then use a wash bottle or dropper to fill the flask to the mark so that the meniscus sits on the mark. Do not overfill or you will not be able to calculate a precise concentration.
6. Stopper the flask and mix again.
7. This is your primary standard. You will use it to standardize other solutions.

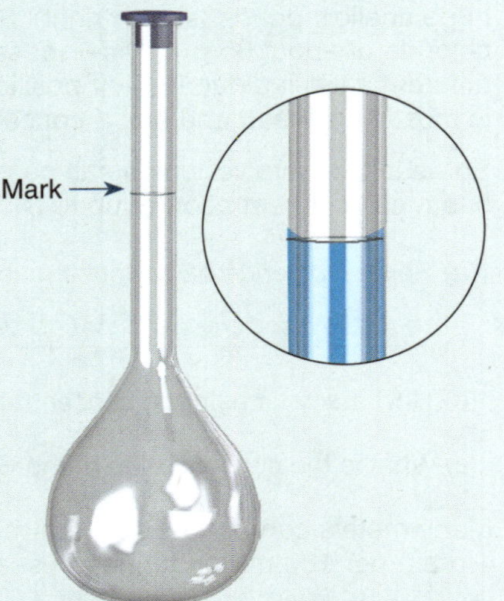

A volumetric flask is designed to have an exact volume (to within +/– 1%) when filled to the mark.

©2020 BIOZONE International
ISBN: 978-1-927309-82-7
Photocopying Prohibited

2. Calculate the exact concentration of your standard flask from the mass you weighed at the start of the investigation:
$n = m \div M = 1.3 \div 106 = 0.012$ moles. $c = n \div V = 0.0122 \div 0.25 = 0.049$ mol/L

3. (a) Why should distilled water be used to make up a standard solution instead of tap water?
Distilled water has had all the impurities taken out of it. Tap water contain a lot of dissolved ions from its treatment that could interfere with any reactions.

(b) What should you do if you overfill your volumetric flask? Start again. If you overfill the flask this will make the solution more dilute than it should be and you will not be able to calculate the true concentration.

Standardizing HCl

- The sodium carbonate solution can now be used to standardize other solutions that are more suitable for testing unknown solutions against. This is done by **titration**.
- Your lab supply of hydrochloric acid (HCl) may be 1 mol/L. This is too concentrated for the volumetric analysis of many solutions. You must produce a 0.1 mol/L solution and standardize it using your sodium carbonate solution.

INVESTIGATION 3.6: Standardizing HCl

See appendix for equipment list.

1. Place 50 mL of 1 mol/L HCl into a clean, dry 100 mL beaker. Using a 25 mL (or similar volume) pipette transfer 25 mL (or similar) to a clean, dry 250 mL volumetric flask.
2. Add distilled water to the flask until it is about half full. Stopper the flask and invert several times to ensure the HCl solution is thoroughly mixed.
3. Fill the flask with distilled water almost to the mark. Then use a wash bottle or dropper to fill the flask to the mark.
4. Rinse a 50 mL burette with the HCl solution, then place the burette in a clamp stand and fill it with the HCl solution.
5. Transfer four 20 mL samples of Na_2CO_3 solution into four clean 100 mL conical flasks. Add two drops of methyl orange indicator to each flask.
6. Record the initial burette volume below. Add the HCl solution from the burette to the Na_2CO_3 while swirling the flask until the indicator **just changes color** (the **equivalence** point).
7. Record the final volume in the table below and calculate the difference. This is the **titre** (the minimum volume of a solution needed to reach the end point in a titration). Do not use this volume in your final calculations (it is a just trial to become familiar with the volume of HCl solution required).

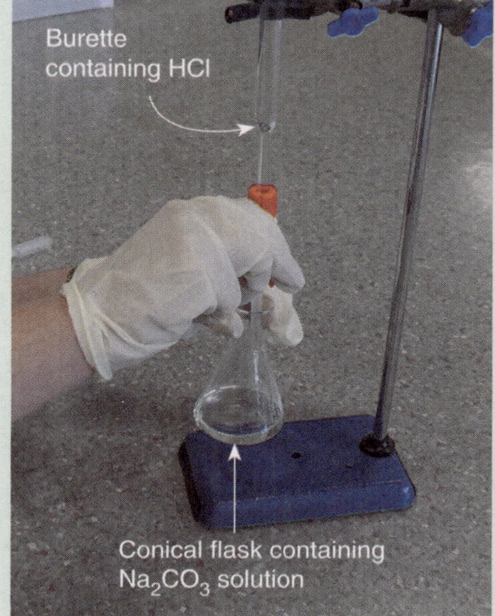

Burette reading	Trial titration	First titration	Second titration	Third titration
Initial reading	50.0 ml	29.0 ml	50.0 ml	30.0 ml
Final reading	29.0 ml	9.0 ml	30.0 ml	10.0 ml
Difference (titre)	21.0 ml	20.0 ml	20.0 ml	20.0 ml

8. Carry out the titration at least three more times and record the volume added for each in the table.

4. Balance the equation for sodium carbonate and hydrochloric acid: $Na_2CO_3 + 2HCl \rightarrow 2NaCl + CO_2 + H_2O$

5. (a) Calculate $n(Na_2CO_3)$ in the conical flask: $n(Na_2CO_3) = c \div V = 0.049 \times 0.02 = 9.81 \times 10^{-4}$

(b) Calculate n(HCl) used: $n(HCl) = 2 \times 9.81 \times 10^{-4} = 1.96 \times 10^{-3}$ moles.

(c) Calculate the average (mean) volume of HCl solution used: $V(HCl) = 0.02$ L

(d) Calculate the concentration of HCl in the volumetric flask: $c(HCl) = n \div V = 1.96 \times 10^{-3} \div 0.02 = 0.098$ mol/L

©2020 **BIOZONE** International
ISBN: 978-1-927309-82-7
Photocopying Prohibited

44 Verifying a Claim

Key Question: How can we use standardized solutions to verify the claims made by the manufacturers of products?

Finding the concentration of ethanoic acid in vinegar

- A standard sodium hydroxide solution is useful for testing the acid concentration of various solutions. It must be standardized whenever it is used as it absorbs carbon dioxide from the air, which can change the concentration.
- In the following investigations, you will standardize a solution of NaOH from the lab supply then immediately use it to calculate the concentration of ethanoic (acetic) acid in a store-bought white vinegar.
- For the following set of investigations use the following molar masses: M(H) = 1.0 g/mol, M(C) = 12.0 g/mol, M(O) = 16.0 g/mol, M(Na)= 23.0 g/mol, M(Cl) = 35.5 g/mol.

INVESTIGATION 3.7: Standardizing NaOH

See appendix for equipment list.

1. Add 50 mL of 1 mol/L NaOH solution to a clean, dry 100 mL beaker. Transfer 25 mL to a 250 mL volumetric flask using a 25 mL pipette.

   Alternatively weigh 1 gram of solid NaOH in a 100 mL beaker and dissolve with distilled water before transferring to a volumetric flask.
2. Following the same procedure as to produce the HCl earlier, dilute the NaOH by filling the volumetric flask up to the mark with distilled water.
3. Rinse a burette with the dilute NaOH solution. Then fill the burette with the NaOH solution.
4. Rinse a pipette with your standardized HCl solution then pipette four 20 mL samples into four clean, dry 100 mL conical flasks.
5. Add two drops of phenolphthalein indicator to the conical flasks. This will turn pink when the HCl/NaOH reaction is complete.
6. Again, you will need to carry out at least three titrations plus a trial run. Use the table below to record your results.
7. Record the initial burette volume. Add the NaOH solution from the burette to the HCl while swirling the flask until the indicator just changes color. Record the final volume and calculate the difference (the titre).
8. Carry out the titration at least three more times and record the volume added for each.

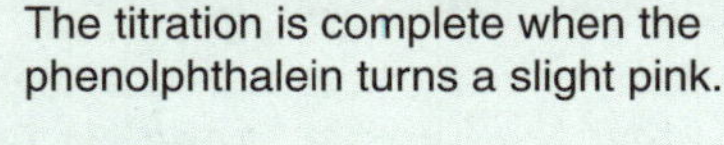

The titration is complete when the phenolphthalein turns a slight pink.

Burette reading	Trial titration	First titration	Second titration	Third titration
Initial reading	50.0 ml	28.0 ml	50.0 ml	30.0 ml
Final reading	28.0 ml	8.0 ml	30.0 ml	10.0 ml
Difference (titre)	22.0 ml	20.0 ml	20.0 ml	20.0 ml

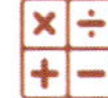

1. Write a balanced equation for sodium hydroxide (NaOH) reacting with hydrochloric acid (HCl):

   $HCl + NaOH \rightarrow NaCl + H_2O$

2. (a) Calculate n(HCl) in the conical flasks: $n(HCl) = c \div V = 0.098 \times 0.02 = 1.96 \times 10^{-3}$

   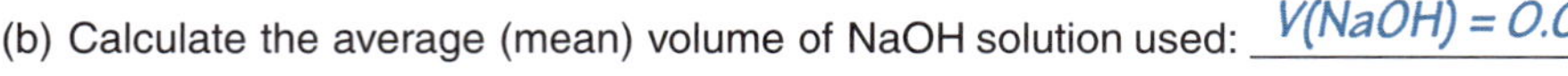

   (b) Calculate the average (mean) volume of NaOH solution used: $V(NaOH) = 0.02$ L

   (c) Calculate n(NaOH) used: 1:1 reaction so $n(NaOH) = 1.96 \times 10^{-3}$

   (d) Calculate the concentration of NaOH in volumetric flask: $c(NaOH) = n \div V$

   $= 1.96 \times 10^{-3} \div 0.02 = 0.098$ mol/L

©2020 BIOZONE International
ISBN: 978-1-927309-82-7
Photocopying Prohibited

▸ You can now use the standardized NaOH to investigate the concentration of ethanoic acid in vinegar.

INVESTIGATION 3.8: Analyzing vinegar See appendix for equipment list.

1. Refill the burette with your standardized NaOH solution.
2. Rinse a 25 mL pipette with vinegar. Transfer 25 mL of vinegar to a 100 mL volumetric flask and fill to the mark with distilled water (to dilute the vinegar).
3. Rinse a 10 mL pipette with distilled water then use it to transfer 10 mL of vinegar solution into each of four 100 mL conical flasks.
4. Add two drops of phenolphthalein indicator to the conical flasks.
5. Again, you will need to carry out at least three titrations plus a trial run. Use the table below to record your results:

Burette reading	Trial titration	First titration	Second titration	Third titration
Initial reading	*70.0 ml*	*70.0 ml*	*70.0 ml*	*70.0 ml*
Final reading	*12.8 ml*	*13.2 ml*	*13.2 ml*	*13.2 ml*
Difference (titre)	*57.2 ml*	*56.8 ml*	*56.8 ml*	*56.8 ml*

Record the initial burette volume. Add the NaOH solution from the burette to the vinegar while swirling the flask until the indicator just changes color. Record the final volume and calculate the difference.

6. Carry out the titration at least three more times and record the volume added for each.

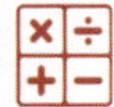

3. If it is required, balance the equation for the reaction of sodium hydroxide and ethanoic acid:

$NaOH + CH_3COOH \rightarrow CH_3COONa + H_2O$ *The equation is already balanced*

4. (a) Calculate the average (mean) volume of NaOH solution used: *V(NaOH) = 0.0568 L*

(b) Calculate n(NaOH) used: *$n(NaOH) = c \times V = 0.098 \times 0.0568 = 5.566 \times 10^{-3}$ moles*

(c) Calculate $n(CH_3COOH)$ in the conical flasks: *1:1 reaction so $n(CH_3COOH) = 5.566 \times 10^{-3}$ moles*

(d) Calculate the concentration of CH_3COOH in the conical flask: *$c(CH_3COOH) = n \div V$*

$5.566 \times 10^{-3} \div 0.025 = 0.223$ mol/L

(e) The CH_3COOH in the vinegar was diluted by how many times in the volumetric flask?

Diluted by 4 times.

(f) What was the original concentration of ethanoic acid in the vinegar?

Original undiluted vinegar = $0.223 \times 4 = 0.89$ mol/L

5. What are some errors that may have arisen in your analysis of the vinegar? *Errors may come from incorrect reading of the burette, incorrect pipetting and production of standards. In a serial set of titrations these little errors can add up to produce large difference in results.*

6. Check your results with others in your class. How close were you to their results? *Students results.*

7. How close were you to the actual concentration of acid in the vinegar (as stated on the bottle)?

Many vinegars are about 5% CH_3CHOOH or around 1 to 0.8 mol/L

8. Your NaOH solution could now be used analyze the concentration of other solutions. Use an electronic format to write and share a method for analyzing the total acid concentration in fresh squeezed orange juice.

See long answer section

©2020 **BIOZONE** International
ISBN: 978-1-927309-82-7
Photocopying Prohibited

45 Reaction Rates

Key Question: During reactions, atoms must collide. How does the way they do this affect a reaction?

Collision theory

- Collision theory describes the mechanism by which chemical reactions occur.
- In order for an atom or molecule to react with another atom or molecule they must collide with each other.
- The collision must be strong enough to break the intramolecular bonds in the reactants (e.g. the covalent bonds) so that the reaction can occur and new products be formed.

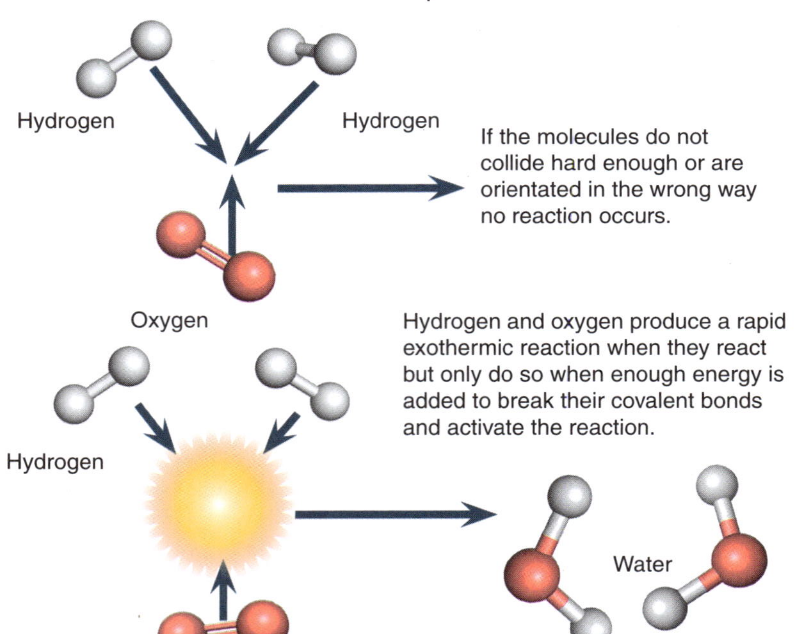

The Hindenburg disaster, 1937. The huge fireball seen in the photo was caused by burning hydrogen gas (used for buoyancy). The reason for the hydrogen's ignition is still not known, but may have been a spark from static electricity or possibly even lightning.

INVESTIGATION 3.9: Reaction rates

See appendix for equipment list.

Caution is required when handling acids in the lab. Avoid contact with skin. Wear protective eyewear and gloves.

1. Place 20 mL of 1 mol/L HCl in a 50 mL beaker. Place 0.5 grams of powdered $CaCO_3$ into the HCl and time how long it takes for the $NaHCO_3$ to fully react.
2. Now place 20 mL of 2 mol/L HCl into a 50 mL beaker and add 0.5 grams of powdered $CaCO_3$ into the HCl. Again time how long it takes for the $CaCO_3$ to fully react.
3. Now place 20 mL of 1 mol/L HCl in a 50 mL beaker. Make an ice bath using a 250 ml beaker filled with about 100 mL of iced tap water. Place the 50 mL beaker with the acid into the ice bath and wait for a minute or so while the acid cools down. Now add 0.5 grams of powdered $CaCO_3$ into the HCl. Again time how long it takes for the $CaCO_3$ to fully react.
4. Finally place 20 mL of 1 mol/L HCl in a 50 mL beaker. This time add 0.5 grams of $CaCO_3$ chips. Again time how long it takes for the $CaCO_3$ to fully react.

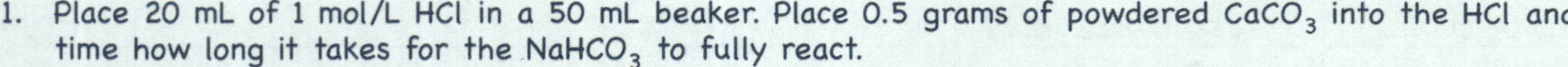

Reaction mix	Time to fully react (s)
1 Mol HCl + $CaCO_3$	
2 Mol HCl + $CaCO_3$	*Student's own response*
1 Mol HCl + $CaCO_3$ 5°C	
1 Mol HCl + $CaCO_3$ chips	

1. What factors affect the rate of a chemical reaction? *Concentration (higher concentration = faster rate), surface area (greater surface area = faster rate), temperature (higher temperature = faster rate)*

©2020 BIOZONE International
ISBN: 978-1-927309-82-7
Photocopying Prohibited

Temperature affects reaction rate

- A substance's temperature (as measured by a thermometer) is a measure of the average kinetic energy of the particles in it. For any substance at any particular temperature, the particles in it will have a range of kinetic energies. As the temperature increases, more particles will be moving fast enough to react when they collide.
- The activation energy is the energy needed for a substance to react. Increasing the temperature increases the number of particles with enough energy to react.

Activation energy
Higher temperature
Moderate temperature
Lower temperature
Number of particles
Kinetic energy

Catalysts

- Catalysts are important in many everyday and industrial reactions. Catalysts are chemicals that increase the rate of a reaction by lowering the activation energy required for the reaction to proceed.
- They do this by offering an alternative reaction pathway or orientating the reactants into the best positions for a reaction.
- Catalysts are not used up in the reaction.
- One of the most common everyday reactions involving a catalyst occurs in **catalytic converters**. These are found in the exhaust systems of gasoline and diesel fueled cars.
- The incomplete combustion of fuel and the reaction of nitrogen at high temperature with oxygen produces harmful gases (e.g. CO, NO). Catalytic converters convert these into less harmful gases (e.g. CO_2).

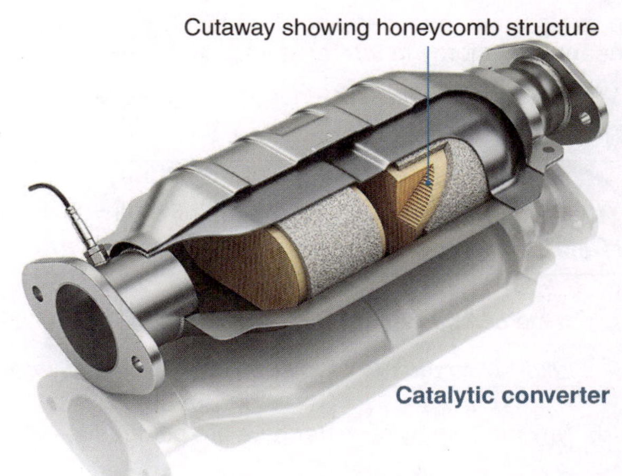

Catalytic converter

2. In terms of collisions between particles, explain why increasing the concentration of reactants in solution would increase the rate of a reaction:

   More particles of reactants means more collisions per unit of time. This means more chance of a reaction are there are more particles moving at the correct speed or with the correct orientation for a reaction to occur.

3. In terms of collisions, explain why decreasing the temperature of reactants would increase the rate of a reaction:

   Decreasing the temperature decreases the speed at which the particles are moving. This means they collide slower and less often. Less particles will be beyond the activation energy and so the reaction will proceed at a slower rate.

4. In terms of collisions, explain why increasing surface area of reactants would increase the rate of a reaction:

   Increasing the surface area increase the number of reactants colliding with increases the chances of a successful reaction and thus increases the reaction rate.

5. Study the image of the catalytic converter shown above right. The catalytic compounds coat the honeycomb like inner structure of the converter. Why do you think the converter has this honeycomb like structure?

   It increases the surface area allowing for more catalyst to be exposed to the reactants and thus increasing the rate of the reaction.

6. Enzymes are special kinds of biological catalysts. Each enzyme in the body catalyzes a specific reaction. It is estimated that there are 75,000 kinds of enzymes in the human body. Why so many enzymes are needed?

   There are many different reactions that occur in the body. Enzymes have a very specific shape and so only catalyze one or a very few types of reaction. Thus many different kinds of enzymes are needed.

©2020 **BIOZONE** International
ISBN: 978-1-927309-82-7
Photocopying Prohibited

46 Energy in Chemical Reactions

Key Question: What is the difference between exothermic and endothermic reactions and where does the energy come from or go to in these reactions?

- The image right shows some wood on fire. Fire is self sustaining chemical reaction. The heat (energy transfer) from the fire (the chemical reaction) delivers enough energy to the reactants (the wood and oxygen in the air) to take them past their activation energy. These then react and release more energy in turn.
- The reaction of wood with oxygen is **exothermic**, meaning it releases energy. A reaction that removes energy from a system (it gets colder) is called **endothermic**.
- In exothermic reactions energy is lost from the reactants as heat. This means the energy in the reactants is less than the energy in the product.

1. When a reaction releases energy, would the temperature of the surrounding environment increase or decrease?

   Increase

2. When a reaction uses energy would the temperature of the surrounding environment increase or decrease?

   Decrease

3. (a) In the combustion of wood shown above, where is the heat coming from?

   From the release of excess energy in the wood as it forms new products (CO_2 and water)

   (b) Why does spraying water on a fire put the fire out? Water removes excess energy when it turns to steam. This energy is then not available to move more wood and oxygen molecules past their activation energy and so the reaction slows and eventually stops.

4. Consider the following reactions: 1: A + B → AB + heat energy
   2: CD + heat energy → C + D

   (a) Is reaction 1 exothermic or endothermic? Endothermic

   (b) Is reaction 2 exothermic or endothermic? Exothermic

   (c) On the axis below sketch a column graph to show energy of the reactants and products for each reaction.

Reaction 1

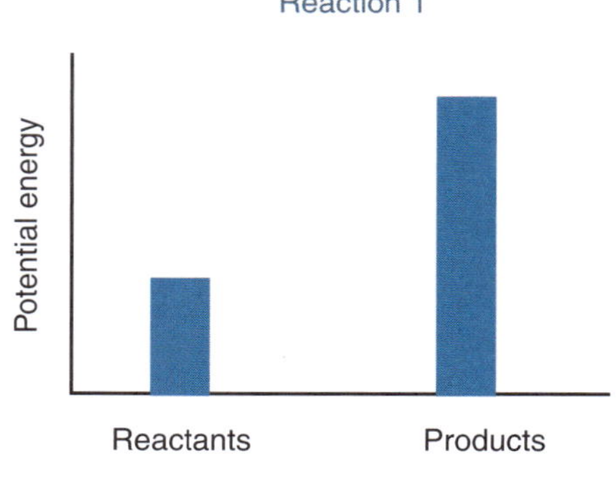

Reaction 2

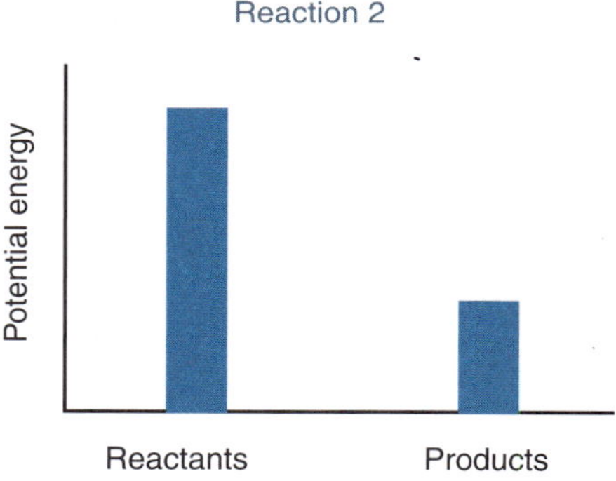

5. (a) Atoms in molecules and compounds are held together by bonds. It takes energy to break these bonds. Is bond breaking therefore exothermic or endothermic?

   Endothermic

   (b) Bond formation releases energy because atoms on their own are less stable than those sharing electrons. Is bond formation exothermic or endothermic?

   Exothermic

   (c) If a reaction is exothermic, what does this tell us about the bonds in the reactants compared to the products?

   It tells us the energy needed to break the bonds was less than the energy released when forming new bonds

©2020 **BIOZONE** International
ISBN: 978-1-927309-82-7
Photocopying Prohibited

INVESTIGATION 3.10: Endothermic and exothermic reactions

See appendix for equipment list.

The chemicals you will be using are all irritants.
Wear eyewear and gloves.

1. For each of the following reactions measure the initial and final temperature and record them in the table below:
2. To a 25 mL beaker add 1 gram of solid sodium hydroxide with 20 mL of water and mix.
3. To a 25 mL beaker add 1 gram of solid ammonium chloride to 20 mL of water and mix.
4. To a 25 mL beaker add 5 mL of 1 mol/L hydrochloric acid to 5 mL 1 mol/L sodium hydroxide.
5. In a beaker mix 1 gram of solid ammonium nitrate with 20 mL of water and mix.

	Initial temperature (°C)	Final temperature (°C)	Temperature change (°C)	Endothermic or exothermic
$NaOH_{(s)} \rightarrow Na^+_{(aq)} + OH^-_{(aq)}$	20°C	33°C	+ 13°C	Exothermic
$NH_4Cl_{(s)} \rightarrow NH_4^+{}_{(aq)} + Cl^-{}_{(aq)}$	20°C	13°C	-7°C	Endothermic
$HCl_{(aq)} + NaOH_{(aq)} \rightarrow NaCl_{(aq)} + H_2O_{(l)}$	20°C	24°C	+ 4°C	Exothermic
$NH_4NO_{3(s)} \rightarrow NH_4^+{}_{(aq)} + NO_3^-{}_{(aq)}$	20 °C	16°C	-4°C	Endothermic

6. Consider the first reaction of solid NaOH forming aqueous NaOH (i.e. dissolving NaOH). Note that NaOH is an ionic substance. When it dissolves, the ionic bonds are broken.

   (a) Does breaking an ionic bond require energy? *Yes. Bond breaking requires energy*

   (b) Hydrogen bonds between water molecules are also broken when the NaOH dissolves. Does breaking these bonds require energy?

   Yes, breaking these bonds also requires energy

   (c) The final step in the dissolving of NaOH is the formation of bonds between the Na^+ and OH^- ions and the polar water molecules. Does forming the bonds between the water molecules and the Na^+ and OH^- ions require energy or release energy?

   Forming bonds releases energy

   (d) The overall process of NaOH dissolving in water is exothermic, it releases energy. What does this tell us about the amount of energy required to break the initial bonds and form new ones in this reaction?

   The amount of energy required to break the bonds is less than the energy that is released when new bonds form. The excess energy is transferred to the water and the water's temperature increases.

   (e) If a reaction releases energy what does this tell us about the energy in the reactants compared to the products?

   There was more stored energy (energy stored in the chemical bonds) in the reactants than there is in the products.

7. Consider the reaction of ammonium chloride dissolving in water. A similar process of bond breaking and forming occurs as with the NaOH. Why does the water temperature decrease in this reaction?

   The water temperature decreases because more energy is required to break the bonds than is released. This energy is taken from the water molecules and thus they slow down, and the water's temperature decreases.

©2020 **BIOZONE** International
ISBN: 978-1-927309-82-7
Photocopying Prohibited

47 Enthalpy and Chemical Reactions

Key Question: What is enthalpy and why is it important in chemical reactions?

Heat, energy, and enthalpy

- **Enthalpy** *(H)* is a term used to describe the heat energy in a system. If there is a change of heat energy then there is a change in enthalpy *(ΔH)*. *ΔH* is measured in kJ.
- During a reaction, the change in enthalpy is equal to the enthalpy of the products – the enthalpy of the reactants (i.e. the final enthalpy – the initial enthalpy) or

$$\Delta H = H(\text{products}) - H(\text{reactants})$$

- It is not possible to measure the internal heat or enthalpy of a system, but it is simple to measure the change in the enthalpy of a system by measuring the change in temperature of water in contact with the system.
- You have already measured the temperature change of water in some chemical reactions. You can work out the change in enthalpy of a reaction by measuring the volume of water as well.
- It takes 4.2 J of energy to raise the temperature of 1 g (= 1 mL) of water by 1°C. Thus if you know the temperature change and the volume of water, then the total amount of energy released or absorbed can be calculated. This can be written as $\Delta Q = m \times \Delta T \times c$ where ΔQ= the change in energy, m = mass, ΔT = change in temperature, and c = the specific heat of water.
- For example if a kettle raises the temperature of 200 mL of water by 20°C then the energy produced by the kettle is $200 \times 20 \times 4.2 = 16.8$ kJ.

INVESTIGATION 3.11: Investigating enthalpy

See appendix for equipment list.

Copper sulfate is an irritant and toxic if ingested. Wear gloves and eyewear.

1. Obtain a clean and dry styrofoam cup.
2. Measure 100 mL of 0.1 mol/L copper sulfate solution in a 100 mL measuring cylinder and add to the cup.
3. Record the initial temperature of the solution. *Example 20 °C*
4. Add about 1 gram of zinc powder and stir rapidly. Record the highest temperature the solution reaches. *Example: 25 °C*

1. (a) What was the exact volume of copper sulfate solution used? *Student's answer, e.g. 100 mL*

(b) Calculate the number of moles of copper sulfate in the solution (c = n/V): *$n(CuSO_4) = c \times V$*
$= 0.1 \times 0.1 = 0.01$ moles

(c) Calculate the mass of copper sulfate in the solution (n=m/M) $M(CuSO_4)$ = 159.6 g/mol:
$m(CuSO_4) = n \times M = 0.01 \times 159.6 = 1.595$ g

(d) Calculate the temperature change for the reaction: *5°C*

(e) Calculate the energy produced by the reaction (Q = m x ΔT x 4.2 J): *$100 \times 5 \times 4.2 = 2100$ J*

(f) Calculate the amount of energy released per mole of copper sulfate that reacts: *0.01/ 1 = 100.*
0.01 moles = 2100 J therefore $2100 \times 100 = 210{,}000$

(g) Write a balanced equation for the reaction of copper sulfate and zinc including the value of ΔH:
$CuSO_4 + Zn \rightarrow ZnSO_4 + Cu$ ΔH −210kJ (actual ΔH −219kJ)

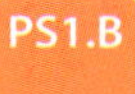

©2020 **BIOZONE** International
ISBN: 978-1-927309-82-7
Photocopying Prohibited

Modeling energy in chemical reactions

- Magnesium, paper, or many other substances often need an ignition source before they will burn. The paper in this book is unlikely to burst into flames on its own. Even a particularly flammable and volatile substance such as gasoline will not ignite and burn unless brought near an ignition source.
- Once they are started, these combustion reactions release a lot of energy. However, they cannot start the burning process by themselves. They need an input of energy, the activation energy.
- The axes below show a graph of the energy changes in the exothermic reaction of substances A and B forming AB.

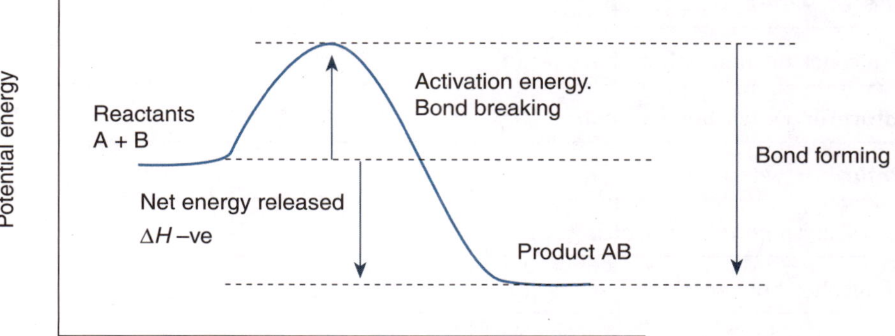

Reaction progress

2. (a) Sketch a graph of the energy changes for the reaction between copper sulfate and zinc metal (page 96):

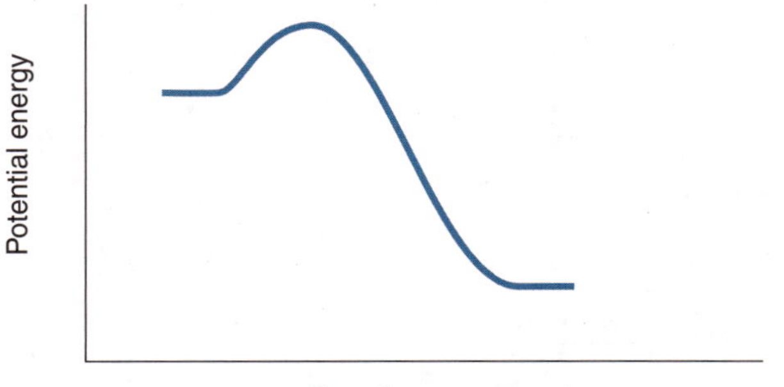

(b) Sketch a graph of the energy changes for the dissolving or ammonium nitrate in water (page 95):

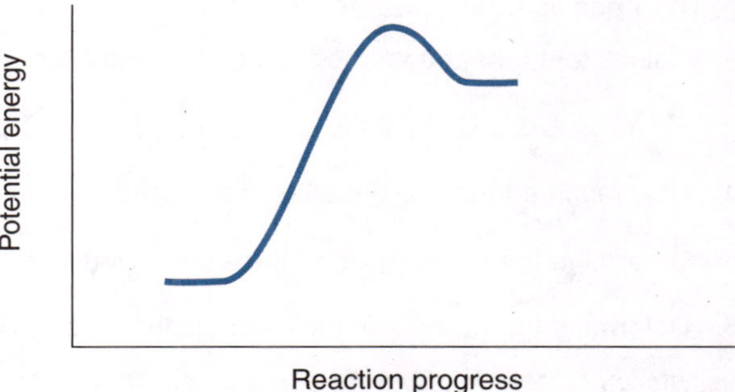

Finding the ΔH of a fuel

- Alcohols can be used as fuels because they have a high energy density and combust easily. By measuring the temperature of water heated by the alcohols the enthalpy of the combustion reaction can be calculated.

INVESTIGATION 3.12: Energy from alcohols

See appendix for equipment list.

Alcohols are highly flammable. Keep them away from an ignition source when not using them. Vapor is also flammable. Keep the lids of the bottles on when not in use. Wear eyewear .

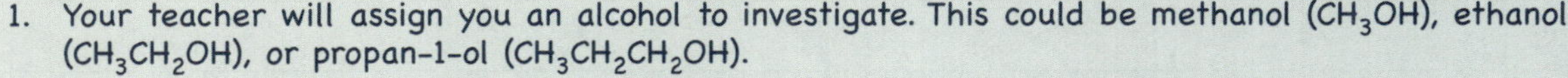

1. Your teacher will assign you an alcohol to investigate. This could be methanol (CH_3OH), ethanol (CH_3CH_2OH), or propan-1-ol ($CH_3CH_2CH_2OH$).
2. Weigh an empty, dry soda can and record the mass in the table on the next page.
3. Fill the soda can with distilled water to about 2-3 cm from the top. Reweigh the soda can and record the mass on the table.
4. Fill the alcohol burner with one of the alcohols and weigh it. Record the mass in the table.
5. Place the soda can on a tripod or on a ring clamp and secure with a second clamp. Adjust the height of the soda can on the clamp stand so that the flame from the alcohol burner will be just touching the bottom of the soda can.
6. Place a thermometer in the water in the soda can, stir gently for a few seconds and record the temperature when the thermometer shows a stable temperature.
7. Light the alcohol burner and immediately start the stop watch. Time 7 minutes.
8. After 7 minutes remove the alcohol burner (and put it out) and immediately record the temperature of the water in the soda can.
9. Record the mass of the alcohol burner and calculate the mass of alcohol used.

©2020 **BIOZONE** International
ISBN: 978-1-927309-82-7
Photocopying Prohibited

Name of alcohol:	e.g. Methanol
Mass of dry, empty soda can (g)	
Mass of soda can with water (g)	
Mass of water in soda can (g)	e.g. 300 g
Mass of alcohol burner before burning (g)	
Temperature of water before heating (°C)	
Temperature of water after heating (°C)	
Temperature change of water (°C)	e.g. 20
Mass of alcohol burner after burning (g)	
Mass of alcohol used (g)	e.g. 1.2 g

▶ The specific heat of water = 4.2 J/g/°C. $\Delta Q = m \times \Delta T \times c$. M(H) = 1.0 g/mol, M(C) = 12.0 g/mol, M(O) = 16.0 g/mol.

3. Calculate the heat absorbed (Q) by the water for the alcohol you investigated: *Answer will depend on student's measurements. For example 4.2 × 300 g × 20°C = 25,200 J = 25.2 kJ*

4. How much energy did the alcohol release? *25.5 kJ (Assuming all heat energy went into the water)*

5. Determine the molar mass of the alcohol you investigated: *Methanol: 32 g/mol. Ethanol: 46 g/mol. Propan-1-ol: 60 g/mol*

6. Determine the number of moles of alcohol used (n=m/M): *Answers will depend on student's measurements and alcohol used. For example, mass methanol used = 1.1 grams. Then n = m/M = 1.2 /32 = 0.038 moles.*

7. Calculate the energy per mole (in kJ/mol) released during combustion (enthalpy of combustion) of your alcohol:

If 0.038 moles of methanol released 25.2 kJ then 1 mole releases = 1/ 0.038 = 26.67 × 25.2 = 672.1 kJ

8. The values for enthalpy of combustion ($\Delta_c H$) for the above alcohols are:
methanol = –726 kJ/mol, ethanol = –1368 kJ/mol, and propan-1-ol = –2020 kJ/mol.
Calculate your percentage error based on the heat of combustion value for your alcohol:

672.1 – 726 / 726 × 100 = 7.4% (for methanol)

9. Explain any difference between your calculated enthalpy of combustion and the actual enthalpy of combustion:
Some energy will inevitably be lost through the walls of the can and lost to the air. Also combustion is unlikely to be complete. Some energy will be lost when soot or carbon monoxide is produced instead of carbon dioxide.

10. Write a balanced equation for your alcohol reacting with oxygen (assume the only products are carbon dioxide and water). Include the ΔH of the reaction.

Student's answer. (see Q 15)

11. Convert your calculated enthalpy of combustion from kJ/mol into kJ/g: *kJ/mol = 672.1. 1 mole methanol = 32 g*
672.1 ÷ 32 = 21.0 kJ per gram.

©2020 **BIOZONE** International
ISBN: 978-1-927309-82-7
Photocopying Prohibited

12. Methanol, ethanol, and propan-1-ol, all have essentially the same density (0.8 g/mL to 1 decimal place). The energy of fuels is often compared by the amount of energy released per gram. Why is this a more useful than using energy per mole?

Energy per mole is a more technical notation and is uncommon outside of chemistry. Grams are a more common measure. Because the fuels all have a similar density, using energy per gram provide a simple and relatively accurate comparison of their potential energy output. It also makes measuring precise energy needs simple (weighing a gram is more common than weighing a mole).

13. Pool the class data and fill in the table below:

Alcohol	Enthalpy of combustion (kJ/g)	Average enthalpy of combustion (kJ/g)
Methanol	*Class data will vary*	
Methanol		
Methanol		
Methanol		
Ethanol		
Ethanol		
Ethanol		
Ethanol		
Propan-1-ol		
Propan-1-ol		
Propan-1-ol		
Propan-1-ol		

14. Using the class data of the enthalpy of combustion for the three different alcohols, if each one could be purchased for the same price, which one would offer you the best value? Explain your choice:

Methanol = 22.7 kJ/g Ethanol = 29.7 kJ/g Propan-1-ol = 33.7 kJ/g.

Propanol is better than ethanol, which is better than methanol.

15. Balance the combustion equation for each of the alcohols below:

(a) Methanol: *2* CH_3OH + *3* $O_2 \rightarrow$ *2* CO_2 + *4* H_2O

(b) Ethanol: __ CH_3CH_2OH + *3* $O_2 \rightarrow$ *2* CO_2 + *3* H_2O

(c) Propan-1-ol: *2* $CH_3CH_2CH_2OH$ + *9* $O_2 \rightarrow$ *6* CO_2 + *8* H_2O

16. Can you explain why there is a difference in the amount of energy released during the combustion of each type of alcohol per gram:

Per gram, more of the larger alcohols are made up of carbon, which when combusted release energy when new bonds are formed.

©2020 **BIOZONE** International
ISBN: 978-1-927309-82-7
Photocopying Prohibited

48 Bond Energies

Key Question: How can we use the energy stored in bonds to predict or calculate the energy in a reaction?

- It takes energy to break a chemical bond. Energy is released when a bond is formed. The difference between the amount of energy used to break a bond and the energy released when new bonds form will determine the ΔH of the reaction.
- **Average bond enthalpy** is the energy required to break one mole of a specified bond type in a gaseous state. By knowing the bond enthalpies of the reactants and products in a reaction we can calculate the overall enthalpy of the reaction without having to carry out any experiments.
- Consider the following reaction:

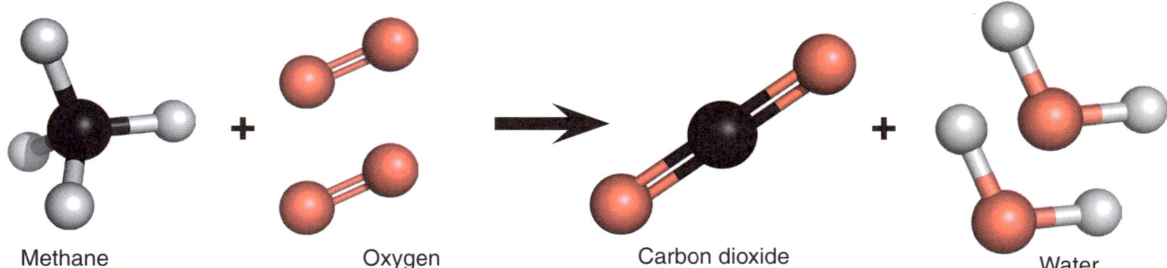

Bond	Average bond enthalpy
C – H	413 kJ/mol
O = O	498 kJ/mol
C = O	805 kJ/mol
O – H	464 kJ/mol
C = C	611 kJ/mol
C – C	349 kJ/mol
H – Br	370 kJ/mol
C – Br	276 kJ/mol
Br – Br	190 kJ/mol

For any chemical reaction, the enthalpy of the reaction (ΔH_r)
ΔH_r = the sum of the bond enthalpies in reactants – the sum of the bond enthalpies in products

- For the reaction above there are (4 × C–H bonds + 2 × O=O bonds) – (2 × C=O bonds + 4 × O–H bonds).
- Thus: (4 × 413 + 2 × 498) – (2 × 805 + 4 × 464) = 2648 – 3466 = –818 kJ/mol for the combustion of CH_4.
- The measured enthalpy of combustion for methane is –882 kJ/mol. If we take account for the fact that water returns to the liquid state and add in the enthalpy of vaporization for water (41 kJ/mol) then we get a more accurate answer: 2648 – (3466 +(2 × 41)) = –900 kJ/mol.
- The final discrepancy is due to the use of average bond enthalpies (for example the enthalpies of C–H vary from 439 kJ/mol to 411 kJ/mol depending on the molecule and its state).

1. Estimate the enthalpy change for the reaction between ethene and oxygen: $CH_2CH_{2(g)} + 3O_{2(g)} \rightarrow 2CO_{2(g)} + 2H_2O_{(l)}$:

   (4 × 413) + 611+ (3 × 498) = 3757   (4 × 805) + (4 × 464) + (2 × 41) = 5158

   3757 - 5158 = -1401 kJ / mol of ethene

2. Estimate the enthalpy change for the reaction between ethene and bromine: $CH_2CH_{2(g)} + Br_{2(g)} \rightarrow CH_2BrCH_2Br_{(g)}$:

   (4 × 413) + 611 + 190 = 2453   (4 × 413) + 349 + (2 × 276) = 2553

   2453 - 2553 = -100 kJ / mol ethene

PS3.D PS1.B EM

©2020 BIOZONE International
ISBN: 978-1-927309-82-7
Photocopying Prohibited

49 Reversible Reactions

Key Question: Some reactions are reversible. Under what conditions do reversible reactions occur?

- Many of the chemical reactions you have studied so far have been ones that go to completion. This means all the reactants combine to form products, with none left over when the reaction finishes. Sometimes reactions are reversible, the products can reform back into reactants.

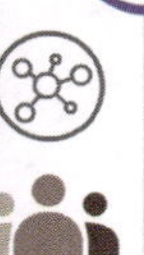

INVESTIGATION 3.13: Reversibility

See appendix for equipment list.

Caution is required when handling HCl and NaOH as they are corrosive. Wear protective eyewear, gloves and clothing.

1. Add 5 mL of dilute HCl to a test tube. Into a separate test tube add 5 mL of dilute NaOH solution.
2. Hold a piece of red litmus paper so 1 cm of the end is in the NaOH solution. Record the color change:
   Changes to blue
3. Now place the same 1 cm end of the litmus paper into the HCl. Record the color change:
   Changes to red
4. Place the end the litmus paper back into the NaOH. Record the color change:
   Changes to blue
5. Place the end of the litmus paper in the HCl one more time. Record the color change:
   Changes to red
6. Now hold blue litmus paper so that 1 cm of the end is in the HCl. Record the color change:
   Changes to red
7. Now place the same 1 cm end of the litmus paper into the NaOH. Record the color change:
   Changes to blue
8. Place the end the litmus paper back into the HCl. Record the color change:
   Changes to red
9. Place the end of the litmus paper in the NaOH one more time. Record the color change:
   Changes to blue

- The chemical used in both the red and blue litmus paper is litmus. Litmus in the red form can be written as HLt and litmus in the blue form can be written as Lt^-.
- The equation for placing red litmus into NaOH is: $HLt_{(aq)}$ (red) + $NaOH_{(aq)} \rightarrow H_2O_{(l)} + Na^+_{(aq)} + Lt^-_{(aq)}$ (blue)
- And placing blue litmus into HCl can be written as: $Lt^-_{(aq)}$ (blue) + $HCl_{(aq)} \rightarrow HLt_{(aq)}$ (red) + $Cl^-_{(aq)}$
- In the equations above the compounds and ions involved in the color change are HLt, H^+, and Lt^-.

1. (a) Write an equation for the color change of red litmus to blue litmus:
   $HLt \rightarrow H^+ + Lt^-$

   (b) Write an equation for the color change of blue litmus to red litmus:
   $H^+ + Lt^- \rightarrow HLt$

2. (a) Under what conditions does HLt form? *Acid conditions*

   (b) Under what conditions does Lt^- form? *Basic conditions*

©2020 **BIOZONE** International
ISBN: 978-1-927309-82-7
Photocopying Prohibited

SC SSM PS1.B

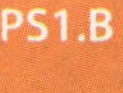

Forwards and backwards

- Some reactions do not go to completion. As the product is made, some of those products break up back into reactants so that there is always some reactant left, even when the reaction appears to have finished.
- This can be represented by the general equation aA+ bB $\rightleftharpoons$ cC + dD.
- The symbol $\rightleftharpoons$ means the reaction goes in both directions. The forward direction is left to right. The reverse direction is right to left.

INVESTIGATION 3.14: Equilibrium

See appendix for equipment list.

1. **Reaction A**: Fill a large beaker or basin (e.g. 1 L) to three quarters full with water. Label this beaker the "reactants". Add a few drops of food coloring (optional) to help see the water more clearly. Beside the full beaker, place an equal sized empty beaker or basin. Label this beaker the "products".
2. Place a small (e.g. 100 mL) beaker into the full reactants basin so that it fills (do not force fill it). Transfer the contents to the products basin.
3. Now place the 100 mL beaker into the products basin, let it fill, and transfer the contents back to the reactants basin.
4. Continue this back and forth process until the volumes of reactants and products remains steady.
5. At this point you have reached an **equilibrium.**
6. **Reaction B**: Now add some more reactant to the "reaction" by filling the reactant basin until it is three quarters full again. Continue the back and forth process as in steps 2 and 3 until the equilibrium is re-established.
7. **Reaction C**: Now remove some product by emptying the products basin to waste until it is one quarter full. Continue the back and forth process as in steps 2 and 3 until the equilibrium is re-established.
8. **Reaction D**: Now empty the products basin and reset the reactants basin to three quarters full. You will now alter the rate of the forward and reverse reactions. Use the 100 mL beaker for the forward reaction, transferring reactants to products. Use a 50 mL beaker to transfer products to the reactants.
9. Continue this until an equilibrium is reached.
10. **Reaction E**: Reset the reactants and products again. You will now change the rate of the backwards reaction. Use the 50 mL beaker for the forwards reaction (from reactants to products) and the 100 mL beaker for the backwards reaction. Continue this until an equilibrium is reached.

3. (a) Describe the rate of the forward reaction compared to the backward reaction at the start of reaction A:

The rate of the forward reaction is faster than the backward reaction

(b) Describe the rate of the forward reaction compared to the backward reaction as reaction A neared and then reached equilibrium:

The rate of the forward reaction slows and becomes equal to that of the backward reaction.

(c) What happened in reaction B when more reactant was added to the reaction?

More product was formed. The rate of the forward reaction increased before slowing and becoming equal to the backward reaction again.

(d) What happened in reaction C when some product was removed from the reaction?

The product was replaced by the forward reaction. The backward reaction initially slowed before increasing and equaling that forward reaction.

(e) Describe the reaction that occurred in reaction D in terms of reactants, products, and the equilibrium that was established:

The equilibrium that was established favored the products. Less of the reactants were "unreacted". More product was needed before the forward and backwards reactions became equal.

(f) Describe the reaction that occurred in reaction E in terms of reactants, products, and the equilibrium that was established:

The equilibrium that was established favored the reactants. Less of the products were "reacted". Less product was needed before the forward and backwards reactions became equal.

©2020 BIOZONE International
ISBN: 978-1-927309-82-7
Photocopying Prohibited

50 Equilibrium Reactions

Key Question: How do the conditions of a dynamic equilibrium affect the outcome?

Dynamic equilibrium

- If something is dynamic, it is forever changing. But when we react two chemicals together the reaction may proceed for a while then appear to remain static.
- How can a system be dynamic and yet remain the same? The previous investigations showed that an equilibrium is established when the rate of the forwards reaction equals the rate of the backwards reaction. Thus the reactions continue but at equal rates, producing no noticeable change. This can be shown on a graph, as shown below left:

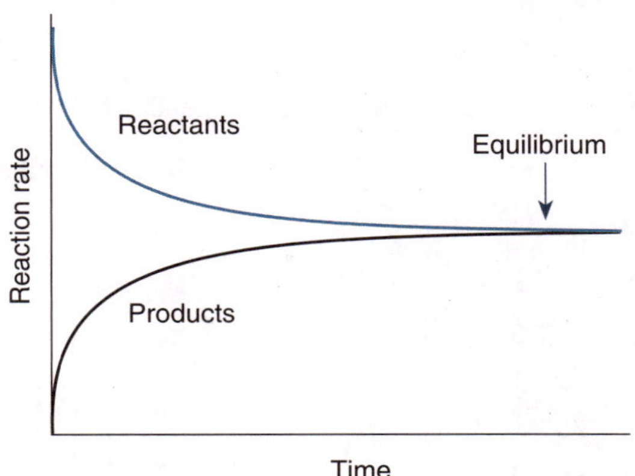

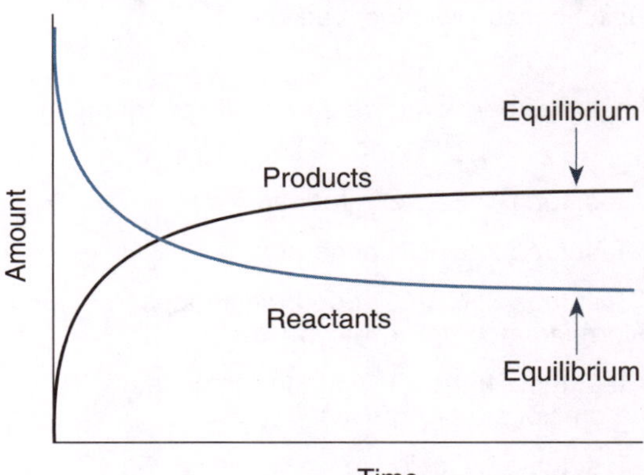

- The amount of products and reactants do not need to be equal at equilibrium, in fact in most cases they are not (above right). But as the amount of product increases the amount of products returning to reactants also increases. So if the rates of the forward and backwards reactions are not equal at the start, they will become equal as reactants turn into more products.
- Study the graphs below and answer the questions:

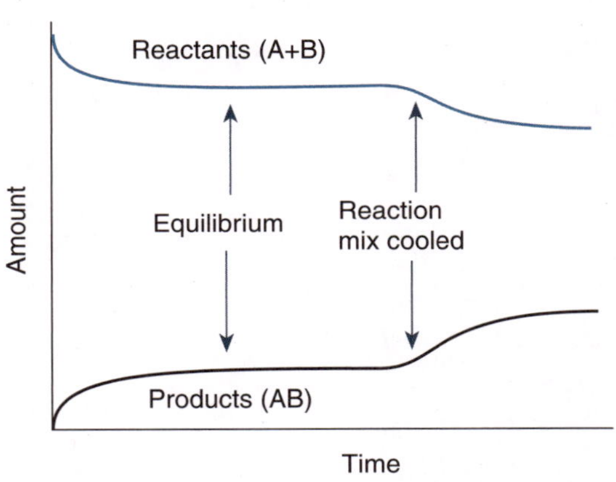

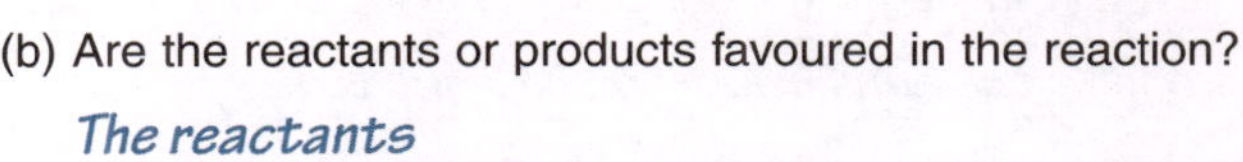

1. For the reaction $A + B \rightleftharpoons AB -\Delta H$ shown left:

   (a) Which side of the equation does the equilibrium lie closest to once equilibrium has been reached (left or right)?

   Left

   (b) Are the reactants or products favoured in the reaction?

   The reactants

   (c) The reaction mixture was cooled. What happen to the amounts of reactants and products?

   Reactants decreased, products increased

   (d) Why did this happen? *The equilibrium moved to the right counteract the reduction in temperature by moving in the direction that releases energy.*

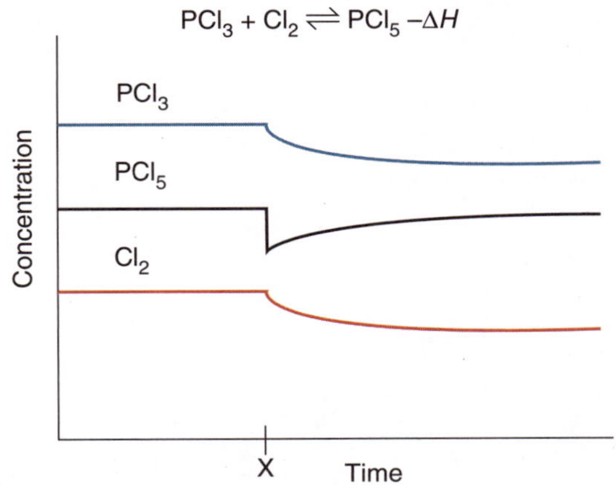

2. The graph left shows the reaction $PCl_{3(g)} + Cl_{2(g)} \rightleftharpoons PCl_{5(g)}$ is at equilibrium. At time X some of the PCl_5 was removed:

   (a) How did the position of the equilibrium respond to the removal of PCl_5?

   The position was moved to the left then moved to the right as the equilibrium was restored.

   (b) Why did this happen? *Removing PCl_5 initially reduced its concentration (the left). The equilibrium moved to the right to increase the concentration of PCl_5*

©2020 BIOZONE International
ISBN: 978-1-927309-82-7
Photocopying Prohibited

PS1.B

3. The image below shows six test tubes containing the complex ion, iron thiocyanate $FeSCN^{2+}_{(aq)}$.

- Iron thiocyanate is a complex ion that forms when potassium thiocyanate (KSCN) solution is added to iron nitrate ($Fe(NO_3)_3$ solution.
- The reaction between Fe^{3+} and SCN^- can be written as:

  $$Fe^{3+}_{(aq)} + SCN^-_{(aq)} \rightleftharpoons FeSCN^{2+}_{(aq)} -\Delta H$$

  The enthalpy for the forward reaction (left to right) is negative (it is exothermic).
- - The iron thiocyanate ion produces a blood red solution.
  - Fe^{3+} has a pale yellow-brown color.
  - SCN^- is colorless.
- By changing the conditions of the equilibrium, the solution can be made lighter or darker.

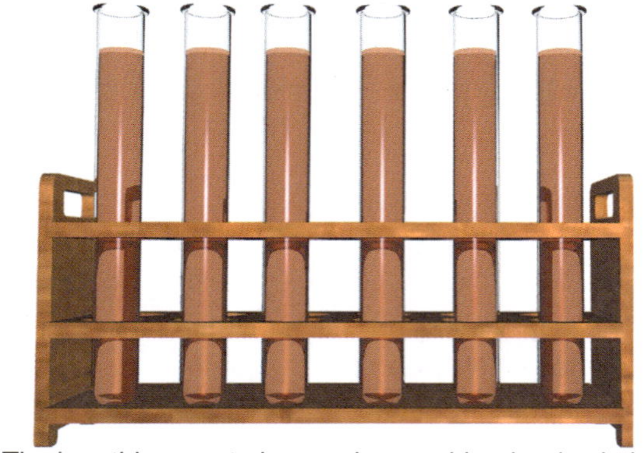

The iron thiocyanate ion produces a blood red solution

The image below shows the results of changing the conditions in the test tubes. The changes are detailed below:

- Test tube 1: $Fe(NO_3)_3$ added (adds Fe^{3+}).
- Test tube 2: KSCN added (adds SCN^-).
- Test tube 3: NH_4Cl added (decreases Fe^{3+} ion by forming a complex with Cl^-).
- Test tube 4: $AgNO_3$ added (decreases SCN^- by forming a precipitate of AgSCN).
- Test tube 5: Cooled in ice water.
- Test tube 6: Heated in hot water bath.

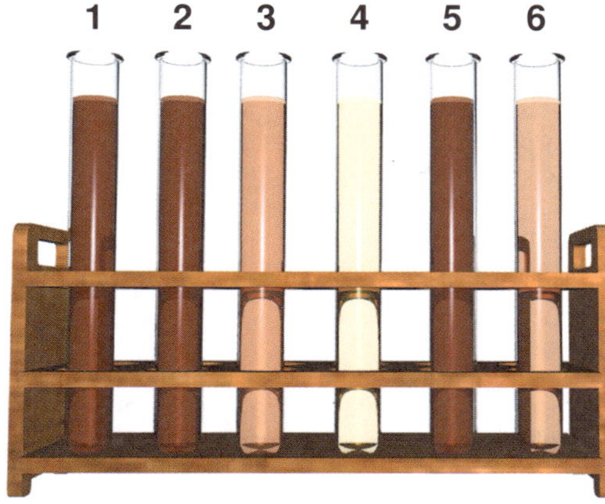

(a) Explain why adding more Fe^{3+} to test tube 1 produces a deeper red color: More Fe^{3+} was added to the solution so more $FeSCN^{2+}$ was formed by the equilibrium shifting to the right to use up the Fe^{3+}.

(b) Explain why adding more SCN^- to test tube 2 produces a deeper red color: Adding more SCN^- to the solution shifts the equilibrium to the right to use up the SCN^-.

(c) Why would removing Fe^{3+} ions (test tube 3) make the solution lighter? Removing Fe^{3+} reduces the amount of $FeSCN^{2+}$. Fe^{3+} is replaced as the equilibrium shifts to the left and SCN^- is produced further adding to the light solution.

(d) In test tube 4, which direction did the reaction shift (forwards or backwards). Explain why it shifted in this direction: The equilibrium shifted in the backwards direction. Removing the SCN^- causes FeSCN to break up and form Fe^{3+} and SCN^- (lighter).

(e) The solution in the ice bath (test tube 5) became darker. This means the reaction moved in the forwards direction. Why did it move in this direction?
The equilibrium shifts in the forward direction which is exothermic. This produces heat to counter act drop in temperature. This produces more $FeSCN^{2+}$ (darker).

(f) Test tube 6 was heated and became lighter in color. This means the backwards endothermic reaction occurred. Can you explain why?
The endothermic reaction uses the heat to drive the backwards reaction, reducing the temperature of the system. This produces more Fe^{3+} and SCN^- and removes $FeSCN^{2+}$ making the solution lighter.

©2020 BIOZONE International
ISBN: 978-1-927309-82-7
Photocopying Prohibited

51 Le Châtelier's Principle

Key Question: What is the principle behind how chemical equilibria behave?

Le Châtelier's principle

- The previous investigations have demonstrated how equilibria behave when conditions change. For example, for the reaction $Fe^{3+} + SCN^- \rightleftharpoons FeSCN^{2+}$, the equilibrium shifts to the right (the forward reaction) if more Fe^{3+} is added (e.g. the concentration of Fe^{3+} is increased). This removes some of the Fe^{3+} from the solution. If the temperature is increased then the equilibrium shifts to the left (the backwards endothermic reaction) to remove some of that heat (i.e. the energy is used to drive the endothermic reaction).
- These observations about equilibria were summarized by Henry Louis Le Châtelier, a French chemist who published many papers on equilibria in the late 19th and early 20th centuries.

Iron thiocyanate equilibrium

- **Le Châtelier's principle** states: When a system at equilibrium is subjected to change (e.g. in temperature, pressure, concentration, or volume) the system changes to a new equilibrium and this change partially counteracts the change applied to the system.
- In other words, an external stress can change the system and the system reacts in a way that reduces or opposes the change as much as possible.
- For example consider the system $A_{2(g)} + B_{2(g)} \rightleftharpoons A_2B_{2(g)}$ (right).

  The system on the left has more gas molecules than on the right. If the pressure is increased, the equilibrium will shift to the right to decrease the pressure. It does this by reducing the number of gas molecules, increasing the number of A_2B_2 molecules and reducing the number of A_2 and B_2 molecules to produce a lower total number of gas molecules.

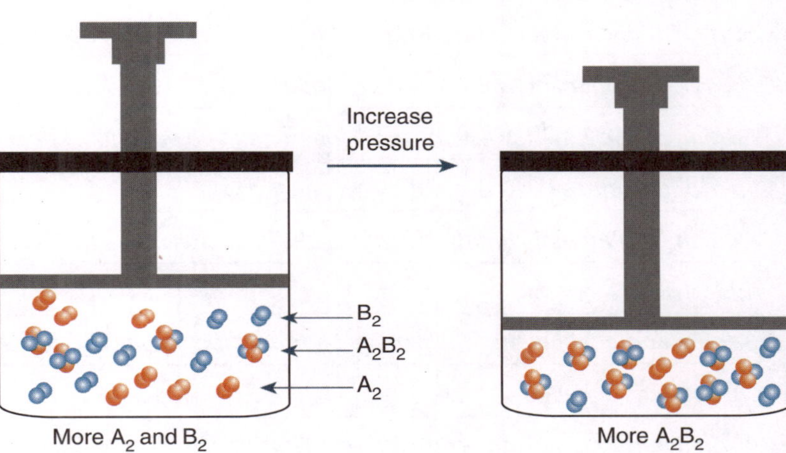

1. (a) Imagine the volume in the system above left was increased. Draw a diagram similar the one above right to show how the system and molecules would react:

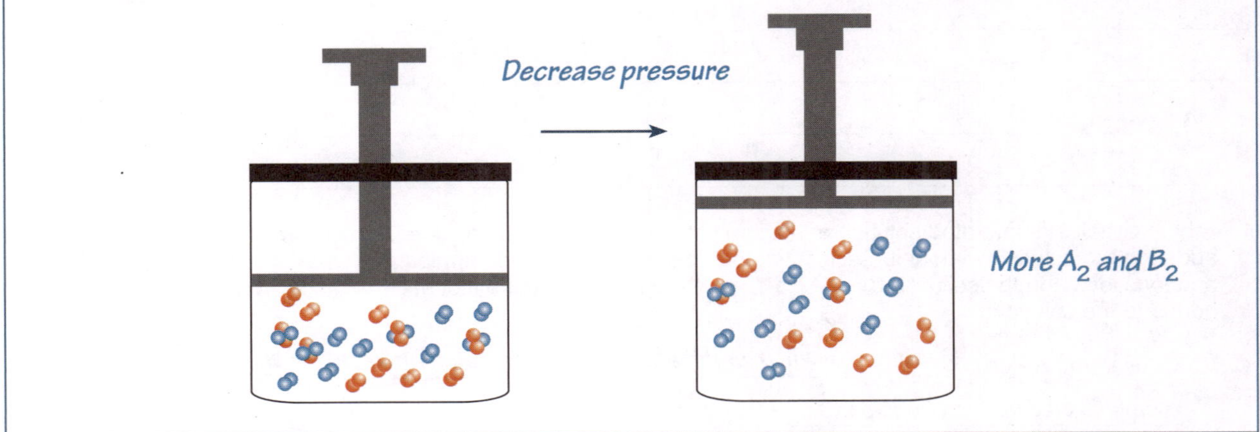

(b) More of gas B_2 was added to the reaction vessel on the left. Explain the response of the equilibrium:

The equilibrium will move to the right (increase the rate of the forward reaction). This removes some of the B_2 and decreases the total number of gas molecules and the associated pressure.

©2020 BIOZONE International
ISBN: 978-1-927309-82-7
Photocopying Prohibited

SC SSM PS1.B

2. The equilibrium between hydrogen and iodine can be written as $H_{2(g)} + I_{2(g)} \rightleftharpoons 2HI_{(g)}$ $-\Delta H$.

(a) Which way would the equilibrium shift if the system's pressure was increased? Explain your answer:

The equilibrium will remain in the same position because the number of molecules on either side of the equilibrium is the same.

(b) A chemist wanted to produce more HI from the system. How would they be able to do this?

Adding more H_2 or I_2, or removing HI will move the reaction in the forward direction, producing more HI.

3. In aqueous solution, the chromate ion ($CrO_4{}^{2-}{}_{(aq)}$) is yellow and the dichromate ion ($Cr_2O_7{}^{2-}{}_{(aq)}$) is orange. An equilibrium is produced when either ion is dissolved in water:

$$CrO_4{}^{2-}{}_{(aq)} + 2H^+ \rightleftharpoons Cr_2O_7{}^{2-}{}_{(aq)} + H_2O_{(l)}$$

A chemist had the following solutions and solids available:
dilute HCl solution, dilute NaOH solution, solid sodium dichromate, solid sodium chromate, and a dilute solution of barium chloride.
Note barium ions react with chromate ions to form a precipitate.

Capaccio CC 3.0

Describe the effect of each of the substances on the chromate/dichromate equilibrium and use Le Châtelier's principle to explain the effect.

Adding dilute HCl add H^+ ions to the equilibrium. The reaction will move to the right (forwards) to remove the H+. The solution will go more orange.

Adding dilute NaOH will remove H^+ ions from the solution. The reaction will move to the left (backwards) to produce more H^+. The solution will go more yellow.

Adding sodium dichromate will move the reaction move to the left (backwards), removing some of the added dichromate.

Adding sodium chromate will move the reaction to the right (forwards), removing some of the added chromate.

Adding barium chloride produces solid barium chromate. This removes chromate from solution. The reaction will to the left (backwards) to replace some of the lost chromate.

4. A beaker holds a saturated solution of sodium chloride. A saturated solution is one in which no more substance can be added to the solution without it precipitating out to form a solid. At saturation, the solution is in equilibrium between the aqueous ions and the solid: $NaCl_{(s)} \rightleftharpoons Na^+{}_{(aq)} + Cl^-{}_{(aq)}$. Explain the effect each of the following would have if they were added to the solution:

(a) Concentrated HCl: Adding HCl will precipitate NaCl out of solution because the reaction will shift to the left to remove the excess Cl^- ions.

(b) Concentrated NaOH: Adding NaOH will precipitate NaCl out of solution because the reaction will shift to the left to remove the excess Na^+ ions.

(c) Solid KCl: Adding solid KCl will precipitate NaCl out of solution because the reaction will shift to the left to remove the excess Cl^- ions.

©2020 **BIOZONE** International
ISBN: 978-1-927309-82-7
Photocopying Prohibited

52 Industrial Equilibria

Key Question: How is equilibrium chemistry used in industry?

Chemists make your stuff

- An understanding of chemistry has allowed humans to produce substances that are entirely new to this universe. What's more, we are able to produce these substances on large scales. Every year we produce more than 300 million tonnes of plastic, a substance that does not exist naturally.
- Almost everything you have or use has in some way made use of or been made by industrial chemistry. From the stainless steel spoon you used to eat your breakfast to the shampoo you used to wash your hair to the artificial leather on your shoes.
- Industrial chemistry takes the principles and ideas of chemists and scales them up to produce the vast quantities of substances needed to keep society working. Without the huge factories producing ammonia or sulfuric acid many common fertilizers would be in short supply, limiting the amount of food we could grow. Without the massive iron smelters we wouldn't have the steel needed to produce virtually anything requiring strong structural support, e.g. cars, bridges, chairs, or buildings.
- Many of the most important industrial processes are based on simple chemical equilibria.

Producing ammonia

- The main industrial procedure for producing ammonia is the Haber process. This process manipulates the equilibrium between nitrogen, hydrogen, and ammonia. It was first demonstrated by Fritz Haber in 1909 and industrial scale production began in 1913.
- Ammonia is an important precursor chemical in the production of nitrates. The Haber process was first developed to replace the need to mine nitrates for use in fertilizer production and other industrial processes. However, most of the nitrates initially produced ended up being used for ammunition in both the First and Second World Wars.
- The reaction for the formation of ammonia from nitrogen and hydrogen is:

$$N_{2(g)} + 3H_{2(g)} \rightleftharpoons 2NH_{3(g)} \quad \Delta H = -92\ kJ$$

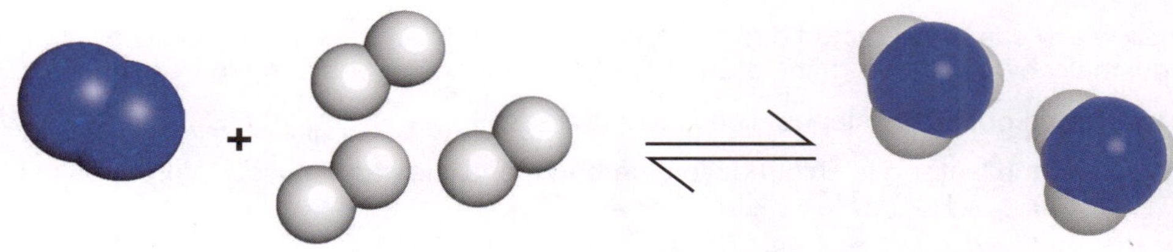

- The rate of the ammonia reaction is very slow (almost zero at room temperature). A catalyst is used to speed it up.
- The catalyst works most efficiently at around 400°C. This shifts the equilibrium to the left. To counteract this, the pressure of the system is increased, although this comes at an energetic (and therefore economic) cost.

1. (a) Is the forward reaction endothermic or exothermic? *Exothermic*

   (b) To produce the greatest amount of NH_3 in which direction should the equilibrium shift? *To the right (forwards)*

   (c) Can this be achieved by raising or lowering the temperature of the reaction? *Lowering the temperature*

   (d) What would the effect of this be on the rate of the reaction? *The rate of the reaction would be reduced (particles move slower and do not have the energy to react).*

2. (a) What is the effect of a catalyst on the rate of the reaction? *It increases the rate of the reaction*

   (b) What would be the effect of using a catalyst on the position of the equilibrium? *No effect*

3. What would the effect of increasing the pressure on the system be? *The equilibrium would shift to the right (forwards) to produce the least number of gas molecules.*

©2020 BIOZONE International
ISBN: 978-1-927309-82-7
Photocopying Prohibited

SC

PS1.B

- The graph right shows the effect of temperature and pressure on the percentage of ammonia formed during the reaction of N_2 and H_2.
- Because the reaction is so slow, the amount of product produced at 200°C is very low even if the percentage of product produced from reactants is high. The Haber process reacts nitrogen and hydrogen over a catalyst at 200 atmospheres and 400°C to produce ammonia. The yield is about 15% but at a much faster rate than at lower temperatures. This is a compromise between speed and cost.

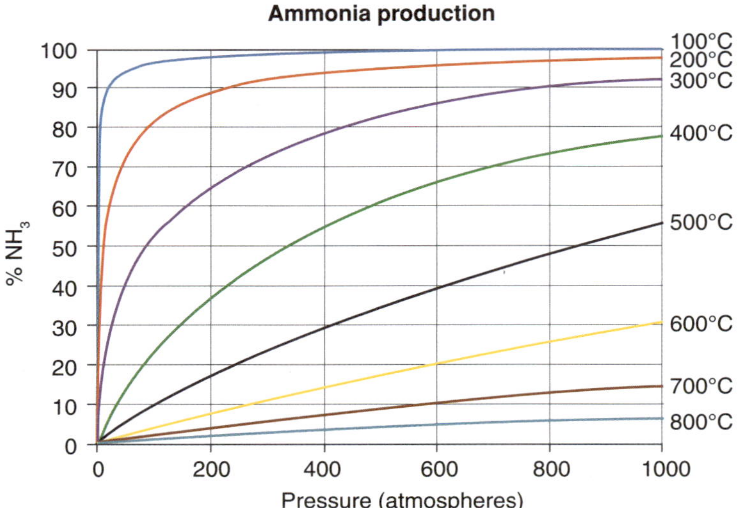

4. From the graph:

(a) What is the effect of increasing the temperature of the reaction?

Increasing the temperature decreases the percentage of NH_3 produced.

(b) What is the effect of increasing the pressure of the reaction?

Increasing the pressure increases the percentage of NH_3 produced.

5. Why is a compromise of 200 atmospheres and 400°C used in the production of ammonia?

400°C is hot enough to increase the reaction to an acceptably high rate but low enough that it does not affect the percentage of NH_3 produced too much. 200 atmospheres is high enough to push the equilibrium to the right but low enough that the economic cost is not prohibitively high (obtaining high pressures and building reaction vessels that can withstand them is costly). Thus, there is a compromise between output and cost.

Getting the hydrogen

- Unlike nitrogen which can be extracted directly from the air, hydrogen gas is not readily available and must be produced industrially by steam reforming of methane. The production of hydrogen is worth over $100 billion a year.
- The reaction is carried out in two stages: The first stage is: $CH_{4(g)} + H_2O_{(g)} \rightleftharpoons CO_{(g)} + 3H_{2(g)}$ ΔH 206 kJ
- In the second stage (the water-gas shift reaction) carbon monoxide can be used to produce more hydrogen: $CO_{(g)} + H_2O_{(g)} \rightleftharpoons CO_{2(g)} + H_{2(g)}$ ΔH –45 kJ.

6. (a) Under what conditions would the first reaction be carried out in order to produce the greatest amount of hydrogen?

Decrease the pressure. The equilibrium will move in the forward direction to increase the number of molecules. Increase the temperature. The equilibrium will move in the forward direction to use up the excess heat. Constantly remove hydrogen from the system to drive the reaction in the forwards direction.

(b) Identify two ways of increasing the products of the second reaction:

Decrease the temperature, constantly remove the products.

Getting the nitrogen

- Some of the hydrogen produced by steam reforming is used to produce pure nitrogen for use in the production of ammonia. Air is fed into a reactor along with the hydrogen: $4N_{2(g)} + O_{2(g)} + 2H_2 \rightleftharpoons 2H_2O_{(g)} + 4N_{2(g)}$

7. The equilibrium lies far to the right. What does this mean in terms of the ease of producing the products?

It means producing the products is relatively easy because the equilibrium is in favor of the forwards reaction and therefore the products

©2020 **BIOZONE** International
ISBN: 978-1-927309-82-7
Photocopying Prohibited

Producing sulfuric acid

- It has been said that you can measure a country's industrial strength by its production of sulfuric acid.
- Sulfuric acid is used in the manufacture of fertilizers, dyes, fabrics, detergents and many other everyday materials. It is also important as an electrolyte in lead-acid batteries used to start car engines. Over 250 million tonnes of sulfuric acid are produced each year.
- The process is carried out in several steps.

Molten sulfur is fed into the sulfur burner along with air (oxygen), forming sulfur dioxide.

$S_{(s)} + O_{2(g)} \rightarrow SO_{2(g)}$

Sulfur dioxide is converted into sulfur trioxide using a vanadium oxide catalyst. The reaction is carried out in several steps beginning at around 450°C and dropping to about 350°C.

$2SO_{2(g)} + O_{2(g)} \rightleftharpoons 2SO_{3(g)}$

Waste gases are reacted with calcium carbonate to prevent their loss to the environment.

The sulfur trioxide is passed into a chamber where it is dissolved in 98% sulfuric acid. There it reacts with water to produce more sulfuric acid.

$SO_{3(g)} + H_2O_{(l)} \rightarrow H_2SO_{4(l)}$

8. The equilibrium reaction between sulfur dioxide and oxygen is: $2SO_{2(g)} + O_{2(g)} \rightleftharpoons 2SO_{3(g)}$ ΔH –196 kJ.

(a) Would this reaction be carried out under high or low pressure? *High pressure*

(b) Explain why constantly removing the SO_3 produces more product: *By removing the SO_3 from the system, the effective concentration or amount of SO_3 on the right hand side of the equilibrium is reduced. This will favor the forwards reaction and produce more SO_3 from the reaction.*

(c) Why is the reaction carried out at a high temperature when the equilibrium suggests this would not be favorable?

High temperatures increase the reaction rate so that reactants actually react to form products and an equilibrium is achieved (lower temperatures used later then take advantage of this equilibrium).

PS1-6 Producing ethanol

- About 93% of all ethanol is production by fermentation in industrial processes. The other 7% is produced by the direct hydration of the hydrocarbon ethene following the equation: $C_2H_{4(g)} + H_2O_{(g)} \rightleftharpoons C_2H_5OH_{(g)}$ ΔH –45 kJ

9. You are an industrial chemist and want to produce ethanol by direct hydration of ethene. Using the equilibrium equation above, specify what changes in conditions would produce the greatest amount of ethanol. Explain how each of these changes would affect the equilibrium and why:

To produce the most amount of ethanol the system should placed under high pressure and relatively low temperature. A catalyst will speed up the reaction. A high temperature will also speed up the reaction.

High pressure will force the equilibrium to the right to produce less molecules. This favors the product ethanol. The enthalpy of the reaction is negative (forward reaction is exothermic). A reduced temperature will cause the equilibrium to move to the right. However reducing the temperature too far will reduce the reaction rate. Thus temperature should be high enough for the reaction to occur. Around 300°C is effective (not too high, not too low).

©2020 **BIOZONE** International
ISBN: 978-1-927309-82-7
Photocopying Prohibited

53 Acids and Bases

Key Question: How is equilibrium involved in the formation of acids and bases?

Acids and bases

- You will be familiar with various acidic substances (such as hydrochloric acid and vinegar) and basic substances (such as sodium hydroxide and ammonia). In this investigation you will test the pH of these and other substances:

INVESTIGATION 3.15: pH scale

See appendix for equipment list.

Caution is required when handling acids and bases as they are corrosive. Wear protective eyewear, gloves and clothing.

1. You will need distilled water and 0.1 mol/L solutions of hydrochloric acid (HCl), nitric acid (HNO_3), ethanoic acid (CH_3COOH), citric acid ($C_6H_8O_7$), sodium hydroxide (NaOH), ammonia solution (NH_3), sodium carbonate (Na_2CO_3), and sodium hydrogen carbonate ($NaHCO_3$).
2. Test the pH of each acid or base by placing a few drops on a watch glass and touching a piece of pH paper to them.
3. Use the color change to identify the pH of the solution and record this in the table below.

Substance	pH
H_2O	~7
HCl	1
HNO_3	1
CH_3COOH	2.88
$C_6H_8O_7$	2.08

Substance	pH
NaOH	13
NH_3	11.1
Na_2CO_3	11.8
$NaHCO_3$	8.1

The pH scale

- The pH scale ranges from 0 to 14. Substance with a pH of 1 are very acidic, whereas substances with a pH of 14 are very basic. Neutral substances have a pH of 7.

1. The diagram below shows the pH scale. Show the position of the substances you tested above on the scale. Then using books, or the internet find out the pH of a range of household substances (e.g. bleach) and add them to the diagram:

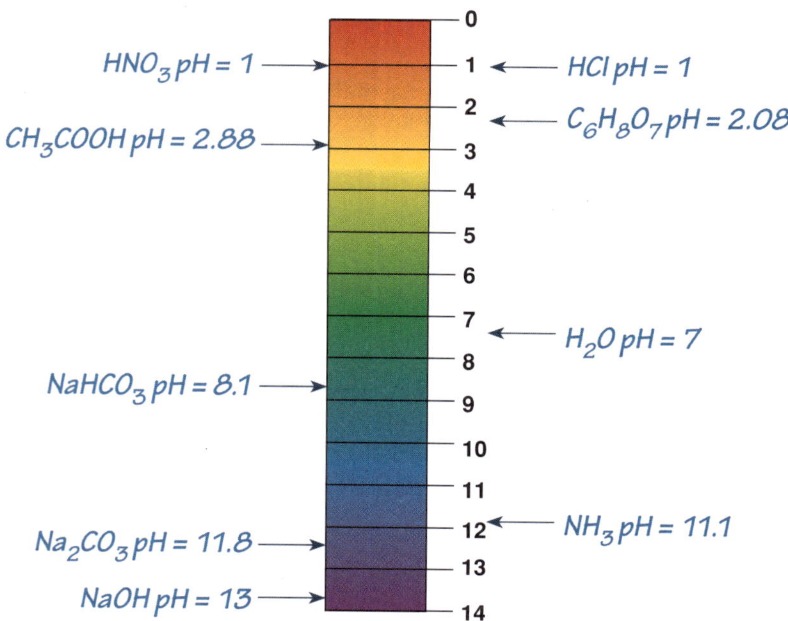

©2020 BIOZONE International
ISBN: 978-1-927309-82-7
Photocopying Prohibited

pH of acids and bases

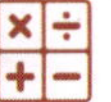

- The pH scale is based on the concentration of H^+ ions in a solution (H^+ ions actually exist as H_3O^+ in solution). It is a logarithmic scale so that the concentration of H^+ changes by ten times from one pH unit to the next.
- pH actually means the hydrogen (H) potential (p) of the solution (p is also stated as power, or *potentia* (capacity)).
- Water exists in an equilibrium with the hydronium ion (H_3O^+) and the hydroxide ion (OH^-):

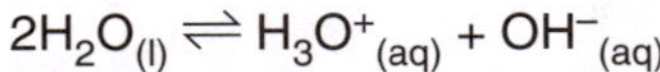

$$2H_2O_{(l)} \rightleftharpoons H_3O^+_{(aq)} + OH^-_{(aq)}$$

What's the $[H^+]$ of a lemon? Find out by testing its pH.

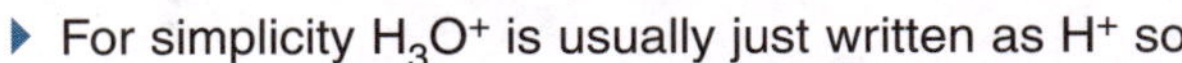

- For simplicity H_3O^+ is usually just written as H^+ so:

$$H_2O_{(l)} \rightleftharpoons H^+_{(aq)} + OH^-_{(aq)}$$

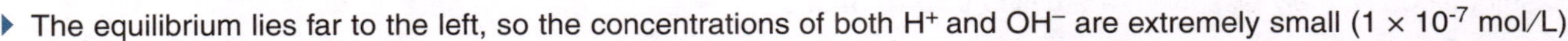

- The equilibrium lies far to the left, so the concentrations of both H^+ and OH^- are extremely small (1×10^{-7} mol/L).

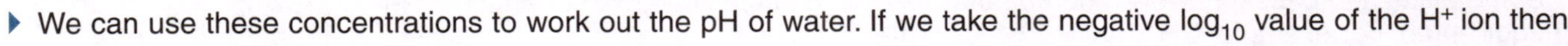

- We can use these concentrations to work out the pH of water. If we take the negative $\log_{10}$ value of the H^+ ion then:

$$-\log_{10} 1 \times 10^{-7} = 7$$

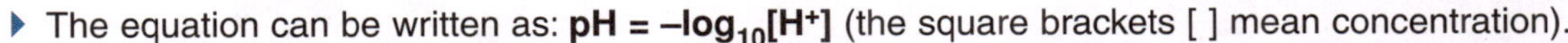

- The equation can be written as: **$pH = -\log_{10}[H^+]$** (the square brackets [ ] mean concentration).
- If we know the pH of a solution we can also work out the H^+ concentration of that solution using the negative inverse $\log_{10}(10^x)$. So if the pH is 5.3 then: $10^{-5.3} = 5.0 \times 10^{-6}$ mol/L (of H^+ ions).
- The equation can be written as **$[H^+] = 10^{-pH}$**
- It is useful to note at this stage that as $[H^+]$ increases $[OH^-]$ decreases at the same rate. This can be seen in the equation $[H^+][OH^-] = 1 \times 10^{-14}$. This value is referred to as K_w and is constant.

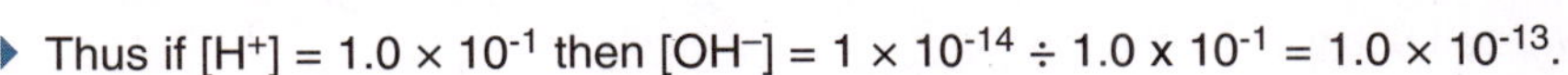

- Thus if $[H^+] = 1.0 \times 10^{-1}$ then $[OH^-] = 1 \times 10^{-14} \div 1.0 \times 10^{-1} = 1.0 \times 10^{-13}$.

2. Calculate the pH of the following H^+ concentrations:

   (a) 1.0×10^{-3} mol/L: *3.0*

   (b) 1.7×10^{-6} mol/L: *8.6*

   (c) 2.3×10^{-9} mol/L: *5.8*

   (d) 6.0×10^{-12} mol/L: *11.2*

3. Calculate the H^+ concentration of a solution with the following pH:

   (a) pH 1.0: *0.1 mol/L*

   (b) pH 3.6: *2.5×10^{-4} mol/L*

   (c) pH 14.0: *1×10^{-14} mol/L*

   (d) pH 11.4: *3.9×10^{-2} mol/L*

4. How many times greater is the H^+ concentration in a pH 3 solution than a pH 4 solution? *10 times greater*

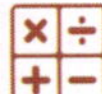

5. A solution of NaOH is made up to 0.25 mol/L. Calculate its pH: *$[OH^-] = 0.25$ mol/L.*
   $[H^+] = K_w \div [OH^-] = 1 \times 10^{-14} \div 0.25 = 4 \times 10^{-14}$. $pH = \log [H^+] = -\log 4 \times 10^{-14} = 13.4$

6. The following substances are acids or bases. The ions they form in solution are shown:

Substance	Acid/base	Ions in solution
HCl	Acid	H^+ Cl^-
HNO_3	Acid	H^+ NO_3^-
CH_3COOH	Acid	H^+ CH_3COO^-
NaOH	Base	Na+ OH^-
KOH	Base	K+ OH^-
NH_3	Base	NH_4^+ OH^-

   (a) What ion(s) do the acids have in common? *H^+ ions*

   (b) What ion(s) do the bases have in common? *OH^- ions*

   (c) Complete the reaction: NH_3 + *H_2O* $\rightleftharpoons NH_4^+ + OH^-$

   (d) The reaction in (c) above is an equilibrium. How might this affect the expected pH of an NH_4OH solution?
   It would be higher than expected due to fewer OH- ions present.

©2020 BIOZONE International
ISBN: 978-1-927309-82-7
Photocopying Prohibited

Strong and weak acids

- Why do some acids have different pH values to others? Why are some very low and others close to 7? Why do some bases have high pH values while others are relatively low?
- Acids and bases dissociate when placed in water. The more strongly they dissociate, the greater the effect on the pH. For example, when HCl is placed in water, it dissociates completely into H^+ and Cl^- ions:
  $HCl_{(aq)} \rightarrow H^+_{(aq)} + Cl^-_{(aq)}$
- A solution of 0.1 mol/L HCl will produce 0.1 mol/L H^+ ions. Acids that completely dissociate are called **strong acids**.
- The same occurs for bases. Bases such as NaOH completely dissociate into ions (Na^+ and OH^-). A 0.1 mol/L solution of NaOH will produce 0.1 mol/L of OH^- ions. Bases that completely dissociate are called **strong bases**.
- Some acids and bases, however, do not fully dissociate but instead exist in an equilibrium. For example: $CH_3COOH_{(aq)} \leftrightharpoons CH_3COO^-_{(aq)} + H^+_{(aq)}$.
  Because the $[H^+]$ is lower than the concentration of the compound, it will have a higher pH than might be expected.
- Acids and bases that do not fully dissociate are called **weak acids** and **weak bases**.

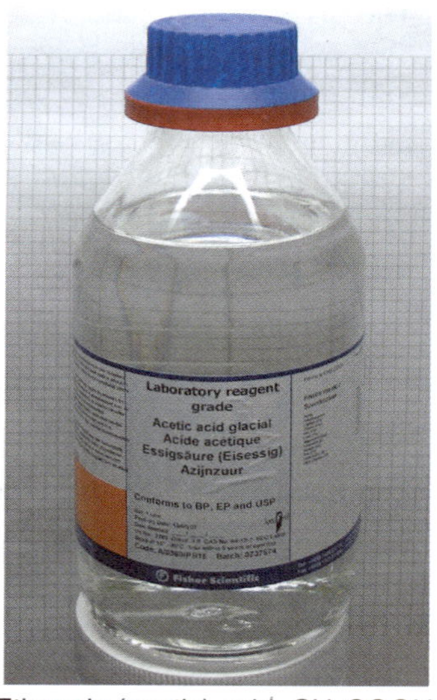

Ethanoic (acetic) acid, CH_3COOH

7. (a) 1 gram of $NaHCO_3$ (sodium hydrogen carbonate) is placed into excess 0.5 mol/L solution of HCl. The reaction is vigorous, with many bubbles of CO_2 appearing. Write an equation for the reaction:

$NaHCO_{3(s)} + HCl_{(aq)} \rightarrow NaCl_{(aq)} + CO_{2(g)} + H_2O_{(l)}$

(b) 1 gram of $NaHCO_3$ sodium hydrogen carbonate is placed into excess 0.5 mol/L solution of CH_3COOH. The reaction is slower that the reaction with HCl. Write an equation for the reaction:

$NaHCO_{3(s)} + CH_3COOH_{(aq)} \rightarrow CH_3COOCl_{(aq)} + CO_{2(g)} + H_2O_{(l)}$

(c) Explain why the rates of reaction for HCl and CH_3COOH are different:

The rates are different because HCl is a strong acid. All H+ ions are in solution and react immediately, producing a higher reaction rate. CH_3COOH is in equilibrium. Not all H+ ions are in solution. Thus the lower concentration of H+ reduces the rate of the reaction.

(d) Explain why both reactions form the same volume of CO_2 gas:

The concentrations of each acid are the same. Therefore there is the same amount of H+ ions available for each reaction. However HCl will produce the gas quickly whereas the CH_3COOH will produce the gas more slowly as the equilibrium slowly shifts to the right as CO_2 is released.

8. (a) A 1 cm strip of Mg ribbon is place in 1 mol/L HCl. The reaction produces H_2 gas. Write an equation for the reaction:

$Mg_{(s)} + 2HCl_{(aq)} \rightarrow MgCl_{2(aq)} + H_{2(g)}$

(b) A 1 cm strip of Mg ribbon is place in 1 mol/L H_2SO_4. The reaction also produces H_2 gas. Write an equation for the reaction:

$Mg_{(s)} + H_2SO_{4(aq)} \rightarrow MgSO_{4(aq)} + H_{2(g)}$

(c) Explain why the reaction between H_2SO_4 and Mg is much more vigorous than HCl and Mg:

H_2SO_4 has twice has twice as many H+ ions in solution as HCl. The the concentration of H+ ions is therefore twice as great in H_2SO_4 as in HCl. As the result the reaction rate will be higher in H_2SO_4 because a higher concentration of reactants increases the reaction rate.

9. Explain how acetic acid can be a concentrated weak acid: *Concentrated means having a lot of something in solution. So CH_3COOH can have many CH_3COOH units in solution (say a concentration of 17 mol/L). However because CH_3COOH does not dissociate fully it is classed as a weak acid. Thus is can be concentrated but still weak.*

©2020 **BIOZONE** International
ISBN: 978-1-927309-82-7
Photocopying Prohibited

54 Review Your Understanding

Key Question: There are many antacid formulas on the market. How do they work?

- At the beginning of this chapter you were introduced to some chemical equations relating to the neutralizing of stomach acid by antacid tablets.
- With the concepts you have learned throughout this chapter you should now be able to answer questions posed about these tablets and devise a way of finding out the mass of the active ingredient in a particular tablet.

1. At the start of the chapter you were asked to complete the word equations below. Redo them now, then check them against your original answers:

   (a) Hydrochloric acid + calcium carbonate → *Water + carbon dioxide + calcium chloride*

   (b) Hydrochloric acid + sodium hydrogen carbonate → *Water + carbon dioxide + sodium chloride*

   (c) Hydrochloric acid + magnesium hydroxide → *Water + magnesium chloride*

2. Complete and balance the chemical equations for each of the following reactions:

   (a) $2HCl + CaCO_3 \rightarrow H_2O_{(l)} + CO_{2(g)} + CaCl_{2(aq)}$

   (b) $HCl + NaHCO_3 \rightarrow NaCl_{(aq)} + H_2O_{(l)} + CO_{2(g)}$

   (c) $2HCl + Mg(OH)_2 \rightarrow MgCl_{2\,(aq)} + 2\,H_2O_{(l)}$

3. For question 2(c) explain the balancing of the reaction in terms of the atoms and ions present. You may wish to draw a diagram to help you:

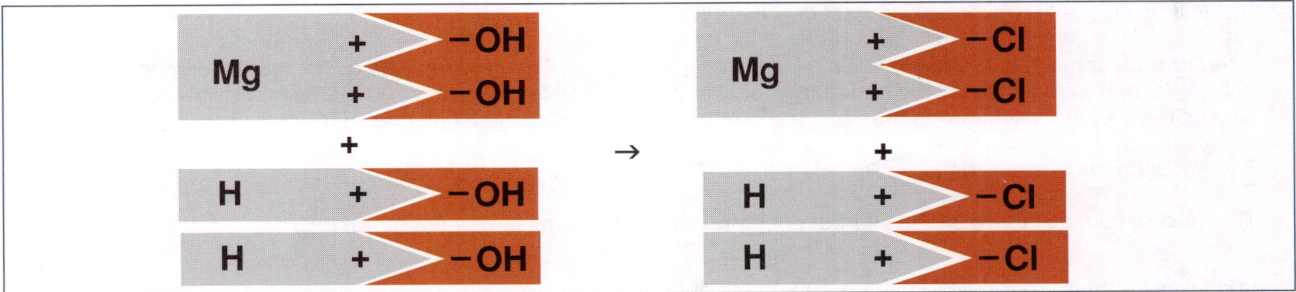

HCl forms H^+ and Cl^- ions. $Mg(OH)_2$ forms Mg^{2+} and OH^- ions. Two Cl^- ions are needed to balance the charge on the Mg^{2+} ion. Therefore there needs to be 2 units of HCl to provide the Cl^- ion. The $2H^+$ ions from the 2 HCl joins with the 2 OH^- ions to form 2 units of H_2O (water).

4. At the start of the chapter you were asked to develop a method for working out the amount of active ingredient in an antacid tablet. You may now want to modify your method. How would you find out the amount (the mass) of calcium carbonate in an antacid tablet with calcium carbonate as the active ingredient? Calcium carbonate doesn't dissolve in water, but it does dissolve in acid.

   - *Dissolve the calcium carbonate in a known excess volume of an known concentration of HCl.*
   - *React the left over acid with a sodium hydroxide solution of known concentration until the solution is neutral.*
   - *From the known volume and concentration of NaOH used, the moles of HCl left over can be calculated.*
   - *The moles of HCl used up to start with can then be calculated.*
   - *The moles of calcium carbonate reacted with the HCl can then be calculated.*
   - *From the moles of calcium carbonate, the mass of calcium carbonate can be calculated.*

©2020 BIOZONE International
ISBN: 978-1-927309-82-7
Photocopying Prohibited

55 Summing Up

Chemical reactions

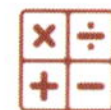

PS1-2 1. Balance the following equations:

(a) Ca + 2HCl → $CaCl_2$ + H_2

(b) 2Li + $2H_2O$ → 2LiOH + H_2

(c) $2C_2H_6$ + $7O_2$ → $4CO_2$ + $6H_2O$ *Also acceptable:* C_2H_6 + 3½ O_2 → $2CO_2$ + $3H_2O$

(d) C_2H_5OH + $2O_2$ → 2CO + $3H_2O$

2. A chemist accidently mislabeled a container holding a silvery metal element. To find out what the metal was, the chemist carried out a series of tests. The results are stated below:

The density of the metal lay between sodium and sulfur
When reacted with acid, the reaction was relatively fast, much faster than iron in the same concentration of acid.
When reacted with water the reaction was relatively slow, much slower than adding sodium to water.
When burned in oxygen, a one gram piece of metal reacted with 0.66 grams of oxygen (M(O)=16 g/mol).

What do you think this metal was? Explain your answer: *Density between Na and S so must be Mg or Al. Fast acid reaction but slow water reaction suggests Mg. Al ordinarily does not react with acid or water due to oxide coat (although removing the coat prompts a fast reaction). 0.66 g of O = 0.041 moles. 1 g of Mg = 0.041 moles therefore 1:1 ratio MgO. 1 g of Al = 0.037 moles. Metal is most probably Mg.*

PS1-7 3. Fertilizers show the different percentage of components in the fertilizer by printing an NPK rating on the bag. This stands for nitrogen – phosphorus – potassium. A rating of 13 – 10 – 10 contains 13% nitrogen by mass, 10% phosphorus (actually P_2O_5), and 10% potassium (actually K_2O) by mass.

A fertilizer company made a fertilizer claiming a rating 15 – 0 – 0. The nitrogen was contained as ammonium sulfate ($(NH_4)_2SO_4$). A student tested the percentage of N in the fertilizer by dissolving 100 grams of fertilizer in 1 L of distilled water. 50 mL of the solution was then reacted with excess 1 mol/L barium chloride ($BaCl_2$) solution.

The equation for the reaction is $BaCl_{2(aq)} + (NH_4)_2SO_{4(aq)} \rightarrow BaSO_{4(s)} + 2NH_4Cl_{(aq)}$.

The solid barium sulfate was then filtered, dried and weighed. Its mass was 6.3 grams.

(a) Calculate the mass of $(NH_4)_2SO_4$ in 100 grams of fertilizer:

In 100 g fertilizer, 15 g is N. n(N) = 15÷14 = 1.07 moles.

From stoichiometry: 2N: 8H: 1S: 4O or 1N: 4H: ½S:2O. So m($(NH_4)_2SO_4$) = (1.07 × 14)+(1,07 × 4 × 1) + (1.07 × 0.5 × 32.1) + (1.07 × 2 × 16) = 70.7 g (Or (14+4) × 1.07) + (32.1 + 16 × 4) x 0.54 = 70.7 g)

(b) Calculate the concentration of the $(NH_4)_2SO_4$ solution: *n($(NH_4)_2SO_4$) = m÷M = 70.7 ÷ 132.1= 0.535 moles.*

c = n ÷ V = 0.535 ÷ 1000 = 5.35 × 10^{-4} mol/L

(c) How many moles of SO_4 are in 50 mL of the $(NH_4)_2SO_4$ solution?

Stoichiometry = 1:1 so 0.535 mol/L $(NH_4)_2SO_4$ produces 0.535 mol/L SO_4.

n(SO_4) = 0.535 ÷ 1000 × 50 = 0.02675

(d) What mass of $BaSO_4$ would be expected to be produced in the reaction?

m(SO_4) n × M = 0.02675 × 96.1 = 2.57 g. SO_4: Ba = 1:1 so 0.02675 × 137.3 = 3.67 g Ba.

2.57 + 3.67 = 6.24 g would be expected.

(e) Was the fertilizer company correct in saying its fertilizer contained 15%N? Give reasons for your answer:

The student obtained 6.3 grams of $BaSO_4$ which is more than expected. We can therefore conclude that the manufacturer is correct in their claim (often manufacturers claims are slightly under actual measures to take into account errors in weighing).

EM

©2020 BIOZONE International
ISBN: 978-1-927309-82-7
Photocopying Prohibited

4. A mining company wanted to mine iron ore. They had the choice of two sites. Site 1 contained ore with approximately 80% hematite (Fe_2O_3). The second site contained ore with approximately 70% magnetite (Fe_3O_4). Determine which site will produce the greatest amount of iron.

M(Fe_2O_3) = 159.6 g/mol. M(2Fe) = 55.8 × 2 = 111.6. % Fe in Fe_2O_3 = 111.6/159.6 × 100 = 69.9%.

In a 100 g sample, 80 g will be Fe_2O_3. So 80 × 69.9 % = 55.9 g Fe.

M(Fe_3O_4) = 231.4 g/mol. M(3Fe) = 55.8 × 3 = 167.4. % Fe in Fe_3O_4 = 167.4/231.4 × 100 = 72.3 %.

In a 100 g sample, 70 g will be Fe_3O_4. So 70 × 72.3 % = 50.6 g Fe. It is better to mine the Fe_2O_3.

S1-4

Thermochemistry

5. (a) Define the term endothermic: *Endothermic is the absorption of energy. An endothermic reaction absorbs energy from its surroundings.*

(b) Sketch a graph showing the enthalpy change for an endothermic reaction:

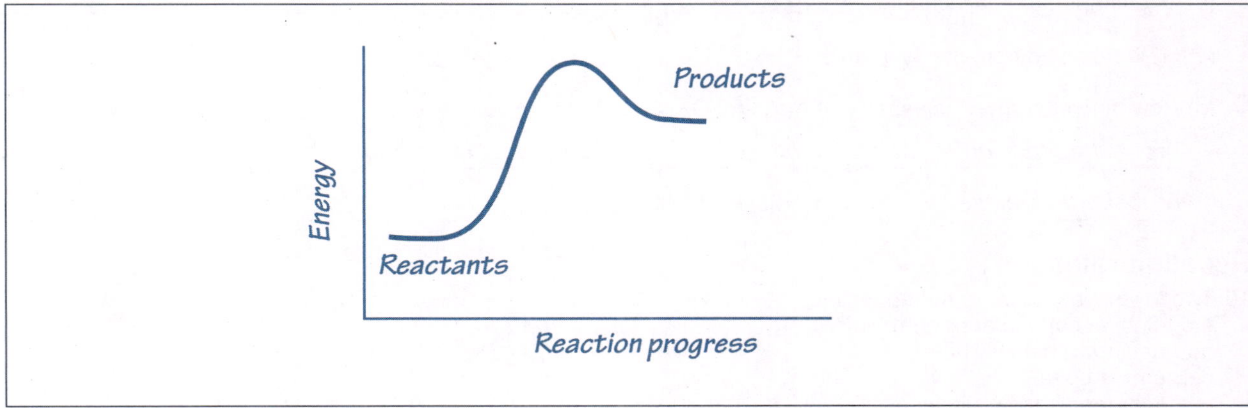

6. Sketch a graph showing the enthalpy change for the reaction Mg + ½O_2 → MgO ΔH –601.6 kJ/mol. Use you graph to help explain the term exothermic:

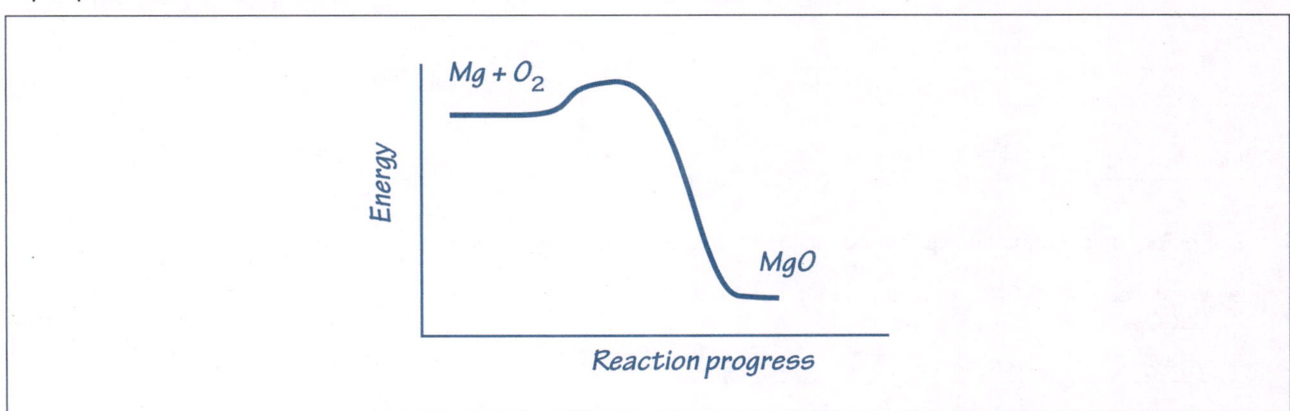

An exothermic reaction is one that releases energy into the environment and so feels hot. The bonds between Mg^{2+} and O^{2-} ions have less energy than the bonds between Mg atoms and between O_2 atoms. The excess energy is released to the environment, increasing its temperature

7. The reaction of carbon with oxygen is $C_{(s)} + O_{2(g)} \rightarrow CO_{2(g)}$ $\Delta_r H$ –393.5 kJ/mol

(a) Explain what the notation for the reaction means:

One mole of solid carbon (C) reacts with one mole of oxygen gas (O_2) to form one mole of carbon dioxide gas (CO_2). The reaction releases 393.5 kJ of energy per mole.

(b) How much energy is released is 2.6 moles of carbon is reacted completely?

2.6 × 393.5 = 1023.1 kJ

©2020 BIOZONE International
ISBN: 978-1-927309-82-7
Photocopying Prohibited

EM PS1.A

PS1-6 Equilibrium

8. When ammonia dissolves in water the following equilibrium occurs: $NH_{3(g)} + H_2O_{(l)} \rightleftharpoons NH_4^+{}_{(aq)} + OH^-{}_{(aq)}$

(a) Describe a simple method for detecting the ammonia: *Adding a hydroxide will push the equilibrium to the left. This will produce more NH_3 gas, will to be able to be detected by its smell.*

(b) How could you show OH^- ions were present? *Test using litmus. If it turns blue then OH^- ions are present*

9. Consider the following equation: $2O_{3(g)} \rightleftharpoons 3O_{2(g)}$ $\Delta H = -285$ kJ. Describe the effect of the following changes in conditions on the number of $O_{2(g)}$ molecules in the system:

(a) The pressure is increased: *Decrease O_2*

(b) The temperature is increased: *Decrease O_2*

(c) O_3 is removed from the system: *Decrease O_2*

(d) How much energy is released is 4 moles of O_3 is reacted completely? *From the reaction equation 2 moles of O_3 reacting will release 285 kJ of energy. Therefore 4 moles of O_3 reacting will release $285 \times 2 = 570$ kJ of energy.*

PS1-5 Reaction rates

10. Two investigations were set up testing the effect of surface area on reaction rate as shown below. The mass of the calcium carbonate chips was kept the same for the two investigations:

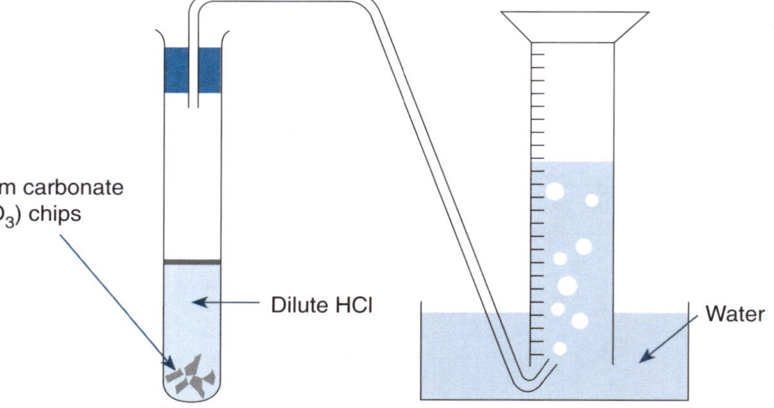

	Size of $CaCO_3$ chips
Investigation 1	Large
Investigation 2	Small

The graph, below right, shows the production of gas for investigation 1 and investigation 2:

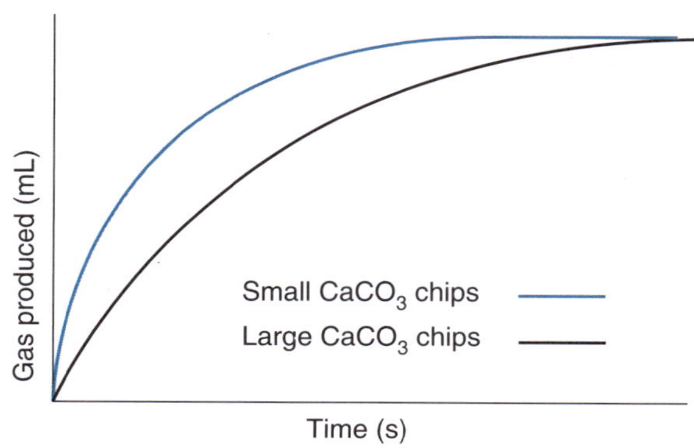

(a) What was the effect of reducing the size of the $CaCO_3$ chips?

It speeds up the reaction rate.

(b) Explain your answer in terms of collision theory:

There is more surface area available in the small calcium carbonate chips for the acid to react with. Thus more reactions occur in a given time, increasing the rate of the reaction overall.

(c) The students decided to increase the concentration of the HCl and run the investigation again using a new set of small $CaCO_3$ chips. How will this affect the rate of the reaction? Explain your answer:

This would increase the rate of the reaction. There are now more acid particles available to react in a given time so the reaction will be faster overall.

PS1.B P

©2020 BIOZONE International
ISBN: 978-1-927309-82-7
Photocopying Prohibited

4: Nuclear Processes

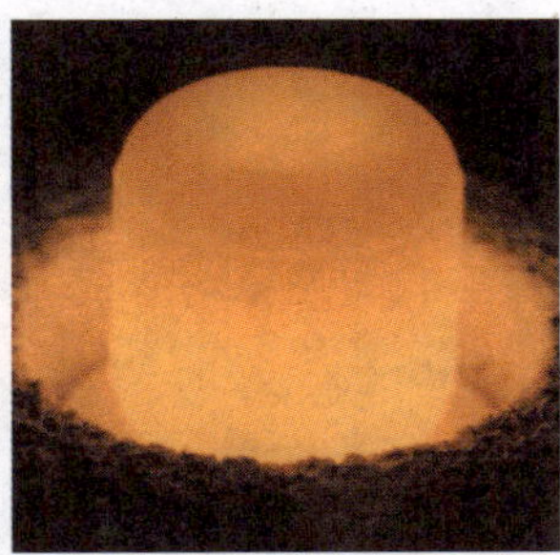

Anchoring Phenomenon

Nuclear power: How do you power a city or destroy it with just a few kg of nuclear material? **56 62**

Modifying our atomic model. What is the strong nuclear force?

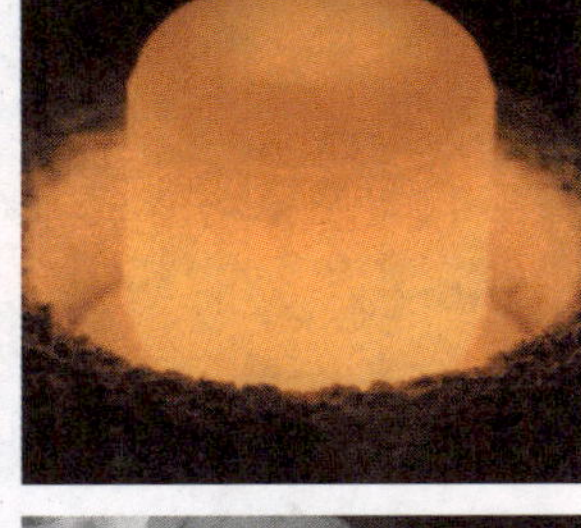

☐ 1 RECALL: In this chapter, we are concerned with processes within the nucleus of an atom and the strong and weak nuclear forces that govern these processes. How do we know about the nuclear structure *[CCC-6]* when its scale *[CCC-3]* is too small to observe directly? Recall how Rutherford and his students use experimentation to modify the model *[CCC-4]* of the atom existing at the time and determine that the nucleus, with its positive charge, was at an atom's center *[SEP-7]* *[PS1.A]*. **19 20**

☐ 2 Recall the atomic models you explore in Chapter 2. Why don't the protons in the nucleus fly apart as a result of the repulsive forces between them? You will now have to modify your model *[SEP-2]* of the atom, including through investigation *[SEP-3]*, to include the elementary particles that make up neutrons and protons (quarks) and the strong nuclear force that holds nuclei together *[PS1.C]* *[CCC-5]* *[HS-PS1-8]*. **21 57**

What is radioactive decay?

☐ 3 Recall what the atomic number of an element tells you about the makeup of the nucleus of an atom of that element *[PS1.A]*. What does the atomic mass tell you and why it is not usually a whole number? Recall how the proton number and nucleon number are expressed as a nuclide number. You have used nuclide number to determine the numbers of subatomic particles in the isotopes of different elements. Why are isotopes important when we model *[SEP-2]* nuclear processes? Explain why isotopes become less stable as the size of the atomic nucleus increases. What determines the stability of the atomic nucleus? *[PS1.C]* *[CCC-5]* *[HS-PS1-8]* **23 36 58**

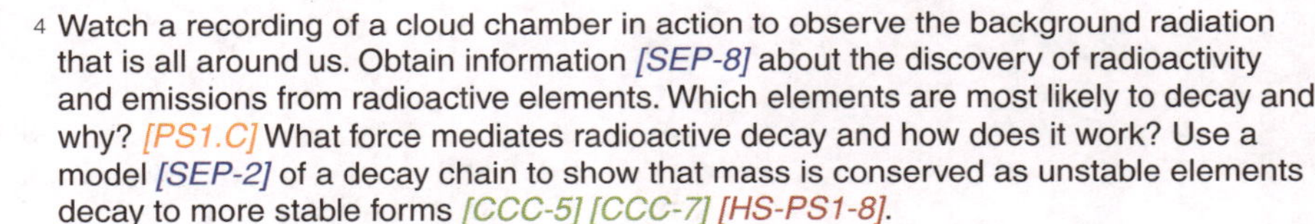

☐ 4 Watch a recording of a cloud chamber in action to observe the background radiation that is all around us. Obtain information *[SEP-8]* about the discovery of radioactivity and emissions from radioactive elements. Which elements are most likely to decay and why? *[PS1.C]* What force mediates radioactive decay and how does it work? Use a model *[SEP-2]* of a decay chain to show that mass is conserved as unstable elements decay to more stable forms *[CCC-5]* *[CCC-7]* *[HS-PS1-8]*. **59 63**

What is mass-energy equivalence and why is it important?

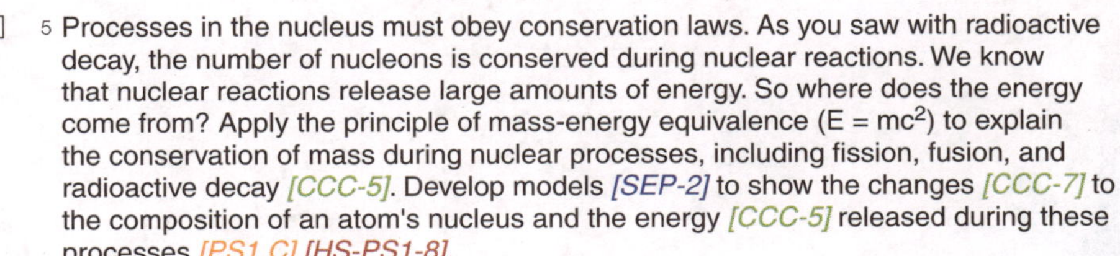

☐ 5 Processes in the nucleus must obey conservation laws. As you saw with radioactive decay, the number of nucleons is conserved during nuclear reactions. We know that nuclear reactions release large amounts of energy. So where does the energy come from? Apply the principle of mass-energy equivalence ($E = mc^2$) to explain the conservation of mass during nuclear processes, including fission, fusion, and radioactive decay *[CCC-5]*. Develop models *[SEP-2]* to show the changes *[CCC-7]* to the composition of an atom's nucleus and the energy *[CCC-5]* released during these processes *[PS1.C]* *[HS-PS1-8]*. **60**

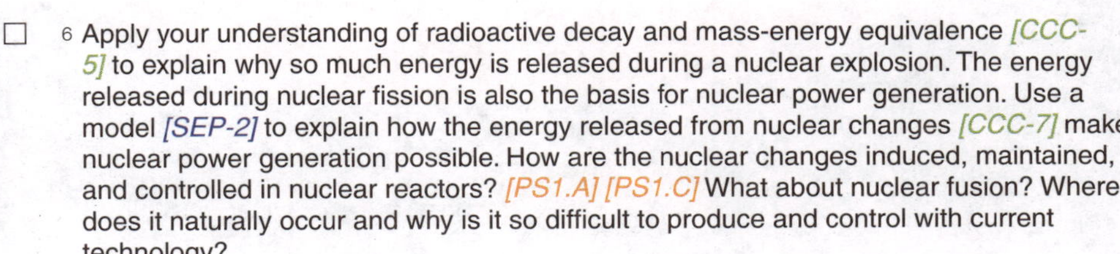

☐ 6 Apply your understanding of radioactive decay and mass-energy equivalence *[CCC-5]* to explain why so much energy is released during a nuclear explosion. The energy released during nuclear fission is also the basis for nuclear power generation. Use a model *[SEP-2]* to explain how the energy released from nuclear changes *[CCC-7]* make nuclear power generation possible. How are the nuclear changes induced, maintained, and controlled in nuclear reactors? *[PS1.A]* *[PS1.C]* What about nuclear fusion? Where does it naturally occur and why is it so difficult to produce and control with current technology? **60 61**

Challenge question:

The Sun is a fusion reactor, transforming lighter elements into heavier elements with the release of energy. Where does the energy come from and how would you calculate the energy released when one atom of helium is produced?

Teacher's Notes

4. Nuclear Processes

Anchoring phenomenon

The anchoring phenomenon introduces nuclear processes through their use in electricity generation and in nuclear weapons. The uncontrolled chain reaction that occurs when a nuclear weapon is detonated is an interesting contrast to how the reaction is controlled in nuclear power plants. Students find it interesting that just a small amount of mass produces an enormous amount of energy. Students may discuss where this energy comes from compared to other fuel types, or why elements with large mass numbers are used in power plants.

Modifying our atomic model. What is the strong nuclear force?

1. It is useful to recall the atomic model of protons and neutrons in the nucleus of the atom. Students may wish to review Rutherford's experiment and the development of the atomic model as background to this chapter.

2. So far students have treated protons and neutrons as fundamental (elementary) particles. Students now learn that this is not the case as they learn about the existence of quarks. Although students do not need to know details about quarks, knowledge of their arrangement and binding inside the proton and neutrons is useful in understanding where mass comes from and how energy is released in a nuclear reaction. It also adds to and modifies their understanding of the atomic model The Nature of Science). The charge of quarks has been mentioned but their colors have been deliberately left out in order to simplify the model. Students can visualize how the strong nuclear force operates by investigating with neodymium magnets (or observing a teacher-led demonstration).

What is radioactive decay?

3. Students has already come across isotopes. In activity 58, they review isotopes and the differences in mass number of different isotopes. They see that the larger an atom's nucleus, the more neutrons are needed to balance the increasing electromagnetic repulsion between protons.

4. Students study the different ways in which radioactive isotopes decay. They also study the effect of having too many or too few neutrons on the type of decay and write a simple decay series for a radioactive isotope, using beta negative decay. Students may like to use the dynamic periodic table https://ptable.com/ to confirm their decay series. A link to the table is included on **BIOZONE's Resource Hub**. It is important students realize that nucleon number is conserved in nuclear reactions.

What is mass-energy equivalence and why is it important?

5. Students compare the energy released in chemical versus nuclear reactions and note that it is far larger in nuclear reactions per gram of material. Students should understand that the energy stored in the bonds between quarks and hadrons (protons and neutrons) has a mass equivalent. Thus when energy is released when these bonds are rearranged, mass is reduced. Following Einstein's equation, a small amount of mass lost produces a huge amount of energy. Students study the neutron induced fission of uranium-235. The reaction produces neutrons that can cause the fission of other nearby uranium-235 atoms. Thus a chain reaction can occur. Students study how this chain reaction can be controlled in nuclear power plants and how the energy produced can be used to power homes and industry.

6. Nuclear explosions demonstrate the immense power contained within the nucleus of an atom. Studying the various events and energy release in nuclear explosions provides students with another way of understanding nuclear processes. A visual model (page 129) helps emphasize the scales involved. It is also important to understand that although nuclear weapons were primarily built as vehicles for mass destruction and warfare, they also served to verify theories of nuclear processes and even changed scientists' understanding of some of these processes (reference the Castle Bravo test that was three times larger than predicted). Students study the difference between fission and fusion and use binding energies to understand why some elements undergo fusion and some are fissionable. Students compare and contrast fission and fusion technologies and explore how these can be harnessed. They can then ask questions about the benefits, drawbacks, and challenges of nuclear technologies

56 Nuclear Power

Key Question: How do you power a city, or destroy it, with just a few kilograms of nuclear material?

- Nuclear power has two sides to its use. One side is for the good of all people. This might be through a nuclear power plant producing electricity or even through radioactive materials being used to destroy cancer cells.
- The other side is for harm. Nuclear weapons are the most destructive devices humans have ever created. Even the smallest of today's nuclear devices are capable of destroying entire cities and contaminating the environment.
- However both nuclear power plants and nuclear weapons use the same principles of either splitting atoms (fission) or sometimes joining atoms (fusion) to produce energy.
- Nuclear reactions produce an enormous amount of energy. As a reference, think of this. The "Little-Boy" atomic bomb dropped on Hiroshima at the end of the Second World War was the equivalent of using 15 thousand tonnes of TNT. However the amount of nuclear material used in the bomb itself was just 64 kilograms.
- As another comparison, a nuclear power plant producing 1000 MW of electricity uses about 27 tonnes of uranium fuel a year. Only about 5% of the fuel is actually usable uranium. A coal fired power plant producing the same amount of electricity uses about 2.5 million tonnes of coal a year. A solar power station producing the same amount of electricity covers about 23 km^2 of land (5680 acres).
- But how are atoms split and where does the energy released come from?
- Why does one process allow us to produce in a controlled way the electricity we need, while the other produces an apparently out-of-control explosion?
- Only certain elements can be used to produce nuclear power, the most common of these are uranium (U) and plutonium (Pu). Why can't other elements be used? What makes U and Pu useful while other elements are not?

United States Department of Energy

The Castle Bravo nuclear test yielded 15.5 megatons. This was nearly three times more than expected due to unexpected nuclear reactions during the explosion. It consisted of two nuclear devices. A fission device was used to detonate a fusion device.

The reactor dome and cooling tower are stereotypical features of nuclear power plants.

1. In your own words define the following:

   (a) Radioactive: The emission of radiation and particles due to the decay (splitting) of an atom.

   (b) Fission: The splitting of an atom.

   (c) Fusion: The combining of two atoms to form one.

2. Use your periodic table to find:

   (a) The relative atomic mass of uranium: 238.0

   (b) The relative atomic mass of plutonium: 244

   (c) How might these numbers explain their use in the nuclear industry? The high atomic mass means many protons and neutrons in the nucleus. This produces instabilities in the nucleus that can result in the nucleus splitting. Splitting the nucleus releases energy. Thus they are used in the nuclear industry to produce energy that can be capture and used to produce electricity.

©2020 BIOZONE International
ISBN: 978-1-927309-82-7
Photocopying Prohibited

57 Inside the Nucleus

Key Question: What is the internal structure of the atomic nucleus?

A new science

- Recall our earlier model of an atom. It describes three types of subatomic particles: protons and neutrons in the nucleus of an atom, surrounded by orbiting electrons.
- However, in the early 1960s, several scientists independently proposed that neutrons and protons (members of a group of subatomic particles called **hadrons**) were not in themselves elementary (indivisible) particles, but were made of smaller particles they called **quarks**. They assigned properties to these particles, including mass, spin, and electrical charge.
- At the time, there was no physical evidence for these elementary particles, and it was not until 1968 that a new technology called deep inelastic scattering (much like Rutherford scattering) provided the first convincing evidence of the existence of quarks.
- Since then, physicists have learned much more about elementary particles through experimentation. So how do you "look" inside an atom? Of course you cannot, but you can find out about it indirectly by looking at what happens when subatomic particles collide.
- The biggest 'piece' of experimental equipment is CERN's **Large Hadron Collider** or LHC (image, top right). The LHC is the world's largest particle accelerator. It lies beneath the France-Switzerland border and consists of a 27 km ring of superconducting magnets (image, bottom right) and particle accelerating structures, all housed within a concrete tunnel.

A small section of the LHC

Maximilien Brice cc 4.0

One of the quadrupole electromagnets, used to direct the particles.

gamsiz cc 2.0

- Inside the accelerator, the strong magnetic field guides two high-energy particle beams in opposite directions at close to the speed of light before they are made to collide. The results (by-products) of these collisions are detected by particle detectors. Analyzing these by-products allows physicists to test the predictions of different theories of particle physics.

Inside the nucleus

- There are two types of elementary particles we need to know about. One type, the **electron**, we know. Recall that it has an electrical charge of -1. The other type is the **quark**. Quarks make up protons and neutrons (nucleons). There are six types ('flavors') of quarks. Two types ('up' and 'down') make up the protons and neutrons in atomic nuclei.
- Each 'up' quark has a charge of $+^{2}/_{3}$. Each 'down' quark has a charge of $-^{1}/_{3}$. The sum of the charges of the quarks that make up a nuclear particle determines its electrical charge.

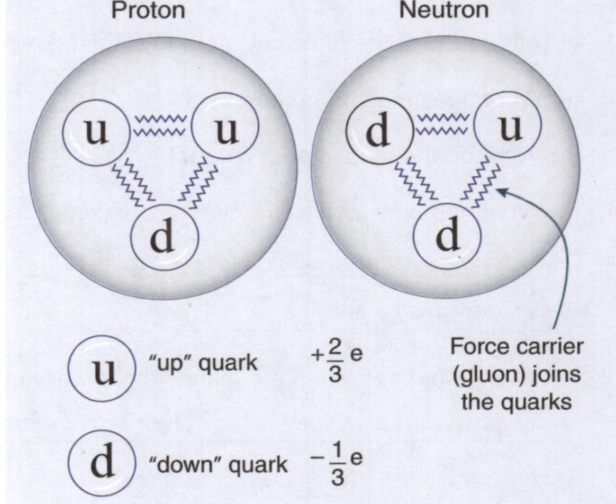

1. Study the diagram of the composition of protons and neutrons:

   (a) What quark 'flavors' make up a proton?

   2 ups and 1 down.

   (b) What quark 'flavors' make up a neutron?

   1 up and 2 downs.

   (c) Use the quark charges to explain why a proton has a charge of +1 and a neutron is neutral (no charge):

   2 x (2/3) + -1/3 = 1.

   2/3 + 2 x (-1/3) = 0.

2. Like charges repel each other. Since the protons in the nucleus are so close together we would expect them to fly apart. But they do not. What does this tell you about the forces present in the nucleus?

   There must be something holding the protons (and neutrons) together with enough force to overcome the repulsive forces between the particles. It likely acts at very short range (because as soon as the protons separate even slightly they repel each other and fly apart).

©2020 BIOZONE International
ISBN: 978-1-927309-82-7
Photocopying Prohibited

The strong nuclear force

- Four fundamental forces govern how the universe and all things in it behave (table right). We will look at gravitational and electromagnetic forces in the following chapters. In this chapter, we examine the strong and weak nuclear forces, which govern nuclear interactions.
- Based on the fact that like charges repel there must be a strong force over-riding this repulsion and holding the positive protons together in the nucleus. This strong nuclear force (or **strong force**) is the strongest force known.

Force	Action	Range	Relative strength
Gravitational	Acts between objects with mass	Infinite	Much weaker
Weak nuclear	Governs particle decay so responsible for radioactivity	Short range –within the diameter of a nucleus	
Electromagnetic	Acts on electrically charged particles	Infinite	
Strong nuclear	Binds quarks together	Short range – within 0.1% of a proton's diameter	Much stronger

- The strong force holds quarks together to form protons and neutrons and counteracts the tendency of the positively-charged protons to repel each other.

INVESTIGATION 4.1: Modeling the strong nuclear force

See appendix for equipment list.

This may be best done as a demonstration by your teacher. Neodymium magnets can be brittle and break with repeated impacts. The activity will work with other magnets as well, but the most dramatic effect is achieved with neodymium magnets, which are the strongest type of readily available permanent magnet.

CAUTION: Hold the magnets firmly as they can cause a painful injury if they pull together suddenly and "pinch" skin.

1. You will need an old click pen and two small neodymium magnets with central holes.
2. Take the old pen apart and remove the thin plastic tube that holds the ink and the spring from the pen case.
3. Place one neodymium magnet on to the ink tube.
4. Place the spring on after the magnet.
5. Place the second magnet after the spring so that it is attracting the first magnet.
6. You now have two magnets separated by the spring.
7. Slowly push the two magnets towards each other and observe the result.

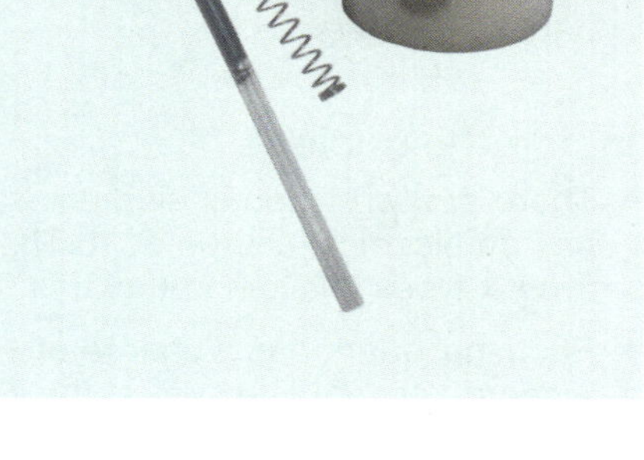

3. In your model of the forces inside a nucleus (above):

(a) What does the spring represent? *Electrostatic repulsive force.*

(b) What do the magnets represent? *Strong nuclear force.*

4. (a) What happens as the two magnets become closer? Can the spring keep the magnets apart?

Once the magnets are close enough together, the attractive force is too strong for the spring to keep the magnets apart.

(b) What happened when the magnets got really close together?

The magnets snapped together very quickly.

5. Describe how this model demonstrates the forces inside the atomic nucleus, even though it is a simplification:

The electrostatic repulsive forces push like charged particles away. When enough energy is added to push the protons close enough together, the strong nuclear force takes over and is strong enough to overcome the electrostatic repulsive forces.

©2020 BIOZONE International
ISBN: 978-1-927309-82-7
Photocopying Prohibited

58 Isotopes

Key Question: Why are some isotopes more common than others?

- You have already seen in Chapter 2 that atoms can have different atomic masses based on the number of neutrons in the nucleus. Recall that these are called isotopes.
- As we saw in Chapter 2, isotopes are important when determining the relative atomic mass of an atom. They are just as important when dealing with nuclear processes.
- It should be noted that some isotopes are not stable. This is why only some isotopes are common in nature. For example you have already come across the isotopes of iron (Fe) in Chapter 3. Recall that most Fe atoms have 30 neutrons, but some have 28, 31 and 32 neutrons. There are others with less or more neutrons but these are so unstable they usually fly apart in less than a second if they are ever produced.
- This instability becomes more pronounced as the atomic nucleus gets larger and as the imbalance between protons and neutrons become larger. It is the ratio of protons to neutrons that is critical to the stability of the nucleus.

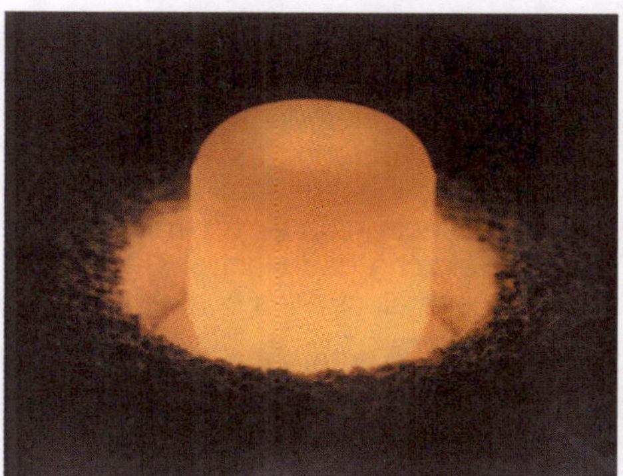

All isotopes of plutonium are unstable. As they decay (break apart) they give off heat making a sample warm to the touch. This heat has been used to power numerous space probes (e.g. Voyagers 1 and 2).

1. Find the element uranium on the periodic table.
   (a) What is the atomic number of uranium? 92
   (b) How many protons in its nucleus? 92
2. Study the graph of neutrons vs protons in an atom's nucleus right.
   (a) What happens to the ratio of protons to neutrons as the number of protons increases?
   It decreases (less protons to neutrons)
   (b) What is the maximum number of protons a stable atom can have (i.e. there is at least one non-radioactive isotope)?
   82
   (c) What is the name of this atom? Lead

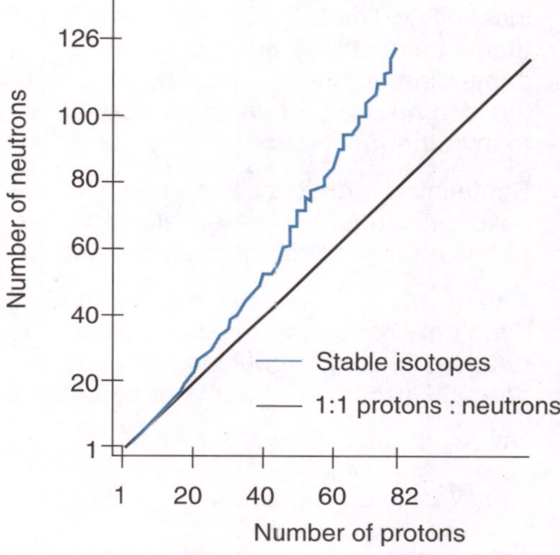

3. Go to **BIOZONE's Resource Hub** and access the **Dynamic Periodic Table** tagged to this activity.
   (a) Click on the "Isotopes" tab. Click on "All". How many isotopes (versions) of uranium are there? 27
   (b) Complete the table for the six isotopes of uranium below:

Element and isotope	Atomic number	Mass number	Number of protons	Number of neutrons	Nuclide notation
Uranium-232	92	232	92	140	$^{232}_{92}U$
Uranium-233	92	233	92	141	$^{233}_{92}U$
Uranium-234	92	234	92	142	$^{234}_{92}U$
Uranium-235	92	235	92	143	$^{235}_{92}U$
Uranium-236	92	236	92	144	$^{236}_{92}U$
Uranium-238	92	238	92	146	$^{238}_{92}U$

   (c) What do you notice about the uranium isotopes?
   There are significantly more neutrons than protons in the nucleus.

©2020 **BIOZONE** International
ISBN: 978-1-927309-82-7
Photocopying Prohibited

59 Radioactive Decay

Key Question: How do unstable atoms decay and release energy?

Radioactivity that keeps you safe

- A common type of smoke detector is the ionization smoke detector. A radioactive element (typically americium – Am) bombards air particles entering the smoke detector with **alpha particles** (essentially a helium nucleus, i.e. two protons and two neutrons with no electrons) turning them into positively charged ions and negatively charged electrons. These electrons and ions are attracted to opposite electrodes completing a circuit between the electrodes.
- When smoke enters the smoke detector, the smoke particles attach themselves to the ions and stop the electric current, blocking the circuit. The alarm sounds when the circuit is broken.

Radioactive atoms and decay

- The strong nuclear force between all nucleons stabilizes a nucleus by offsetting the electrostatic repulsion between the protons. However, if there are too many neutrons, or too few neutrons for a given number of protons, then the nucleus will be unstable. The stability of a nucleus requires the neutron to proton ratio to be about 1:1 for smaller nuclei up to 1:1.5 for the larger nuclei.
- In unstable elements, the protons and neutrons do not stay together indefinitely. The nucleus of the atoms in these elements decay into different atoms by emitting particles. Radioactive decay occurs at varying rates depending on the isotope. But each isotope decays at a specific rate and this can be used for various applications including dating rocks, fossils, and many other mineral-based objects.
- Sometimes a radioactive element will spilt to form an **alpha particle** (a helium nucleus) and a new element. At other times, it will produce a **beta particle** (an electron or positron) and form a new element or isotope.
- A third way a radioactive isotope can decay is by **gamma radiation**. This does not change the mass or the atomic number of the isotope but it removes energy, producing a slightly more stable atom. Gamma decay normally occurs after alpha or beta decay.

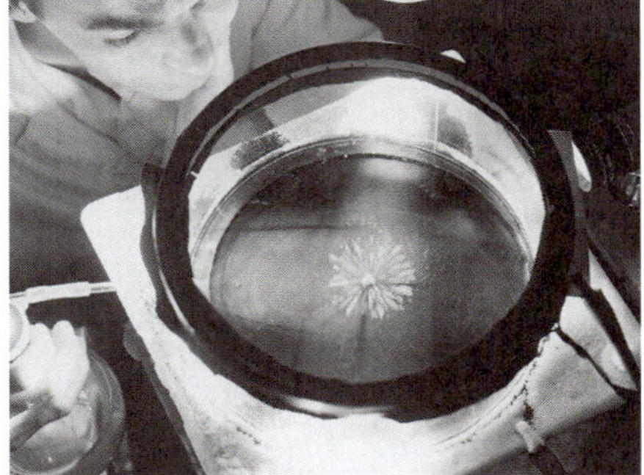

NASA

Traces left by alpha particles in a cloud chamber

1. (a) What is an alpha particle? *A helium nucleus (2 protons and 2 neutrons only).*

   (b) What is a beta particle? *An electron (or positron) emitted from the nucleus.*

2. Study the table below of radioactive elements below the atomic number of lead:

Element isotope	Protons	Neutrons	Type of radioactive decay	Ratio protons : neutrons
Oxygen	8	11	beta (electron emission)	*1 : 1.38*
Oxygen	8	6	beta (positron emission)	*1 : 0.75*
Copper	29	38	beta (electron emission)	*1 : 1.31*
Copper	29	30	beta (positron emission)	*1 : 1.03*
Cadmium	48	79	beta (electron emission)	*1 : 1.65*
Cadmium	48	55	beta (positron emission)	*1 : 1.14*
Gold	79	124	beta (electron emission)	*1 : 1.57*
Gold	79	99	beta (positron emission)	*1 : 1.25*

(a) Complete the table by calculating the ratio of protons to neutrons:

(b) Now calculate the ratio of protons to neutrons for the stable isotopes of the elements above (use the mass number to the nearest whole number from you periodic table to estimate the number of neutrons).

Oxygen 1:1, Copper 1:1.20, Cadmium 1:1.33, Gold 1:1.49

(c) What does this tell you about the beta decay an isotope will undergo? *Any isotope with a proton : neutron ratio lower than the ratio of the stable atom will emit a positron (positive beta decay). Any isotope with a proton : neutron ratio greater than the ratio of the stable atom will emit an electron (negative beta decay).*

PS1.C EM

©2020 **BIOZONE** International
ISBN: 978-1-927309-82-7
Photocopying Prohibited

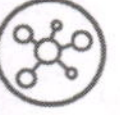

Types of radioactive decay

- Six types of radioactive decay are pictured below:

Decay type	Nuclear equation	Model	Change in atomic/mass number
Alpha decay	$^{A}_{Z}\mathbf{X} \rightarrow {}^{4}_{2}\mathbf{He} + {}^{A-4}_{Z-2}\mathbf{Y}$		A: Decrease by 4 Z: Decrease by 2
Beta decay	$^{A}_{Z}\mathbf{X} \rightarrow {}^{0}_{-1}\mathbf{e} + {}^{A}_{Z+1}\mathbf{Y}$		A: Unchanged Z: Increase by 1
Gamma decay	$^{A}_{Z}\mathbf{X} \rightarrow {}^{0}_{0}\gamma + {}^{A}_{Z}\mathbf{Y}$	γ	A: Unchanged Z: Unchanged
Positron emission (beta positive decay)	$^{A}_{Z}\mathbf{X} \rightarrow {}^{0}_{+1}\mathbf{e} + {}^{A}_{Z-1}\mathbf{Y}$		A: Unchanged Z: Decrease by 1
Electron capture	$^{A}_{Z}\mathbf{X} + {}^{0}_{-1}\mathbf{e} \rightarrow {}^{A}_{Z-1}\mathbf{Y}$		A: Unchanged Z: Decrease by 1
Spontaneous fission	$^{A^1+A^2+A^3}_{Z^1+Z^2}\mathbf{X} \rightarrow {}^{A^1}_{Z^1}\mathbf{Y} + {}^{A^2}_{Z^2}\mathbf{W} + A^3\,{}^{1}_{0}\mathbf{n}$		A (total): Unchanged Z (total): Unchanged

- **Alpha decay** is the only type of radioactive decay that results in an appreciable change in an atom's atomic mass. Two protons and two neutrons are ejected from an atom as a helium nucleus (an alpha particle).
- **Beta decay** includes the emission of an electron (**negative beta decay**) or a positron (**positive beta decay**).
  - When an electron is emitted, the mass number (A) remains the same. A neutron is converted into a proton, raising the atomic number (Z) by one.
  - When a positron is emitted, the mass number remains the same. A proton is converted to a neutron, lowering the atomic number by one.
  - Electron capture has the same effect on the number of protons and neutrons in a nucleus as positron emission.
- In **gamma** decay, an excited nucleus emits gamma rays, but its proton (Z) and neutron count (A–Z) stay the same.
- Most commonly, decay occurs as alpha decay, beta decay via emission of an electron, and gamma decay.

3. (a) Uranium-238 decays by alpha decay. What is the name, atomic number, and mass number of the element it decays to?

   Thorium. Atomic number = 90. Mass number = 234

   (b) Bismuth-212 decays via beta decay (emission of a electron). What is the name, atomic mass, and mass number of the element it decays to?

   Polonium. Atomic number = 84. Mass number = 212

4. Using information on this page write down the possible decay series of the radioactive sodium atom. What is the first stable atom the series will arrive at? You may need to use your periodic table to help you.

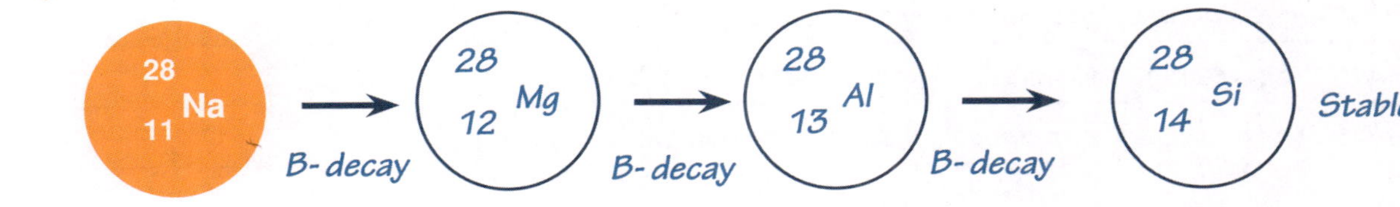

©2020 **BIOZONE** International
ISBN: 978-1-927309-82-7
Photocopying Prohibited

Conservation of nucleons in a nuclear reaction

- In chemistry and physics, we have numerous conservation laws. For chemical reactions, we have:
  - The law of conservation of mass, which states that the mass of the products in a chemical reaction must equal the mass of the reactants.
  - The law of conservation of energy, which states that energy can only transform from one form to another (i.e. chemical → thermal → mechanical → electrical → and so on).
- With nuclear reactions, we now have a new law of conservation: **The law of conservation of nucleon number**. This law states that the total number of nucleons (protons and neutrons) does not change in a nuclear reaction.

5. How is the law of conservation of nucleon number similar to the law of conservation of mass? *Just like the masses of the products equals the masses of the reactants in the conservation of mass, the number of nucleons after a nuclear reaction equals the number of nucleons present before the reaction.*

6. Explain why, during beta decay, the atomic number can change but the nucleon number remains the same: *During beta decay, either a proton changes to a neutron or a neutron changes to a proton. In both instances the number of nucleons remains the same. Mass number is the number of nucleons. Since the number of nucleons does not change, the mass number does not change.*

A brief history of our understanding of radioactivity

- Henri Becquerel discovered radioactivity in 1896, when experimenting with uranium salts. The SI unit for radioactivity, the becquerel (Bq), is named after him.
- The term radioactivity was coined by Becquerel's doctoral student, Marie Curie. She was the first woman to win a Nobel prize (in physics), the only woman to win it twice, and the only person to win it in two different categories (physics and chemistry). Marie Curie died in 1934 of radiation poisoning. The dangers of radiation were not known at that time. The papers and workbooks Curie used are still stored in lead shielded boxes today due to their high levels of radioactivity.
- Ernest Rutherford carried out many experiments in radioactivity. From 1898 until his death in 1937 he made so many discoveries he is often called the "father of nuclear physics".
- The discovery of radioactivity led to attempts to find uses for it. Eventually this led to building nuclear reactors to provide cheap electricity. It also led to the development of nuclear weapons.

Henri Becquerel

Marie Curie

Ernest Rutherford

7. Form a group of three to research Henri Becquerel, Marie Curie, and Ernest Rutherford (choose one scientist each). What were the experiments they did to prove and explain radioactivity? Summarize what you have found out for your choice below. As a group, make a short presentation to the class to summarize the research of all three.

Henri Becquerel experimented with uranium to determine if there was any connection between x-rays and naturally occurring phosphorescence. He proposed that uranium absorbed the Sun's energy and emitted it as x-rays. The experiment involved exposing potassium uranyl sulfate (a sulfate of uranium) to sunlight and placing it on photographic plates wrapped in opaque paper. The plate turned "foggy" when the uranium was exposed to the Sun. However, his idea that the uranium was absorbing energy from the Sun was disproved when the same result was achieved even in the dark. This lead to the discovery that uranium emitted radiation without an external source of energy (such as the Sun). Becquerel also showed that the rays emitted by uranium caused gases to ionize and that they could be deflected by electric or magnetic fields (unlike x-rays).

For other responses see the long answer section.

©2020 **BIOZONE** International
ISBN: 978-1-927309-82-7
Photocopying Prohibited

60 Fission and Chain Reactions

Key Question: How does nuclear fission release energy, and how can we use this to produce electricity?

Where does the energy come from in nuclear reactions?

Chemical vs nuclear explosions

- Explosions release energy very quickly. In traditional chemical explosions (e.g. a stick of dynamite, C4, TNT) the energy released is associated with the electrostatic forces between atoms.
- Chemical bonds between atoms are being broken to make more stable, lower-energy bonds. The excess energy that was stored (potential energy) in the chemical bond is released as kinetic energy, light, heat, etc. These kinds of explosions are, of course, exothermic reactions (chemical reactions that release heat).
- In nuclear fission reactions, the energy comes from breaking the **bonds** between nucleons in the nucleus created by the **strong nuclear force**. Without the strong nuclear force holding them together, the electromagnetic repulsive forces between the protons causes them to separate very quickly. The binding energy (energy needed to hold the nucleons together) is released.

A chemical based (conventional) explosion

Nuclear energy and Einstein's equation

- You may have seen Einstein's famous equation **$E = mc^2$**: Energy equals mass times the speed of light squared. But what does that actually mean?
- This equation combines **mass** and **energy** in one conservation law.
- We know that, under normal circumstances, energy cannot be created nor destroyed, only transformed from one form to another. Likewise, when we perform chemical reactions, the mass of the products is equal to the mass of the reactants in the reaction.
- For nuclear reactions, and within the nucleus, things are quite different.
- When small nuclei are forced together (fusion) the total mass is reduced slightly and the "lost" mass appears as an equivalent release of energy. When a large nucleus splits up (fission) the total mass is also reduced and the "lost" mass again appears as an equivalent release of energy. Einstein's equation predicts exactly this result.
- In general, the quantity of energy released from a nuclear reaction compared with the energy released by a chemical reaction (using suitable samples of similar size) is of the order of a million times greater. Why? It is because the presence of the strong nuclear force in nuclei allows enormous amounts of energy to be stored and/or released. The electrostatic forces holding electrons in an atom and holding atoms in molecules are nowhere near as powerful and consequently store and/or release much less energy.
- One feature of nuclei is that the quarks within them do not contribute significantly to the mass of a proton or neutron. Most of the mass is the mass-equivalent of energy stored by the strong nuclear force holding the quarks together and the nucleons themselves together (right).

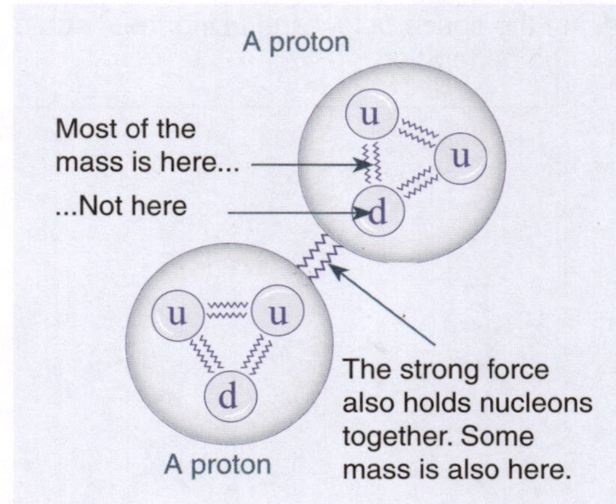

1. Recall that, at extremely short distances, the strong nuclear force is much stronger than the electromagnetic force. How do we know the strong nuclear force is stronger than the electromagnetic force?

   EM forces are stronger the less distance between the charges. As close as the protons are in the nucleus, the repulsive force is incredibly strong. The strong nuclear force is able to keep multiple protons together in a single nucleus in spite of EM forces. This shows it is much stronger.

2. Why is more energy released when breaking the bonds formed by the strong nuclear force than the chemical bonds formed from electrostatic forces?

   Because strong nuclear force is stronger, there is more energy stored in the bonds holding the nucleons together to be released, than in the chemical bonds holding atoms together.

©2020 **BIOZONE** International
ISBN: 978-1-927309-82-7
Photocopying Prohibited

SC EM PS1.C

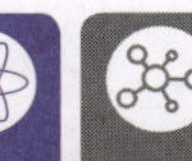

Nuclear fission

- Nuclear fission is a type of radioactive decay and can be spontaneous or induced (as shown below). In this reaction, instead of simply ejecting a small alpha particle or electron, the nucleus of a large atom splits in two, forming two similar sized nuclei.
- This usually occurs as a result of a neutron colliding with a parent nucleus in the controlled conditions of a nuclear reactor or the somewhat less controlled conditions of a nuclear weapon.
- Consider the following reaction:

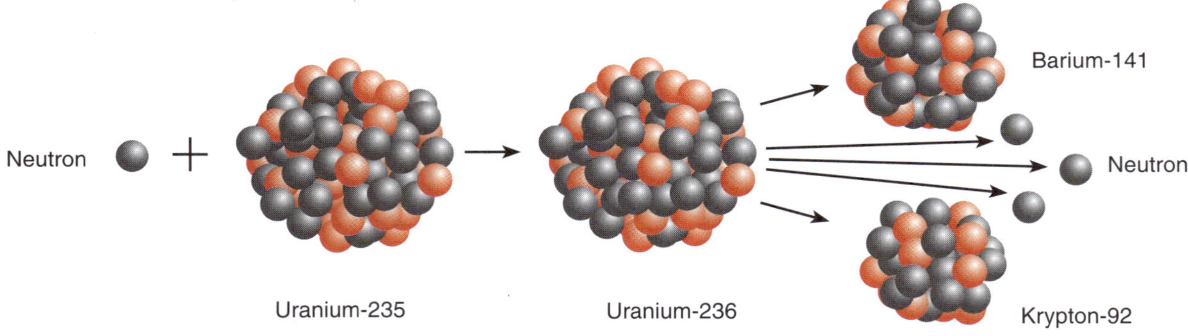

The start of a nuclear chain reaction

- If ^{235}U is bombarded with a neutron, the resulting ^{236}U produced is unstable and undergoes **fission**. The resulting elements ^{92}Kr and ^{141}Ba do not contain as many nucleons as ^{236}U.
- The remaining three neutrons are released as high energy particles, able to bombard other nearby ^{235}U atoms and produce a chain reaction.
- $1n + {}^{235}U \rightarrow {}^{236}U \rightarrow {}^{141}Ba + {}^{92}Kr + 3n$
- In each step, the total nucleon number is a constant value of 236. This is the same in all fission reactions.

3. It is worth pausing now to review what you understand about mass, energy, and reactions. You have seen in the previous chapters(*Structure and Properties of Matter* and *Chemical Reactions*) that in a chemical reaction the mass of the reactants must equal the mass of the products. This is not the case in a nuclear reaction. Given that Einstein's equation tells us mass is just highly concentrated energy write a law about mass and energy in a nuclear reaction:

   In a nuclear reaction the total energy (mass and energy because mass is concentrated energy) will be the same before and after the reaction.

4. In the space below and using the diagram above, draw a diagram to show how a nuclear chain reaction occurs. Label the generations:

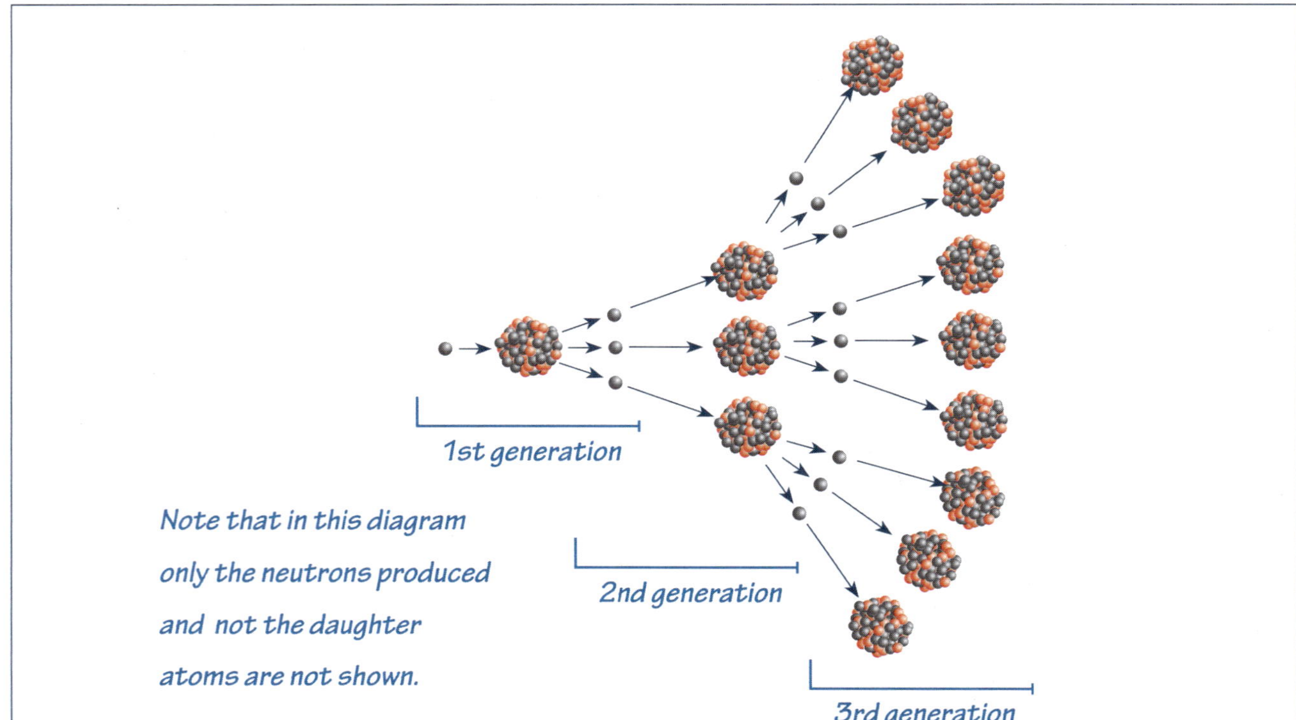

©2020 **BIOZONE** International
ISBN: 978-1-927309-82-7
Photocopying Prohibited

Controlling the chain reaction

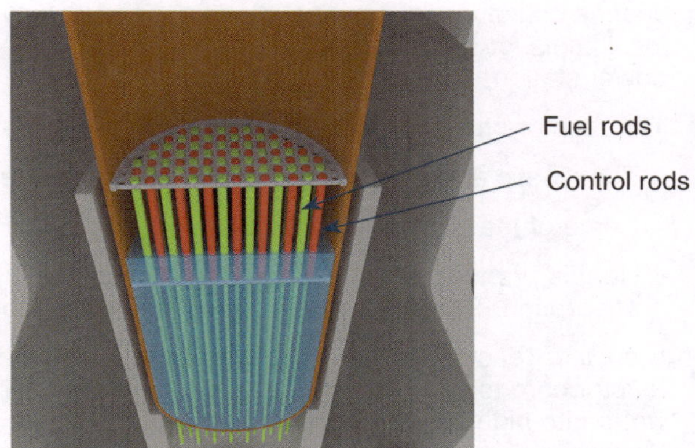

- Recall that when ^{236}U splits it releases neutrons. These can be absorbed by nearby ^{235}U atoms which then split and so on in a chain reaction. A nuclear weapon produces an uncontrolled chain reaction. To maintain a sustained but controlled nuclear reaction, for every 2 or 3 neutrons released, only one must be allowed to strike another uranium nucleus.
- Most nuclear reactors are controlled by using neutron poisons, which absorb neutrons and so reduce the rate at which the chain reaction can proceed. They also use neutron moderators (such as heavy water) which slow down neutrons so that they are more easily "captured" by the nuclei and allow the chain reaction to proceed more rapidly. Nuclear reactors generally have automatic and manual systems to shut the fission reaction down if conditions are unsafe.

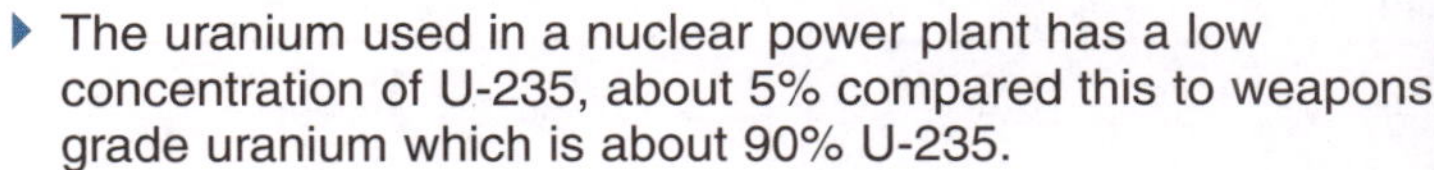

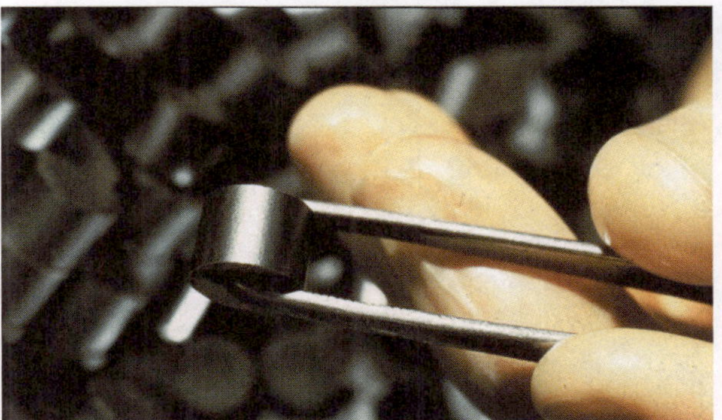

- The uranium used in a nuclear power plant has a low concentration of U-235, about 5% compared this to weapons grade uranium which is about 90% U-235.
- The power output of the reactor is adjusted by controlling how many neutrons are able to produce more fission reactions. Control rods that are made of a neutron poison are used to absorb neutrons. Absorbing more neutrons in a control rod means that there are fewer neutrons available to cause fission, so pushing the control rod deeper into the reactor will reduce the reactor's power output and extracting the control rod will increase it.
- A nuclear power plant produces electricity by using the heat from the reactor (released during the fission reactions) to boil water to produce steam to drive a turbine.

5. A nuclear chain reaction can get out of control very quickly. The nuclear chain reactions produced by the nuclear bombs used at the end of World War II both took about 0.8 microseconds. This was enough time for just 80 generations of reactions. After this time the nuclear material was no longer dense enough (close enough together) to sustain the nuclear chain reaction.

   The chain reaction you drew follows a simple exponential growth rate, tripling at each step. In reality the chain reaction will not follow a perfect exponential growth rate. This can be accounted for by instead doubling the reaction instead of tripling it. This allows for neutrons that do not participate in the chain reaction. The equation can be written as $\mathbf{y = C \times 2^t}$ where y is the number of atoms, C is the initial number of atoms (in this case 1) and *t* the number of generations.

   (a) Use the equation above to calculate the number of atoms in 80 generations of the nuclear chain reaction you drew:

   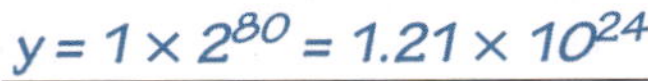

   $y = 1 \times 2^{80} = 1.21 \times 10^{24}$

   (b) Use a spreadsheet to calculate the entire number of atoms that under went a fission reaction in a fission bomb:

   2.42×10^{24} (simply summing all reactions)

   (c) What mass (in grams) of Uranium-235 undergoes fission in the 0.8 microseconds of the reaction?

   $2.42 \times 10^{24} \times 235$ g/mol = ~945 grams

6. What would happen to the nuclear chain reaction if, for every 2 or 3 neutrons released:

   (a) Less than 1 was allowed to strike another uranium nucleus? *The reaction would slow down.*

   (b) More than 1 was allowed to strike another uranium nucleus? *The reaction would increase.*

7. (a) How do the control rods regulate the nuclear chain reaction?

   The control rods absorb neutrons so fewer neutrons are available to continue the nuclear chain reaction.

   (b) Describe the position of the reactor control rods when the reactor is effectively shut down:

   The reactor control rods would be completely inserted into the reactor core.

   (c) Describe the position of the reactor control rods when the reactor is at full power:

   The reactor control rods would be completely retracted from the reactor core.

©2020 **BIOZONE** International
ISBN: 978-1-927309-82-7
Photocopying Prohibited

Using nuclear fission

- Just as conventional power stations generate electricity by harnessing the thermal energy released from burning fossil fuels, the thermal energy released from nuclear fission can be used to produce electricity in a nuclear power station.
- The reactor core generates heat in a number of ways.
  - The kinetic energy of fission products is converted to thermal energy when the nuclei collide with nearby atoms.
  - Some of the gamma rays produced during fission are absorbed by the reactor. Their energy is converted to heat.
  - Heat is produced by the radioactive decay of fission products and materials that have been activated by neutron absorption. This decay heat source will remain for some time even after the reactor is shut down.
- A nuclear reactor coolant – usually water, but sometimes a gas, liquid metal, or molten salt – is circulated past the reactor core to absorb the heat that it generates. The heat is carried away from the reactor and is then used to turn water into high pressure steam (below).

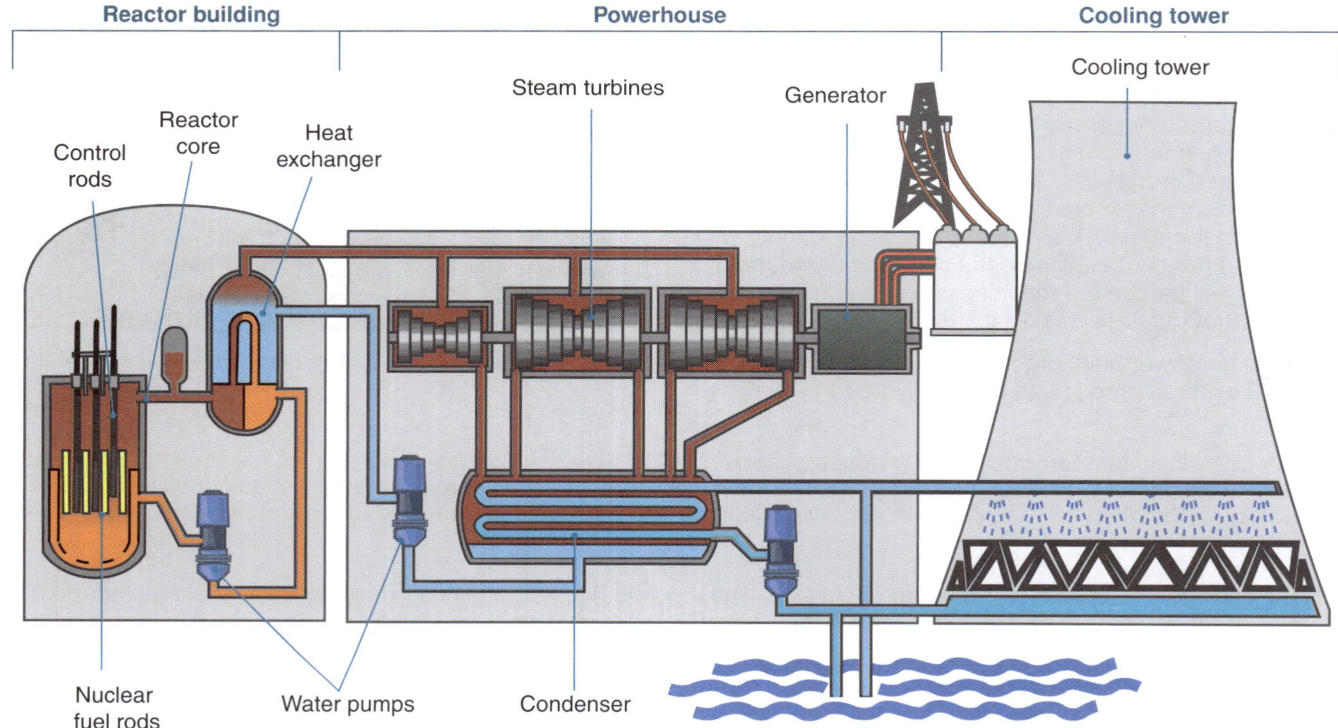

8. (a) How is the heat used in a nuclear power plant generated? The heat comes from the fission reactions of the uranium -235 fuel. This can be from the kinetic energy of ejected neutrons being turned into thermal energy during collisions with other atoms, or energy absorbed from gamma rays.

   (b) Describe how this heat is used to produce electricity: The nuclear reaction releases heat which is used to heat water into steam (either by a heat exchanger or by water flowing directly around the core). This steam is used to turn turbines connected to generators to produce electricity. This is similar to the way combustion power plants produce electricity.

9. Nuclear power stations have a risk of the fuel melting down, but never a nuclear explosion. Using information in this activity and your knowledge of nuclear processes explain why nuclear fuel might melt under extreme circumstances, but not explode like a nuclear bomb:

   Nuclear fuel pellets contain only about 5% of fissile uranium. This is not dense enough for enough neutrons to be able to strike other nuclei for a chain reaction to escalate into a nuclear explosion. The chain reaction will however produce enough heat to melt the fuel (hence meltdown). This heat can cause steam explosions or the melted metal can react with other substances which could explode.

©2020 **BIOZONE** International
ISBN: 978-1-927309-82-7
Photocopying Prohibited

61 Fission vs Fusion

Key Question: How does fission compare to fusion? How does the energy produced by each compare?

- Having seen where the energy in a nucleus is stored, and how the nuclear chain reaction can occur, it is now easier to see how a nuclear explosion produces such massive effects. The splitting of large atoms releases the binding energy of the strong nuclear force. Doing this on a massive scale results in the largest explosions humanity have created.
- When describing the power of a nuclear explosion, it is typical to hear it described as equivalent to a certain amount of TNT (trinitrotoluene).
- The photo of the conventional explosion on the page 125 is the equivalent of 16 tonnes of TNT. The photo of the nuclear explosion (right) is the equivalent of 23,000 tonnes (23 kilotonnes) of TNT.
- Modern nuclear weapons can have extraordinarily high explosive yields. The most powerful nuclear weapon ever tested was the Soviet Tsar Bomba, a hydrogen bomb that used a small fission trigger to detonate the larger fusion device. It had a yield of 50 megatons of TNT (scaled down from the original design of 100 megatons). The explosion was so large the fireball was more than 8 km across and the mushroom cloud formed reached at least 65 km high. Everything within a radius of 55 km was flattened. The blast wave broke window panes up to 900 km away from the test site.

National Nuclear Security Administration

- The diagram below indicates the immense power of these kinds of explosions:

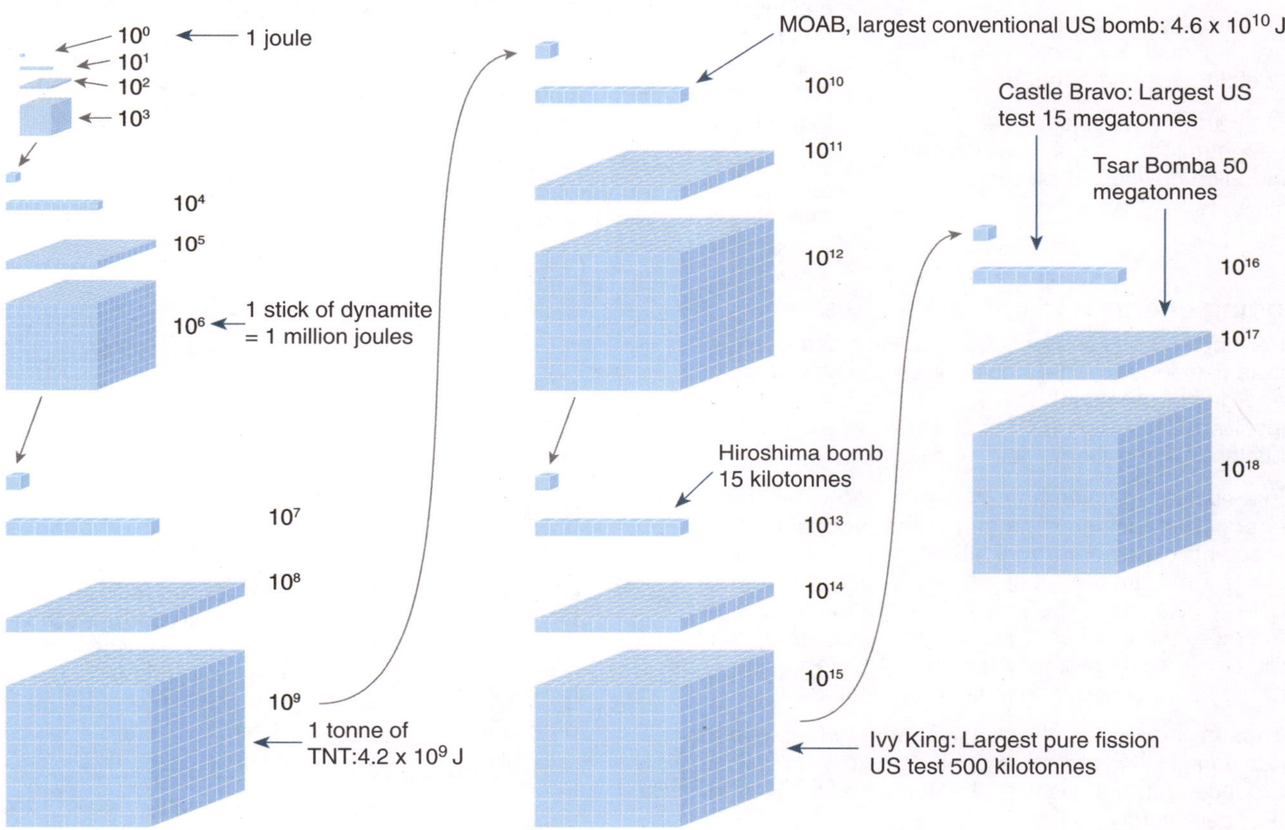

1. (a) TNT and dynamite are common but quite different explosives. A stick of dynamite weighs about 200 grams and produces about 1 megajoule (1 MJ or 1,000,000 J) of energy when detonated. One tonne of TNT produces 4.2×10^9 joules. How many sticks of dynamite would need to be detonated to produce the equivalent explosion of the 23 kilotonne explosion shown?

   $23{,}000 \times 4.2 \times 10^9 = 9.66 \times 10^{13}$ joules

   9.66×10^{13} = 96,600,000 sticks of dynamite

   (b) What mass of dynamite would this be? *96,600,000 x 0.2 kg = 19,320,000 kg = 19.32 kilotonnes*

©2020 BIOZONE International
ISBN: 978-1-927309-82-7
Photocopying Prohibited

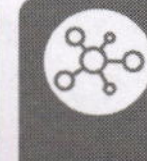

2. The Tsar Bomba released ~ 2.1×10^{17} J of energy. How many times more energy was released by the Tsar Bomba than:

(a) 1 tonne of TNT: *2.1×10^{17} J ÷ 4.2×10^{9} J = 50,000,000 times more energy.*

(b) MOAB: *2.1×10^{17} J ÷ 4.6×10^{10} J = 4,565,000 times more energy.*

(c) Castle Bravo: *50 megatonnes ÷ 15 megatonnes = 3.3 times more energy.*

Nuclear fusion

- The simplest nuclear fusion reaction takes two hydrogen nuclei and fuses them to form helium. The specific way this is done depends on where the reaction occurs.

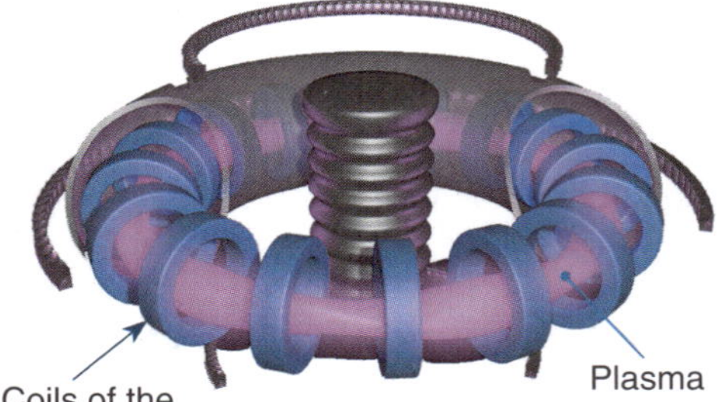

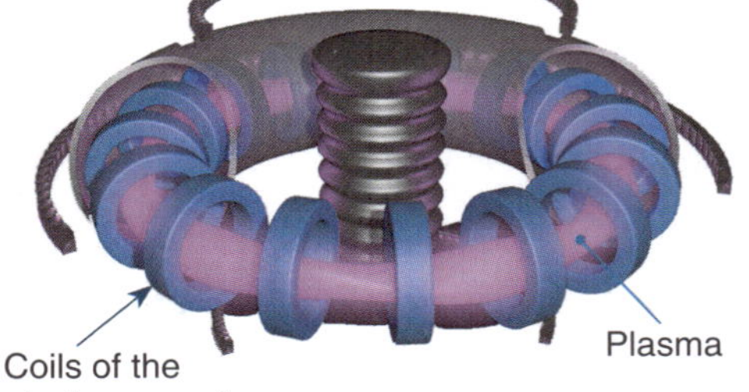

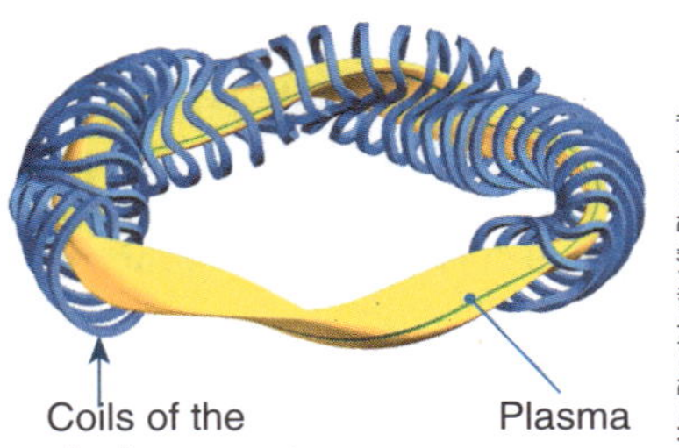

- In stars, fusion occurs naturally by the proton-proton chain reaction. This ends in the reaction: $^{3}He + {}^{3}He \rightarrow {}^{4}He + 2n$.
- Other fusion reactions also occur in stars, such as fusing helium nuclei to form carbon.
- On Earth the simplest (and uncontrolled) form of fusion occurs in fusion devices. The reaction is: $^{2}H + {}^{3}H \rightarrow {}^{4}He + 1n$.
- Fusion reactions produce no radioactive waste and as a result there is much study into developing fusion reactors to produce electricity. The major problems that have yet to be overcome are the enormous pressures and temperatures needed to force hydrogen atoms together.
- Currently two main designs for fusion reactors are being tested. The simplest is called a tokamak, which has a torus shaped reaction chamber (above center). A more complex design (but theoretically more likely to work) is called a stellarator (above). Instead of simple ring shaped electromagnets, it has individually designed twisted electromagnets. The superconducting electromagnets are usually cooled by liquid helium and produce a magnetic field that confines the superheated plasma.

Binding energy and energy release

- Per nucleon fusion reactions produce more energy than fission reactions. However, fission produces more energy per reaction because of the many more nucleons involved. The graph right shows the binding energy per nucleon for the elements.
- In a nuclear reaction energy is released when the binding energy per nucleon is greater in the products than the reactants. Consider a rock above the ground. If the rock starts 1 m from the ground it has a certain potential energy. The closer the rock is brought to the ground the more potential energy must be released. Similarly the closer or more tightly bound the nucleons the more energy is released in their coming together.
- Thus fusing ^{2}H and ^{3}H produces energy because the nucleons in ^{4}He are more tightly bound. And splitting ^{236}U produces energy because the nucleons in ^{92}Kr and ^{141}Ba are more tightly bound than in ^{236}U.

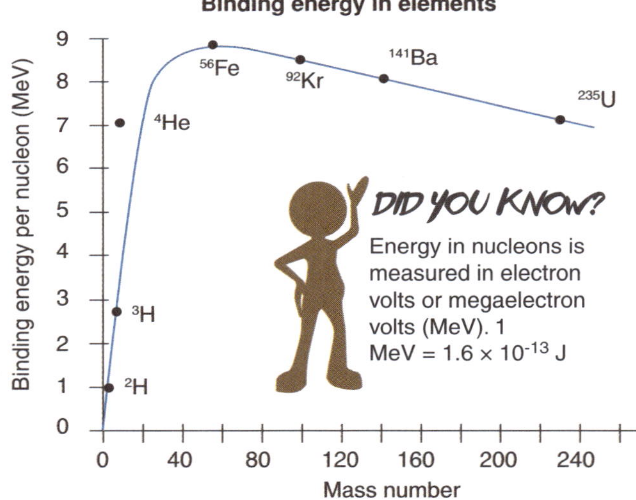

3. Explain why elements below Fe will not undergo fission, while elements above Fe will not fuse:

Elements below Fe on the PT will not fission it would take more energy to pull the nucleons apart than would be produce from actually doing it. Thus they will not spontaneously break apart. Elements above Fe become too unstable when fused together, and they fly apart. More energy is needed to force them together than is available (unless in an exploding star).

4. What is special about the Fe nuclei in terms of nuclear reactions? *Fe is the point that divides elements than can naturally fission from those that can naturally fuse. In Fe the nucleons are bound as tightly together as they can be. Thus no more energy can be released by binding them tighter. Fusing or fissioning Fe nuclei actually requires energy to pull the nucleons further apart (similarly lifting a rock requires energy).*

5. If the binding energy per nucleon for ^2H is 1.1 MeV (megaelectron volts), ^3H is 2.8 MeV, and ^4He is 7.07, how much energy (in MeV) is released in the reaction: $^2H + {}^3H \rightarrow {}^4He + 1n$?

$(2 \times 1.1) + (3 \times 2.8) = 10.6$ MeV. $4 \times 7.07 = 28.28$ MeV energy released $= 28.28 - 10.6 = 17.68$ MeV

6. Why are fission reactors much simpler to build than fusion reactors? *Fission can occur spontaneously or it can be easily induced in large nuclei by the addition of a neutron from another radioactive source. Fusion requires nuclei to be smashed together at very high temperature and pressure, which is difficult to produce and control.*

Thermonuclear weapons

- The Tsar Bomba was the biggest nuclear device ever built. However theoretically nuclear weapons can be many times larger.
- In a simple thermonuclear weapon a sphere of fissile material is surrounded by a sphere of conventional explosives. Below this is a cylinder of material that can undergo fusion with a core of fissile material. Both devices are surrounded by material that helps reflect x-rays and γ-rays.
- The conventional explosives compress the fissile material allowing a chain reaction to occur. X-rays and γ -rays released by the reaction are reflected towards the lower cylinder and compress it. The fission material within is compressed and undergoes a fission reaction. The neutrons released by this strike the fusion material resulting in a fusion reaction.
- Extra neutrons produced by the fusion reaction increase the efficiency of the initial fission reaction. The combined reactions thus produce a much bigger explosion. The reactions take just 0.6 microseconds.

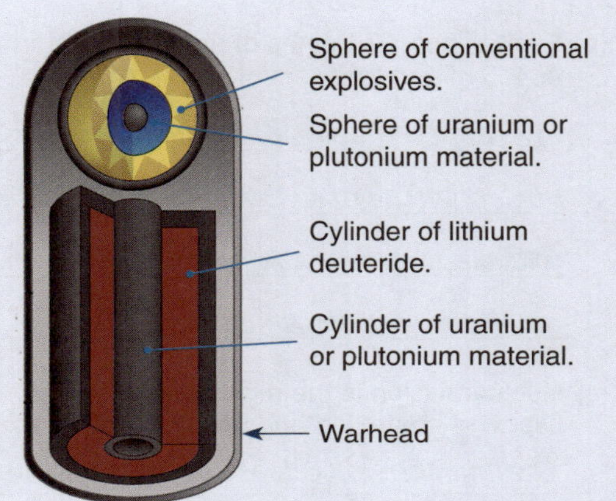

- Fusion weapons are often called hydrogen bombs, but they do not carry hydrogen in them. Instead the fusion device is made of lithium deuteride ($^6Li^2H$). Undergoes two reactions:
- $^6Li + n \rightarrow {}^3He + {}^3H$ then $^2H + {}^3H \rightarrow {}^4He + n$

7. Why does the reaction in thermonuclear weapon only last about 0.8 microseconds? *The reaction only lasts this long because by this time so much heat has been generated the material literally blows apart. This means the nuclear material is not close enough to continue the chain reaction.*

8. Where does the energy come from in these atomic explosions? *Energy comes from the loss of mass in the products of the reaction. This is linked to the strong nuclear force and the binding energy of nucleons (the energy-mass equivalence). The energy in the explosion comes from the release of this binding energy (and thus a reduction of mass).*

9. In the Castle Bravo atomic test (1954) the blast yield was three times greater than expected. Lithium-7 made up 60% of the lithium deuteride fuel. This was though to be inert but when absorbing a neutron split to form $^3H + {}^4He + n$. Why would this lead to a bigger explosion than expected?

The bigger than expected explosion comes from the fact that the ^3H (tritium) produced from the split of lithium 7 can fuse with the ^2H (deuterium). Thus the lithium 7 actually acts as a tritium enricher, adding fuel for further fusion.

©2020 **BIOZONE** International
ISBN: 978-1-927309-82-7
Photocopying Prohibited

62 Review Your Understanding

Key Question: How do you power a city, or destroy it, with just a few kilograms of nuclear material?

- At the beginning of this chapter you were asked questions of how to power a city (or destroy it) using just a few kilograms of material. You should now be in a position to answer those questions.

1. (a) Distinguish between nuclear decay, nuclear fission, and nuclear fusion:

   Nuclear fusion combines two smaller nuclei into a larger one resulting in the release of energy. Nuclear fission works in reverse, splitting large atomic nuclei into smaller daughter nuclei, energy, and radiation. Nuclear decay is similar to nuclear fission in that it deals with large nuclei, however, instead of splitting into smaller nuclei, nuclear decay results in the release of radiation to put the nucleus into a more stable state.

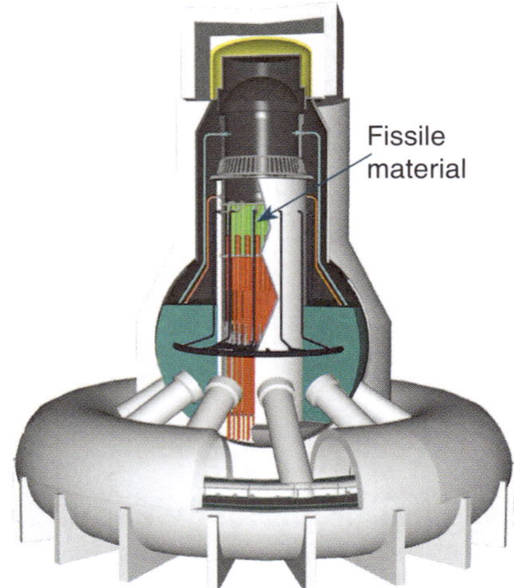

MK1 reactor core

   (b) Explain how 27 tonnes of uranium reactor fuel produces as much energy as 2.5 million tonnes of coal:

   In a reactor fission of Uranium releases energy stored between the nucleons (the strong nuclear force). Burning coal releases energy stored in the bonds between atoms (the electromagnetic force). The strong force is much greater than the electromagnetic force. Thus a small amount of uranium fuel releases as much energy than a large amount of coal.

   (c) Nuclear fission is the most reliable nuclear reaction to provide energy for producing electricity. Explain how nuclear fission is controlled and used to provide electricity:

   Uranium fission releases neutrons that can be absorbed by other uranium nuclei. This makes them unstable so that they may then fission. This can lead to a chain reaction of continuing fission of uranium atoms. In a reactor the chain reaction is controlled using neutron absorbing elements (e.g. boron) as a way of controlling the amount of neutrons available to continue the fission chain reaction. Energy released during the fission reaction is absorbed by water. This turns to steam which can be used to turn a turbine connected to an electricity generator.

2. (a) The Little Boy nuclear bomb (right) converted 0.7 grams of mass into energy. Use $E = mc^2$ to calculate the energy in joules that was released by this small amount (speed of light = 300,000,000 m/s).

   $0.7 \div 1000 = 0.0007$ kg.

   $0.0007 \text{ kg} \times 300{,}000{,}000^2 = 6.3 \times 10^{13}$ J

   (b) If the explosion of one tonne of TNT releases 4.2×10^9 J, what was the kilotonne rating of the Little Boy bomb (show your working)?

   $6.3 \times 10^{13} \text{ J} \div 4.2 \times 10^9 = 15\,000$ tonnes = 15 kilotonnes

3. The combustion of 1 L of gasoline releases about 7.3×10^4 kJ of energy. The mass of the reactants (gasoline and O_2) weigh about a millionth of a gram more than the mass of the products (CO_2 and water). Where has this mass gone?

   The mass has been converted to energy. Just as the mass of the nucleus changes in nuclear reactions due to the release of energy stored in the strong force, molecular mass is affected when energy stored in the electromagnetic (EM) force between bonds is released. The EM force is much weaker than the strong force, so the mass is only fractionally affected,

©2020 BIOZONE International
ISBN: 978-1-927309-82-7
Photocopying Prohibited

63 Summing Up

Nuclear processes

PS1-8

1. ^{238}U decays to ^{206}Pb via a series of decays. The first four decays are shown below:

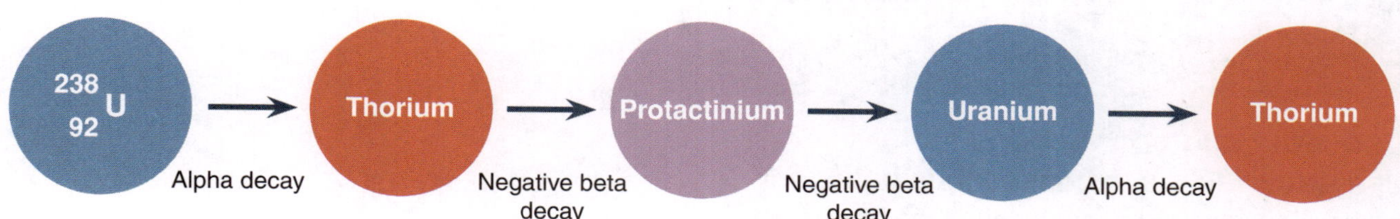

(a) For each of the elements shown, use the type of decay shown to calculate the atomic number and atomic mass of each of the elements shown:

Thorium: *Atomic mass: 234. Atomic number: 90.*

Protactinium: *Atomic mass: 234. Atomic number: 91.*

Uranium: *Atomic mass: 234. Atomic number 92.*

Thorium: *Atomic mass: 230. Atomic number: 90.*

(b) Explain why uranium and thorium appear twice in the series: *Uranium 238 decays via alpha decay losing two protons and two neutrons, thus transforming into thorium. The release of an electron via beta decay transforms a neutron into a proton. This happens twice so uranium 238 regains the two protons over two decay steps. It then decays via alpha decay again, reforming thorium once more.*

2. The diagram below shows materials needed to stop certain types of radiation produced by a radioactive source:

Relative penetrating power of alpha, beta, and gamma radiation

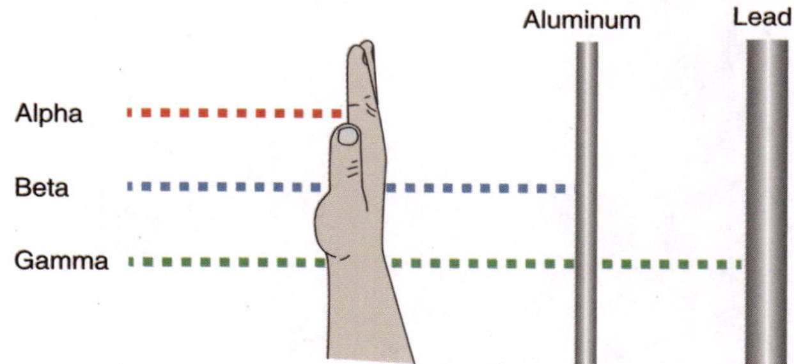

(a) Which type of radiation has the greatest amount of energy? *Gamma radiation.*

(b) Explain the difference in penetrating power of alpha particles and gamma rays: *Alpha particles have relatively large masses and a strong positive charge. They collide with molecules quickly when striking matter. Gamma rays are energy without mass or charge. They are able to pass through material without interacting with anything.*

3. Use a diagram to show the alpha decay of Bismuth ($^{184}_{83}$Bi) and the product formed:

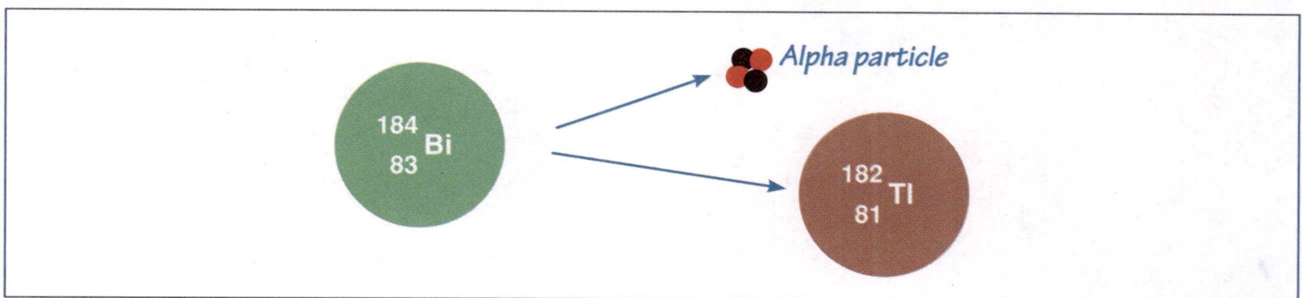

©2020 BIOZONE International
ISBN: 978-1-927309-82-7
Photocopying Prohibited

Motion and Stability: Forces and Interactions

Concepts and connections
Use arrows to make connections between related concepts in this section of the book

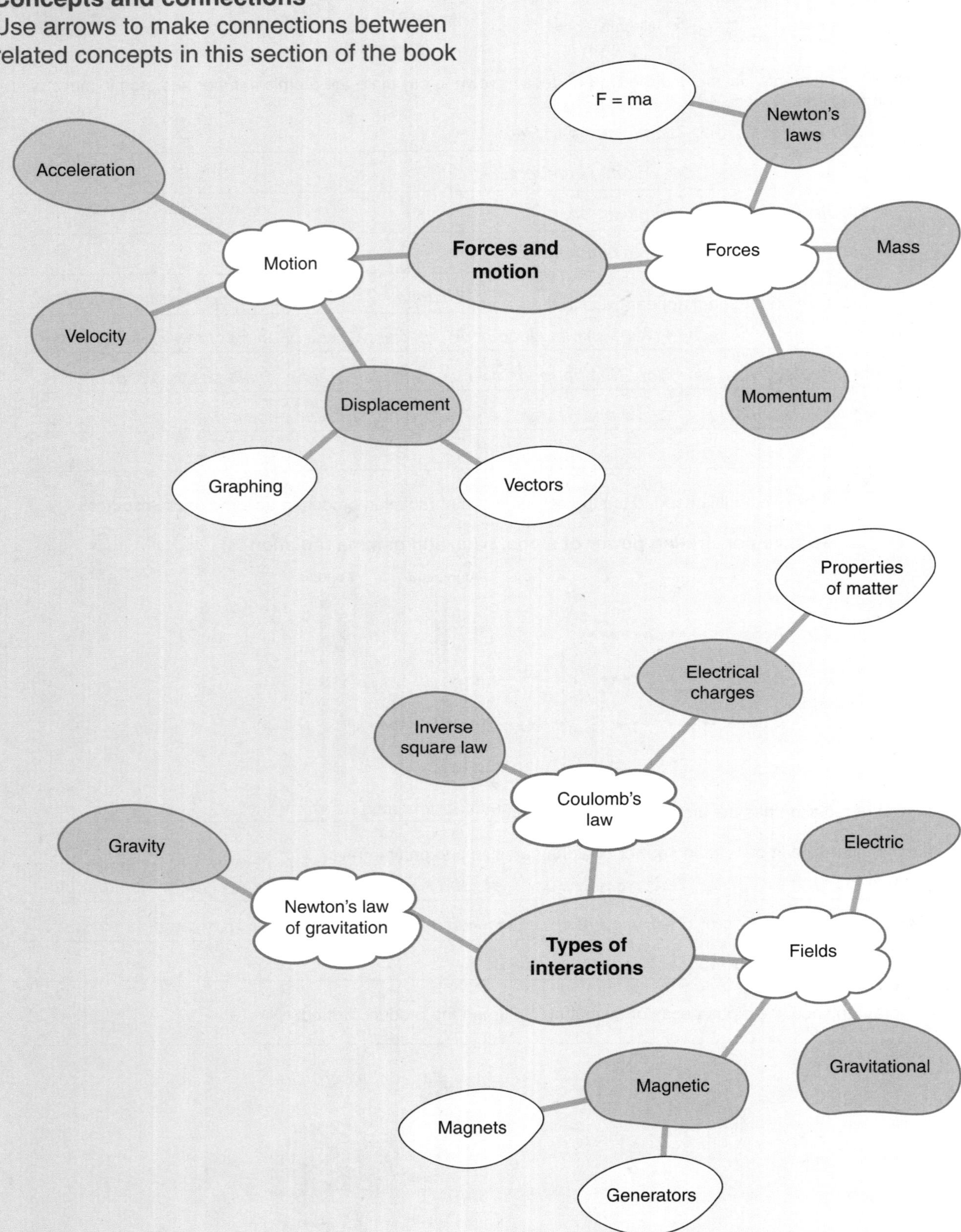

PS2.A

5: Forces and Motion

Activity number

Anchoring Phenomenon

Breaking bricks: How is it possible to break many bricks, blocks of ice, or boards of wood stacked on top of one another? What physical principles explain these feats of strength? 64 76

How can Newton's laws explain how and why things move?

☐ 1 The study of motion is an important part of physics. How do you describe the motion of objects around you? Identify the three main aspects of motion and construct diagrams [SEP-2] to describe these different aspects of movement. Understand the difference between speed and velocity. Use and interpret mathematical representations of the relationship between distance and time to determine velocity [SEP-4]. Working in groups, investigate this relationship for yourselves and plot your results [SEP-3] [PS2.A] [CCC-2]. 65 66

☐ 2 What do we mean when we say something is accelerating? Use a mathematical expression (equation) to calculate the acceleration for an object traveling in a particular direction. Given a known acceleration, time, and final velocity, how would you calculate distance traveled? [SEP-5] Draw and interpret graphs of displacement versus time and velocity versus time [SEP-4] [PS2.A] [CCC-2] [HS-PS2-1]. 66 67

NASA

☐ 3 What do you understand by the term force? What happens when unbalanced forces are applied to an object? Interpret data on the effect of friction on motion [SEP-4]. Explain why friction is an important consideration in the design and operation of moving objects. Weight is a specific kind of force due to gravity but forces can act on objects from any direction. Draw a simple model [SEP-2] to illustrate the effect of a constant force being applied to objects of increasing mass [PS2.A] [CCC-2] [HS-PS2-1]. 68 69 77

☐ 4 You now understand the difference between mass and weight and can analyze the motion of objects subjected to a constant force (gravity). Interpret data [SEP-4] [PS2.A] of the effect [CCC-2] of increasing mass on acceleration and explain the relationship between the force on an object, its mass, and its acceleration. Examples could include a falling object, an object rolling down a ramp, or a moving object being pulled by a constant force [HS-PS2-1]. 69 77

NASA

☐ 5 Understand that forces come in action-reaction pairs which are always equal and opposite. Why don't these action-reaction pairs cancel each other out? Draw and analyze free body diagrams [SEP-2] [SEP-4] to calculate [SEP-5] the net forces on known masses and thus determine their acceleration. 69 70

☐ 6 Extend your study of forces and motion to include an exploration of collisions and momentum. Investigate [SEP-3] the momentum of objects of different mass using marbles on a ramp. Use your results [SEP-5] to explain how momentum affects [CCC-2] the distance the marbles roll [PS2.A]. 71

☐ 7 Analyze the results [SEP-5] of investigations of the effect of collisions on the momentum of colliding objects in an isolated system [PS2.A] [CCC-4]. Explain the results in terms of conservation of momentum [HS-PS2-2]. 72 73 77

☐ 8 Extend your analysis to explain the effects [CCC-2] of collisions between real objects, such as motor vehicles. Investigate [SEP-3] what happens when the force of a colliding object is spread out over a longer (or shorter) period of time. Use math [SEP-5] to demonstrate this and explain [SEP-6] its significance to how much damage occurs during a collision [PS2.A] [HS-PS2-2]. 74 75 77

How can mathematical models of Newton's laws be used to test and improve engineering designs?

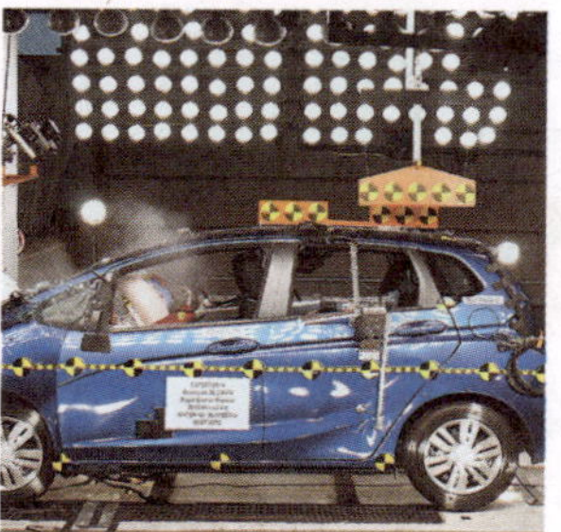

☐ 9 Use your understanding of changes in momentum and impulse to analyze the design of modern safety equipment, e.g. crash helmets for different sports and air bags and crumple zones in cars. Use math [SEP-5] to explain how these safety devices reduce the force delivered during a collision [PS2.A] [CCC-2] [HS-PS2-2]. 75

☐ 10 Demonstrate your understanding of forces and momentum by carrying out an investigation [SEP-3] to design a landing device to protect a raw egg from a fall of at least 4 m. Working in groups, discuss how you will make the lander and the different designs [SEP-6] you could use to cushion the egg at impact and/or slow its descent [CCC-2]. Justify your design choice in terms of Newton's laws of motion [PS2.A] [ETS1.B] [ETS1.C] [HS-PS2-3] [HS-ETS1-2] [HS-ETS1-3]. 75

ADVANCED LEARNING

★ Challenge question:

Does Newton's cradle demonstrate conservation of momentum? Explain your answer accounting for the forces involved.

Download this page from BIOZONE's Resource Hub

5. Forces and Motion

Anchoring phenomenon

The anchoring phenomenon *"Breaking Bricks"* introduces students to the concept of forces and their applications. Students can discuss forces involved and how it is possible to break a brick without breaking your hand (or head). Students could look for some of the more improbable feats of brick breaking on the internet (e.g. breaking 30 or more bricks at once) and discuss if they believe them or not and what forces are involved.

How can Newton's laws explain how and why things move?

1. Studying motion helps review important ideas that underlie many concepts in physics. Terms such as distance and displacement, speed and velocity are covered and used to introduce the concept of scalar and vector quantities. Students use and construct diagrams to illustrate these concepts. Students are introduced to the use of mathematical formulas to describe motion. There is opportunity to work in groups to carry out practical tasks that demonstrate some of the concepts covered.

2. Acceleration is an important part of motion as it is part of the link between the motion of an object and any forces involved (F = *ma*). It is therefore important that students understand the concept of acceleration and are able to use mathematical formulas and diagrams to describe it and relate it to distance and velocity. In many cases, students are shown how a formula can be derived from known information by simple substitution and rearrangement of equations. Understanding how to use these formulas is important when calculating forces involved in collisions (see #10). An important task here is for students to derive an equation for displacement given acceleration and velocity, but not time. Using the equations they already know, students should be able to derive the final equation $Dx = (v_f^2 - v_i^2) \div 2a$.

3. Students are introduced to the concept of forces via Newton's first law. Simple tasks, such as thinking about the motion of a tennis ball, help students build the concept that a change in motion requires a force. Students are introduced to the concept of balanced and unbalanced forces by using and constructing diagrams. Students analyze data on friction and learn that friction is a force that resists motion. The concept of the coefficient of friction is introduced and returned to when studying forces on an incline.

4. It is important that students understand the difference between weight and mass and use these terms correctly. Students can analyze experimental data to show the relationship between force, mass, and acceleration, and thus develop an understanding of Newton's second law. A method for this experiment is given, so students should be able to carry out their own investigation if time permits. Graphing the results helps students visualize the relationships being investigated and how to find an unknown variable in a relationship. Students should be able to use free body diagrams to analyze and describe the effect of forces on objects. Students are then able to apply these ideas to more complex problems, such as free fall and objects on an incline. The Dawn spacecraft is a good example of the long term effect of even a small unbalanced force being constantly applied to an object.

5. Students complete their study of forces with a review of Newton's third law and force pairs. Students compare a diagram showing all the force pairs involved and a free body diagram, showing a summary of the net forces. They use this to explain how a plane increase its altitude.

6. Momentum is introduced using a description of a pool game and balls on a pool table. Students can work in groups to carry out a simple investigation of the effect of mass on momentum using marbles or ball bearings rolling down a ramp.

7. The law of conservation of momentum is an important concept as it explains the way objects behave in collisions and explosions. Students can analyze data from before and after a collision to describe how momentum is affected in a collision and explain the results. Students can also use on-line simulations or a spreadsheet (see **BIOZONE's Resource Hub**) to carry out their own investigations or use the steps in the activity to carry out a lab based investigation using dynamic trolleys or an air-track time if time permits.

8. Students will extend their concepts of momentum to include impulse. Students can work in groups to investigate impulse by investigating the forces experienced by a bungee jumper using elastic and non-elastic materials. In this case, the elastic material spreads out the time of the collision (between the bungee jumper's feet and the cord wrapped around them). Students should be able to use equations from earlier in the chapter to show the effect of extending collision time on the force experienced and explain the significance of this.

How can mathematical models of Newton's laws be used to test and improve engineering designs?

9. Students are given real world collision data and situations. Here they can clearly see the effect of increasing the time over which the force is felt by studying how NASA lands Mars probes and the differences in G forces (multiples of gravity) felt in crashes in cars built four decades apart. This crash data is available on-line at the NHTSA Vehicle Crash Test Database (see **BIOZONE's Resource Hub**) along with photos and reports of test crashes for many different models of car.

10. Students have the opportunity to break into groups and apply their understanding of Newton's laws and collisions and impulse to build a device that will protect an egg from a 4 m fall. This is a good opportunity to review the concepts covered in the chapter and discuss how they might apply to landing space probes on rocky surfaces, or protecting people in car crashes. Students could discuss the differences between landing through an atmosphere, such as on Mars, or on an airless planet, such as Mercury. Students could apply equations for acceleration, velocity, and force to calculate the final velocity of the egg from free fall to the ground and the force it will experience with hitting the ground. They could use this information in the design of their lander, realizing that the egg will experience an uneven force over a small area when it hits the ground.

64 Breaking Bricks

Key Question: How is it possible to break many bricks, blocks of ice, or boards of wood stacked on top of one other? What physical principles explain these feats of strength?

- Have you ever seen a martial arts demonstration where the demonstrators break boards with their hands? These demonstrations usually start with the novices breaking one board and move on to veterans showing off their strength and skill by breaking multiple boards in one strike. Some demonstrations even include brick breaking.
- The karate chop is a well-known technique used to break boards with the hand. However, when dealing with large stacks of hard objects like bricks, this technique is often replaced by a vertical punch or elbow thrust. Can you think of any reason why?
- While hands are the most common body part used to break objects, they are not exclusive. Feet, elbows, knees, and even heads are used to break objects such as boards, bricks, and blocks of ice. Some of the best performers break large stacks of these objects to impress the audience.
- If you have ever hit a solid object, you will know that it can be painful and could even cause injury. How are these demonstrators able to break stacks of these objects without injuring themselves?

1. It is not difficult to find these kinds of videos online. Find one where multiple bricks are broken by hand.

   (a) Is there a specific way they hit the objects? Describe what you observe: *If punching, the performer raises their center of gravity, put as much of their mass over the bricks as possible and punch straight down. If an elbow is used the performer hits with the elbow vertically in line with their shoulder.*

   (b) Explain how technique may aid in breaking bricks: *Proper technique allows performers to increase momentum and thus the impact on the bricks.*

   (c) Why do the bricks break but not the bones in the performer's hands? *Bricks are hard so there is nothing to reduce the impulse of the blow. Hands have some cushioning and flexibility to reduce to force felt from the impact.*

   (d) Explain what you think would happen if the performer hit the bricks slowly instead of quickly? Why? *The bricks are less likely to break and the performer may injure their hand. Less speed means less impulse so the bricks will be affected less. The hand will stop more quickly because the bricks don't break and injuring is possible.*

2. When it comes to breaking things with a head instead of by hand, it is common to see padding, such as folded towels, used for some added protection.

   (a) How does the padding protect the head from injury when striking the bricks? *The padding adds an extra layer of cushioning to reduce the impulse to the head (and importantly the brain).*

   (b) Is there any similarity between how the towels protect the head and what happens when a hand strikes a stack of bricks? Explain: *There is less padding in the head. The towel gives extra protection. The hand has extra layers of muscle, tendons, and joints to reduce impulse from the impact.*

©2020 **BIOZONE** International
ISBN: 978-1-927309-82-7
Photocopying Prohibited

65 Distance and Displacement

Key Question: What do distance and displacement mean, how are they different, and why is it important to have different ways to describe an object's final position?

Describing motion

- What is motion? How do we describe it?
- Take a moment to observe the movement of objects around you. The movement could be of the tip of your finger, or your classmate wandering around the room, or your teacher walking to the other side of the desk. How would you describe the motion of these things?

1. Describe the motion or movement of three things in your classroom:

   (a) *Student's answer.*

   (b) *Student's answer.*

   (c) *Student's answer.*

2. The image below shows two trucks racing. Write down how the motion of these trucks could be described:

Student's answers may may include: motion is linear or in one dimension. Trucks appear to moving at high speed or velocity. Motion could be described as left to right across the image (i.e. direction is included).

3. Looking at your descriptions of movement from the previous questions:

   (a) Identify specific words you used to describe the movement of these things and write them in a list:
   Student's answer.

   (b) Compare your list of words with a classmate and add any additional words they used to your list:
   Student's answer.

   (c) From your list, identify and highlight any words which can be associated with a specific numerical value (i.e. **fast** is a common word to describe an object in motion, but it is both relative and unspecific. However, **speed** is a word which you can attach a specific value to: "the speed limit was 55 mph"). *Student's answer.*

©2020 **BIOZONE** International
ISBN: 978-1-927309-82-7
Photocopying Prohibited

Scalars and vectors

- Physics is a mathematical science. The underlying concepts and principles have a mathematical basis. When we describe motion we use words associated with mathematical quantities that have strict definitions: **distance**, **displacement**, **speed**, **velocity**, **acceleration**, etc.

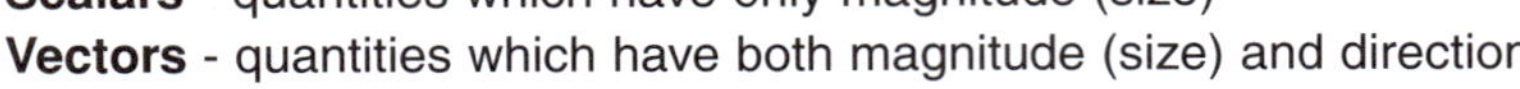

- The mathematical quantities that we use to describe motion can be divided into 2 categories:
  Scalars - quantities which have only magnitude (size)
  Vectors - quantities which have both magnitude (size) and direction

Displacement and distance

- An important part of physics is the study of motion. Motion is the movement of an object from one point to another. Two important aspects of motion are the distance moved and the direction traveled.
- Distance is a scalar quantity. It refers to total distance covered during an object's motion. Distance is measured in meters (m).
- Displacement is a vector quantity. It refers to the straight–line distance from the initial position to the final position and includes direction. In other words, displacement is an object's overall change in position.

4. Identify the following quantities as being either a vector or a scalar (circle one)

(a) 7 km (scalar / vector) — *scalar circled*

(b) 0 degrees Celsius (scalar / vector) — *scalar circled*

(c) 2000 Calories (scalar / vector) — *scalar circled*

(d) 1 Terrabyte (scalar / vector) — *scalar circled*

(e) 75 km/h, North (scalar / vector) — *vector circled*

(f) Falling at a rate of 9.81 m/s^2 (scalar / vector) — *vector circled*

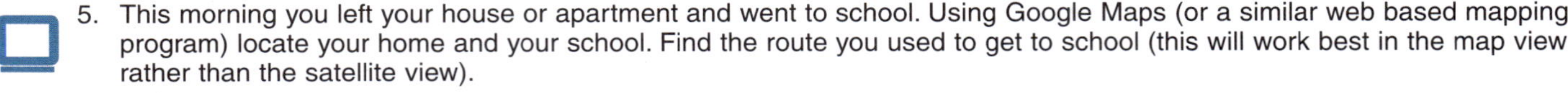

5. This morning you left your house or apartment and went to school. Using Google Maps (or a similar web based mapping program) locate your home and your school. Find the route you used to get to school (this will work best in the map view rather than the satellite view).

(a) On graph paper set a grid with x and y axes extending in both positive and negative directions. The point where the axes cross defines your home. Call this the origin, the point at which your journey began. *Student's answer*

(b) Map your journey on the grid, starting from the origin. Make sure you use a uniform scale throughout. *Student's answer*

(c) Use the mapping program to find out the total distance of your journey. *Student's answer*

(d) Draw a straight line from the origin to the end point of your journey (the school). Using the measurements of your x and y axes, you can work out the length of this line. This is the actual distance from your house to your school. How much further is your total distance traveled compared to the actual distance to school?

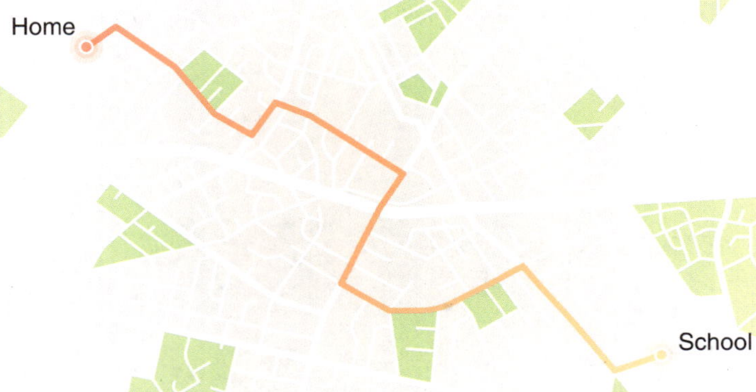

Unless a student lives next to the school, they will travel further to get to school than the distance.

(e) Would the total distance you traveled to school be a scalar or vector value? Justify your answer:

Total distance would be a scalar value because it involves multiple changes in direction but has no direction itself.

(f) What would be a vector value on your map of your travels to school? Explain why:

The straight line between home and the school would be a vector because it has both distance and a specific direction.

6. A space shuttle takes off from the Kennedy Space Center in Cape Canaveral, Florida, completes 8 orbits around the Earth, and then lands at Edwards Air Force Base in California. At the same time, a photographer flies from Florida to California to take pictures of the shuttle. Who undergoes the greater displacement? Justify your answer.

Both the space shuttle and the photographer have the same displacement. Even though the space shuttle traveled much further, the final destination was the same, therefore displacement was the same.

©2020 **BIOZONE** International
ISBN: 978-1-927309-82-7
Photocopying Prohibited

Visualizing motion

- We live in a physical world - a world we can see. Even though physics is a mathematical science, we rely on representing physical concepts in a visual manner. Visualizations allow us to better understand concepts and apply them to the real world.
- In physics, we use multiple ways to describe physical concepts. These include words, graphs, numbers, equations, and diagrams and are often converted from one form to another (i.e. a word problem can be used to create a graph, which can then be translated into an equation).
- In order to convert between multiple means of describing physical concepts we must be very specific about the words we use (this is why science has a lot of technical terms). Without specific terminology concepts can be misinterpreted and translated incorrectly.
- In physics, we typically deal with vectors - we want to know where something is going. For simplicity, when dealing with linear (1-dimensional) motion, we identify direction with positive and negative numbers. Positive numbers indicate motion to the right. Negative numbers indicate motion to the left.

7. Imagine a man walking along a flat straight street. He starts at home (0) and walks 10 m to the right. Realizing he's walking in the wrong direction he turns around and walks 50 m in the opposite direction to the store. Plot the two stages of the walk on the number line below:

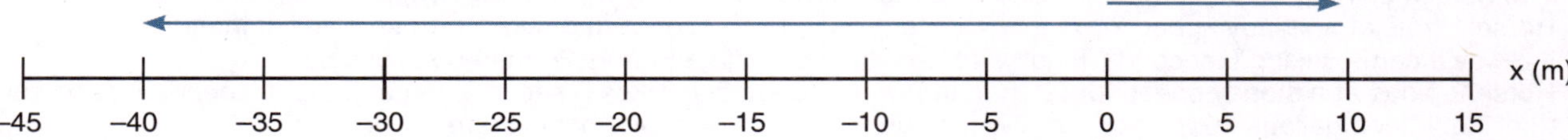

(a) What is the total distance the man walked? *10 m + 50 m = 60 m*

(b) What is man's the displacement? *10 m + -50 m = -40 m*

8. (a) On the grid below plot the following motion. A person walks 10 m west, then 10 m north, then 20 m east, and finally 20 m south.

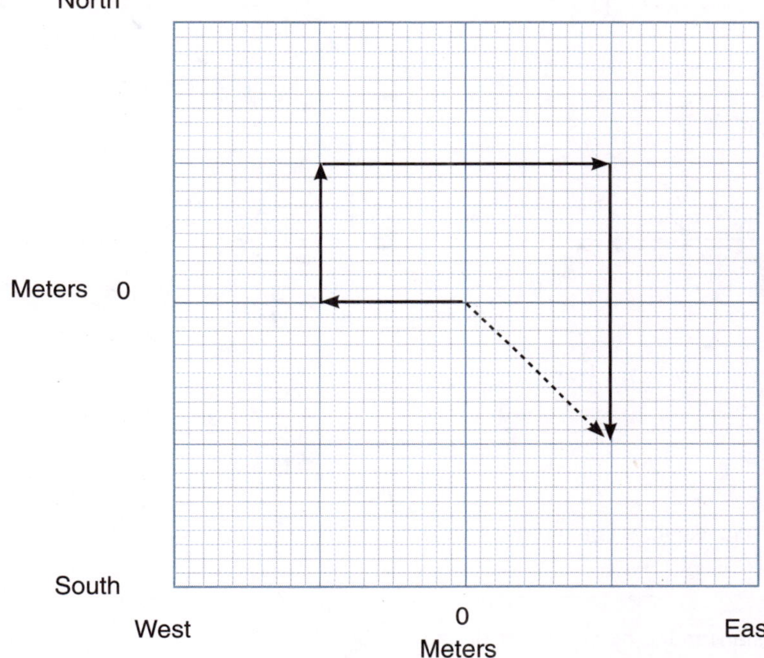

(b) What is the total distance the person walked? *10 m + 10 m + 20 m + 20 m = 60 m*

(c) What is the person's displacement at the end of the journey?
Hint: You might need to use **Pythagoras $a^2 + b^2 = c^2$** to find the displacement.

$c = \sqrt{(10^2 + 10^2)} = \sqrt{200} = 14$ m

9. Describe two situations, one in which distance traveled is the most important factor and one in which displacement is the most important factor.

Total distance traveled would matter when calculating fuel consumption on a trip. Displacement would matter when traveling to or from a specific place (e.g. how far from home are we?)

©2020 BIOZONE International
ISBN: 978-1-927309-82-7
Photocopying Prohibited

66 Speed and Velocity

Key Question: In horizontal motion, how do we distinguish between speed and velocity and how do we represent these concepts mathematically?

The role of time

- Time is important when describing motion. If another person traveled the same route in the journeys on the previous page, but took more or less time, the motion would be different because their speed would be different.
- A number line and a grid simply show an object's position in one dimension and two dimensions respectively. We can show the effect of time by plotting a distance versus time graph or a displacement versus time graph. Distance and displacement are usually plotted along the y-axis whereas time is plotted along the x-axis.
- For motion in one-dimension, distance traveled will always be positive as it is cumulative. Displacement can be positive or negative depending on whether the object is going away from the starting point or going back towards the starting point. Time always runs forward and so is always positive.

1. Two horses starting at the same point and time move towards a newly filled water trough.
Horse 1 trots at a steady speed 20 m to the water trough in 4 seconds. It drinks for 10 seconds. It then trots 30 m back the way it came, taking 6 seconds. Finally, it turns around again and trots 20 m in 5 seconds.
Horse 2 plods at a steady speed for 20 m to the water trough and takes 8 seconds to get there. It spends 6 seconds drinking. It then gallops 30 m back the way it came in 3 seconds and remains there eating grass.

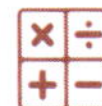

(a) In two different colors, plot the movement of the horses on the displacement-time graph below:

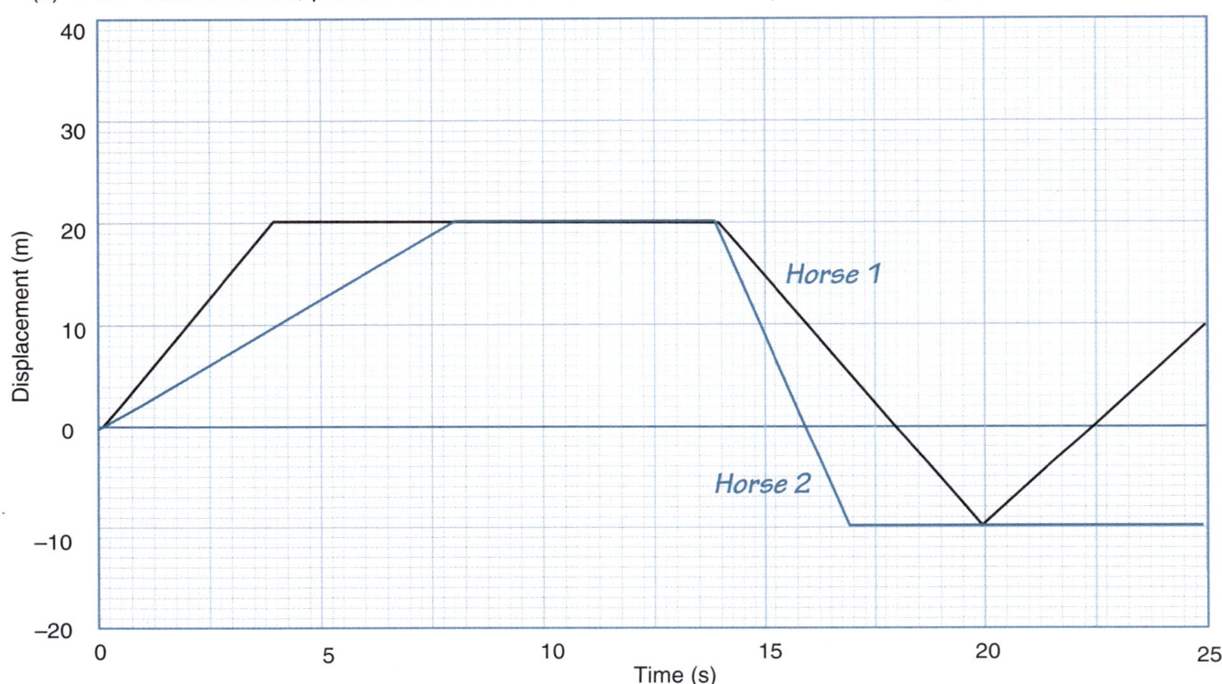

(b) What are the horses' displacements after 25 seconds?

Horse 1: 10 m   Horse 2: -10 m

(c) What are the total distances the horses move during this time?

Horse 1: 20 + 30 + 20 = 70 m   Horse 2: 20 + 30 = 50 m

(d) Describe the difference between the lines created by the displacement of the two horses in the first 4 seconds:

The line for horse 1's displacement over time is steeper for the first 4 seconds

(e) Describe the difference between the lines created by the displacement of the 2 horses between the 14 and 17 second marks of the graph:

The line for horse 2's displacement over time is steeper on the downwards gradient between 14 and 17 seconds.

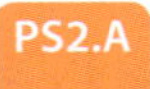

©2020 BIOZONE International
ISBN: 978-1-927309-82-7
Photocopying Prohibited

2. (a) Draw a displacement-time graph for the following:
A person walks at a steady speed 10 m along a straight footpath in 5 seconds. He then stops for 3 seconds before walking at a steady speed another 5 m in 3 seconds. He then turns around and jogs at a steady speed 18 m in the opposite direction in 4 seconds.

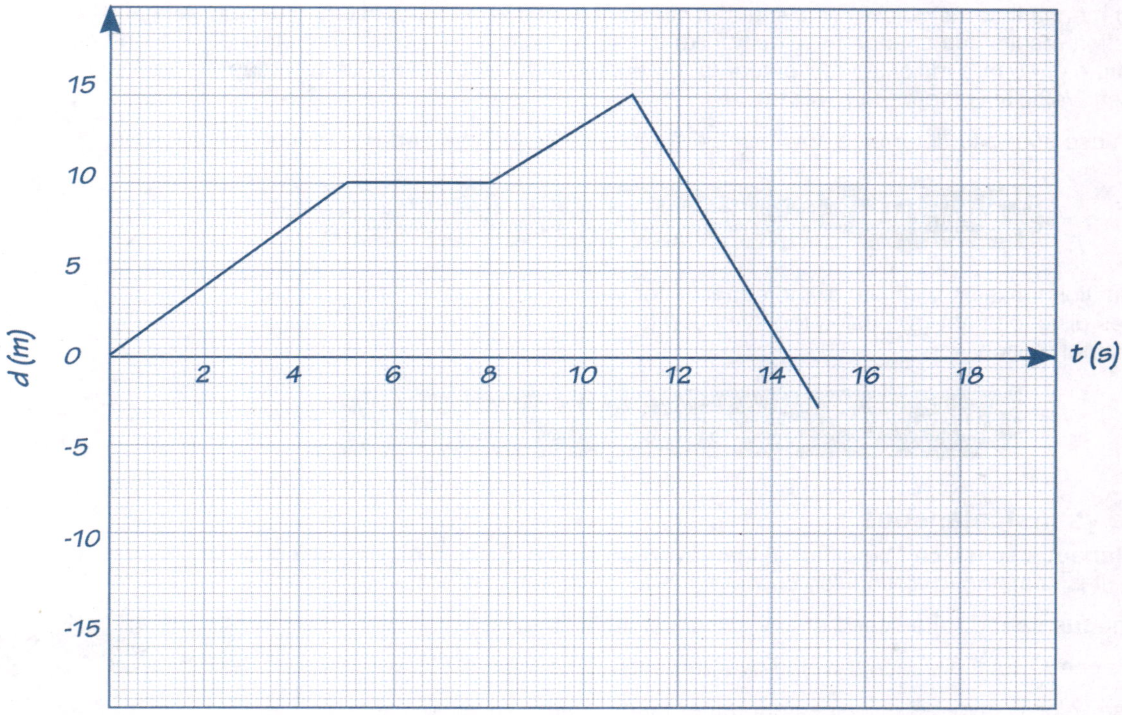

(b) Draw a velocity-time graph for the person's walk in 2. (a):

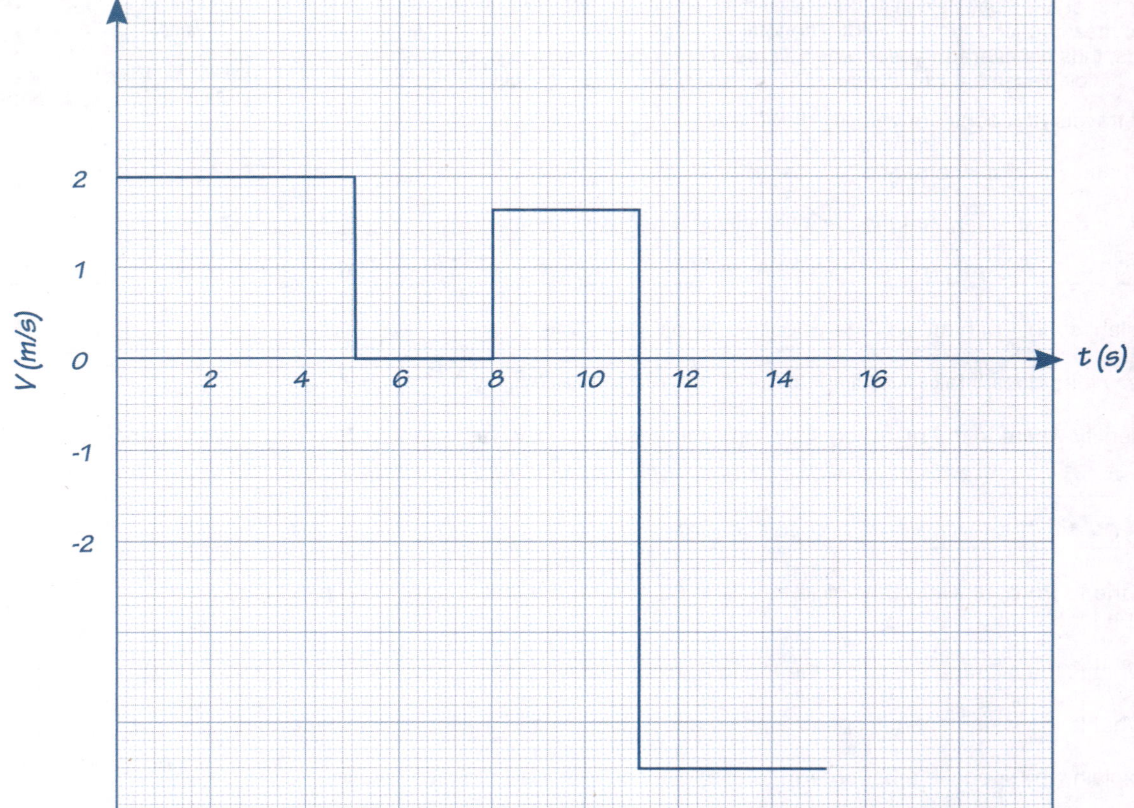

©2020 BIOZONE International
ISBN: 978-1-927309-82-7
Photocopying Prohibited

Speed and velocity

- **Speed** describes how fast something is moving. It is the rate at which an object covers a certain distance. In physics, speed is measured in meters per second (m/s).
- Like distance, speed is a **scalar** value. It is only described by its magnitude and no direction is included.
- **Velocity** takes into account an object's speed and direction. Thus, like displacement, it is a vector. Constant velocity means that the object is moving at a constant speed and in a constant direction. A change in speed or direction results in a change in velocity. Like speed, velocity is measured in meters per second but will have direction, e.g. +5 m/s indicates movement to the right in a frame of reference whereas -5 m/s indicates movement to the left. Velocity is therefore a **vector**.
- We can use speed to determine how much distance is traveled in a certain amount of time using the equation:

Change in distance (Δd) = speed (v) × change in time (Δt)

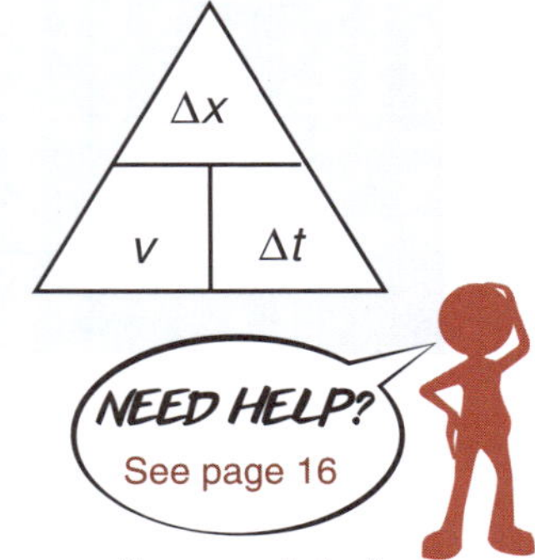

- The equation looks very similar to the equation for velocity. Change in distance ($\Delta \boldsymbol{d}$) becomes displacement ($\Delta \boldsymbol{x}$) using the symbol $\boldsymbol{x}$ to indicate movement along the x-axis in horizontal motion.

Displacement (Δx) = velocity (v) × change in time (Δt)

NEED HELP? See page 16

Average vs instantaneous

- **Instantaneous** speed can be different than **average** speed. Since a moving object often changes its speed during motion, it is common to distinguish between the average and instantaneous speed.

Instantaneous speed - the speed at any given instant in time.

Average speed - total distance traveled over the entire time interval.

- Because velocity involves displacement rather than distance traveled, average velocity can be very different than average speed.

- Example: A football coach instructs the team to sprint the length of the field and jog the width for 3 laps around the field as a training exercise. Jerome finishes the exercise first with a time of 3 minutes and 4 seconds.
- The length of a football field is 100 yards and its width is 53 1/3 yards (Note: yards is the specific measure for a football field). The total distance Jerome ran was 920 yards in 184 seconds; thus his **average** speed was 5 yards/second. However, since he ended where he began, his displacement was 0 yards. This means his **average** velocity is 0 yards/second.

3. A car travels at a constant velocity of -20 m/s

(a) What is the displacement of the car after 10 s? *-200 m* After 15 s? *-300 m*

(b) What does it mean that the car's velocity is a negative number? *A negative number could indicate the car is traveling to the left from the frame of reference or that it is driving in reverse.*

4. Abdullahi sprints with an average velocity of 8 m/s for 100 m.

(a) How long does it take Abdullahi to sprint the 100 m race? *100 ÷ 8 = 12.5 s*

(b) Identify where in the race Abdullahi's instantaneous velocity was less than 8 m/s. Justify your answer:

Instantaneous velocity was less than 8 m/s at the beginning of the race. The race starts from rest so Abdulahi must accelerate to gain speed.

5. On a trip to Punxsutawney, a vehicle travels at 60 mph for the first hour of the trip, 66 mph for the second hour, and 62 mph for the third.

(a) What is the average speed for the trip to Punxsutawney? *(60 + 66 + 62) ÷ 3 = 62.7 mph*

(b) What is the instantaneous speed 2 hours and 3 minutes into the trip to Punxsutawney? *62 mph*

(c) Explain why speed is a better term to use to describe a road trip than velocity: *During a road trip, cars must turn with the road to stay on it. The turns mean velocity is constantly changing.*

©2020 BIOZONE International
ISBN: 978-1-927309-82-7
Photocopying Prohibited

Position-time graphs

- We can learn a lot about the movement of an object through a position-time graph (or displacement-time graph). The graph will have different properties depending on the type of movement the object undergoes.
- An object moving with constant velocity will show a straight line when graphed.
- Positive velocity is depicted by a line going up and to the right.
- Negative velocity is depicted by a line going down and to the right.

Constant positive velocity

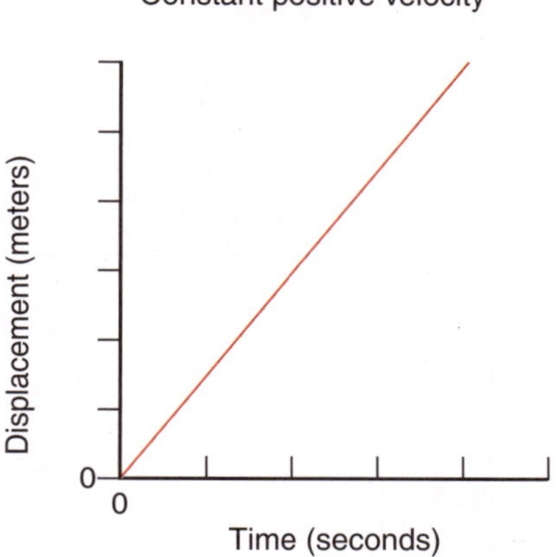

Constant negative velocity

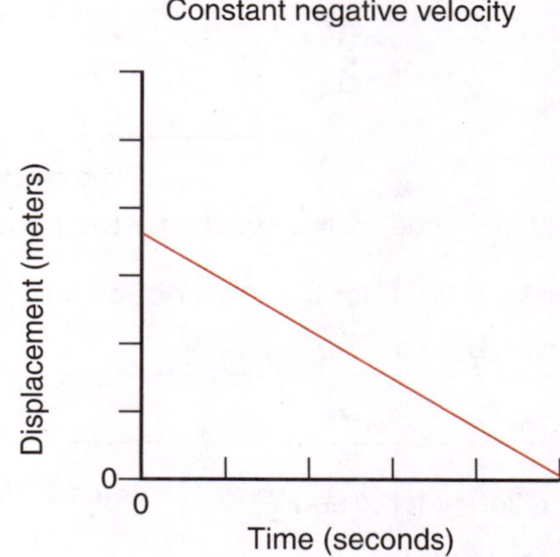

- When the velocity of an object changes, it is undergoing **acceleration** and will show a curve when graphed.
- Positive changing velocity (acceleration) is depicted by a line curving up and to the right.
- Negative changing velocity (deceleration) is depicted by a line curving down and to the right.

Positive changing velocity
Positive acceleration

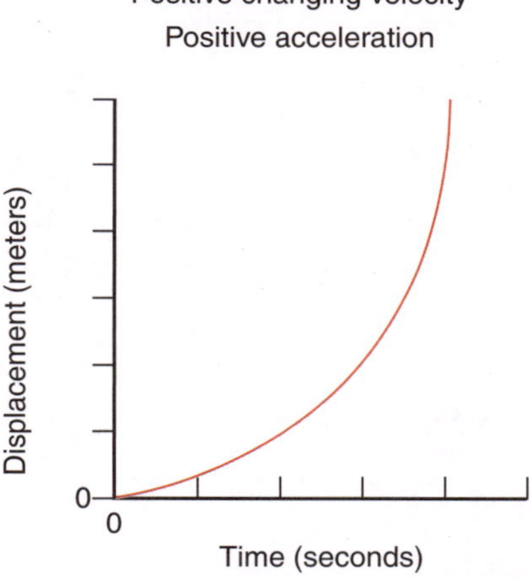

Negative changing velocity
Negative acceleration (deceleration)

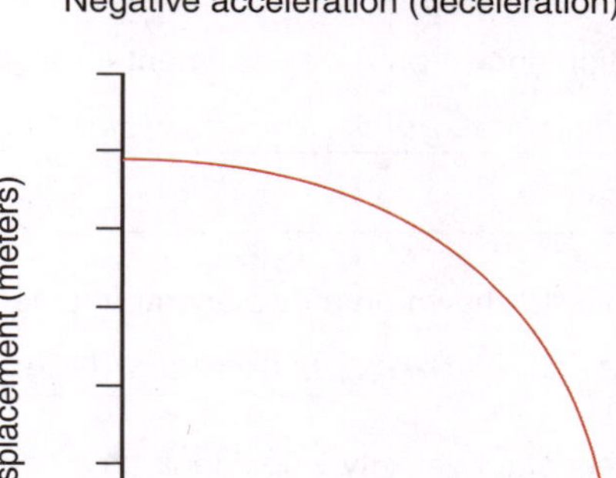

- Velocity can be found on a position-time graph from the gradient, since gradient is rise/run and rise/run = $\Delta \boldsymbol{x}/\Delta \boldsymbol{t}$, and $\boldsymbol{v} = \Delta \boldsymbol{x}/\Delta \boldsymbol{t}$. The steeper the gradient the higher the velocity of the object.

Fast constant velocity

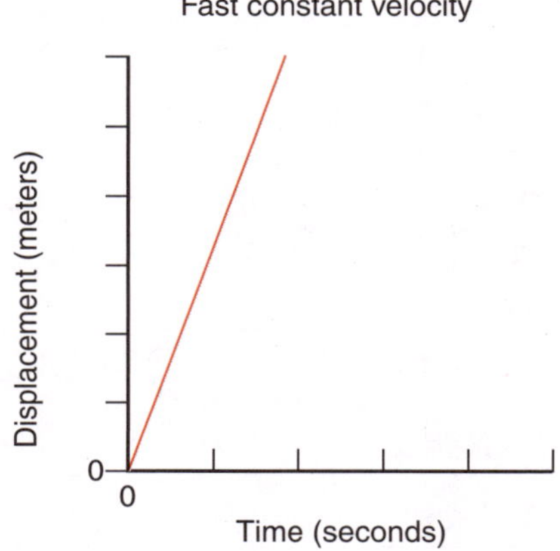

Slow constant velocity

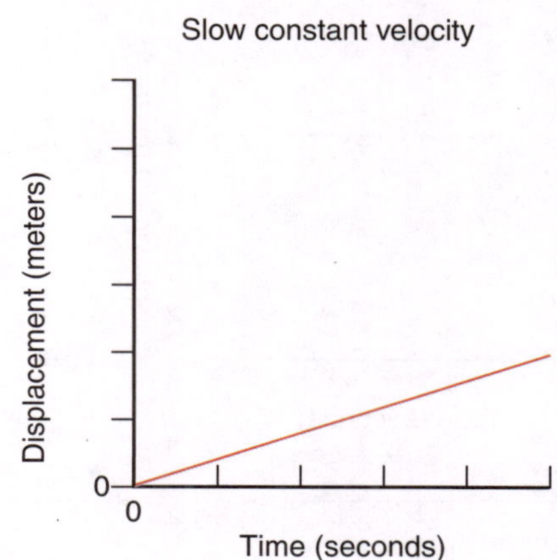

©2020 **BIOZONE** International
ISBN: 978-1-927309-82-7
Photocopying Prohibited

- A straight horizontal line indicates an object is not moving. The velocity of an object at rest is 0 m/s.

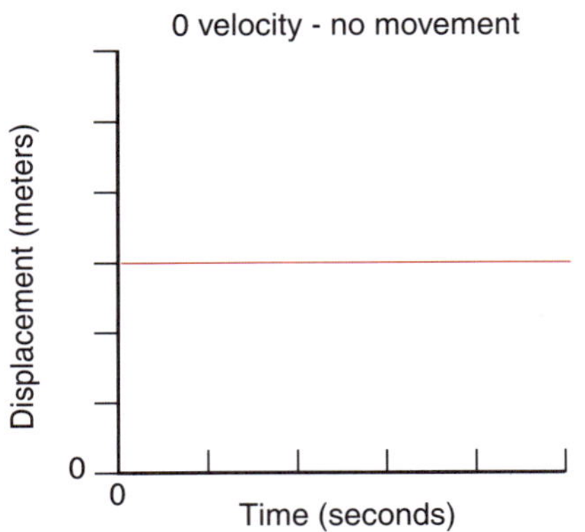

The displacement (from the origin) remains the same so the velocity must be zero.

6. The axes right show the motion graphs for two people, X and Y:

   (a) Calculate $\Delta x/\Delta t$ (= v) for X and Y (include units):

   x = 30 m ÷ 3 s = 10 m/s

   y = 15 m ÷ 5 s = 3 m/s

   (b) How is v different for X and Y? *The velocity of x is greater than the velocity of y.*

7. How is constant velocity shown on a distance-time or displacement-time graph?

   Constant velocity is represented by a straight line.

8. How is direction shown on a displacement-time graph? *Lines going up and to the right indicate forward (right) velocities while lines going downward right indicate backwards (left) velocities.*

9. How is magnitude shown on a displacement-time graph? *The steepness of the lines indicate magnitude. The steeper the line, the greater the magnitude.*

10. Predict how constant velocity would look on a velocity-time graph. How is this different from a displacement-time graph?

    Constant velocity would be a flat line on a velocity-time graph compared to a straight rising (or descending) line on a displacement time graph.

11. Study the graphs below and state the type of motion they are showing:

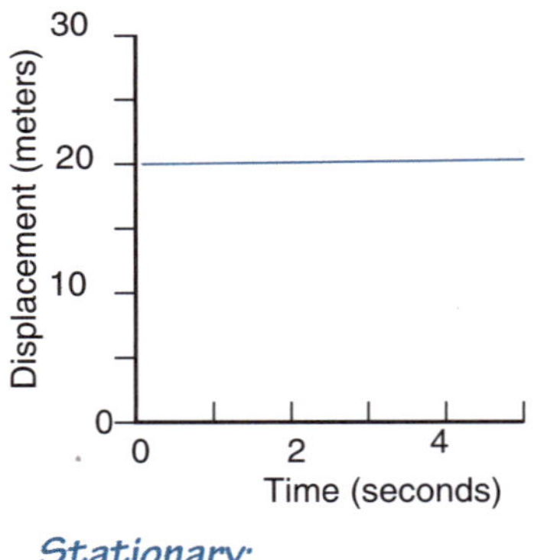

Stationary: The person or object is not moving

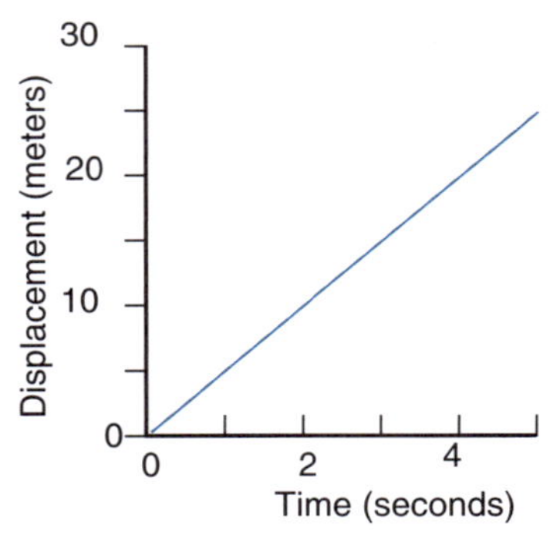

Constant velocity (forwards)

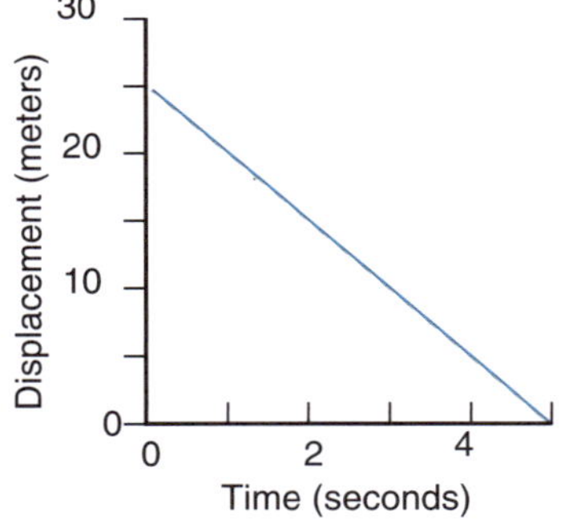

Constant velocity (backwards or reverse)

©2020 **BIOZONE** International
ISBN: 978-1-927309-82-7
Photocopying Prohibited

INVESTIGATION 5.1: Distance, displacement, and velocity

See appendix for equipment list.

1. This investigation involves the whole class. Set out a straight line (one-dimensional) course with 9 markers at 10 m intervals as shown below. The starter and the runner are at the middle of the course.
2. On the right side of the course, each marker has 2 timers, one for the outward journey and one for the return journey.

3. On the starter's signal (they should use a large visual signal like a flag):
   (a) The runner runs from the middle of the course to one end and then back to the opposite end.
   (b) All the timers start timing at the starter's signal and stop when the runner reaches their marker.
4. Complete the tables below:

Table 1

d = distance traveled (m)	t = time elapsed (s)
0	0
10	
20	
30	
40	
50	
60	
70	
80	
90	
100	
110	
120	

Table 2

d = displacement (m)	t = time elapsed (s)
0	0
10	
20	
30	
40	
30	
20	
10	
0	
-10	
-20	
-30	
-40	

Table 3

$\Delta d = d_{final} - d_{initial}$ (m)	$\Delta t = t_{final} - t_{initial}$ (s)	$v = \Delta d/\Delta t$ (m/s)	t = mid interval time (s)
10 – 0 = 10			
20 – 10 = 10			
20 – 10 = 10			
30 – 20 = 10			
40 – 30 = 10			
30 – 40 = -10			
20 – 30 = -10			
10 – 20 = -10			
0 – 10 = -10			
-10 – 0 = -10			
-20 – -10 = -10			
-30 – -20 = -10			
-40 – -30 = -10			
	NOTE: $t_{mid\ interval} = (t_{final} + t_{initial}) \div 2$		

5. Use Table 1 to plot a graph of distance versus time on grid 1 opposite. Use the plotted points to make a smooth trendline.
6. Use Table 2 to plot a graph of displacement versus time on grid 2 opposite. Use the plotted points to make a smooth trendline.
7. Use the last two columns of Table 3 to plot a graph of velocity versus time on grid 3 opposite. Make a smooth trendline.

12. Why is it necessary for the starter to use a visual signal?

Sound travels more slowly than light, so you will see the signal before you will be able to hear the starter say 'go'.

13. Why was it necessary to use the mid-interval times for the third table and third graph?

Since velocity is changing, the velocity calculated is the average velocity over that interval. Assuming constant acceleration, the average velocity would be the velocity at the half-way point of the interval.

©2020 BIOZONE International
ISBN: 978-1-927309-82-7
Photocopying Prohibited

Graph 1: Distance versus time

Graph 2: Displacement versus time

Graph 3: Velocity versus time

©2020 **BIOZONE** International
ISBN: 978-1-927309-82-7
Photocopying Prohibited

14. Describe the shape of the distance-time graph: *Answers may vary based on student's results but will generally show a broken line going up and to the right*

15. Describe the shape of the displacement-time graph: *Answers may vary based on student results but will generally show a broken line going up and to the right to a peak at 40 m from the origin followed by a broken line going down and to the right continuing to a distance of -40 m from the origin.*

16. Describe the shape of the velocity-time graph: *Answers may vary based on the student's results but may show a broken line going up and to the right followed by a broken line going down and to the right. The lines may become horizontal at intervals along the graph.*

17. Racing karts can reach very high speeds but speeds and engine sizes are limited at younger age groups. The graph below shows the displacement-time graph of a racing kart driving in a straight line from a stationary start.

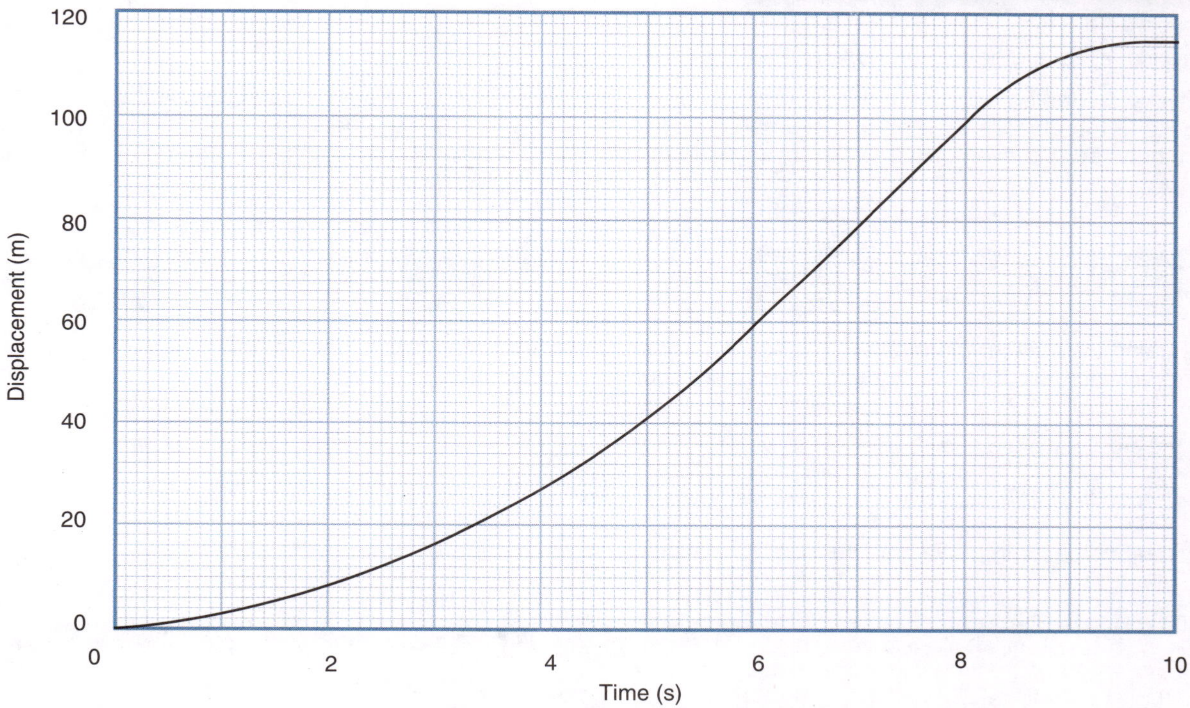

(a) Describe the motion of the racing kart: *It accelerates for the first 6 seconds. It then reaches a constant velocity for 2 seconds before rapidly decelerating in about 1.5 seconds and coming to a stop.*

(b) At what time did the kart reach its top speed? *6 seconds*

(c) Calculate the top speed of the kart: *Distance between 6s and 8 s = 40 m. 40 m ÷ 2s = 20 m/s*

(d) How long did it take the kart to come to a complete stop? *About 1.5 s*

(e) At what rate did the kart slow down? *From 20 m/s to 0 m/s in 1.5 seconds. Slowed at a rate of 13.3 m/s/s*

©2020 **BIOZONE** International
ISBN: 978-1-927309-82-7
Photocopying Prohibited

67 Acceleration

Key Question: How do we use what we know about constant acceleration to solve unknown values of displacement, time, and velocity?

Acceleration

- **Acceleration** occurs when velocity changes. Acceleration can be changed by altering speed or direction (or both). It is defined as the change in velocity over the time elapsed.

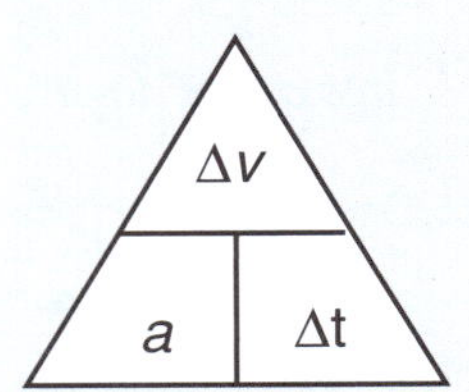

Acceleration (a) = change in velocity (Δv) ÷ change in time (Δt)

- In everyday language, we talk about accelerating (speeding up) and decelerating (slowing down) as would describe the skiers below.

- In physics, acceleration can be positive and negative.
- Positive acceleration acts in the direction of an object's movement. Negative acceleration acts in the direction opposite to the object's movement. Thus negative acceleration, if it persists, means that an object will not only slow down, but stop and eventually travel backwards in the opposite direction as shown in the diagrams (1-4) below.

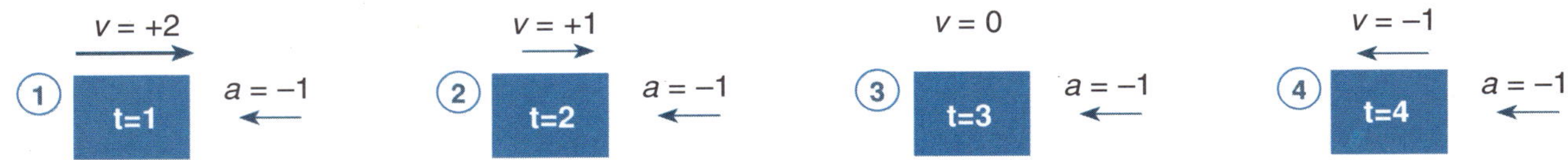

- Acceleration is measured in meters per second per second (m/s^2). A car accelerating from a stationary start at 5 m/s^2 will increase its velocity by 5 meters per second every second.

1. For the car mentioned above, what will its velocity be after:

   (a) 1 second: $5\ m/s^2$ (b) 2 seconds: $10\ m/s^2$ (c) 3 seconds: $15\ m/s^2$

2. Two cars compete in a straight-line race. The velocities of each car are shown in the table below:

Time (s)	Velocity of car 1 (m/s)	Velocity of car 2 (m/s)
0	0	0
1	10	7.5
2	20	15.0
3	30	22.5
4	40	30.0

(a) Calculate the average acceleration of car 1: $a_{avg} = (v_f - v_i) \div \Delta t = (40\ m/s - 0\ m/s) \div 4\ s = 10\ m/s^2$

(b) Calculate the average acceleration of car 2: $a_{avg} = (v_f - v_i) \div \Delta t = (30\ m/s - 0\ m/s) \div 4\ s = 7.5\ m/s^2$

(c) Calculate the average velocity of car 1: $v_{avg} = 0.5(v_f + v_i) = 0.5(40\ m/s) = 20\ m/s$

(d) Calculate the average velocity of car 2: $v_{avg} = 0.5(v_f + v_i) = 0.5(30\ m/s) = 15\ m/s$

(e) How far did car 1 travel in the 4 second race? $d = v_{avg} \times \Delta t = 20\ m/s \times 4\ s = 80\ m$

(f) How far did car 2 travel in the 4 second race? $d = v_{avg} \times \Delta t = 15\ m/s \times 4\ s = 60\ m$

©2020 BIOZONE International
ISBN: 978-1-927309-82-7
Photocopying Prohibited

Velocity-time graphs

- Similar to displacement-time graphs, we can learn a lot about the movement of an object through a velocity-time graph. The graph will have different properties depending on the type of movement the object undergoes.
- An object moving with constant acceleration will show a straight line when graphed.
- Positive acceleration is depicted by a line going up and to the right.
- Negative acceleration is depicted by a line going down and to the right.
- An object moving with constant velocity (no acceleration) is depicted by a horizontal line.

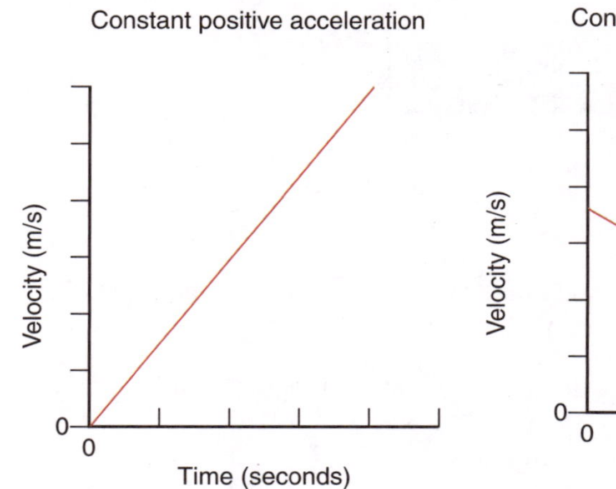

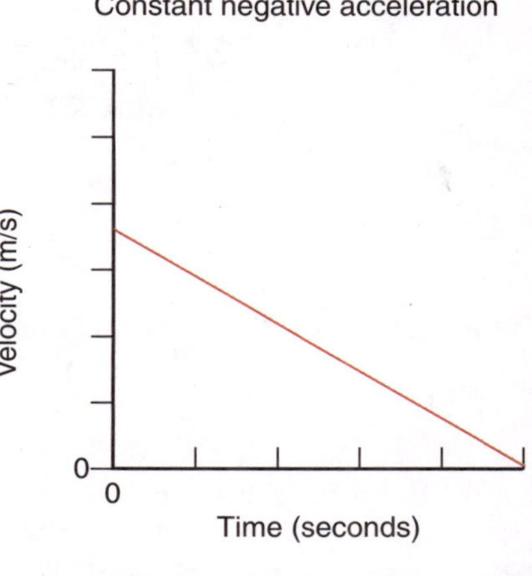

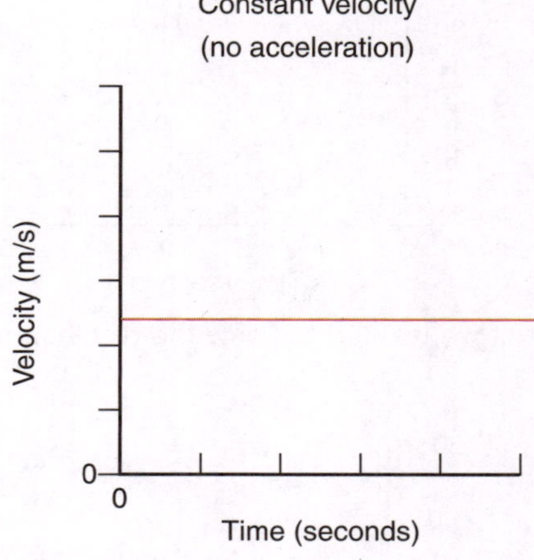

- Acceleration can be found on a velocity-time graph from the gradient, since gradient is rise/run and rise/run = $\Delta \boldsymbol{v}/\Delta \boldsymbol{t}$, and $\boldsymbol{a} = \Delta \boldsymbol{v}/\Delta \boldsymbol{t}$. The steeper the gradient the higher the acceleration of the object.

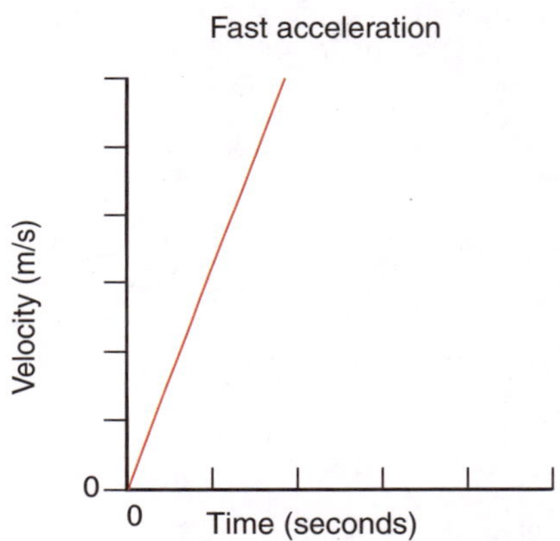

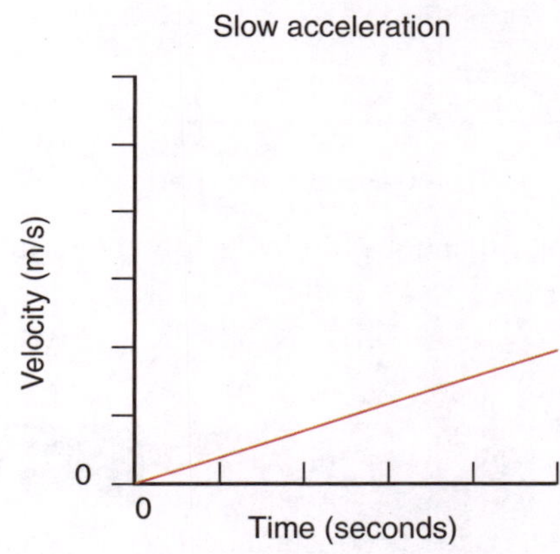

- Because displacement (Δx) is velocity (v) multiplied by elapsed time (Δt), an object's displacement can be determined by calculating the area under a velocity-time graph.

3. Calculate the displacement of the objects depicted in the following velocity-time graphs:

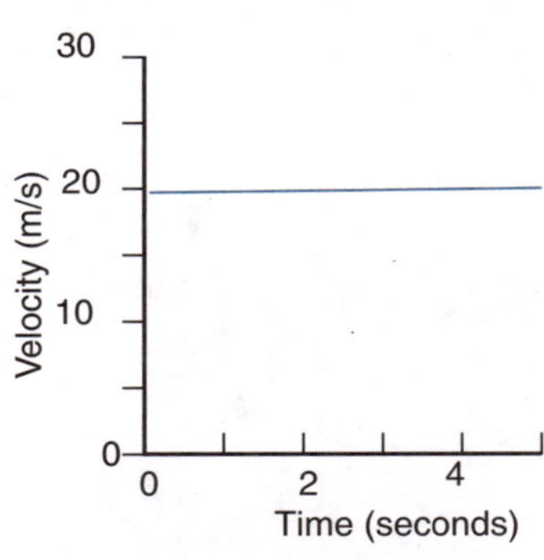

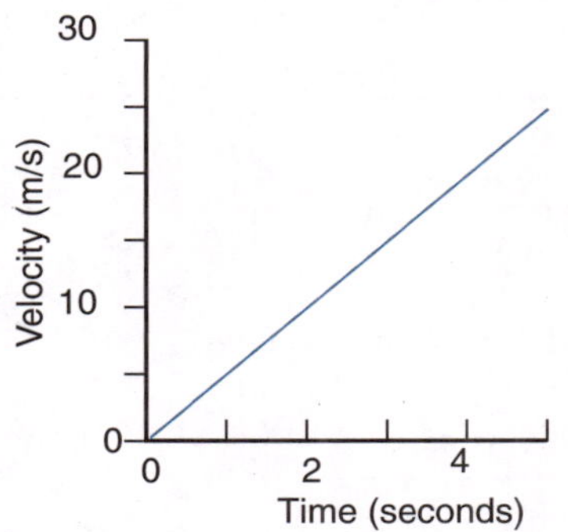

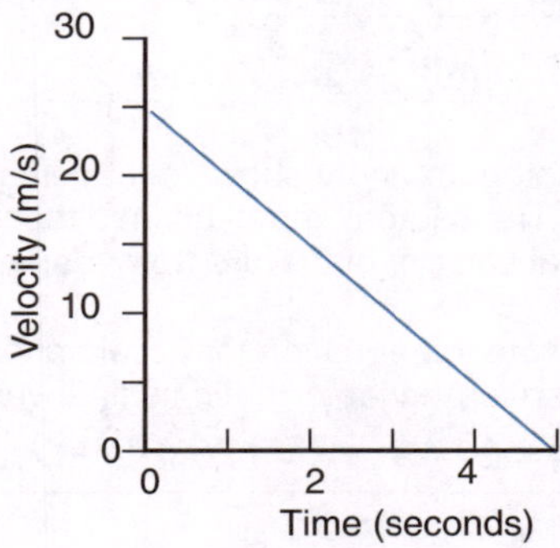

20 × 5 = 100 m

0.5 × 25 × 5 = 62.5 m

0.5 × -25 × 5 = -62.5 m

©2020 **BIOZONE** International
ISBN: 978-1-927309-82-7
Photocopying Prohibited

Displacement and velocity with constant acceleration

- Given constant acceleration, it is intuitive to be able to calculate velocity at any given point in time.

Let's consider a car traveling at a constant velocity of 10 m/s. The car then accelerates at a constant rate of 10 m/s^2 for 5 seconds. The velocity-time graph for this movement is plotted below:

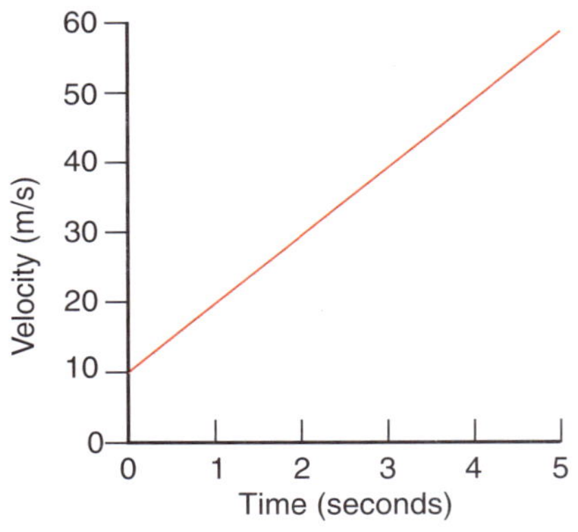

- Starting at 10 m/s, after 1 second of constant acceleration has gone by the car's velocity will have increased by 10 m/s for 20 m/s. After 2 seconds, the car will be traveling at 30 m/s.
- We can calculate the velocity after acceleration (v_f) with the following formula:

$$v_f = v_i + a\Delta t$$

- If we wanted to know the displacement of the car during this time, we would simply find the area under the line.

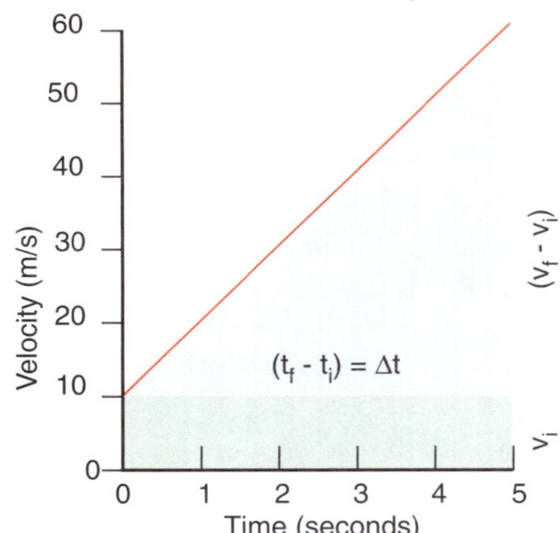

- We need to find the area of two shapes: a rectangle and a triangle.
- The area of the rectangle is:
  10 m/s (v_i) × 5 s (Δt) = 50 m
- The area of the triangle is:
  ½ × (60 m/s (v_f) - 10 m/s (v_i)) × 5 s (Δt) = 125 m
- Add the 2 areas together and we get a displacement (Δx) of 175 m.
- If we ignore the numerical values, we can see we also get the equation for displacement with constant acceleration, where ½ $(v_f - v_i)\,\Delta t$ is the area of the triangle and $v_i\,\Delta t$ is the area of the rectangle:
  $\Delta x = v_i \Delta t + \frac{1}{2}(v_f - v_i)\,\Delta t$

- We can simplify this equation by factoring out Δt, distributing the 1/2, and combining like terms:

$$\Delta x = (v_i + \tfrac{1}{2}(v_f - v_i))\,\Delta t$$
$$\Delta x = (v_i + \tfrac{1}{2}v_f - \tfrac{1}{2}v_i)\,\Delta t$$
$$\Delta x = (\tfrac{1}{2}v_i + \tfrac{1}{2}v_f)\,\Delta t$$

$$\Delta x = \tfrac{1}{2}(v_i + v_f)\,\Delta t$$

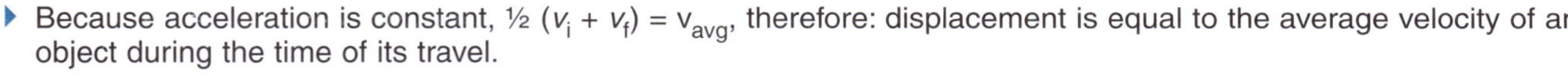

- Because acceleration is constant, ½ $(v_i + v_f) = v_{avg}$, therefore: displacement is equal to the average velocity of an object during the time of its travel.

4. In question 2(e) and (f) you were asked to calculate the displacement of two dragsters. Use the equation above to calculate their displacement. Do these answers agree with your initial calculations?

   $d = (40\ m/s)^2 \div (2 \times 10\ m/s^2) = 1600\ m^2/s^2 \div 20\ m/s^2 = 80\ m$;

   $d = (30\ m/s)^2 \div (2 \times 7.5\ m/s^2) = 900\ m^2/s^2 \div 15\ m/s^2 = 60\ m$ The answers are the same.

5. The photograph right shows a jet being launched from an aircraft carrier. The catapult that launches the jet accelerates it at 33 m/s^2 to a final velocity of 70 m/s from a standing start.

   (a) Rewrite the equation for acceleration to find Δt and find the time taken for the jet to reach its final velocity:

   $\Delta t = \Delta v \div a$ $\Delta t = (70\ m/s - 0\ m/s) \div 33\ m/s^2 =$

   2.1 s

   (b) What is the distance covered by the jet as it launches?

   $\Delta x = 0.5\,(70\ m/s + 0\ m/s) \times 2.1\ s = 73.5\ m$

©2020 BIOZONE International
ISBN: 978-1-927309-82-7
Photocopying Prohibited

Displacement under constant acceleration with missing variables

- If acceleration is constant, we are able to calculate for displacement even when we are missing a value.
- We need to use two formulas:

$$a = (v_f - v_i) / \Delta t$$

$$\Delta x = \frac{1}{2}(v_i + v_f)\Delta t$$

- What if we do not have an initial velocity (v_i)?
- First rearrange the formula for acceleration to solve for v_i:

$$a = (v_f - v_i) / \Delta t$$
$$a\,\Delta t = v_f - v_i$$
$$v_i = v_f - a\,\Delta t$$

- Substitute this value for v_i into the displacement formula and simplify:

$$\Delta x = \frac{1}{2}(v_i + v_f)\Delta t$$
$$\Delta x = \frac{1}{2}((v_f - a\,\Delta t) + v_f)\Delta t$$
$$\Delta x = (\frac{1}{2}(v_f - a\,\Delta t) + \frac{1}{2}v_f)\Delta t$$
$$\Delta x = (\frac{1}{2}v_f - \frac{1}{2}a\,\Delta t + \frac{1}{2}v_f)\Delta t$$
$$\Delta x = (v_f - \frac{1}{2}a\,\Delta t)\Delta t$$

$$\Delta x = v_f\,\Delta t - \frac{1}{2}a\,\Delta t^2$$

- Note: These equations and the ones you will derive in questions 6 and 7 are important as they will surface in later chapters as well.

6. What if we do not have a final velocity (v_f)? Using the two equations at the top of the page, and a similar procedure used above, create an equation to solve for displacement without v_f:

$a = (v_f - v_i)/\Delta t \rightarrow a\Delta t = v_f - v_i \rightarrow v_f = v_i + a\Delta t$. Substitute into displacement formula

$\Delta x = 0.5(v_i + v_f)\Delta t \rightarrow \Delta x = 0.5(v_i + v_i + a\Delta t)\Delta t \rightarrow \Delta x = v_i\Delta t + 0.5a\Delta t^2$

7. What if we do not have time (Δt)? Again using the two equations at the top of the page, and a similar procedure used above, create an equation to solve for displacement without Δt:

$a = (v_f - v_i)/\Delta t \rightarrow a\Delta t = (v_f - v_i) \rightarrow \Delta t = (v_f - v_i)/a$.

Substitute into displacement formula

$\Delta x = 0.5(v_i + v_f)\Delta t \rightarrow \Delta x = 0.5(v_i + v_f)(v_f - v_i)/a \rightarrow \Delta x = (v_iv_f + v_f^2 - v_i^2 - v_iv_f)/2a \rightarrow \Delta x = (v_f^2 - v_i^2)/2a$.

8. A land-speed car can decelerate at 9.8 m/s^2.

(a) If it decelerates, what is its velocity after the first 10 s of deceleration if its initial velocity was 885 km/h (245.8 m/s)?

$v_f = v_i + a\Delta t \rightarrow v_f = 245.8\ m/s + -9.8\ m/s^2 \times 10s =$

$147.8\ m/s$

(b) How much ground did the car cover in those first 10 s?

$\Delta x = 0.5(v_i + v_f)\Delta t \rightarrow \Delta x = 0.5\,(245.8\ m/s + 147.8\ m/s) \times 10s = 1376.8\ m$

(c) If the car comes to a complete stop, what was the car's total displacement during deceleration?

$\Delta x = (v_f^2 - v_i^2)/2a \rightarrow \Delta x = -(245.8\ m/s)^2 \div (2 \times -9.8\ m/s^2)$

$= 60{,}417.64\ m^2/s^2 \div -19.6\ m/s^2 = 3082.5\ m$

©2020 **BIOZONE** International
ISBN: 978-1-927309-82-7
Photocopying Prohibited

68 Introduction to Forces

Key Question: What are forces and how do they affect the motion of objects?

Force

- Force is an important term in physics. It refers to a push, pull, or twist acting on an object. Forces can cause objects to move, speed up, slow down, change direction, change shape, etc.
- Forces are measured in **newtons** or kg m/s^2. They are vectors as they have size and direction.
- A force requires an "agent" to provide the push, pull, or twist, e.g. a soccer player's boot kicking a ball or a bungee rope pulling a bungee jumper up from the lowest point of the jump.
- Contact forces require contact between the agent and an object, e.g. a bat hitting a ball.
- Long-range forces act without physical contact between the agent and the object, e.g. magnetic, electrostatic and gravitational forces.

The force applied to the ball by kicking it has both size and direction.

Forces change an object's motion

- Ever had this happen to you? You sit on a sled or a trolley. Your friend grabs the sled's rope and takes off running, intending to give the sled a mighty pull. The sled suddenly moves in the direction your friend pulls but you stay exactly where you are as the sled flies out from under you.
- It is the natural tendency of an object to resist changes in its state of motion. This resistance to change is called **inertia**. It is the reason you stayed where you were while the sled moved from under you. You were not attached to the sled, so it moved and you stayed in your current state of motion, that is, not moving.
- **Newton's 1st law** states an object at rest stays at rest and an object in motion stays in motion (same speed and direction - velocity) unless acted upon by an unbalanced **force**. This law describes an object's resistance to change, which is why it is often referred to as the **law of inertia**.
- It also implies that to change an object's motion a force must act on that object.

1. A ball is placed on flat ground and then left there with no one or thing touching it other than the ground.

   (a) Is the ball likely to move? *No*

   (b) Why? *There is nothing to make it move. No force is pushing or pulling it.*

   (c) How could someone get the ball to move? *They would need to apply a force in one direction (but not downwards)*

2. (a) A single ball bearing rolls along a smooth metal surface. What is needed to bring the ball to a stop?

   A force acting in the opposite direction to the ball bearing's travel. The force only acts until the ball stops.

   (b) What is needed to change the direction of the ball bearing's motion? *A force acting in any direction other than the direction of the ball bearing's travel (but not downwards).*

3. What happens to a stationary object when a force is applied to it in one direction?

   It accelerates in the direction the force is applied.

4. What happens if a greater force is applied to the object?

   The object accelerates at a greater rate.

©2020 **BIOZONE** International
ISBN: 978-1-927309-82-7
Photocopying Prohibited

Normal force

- When an object comes in contact with a surface (the ground or another object), that surface exerts a force on the object which resists changes in the surface. This force is called the **normal force** (F_N) and it is always perpendicular to the surface.
- Normal force is a force of resistance. It prevents the surface from deforming or breaking due to force applied through contact with an object.
- When an object is placed on the ground, the ground exerts a normal force on the object equal to the force of gravity acting on the object (its weight).
- Without an equal normal force to counter the weight of the object, the object would pass through the surface. Think about what would happen if you were to step onto a surface of water like in a pool. Water does not resist deformation well and therefore does not provide enough normal force to support your weight.
- Normal force is not only resistance to weight. It is exerted whenever an object comes in contact with a surface. If you were to push on a wall, the wall will exert a normal force equal to the force of your push to resist change.
- When there is no change in the surface, the normal force is equal the force applied by the object. If the normal force is less than the applied force, the object will pass through the surface.
- The normal force is always in the opposite direction of the force applied. If the force is a pull, the direction of the normal force is away from the surface.

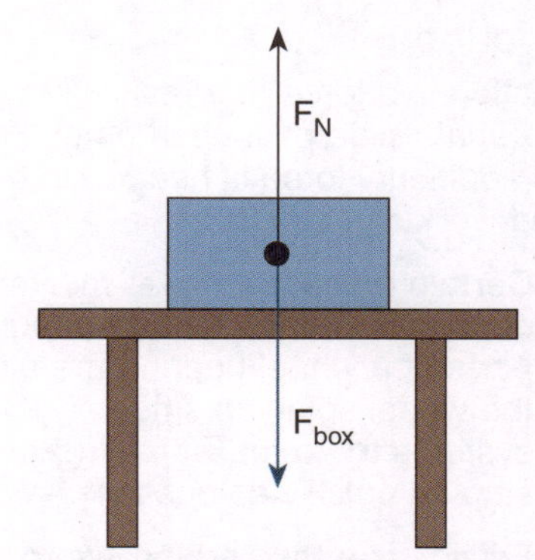

The normal force is equal and opposite to the force applied by the box

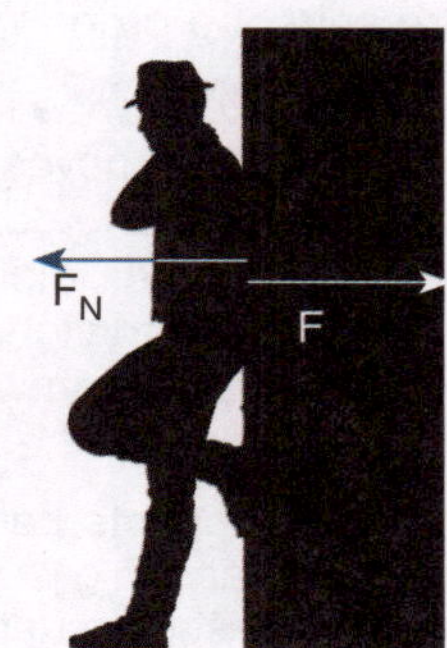

5. On the diagram below draw arrows to show the direction of the following forces:

   (a) The normal force.

   (b) The force produced by the block due to gravity.

   (c) The force in the direction the block would accelerate due to the incline.

   (d) The force of friction (resistance to motion) between the block and surface.

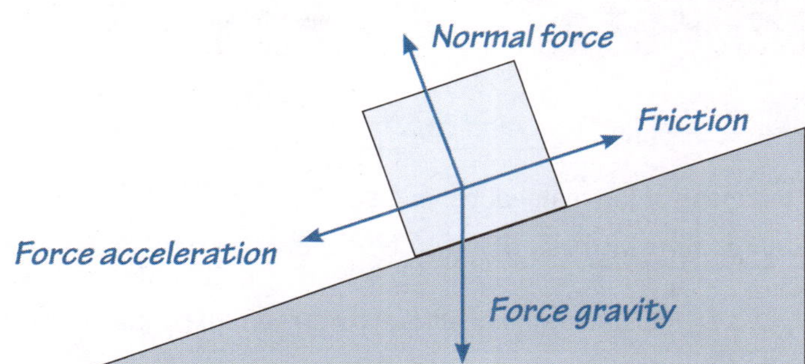

   (e) From your diagram comment on the relationship between the normal force and the force due to gravity for the ball on a sloped surface and contrast this with these force on a horizontal surface:

   On a horizontal surface the normal force and the force of gravity act in opposite directions. This object therefore does not move. On a sloped surface the normal force acts at an angle compared to the force of gravity. This means the object will move because the forces acting on it are not equal and opposite.

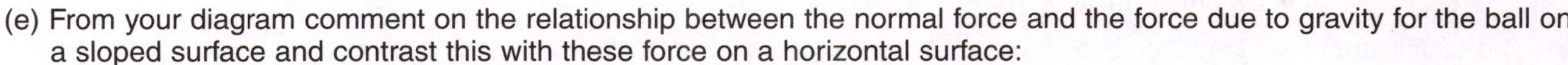

   (f) As the slope became steeper what force would have to increase in order for the ball to remain stationary:

   Friction

   (g) What force would this have to equal? The force causing the acceleration down the slope.

6. What happens if the force applied to a surface by an object is greater than the normal force?

   The surface with deform or break.

©2020 BIOZONE International
ISBN: 978-1-927309-82-7
Photocopying Prohibited

Friction

Friction is a force that resists motion. It is caused when materials slide past each other. Friction is industrially important as additional force is needed to overcome it before useful motion and work can occur.

- Certain processes try to minimize or maximize friction depending on the outcome required. For example, the moving parts of a wheel bearing are oiled to reduce friction and keep the wheel spinning smoothly. On the other hand, braking systems try to maximize friction to bring a moving object to a stop as quickly as possible.
- Friction is a "two-edged sword" in motion. Without it, there would be no stopping a moving object and no grip (try running on ice in socks). However, friction opposes the rolling of a tire or the spin of a propeller, and even movement through air.
- Jet planes experience high friction as they move through the air. As the plane moves through the air it hits air molecules. The faster the plane goes the more air molecules are hit per unit of time. The air molecules are also hit harder. These two factors increase friction which acts to slow the plane down. Eventually the friction will equal the force produced by the plane's engine and the plane will reach a constant velocity.

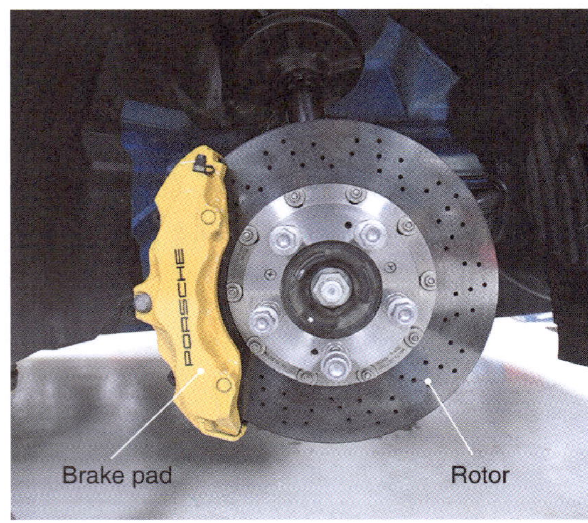

In a car, two high friction brake pads engage the rotor (attached to the wheel) to stop it spinning. The rotational energy of the rotor is converted to heat. The more quickly the rotor can heat up and disperse the heat, the faster it can stop.

7. A group of students wanted to study the friction produced by various materials. Each material used was loaded with additional mass to give the same total mass each time. The force meter was attached to the material and pulled until the material just started to move. The force required for this was recorded. This step was repeated several times for each material and the results averaged. The setup and results are shown below:

Material being tested

Force meter

Smooth, level bench top

Material	Average force required to move object
Wood (oak)	2
Paper	1.5
Sand paper	3
Plastic	1.5

(a) Which material produced the most friction? *The sand paper*

(b) Why did the students adjust the mass of the material being tested? *Friction is a force acting in opposition to applied force. Mass is a component of force (F = ma) and should be constant for a fair test.*

8. The force of friction (F_f) can be found by multiplying the coefficient of friction (μ) by the normal force (F_N). The coefficient of friction can be two different things:

- The coefficient of static friction μ_s shows the relationship between F_f and F_N when two objects are not moving.
- The coefficient of dynamic friction μ_k shows the relationship between F_f and F_N when either of the objects are moving.

The coefficient of static friction is always greater than for dynamic friction (it is easier to keep an object moving than to get it to start moving).

(a) The coefficient of static friction for oak wood is 0.62. What was the force exerted by the wood on the bench top in question 7?

$2 \div 0.62 = 3.22$ N

(b) The mass of the wood was 330 grams. How does this mass and the force the wood exerts compare?

$F = ma = 0.33 \times 9.8 = 3.2$ The mass exerts the same force as calculated above.

9. (a) Describe some occasions where friction needs to be minimized: *Student's answers may vary. Possible answers may include: engines, moving machine parts, skiing.*

(b) List some occasions when friction needs to be increased: *Student's answers may vary. Possible answers may include: brakes, grip, parachutes.*

©2020 **BIOZONE** International
ISBN: 978-1-927309-82-7
Photocopying Prohibited

69 Newton's Second Law

Key Question: What happens to an object when the forces acting on it are not balanced and how do we use what we know about motion to solve force equations?

- Newton's 1st law of motion predicts that only an unbalanced force will cause an object to accelerate by changing its speed, direction, or both speed and direction.
- Newton's 2nd law applies to the behavior of objects for which all existing forces are not balanced and states the acceleration of an object is dependent upon the net force acting upon the object and the mass of the object.
- The acceleration of the object is directly proportional to the net force applied to it and inversely proportional to the mass of the object. In other words, the greater the force applied, the greater the acceleration and the more massive the object, the slower the acceleration.

Weight is not your mass

Everyday language often gives meanings to words that are different to what they have in physics. Mass and weight are good examples. When most people talk about weight they actually mean mass.

- **Mass** (*m*) measures the amount of matter (atoms) making up an object. The SI unit for mass is kilogram (kg). Take the stone on the right as an example. If you placed it on an electronic balance, you might get a readout of 50 grams. This is the stone's mass.
- But how does the scale know how much matter is in the stone? It doesn't count every atom, so what is it doing? It is measuring the force due to gravity (weight) in newtons acting on the stone. Near the surface of the Earth, scales or balances are calibrated so that the force due to gravity they are measuring is divided by 9.8, which gives the result on the readout as the mass in kilograms.
- **Weight** is the measurement of the pull of gravity on an object, so it is also called the **force due to gravity**. It is a force, so it is measured in newtons (N). The mass of an object remains the same (unless it gains or loses atoms) but its weight will depend the strength of gravity at its location. Thus, your weight on Earth is your mass multiplied by the strength of gravity.

1. Write an equation for calculating weight: *W = mg (weight = mass x gravity)*

2. Complete the table below:

Place	Strength of gravity (N/kg)	Mass (kg)	Weight (N)	Mass (kg) on Earth that has the same weight
Earth	9.8	70	*686.7*	*70.0*
Mars	3.7	70	*259.7*	*26.5*
Moon	1.6	70	*113.4*	*11.6*
Jupiter	24.8	70	*1735.3*	*176.9*

3. If newtons are measured in kg m/s^2 and mass is measured in kg what are the actual units for the strength of gravity and what is this called?

g = W/m = kg m/s^2 / kg = m/s^2 gravity is therefore m/s^2 which is acceleration.

4. Rewrite you equation from question 1 to substitute in force (F) and the property from question 3:

F = ma

5. Determine the accelerations that result when a 10 N force is applied to a 3 kg and 6 kg object respectively:

a = F/m = 10 N ÷ 3 kg = 3.3 m/s^2

a = 10N ÷ 6 kg = 1.7 m/s^2

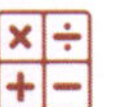

6. A force of 15 N is applied on a brick causing it to accelerate at a rate of 5 m/s^2. Determine the mass of the brick:

m = F/a = 15 N ÷ 5 m/s^2 = 3 kg

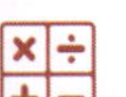

7. A skateboard is accelerating at a rate of 2 m/s^2. If the net force is tripled and the mass doubled, what is the new acceleration of the skateboard?

a = F/m, 2 m/s^2 = F/m, 3F/2m = 3/2 F/m = 3/2 × 2 = 3 m/s^2

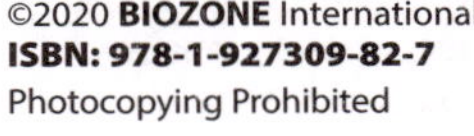

©2020 BIOZONE International
ISBN: 978-1-927309-82-7
Photocopying Prohibited

CE  PS2.A 

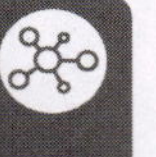

- Newton's 2nd law can be expressed as an equation: $a = F/m$
- This equation is more commonly written:

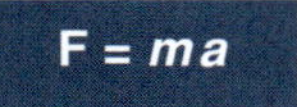

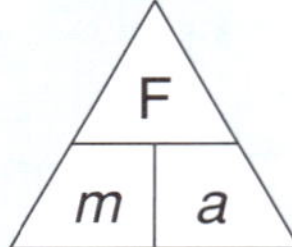

Force, mass, and acceleration

Weight is the effect of the acceleration of gravity on an object. By using string and a pulley, the force of gravity accelerating a falling object can be used to accelerate a trolley horizontally along a bench at the same time.

- Two students were investigating acceleration. If they keep the force constant, they wondered what effect changing the mass would have on the acceleration:

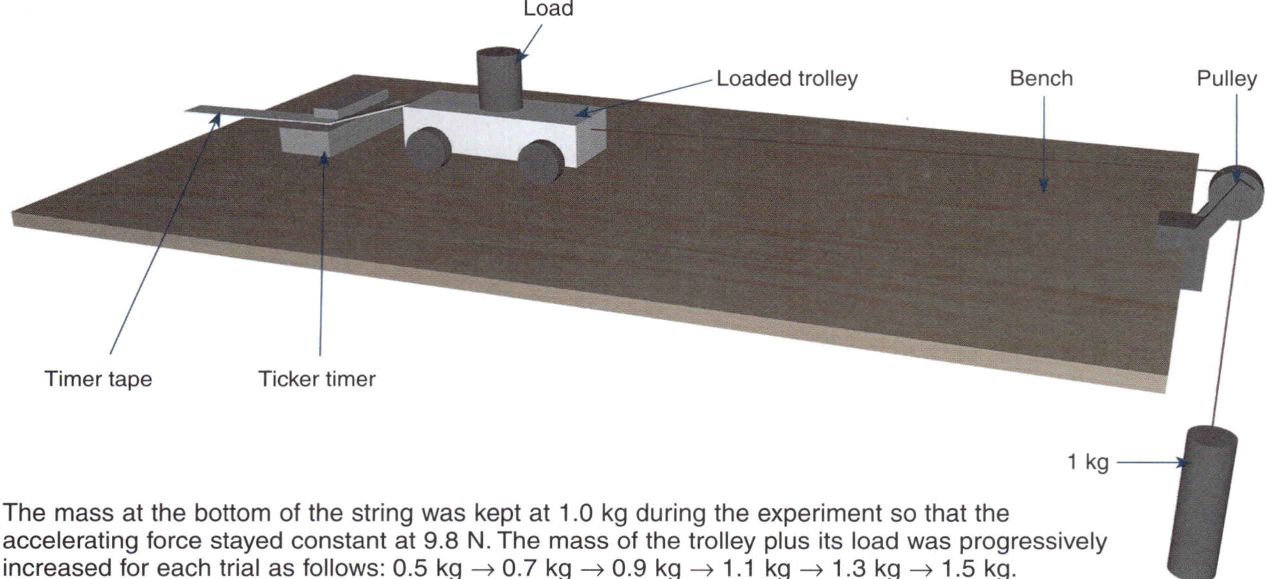

- The mass at the bottom of the string was kept at 1.0 kg during the experiment so that the accelerating force stayed constant at 9.8 N. The mass of the trolley plus its load was progressively increased for each trial as follows: 0.5 kg → 0.7 kg → 0.9 kg → 1.1 kg → 1.3 kg → 1.5 kg.

  The corresponding increase in the total mass of the system (loaded trolley mass + 1 kg) for each trial is therefore: 1.5 kg → 1.7 kg → 1.9 kg → 2.1 kg → 2.3 kg → 2.5 kg.
- Before the 1 kg mass was attached to the string, they compensated for friction by squeezing a suitably sized piece of plasticine onto the string, just below the pulley. If the system (loaded trolley, pulley, string, and ticker tape with timer running) moves at near constant speed after being given a slight nudge then, as near as possible, friction within the system has been balanced out.
- They attached timer tape to the trolley and threaded the tape through a ticker timer.
- They then started the timer and let the trolley go. The trolley rolled forward as the 1 kg fell to the floor. When it hit the floor, they stopped the ticker timer.
- The students reset the equipment and repeated the procedure to produce a tape for each different total mass.
- By counting the spaces between the dots on the timer tape the students were able to calculate the exact time the 1 kg was falling. Each space is equal to 0.02 seconds (the timer makes 50 dots per second). They multiplied the number of spaces by 0.02 to determine the exact time the 1 kg was falling.
- The distance the trolley rolled was determined by measuring the distance between the first and last dot.
- Their results are shown in the table below:

Trial	System mass (kg)	$\frac{1}{\text{System mass}}$	Force causing acceleration (N)	Total time (s)	Distance (m)	Acceleration (m/s^2)
1	1.5	0.667	9.8	0.61	1.2	6.45
2	1.7	0.588	9.8	0.64	1.2	5.86
3	1.9	0.526	9.8	0.68	1.2	5.19
4	2.1	0.476	9.8	0.72	1.2	4.63
5	2.3	0.435	9.8	0.75	1.2	4.27
6	2.5	0.4	9.8	0.78	1.2	3.94

8. (a) Complete the column for the 1/system mass during each trial:

   (b) Complete the acceleration column using the equation $\mathbf{a = 2d \div t^2}$ to calculate the acceleration of the trolley:

©2020 **BIOZONE** International
ISBN: 978-1-927309-82-7
Photocopying Prohibited

9. Explain the physics behind the technique used to compensate for friction:
The mass of plasticine on the string allowed gravity to work on it. If the system moved at constant velocity, there was no acceleration, so the forces of gravity and friction were equal.

 10. Plot acceleration vs 1/system mass with a line of best fit below. We use 1/system mass so the slope is positive:

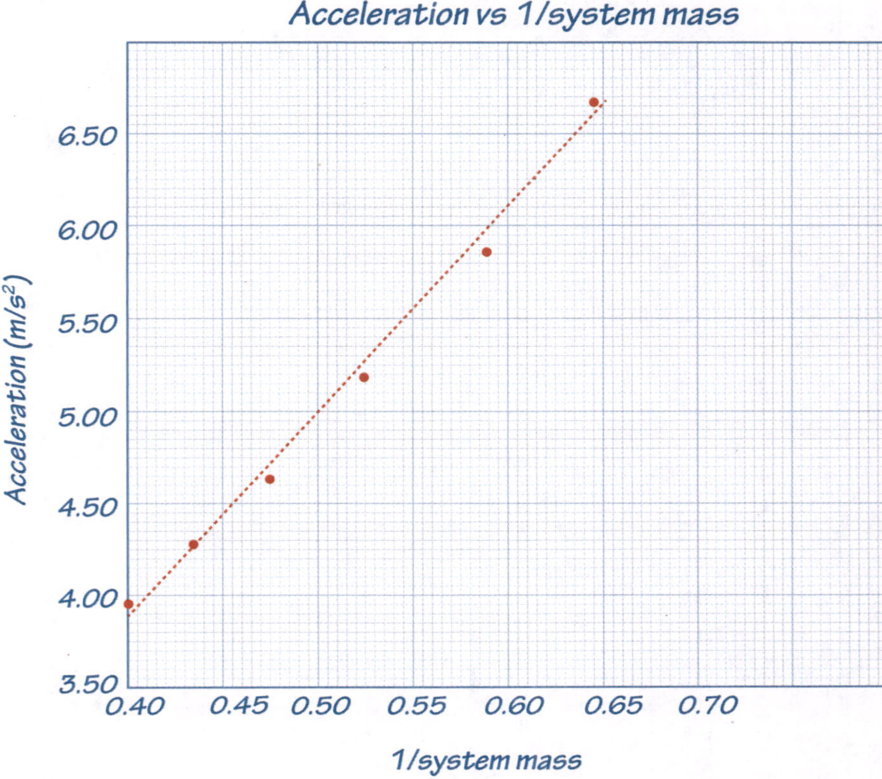

11. (a) What was the general effect on the acceleration when more mass was added to the system?
The higher the mass, the less the acceleration.

(b) Calculate the gradient of the graph: *Gradient = $\Delta y \div \Delta x = 2.51 \div 0.267 = 9.4$*

(c) The graph is a straight line, so it has the equation **a = gradient × 1/m + intercept**. From **F = m × a** we have the equation **a = F × 1/m + 0**. By comparing these two equations what is the value of F? Comment on its significance.
F would be equal to the gradient = 9.4. Since the graph shows a straight line of best fit, the gradient is constant. This means F is the net force after friction has acted against the force causing acceleration.

(d) Explain your answer from 11(a): *Acceleration is less as mass increases because the force acting on it has been kept constant. This means the force is proportionally less each time and so its effect is reduced according to a = F/m.*

(e) From the table of data, what would the velocity of the trolley from trial 1 be after 3 seconds?
19.35 m/s (after 1st second 6.45 m/s, 2nd 12.9 m/s, 3rd 19.35 m/s)

(f) From the table of data, what would the velocity of the trolly from trial 6 be after 3 seconds?
11.82 m/s (after 1st second 3.94 m/s, 2nd 7.88 m/s, 3rd 111.82 m/s)

©2020 **BIOZONE** International
ISBN: 978-1-927309-82-7
Photocopying Prohibited

- A second group of students used the same equipment to carry out a slightly different investigation. They wanted to keep the mass of the system constant while increasing the force acting on the system. They achieved this by moving masses from the loaded trolley and adding them to the 1 kg at the bottom of the string.
- Their results are shown in the table below:

Trial	System mass (kg)	Mass causing the acceleration (kg)	Force causing the acceleration (N)	Total time (s)	Distance (m)	Acceleration (m/s^2)
1	2.6	1.0	9.8	0.80	1.2	3.8
2	2.6	1.2	11.8	0.73	1.2	4.5
3	2.6	1.4	13.7	0.67	1.2	5.3
4	2.6	1.6	15.7	0.63	1.2	6.0
5	2.6	1.8	17.6	0.59	1.2	6.9
6	2.6	2.0	19.6	0.56	1.2	7.7

12. For the table above:

(a) Complete the column for the force causing the acceleration during each trial (in newtons).

(b) Complete the column for the acceleration during each trial using the equation $\mathbf{a = 2d \div t^2}$

13. Plot a graph of acceleration versus force, with a line of best fit, on the grid below:

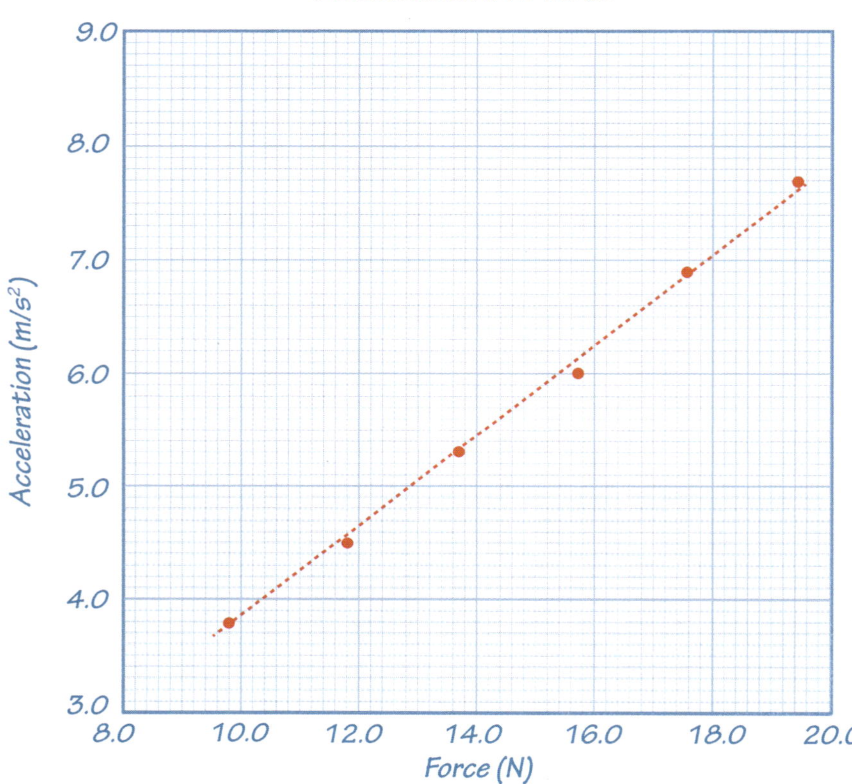

14. Why was the force acting on the system increasing? The hanging mass was increasing. This has the effect of increasing the force acting on the system via F = ma.

15. What was the general effect on the acceleration when more force was used?
As the force increased, the acceleration increased.

16. (a) Calculate the gradient of the graph: Gradient = $\Delta y \div \Delta x = 3.9 \div 9.8 = 0.40$

(b) The graph is a straight line, so it has the equation **a = gradient × F + intercept**. From **F = m × a** we have the equation **a = 1/m × F + 0**. By comparing these two equations, calculate a value for m. Comment on its significance:

Gradient = 1/m Therefore, m = 1 / 0.40 = 2.5 (2.6 before earlier rounding) kg. This is the mass of the system. If the total mass of the system is constant, acceleration is proportional to force.

©2020 BIOZONE International
ISBN: 978-1-927309-82-7
Photocopying Prohibited

Free body diagrams

Free body diagrams are used to show the forces acting on an object. Consider the photograph of the jet fighter flying in level flight and at a constant velocity below:

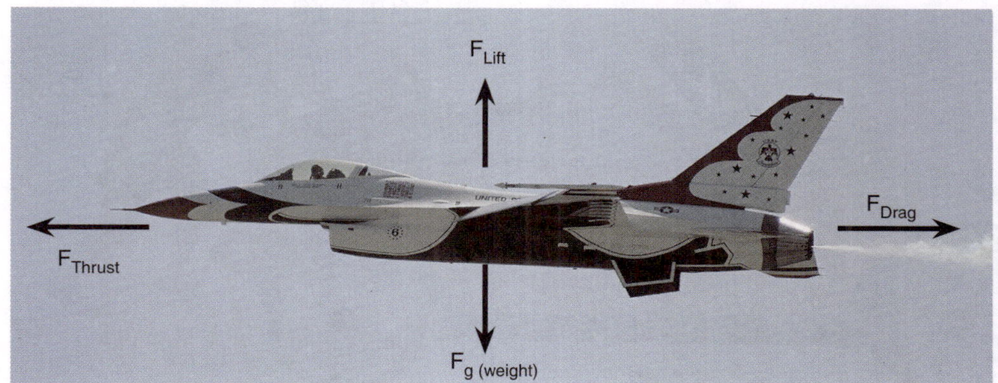

- This **free body diagram** shows the forces acting on the jet plane. If it is flying horizontally at constant velocity, then the forces will be balanced. The force due to gravity acting on the aircraft is balanced by lift force generated by the wings. The thrust force of the engine is balanced by the drag force caused by air resistance.

17. The plane shown above is traveling at a constant speed. If the thrust from the engine is increased from say 89,000 N (about 9,000 kg) of thrust to 96,000 N (about 9,800 kg) of thrust, how would the motion of the plane be affected?

The plane would accelerate until the friction from air resistance equaled the new force of the thrust. At that point the plane would reach its new higher velocity and velocity would remain constant (forces are balanced).

18. The diagrams below show the forces acting on a trolley of 0.5 kg at different times:

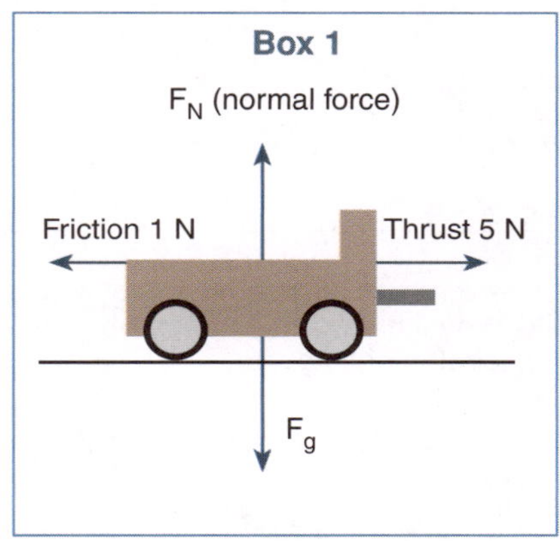

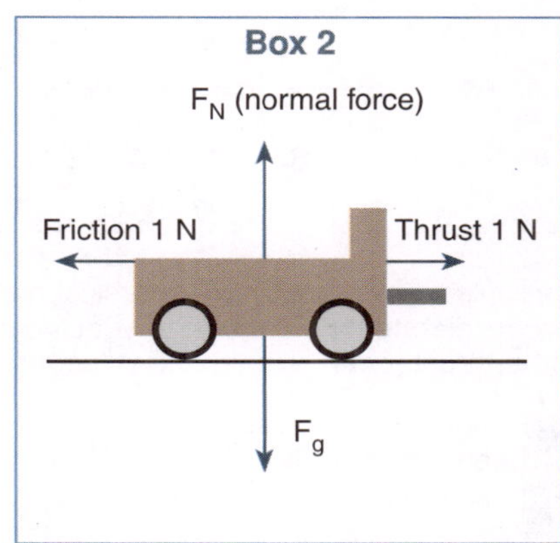

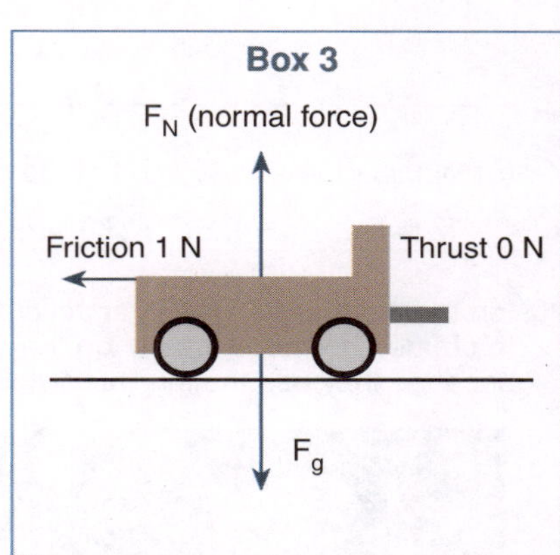

(a) Complete the table below:

	Box 1
$F_{(net)}$	*4 N to the right*
m	*0.5 kg*
a	*8 m/s²*

	Box 2
$F_{(net)}$	*0 N*
m	*0.5 kg*
a	*0 m/s²*

	Box 3
$F_{(net)}$	*1 N to the left*
m	*0.5 kg*
a	*-2 m/s²*

(b) Describe the motion of the trolley in box 1: *Rapid acceleration to the right.*

(c) Describe the motion of the trolley in box 2: *No acceleration - constant speed.*

(d) Describe the motion of the trolley in box 3: *Deceleration.*

©2020 **BIOZONE** International
ISBN: 978-1-927309-82-7
Photocopying Prohibited

19. The battery in Rebekah's car has gone flat. Michael and Eddie are push-starting the car. The diagram below shows the forces applied to the car. The men need to apply enough force to overcome the forces opposing the movement of the car. The mass of the car is 630 kg. Rebekah has a mass of 62 kg.

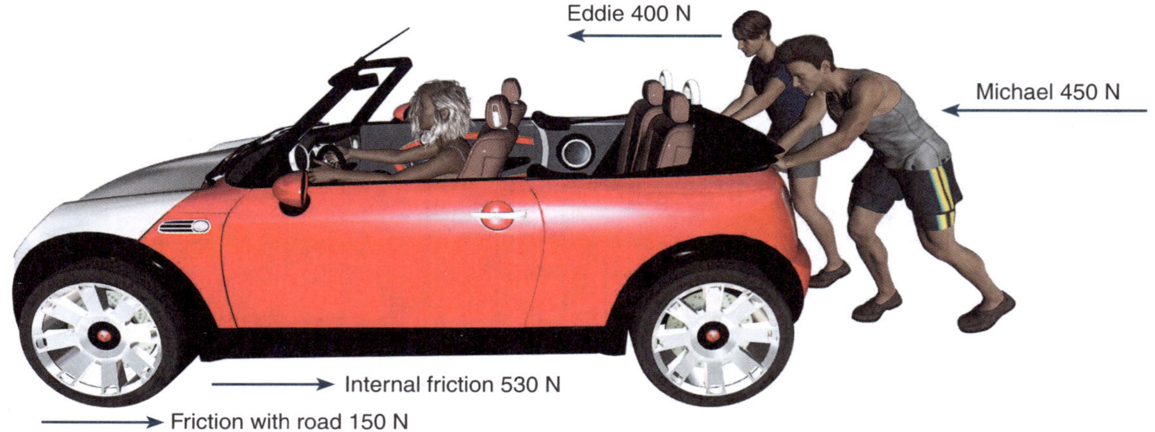

(a) Draw a free body diagram showing the net forces applying to the car:

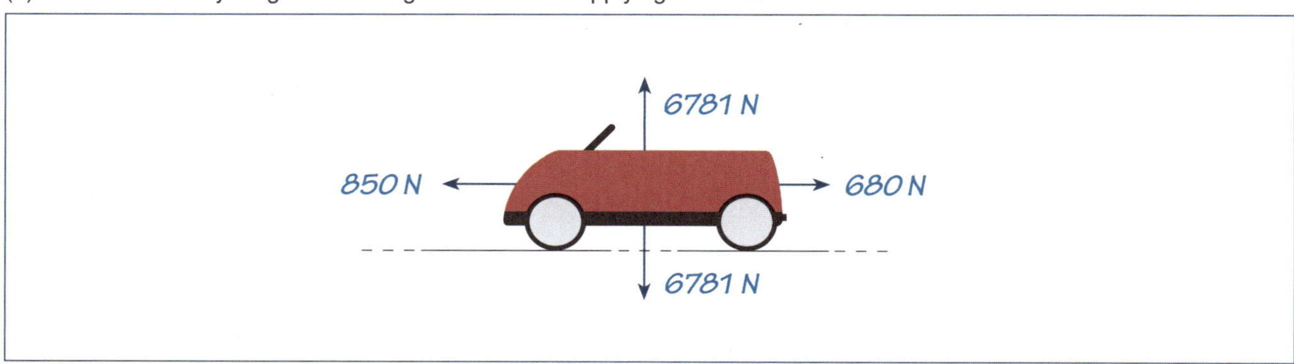

(b) What is the acceleration of the car as Michael and Eddie push it? *Net thrust = 850 N - 680 N = 170 N*

Total mass = 630 + 62 = 692 kg   a = F ÷ m = 170 ÷ 692 = 0.25 m/s^2

20. Ion thrusters are a type of spacecraft propulsion system. Unlike chemical rockets, which burn large amounts of fuel to create enormous thrust, ion thrusters accelerate ions through an electrical field, creating very little thrust. However, because they use the ion "fuel" slowly the thruster can remain active for very long lengths of time.

NSTAR ion thruster on the Deep Space 1 spacecraft

The Dawn spacecraft was launched in 2007 and arrived at Ceres in 2015. Its main propulsion systems was a xenon based ion thruster.

(a) Ion thrusters produce only small amounts of thrust. It took 96 hours to accelerate from 0 to 96 kilometers per hour. What was the acceleration of the space probe in m/s^2?

$a = \Delta v / \Delta t$ = (96,000 m ÷ 3600 s) ÷ 345,600 s = 7.72×10^{-5} m/s^2

(b) The mass of the dawn spacecraft was 1.2177×10^3 kg. Calculate the force the ion thruster produced:

F = ma = 1.2177×10^3 kg x 7.72×10^{-5} m/s^2 = 0.094 N

(c) Over the course of its mission, the ion thruster on Dawn managed a velocity change of 11.5 km/s. How long would it need to be on (at full thrust) to reach change in velocity?

$\Delta t = \Delta v \div a$ 11.5 x 1000 = 11,500. 11500 m/s / 7.72×10^{-5} m/s^2 = 148,963,730 s = 4.7 years.

(Total run time of the engine was 5.9 years so not on full thrust constantly)

©2020 **BIOZONE** International
ISBN: 978-1-927309-82-7
Photocopying Prohibited

Free fall

- **Gravity** is a property of all masses that causes them to pull together. This pull is almost insignificant unless at least one of the masses is extremely large, e.g. Earth's gravity pulls all smaller objects near it towards its center.
- **Free fall** is a special type of motion in which gravity is the only force acting on an object. Objects in free fall are not encountering a significant force of air resistance, they are falling under the sole influence of gravity. Under such conditions, all objects fall with the same acceleration, regardless of mass.
- On Earth, objects in free fall accelerate at a rate of 9.8 m/s^2. This specific acceleration value is so important in physics that it has its own specific name (**acceleration of gravity**) and symbol (**g**).
- When solving free fall problems, we simply substitute **g** for acceleration (**a**) in our acceleration equations.
- Because the acceleration towards the center of the Earth due to the effects of gravity is always down (from our perspective) and we care about direction in physics, **g** = -9.8 m/s^2. The negative sign indicates the acceleration is in the downwards direction.
- If you were to drop a rock from the top of a cliff, in 1 second it will reach a speed of 9.8 m/s (a velocity of -9.8 m/s because it is motion in the down direction). After 2 seconds, it will be traveling at 19.6 m/s. After 3 seconds, it has reached 29.4 m/s. And so on.

Air resistance

- In the late 1500s, Galileo conducted a series of experiments where he demonstrated that heavy objects and light objects fall at the same speed.
- In spite of popular belief, Galileo did not state that the objects would hit the ground at the same time - he understood **air resistance**. He did reason that without air resistance, the objects would fall at the same rate.
- Air resistance or **drag** is the force opposing the downward acceleration due to gravity. This force is due to contact with air particles as the object pushes through them. Inertia makes these air particles resist changes in their states of motion resulting in drag.
- The faster an object's velocity, the greater the opposing air resistance. This is due to contact with a greater number of particles in the same time-frame.

Galileo

Terminal velocity

- During free fall a point is reached at which gravitational force is equal to the opposing force of air resistance. At this point the object stops accelerating and falls at a constant velocity. This is called **terminal velocity**
- Surface area is a very important factor when determining terminal velocity. The greater the surface area, the greater the air resistance, and subsequently the slower the terminal velocity.
- Skydivers take advantage of surface area to safely jump out of aircraft. Using a parachute, they significantly increase their surface area, thus reducing their terminal velocity to a safe speed.

F_g > air resistance
Skydiver accelerates

F_g = air resistance
Skydiver falls at constant velocity
Terminal velocity

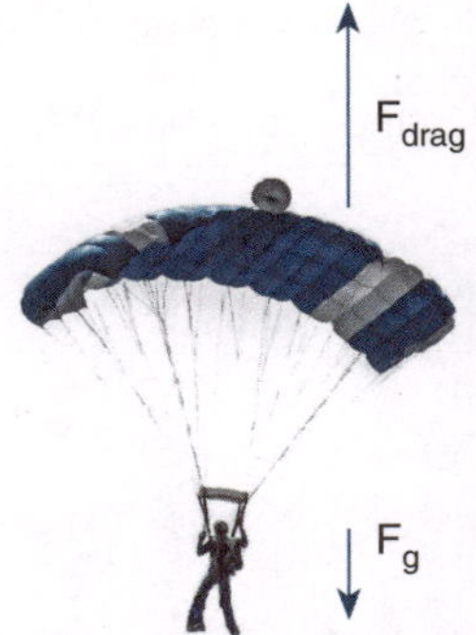

F_g < air resistance
Skydiver decelerates

21. (a) Complete the table below of the velocity of a falling object in a vacuum (recall $\Delta x = v_f \Delta t - ½ a \Delta t^2$):

Time	Velocity	Displacement
1 s	-9.8 m/s	-4.6 m
2 s	-19.6 m/s	- 19.7 m
3 s	- 29.4 m/s	-44.1 m
4 s	-39.2 m/s	-78.4 m
5 s	-49.0 m/s	-122.5 m
6 s	-58.8 m/s	-176.4 m

(b) Describe the shape of the velocity-time and displacement-time graphs this data would produce:

The velocity time graph is a straight line diagonally upwards to the right. The displacement time graph is a curve line diagonally upwards to the right.

©2020 **BIOZONE** International
ISBN: 978-1-927309-82-7
Photocopying Prohibited

Problems in free fall

- Like mathematics, solving problems in physics requires practice to acquire proficiency. Many problems require careful thought and planning before you are able to calculate.
  - If not already provided, draw a picture or diagram to visualize the problem.
  - Identify known and unknown variables and label them on your diagram.
  - Determine the equation(s) you need to use (you may need to use more than one).
  - Include units and check they agree with the answer you are looking for.
- For short drops with dense objects (typical of the problems you will be solving) we can reasonably ignore the effects of air resistance.

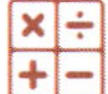

22. You are on top of a building that is 55.0 m tall. You toss a ball straight up. It travels 35.0 m up before it stops and begins to fall back down.

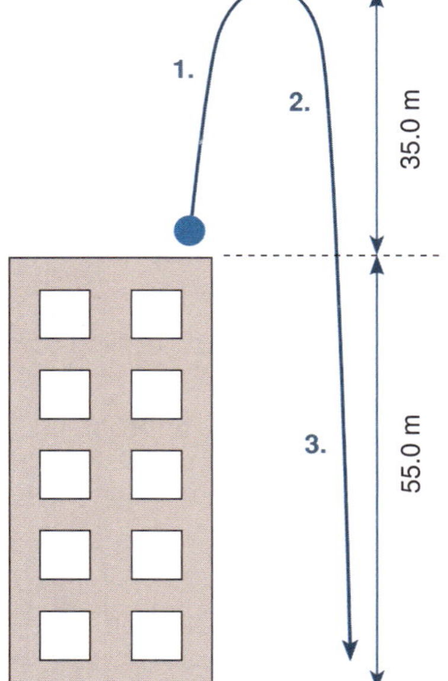

- $a = g = -9.8\ m/s^2$
- $V_{i\,(1.)} = (?)$
- $V_{f\,(1.)} = 0\ m/s$
- $\Delta_{x(1.)}$ 35 m

(a) What was the ball's initial velocity (how fast was the ball when it left your hand)?

$\Delta x = (v_f^2 - v_i^2) \div 2a \rightarrow v_i = \sqrt{(v_f^2 - 2a\Delta x}$

$= \sqrt{(0 - 2 \times -9.8\ m/s^2 \times 35\ m} =$

$= \sqrt{686\ m^2/s^2} = 686\ m/s$

(b) It goes up and then falls down to the ground below. How much time was the ball in the air? Hint: you will need multiple steps to solve this problem.

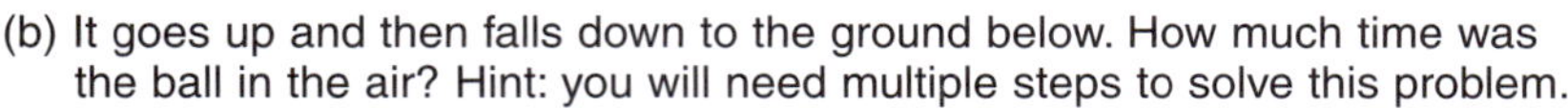

1. $a - (v_f - v_i) \div \Delta t_1 \rightarrow \Delta t_1 = (v_f - v_i) \div a = -26.2\ m/s \div -9.8\ m/s^2$

$= 2.7\ s$

2. $\Delta t_2 = -26.2\ m/s \div -9.8\ m/s^2 = 2.7 s$

3. $vf = \sqrt{(v_i^2 + 2a\Delta t)} = \sqrt{-26.2\ m/s\ 2 + 2 \times -9.8 m/s2 \times 55\ m}$

$= \sqrt{1764.44\ m^2/s^2} = 42\ m/s$

$\Delta t_3 = (-42\ m/s - -26.2\ m/s) / -9.8\ m/s^2 = 1.6\ s$

$\Delta t_{total} = \Delta t_1 + \Delta t_2 + \Delta t_3 = 2.7 s + 2.7 s + 1.6\ s = 7 s$

23. A ball rolls down a ramp from rest and travels a distance of 3.00 m in 2.05 s. We can ask questions about this motion:

(a) Use the space below to draw and label your own diagram to visualize the problem:

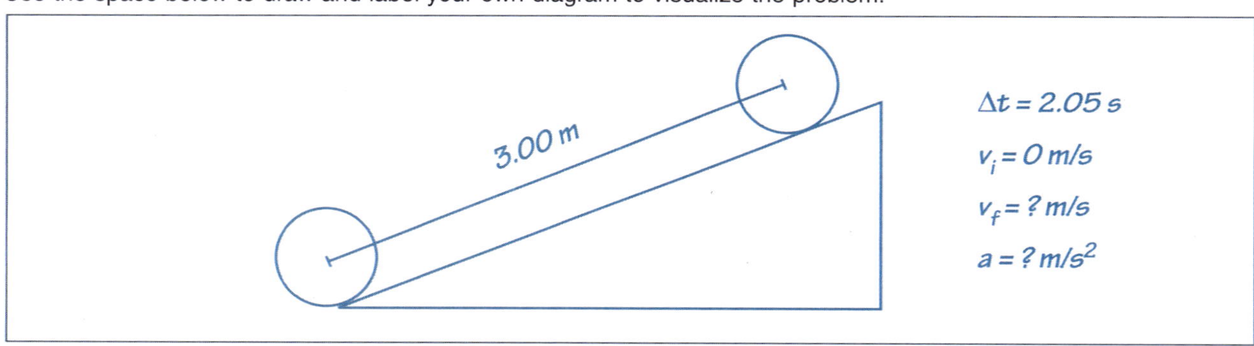

(b) What is the ball's acceleration? $\Delta x = v_i \Delta t + 0.5a\Delta t^2 = 0.5a\Delta t^2 \rightarrow a = 2\Delta x \div \Delta t^2$

$(2 \times 3.00\ m) \div 2.05\ s^2 = 1.43\ m/s^2$

(c) What is the ball's final velocity at the bottom of the ramp?

$v_f = v_i + a\Delta t = a\Delta t = 1.43\ m/s^2 \times 2.05\ s = 2.93\ m/s$

(d) What is the ball's average speed going down the ramp?

$v_{avg} = 0.5(v_f + v_i) = 0.5 \times 2.93\ m/s = 1.47\ m/s$

©2020 BIOZONE International
ISBN: 978-1-927309-82-7
Photocopying Prohibited

g on an incline

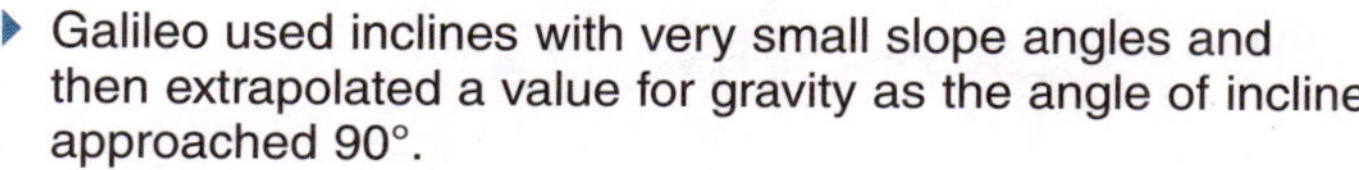

- In Galileo's time, it was difficult to measure short increments of time accurately, so to study gravity, Galileo had to slow it down. He did this by using inclined planes. Gravity still caused the motion, but its effect was decreased enough to allow Galileo to gather usable data.
- Galileo used inclines with very small slope angles and then extrapolated a value for gravity as the angle of incline approached 90°.
- Galileo found the acceleration due to gravity on an incline was directly proportional to the sine of the incline angle (sin θ).
- To solve problems involving gravity on an incline, it helps to visual the problem by sketching a free body diagram (right).

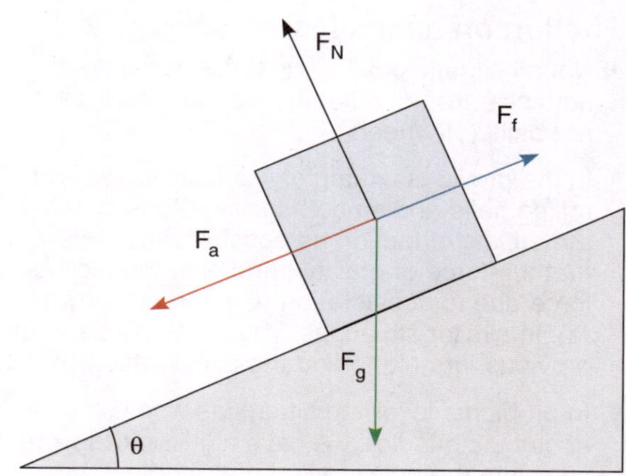

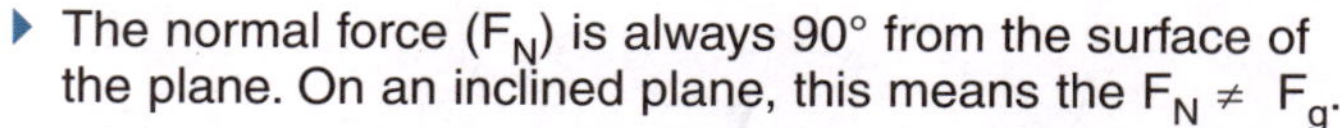

- The normal force (F_N) is always 90° from the surface of the plane. On an inclined plane, this means the $F_N \neq F_g$.
- We need to break F_g up into two components: the force of gravity perpendicular to the inclined plane and the force of gravity parallel to the inclined plane. To do this, we will create a right angled triangle as shown on the right.
- Note: the force of gravity parallel to the inclined plane is the component of gravity causing the box to accelerate down the inclined plane. We will call it F_a and this is the value we are solving for. The force of gravity perpendicular to the inclined plane is the component of gravity exerting force on the surface of the inclined plane and is equal to the normal force. We will call this component $-F_N$.

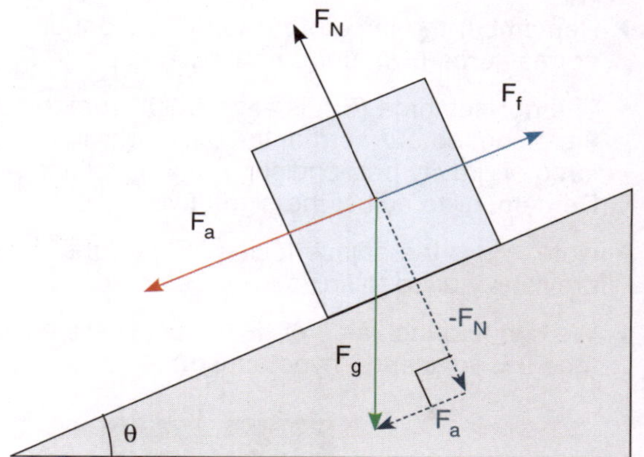

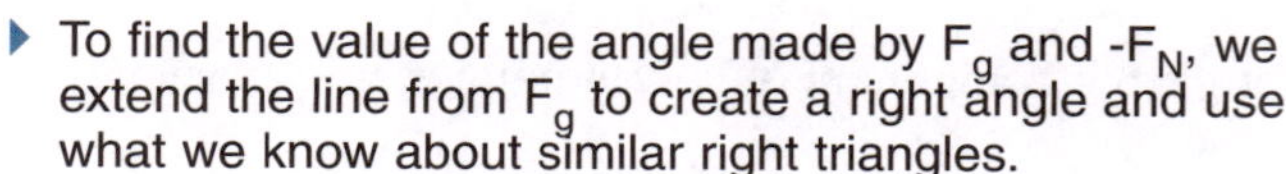

- To find the value of the angle made by F_g and $-F_N$, we extend the line from F_g to create a right angle and use what we know about similar right triangles.
- The measure of this angle is the same as the angle of incline of the plane (θ).
- sin θ = opposite ÷ hypotenuse = $F_a \div F_g$, therefore:

$$F_a = F_g \sin \theta$$

- Dividing both sides of the equation by m, we get:

$$a = g \sin \theta$$

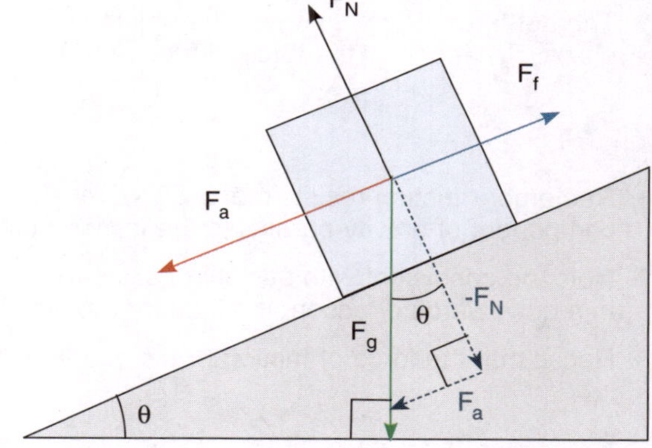

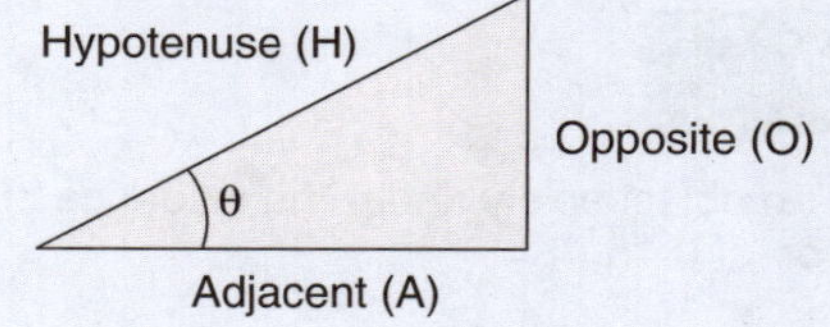

H is always the longest side of a right angle triangle and is opposite the right angle. O is opposite theta. The sine of angle theta (θ) can be calculated by measuring the value of O and dividing it by the value of H. Theta can be obtained from the inverse function of sine theta ($\sin^{-1}$).

24. (a) Using the derived equation above, calculate the acceleration of a box sliding down a frictionless inclined plane angled at 35°.

$a = g \sin\theta = -9.8\ m/s^2 \times \sin 35° = -5.62\ m/s^2$

(b) Do you think this acceleration is slow enough for Galileo to make accurate measurements of time? Explain:

No. The acceleration is more than half the acceleration of gravity. Objects will still be moving too fast to observe and record measurements clearly. Acceleration would need to be slower to make accurate measurements (smaller angle needed).

©2020 **BIOZONE** International
ISBN: 978-1-927309-82-7
Photocopying Prohibited

Friction on an incline

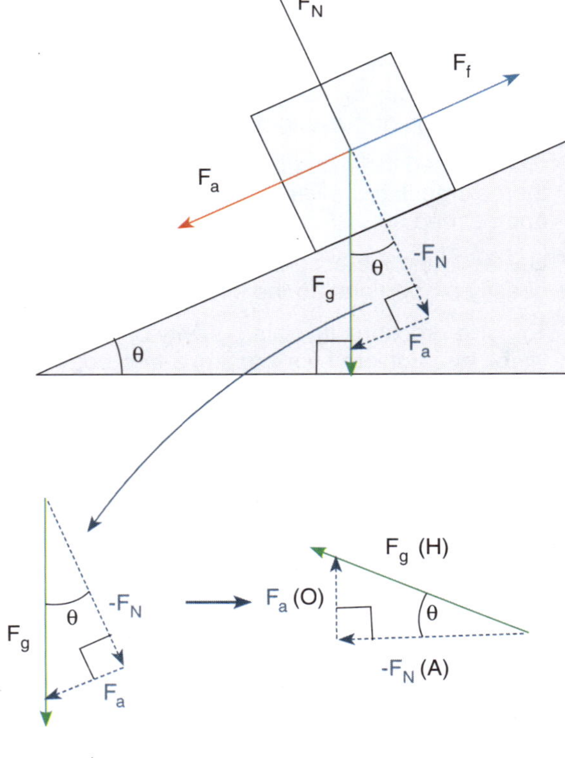

- Not all situations allow us to treat friction as negligible. Most surfaces are not smooth and most objects on the incline are not balls (or wheels).
- In the previous example, like Galileo, we were looking at rolling balls and smooth inclines to eliminate as much of the effects of friction as possible. Thus, we were able to say that the force of gravity parallel to the inclined plane was the force due to acceleration (F_a) and subsequently acceleration (a). In similar situations, you are still able to use the previous formula to find the acceleration of the object.
- In problems involving situations where friction is a contributing factor, we have a little more work to do to calculate the acceleration of the object.

- Remember the force of friction (F_f) is dependent on the normal force (F_N), but on an inclined plane $F_N \neq F_g$.
- The normal force (F_N) is always 90° from the surface of the plane, so our normal force is equal in magnitude to the force of gravity perpendicular to the inclined plane ($-F_N$). Remember to adjust the sign of your value.
- Note: $-F_N$ is the adjacent side of the right triangle we previously used to solve for acceleration (a).
- We can find the value of $-F_N$ by using the cosine of theta ($\cos\theta$ = adjacent ÷ hypotenuse):

$$F_N = -F_g \cos\theta$$

- The net force (F_{net}) acting on the object (causing acceleration) is the component of gravity parallel to the inclined plane (F_a) added to the force of friction (F_f):

$$F_{net} = F_a + F_f$$

- Remember friction resists motion. This means the value of the force friction (F_f) will be opposite in value to the component of gravity parallel to the inclined plane (F_a).
- Note: be consistent with assigning signs to indicate direction. If your acceleration is negative due to gravity acting in a downwards direction, the sign for friction will be positive.
- Recall that the force of friction (F_f) is equal to the coefficient of friction (μ) multiplied by the normal force (F_N):

$$F_f = \mu F_N$$

- Combining our equations, we get the acceleration of an object on an inclined plane:

$$a = g(\sin\theta - \mu\cos\theta)$$

25. (a) Using the equation derived above, what would the acceleration of an object solely under the influence of gravity be with a mass of 30 kg on an incline of 30° with a coefficient of friction of 0.15.

$a = -9.8\ m/s^2 \times (\sin 30° - 0.15 \cos 30°) = -3.6\ m/s^2$

(b) The same object is set up on the same inclined plane but on a different planet. Calculate the force of gravity (F_g) on this planet if the acceleration of the object going down the ramp was measured to be -2 m/s².

$-2\ m/s^2 = g \times (\sin 30° - 0.5 \cos 30°) \rightarrow g = 2\ m/s^2 \div 0.37$

$g = -5.4\ m/s^2$

(c) The object is placed on a new 30° incline (on Earth) with an unknown coefficient of friction. Find the coefficient of friction if the acceleration of the object was measured to be -4.05 m/s².

$-4.05\ m/s^2 = -9.8\ m/s^2 (\sin 30° - \mu\cos 30°) \quad 0.413 = 0.5 - \mu 0.846$

$-0.087 = -\mu 0.846 \quad -\mu = -0.087 \div 0.846 \quad \mu = 0.1$

©2020 **BIOZONE** International
ISBN: 978-1-927309-82-7
Photocopying Prohibited

70 Newton's Third Law

Key Question: What are reaction pairs and why do they not cancel each other out?

Action-reaction pairs

The picture on the near right shows an artist's concept of NASA's Space Launch System (still in production) on the launch pad, before launch. What are the forces acting on the stationary SLS? The SLS presses down onto the Earth because of gravity (we'll ignore the launch pad for simplicity). But the SLS isn't going down through the ground. Something is stopping it. What?

- Forces come in pairs. The force of the SLS pressing down on the Earth is balanced by the reaction force of the Earth pressing back against the SLS. These forces are always equal and opposite.
- It is important to note that the action and reaction forces act on different objects. This is why the opposing forces do not cancel each other out.
- If the SLS had equal-sized booster rockets on both ends pointing in opposite directions and fired them at the same time, the SLS would not go anywhere. This is because the booster rockets would exert equal and opposite forces on the shuttle and they cancel each other out. So this would *not* be an example of a reaction pair.

Both images: NASA

- However, when the booster rockets do fire they provide a directional thrust on the SLS by pushing against the Earth and then the air. Because the Earth is so massive, it does not move but the SLS does. By contrast, an incredible quantity of air is displaced by the expanding gas from the ignition of the rocket fuel to allow the SLS to ascend. Both the Earth and the SLS exert the same amount of force, but the difference is the effect the forces have on different objects.

1. Draw arrows showing the two forces mentioned above onto the photo of the SLS on the left. Label them F_{SE} (for force of SLS on Earth) and F_{ES} (for force of Earth on SLS).

2. At launch, the SLS's engines provide a force (thrust) to lift the SLS off the launchpad. What forces are acting on the SLS now? Draw arrows on the photo on the right to show the forces on the SLS. Label the forces appropriately.

3. (a) The image below shows a ferry being driven forward. The engines are providing the action force (thrust). The hull provides another action force (buoyancy). On the photo, draw and label the action and reaction forces on the ferry.

(b) Describe how the action and reaction forces act on the ferry: *The propellers push on the water. The water's resistance to change pushes back on the propellers, providing thrust. Thrust is resisted by drag (friction through a fluid, like water). The gravitational force of the Earth on the ferry and the ferry on the Earth is weight. The surface of the water resists the weight of the ferry providing support as buoyancy.*

©2020 BIOZONE International
ISBN: 978-1-927309-82-7
Photocopying Prohibited

CE PS2.A

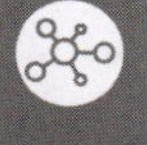

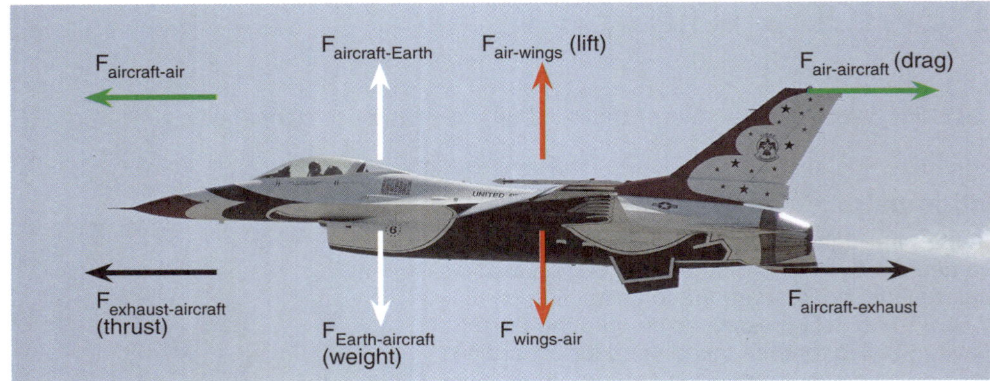

4. (a) The diagram above shows the forces-pairs acting on an aircraft in flight. If the aircraft increases its altitude, what are the main forces involved and explain what needs to happen to them:

Weight and lift. The force of lift (the force of the air acting on the wings) needs to be greater than the weight (the force of gravity acting on the aircraft) for the aircraft to gain altitude.

(b) In the left box below show the force pairs operating as the aircraft gains altitude and in the right box below show the free body forces operating as the aircraft gains altitude:

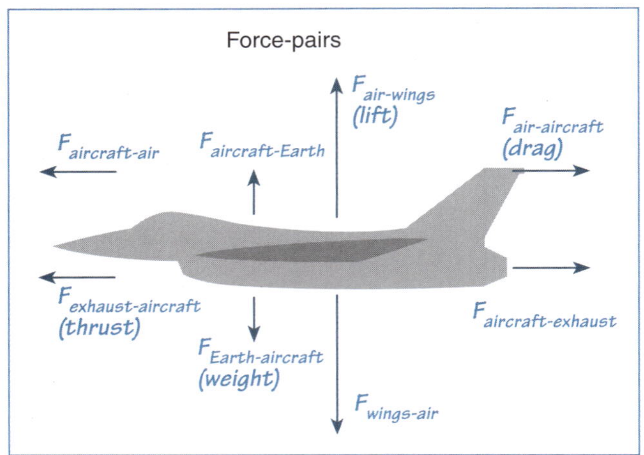

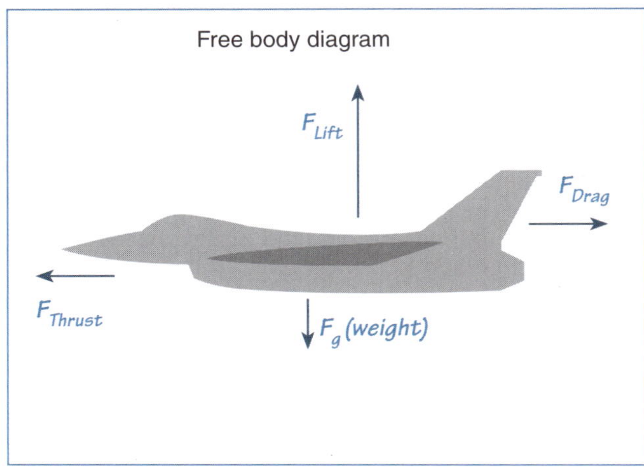

5. (a) Form small groups and use at least two reference sources to research one of the laws of motion identified below. Make a visual presentation to the class to include a general description of the law and an example demonstrating application of the principle.

(b) Using what you have learned in this activity and from the presentations of your classmates, describe an example of each of Newton's three laws below.

Law	Description	Example
First law: **Law of inertia**	Every object in a state of uniform motion tends to remain in that state of motion unless it is subjected to an unbalanced external force.	(a) *Student's answers will vary but could include a bat hitting a ball.*
Second law: **Definition of force**	$F = ma$ An object's acceleration (a) depends on its mass (m) and the applied force (F).	(b) *Student's answers will vary but could include a baseball is easier to throw than a shot put.*
Third law: **Law of reciprocity**	For every action, there is an opposite and equal reaction. When one body exerts a force on a second body, the second body simultaneously exerts a force equal in magnitude and opposite in direction on the first body.	(c) *Student's answers will vary but could include a gun recoiling when fired.*

©2020 **BIOZONE** International
ISBN: 978-1-927309-82-7
Photocopying Prohibited

71 Introduction to Momentum

Key Question: How do we describe the quantity of motion in an object and how do we use it to explain why some objects are more difficult to stop than others?

Momentum

Ever played pool or gone bowling? These games operate on the same principle: using the movement of one ball to move or knock over other objects.

- In the game of pool (right) the white ball is used to maneuver the others around the table. Sometimes when it hits a target ball, the target will move away while the white ball stops moving. Other times, the white ball continues to move but more slowly after hitting the target ball, which also moves but more slowly.
- A important part of pool or bowling, is the concept of **momentum**. In pool, momentum for the white ball is transferred to the target ball to make it move. How this is done will determine the way the target ball and the white ball move after colliding.
- Momentum is a physics term referring to the quantity of motion an object has.
- Momentum can be defined as "mass in motion". All objects have mass, so if an object is moving, it has momentum.
- Momentum is dependent on two variables: mass and velocity.
- Momentum (p) is equal to the mass of an object (m) multiplied by the velocity of the object (v):

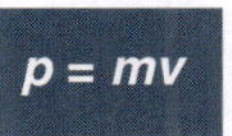

$$p = mv$$

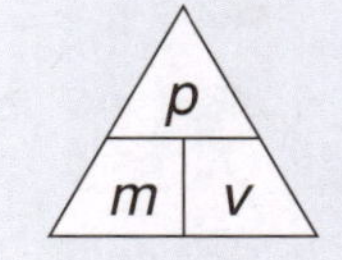

- Momentum has units of kilogram meters per second (kg m/s).
- Because momentum has a velocity component, it is a vector with the same direction as the velocity.

1. Determine the momentum of:

   (a) a 60 kg running back running east at 10 m/s: $p = 60 \text{ kg} \times 10 \text{ m/s east} = 600 \text{ kg m/s east}$

   (b) a 36,000 kg semi traveling north on interstate 95 at 30 m/s: $p = 36{,}000 \text{ kg} \times 30 \text{ m/s north}$ $= 1{,}080{,}000 \text{ kg m/s north}$

   (c) a 5.0 gram snail moving south at 2.0 meters per hour: $p = 0.005 \text{ kg} \times 0.00056 \text{ m/s south}$ $= 2.8 \times 10^{-6} \text{ kg m/s south}$

2. A car has 30,000 units of momentum. What would the car's new momentum be if:

   (a) its velocity was halved: 15,000 units

   (b) its velocity was doubled: 60,000 units

   (c) its mass was doubled: 60,000 units

   (d) its velocity was doubled and its mass halved: 30,000 units

3. Which has the greater momentum: a 25,000 kg truck moving at 5 m/s, or a 1200 kg car moving at 21 m/s?

   $p_{truck} = 25{,}000 \text{ kg} \times 5 \text{ m/s} = 125{,}000 \text{ kg m/s}$

   $p_{car} = 1200 \text{ kg} \times 21 \text{ m/s} = 25{,}200 \text{ kg m/s}$

4. A 990 kg car reduces its velocity from 22 m/s to 13 m/s. Calculate the change in the car's momentum:

   $\Delta p = p_f - p_i = (m \times v_f) - (m \times v_i) = m(v_f - v_i)$

   $= 990 \text{ kg} \times (13 \text{ m/s} - 22 \text{ m/s}) = 990 \text{ kg} \times -9 \text{ m/s} = -8910 \text{ kg m/s}$

©2020 **BIOZONE** International
ISBN: 978-1-927309-82-7
Photocopying Prohibited

Marble momentum

- Marbles (or ball bearings) are great for exploring motion. Because they are made of glass or steel, they are smooth and hard. This means they roll with very little friction on smooth, hard surfaces and convert very little kinetic energy to heat and sound during collisions. When collisions are like this, they are said to be **elastic collisions**.

INVESTIGATION 5.2: Investigating momentum

See appendix for equipment list.

You can work in pairs or small groups for this investigation.

1. You will need a marble and a ball bearing of similar size but different mass. Set up the following:

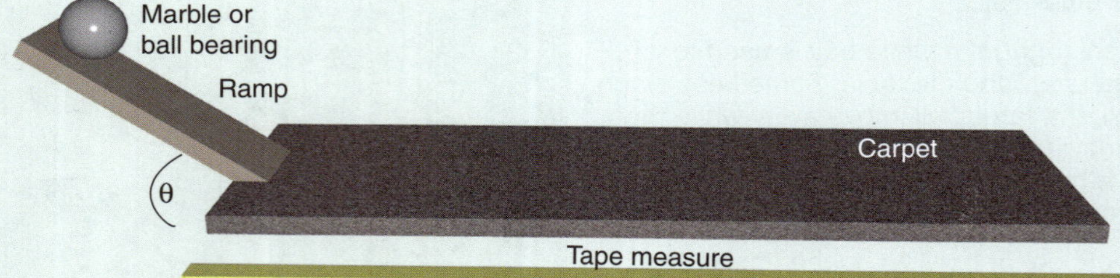

2. Measure the mass of marble.
3. Measure the length of the ramp and place the marble at the top.
4. The velocity of the marble at the bottom of the ramp can be calculated by combining the equation you derived in question 7 of page 151 and using right angle trigonometry to find the acceleration:

$$v = \sqrt{2g\sin\theta\Delta x}$$

where $g = 9.8\ m/s^2$, θ is the angle of incline of the ramp and Δx is the length of the ramp in meters.

5. Roll the marble down the ramp four times and record the distance that it runs along the carpet each time on the table below.
6. Calculate the average distance and record it in the appropriate table below.
7. Measure the mass of the ball bearing.
8. Replace the marble with the ball bearing and repeat steps 6 and 7.
9. Calculate the velocity of the marble and ball bearing as they leave the ramp and add to the table.

Test 1	Mass of marble (kg)	Length of ramp (m)	Velocity (m/s)	Distance rolled (m)
1	*Student's own answer*			
2				
3				
4				
Average				

Test 2	Mass of ball bearing (kg)	Length of ramp (m)	Velocity (m/s)	Distance rolled (m)
1				
2				
3				
4				
Average				

5. Describe the investigation in terms of momentum and describe the effect that change in momentum had on the distance rolled by the marble and ball bearing:

Gravity acted on the marble and the ball bearing causing them to gain momentum. Gravity acted on both objects with the same acceleration so the velocity of each was the same at the bottom of the ramp. The ball bearing had more momentum because it had more mass and traveled a further distance than the marble.

©2020 **BIOZONE** International
ISBN: 978-1-927309-82-7
Photocopying Prohibited

72 Investigating Collisions

Key Question: How do collisions influence the momentum of objects and how do we account for their relative influences mathematically?

Collisions

Moving objects have momentum. But what happens when two objects moving in opposite directions crash into each other? Where does the momentum go?

- Some students wanted to investigate this problem. They set up two carts on an air track to minimize friction. The mass of each cart was measured. Cart 1 had a mass of 0.75 kg, cart 2 had a mass of 0.73 kg
- The students noted that sometimes objects stick together when they collide (**completely inelastic collision**), and other times they rebound (**elastic collision**). To simplify their investigation, they fitted magnets to the carts so that they would stick together when they collided. A radar speed gun was available to measure velocity before and after the collisions.
- For the first investigation, cart 2 was made stationary in the center of the air track. Cart 1 was given a push in the direction of cart 2. The results of four trials that they carried out are shown in Table 1.

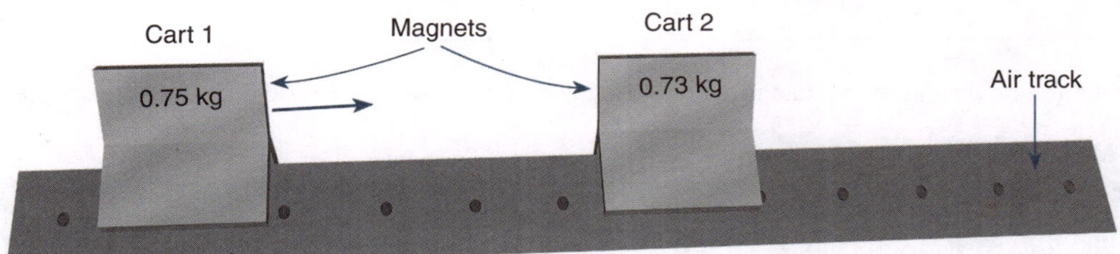

Table 1

Velocity cart 1 (m/s)	Momentum cart 1 (kg m/s)	Velocity cart 2 (m/s)	Momentum cart 2 (kg m/s)	Velocity cart 1&2 after collision (m/s)	Total momentum of carts (kg m/s)
0.52	0.39	0.00	0.00	0.26	0.385
0.64	0.48	0.00	0.00	0.33	0.49
0.32	0.24	0.00	0.00	0.16	0.24
0.13	0.098	0.00	0.00	0.07	0.10

- The students then decided to investigate both carts in motion. They pushed the carts in opposite directions towards each other. They kept the magnets in place so that the carts would stick together after the collision.
- Because cart 2 was moving in the opposite direction of cart 1, the students recorded this as a negative velocity. The results of four trials that they carried out are shown in Table 2.

Table 2

Velocity cart 1 (m/s)	Momentum cart 1 (kg m/s)	Velocity cart 2 (m/s)	Momentum cart 2 (kg m/s)	Velocity cart 1&2 after collision (m/s)	Total momentum of carts (kg m/s)
0.41	0.31	–0.11	-0.80	0.15	0.222
0.12	0.09	–0.35	-0.256	–0.11	-0.1628
0.37	0.278	–0.36	-0.263	0.01	0.0148
0.29	0.218	–0.30	-0.219	0.00	0

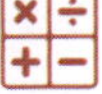

1. (a) Complete Tables 1 and 2 (above) by calculating the momentum of each cart:

   (b) Explain the students' results: *The momentum of cart 1 plus the momentum of cart 2 should equal total momentum. There will be small differences due to rounding and accuracy of measurements.*

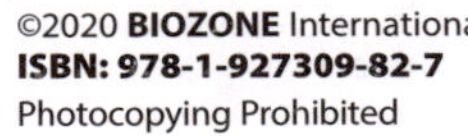

©2020 BIOZONE International
ISBN: 978-1-927309-82-7
Photocopying Prohibited

Elastic collisions

- The students from the previous page decided to investigate collisions in which the objects rebound. They reversed the orientation of one set of magnets so that the magnets would repel each other as the carts came together. The carts would then "rebound" after the collision.
- In the first two trials cart 2 was stationary. In trials 3-5, cart 2 was moving. The results of the five trials that they carried out are shown in Table 3:

Table 3

Trial	Velocity cart 1 (m/s)	Velocity cart 2 (m/s)	Velocity cart 1 after collision (m/s)	Velocity cart 2 after collision (m/s)
1	0.49	0.00	0.01	0.50
2	0.55	0.00	0.01	0.56
3	0.21	–0.56	–0.55	0.22
4	0.45	–0.43	–0.42	0.46
5	0.40	0.10	0.10	0.40

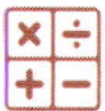

2. Complete Table 4 (below) using the results in Table 3. Recall that mass of cart 1 = 0.75 kg and mass of cart 2 = 0.73 kg.

Table 4. Momentum from Table 3 results

Momentum cart 1 (kg m/s)	Momentum cart 2 (kg m/s)	Momentum cart 1 after collision (kg m/s)	Momentum cart 2 after collision (kg m/s)
0.368	0.00	0.0075	0.365
0.413	0.00	0.0075	0.409
0.158	-0.409	-0.4125	0.161
0.338	-0.314	-0.315	0.336
0.3	0.073	0.075	0.292

3. Explain the students' results from Tables 3 and 4: The momentums of cart 1 and 2 together after the collisions is equal to the momentums of cart 1 and 2 together before the collision.

4. Looking at the results from Tables 1 and 2 from the previous page and tables 3 and 4 above, write a mathematical expression to model the results.

$P_{1i} + P_{2i} = P_{1f} + P_{2f}$

5. A car with a mass of 1000 kg is traveling east with a speed of 20 m/s. A 2000 kg pickup truck traveling west at 15 m/s crosses the median and the two vehicles collide.

(a) If the two vehicles stuck together upon impact like the carts in the students' investigation on the previous page, what would the resulting velocity of the vehicles be immediately after impact?

$P_{car} + P_{truck} = P_{car + truck}$, $(1000\ kg \times 20\ m/s) + (2000\ kg \times -15\ m/s) = (1000\ kg + 2000\ kg) \times v_f$,

$-10{,}000 = 3000 \times v_f$, $v_f = -10{,}000 \div 3000 = 3.33$ m/s west.

(b) Perfectly elastic collisions (like the carts in the students' investigation above), don't usually happen in car crashes because energy is lost as heat and sound, and in the deformation of the vehicles. These means they rebound more slowly than might be expected. What would the resulting velocity of the car be if the velocity of the truck after the impact was 3.5 m/s to the east?

$20{,}000\ kg\ m/s + -30{,}000\ kg\ m/s = -10{,}000\ kg\ m/s = (1000\ kg \times v_{car}) + (7000\ kg\ m/s)$,

$(1000\ kg \times v_{car}) = -10{,}000\ kg\ m/s - 7{,}000\ kg\ m/s = -17{,}000\ kg\ m/s$,

$v_{car} = -17{,}000\ kg\ m/s \div 1000\ kg = -17\ m/s$ (17 m/s west)

©2020 BIOZONE International
ISBN: 978-1-927309-82-7
Photocopying Prohibited

73 Law of Conservation of Momentum

Key Question: What is the law of conservation of momentum and how does it explain why force pairs do not cancel each other out?

PS2-2

Conservation of momentum

Momentum (p) is always conserved (i.e. total momentum is the same before and after a collision or interaction) as long as there are no forces outside the immediate situation affecting the colliding objects (i.e. causing additional accelerations). The following examples illustrate these ideas.

	Before colliding	After colliding	Explanation
A	m_1 = 1500 kg, v_1 = 30 m/s; m_2 = 1000 kg, v_2 = 20 m/s. Relatively smooth horizontal surface	m_1 = 1500 kg, v_3 = 20 m/s; m_2 = 1000 kg, v_4 = ?? m/s. Relatively smooth horizontal surface	No forces outside the immediate interaction causing significant accelerations of the colliding objects so… **p after = p before** $m_1v_1 + m_2v_2 = m_1v_3 + m_2v_4$
B	m_1 = 1500 kg, v_1 = 30 m/s; m_2 = 1000 kg, v_2 = 20 m/s. Rough sloping surface	m_1 = 1500 kg, v_3 = ?? m/s; m_2 = 1000 kg, v_4 = ?? m/s. Rough sloping surface	There are at least two forces outside the immediate interaction that will cause significant extra accelerations **Momentum (p) will not be conserved!**

Explosions

- Explosions throw objects in all directions, but they still obey the law of conservation of momentum. Imagine the firework in the photograph has been fired straight up and has reached the highest point of its flight. Its momentum in that instant is zero.
- At that exact moment, it explodes. What is the combined momentum of all the fragments now? Conservation of momentum states that they must add up to zero.
- Consider the simplified diagram below:

A= 2 kg, –2 m/s — AB 4 kg, 0 m/s — B= 2 kg, 2 m/s

- The momentum of fragment A is exactly opposite the momentum of fragment B.

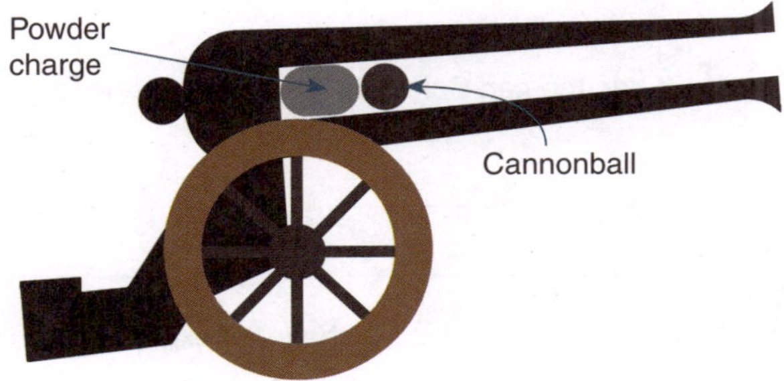

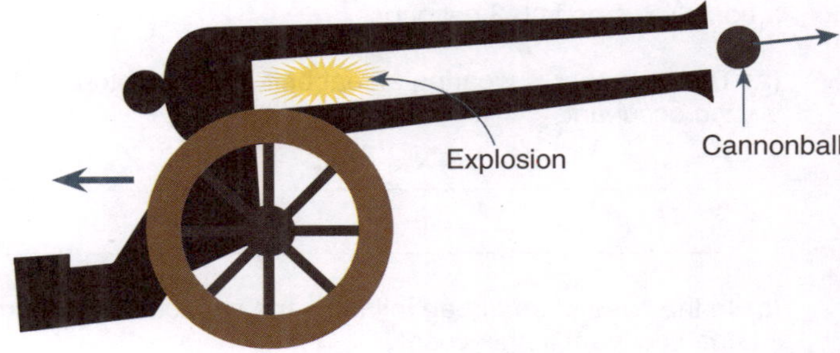

1. Consider the cannon above. The cannon has a mass of 900 kg and is at rest before firing. The cannonball weighs 5 kg. When fired, the cannonball exits the barrel at 230 m/s.

   Calculate the velocity of the cannon after it is fired: $p_b = -p_c$ Therefore $m_b v_b = m_c \cdot -v_c$ Therefore $-v_c = (m_b v_b) \div m_c$ = (5 kg x 230 m/s) ÷ 900 kg = -1.28 m/s

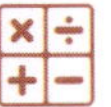

2. An object at rest explodes into two equal parts, A and B. Part A flies off at 20 m/s. What is the velocity of part B?

   -20 m/s

©2020 BIOZONE International
ISBN: 978-1-927309-82-7
Photocopying Prohibited

SSM

PS2.A

74 Impulse

Key Question: How does changing the time over which a force is applied to an object influence the object?

Impulse

Collisions can be instantaneous, or they can occur slowly. Imagine falling out of a plane at an altitude of 5500 meters. You would expect that when you hit the ground, the force of the impact would be fatal. True? Not necessarily.

- In 1944 Royal Airforce tail gunner Nicholas Alkemade jumped from a flaming Lancaster Bomber at 5500 meters without a parachute after it was destroyed over Germany. He survived with only a few bruises and a sprained ankle because his fall was cushioned by thick tree branches and nearly 0.5 m of snow.
- Because the force of the impact was spread out over a long period, Alkemade was able to survive the fall. Vesna Vulovic, an air hostess in Yugosalvia, holds the record for surviving the highest fall, from a DC-9 that blew up at 10,160 m in 1972.
- • Recall F=*ma*. We know that $a = \Delta v/\Delta t$ so we can substitute *a* in the equation to give $F = \Delta(mv)/\Delta t$.
  • Multiplying both sides by Δt gives $F\Delta t = \Delta mv$.
  • Thus the force multiplied by the change in time (or **impulse J**) equals the change in momentum (*p*) of the object.
  • It has the unit of Ns (newton seconds).

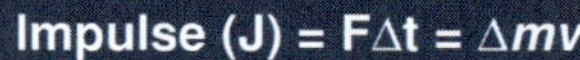

Impulse (J) = $F\Delta t = \Delta mv$

1. Consider an object with a mass of 100 kg moving at 10 m/s. Its momentum is therefore 1000 kg m/s. How can this object be stopped? It can be stopped quickly by applying a large force or slowly by applying a small force.

   (a) The object mentioned above has a momentum of 1000 kg m/s. A force of 200N is applied to the object. How quickly will the object stop?

   $\Delta t = \Delta p \div F = 1000$ kg m/s $\div$ 200 N = 5 s

   (b) The object has a force of 10 N applied to it. How quickly does the object stop now? $\Delta t = \Delta p \div F$

   = 1000 kg m/s $\div$ 10 N =100s

2. A 3 kg object traveling with a velocity of 20 m/s experiences a force of 15 N for 5 seconds in the same direction of travel. Calculate the object's new velocity:

   $\Delta t = \Delta p \div F = m(v_f - v_i) \div F$ Therefore $v_f = (F\Delta t) \div m + v_i$

   = (15 N x 5 s) $\div$ 3 kg + 20 m/s = 45 m/s

3. A car with an occupant weighing 80 kg is moving at 27 m/s (about 100 kmph). The car hits a tree head on and comes to a complete stop in 0.3 seconds.

   (a) The occupant is wearing a seat belt and thus stops at the same rate as the car. Calculate the force experienced by the occupant:

   $F = \Delta p \div \Delta t$ = (80 kg x 27 m/s) $\div$ 0.3 s = 7,200 N

   (b) In the case of an airbag inflating, the impact time is increased by 0.1 seconds. Calculate the force experienced by the occupant in this case:

   $F = \Delta p \div \Delta t$ = (80 kg x 27 m/s) $\div$ 0.4 s = 5,400 N

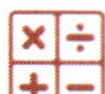

4. A tennis racket hits a stationary ball (mass 60 g) at 40.0 m/s. The impact time is 0.001 seconds. Calculate the force exerted on the ball:

   $F = \Delta p \div \Delta t$ = (0.060 kg x 40.0 m/s) $\div$ 0.001 s = 2400 N

©2020 **BIOZONE** International
ISBN: 978-1-927309-82-7
Photocopying Prohibited

Impulse in bungee jumping

- To stop an object in motion, a large force can be applied to stop the object quickly or a small force can be applied to stop the object slowly.
- In both cases the object comes to a stop, but the effects of the stop on the object are very different.

INVESTIGATION 5.3: Investigating impulse

See appendix for equipment list.

1. Work in pairs. You will need an old doll or small stuffed animal. Measure the mass of the "bungee jumper" in kilograms.
2. Set up the following equipment as shown right (make sure the stand is securely clamped to the tabletop)
3. Create a bungee cord using elastic bands.
4. Attach one end of the cord to the "jumper" and the other end to a dual range force sensor attached to the stand.
5. Allow the bungee jumper to fall while you collect force time data using a program such as Logger Pro.
6. Sketch a graph of force vs time below. Take note of the peak force value.
7. Replace the elastic band bungee cord with a length of inelastic string or twine the same length as the elastic bungee cord.
8. Allow the bungee jumper to fall again and sketch the graph of force vs time below. Again note the peak force value.

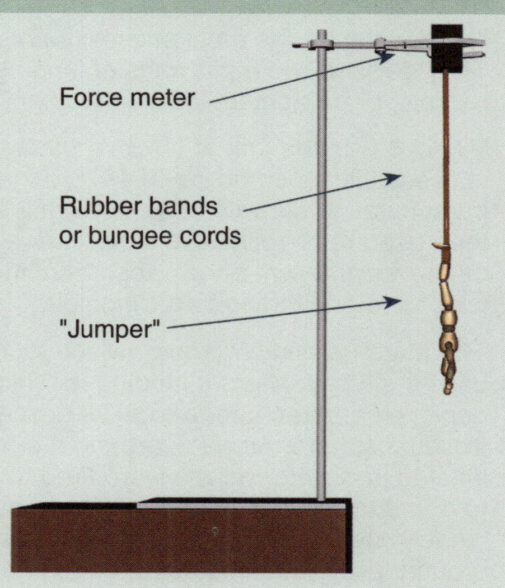

With elastic band

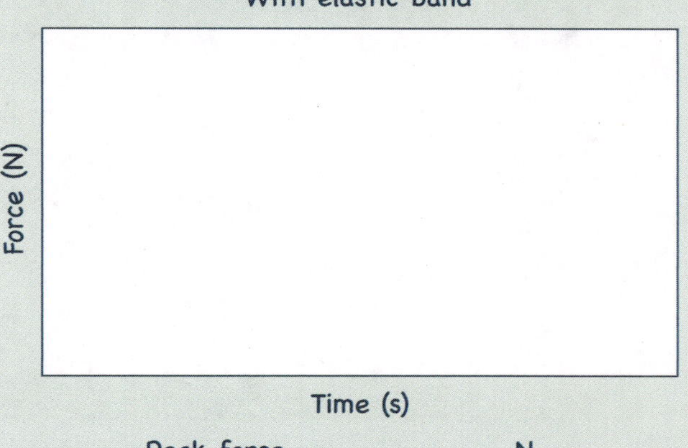

Peak force = ____________ N

Without elastic band

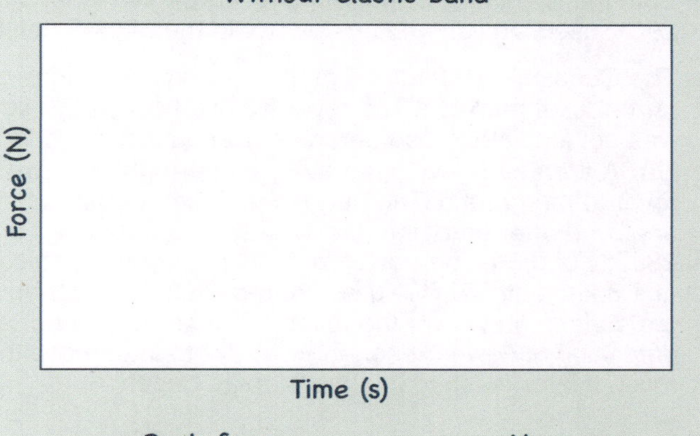

Peak force = ____________ N

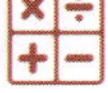

5. Measure the length of the elastic band bungee cord (or the length of inelastic string as they were the same length) in meters and the mass of the bungee jumper in kilograms.
   (a) Calculate the velocity of the jumper when it reached the end of the cord (before the cord applied a force to stop it):
   Student's answer based on length of cord.
   Students can find time from the data logger / force time graph or $t = \sqrt{(2d/a)}$ *and* $v_f = (\Delta x \div \Delta t) - v_i$
   (b) What is the magnitude and direction of the impulse required to stop the bungee jumper: ____________
   Student's answer based on results
   (c) Compare the peak forces identified in the investigation and calculate the time it took the cords to stop the jumper:
   Student's answer based on results

6. Using the data you collected from the investigation and what you know about impulse, explain why bungee cords are made out of elastic materials rather than something like rope:
   Elastic materials spread the momentum of the person over a longer time frame, so reducing the impulse felt by the person. An inelastic material like rope would simply stop the person. The high impulse of the sudden stop would probably make the jump fatal.

©2020 **BIOZONE** International
ISBN: 978-1-927309-82-7
Photocopying Prohibited

75 Crumple Zones and Crash Helmets

Key Question: How are the principles of momentum and impulse applied to limit damage to sensitive objects and save lives?

Landing on Mars

- NASA is by far the most successful of all the space agencies that have tried to land probes and rovers on Mars. They devised two main ways of landing probes and rovers on to the surface safely: using parachutes and retro-rockets, or parachutes and air bags.
- Viking 1 was the first of NASA's Mars landers to successfully touch down (in 1976). The lander used parachutes to slow its descent from 250 m/s to 60 m/s. It then used retro-rockets to slow its descent below 2.4 m/s before touchdown. Shock absorbers in the legs reduced the final force on landing to a slight jolt.
- The Mars Pathfinder lander, which touched down in 1997, used a slightly different landing technique. The lander also deployed a parachute to slow its descent after entering the atmosphere. Air bags around the lander's frame were inflated and retro-rockets were fired to bring the lander to a sudden halt at just 98 meters above the ground. The lander was then cut loose from the parachute and fell to the ground, using the inflated air bags to cushion its landing. When it hit the ground, it bounced up to 15 m high and experienced a maximum force of 18 G.

- Both the Spirit and Opportunity rovers, which landed in 2004, used similar landing devices to the Mars Pathfinder.

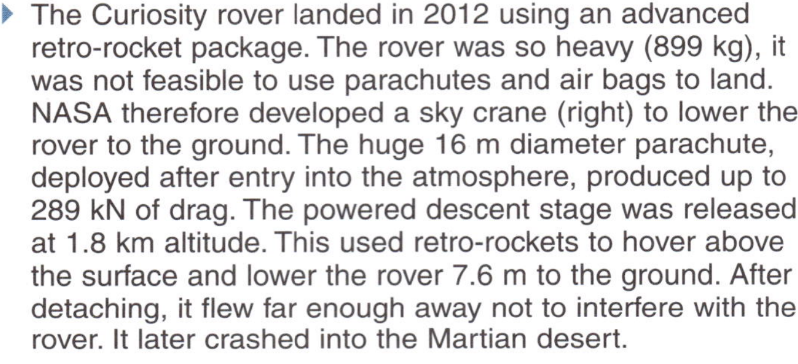

- The Curiosity rover landed in 2012 using an advanced retro-rocket package. The rover was so heavy (899 kg), it was not feasible to use parachutes and air bags to land. NASA therefore developed a sky crane (right) to lower the rover to the ground. The huge 16 m diameter parachute, deployed after entry into the atmosphere, produced up to 289 kN of drag. The powered descent stage was released at 1.8 km altitude. This used retro-rockets to hover above the surface and lower the rover 7.6 m to the ground. After detaching, it flew far enough away not to interfere with the rover. It later crashed into the Martian desert.

Mars Pathfinder lander

Curiosity rover and sky crane

NASA

1. (a) What are the two main methods used by NASA to land its rovers or landers on Mars?

   Parachutes and retro-rockets, and parachutes and airbags.

   (b) Would these methods work on the Moon or Mercury? *In part. Parachutes cannot be deployed on the Moon and Mercury because there is no atmosphere. Airbags and retro-rockets could be used because the gravity on the Moon and Mercury is less than on Mars.*

2. (a) Explain why the Curiosity rover used a sky crane to lower it to the surface: *The Curiosity rover was too heavy for the airbags and it would have been damaged if not lowered slowly and with control.*

   (b) The acceleration due to gravity of Mars is 3.71 m/s^2. What is the weight of the Curiosity rover on Mars?

   F (weight) = ma = 899 kg x 3.71 m/s^2 = 3,335 N

   (c) What equivalent mass on Earth would produce this weight? ______

3. Why was it so important for these rovers to reduce the force experienced during landing?

   To reduce the damage to the rovers.

PS2.A CE

©2020 **BIOZONE** International
ISBN: 978-1-927309-82-7
Photocopying Prohibited

Crumple zones

"Cars today, they're just not built tough. A small crash and they're just wrecked". Have you heard this before? Or how many times have you heard someone say, "I like being in a SUV or four-wheel drive, they're solid, tough, and can take a crash."

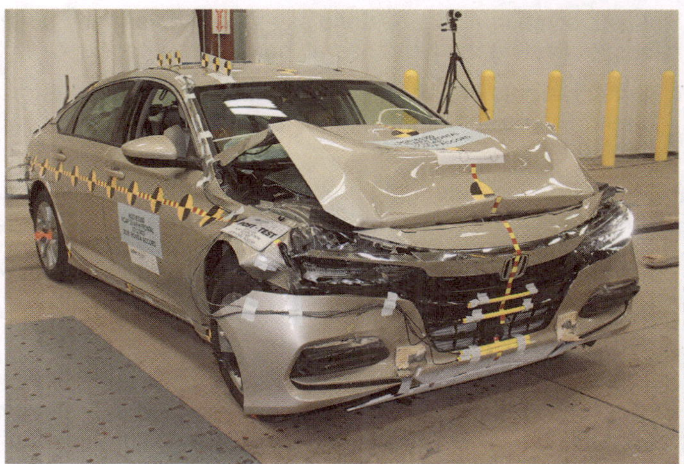
NHTSA: National Highway Traffic Safety Administration

- Is this thinking logical when it comes to a crash? From your studies of force, momentum, and impulse, would you rather be in a car that crumpled when it had a crash or was rigid and kept its shape?
- Cars today have sophisticated safety systems that often allow the car to avoid a collision. However, these do not prevent all crashes. In the case of a crash, the frame of the car is designed to crumple up.
- The crumpling effect increases the time the force of the crash is applied, reducing the force felt by the occupants.

4. Why do cars have crumple zones? *Crumple zones increase the time it takes for the car to come to a stop. The force is applied over a longer time, reducing impulse. Occupants of the car will experience less impact force and will sustain less damage.*

5. What other safety devices do modern cars regularly have installed to protect the occupants in the event of a crash?

Student's answers may include seat belts, front and side air bags, and proximity sensors.

6. Motorbikes have little capacity for crumple zones (although this is performed to a degree by the front wheel and steering system). Bicycles have virtually no capacity for a crumple zone. How do riders protect themselves in a crash?

Student's answers may include helmets, knee and elbow pads, and protective riding clothing.

7. The graphs below show acceleration data from crash tests performed on a modern 2018 vehicle and a vehicle from 1980.

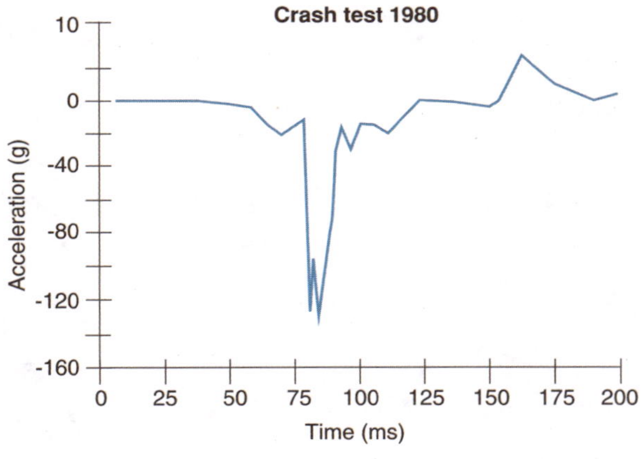

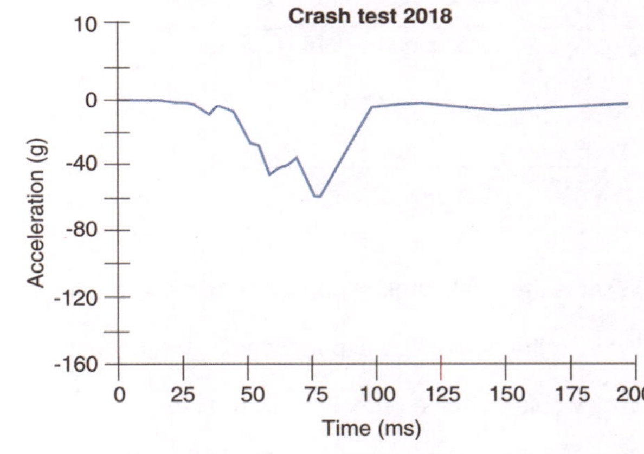

(a) If the masses of each car were 1000 kg, calculate the peak forces the occupants would experience in both crashes:

F_{1980} = 1000 kg × -140 × 9.8 m/s^2 = -1,372,000 N

F_{2018} = 1000 kg × - 60 × 9.8 m/s^2 = -588,000 N

(b) Which car do you feel would keep you safer in a crash? Explain your reasoning using ideas of impulse and the forces experienced during the crash tests:

The 2018 car would keep the occupants safer in a crash. The crumple zones of the 2018 car spread the force of the crash over a longer period of time, reducing the impulse on the occupants. The peak force experienced in the 2018 car was less than half the force experience in the 1980 car.

©2020 **BIOZONE** International
ISBN: 978-1-927309-82-7
Photocopying Prohibited

Crash helmets

Helmets come in all shapes and sizes and serve many purposes. In virtually all cases, protecting the head has at least some importance in the helmet design.

- Crash helmets are specifically designed to reduce the force of impact during a crash. The design of the helmet is constrained by the conditions under which it will be used.
- Motorcycle helmets (right) are normally designed to protect the entire skull and face. For a pedal cyclist this design would be very heavy and would quickly become much too hot. The much lower speeds involved allow cycle helmets to be designed with air gaps and protect only the skull.

Mats

Mats are used for a variety of reasons and serve many purposes. On the workfloor, they are used to reduce the stress on the joints of workers who spend a lot of time walking or standing on hard surfaces. On the gym floor, they are used to lessen impact as cheerleaders, gymnasts, and martial artists practice their tumbles, jumps and flips. In track and field, they protect high jumpers and pole vaulters when they fall.

- During a routine, gymnasts can leap to incredible heights before landing on the floor. Over time repeated jumps can lead to damage to the joints. Floor mats reduce these injuries by increasing the time it takes for the gymnast to come to a stop after landing.
- The mats used by individuals doing yoga are much thinner than those used by gymnasts. Yoga is a low-impact activity and does not require the same level of joint protection. This means that yoga mats are easier to transport.

8. The graphs below show the simulated force experienced on the head during a crash with and without a helmet.

Without helmet

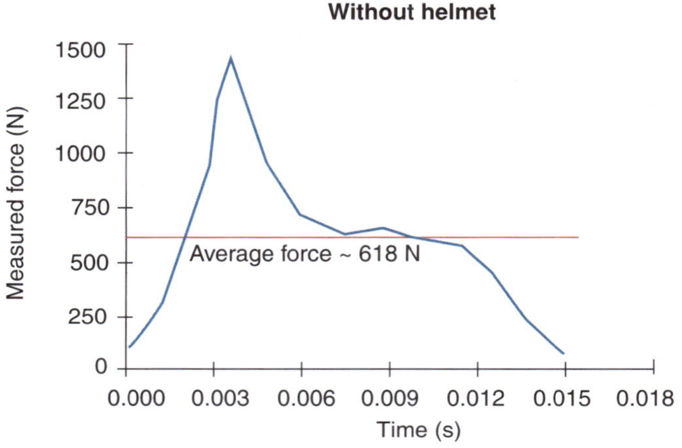

With helmet

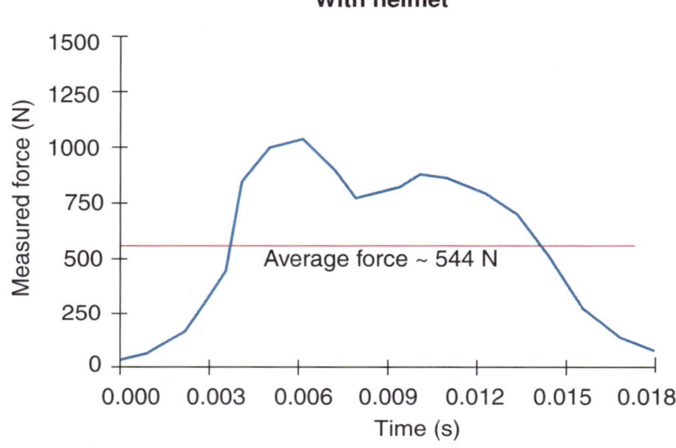

(a) What is the peak force experienced without a helmet? *Almost 1500 N.*

(b) What is the peak force experienced with a helmet? *A little over 1000 N.*

(c) Why would both the peak and average force experienced be lower when wearing a helmet?

The padding in the helmet allows the impact to occur over a longer time period.

9. A 40 kg gymnast is practicing her routine. On her final jump, she reaches a height of 2.0 meters before landing on the gym mat. The mat stops the gymnast in 0.05 seconds.

(a) What is the magnitude and direction of the impulse required to stop the gymnast? $J = m\Delta v = m(v_f - v_i) = mv_f$

$v_f = \sqrt{(2 \times g \times \Delta x)} = \sqrt{2 \times 9.8 \times 2} = 6.25$ m/s. $J = 40$ kg $\times 6.26 = 250$ kg m/s (up)

(b) What is the average force the mat exerts on the gymnast?

$F = J \div \Delta t = 250$ kg m/s $\div 0.05$ s $= 5000$ N

(c) Our gymnast decides to practice her routine at home on the driveway and sprains her ankle. The driveway stops the gymnast in 0.01 seconds. Explain why the gymnast injured herself at home (use calculations to justify your answer):

$F = J \div \Delta t = 250$ kg m/s 0.01s = 25,000 N. *Because the driveway stopped the gymnast more quickly she experienced a force 5 times greater than on the mat.*

©2020 **BIOZONE** International
ISBN: 978-1-927309-82-7
Photocopying Prohibited

INVESTIGATION 5.4: Building a lander

See appendix for equipment list.

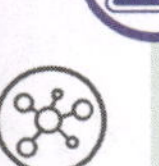

PS2-3

TS1-2

1. You will now have the opportunity to put your knowledge of forces and momentum to the test. The objective of this investigation is to build a device that will protect a lander from descent of at least four meters.
2. The conditions: You will build a landing device that will protect a raw egg from a fall of at least four meters. Your are free to use the equipment (below) in any way but there may be no platform for the lander to fall onto (landers don't have convenient foam landing pads to fall onto when they reach their destination).
3. The equipment: 1 egg, 60 cm of tape, 5 rubber bands, 1 small garbage or plastic bag, 10 paper clips, 1 m string, 20 plastic or paper straws, 1 plastic egg or similar sized object for testing. Your teacher may modify this equipment as they wish.
4. Building time will be determined by your teacher.
5. Before you begin, discuss with your group how you will construct the lander and in what ways you could cushion the egg at impact or slow its descent to reduce the shock of landing. You could research lander design online to gather ideas before you begin construction.

TS1-2

Lander design

10. Now that you have built and tested your egg lander you must evaluate its structure and performance.

(a) Describe the structure of your lander: *Student's answer.*

(b) On a scale of 1-5, how did your egg survive the fall? (1= completely scrambled, 5 = safe and sound): *Student's answer.*

(c) If your egg cracked, what could have been done to keep it from cracking if you repeated the test?

Student's answer.

(d) Recall the landing devices designed used by other groups in your class. Which of these were the most effective? Can you explain why?

Student's answer.

(e) Given enough material, it would be easily possible to design a lander that would protect the egg from a much higher fall than 5 m. The same applies to planetary landers. However, there are numerous constraints on the development of these devices, one of which is cost. Discuss with your group what other constraints there might be on the design on a lander for Mars:

Student's answers may include: availability of materials (some necessary materials may not exist yet), time (scheduling deadlines), weight (heavier objects take more effort to launch), and Martian atmosphere (different density than Earth) and physical conditions of Mars.

©2020 **BIOZONE** International
ISBN: 978-1-927309-82-7
Photocopying Prohibited

76 Review Your Understanding

Key Question: How is it possible to break many bricks, blocks of ice, or boards of wood stacked on top of each other? What physical principles explain these feats of strength?

- Recall the beginning of the chapter when you were asked questions involving breaking bricks. With what you have learned, think about how the concepts of forces, momentum, and impulse are involved in breaking bricks.
- One very important principle involved in brick-breaking is momentum. The more momentum an object has, the more force it can generate when it comes into contact with another object. This is why the velocity of the strike to the brick is important. The faster the strike, the greater the momentum. When breaking bricks, the downward punch increases the momentum available by involving some of the mass of the individual to the strike. Increased velocity and increased mass both contribute to increased momentum.

1. Assume a hand comes to a complete stop when it strikes a brick. Using a downward punch, the performer is able to put 20 kg of mass behind the strike. The brick requires an impulse of 250 kg m/s to break.

   (a) Calculate the minimum velocity of the performer's hand must strike the brick at in order to break it:

   $v = J \div m = 250$ kg m/s $\div$ 20 kg = 12.5 m/s

   (b) If the performer's hand comes to a complete stop in 0.1 s, what was the force applied to the brick and the hand?

   $F = J \div \Delta t = 250$ m/s $\div$ 0.1 s = 2500 N

2. When breaking bricks, performers punch through the brick instead of stopping as soon as they make contact. Because the punch continues after the brick breaks, it takes longer for the hand to come to a complete stop.

   (a) Using the same information from question 1, what was the force applied if the performer's hand comes to a complete stop in 1.0 s instead?

   $F = J \div \Delta t$ = 250 kg m/s $\div$ 1.0 s = 250 N

   (b) Explain why the impulse remains the same: *Impulse is a measure of change in momentum, which is simply the change in the velocity of a mass. This change in momentum can be accomplished by a large force in a short time or by a small force over a long time*

3. When a fist punches a brick, every arm joint lined up with the strike, from the metacarpals of the hand, the carpals of the wrist, to the elbow and shoulder, act as cushions, extending the time over which the force is experienced.

   (a) Explain how this increased time protects the bones in the hand: *Since impulse remains the same (Question 2b), extending the time reduces the force experienced by the bones in the hand making it less likely they will break.*

   (b) Why does the brick still break even if the bones do not? *Because the brick is rigid it has no way to extend the force applied to it.*

4. When the head is used to break bricks, padding is added to the stack where the head will come in contact.

   (a) Explain why more padding is used when using the head to break bricks: *The padding extends the time, therefore lowers the force experienced during impulse. The head has less ability to reduce time on its own so padding is used.*

   (b) Why do you think it is less common to see performers breaking bricks with their heads than with their hands? *If a performer were to make a mistake breaking bricks with their hands they would break their bones and require time to heal. Because of the importance of the brain (in your head) if the performer were to make a mistake, brain damage/concussion are likely. There is less long term risk in breaking breaks by hand.*

©2020 **BIOZONE** International
ISBN: 978-1-927309-82-7
Photocopying Prohibited

77 Summing Up

Forces

PS2-1

1. A block with a mass of 2 kg is at rest on a frictionless surface. Read the descriptions above the diagrams then add labels and arrows to the diagrams to show the unbalanced forces involved and complete the table under the diagrams:

The block is pushed from its left with a force of 10 N for 1 second.

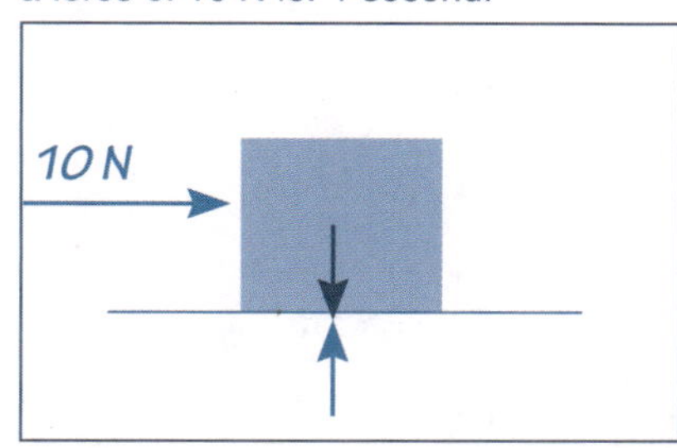

The block is no longer pushed. It is left to move for 3 seconds.

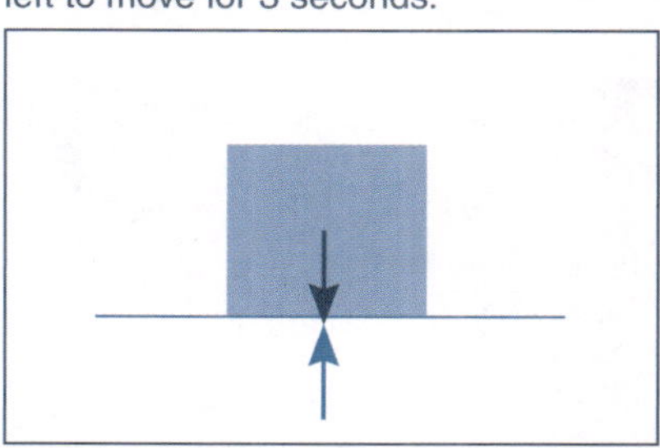

The block is brought to a stop by applying a force of 5 N.

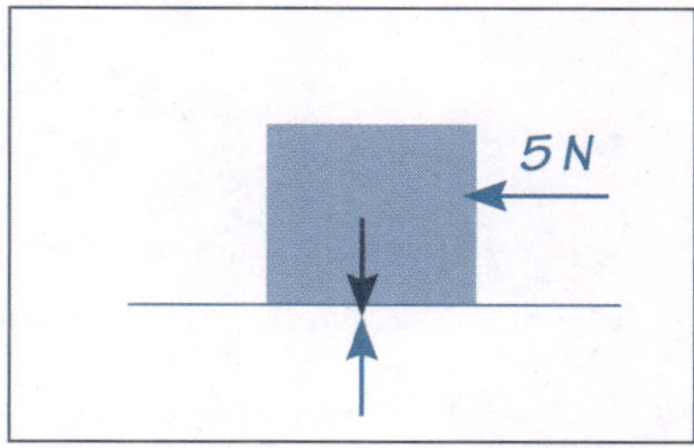

Acceleration	Velocity	Acceleration
$a = F \div m = 10\ N \div 2\ kg = 5\ m/s^2$	$\Delta v = a \Delta t = 5\ m/s^2 \times 1\ s = 5\ m/s$	$a = F \div m = -5\ N \div 2\ kg = -2.5\ m/s^2$

2. The following method is sometimes proposed for long distance space travel. A spaceship fires its engines at full thrust at its point of origin. It continues traveling with engines on full thrust for half of its journey. It then switches off its engines, turns around (180°) and restarts its engines at full thrust for the second half of the journey facing back the way it came. Explain why this would produce the shortest travel time and would bring the ship to a rest at the end of the journey.

Direction of travel

First half of journey

Second half of journey

At full thrust, the spaceship will be accelerating at its maximum rate for the first half of its journey, reaching maximum velocity at the halfway point. Because no other forces are acting on the ship its only way of slowing down is to turn around and fire its engines at full thrust in the opposite direction, producing an equal but opposite force to the original thrust that will bring the ship to a stop by the end point of the journey.

3. Two skydivers jump out of a plane. They both adopt the same body orientation while falling (horizontal star position). Skydiver A has a mass of 75 kg. Skydiver B has a mass of 85 kg.

(a) What is the magnitude of the force on skydiver A?

$F\ (weight) = ma\ (a = g = 9.8\ m/s^2) =$

$75\ kg \times 9.8\ m/s^2 = 735\ N$

(b) What is the magnitude of the force on skydiver B?

$F\ (weight) = ma\ (a = g = 9.8\ m/s^2) =$

$85\ kg \times 9.8\ m/s^2 = 833\ N$

(c) Both skydivers reach terminal velocity (acceleration is zero). This is the point at which the force of air resistance equals the weight. Explain why the terminal velocity of skydiver A is less than that of skydiver B:

As the skydivers accelerate down, air resistance increases. The force due to air resistance will match the smaller weight of skydiver A before that of skydiver B.

(d) They open their parachutes. Explain why their velocity decreases until they reach a constant velocity of ~25 kmph:

Parachutes increase the surface area the skydivers' weight is spread over causing greater air resistance. Therefore both skydivers decelerate until a new, lower terminal velocity is reached.

©2020 BIOZONE International
ISBN: 978-1-927309-82-7
Photocopying Prohibited

CE PS2.A

4. A student builds a cart powered by a fan. When the fan is turned on, the sail will catch the air moved by the fan.

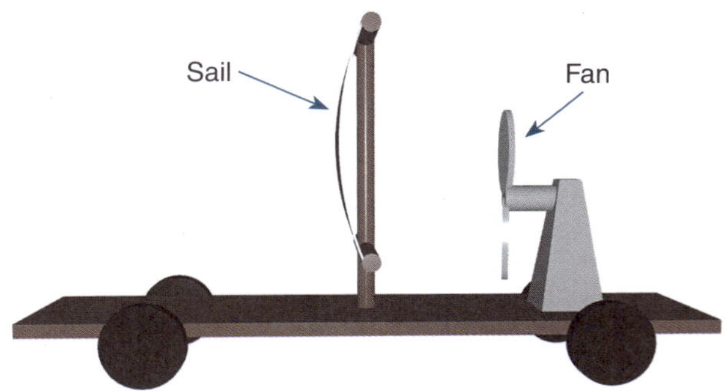

Explain why the cart will not move when the fan is turned on:

When the fan is on, it pushes against the air which would cause the cart to move to the right. However, the air that the fan has acted on will push against the sail to the left with the same force as the fan blows. The forces cancel out and there will be no movement.

PS2-2 Momentum

5. An astronaut floating in space holds a ball. The astronaut-ball system has a velocity of 0 m/s. The 70 kg astronaut pushes the 1.3 kg ball directly away from her so that the ball attains a velocity of 1.5 m/s. Calculate the velocity of the astronaut after she releases the ball:

   $m_a v_a = m_b v_b$ Therefore $v_a = (m_b v_b) \div m_a = (1.3 \text{ kg} \times 1.5 \text{ m/s}) \div 70 \text{ kg} = 0.028 \text{ m/s}$

6. Two astronauts (each with mass 70 kg) push off each other in space. One carries 46.7 kg of equipment and moves away with a speed 0.3 m/s.

   (a) Calculate the velocity of the 'unloaded' astronaut:

   $m_{a1} v_{a1} = m_{a2} v_{a2}$ and $m_{a1} = 70 \text{ kg} + 46.7 \text{ kg}$

   Therefore, $v_{a2} = (116.7 \text{ kg} \times 0.3 \text{ m/s}) \div 70 \text{ kg} = 0.5 \text{ m/s}$

   (b) Determine how far apart the astronauts are after 5 s. Show your working:

   $d = v\Delta t = (0.3 \text{ m/s} + 0.5 \text{ m/s}) \times 5 \text{ s} = 4 \text{ m}$

Drawing F.Hicks

7. A truck with a mass of 15,000 kg moving at 25 m/s crashes into the back of a small 1100 kg car moving at 15 m/s. They stick together after the collision.

   (a) Calculate the momentum of the system before the collision and after the collision: $p_i = p_f$

   Therefore, $(15{,}000 \text{ kg} \times 25 \text{ m/s}) + (1{,}100 \text{ kg} \times 15 \text{ m/s}) = 375{,}000 + 16{,}500 = 391{,}500 \text{ kg m/s}$

   (b) Calculate the velocity of the truck-car system after the collision: $p = mv$

   Therefore $v = p \div m = 391{,}500 \text{ kgm/s} \div (15{,}000 \text{ kg} + 1{,}100 \text{ kg}) = 24.3 \text{ m/s}$

8. A golf club swings and hits a golf ball from a tee. The golf club makes contact with the golf ball for 0.5 milliseconds (5.00 x 10^{-4} seconds). The mass of the golf ball is 0.045 kilograms. The velocity of the ball off the tee is 78 m/s.

   (a) What is the impulse experienced by the golf ball? Impulse $= \Delta p = m\Delta v =$

   $0.045 \text{ kg} \times 78 \text{ m/s} = 3.51 \text{ kgm/s}$

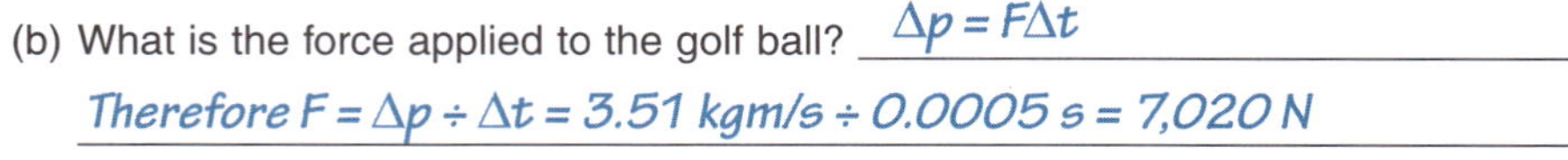

   (b) What is the force applied to the golf ball? $\Delta p = F\Delta t$

   Therefore $F = \Delta p \div \Delta t = 3.51 \text{ kgm/s} \div 0.0005 \text{ s} = 7{,}020 \text{ N}$

   (c) Use two different methods to show the acceleration of the golf ball off the tee is 156,000 m/s^2.

   $F = ma$ Therefore, $a = F \div m = 7{,}020 \text{ N} \div 0.045 \text{ kg} = 156{,}000 \text{ m/s}^2$

   $a = \Delta v \div \Delta t = 78 \text{ m/s} \div 0.0005 \text{ s} = 156{,}000 \text{ m/s}^2$

©2020 **BIOZONE** International
ISBN: 978-1-927309-82-7
Photocopying Prohibited

PS2.B
PS3.C

6: Types of Interactions

Anchoring Phenomenon

Force field: Why does a swinging magnetic pendulum not hit against a copper plate? 78 90

How can different objects interact when they are not touching?

☐ 1 Recall your understanding of weight from the previous chapter. Predict what would happen to your weight if your mass doubled, or if the Earth's mass doubled. You should now understand gravity as an attractive force between objects with mass. Calculate *[SEP-5]* the gravitational force between objects of different mass and explain why we don't notice the acceleration of massive objects, such as the Earth *[PS2.B] [CCC-1]*. 79

☐ 2 Recognize patterns *[CCC-1]* in the distance of objects from the Sun and their orbital period (Kepler's ratio). Use this pattern to calculate *[SEP-5]* orbital parameters. Newton derived his law of gravitation from Kepler's ratio and his understanding of the centripetal force pulling planets towards the Sun *[PS2.B]*. Use Newton's law of gravitation *[SEP-2]* to calculate *[SEP-5]* the effect of changing variables such as mass or distance between celestial objects. Explain the significance of the inverse square law described in Newton's law of gravitation to the weight force of objects of the same mass at different locations, e.g. on Earth and in space *[SEP-6] [PS2.B] [CCC-1] [HS-PS2-4]*. 80 91

☐ 3 Identify examples of electrostatic phenomena. Investigate electrostatics using everyday objects *[SEP-3]* and explain your observations. Electrostatic forces obey the inverse square law in the same way as gravity *[PS2.B] [CCC-1]* but, unlike gravity, they can attract or repel. What are the properties of electric charge and how do you distinguish charging from polarization? Classify materials based on their conductivity and calculate *[SEP-5]* the charge (in C) for objects based on the number of excess protons or electrons? 81

☐ 4 Apply a mathematical model *[SEP-2]* to demonstrate the inverse square law for gravity (Newton's law of gravitation) and electric charges (Coloumb's law) *[SEP-5]*. What happens to the magnitude of the force when we change the mass (or charge) of one object or the distance between two objects in a system? *[PS2.B] [CCC-1] [CCC-4] [HS-PS2-4]* 82

How are electricity and magnetism connected?

☐ 5 How do the electric fields of charged particles affect their interactions? Draw electric field lines to show how charged particles interact when they influence one another. Develop and use a model *[SEP-2]* of two objects interacting through electric or magnetic fields to illustrate the forces between objects and the changes in the energy of the objects due to the interaction *[PS3.C] [CCC-2] [HS-PS3-5]*. 83 91

☐ 6 Electromagnetic theory links magnetism and electricity *[PS2.B] [PS3.C]*. Like electric fields and gravitational fields, magnetic field enable forces to be exerted from a distance *[CCC-2]*. Investigate magnetic fields using bar magnets and iron filings or powder *[SEP-3]*. How do you explain the patterns *[CCC-1]* ? Use your investigations to make a predictive model *[SEP-2]* of the strength of a magnetic field a the distance from the magnet increases. 84

☐ 7 Where does a magnetic field come from? What makes an object magnetic and what sorts of objects are attracted to magnets? Investigate *[SEP-3]* the properties of ferromagnetic materials by making a magnet *[PS2.B] [PS3.C]*. Use your understanding of atomic structure *[SEP-2] [PS1.A]* and electrostatic forces to explain what is happening *[CCC-2]*. 85

☐ 8 Investigate *[SEP-3]* the influence *[CCC-2]* of current electricity on magnets (and vice versa) *[PS2.B]*. Use a model *[SEP-2]* and your understanding of the relationship between current electricity and magnetism to explain the effect *[CCC-2]* of magnetic fields on conductors (such as metal projectiles) *[PS3.C] [HS-PS3-5]*. 86 87 91

☐ 9 Electric generators are essential for nearly all types of electricity generation. You can now use your understanding of electricity and magnetism to induce *[SEP-3] [CCC-2]* directional electron flow *[HS-PS2-5]* and to design *[SEP-6]* and build a generator and/or a motor *[PS3.A] [CCC-5] [HS-PS3-3]*. 88

How do forces determine the properties of materials?

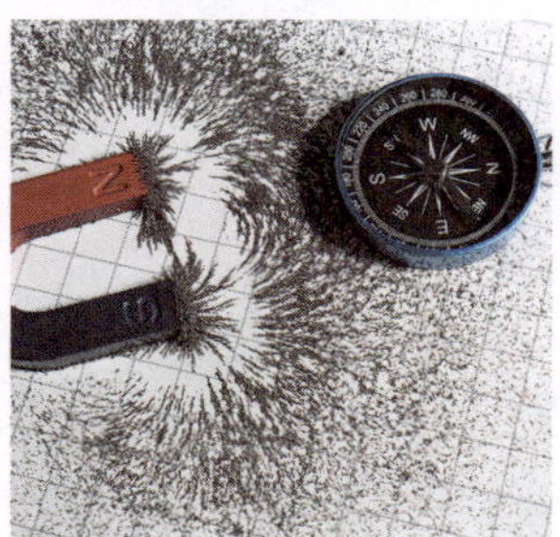

☐ 10 Use your understanding of atomic structure and electrostatic forces to explain the properties of substances in which the intramolecular bonds of the molecules are ionic, covalent, or metallic *[PS1.A]*. How do intermolecular forces help us to explain *[SEP-6]* the properties of substances *[CCC-6]* and their interactions with other substances? *[PS2.B]* Investigate *[SEP-3]* the properties of a wide range of everyday materials and relate these properties to their use. Obtain information about the molecular level interactions of various materials (conductors, semiconductors, and insulators). How does the molecular structure relate to the material's properties at the macroscopic level and therefore its various industrial and everyday applications *[SEP-8] [PS2.B] [CCC-6] [HS-PS2-6]*. 89 91

Download this page from BIOZONE's Resource Hub

6. Types of Interactions

Anchoring phenomenon

Magnetic braking is demonstrated by the movement of a strong magnet near a copper plate. It is fascinating to watch and seems almost like a magic trick. It is used as an anchoring phenomenon as it demonstrates many of the interactions occurring over distance, including gravity (the falling magnet) and electromagnetism. Students recall what they know of momentum to try to explain what is happening when the magnet stops swinging. They may realize that a force is acting on the system, which changes its momentum and thus it does not violate the law of conservation of momentum.

How can different objects interact when they are not touching?

1. Mass, weight, and gravity are interconnected, so reviewing mass and weight is a good starting point from which to look more closely at types of interactions. Simple questions such as *"what would happen if the mass of the Earth doubled?"* prompts students to think about the relationship between the Earth's gravity and its mass, and the Earth's gravity and its effect on nearby objects. Students need to realize the gravitational force between two objects is equal and opposite and depends on the mass of both objects.

2. Newton's law of gravitation ($F = GM_1M_2/r^2$) is one of the most important equations in physics and many other laws we use can be derived from it, including $F = ma$. In activity #80, students are stepped through the derivation of this law, starting with Kepler's third law. This piece of mathematical derivation will seem difficult for some students. What they should understand is that by following a series of logical steps and by combining several related pieces of mathematics (Kepler's law, centripetal force, and acceleration) new mathematical revelations can be made. This can be related to students rearranging an equation to find a previously unknown variable. Students should also understand the importance of the inverse square law as it also relates to Coulomb's law and many other physical phenomena.

3. Electrostatics are something we experience often, especially static shocks. This is a simple way to introduce the topic. Using a simple balloon or plastic comb to pick up small pieces of paper, sticking a balloon to the ceiling, or using a balloon to levitate another balloon are all simple and fun ways to demonstrate electrostatics and forces at a distance. A Van-de-Graaff machine can produce interesting results on student's hair (or watch a video of one in use). Students need to understand that the difference in charge comes from the movement of electrons to or from an object and that some objects tend to accumulate electrons and others to lose them.

4. Students can recognize that Coulomb's law is similar to Newton's law except that the value produced can be either positive or negative. Using a simple calculation of the force experienced by electrons at different distances from a nucleus and under different charges allows students to explore how Coulomb's law works. The effect of the inverse square law is demonstrated using Newton's law. Students then have to opportunity to apply this to Coulomb's law.

How are electricity and magnetism connected?

5. All charges produce electric fields. Students should understand how these fields interact. Electrical fields point from positive charge to negative charge and thus opposite charges attract and like charges repel. Students can see that electric fields can store energy (PS3.A secondary) in the same way as a gravitational field. A slight modification of Coulomb's law can be used to determine the strength of an electric field at any distance from a charge.

6. Students are introduced to magnetism by investigating three aspects of magnetism. They can see that a magnetic field points from the north pole to the south pole and that the strength of the field depends on the square of the distance. This is similar to electric fields. They may expect an interaction between a charged ruler and a magnet, but because the charge is not moving there is no interaction.

7. Students draw on their knowledge of atomic theory and electron structure to explain where magnetism comes from. However, to explain magnetism it is necessary to extend our model of electron structure to include sub orbitals and electron pairing. Paired electrons spin in opposite directions and cancel each other out. Unpaired electrons all spin in the same direction and this unpaired spin is important in magnetism. Students also learn that the magnetic field of groups of atoms in a crystal (domains) must be induced to align in order for the crystal to become a magnet. Thus iron can be magnetic due to the many unpaired electrons but not necessarily a magnet because the magnetic fields of the domains are pointing in random directions.

8. Electricity flowing through a wire produces a circular magnetic field at 90° to the direction of the current in the wire. Students investigate this through a series of simple experiments using compasses and battery cells. It is recommended that if using cells, standard 1.5v alkaline cells are used rather than higher power lithium batteries. You could ask your students why we make this recommendation! (Lithium batteries have a greater power density and lithium is flammable so a short circuit may cause them to explode). The relationship between the direction of the magnetic and the direction of the flow of electricity can be modeled using the right hand grip rule. An extension of this is the effect of placing a current-carrying object in a magnetic field. The wire experiences a force that can be modeled by the right hand rule. This is the basis of electric motors and railguns.

9. Students have seen that a current-carrying wire produces a magnetic field. Now they see that the opposite can happen; a moving magnetic field can produce an electric current. Students apply their knowledge of magnetism and electricity to build a motor. The stronger the magnets and the more coils of wire used, the better the results will be. A low voltage bulb or LED is best used to produce light, or a voltmeter or ampmeter can be used to show voltage and current. There are many videos on the internet about how to build a simple generator and a motor and these may be useful to show the students if they are having difficulty with the investigations. Visit **BIOZONE's Resource Hub** for some examples.

How do forces determine the properties of materials?

10. Students return to properties of matter (PS1.A secondary) using their new knowledge of electrostatics. The properties of materials depend on the interactions of the atoms and molecules that make them up. Students should be able to use their understanding of electrostatics to explain the behavior of certain materials and how this dictates their use. Students should be able to explain, for example, that the boiling point of alkanes increases with the number of carbon atoms due to the larger number of interactions (Van de Waals forces) between the greater number of atoms in the molecule, or that network solids have high melting points because every atom is covalently bonded to every neighboring atom in the network. Students can extend their understanding of the properties of matter by surveying what kinds of materials are used for various purposes and analyzing any patterns of use.

78 Force Field

Key Question: Why does a swinging magnetic pendulum not hit against a copper plate?

- We know that when we aim a mass on a string at a vertical wall or barrier and release it from a height the mass strikes the wall. Momentum from the mass is transferred to the wall. The wall may be damaged or the pendulum may rebound, but the pendulum always hits the wall.
- The law of conservation of momentum is one of the basic principles in the physics of motion and it is used in a variety of ways to do work. Swinging wrecking balls, golf clubs, and sledge hammers all transfer momentum to other objects through impact to accomplish specific tasks. All follow the conservation of momentum.
- Something strange seems to happen when a strong magnet is swung at a copper plate (below). There are many videos online where you can see this demonstrated. View one on the **BIOZONE Resource Hub**. When the magnet is swung at the copper plate, it quite suddenly slows down and comes to a complete stop, never hitting the copper plate. It is as if the magnet hits an invisible shield or force field. What is happening here?

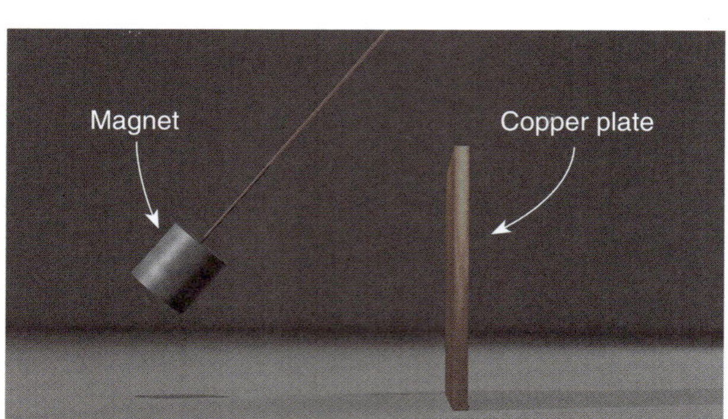

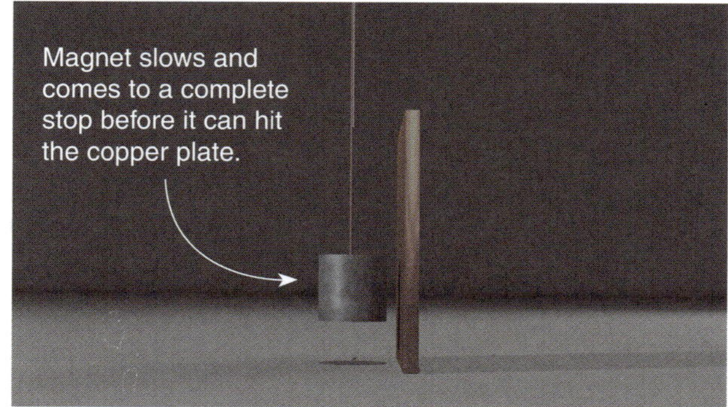

1. When the magnet suddenly stops, it does not hit the copper plate and there does not appear to be any transfer of momentum. Do you think the law of conservation of momentum is still being followed? Explain:

   The law of conservation of momentum states that momentum is conserved in a collision, but a collision does not occur between the copperplate and the magnet. However, a force acting on the system can change the momentum of the system without a collision. The magnetic field does work on the system changing the momentum while still following the conservation law.

2. Newton's first law states that an object in motion will continue moving in the same direction unless acted upon by another force. The magnet does not hit the copper plate, so what caused it to stop moving? Explain:

   The magnet produces a magnetic field due to the arrangement of atoms within the magnet. Even though the copper is not magnetic its electrons are still influence by the field and the movement exerts a net force on the magnet-copper plate system.

3. In groups, discuss how you think the magnet and the copper plate are interacting. Explain why you think the strong magnet stopped when it was swung at the copper plate:

   Copper is not magnetic so there is no force of attraction to a magnet however the electrons in the copper are still influenced by the electromagnetic field produced by the moving magnet. The electrons in the copper react to the magnetic field and move temporarily making the copper plate an electromagnet.

©2020 **BIOZONE** International
ISBN: 978-1-927309-82-7
Photocopying Prohibited

79 Gravity

Key Question: What is gravity and what does it do?

Gravity

- Gravity is a property of mass which acts as an attractive force between all objects with mass. The greater the mass of the object, the greater the gravity it possesses.
- All masses have gravity, so there is an attractive force between all objects – you and your friend, you and your desk, you and the Earth. We do not feel the attractive forces between us and all the objects around us because gravitational force is so weak that the forces involved are insignificant unless at least one of the objects is very large, e.g. the Earth, the Moon, or the Sun.
- The Earth is massive, so the gravitational force of attraction is large enough to pull us towards its center of mass. When we jump, we exert a force on the Earth that propels us away from the ground. This force was instantaneous, so as soon as we left the ground, we experienced an unbalanced force due to gravity and started accelerating back towards the center of the Earth.
- Even when we are on the ground, gravity is still acting on us. Gravitational force is pulling us down (towards the Earth's center of mass). We describe gravitational force acting on an object as **weight**. Weight (measured in N) is based on the mass of the object and multiplied by the strength of gravity. The strength of gravity is directly related to the mass of the large body. On Earth, the value for the strength of gravity is 9.8 N/kg.

You can jump much higher on the Moon than on Earth because of the low gravity, even when wearing an 80 kg space suit.

- When you fall towards the Earth the Earth also falls a little towards you. This may seem strange, but recall that an object's acceleration is related to the force applied to it and its mass. The force applied to you is the same force applied to the Earth, but the Earth is so massive, its acceleration is undetectable.

1. Using the equation **W = mg** where **W** is your weight, **m** is your mass and **g** is the strength of gravity:

   (a) Describe what would happen to your weight if your mass doubled: Gravity would have twice as much mass to affect. Weight would double, e.g. $W = mg$: $19.6 = 2 \times 9.8$, $39.2 = 4 \times 9.8$

   (b) Now, instead, what would happen to your weight if the same sized Earth doubled its mass (which doubles its gravity): Gravity would pull on you twice as hard. Weight would double.

   (c) What do you notice about your weight when you compare the values in 1(a) and (b)? In both instances, weight is doubled.

2. (a) If you had the same mass as the Earth would you now fall towards the Earth or would the Earth fall towards you? Explain: Both you and the Earth would fall towards each other equally as your masses will be equal and you will accelerate at the same rate.

   (b) What if you now had double the mass of the Earth? How would the falling situation now change? You and the Earth would still attract each other, but now it would appear the Earth is falling because it would be moving faster than you (because of your greater mass)

3. Why do you think we do not feel the attractive force of gravity between ourselves and other objects on Earth? We only feel attracted to Earth because of how large Earth is. Other objects are too small for gravity to be significant.

©2020 BIOZONE International
ISBN: 978-1-927309-82-7
Photocopying Prohibited

Gravitational attraction between objects

- There is a symmetrical gravitational force of attraction between all pairs of objects that have mass. Consider object A in the gravitational field of object B: the strength of the field where A is depends on B's mass. The force of gravity on A depends on its own mass and the strength of B's gravity at that position. Thus the size of the gravitational force on A depends on the mass of both objects. Logically, this situation can be viewed totally in reverse. Hence A and B exert exactly the same amount of gravitational force on each other but in opposite directions.
- Remember everyday objects (pens, cups, tables, people, cars) each have gravity too, but their masses are minuscule when compared with moons, planets, and stars. The gravitational force they exert on each other is so extremely small we do not notice it.
- In 1798, Henry Cavendish demonstrated it is possible to observe and measure the gravitational attraction between objects. A modern representation of his experiment is shown below. Cavendish was able to measure the force of gravity between masses by precisely measuring the twisting force on the wire thread used to suspend the smaller masses. The twisting was produced by the movement of the smaller masses towards the larger fixed masses.

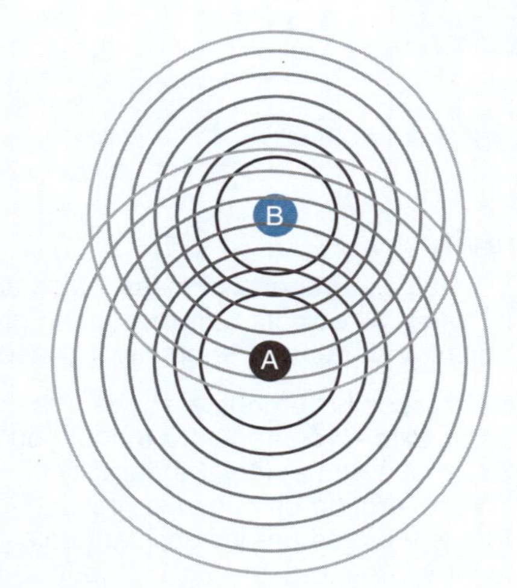

A pulls on B and B pulls on A. The strength of the gravitational field looks the same from either object's perspective.

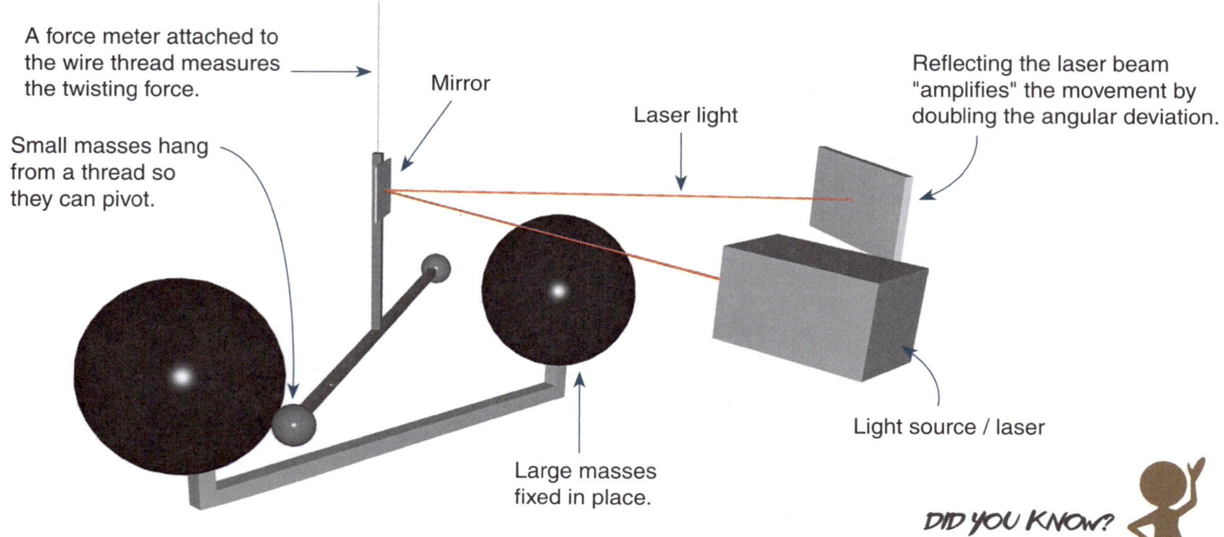

DID YOU KNOW?

Gravity is the weakest of the universe's four fundamental forces. Its effects are only noticeable when there is a huge amount of mass in one place.

4. Cavendish's experiment was designed to find a value called the **gravitational constant (G)** (see next activity). It is part of the equation $\mathbf{F = GM_1M_2/r^2}$, where $\mathbf{M_1}$ and $\mathbf{M_2}$ are masses and **r** is the distance between them. How could Cavendish's experiment find G?

Cavendish knew the masses of mass 1 and 2 (by weighing them) and the distance between them. Measuring the force gave him F. He could then use these known variables to calculate G by rearranging the original equation.

5. Consider two objects. Object A is a million times more massive than object B and therefore produces a gravitational force proportionally larger than object B. Object B is placed inside the gravitational field of object A. Explain the force experienced from the point of view of both objects and how this can explain the fact the gravitational force they exert on each other is opposite but equal.

The gravitational force experienced by B depends on how hard it pulls on A and how hard A pulls on it. From A's point of view the gravitational force depends of how hard it pulls on B and how hard B pulls on it. These experiences are therefore the equal (i.e. A+B = B + A) But A is pulling in one direction and B in the other. They are therefore opposite. Much like a tug of war, the force exerted on the rope depends on the strength of both teams, but the large team will stay still, while the smaller team will be pulled towards the larger team.

©2020 BIOZONE International
ISBN: 978-1-927309-82-7
Photocopying Prohibited

80 Newton's Law of Gravitation

Key Question: How do we calculate the strength of the force of attraction between objects?

PS2-4 Gravity and orbits

- The Sun is the overwhelming source of gravity in our solar system. Its gravity reaches far into space and all the objects in the solar system orbit around it.
- But what is an orbit? The planets are all falling towards the Sun. However, they are also moving sideways relative to the Sun. The effect is that they trace elliptical orbits through space.
- It is important to remember that the gravitational pull the planets exert on the Sun is equal to the Sun's gravitational pull on the planets, but because the Sun is so much bigger, it barely moves, instead it wobbles around its axis on a tiny orbit of its own.
- The closer the match between the mass of the orbiting objects the more pronounced this wobble is. The center of the Earth - Moon system for example is 4670 km from the center of the Earth (i.e. within the radius of the Earth).

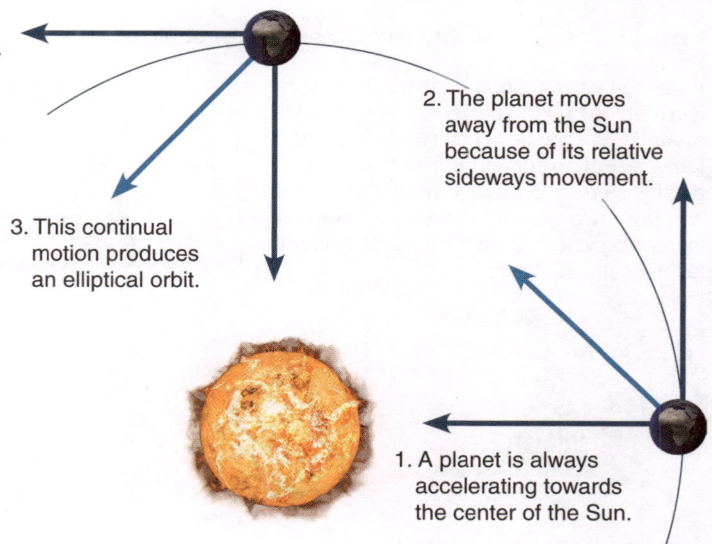

Kepler and orbits

- Between 1609 and 1619 Johannes Kepler used data collected by his mentor Tycho Brahe to produce three important laws of planetary motion.
- His third law is very important in deriving the Newton's law of gravitation and states that the ratio of the square of the period (**T**) to the cube of the average distance from the Sun (**a**) is the same for any planet in the solar system.
- This can be represented through the equation:

$$\frac{T_1^2}{a_1^3} = \frac{T_2^2}{a_2^3}$$

- The ratio for a system is constant (e.g. T^2/a^3 for the Earth-Sun orbit is the same as T^2/a^3 for Mars-Sun orbit).
- In our solar system, this equation can be simplified to astronomical units (AU: the distance from the Earth to the Sun = 150 million km) and years, such that: $\mathbf{a^3 = T^2}$

Examples of T and a

Planet	a (AU)	T (Earth years)	a^3	T^2
Venus	0.72	0.62	0.37	0.38
Earth	1	1	1	1
Jupiter	5.203	11.86	140.9	140.7
Saturn	9.537	29.46	867.4	867.9

1. Calculate the following:

(a) Mercury's distance to the Sun is 0.39 AU. Calculate its orbital period in years: $a^3 = T^2$ Therefore, $T = 0.24$ years.

(b) Neptune takes 164.79 years to orbit the Sun. What is its distance from the Sun in AUs? $a = 30.06$ AU

(c) How far is your answer to (b) in km? $a = 4.509 \times 10^9$ km

(d) Calculate the orbital period of a new planet found to be orbiting 62 AU from the Sun: $T = 488.2$ years

2. (a) Use $T_1^2/a_1^3 = T_2^2/a_2^3$ to solve the following: Jupiter's moon Ganymede, takes 7.15 Earth days to orbit Jupiter. Ganymede is measured to be 1,070,000 km from Jupiter's center. A second moon of Jupiter, Callisto, takes 16.69 Earth days to orbit Jupiter. How far away is Callisto from the center of Jupiter?

Variables a and T need to be converted to AUs and years. $a_C^3 = (T_C^2 \times a_G^3) \div T_G^2$

$= (0.0457^2 \times 0.0071^3) \div 0.02^2 = 1.89 \times 10^{-6}$ Therefore, $a_C = 0.012$ AU or 1.8 million km.

(b) Jupiter's moon Io takes just 1.77 days to orbit the planet. How far away from Jupiter is Io?

$a_I^3 = (T_I^2 \times a_G^3) \div T_G^2 = (0.0048^2 \times 0.0071^3) \div 0.02^2 = 2.07 \times 10^{-8}$

Therefore $a_I = 0.00275$ AU or 412,500 km

©2020 BIOZONE International
ISBN: 978-1-927309-82-7
Photocopying Prohibited

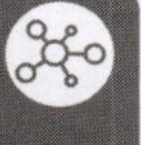

The cannon conundrum

- Isaac Newton also studied the orbits of the planets and knew that a force was required to keep the planets in those orbits. Without this force the planets would fly out of their orbit and travel in a straight line into space.
- Newton realized that the force that caused objects on Earth to fall (gravity) was the same force that held cosmic bodies in their orbits.
- He proposed a thought experiment, shown below.

A cannon sits on top of a tall hill. The cannon is aimed so that the barrel is horizontal. All around it is a flat level plain that stretches forever, with no curvature. The force of gravity is equal to that on Earth. There is no air resistance. As cannonballs are fired with increasing velocity, the balls travel further but fall the same distance.

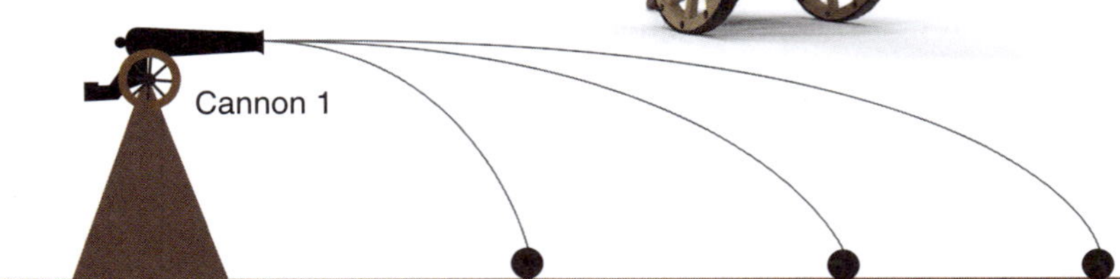

Cannon 2

Newton reasoned that with a great enough horizontal velocity, the cannon ball would never hit the ground because the ground would always curve away from it. The cannon ball would instead orbit the Earth, traveling at the same speed but perpetually falling towards the center of the Earth.

A second cannon also sits on top of a tall hill. The barrel of the second canon is parallel with the barrel of the first canon. However, the "flat plain" this time is the curved surface of a planet. The force of gravity is again equal to Earth's gravity. As cannonballs are fired with increasing velocity in this cannon, the balls travel further and also fall greater distances as the surface of the planet curves away from them.

- Newton also realized that the force (caused by gravity) pulling the planets towards the Sun was a **centripetal force**.
- Newton used this realization and Kepler's third law to derive his law of gravitation.

What is centripetal force?

- Objects moving in a circle, such as in the fair ground ride shown right experience two important forces.
- You may have ridden a fair ground ride and experienced a force trying to push you away from the center of the ride into your seat. At the same time you will have felt the seat you were in pushing back on you with the same force.
- The first force is the centrifugal force, it pulls objects away from the center of motion. The **centripetal force** pulls objects towards the center of motion.
- Because these forces are caused by the changing motion of objects rather than a property of the objects themselves they are often called pseudo-forces.
- Centripetal force is what stops you falling out of your seat when a car goes around a corner.

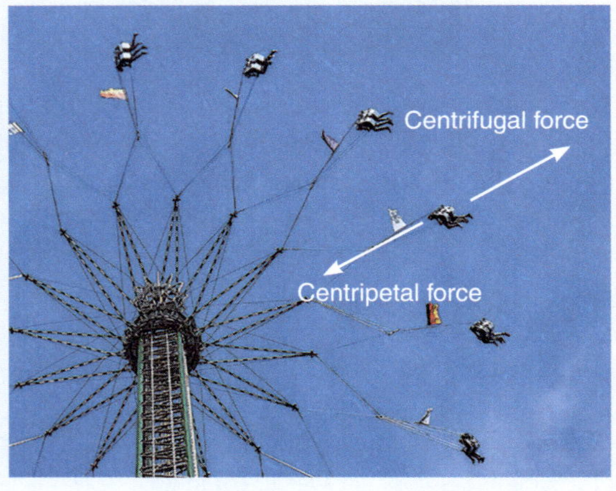

3. Why would the third cannon ball in the second example of Newton's cannon never reach the ground?
The gravity pulls the cannonball towards its center of mass, however the speed of the cannonball will carry it far enough that the ground will curve away from it before it hits the ground.

4. How are gravity and the centripetal force similar and different? *Centripetal force is the force directed towards the center of the circular motion. Gravity can be a centripetal force in orbits but gravity acts on objects whether they are in orbit or not. Many centripetal forces act only when another force is exerted such as the tension force in the cords in the ride above. Gravity is the result of mass. Centripetal force is the result of motion.*

©2020 **BIOZONE** International
ISBN: 978-1-927309-82-7
Photocopying Prohibited

Centripetal acceleration

- Can an object accelerate if it is moving at a constant speed? Yes it can. Remember acceleration (a vector) is a change in velocity and velocity changes by either changing its magnitude (speed), its direction, or both.
- An object traveling at constant speed in a circular path experiences acceleration because the direction of the velocity is constantly changing. This acceleration can be felt when sitting in a turning a car. What you feel is the sideways acceleration due to your changing direction. Acceleration is in the direction of the change in velocity. This is in the direction of the center of rotation or center of a circular path.

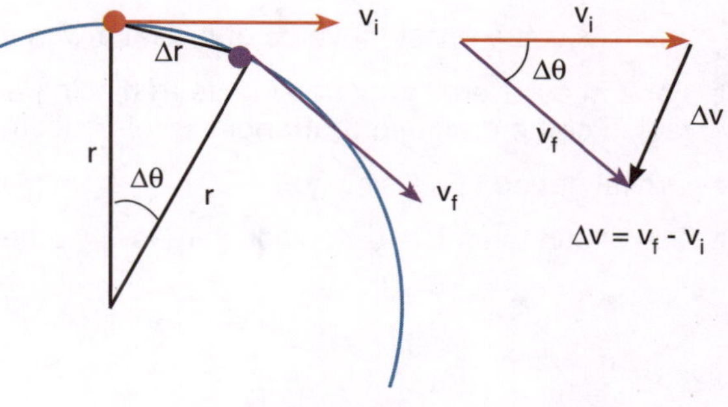

- Consider the diagram near right. The direction of the instantaneous velocity is shown at two points along a circular path.
- The vector diagram far right shows the direction of the change in velocity (Δv). Acceleration is in the direction of the change in velocity, which points directly to the center of the circular path.
- To find the magnitude of centripetal acceleration, we use what we know of similar triangles. Because speed is constant, the magnitude of v_i and v_f is the same (v), creating an isosceles triangle, (far right). The triangle created in the circle (near right) is also an isosceles triangle, with the length of two sides equal to the radius of the circle.
- The measure of the angle between the similar sides for both triangles is the same (Δθ) making them similar triangles.

Δv : v = Δr : r

- Knowing the properties of similar triangles, the ratio of $\Delta v : v$ is equal to the ratio of $\Delta r : r$, therefore:

$$\frac{\Delta v}{v} = \frac{\Delta r}{r}$$ **Equation 1**

- Remember acceleration is $\Delta v \div \Delta t$, so if we rearrange equation 1 above to solve for Δv then we can substitute that equation into the equation for acceleration. We do this by multiplying each side by v:

$$\Delta v = \frac{v\Delta r}{r}$$ **Equation 2**

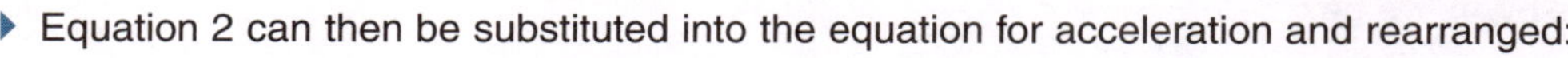

- Equation 2 can then be substituted into the equation for acceleration and rearranged:

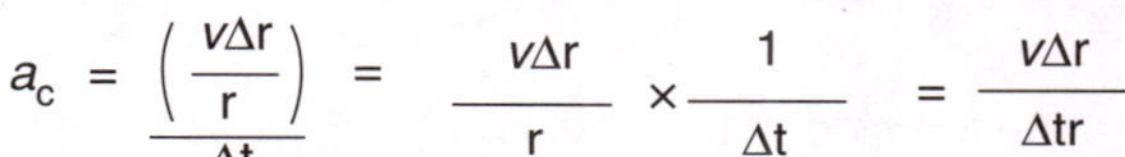

$$a_c = \frac{\left(\frac{v\Delta r}{r}\right)}{\Delta t} = \frac{v\Delta r}{r} \times \frac{1}{\Delta t} = \frac{v\Delta r}{\Delta t r}$$

- Since Δr is the change in distance, we see that $\Delta r \div \Delta t$ is speed (v), which is constant, therefore: $a_c = v^2 \div r$:

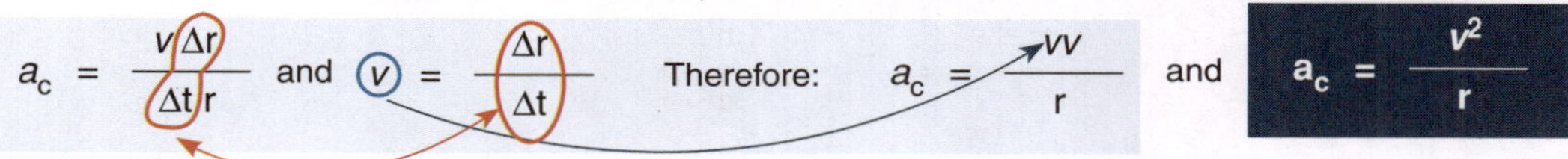

$$a_c = \frac{v\Delta r}{\Delta t r} \text{ and } v = \frac{\Delta r}{\Delta t} \text{ Therefore: } a_c = \frac{vv}{r} \text{ and } a_c = \frac{v^2}{r}$$

Centripetal force

- From Newton's first law, we know that an object will continue moving in a straight path unless acted on by an external force. When looking at planetary bodies orbiting the Sun, this force is the gravitational force. When considering a twirling tethered ball, this force is the tension force in the string.
- Centripetal force is the name given to the net force causing an object to move in a circular path. Multiple individual forces can be involved as long as they add up to give a net force towards the center of the circular path.
- From Newton's second law: $a = F \div m$
- Substituting $v^2 \div r$ from centripetal acceleration (above) in for a we get:

$$\frac{v^2}{r} = \frac{F}{m}$$

- Rearranging the equation to solve for force, we can show that centripetal force (F_c) is:

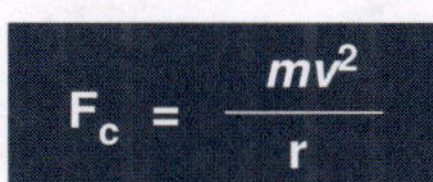

$$F_c = \frac{mv^2}{r}$$

5. A 1000 kg car makes a right hand turn at a constant velocity of 15 m/s. The radius of the arc turned is 25.0 m.

   (a) Determine the acceleration of the car: $a_c = v^2/r = (15\ m/s)^2 \div 25\ m = 9\ m/s^2$

   (b) Calculate the centripetal force acting on the car: $F_c = (mv^2) \div r = (1000\ kg \times 15^2) \div 25\ m = 9000\ N$

   Alternatively $F = ma = 1000\ kg \times 9\ m/s^2 = 9000\ N$

©2020 **BIOZONE** International
ISBN: 978-1-927309-82-7
Photocopying Prohibited

Deriving the law of gravitation

We now have all the components necessary for deriving Newton's law of gravitation:

- Consider a planet (with mass *m*) orbiting the Sun (with mass *M*) on a circular orbit (planets orbit in ellipses but using a circle makes the equations simpler).
- Kepler's third law states that the ratio T^2 to a^3 is constant for every planetary body orbiting the same star (or planet). Therefore,

$$\frac{T^2}{a^3} = k$$

  where *k* is the constant value of the ratio with units of $s^2 \div m^3$.

- For a circular orbit in general units (replacing *a* as the average distance the planet is from its star with *r* as the radius of the orbit and rearranging), this becomes $T^2 = kr^3$.
- Orbital period (T) is simply: $T = 2\pi r \div v$ (simply the circumference of a circle divided by the speed of the orbit).
- Substituting this T into Kepler's third law, we get: $4\pi^2 r^2 \div v^2 = kr^3$ (below):

$$T^2 = \left(\frac{2\pi r}{v}\right)^2 = \frac{(2\pi r)^2}{v^2} = \frac{4\pi^2 r^2}{v^2} = kr^3$$

- Algebraically, we can rearrange this to look like: $v^2 \div r = 4\pi^2 \div kr^2$ (below):

$$\frac{4\pi^2 r^2}{v^2} = kr^3 \rightarrow 4\pi^2 r^2 = kr^3 v^2 \rightarrow \text{Divide each side by r} \rightarrow 4\pi^2 r = kr^2 v^2 \rightarrow \text{Divide each side by r again}$$

$$\rightarrow 4\pi^2 = \frac{kr^2 v^2}{r} \quad \text{Divide each side by } kr^2 \rightarrow \frac{4\pi^2}{kr^2} = \frac{v^2}{r}$$

NEED HELP?
See page 16

- If we then multiply both sides by *m*: we get: $$\frac{mv^2}{r} = \frac{4\pi^2 m}{kr^2}$$

- This arrangement is important because the left side of the equation ($mv^2 \div r$) is the equation for centripetal force derived earlier on the opposite page. This is the force that holds the planet in circular motion. This must then be the gravitational force that holds the planet in orbit, therefore:

$$\frac{mv^2}{r} = F = \frac{4\pi^2 m}{kr^2}$$

- Remember *k* was a constant value for any object orbiting the same star or planet. This number is dependent on the mass of the star or planet (*M*). $k = 4\pi^2 \div GM$, where G is the gravitational constant.
- Substituting this value of *k* into the equation and simplifying we get: $F = GMm \div r^2$

$$F = \frac{4\pi^2 m}{kr^2} = \frac{4\pi^2 m}{\left(\frac{4\pi^2}{GM}\right) r^2} = \frac{4\pi^2 m}{r^2} \times \frac{GM}{4\pi^2} \rightarrow 4\pi^2 \text{ cancels} \rightarrow \frac{m}{r^2} \times GM = F = \frac{GMm}{r^2}$$

- Finally we have arrived at Newton's law of gravitation (using M and m for M_1 and M_2).
- We derived this for objects following circular orbits, but it is true for any two objects in any orbit.

Newton and gravity

- In 1687 Isaac Newton published his law of gravitation. It included the equation we derived above:

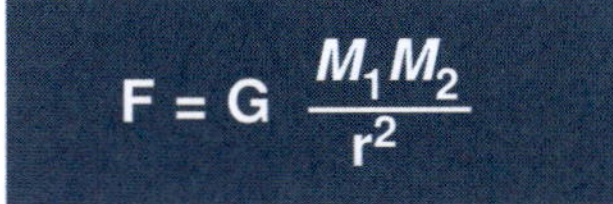

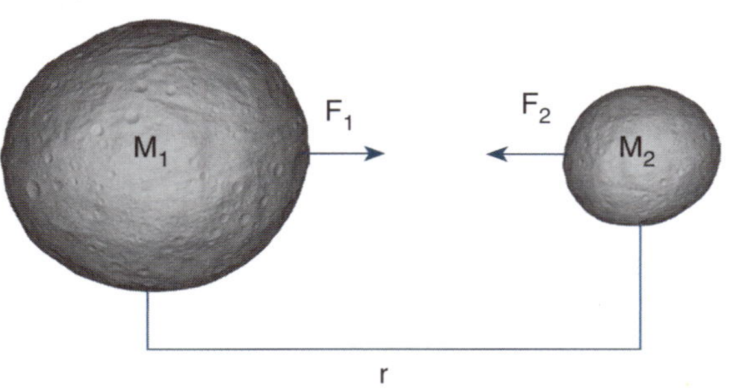

* Remember F_1 and F_2 are equal.

- The equation allows us to calculate the force of gravity acting between two objects, where M_1 and M_2 are the masses of the two objects (in kg), r is the distance between their centers of mass (in meters), and G is the **gravitational constant 6.673 x 10^{-11} $m^3/kg/s^2$ (Nm^2/kg^2)**
- The gravitational constant was unknown at the time. Cavendish was the first to calculate it from his experiments with his torsion balance 71 years after Newton's death (see page 184).

©2020 **BIOZONE** International
ISBN: 978-1-927309-82-7
Photocopying Prohibited

Inverse square law

- **The inverse square** is seen in many areas of physics. For gravity, the force of gravity felt by an object reduces with the square of the distance from the gravitational source, i.e. the further away an object is from a gravitational body, the less the influence of gravity.
- The inverse square law models the behavior of many forces that act over a distance. Because the distance is squared, the strength of the field of force decreases rapidly.

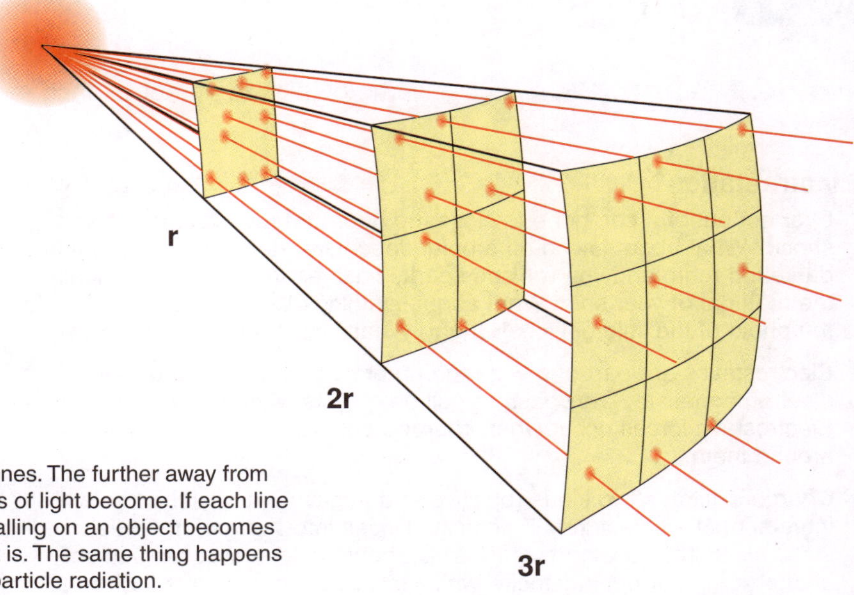

Light traveling from a star moves in straight lines. The further away from the star we get, the more spread out the lines of light become. If each line creates a point of light, the intensity of light falling on an object becomes less the further away from the star the object is. The same thing happens with gravity, electrostatic forces, sound and particle radiation.

- This relationship is described in Newton's law of gravitation and implies that a person on the surface of the Earth will have a different weight from a person in the International Space Station (ISS) because the person in the ISS is further away from the center of the Earth (greater r) than the person on the surface.

6. Using Newton's equation for the law of gravitation, what would be the effect on the force between the Earth and the Moon of changing the following:

   (a) Doubling the distance between the Earth and the Moon: F would (increase / decrease / remain the same) (circle one).

   (b) Doubling the mass of the Moon: F would (increase / decrease / remain the same) (circle one).

   (c) Halving the mass of the Earth: F would (increase / decrease / remain the same) (circle one).

7. Imagine the asteroids shown on the previous page are orbiting each other in space. The mass of asteroid M_1 is 10 billion kilograms. The mass of M_2 is 4 billion kilograms. Their centers are 3600 m apart. Calculate the force F that is acting between them:

   $F = GM_1M_2 \div r^2 = (6.673 \times 10^{-11}\ Nm^2/kg^2 \times 1.0 \times 10^{10}\ kg \times 4.0 \times 10^9\ kg) \div 1.296 \times 10^7 m^2$

   $= 206\ N$

8. How much would a 60 kg person weigh on the ISS? The mass of the Earth is 5.972×10^{24} kg, the radius of the Earth is 6.371×10^6 m, and the average altitude of the ISS is 4.08×10^5 m.

   $W(F) = 6.673 \times 10^{-11}\ Nm^2/kg^2 \times 5.972 \times 1024\ kg \times 60\ kg \div (6.371 \times 10^6\ m + 4.08 \times 10^5\ m)^2 = 520\ N$

9. Calculate the weight for a 60 kg individual at the Moon's distance from the Earth (3.84×10^8 m):

   $W = (6.673 \times 10^{-11}\ Nm^2/kg^2 \times 5.972 \times 10^{24}\ kg \times 60\ kg) \div (3.84 \times 10^8\ m)^2 = 0.162\ N$

10. Use Newton's equation to calculate the force of attraction between the Earth and the Moon ($M_e = 5.972 \times 10^{24}$, $M_{moon} = 7.35 \times 10^{22}$, $r = 3.84 \times 10^8$ m):

    $F = (6.673 \times 10^{-11}\ Nm^2/kg^2 \times 5.972 \times 10^{24}\ kg \times 7.35 \times 10^{22}\ kg) \div (3.84 \times 10^8\ m)^2 = 1.986 \times 10^{20}\ N$

11. It is thought the Sun's gravity could extend out to two light years from its center (1.89×10^{13} km). The Sun's mass is 1.99×10^{30} kg. Calculate the force felt by a 60 kg person at this distance from the Sun.

    $F = (6.673 \times 10\text{-}11\ Nm^2/kg^2 = 1.99 \times 10^{30}\ kg \times 60\ kg) \div (1.892 \times 10^{16} m)^2 = 2.2258 \times 10^{-11} N$

©2020 **BIOZONE** International
ISBN: 978-1-927309-82-7
Photocopying Prohibited

81 Objects of Attraction

Key Question: What causes some objects to attract to each other and others to repel?

Electrostatics

- Ever got out of a car, gone to close the door and received an electric shock? What about taking off a polar fleece sweater or jacket? Try it in a darkened room and you will see sparks flash as the jersey rubs against the material of your shirt. What about lightning? What causes that? Study the photo of the little girl's hair (right) What's causing that to happen?
- Electrostatics are part of a wider group of non-contact forces called electromagnetism. They push or pull on objects without touching them. Electrostatic forces occur when charged objects interact with objects around them.
- Charge exists as two kinds (positive and negative) and is the basic ingredient of all electricity. Electricity divides into two general areas: (i) static electricity (or **electrostatics**) is about stationary accumulations of charge (ii) current electricity, which involves a continuous flow of negatively charged particles (usually electrons).
- Charge is present in all the materials around us, but we are usually unaware of it because most of the positive and negative charges are balanced and evenly spread out. The balance can be "disturbed", e.g. by rubbing materials together. This often results in electrons being transferred from one material to another. The material gaining the electrons is then said to be negatively charged while the material losing the electrons is said to be positively charged.
- Objects with like charges repel each other while objects with opposite charges attract each other.

Electrostatic investigations

- Balloons are well known for producing some interesting electrostatic effects:

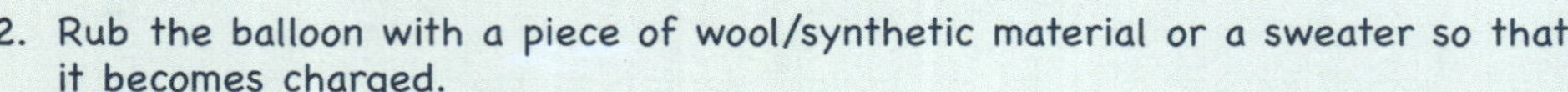

INVESTIGATION 6.1: Balloon electrostatics See appendix for equipment list.

1. In a still, warm room, fully inflate a balloon and hang it from the ceiling or an insulated support with nylon thread or fishing line.
2. Rub the balloon with a piece of wool/synthetic material or a sweater so that it becomes charged.
3. Predict what will happen if you bring the material or sweater used to rub the balloon near the balloon.

   Student's answer.
4. Carry out step 3 and record your observations:

   Students should observe an attraction between the material and the balloon.
5. Fully inflate a second balloon and hang it from the ceiling with more nylon fishing line near the first balloon.
6. Rub both balloons with the same material (wool/synthetic fabric or a sweater). This should give the balloons a charge of the same sign and a similar amount.
7. Predict what will happen to these similarly charged balloons as they hang near each other.

   Student's answer.
8. Carry out step 7 and record your observations:

   Students should observe the balloons repelling each other.

©2020 **BIOZONE** International
ISBN: 978-1-927309-82-7
Photocopying Prohibited

INVESTIGATION 6.2: Threading the needle

See appendix for equipment list.

1. Try threading a standard sewing needle with thin plastic thread (such as from a thread of plastic string). Keep trying!
2. Now thread the needle with normal cotton thread of the same thickness. Was it easier?

1. (a) What happened when you tried to thread the needle with plastic thread? *Student's answer.*
Students may notice the plastic thread is deflected around the metal needle.

(b) What happened when you tried to thread the needle with cotton thread? *Student's answer.*
Students may note it easier to thread the needle with the cotton thread.

(c) Suggest why there was a difference in how the threads behaved while being threaded: *Student's answer.*
Students may describe a difference in charge between the threads and a similar charge between the plastic thread and the needle.

2. For each situation below, state whether the balloons will repel or attract each other:

(a)

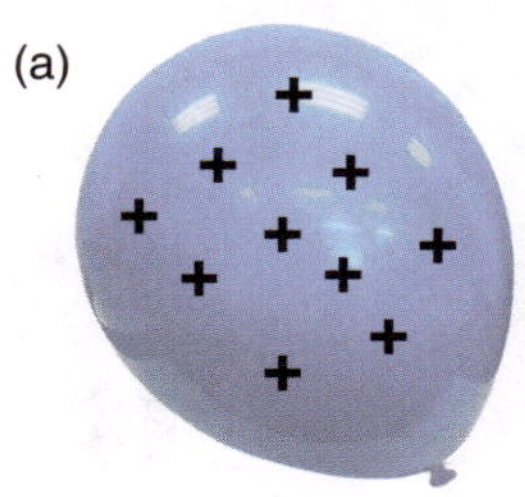

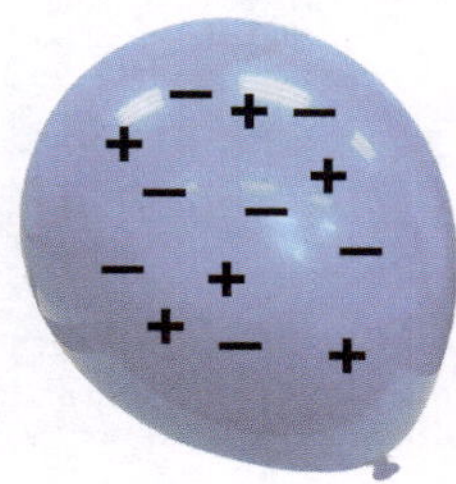

Attract.

(c)

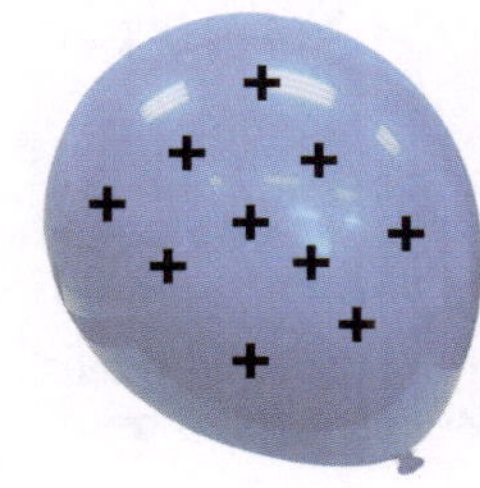

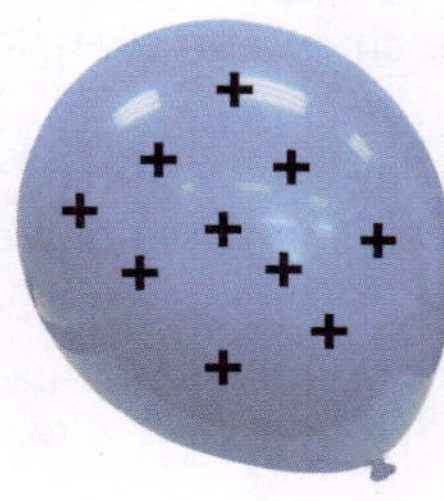

Repel.

(b)

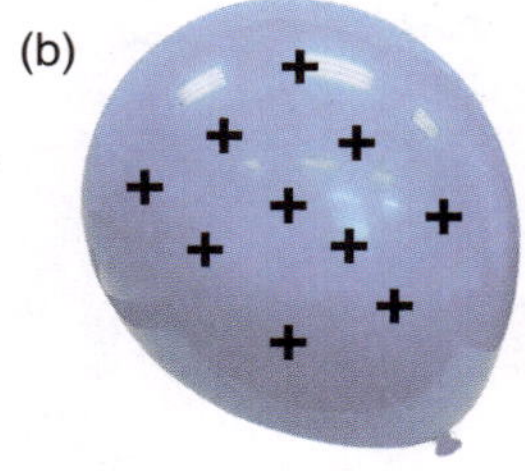

Attract.

(d)

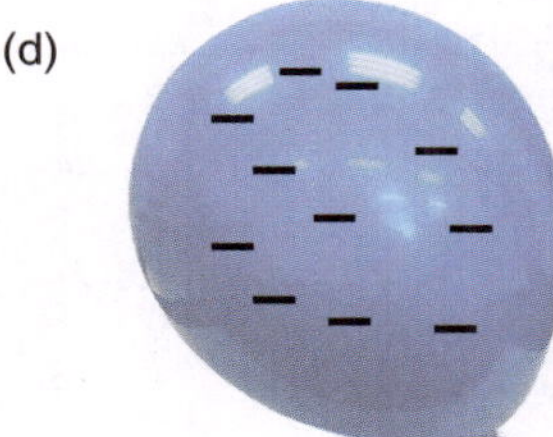

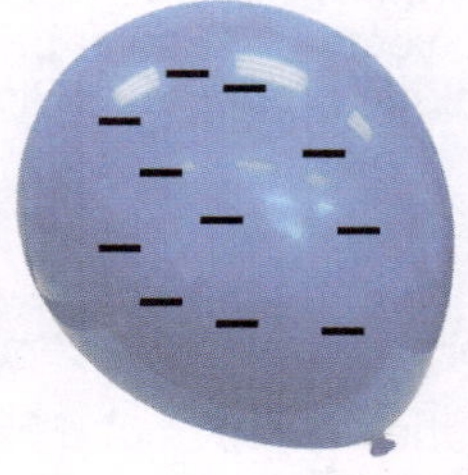

Repel.

(e) Compare and contrast the strength of force between the balloons in (a) and (b) and explain your rationale:
The force experience by (a) will be less than that experienced by (b). It will also be in the opposite direction. The force will be less because the charges are less. The charges are opposite in (a) and the same in (b) thus the forces involved are opposite.

3. A student rubs a balloon with silk to create a charge on the balloon's surface, then brings the balloon near some small pieces of tissue paper. Predict what will happen to the tissue paper and explain your prediction. Try it and find out:
Students should recognize they are adding charge to the balloon which will lead to an attractive or the repulsive forces on the pieces of the tissue paper.

4. What you have been investigating are charges of static electricity. Describe any situations you can think of where you have observed static electricity in your daily life:
Students might suggest static shocks when touching metal or sparks seen on fabric in the dark are examples of static electricity.

©2020 **BIOZONE** International
ISBN: 978-1-927309-82-7
Photocopying Prohibited

Properties of electric charge

- Recall that in atoms protons carry positive charge and electrons carry negative charge.
- Charges are the result of an imbalance of protons and electrons.
  - Negative charges = more electrons in an object than protons.
  - Positive charges = fewer electrons in an object than protons.
- Objects become charged because electrons (negative charges) are transferred from one object to another. Protons are tightly held in the nucleus and so are not exchanged to produce charge.
- Like mass, the charge of an object is a measurable quantity. Charge strength is determined by the difference in quantity between electrons and protons in an object.
- All charge is a multiple of a fundamental unit of charge symbolized by *e*. Electrons have a charge of -*e*. Protons have a charge of +*e*.
- Charge is measured using the unit **coulomb** (**C**). Like the mole, the coulomb is a dimensionless quantity based on the electric charge found in a single electron (-*e*) or proton (+*e*). 1 C is equal to approximately 6.24×10^{18} units of *e*.
- Charge is always conserved. It is not created, only exchanged.
- The charge on a single electron is -1.6×10^{-19} C. The charge on a single proton is $+1.6 \times 10^{-19}$ C.

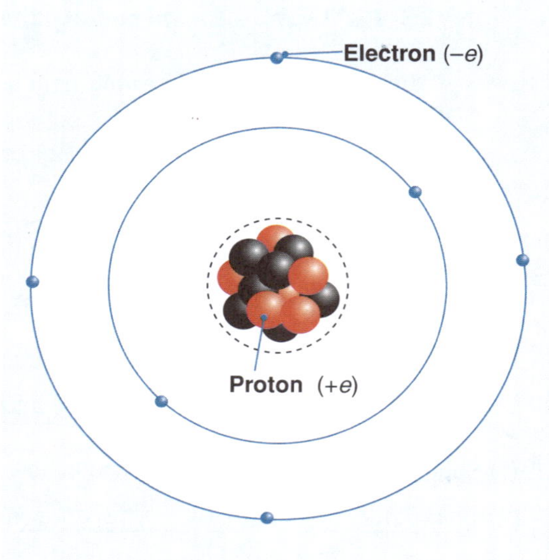

Conductors and insulators

- Materials can be divided into categories based on how well charges move through the material (**conductivity**).
- In **conductors**, the electric charges move freely and distribute evenly over the entire surface. Metals tend to be good conductors.
- In **insulators**, the electric charges do not move freely. They have low conductivity. Plastic, glass, and rubber are good insulators.
- **Semiconductors** are moderate conductors and used extensively in electronics, solar panels, and other technologies for their unique properties. Silicon crystals are an example.

Conducting copper wire surrounded by insulating plastic

Polarization

- In most neutral atoms or molecules, the center of positive charge coincides with the center of negative charge.
- In the presence of a charged object, these centers may separate slightly. This results in more positive charge on one side of the molecule than on the other. This realignment of charge on the surface of an insulator is known as polarization.
- Polarization is not charging. A charge is caused by an imbalance of charge, whereas polarization is simply a separation of charge within the object. When neutral objects are polarized, they are still neutral.

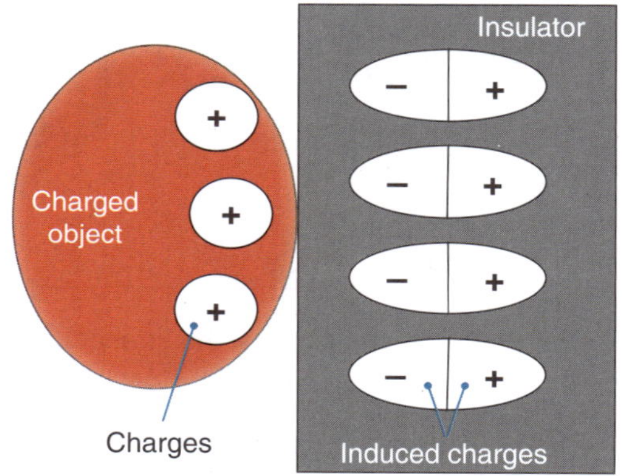

5. Complete the following table with the number of excess protons/electrons and charge in coulombs:

Object	Number of excess protons/electrons	Charge in Coulombs (C)
A	1.0×10^{8} excess electrons	(a) $-1.6 \times 10^{-19} C \times 1 \times 10^{8} e = -1.6 \times 10^{-11} C$
B	1.0×10^{8} excess protons	(b) $-1.6 \times 10^{-11} C$
C	(c) 3.0×10^{8} excess protons	4.8×10^{-9} C
D	3.0×10^{10} excess electrons	(d) $-4.8 \times 10^{-9} C$
E	(e) 3.625×10^{8} excess protons	5.8×10^{-11} C
F	(f) 5.3125×10^{8} excess electrons	-8.5×10^{-9} C

6. The average charge carried by a bolt of lighting is 15 C. What quantity of excess electrons is carried by the lightning bolt?

$15 C \times 6.24 \times 10^{18} e/C = 9.36 \times 10^{19}$ excess electrons.

©2020 **BIOZONE** International
ISBN: 978-1-927309-82-7
Photocopying Prohibited

82 Coulomb's Law

Key Question: How is force influenced by the magnitude of charge on particles and the distance between them?

Coulomb's force

PS2-4

- In 1785, French physicist Charles-Augustin de Coulomb published an equation to explain the force between electric charges (known as Coulomb's law). It stated the magnitude of the electrostatic force between two charges is proportional to the product of the charge magnitudes (their size) and inversely proportional to the square of the distance between them.

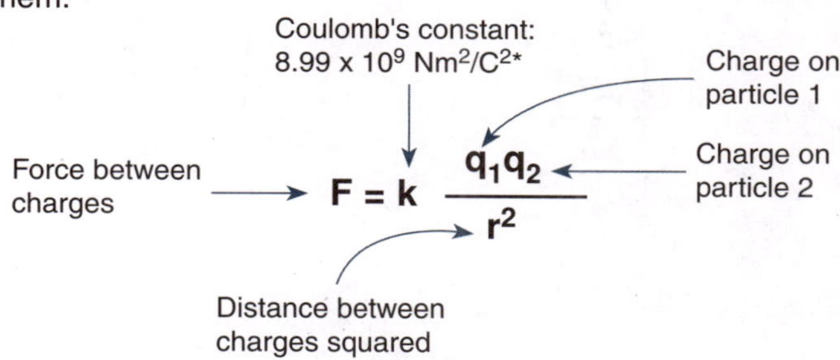

*Newton meters squared per coulomb squared.

- Put simply, the force acting between two charged particles increases as the magnitude (size) of the charges increases and decreases as the distance between them increases.
- We can verify this using simplified quantities placed into the equation and making hydrogen our reference. We can give the proton a charge of 1 and the electron a charge of -1. k is constant so can be ignored for simplicity:
- For a hydrogen atom, the force acting between the proton and the electron will be:

$$F = \frac{1 \times -1}{1^2} = -1$$

← A negative force indicates an attraction between particles.

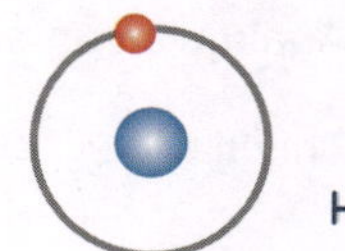

- For a lithium atom, the charge in the nucleus increases to 3 and the distance to the valence electron is roughly 3 times the distance between the hydrogen proton and its electron:

$$F = \frac{3 \times -1}{3^2} = -0.33$$

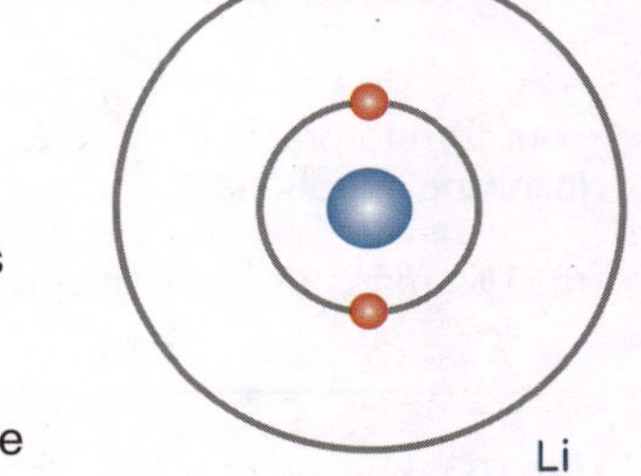

- Thus we can see although the charge in an atom's nucleus may increase as protons are added, the distance between the protons and the electrons can make a big difference in the force acting between the particles.
- Unlike gravity, which is always attractive, electrostatic force forces are either attractive (if the charges are opposite) or repulsive (if the charges are the same).
- Note the similarity between Coulomb's law and Newton's law of gravitation:

$$F = G\frac{M_1 M_2}{r^2}$$

- Recall that the effects of gravity decrease over distance, where the strength of gravity is inversely proportional to the square of the distance between the objects. We can now see that same relationship for electromagnetic forces in Coulomb's equation (i.e. the inverse square law).

1. Two balloons are identically charged at -6.25×10^{-9} C. They are held apart at a distance of 0.50 meters. Determine the magnitude and direction of the electrical force between them.

   $F = k \times (q_1 \times q_2) \div r^2 = (8.99 \times 10^9\ Nm^2/C^2 \times -6.25 \times 10^{-9}\ C \times 6.25 \times 10^{-9}\ C) \div 0.5^2$

   $= 1.4 \times 10^{-6}$ N. The force is positive and therefore repulsive.

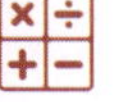

2. Two balloons with charges of $+3.21 \times 10^{-6}$ C and -8.37×10^{-6} C attract each other with a force of 0.604 N. Determine the distance between the two balloons.

   $F = k \times (q_1 \times q_2) \div r^2,\ r = \sqrt{((q_1 \times q_2) \div F)}$

   $= \sqrt{(8.99 \times 10^9\ Nm^2/C^2 \times 3.21 \times 10^{-6}\ C \times -8.37 \times 10^{-6})} \div -0.604\ N = 0.63\ m$

©2020 BIOZONE International
ISBN: 978-1-927309-82-7
Photocopying Prohibited

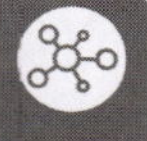

More on the inverse square law

- When using equations involving the inverse square law, it is important to understand what effect changes to any variable in the equation have on the final force.
- Let us return to Newton's law of gravitation and see how the inverse square law applies to it before further studying Coulomb's law.
- What happens to the magnitude of the force between two objects when we change the mass of one of the objects? Let's double the mass of one of the objects:

- We can show what happens in this situation without actually doing a numerical calculation. Instead we can use algebra to prove the result:
- First, we add a subscript to the force (F), to designate a specific value: $F_1 = G\frac{M_1 M_2}{r^2}$
- Double the mass for the 2nd object: $F_2 = G\frac{M_1 2M_2}{r^2}$ (note: the subscript 2 on the force indicates a different value than from the first equation).
- Rearranging the equation, we get: $F_2 = 2G\frac{M_1 M_2}{r^2}$
- Substituting F_1 for $G\frac{M_1 M_2}{r^2}$ since $F_1 = G\frac{M_1 M_2}{r^2}$ we get: $F_2 = 2F_1$.
- The resulting force from doubling the mass of one object doubles the force acting between the objects.

- What if we manipulate the distance between the two objects instead. Let's double the distance between the objects.

- Again we can show the effect calculated above using some algebra and rearrangement of the equation.

  This time: $F_2 = G\frac{M_1 M_2}{(2r)^2}$
- First we must expand the brackets involving the radius: $F_2 = G\frac{M_1 M_2}{2^2 r^2} = G\frac{M_1 M_2}{4r^2}$
- And rearrange to solve in units of G: $F_2 = \frac{1}{4}G\frac{M_1 M_2}{r^2}$. Substituting F_1 for $G\frac{M_1 M_2}{r^2}$ we get: $F_2 = \frac{1}{4}F_1$.

3. Your turn to practice some algebraic manipulations. Demonstrate what happens to the force between two objects in the following situations:

   (a) The mass of one object is tripled. $F_1 = GM_1M_2 \div r^2$, $F_2 = G\,3M_1M_2 \div r^2 = 3(GM_1M_2 \div r^2)$
   $F_2 = 3F_1$

   (b) The masses of both objects is doubled. $F_1 = GM_1M_2 \div r^2$, $F_2 = G\,2M_1 2M_2 \div r^2 = 4(GM_1M_2 \div r^2)$
   $F_2 = 4F_1$

   (c) The distance between the two objects is tripled. $F_1 = GM_1M_2 \div r^2$, $F_2 = GM_1M_2 \div (3r)^2 = GM_1M_2 \div 9r^2$
   $= 1/9\,(GM_1M_2 \div r^2)$, $F_2 = {}^1/_9 F_1$

   (d) The distance between the two objects is halved $F_1 = GM_1M_2 \div r^2$, $F_2 = GM_1M_2 \div (0.5r)^2$
   $= GM_1M_2 \div 0.25r^2 = 1/0.25 \times (GM_1M_2 \div r^2)$, $= 4\,(GM_1M_2 \div r^2)$, $F_2 = 4F_1$

4. The inverse square law also applies to the amount of illumination provided by a source that is a "single" point of light. The Earth is approximately 150 million km from the Sun, whereas Mars is approximately 228 million km from the Sun.

   (a) How would the light intensity of the Sun differ on Mars compared to Earth?
   Mars is approximately 1.5 times further from the Sun than Earth. The intensity of light from the Sun would be significantly less for Mars than for Earth.

   (b) Calculate the approximate difference in the Sun's light intensity on Mars compared to Earth: Intensity is inversely proportional to distance. Setting Earth's distance from the Sun at 1 AU for simplicity, Mars would be approximately 1.5 AUs from the Sun. $1/1.5^2 = 1/2.25 = 4/9 = 0.44 = 44\%$ of the intensity on Earth.

©2020 BIOZONE International
ISBN: 978-1-927309-82-7
Photocopying Prohibited

Coulomb's force and the inverse square law

- The rearrangement of the equations for gravity on the previous page apply in the same way to Coulomb's law, we simply use different notation (k instead of G for example).
- Unlike the equation for gravity, which only ever deals with positive numbers, Coulomb's law may deal with positive and negative numbers (i.e. positive charge and negative charges). This means we may get answers that are positive or negative.
- Coulomb's law is important in understanding how charged particles interact. For example, why does the compound sodium chloride have a different melting point to lithium fluoride? We can answer this if we know a little about the charge the particles carry and the distance they are apart from each other.

5. Using the same algebraic manipulations we used on the previous page, demonstrate what happens to the force between two charges in the following situations:

(a) The magnitude of one of the charges is doubled. $F_1 = kq_1q_2 \div r^2$, $F_2 = k\,2q_1q_2 \div r^2 = 2(kq_1q_2 \div r^2)$

$F_2 = 2F_1$

(b) The distance between the charges is doubled. $F_1 = kq_1q_2 \div r^2$, $F_2 = kq_1q_2 \div (2r)^2 = 1/4(kq_1q_2 \div r^2)$

$F_2 = 1/4 F_1$

(c) Both charges are doubled AND the distance between the charges is doubled.

$F_1 = kq_1q_2 \div r^2$, $F_2 = k\,2q_1 2q_2 \div (2r)^2 = (kq_1q_2 \div r^2)$, $F_2 = F_1$

(d) One of the charges is quadrupled AND the distance between the charges is doubled.

$F_1 = kq_1q_2 \div r^2$, $F_2 = k\,4q_1q_2 \div (2r)^2 = (kq_1q_2 \div r^2)$, $F_2 = F_1$

6. What can you say about the way in which Newton's law of gravitation and Coulomb's law work?

Both laws show force between objects/charges is proportional to the mass/charge but inversely proportional to the distance between them.

7. Use Coloumb's law to answer the following:

(a) Is the force produced positive or negative when two positive charges are brought together? *Positive.*

(b) Would these charges attract or repel each other? *Repel.*

(c) Is the force produced positive or negative when two negative charges are brought together? *Positive.*

(d) Would these charges attract or repel each other? *Repel.*

(e) Is the force produced positive or negative when a positive charge and negative charge are brought together?

Negative.

(f) Would these charges attract or repel each other? ____________

8. What can be said about the sign (positive or negative) in front of the force value in relation to charges attracting or repelling each other?

Negative values for force indicate an attractive force but positive values indicate a repulsive force.

9. An object with a charge of 8.5×10^{-9} C is used to balance the force of gravity on a 1.8 gram balloon with a charge of -6.3×10^{-9} C. If the balloon is perfectly balanced, how high above the balloon is the object?

$F_g = 0.0018\ kg \times 9.8\ m/s^2 = -0.01764\ N$, $F_e = -F_g$, $F_e = kq_1q_2 \div r^2$, $r = \sqrt{(kq_1q_2 \div F)}$

$= \sqrt{(8.99 \times 10^9\ N\,m^2/C^2 \times 8.5 \times 10^{-9} C \times -6.3 \times 10^{-9}) \div 0.01764\ N} = 0.005\ m\ (5\ mm)$

10. Two balloons are charged and each acquires 15 trillion excess electrons. If the mass of each balloon is 1.8 grams, how far below one balloon must the other be in order to lift it?

$F_g = -0.01764\ N$, $r = \sqrt{(8.99 \times 10^9\ N\,m^2/C^2 \times -2.4 \times 10^{-6}\ C \times 2.4 \times 10^{-6}\ C \div 0.01764\ N)}$

$= \sqrt{2.9} = 1.7\ m$

©2020 **BIOZONE** International
ISBN: 978-1-927309-82-7
Photocopying Prohibited

83 Electric Fields

Key Question: How do the electric fields of charged particles affect their interactions?

PS3-5

Electric fields around charged particles

- Electrical forces are field forces. Electricity falls in the category of non-contact forces. An object with charge is able to affect another object even when there is no physical contact between them. Charges influence each other without contact through **electric fields**.
- Electric fields exist in the region of space around a charged object. Conceptually, this field is infinite, but as we learned from Coulomb's law in the previous activity, the strength of the field significantly weakens the further away you are from the field. Unless charges are close together the field force is very, very small. An electric field is considered the immediate region of space around a charged object which is strong enough to influence other charges.
- When another charged object enters an electric field, the field exerts a force on the second charged object. The direction of the field is defined to be the **direction of the electric force that would be exerted on a small positive charge**. Thus, electric field direction is always directed away from positive source charges and towards negative source charges.
- Recall that protons and neutrons are locked in the nucleus so charges are a result of the movement of electrons. Positive charge sources are the result of a net loss of electrons and negative charge sources are the result of a net gain of electrons.

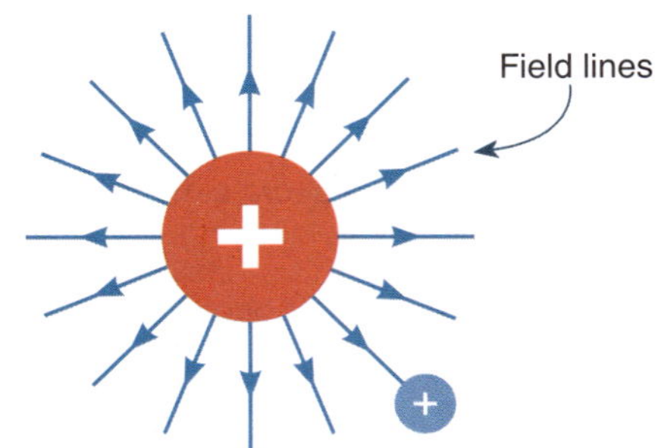

A small positive charge moves away from the larger positively charged sphere.

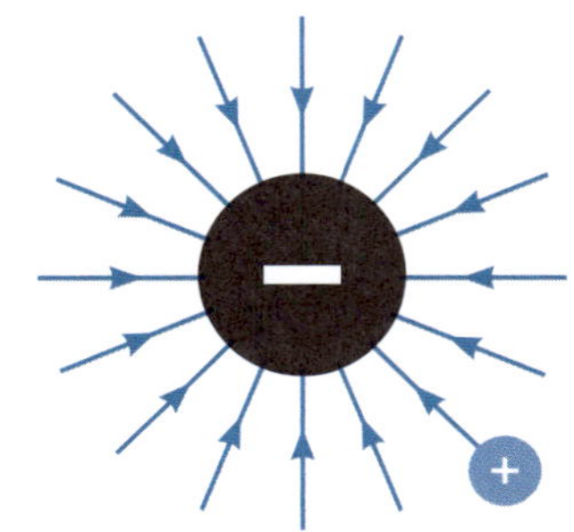

A small positive charge moves towards the larger negatively charged sphere.

1. Draw electric field lines (as shown above right) between the charges below. Charges have equal magnitude:

(a)

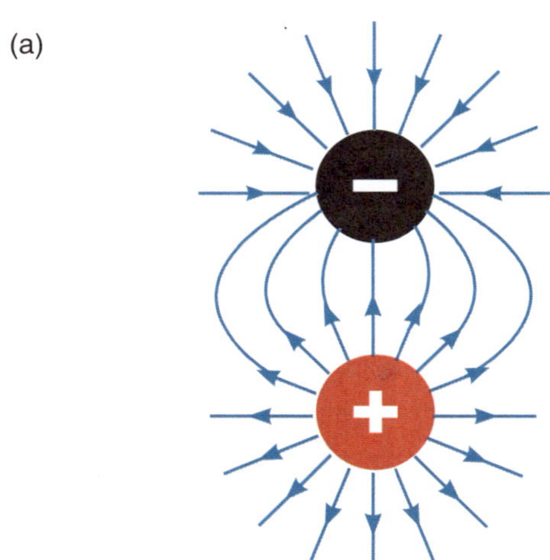

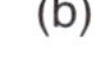

(b)

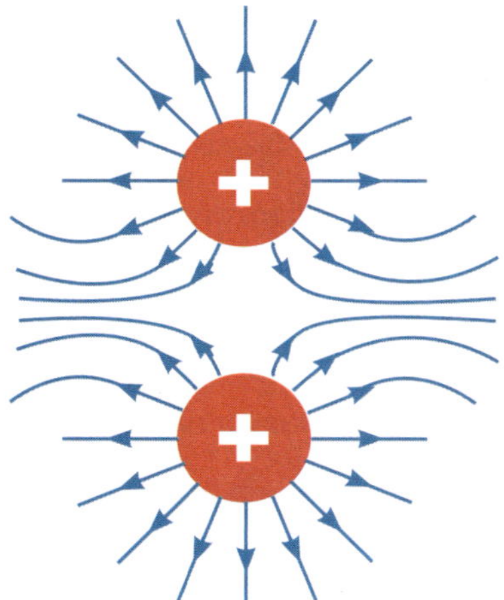

2. The diagram below shows an electric field between two charged plates. Sketch onto the diagram the pathway the positively and the negatively charged particles would take through the field:

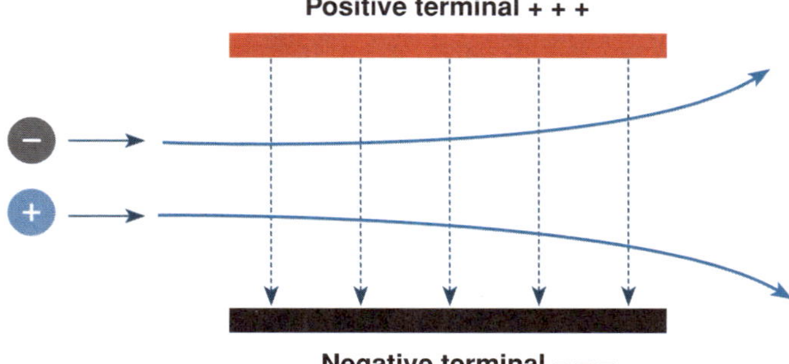

PS3.C CE

©2020 **BIOZONE** International
ISBN: 978-1-927309-82-7
Photocopying Prohibited

Energy and fields

- A planet produces a large gravitational field (top right). An object in the gravitational field possesses energy (measured in joules, J) due to its height above the ground. While it remains at one height above the ground the energy remains stored but has the potential to be used. Because of this we call the energy potential energy (E_p). If the object falls from a height its potential energy is converted into the energy in a moving object (called kinetic energy, E_k) as it falls. The amount of energy available is dependent on the mass, the height, and the strength of gravity.
- The separation of charges produces an electric field (bottom right). In the same way that a mass at a certain height in a gravitational field has a certain potential energy, so the charged particle in an electric field has a certain potential energy. This depends on its charge (q, in Coulombs), its distance from the terminals or charged particles (d, in meters), and the strength of the electric field (E, newton/coulombs). Thus;

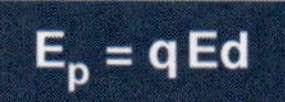

$$E_p = qEd$$

- To move the charged particle in the field requires energy (called work, W). The work done depends on the distance the charged particle moved in the electric field:

$$W = qE\Delta d$$

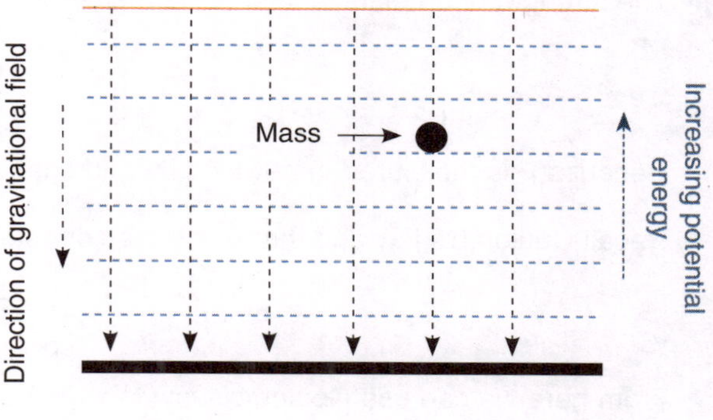

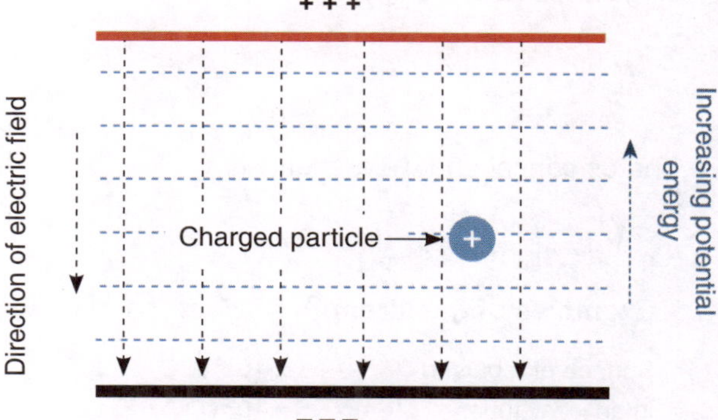

Gravitational and electric fields can be represented in similar ways.

3. Consider the electric field below produced by two charged plates:

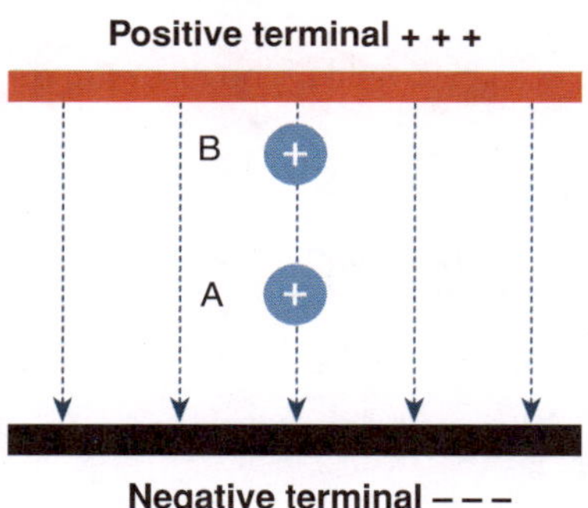

(a) Is the energy stored in the charged particle at point A greater or less than the energy stored in the charged particle at point B? Explain your answer:

The energy stored at point A is less than is than at point B. This is because the position has changed relative to the positive and negative terminals. It has taken energy to move from B towards the + terminal and away from the – terminal (against the electric field). That energy is now stored in the particle.

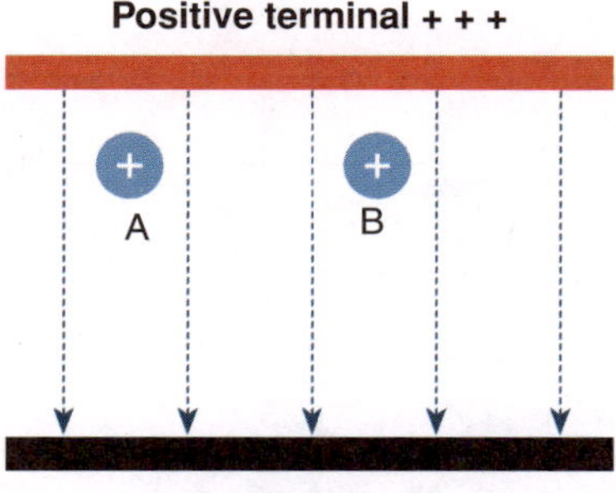

(b) Is the energy stored in the charged particle at point A greater or less than the energy stored in the charged particle at point B? Explain your answer:

The energy is the same at point A and B because the particle's position has not changed relative to the terminals.

4. Electrons has a charge of -1.6×10^{-19} C and protons have a charge of 1.6×10^{-19} C. Explain why a charged particle cannot have a charge of 2.4×10^{-19} C:

2.4×10^{-19} is not a multiple of -1.6×10^{-19}.

It would require the charge of half a proton or electron, which is not possible.

©2020 BIOZONE International
ISBN: 978-1-927309-82-7
Photocopying Prohibited

Strength of an electric field

- The strength of an electric field (E) is measured in newtons per coulomb (N/C), so:

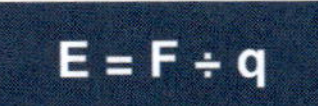

because F is measured in newtons, N and charge is measured in coulombs, C.

- Recall Coulomb's law and that F is the electrostatic force (the force need to separate two charged particles).

$$F = k\,\frac{q_1 q_2}{r^2}$$

- From here we can use Coulomb's law to work out the strength of an electric field produced by a point of charge at any distance from that point of charge. So from the equations above the electric field strength E around any singular point of charge is:

$$E = k\,\frac{q_1 q_2}{r^2} \div q$$

- The qs cancel and we are left with:

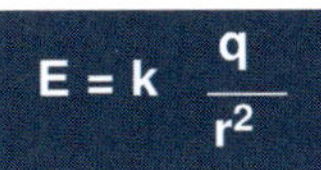

Examples of the strength of electric fields: Electric fields are everywhere. Some examples include:

Surface of a pulsar:	1×10^{14} N/C
In an x-ray tube:	5×10^{6} N/C
In a lightning bolt:	1×10^{4} N/C
In sunlight:	1×10^{3} N/C
Small red light laser:	1×10^{2} N/C
Radiowave:	1×10^{-1} N/C

5. (a) A point of charge has a charge of 3.2×10^{-19} C. Calculate the strength of the electric field at the following distances. Recall that $k = 8.99 \times 10^{9}$ Nm²/C²):

   (i) 2 mm: $(8.99 \times 10^{9}\ Nm^2/C^2 \times 3.2 \times 10^{-19}\ C) \div (0.002\ m)^2 = 7.19 \times 10^{-4}$ N/C

   (ii) 4 mm: $(8.99 \times 10^{9}\ Nm^2/C^2 \times 3.2 \times 10^{-19}\ C) \div (0.004 m)^2 = 1.80 \times 10^{-4}$ N/C

   (b) Calculate the magnitude of the force experienced by a point of charge with a charge of 3.2×10^{-19} C at each of these locations:

   (i) 2 mm: $F = 7.19 \times 10^{-4}\ N/C \times \times 3.2 \times 10^{-19}\ C = 2.30 \times 10^{-22}$ N

   (ii) 4 mm: $F = 1.80 \times 10^{-4}\ N/C \times 3.2 \times 10^{-19}\ C = 5.76 \times 10^{-23}$ N

6. Show that $F = k(q_1 q_2 \div r^2)$ is the same as $F = qE$: $F = k(q_1 q_2 \div r^2)$ and $E = k(q_1 q_2 \div r^2) \div q$ therefore $E = F \div q$.
   Rearrange the equation : $F = qE$.

7. (a) A proton has a radius of about 1.0×10^{-15} m and has a charge of 1.6×10^{-19} C. What is the electric field strength at the surface of the proton?

   $E = k(q \div r^2) = 8.99 \times 10^{9}\ Nm^2/C^2 \times (1.6 \times 10^{-19}\ C \div (1 \times 10^{-15}\ m)^2) = 1.44 \times 10^{21}$ N/C

   (b) Another proton sits in touching distance of the first (side by side). What is the force exerted by one proton on the other?

   $r = 1 \times 10^{-15}\ m \times 2 = 2 \times 10^{-15}$ m,

   $F = k(q_1 q_2 \div r^2) = 8.99 \times 10^{9}\ Nm^2/C^2 \times ((1.6 \times 10^{-19}\ C)^2 \div (2 \times 10^{-15})^2) = 57.5$ N

   (c) Recall the chapter on nuclear process. Why do these protons not fly apart?

   The strong nuclear force holds the protons together.

©2020 **BIOZONE** International
ISBN: 978-1-927309-82-7
Photocopying Prohibited

84 Magnetism

Key Question: How do magnets work and what types of materials do magnets attract?

Magnetism

- Like gravitational and electrostatic forces, magnetic forces also act at a distance. You will probably have some idea about what magnets are. It is likely you have some on your home refrigerator holding up pictures and lists.
- Magnets are interesting in that they affect certain materials without actually touching them. Other objects, such as the Sun also affect objects without touching them. In the case of the Sun, gravitational force affects another object. In the case of magnets, it is magnetic force that affects other objects.
- Like electrostatic charges, magnets can attract some objects but not others.
- Unlike electrostatic charges, magnets do not have a single positive or negative charge. Magnets possess poles which always come in pairs - a north (N) and a south (S) pole.
- Similar to electrostatic charges, like magnetic poles repel and unlike poles attract each other.
- Given the similarities in the way magnets and electrostatic charges behave, can electrostatic charges interact with magnets in the same way they interact with each other?

INVESTIGATION 6.3: Charges and magnets See appendix for equipment list.

1. Rub a plastic ruler with wool and balance it on a small bottle lid.
2. Now rub a second ruler and bring the rubbed end of this second ruler near to the rubbed end of the first ruler. Observe the result.
3. Rub the first ruler with wool again, and again balance it on the small bottle lid. Bring the wool used to rub the ruler near the rubbed end of the ruler. Observe the result.
4. Now bring a magnet near the rubbed end of the ruler balancing on the bottle lid and observe the result. Try both ends of the magnet.
5. Try rubbing the ruler again with the wool and reintroduce the magnet.

1. (a) What happened when the rubbed end of the second ruler was brought close to the rubbed end of the ruler balancing on the bottle lid?

   Students should observe the ruler on the lid move away from the second ruler (repelling)

   (b) What can you say about the two ends of the rulers based on this observation?

   The rules both have the same charge.

   (c) What happened when the piece of wool was brought close to the rubbed end of the ruler balancing on the bottle lid?

   Students should observe the ruler on the lid move towards the wool (attraction)

   (d) How did the rubbed end of the ruler and the magnet interact?

   There was no reaction by the ruler to the magnet.

2. Magnets have a north (N) and south (S) pole. The diagrams below show various alignments of magnets. Predict how these magnets will behave:

(a) *The magnets will attract each other.*

(b) *The magnets will repel each other.*

(c) *The magnets will repel each other.*

©2020 BIOZONE International
ISBN: 978-1-927309-82-7
Photocopying Prohibited

CE P PS3.C PS2.B

INVESTIGATION 6.4: Magnetic fields

See appendix for equipment list.

1. Place an index card (or ruled card) in a clear sandwich bag and then place a teaspoon of iron filings into the bag. For #2-5, you can either sketch what you see or attach a photograph.
2. Lay the sandwich bag down on the bench and gently shake the bag back and forth so there is a thin layer of iron filings on the index card. Sketch what you see in the appropriate box below.
3. Carefully lower the center of the sandwich bag on top of a bar magnet. Sketch what you see in the appropriate box below.
4. Place a second bar magnet under the sandwich bag, lining up the two like poles opposite each other and sketch what you see in the appropriate box below.
5. Finally, reverse the second bar magnet, lining the two opposite poles and again sketch what you see.

Iron filings with no magnet

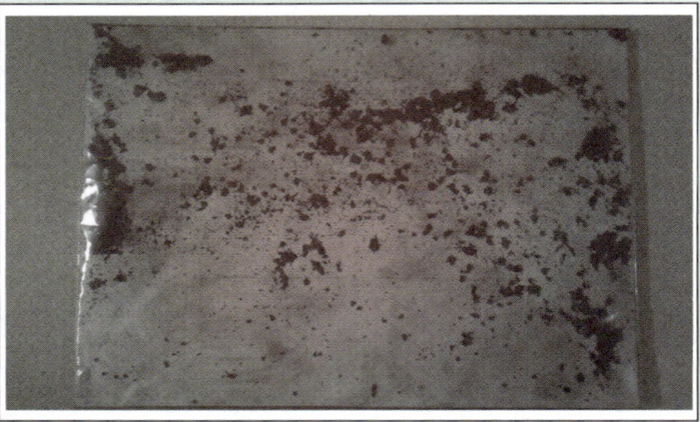

Iron filings with one magnet

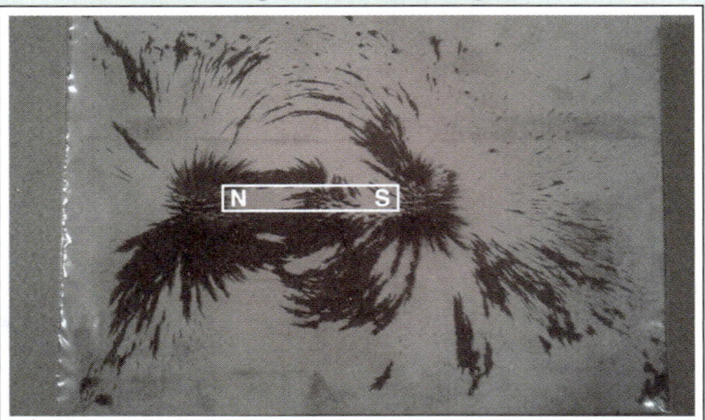

Iron filings with 2 magnets (like poles aligned)

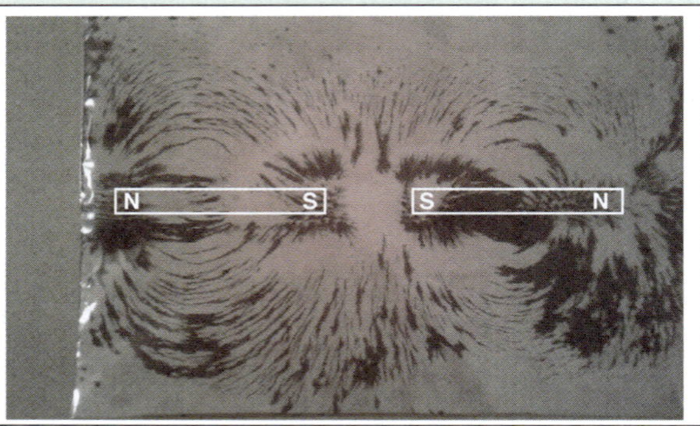

Iron filings with 2 magnets (opposite poles aligned)

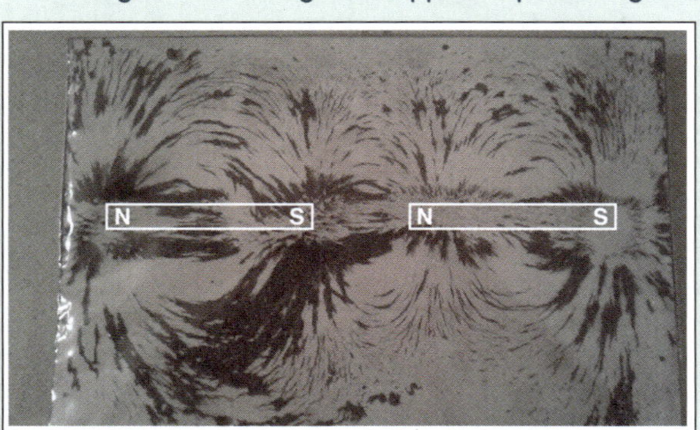

3. (a) How would you explain the patterns of iron filings you see when there is a magnet in place?
   When there is a magnet, the iron filings arrange along the magnetic fields produced by the bar magnets.

   (b) What do the patterns reveal about attraction and repulsion between the magnets? *The filings arrange themselves between ends that attract each other. They filings do not align in the gap between similar ends.*

   (c) How do the patterns of the iron filings relate to the magnetic field of the bar magnet?
   The filings follow the magnetic field and connect one pole to the other.

   (d) What happens to the pattern of iron filings when two like poles are placed next to each other?
   The iron filings are pushed out of the gap between the two like poles.

   (e) What happens to the pattern of iron filings when opposite poles are placed next to each other?
   The filings between the poles arrange themselves in lines between the two poles.

   (f) Can you tell by the patterns where the force of the magnetic field is strongest? Weakest?
   The magnetic field is strongest where the concentration of the iron filings is the greatest.

©2020 **BIOZONE** International
ISBN: 978-1-927309-82-7
Photocopying Prohibited

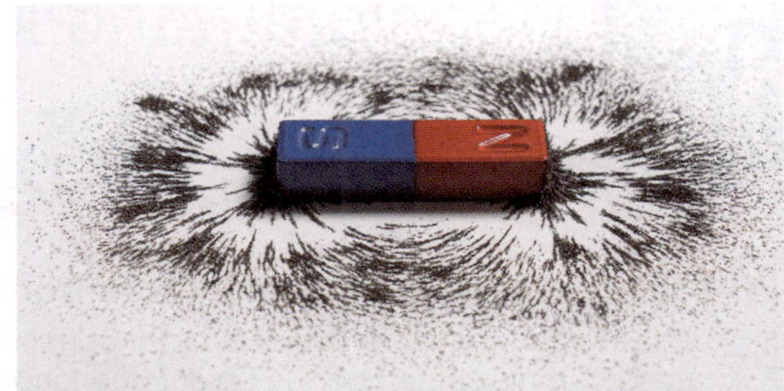

- You have just observed magnetic field lines produced by a magnet acting on iron filings through a plastic sandwich bag and index card. Like gravitational fields, magnetic fields can pass right through materials and affect objects on the other side.
- How is the strength of a magnetic field affected by the distance from the magnet? Is it the same at any distance? Does it get less or more?
- From the result of placing the iron filings around a magnet earlier, you should be able to guess the answers.

INVESTIGATION 6.5: Strength of a magnetic field

See appendix for equipment list.

1. Place a piece of paper between a magnet and a metal paper clip. One at a time, continue placing additional pieces of paper in between the magnet and the paper clip.
2. Record how many pieces of paper can be placed between the magnet and the paper clip before the magnet no longer affects the paper clip.

Student's result.

4. (a) Why does the magnet attract the paper clip even though it is not in direct contact with the paper clip?

Magnetism is a force that acts over a distance. It does not require contact to influence other magnetic materials.

(b) How does the magnetic force acting on the paper clip change as you increase the number of pieces of paper between the magnet and the paper clip (what can you say about the effort it takes to pull the magnet and paper clip apart as the number of pieces of paper increased)?

It took less effort to remove the paper clip from the paper when there was more paper between the magnet and the paper clip. The magnetic field was getting weaker as the distance between increased.

(c) Would the force on the paper clip change if the paper was removed but the distance between the magnet and paper clip stays the same?

The force of attraction would remain the same.

(d) Using your sketches of magnetic fields as a guide, make a predictive sketch below of the strength of the magnetic field as the distance from the magnet increases:

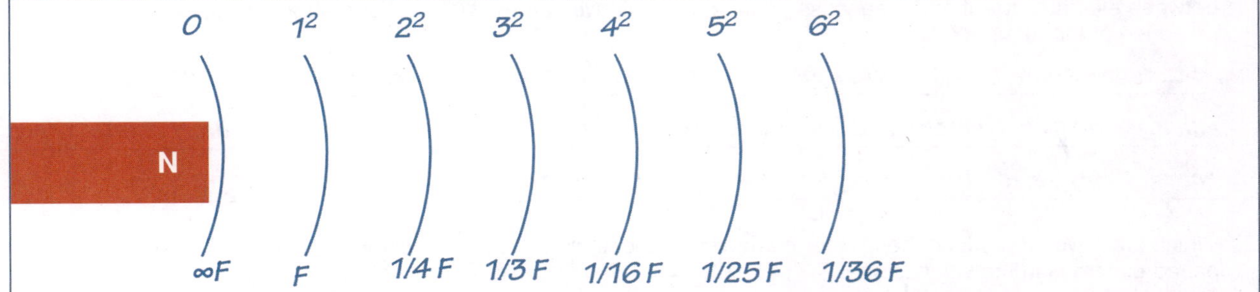

5. The force of attraction was always present between the paper clip and the magnet. When there was too much distance between the paper clip and the magnet, there was not enough magnetic force to hold on to the paper clip and it fell.

(a) Why was the paper clip attracted to the magnet but not the pieces of paper? *The paper clip is metal and many metals are or can be magnetic and be attracted to a magnet. Paper cannot (has no magnetic properties).*

(b) Is there a way to predict what objects will be attracted to a magnet and which will not? Explain:

Certain metals such as iron are attracted to magnets. Materials made of these metals will also be attracted to magnets.

(c) List some objects that you could have substituted for the paper clip in this investigation:

Steel nail, refrigerator door etc.

©2020 **BIOZONE** International
ISBN: 978-1-927309-82-7
Photocopying Prohibited

85 Understanding Magnetism

Key Question: Why are some materials able to produce magnetic fields while others are not?

Where does a magnetic field come from?

- You have seen that when objects carry opposing charges, they are attracted to each other. However, over time, the charges can dissipate and the objects no longer show attraction to each other.
- This occurs because there is a gradual flow of electrons from the area of high concentration to the area of low concentration, which continues until the charges are balanced. This means that the net charge on each object gets smaller and (from Coulomb's law) if the charges are smaller, then the force will be weaker.
- We call a flow of electrons **electric current**. In nature, most electron flows occur rapidly, briefly, (static shock, left) and sometimes violently (lightning).
- When we think of electricity, we often think of the electricity we obtain through electric wires to run our everyday appliances (e.g. lights). This electricity is a current (current electricity) and is a steady flow of electrons. This is normally generated a power plant, but can come from batteries as well.
- Magnetic fields arise from the movement of any charged particles. But what in a magnet is moving to create the magnetic field?
- The model of the atom you studied earlier in this book shows several important things:
  - Electrons fill an atom's energy shells with electron shells near the nucleus filled first.
  - Electrons have a property called spin, either up or down.
  - When orbitals are more than half full, electrons pair up. The spin of the electrons in these pairs is always opposite (i.e. one up one down). It is impossible for both electrons in a pair to have the same spin.
  - Electrons are in constant motion around the nucleus.

Z22 CC4.0

Electron spin was first imaged in 2010 using a scanning tunneling microscope. Cobalt atoms of known electron spin were imaged. Those with a net up spin produced a single peak on the image. Those with a net down spin produced a peak with a cone or flattened peak (above).

- Spinning electrons produce tiny magnetic fields. Paired electrons, spinning in opposite directions, cancel out these magnetic fields.

1. Recall in a previous activity a magnet was placed near the charged end of a ruler. Knowing that there is a relationship between electric current and magnetism, can you explain why there was no observable reaction from the ruler to the presence of the magnet?

   Student's answers will vary. Students should observe that the charge on the ruler was static and static charges do not produce magnetic fields. Students might also note the ruler is made of plastic therefore cannot carry electricity so cannot have a magnetic field.

2. Which atom would produce the greater magnetic field: one with all paired electrons in the valance shell, or one with no paired electrons in the valence shell?

   The atom with no paired electrons in its valence shell would produce the greater magnetic field.

3. (a) The metals iron, nickel, and cobalt can all be magnetized. What does this tell us about their valence electrons?

   These elements tend to have multiple unpaired electrons in their valence shell.

   (b) Substances made of these elements are rarely magnetic on their own (not already a magnet). Suggest why:

   Student's answers will vary. Students should note direction of magnetic field produced by an atom is random due to position. Magnetic fields in opposing directions will cancel each other out leaving a substance with a net magnetic field of zero.

©2020 BIOZONE International
ISBN: 978-1-927309-82-7
Photocopying Prohibited

Ferromagnetic materials

- Ferromagnetic is a term used to describe materials which are attracted by magnets and can become magnets themselves. Iron, cobalt, and nickel are not the only ferromagnetic elements. They are simply the most common elements to demonstrate this property and are easily identified. The strongest magnets are actually made from elements such as neodymium (Nd) and other rare-earth metals.
- In ferromagnetic materials, the magnetic fields of individual atoms can face in any direction. For this reason, even ferromagnetic materials may not show magnetic properties until exposed to an existing magnetic field (from another magnet or an electric current in a coil).

Neodymium magnets (right) are an alloy of neodymium, iron, and boron ($Nd_2Fe_{14}B$). They are the strongest commercially available permanent magnets.

- The magnetic field produced by the magnet causes the ferromagnetic material with an opposite field direction (north attracts south and south attracts north) to line up, instead of being oriented in random directions. This means there is always an attractive force between the ferromagnetic material and a magnet.

Horseshoe magnets are bent so the north and south poles are close together, creating a strong magnetic field.

Magnets are used in the storage of information, from magnetic tapes to computer hard drives.

The Earth produces a magnetic field, which can be detected by simple compasses.

INVESTIGATION 6.6: Making a magnet

See appendix for equipment list.

1. You will need a large nail, needle (or other iron object) and a bar magnet.
2. Mark a center point on the nail.
3. Using one pole of the magnet, stroke the nail from the center out to one end. Repeat this process several times, moving the pole well away from the nail and then to the center to begin each stroke.
4. Repeat step 3 but use the opposite pole of the same magnet and stroke from the middle to the opposite end.
5. Test your nail on other iron objects (or a compass) to see if you have created a magnet. If it is not magnetized yet, repeat steps 3 and 4 several more times and test again.

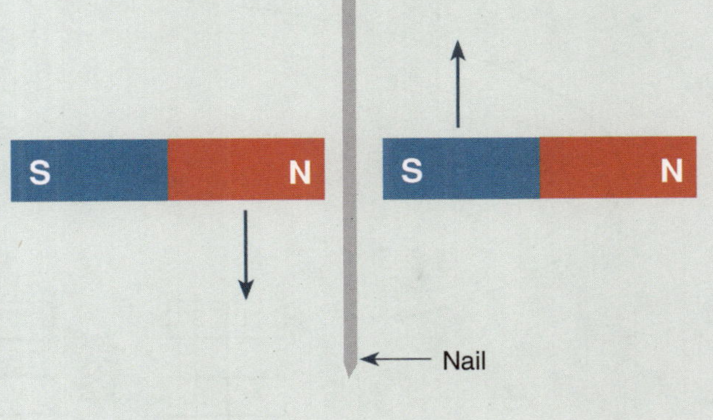

- Atoms in the iron object have magnetic fields, but they are not arranged in any particular order. The random arrangement of the magnetic field directions results in an object that does not appear to have a magnetic field.
- Dragging the north end of a magnet over the surface of the object encourages the atoms with opposite alignment to arrange themselves towards the north magnet. The same for the south end of the magnet.

4. Use electromagnetic forces to explain how this method produces a magnet out of a ferromagnetic material such as iron:

Some atoms will be attracted to the pole of the magnet due to arrangement. Pulling the magnet in one direction will encourage attracted atoms to follow the magnet. Doing this over and over will result in enough atoms arranging themselves in a specific direction to produce a magnetic field.

5. Notice that the materials that make magnets are the same as the materials magnets are attracted to. Why is this?

These materials all have the large numbers of unpaired valence electrons necessary to produce measurable magnetic fields. Materials made of atoms with paired electrons will not be attracted to or become magnets.

©2020 BIOZONE International
ISBN: 978-1-927309-82-7
Photocopying Prohibited

Magnetism in a magnet

- As you learned in Chapter 2 *Structure and Properties of Matter*, electrons are negatively charged particles that orbit the nucleus of the atom in variously shaped "shells" or energy levels.
- Electrons' orbital motion as well as quantum mechanical spin produce an electric current and consequently a magnetic field in the atom.
- In most atoms, the magnetic fields generated by each electron cancel each other out.
- In an atom, two electrons can pair up and occupy an energy level, but their spins are opposite of each other, canceling their magnetic field.
- In a few atoms (like Fe, Co, and Ni) there are unpaired electrons in different energy levels whose spins can align and give the atoms an overall magnetic field.
- When elements like Fe form solids, they form crystalline structures with separate areas (called domains) of many atoms all aligned. Each domain is like a small bar magnet with a north and south pole.

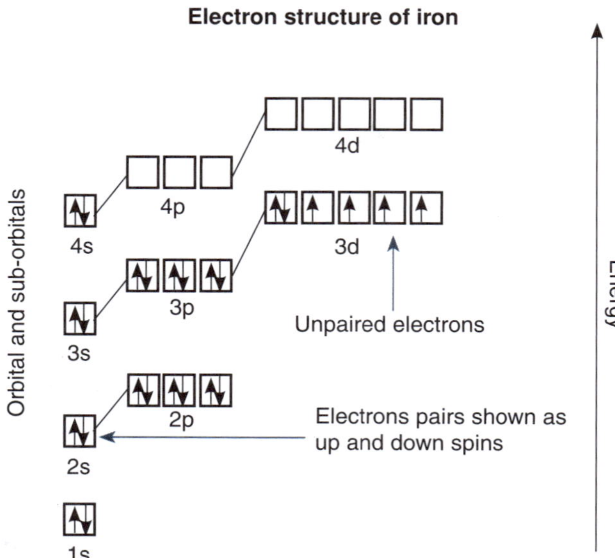

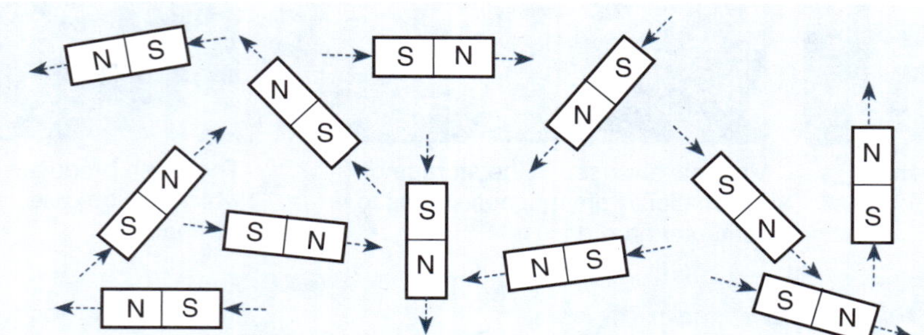

- When all the domains are aligned, the whole object takes on a large-scale magnetic field.

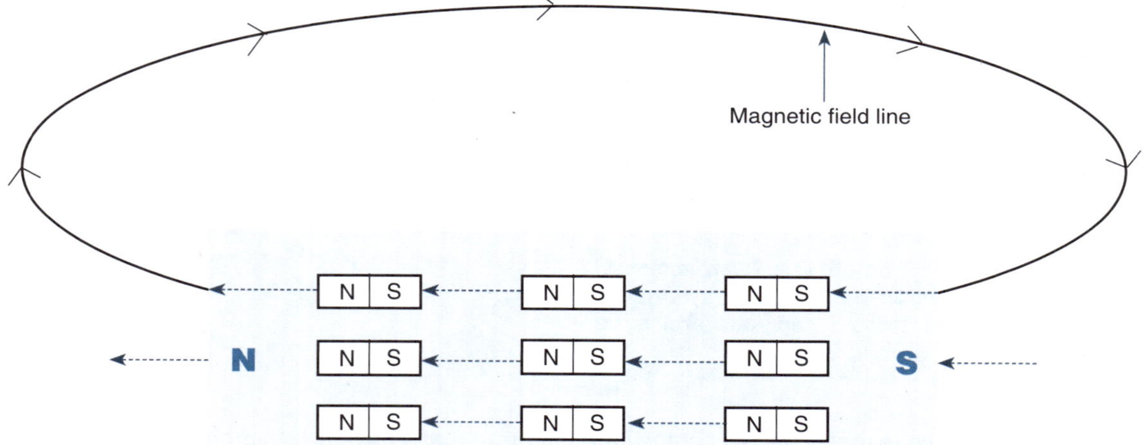

6. Many elements have unpaired electrons. Why is it that Iron (Fe) and similar ferromagnetic materials are able to become magnets while so many other elements are not?

   Because of the way electrons pair up, very few elements possess multiple unpaired electrons. With only 1 or 2 unpaired electrons most elements with unpaired electrons do not create a strong enough field to affect other materials and become magnets.

7. Explain why Iron (Fe) is attracted by magnetic fields even when it is not a magnet itself:

   In raw iron, magnetic domains are pointing in every direction. Because of this, there are always domains orientated so that there is attraction to a magnet.

©2020 BIOZONE International
ISBN: 978-1-927309-82-7
Photocopying Prohibited

86 Electricity and Magnetism

Key Question: What is the relationship between current electricity and magnetism?

Creating a magnetic field from an electric current

- In an earlier activity, you investigated whether electrostatic charges and magnets could interact. While there was no reaction between the charged plastic ruler and the magnet, there is a relationship between current electricity and magnetism.
- While static charges do not interact with magnets due to not producing strong electric fields, current electricity creates magnetic fields which can influence magnets. Let's investigate how this works.

INVESTIGATION 6.7: Electricity affects a compass part 1

See appendix for equipment list.

Caution is required when doing anything with electricity. We recommend school laboratory power supplies and common alkaline 1.5 V cells are used.

PS2-5

1. Create a loop circuit using 1.5 volt AA single cell, insulated wire, and electrical tape. **Do not** attach the second end of the wire to the cell yet.
2. The circuit you will create is essentially a short circuit of the battery so can only be turned on for a short time. The short circuit produces maximum electromagnetic effects but also causes the wire to become hot. Because of the short circuit, the cell should only be connected for the minimum amount of time as it will go flat quickly.
3. Tape the wire to a piece of cardboard in a circular loop (right).
4. Connect the other open end of the wire to the cell so current begins to flow.
5. Bring the compass into the center of the loop.
6. Move the compass around, closer to the wire and away from the wire.
7. Sit the compass right on top of the wire, then move it along the top of the wire and observe the direction the arrow points.
8. Disconnect the wire from the cell. Careful, the wire may be hot.
9. For the second part of the experiment, reverse the cell and repeat the investigation.

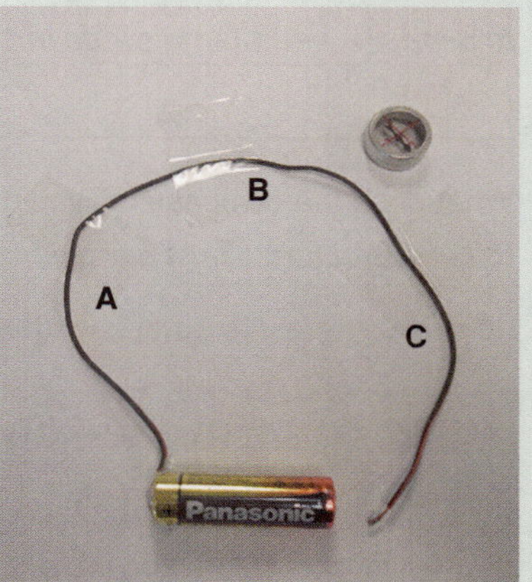

1. For the first part of the experiment, describe the direction the compass arrow pointed at points A, B, and C:

   A: *Arrow will point towards or away from the center of the wire loop depending on direction of current.*

   B: *Will have the same result as A.*

   C: *Will have the same result as A.*

2. For the second part of the experiment, describe the direction the compass arrow pointed at points A, B, and C:

   A: *The arrow will point in the opposite direction of Q1 a,b, and c.*

   B: *Will have the same result as A.*

   C: *Will have the same result as A.*

3. What can be said about the relationship between the direction of the flow of electricity and the direction of the magnetic field produced? Make a statement that predicts the direction a compass will point when near an electric current:

   When a current flows in one direction, the compass will point towards the center of the circuit. When current flows in the opposite direction the compass also points in the opposite direction. The compass will point at right angles to the wire

©2020 BIOZONE International
ISBN: 978-1-927309-82-7
Photocopying Prohibited

INVESTIGATION 6.8: Electricity affects a compass part 2

See appendix for equipment list.

1. Take a compass and wrap three loops of wire around it as shown in the photograph. Secure the wire with clear tape.
2. Connect the cell and observe the direction the arrow points.
3. Reverse the cell and repeat.

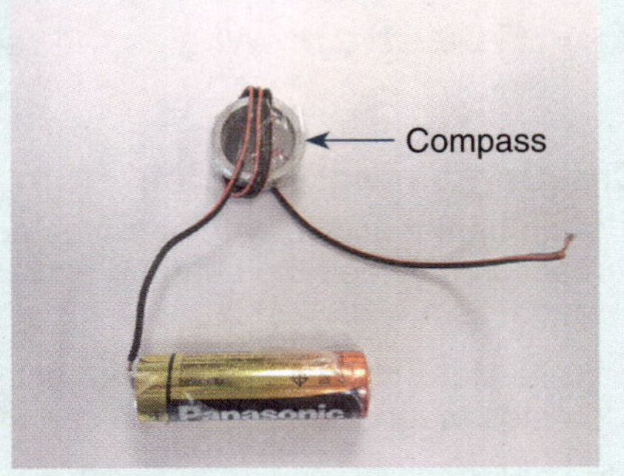

4. Describe the direction the compass arrow points with respect to the wire:
Compass will point at a right angle to the coil. Direction will depend on the direction of the current.

5. How does the direction the compass arrow points compare to the results from the previous investigation where the compass was beside the wire instead of wrapped inside it?
It produces similar results in that the compass points at right angles to the wire.

6. What does the compass tell you about the relationship between the magnetic field and the flow of electricity?
It tells us that there is obviously a magnetic field being created by the flow of electricity. This appears to be at right angles to the flow of electricity through the wire.

INVESTIGATION 6.9: Electricity affects a compass part 3

See appendix for equipment list.

1. Set up a circuit as shown in photograph 1 by cutting a piece of card to a square of about 10 cm by 10 cm and threading a wire through the middle of the card.

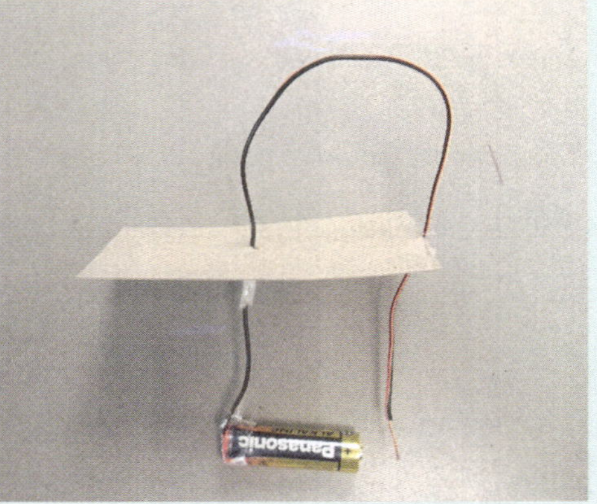

Photograph 1

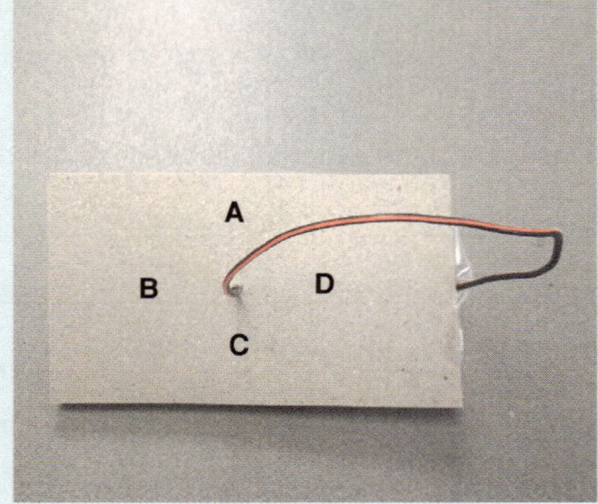

Photograph 2

2. Fold the wire over so that about 10 cm of the wire is perpendicular to the cardboard. Fix the wire in place with tape.
3. The cardboard produces a space between the two parts of the wire and also creates a shelf for a small compass to rest on.
4. Connect the 1.5 volt cell and hold the wire upright with the cell at the bottom. You might need the help of another student to hold the circuit upright and connect the cell while you obtain compass readings in the next step.

©2020 **BIOZONE** International
ISBN: 978-1-927309-82-7
Photocopying Prohibited

5. Place the compass at points A, B, C, and D as shown in photograph 2. Draw a diagram in the space below left to show your results:

6. Reverse the cell and repeat. Draw a diagram in the space below right to show your results.

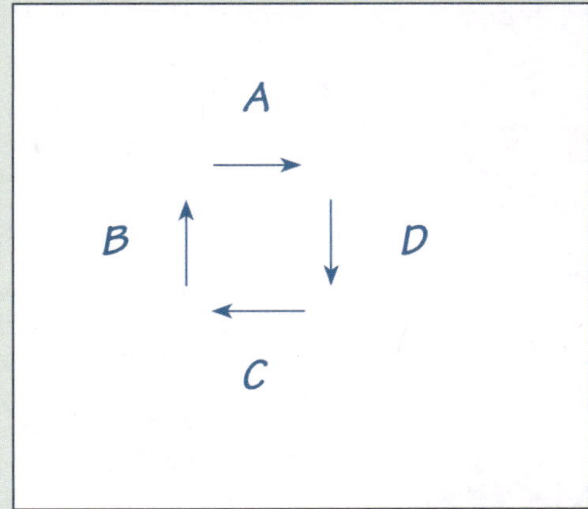

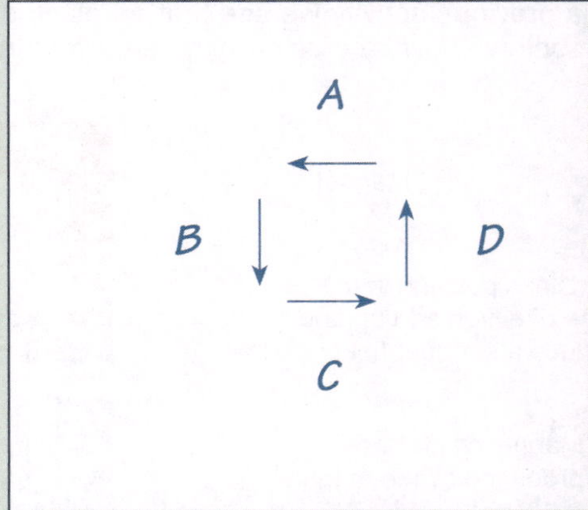

7. (a) What happens to the direction the arrow points when the cell and therefore current is reversed?

The arrows point in the opposite direction.

(b) What kind of field is the electricity in the wire producing that is influencing the compass? *Magnetic field.*

(c) Conventional current flows from the positive terminal to the negative terminal of a cell. What can be said about the direction of the field in Q 7(b) around a wire?

The magnetic field direction circles the flows of electricity along the wire (at a right angle along the length of the wire). This explains why the compass points at right angles to the wire when held above the wire.

Ampère's right hand grip rule

- André-Marie Ampère did much of the early work in electromagnetism. Upon discovering that a compass needle was deflected by an electric current (as you were investigating previously), Ampère developed a mathematical model to understand the relationship between electricity and magnetism.
- One outcome of Ampère's work is **Ampère's right hand grip rule**, which is a simplified method to determine the direction of magnetic deflection without a compass. The rule states that when you know the direction of flow of current electricity, if you grip the wire (phantom grip the wire since we do not recommend grabbing actual current conducting wires with your hand) such that your thumb points in the direction of the current flow, the fingers circling the wire will point in the direction of the magnetic field lines (see right).

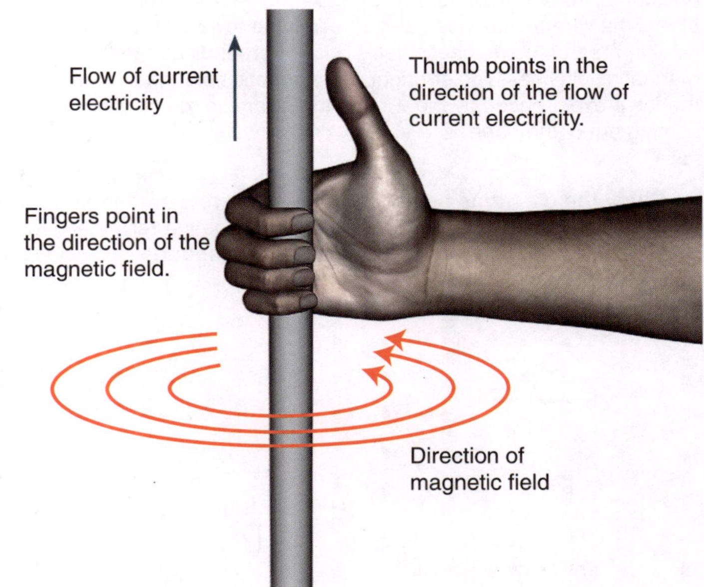

8. Does Ampère's right hand grip rule (above) agree with your diagrams in investigation 6.9 and the direction of the positive and negative end of the battery or power source you used?

Yes the diagram should agree with the students observation. (Student's might like to repeat the investigation, this time taking note of the direction of the current from the battery).

©2020 **BIOZONE** International
ISBN: 978-1-927309-82-7
Photocopying Prohibited

87 Magnetic Fields

Key Question: How does a magnetic field affect the motion of a conductor carrying a current?

PS3-5 ▶ From the previous activity we see that an electric current flowing through a wire produces a magnetic field. The diagram below shows a wire carrying a current inside a magnetic field.

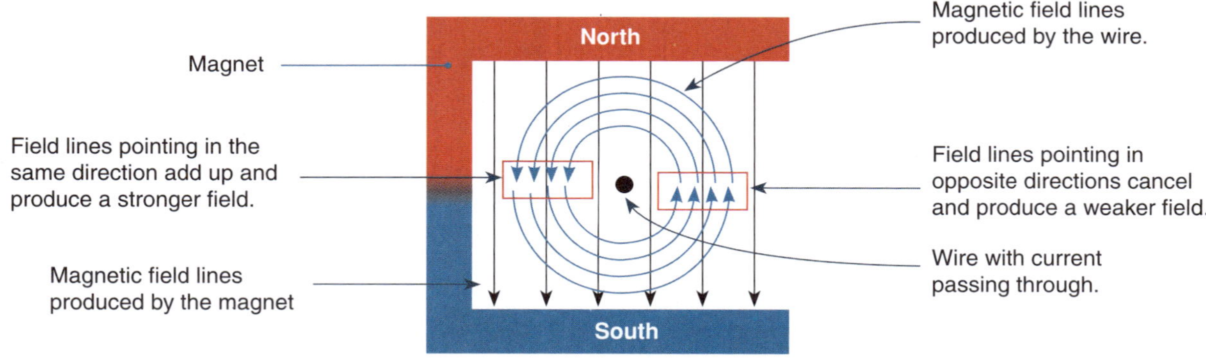

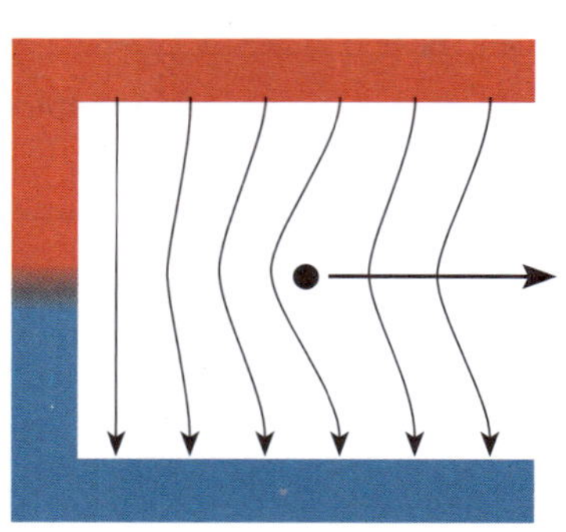

Current

Magnetic field lines

Direction of force

Due to the unequal effect on the magnetic field the field lines become distorted, stronger on one side and weaker on the other. The wire experiences a magnetic force towards the weaker area of the magnetic field. This is perpendicular to the magnetic field lines. The stronger the magnetic field the stronger the force experienced by the wire. In this example the current is coming out of the page.

The direction of the force applied to the wire can be identified using the **right hand rule**. The fingers point in the direction of the magnetic field lines. The thumb points in the direction of the current in the wire. The palm points in the direction of the force experienced by the wire.

1. Study the diagrams below and draw an arrow onto them to show the direction the gray bar would move. Remember current moves from the positive to the negative terminal. Here the magnetic field is independent of the electric field in the wire (it is produced by a separate large magnet):

(a)

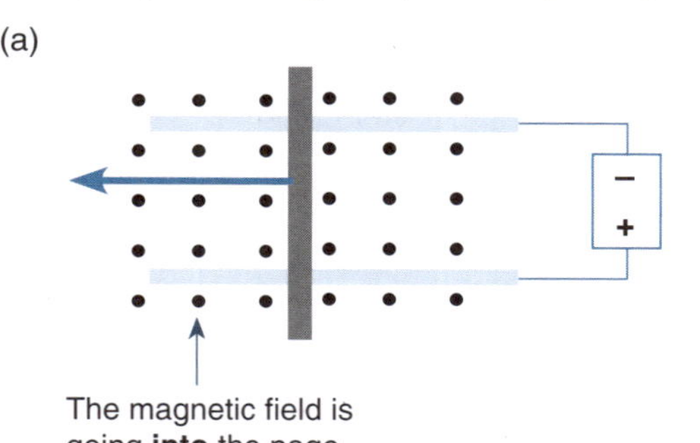

(b)

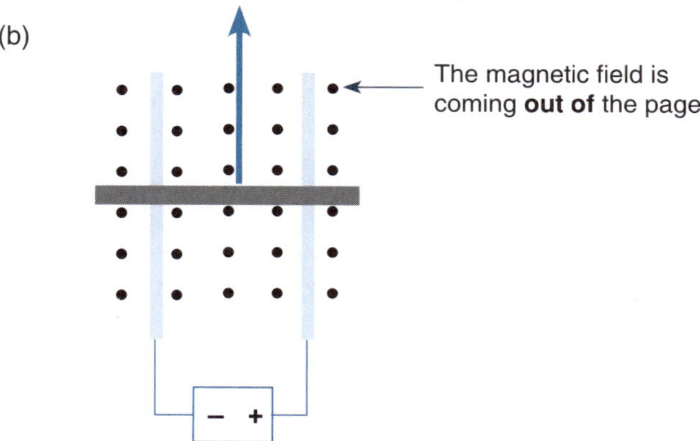

2. For each of the situations above the current is reversed. Which way will the gray bar move now?

(a) *The metal bar would move in the opposite direction (to the right)*

(b) *The metal bar would move in the opposite direction (down)*

   PS3.C  CE

©2020 **BIOZONE** International
ISBN: 978-1-927309-82-7
Photocopying Prohibited

The potential energy in a magnetic field

The railgun is a rather spectacular demonstration of the effect of magnetic fields on conductors carrying a current. The gun consists of two parallel metal rails. The metal projectile sits between the metal rails and is in contact with them. Electricity is passed through the rails and so through the projectile. The rails produce a large magnetic field in the area of the projectile. Because the projectile is carrying current it experiences a force that accelerates it down the rails. Railguns require a huge amount of energy to power. Test guns produced by the US Navy use 25 megawatts of power (enough electricity to power around 19,000 homes).

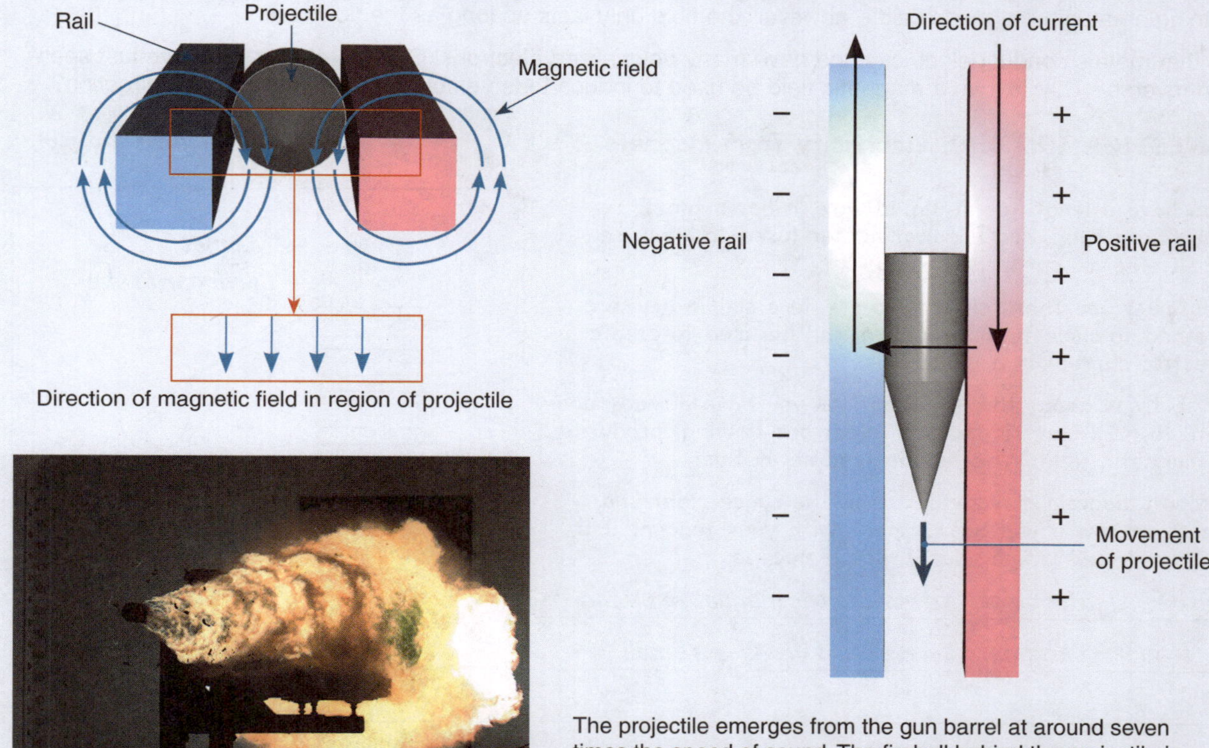

The projectile emerges from the gun barrel at around seven times the speed of sound. The fireball behind the projectile is plasma generated by friction between the projectile and the rails, not explosive powder as in a conventional gun.

3. (a) The energy (in joules, J) in a moving object can be calculated using the $E_k = ½mv^2$. A railgun launches a 12 kg projectile at 2500 m/s. What was the energy provided by the electric current passing through the rails of the gun?

$0.5 \times 12 \times 2500^2 = 37{,}500{,}000$ J = 35.5 MJ

(b) The distance the projectile was accelerated in the railgun above was 6.45 m (the length of the barrel). Calculate the force generated by the gun ($a = v^2 \div 2d$):

$a = (2500 \text{ m/s})^2 \div (2 \times 6.45 \text{ m}) = 6{,}250{,}000 \div 12.9 = 484{,}496.12 \text{ m/s}^2$.

OR: $\Delta x = 0.5(v_i + v_f) \times \Delta t$. $\Delta t = \Delta x \div 0.5(v_i + v_f) = 6.45 \text{ m} \div 0.5 (0 \text{ m/s} + 2500 \text{ m/s}) = 5.16 \times 10^{-3}$ s.

$a = (v_f - v_i) \div \Delta t = (2500 \text{ m} - 0) \div 5.16 \times 10^{-3} \text{ s} = 484{,}496.12 \text{ m/s}^2$

$F = ma = 12 \text{ kg} \times 484{,}496.12 \text{ m/s}^2 = 5{,}813{,}953$ N

(c) Although railguns have real benefits over the use of conventional artillery shells they are unlikely to ever be installed (at least in the perceived future) except on the largest warships or fixed land based stations. Suggest why this is:

Railguns require a lot of electricity to generate the force needed to make them work. This can only (currently) be produced by large electricity generators. Only a ship or power plant would be able to do this.

(d) One of the disadvantages of railguns is the large amount of friction produced by the rapid movement of the projectile. Explain why this is a problem. (Hint: look at the photograph above):

The friction causes so much heat the projectile literally rips the atoms of the surface of the gun barrel. Over time (a relatively small amount) this will severely damage the gun and so it will need replacing.

©2020 **BIOZONE** International
ISBN: 978-1-927309-82-7
Photocopying Prohibited

88 Generators and Motors

Key Question: How do we use the relationship between magnetism and electricity to generate usable electricity?

Generating electricity from magnets

- Flows of electrons (electric currents) create magnetic fields (as seen in investigations 6.7, 6.8, and 6.9). Even a bolt of lightning creates a magnetic field – however, the field only lasts as long as the flow lasts.
- Recall that metals conduct electricity and have many delocalized electrons. Can the effects you have just seen work in reverse? That is, can a magnetic field be used to induce these electrons to all move in one direction?

INVESTIGATION 6.10: Electricity from magnets See appendix for equipment list.

1. You have a length of insulated wire, a bar magnet, a cardboard tube, and a galvanometer (used to measure small electric currents).
2. You must use these objects to create a simple device or method to show that magnetism can be used to create electric current in a wire.
3. First, think about the investigations you have already done (6.7 to 6.9). Electric current (electrons flowing) produced a magnetic field. What is the reverse of this?
4. Draw a diagram of your device in the space right and describe how it will work. Check with your teacher if it will work and refine your device if needed.

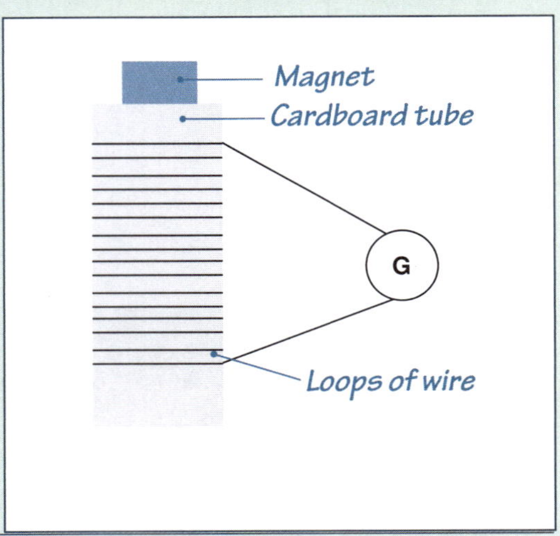

With enough loops of wire moving the magnet back and forth in the cardboard tube should produce a small voltage.

5. Test out your device and record any observations: *Student's observations.*

1. A galvanometer is a device used to detect and measure small electric currents:
   (a) What did the galvanometer show before the magnet was introduced to the wire coil?
   No current.
   (b) What did the galvanometer show when the magnet was being moved back and forth between the coils?
   A current was detected.

PS3-3

2. With some modifications you can turn your device in investigation 6.10 into a fully functioning generator. To do so, you only need strong bar magnets, a spool of enamelled copper wire, cardboard, a nail, and a galvanometer or low voltage light bulb or diode.

   Now consider the following questions:
   – What will you need to do to produce a continuously changing magnetic field?
   – How will the effects of this changing magnetic field be transferred to the copper wire?
   – How will you know if electricity is being produced in the wire?

   Design a generator that could provide continuous electrical power. Draw and label a diagram of your design (right) and describe how it would work (below):

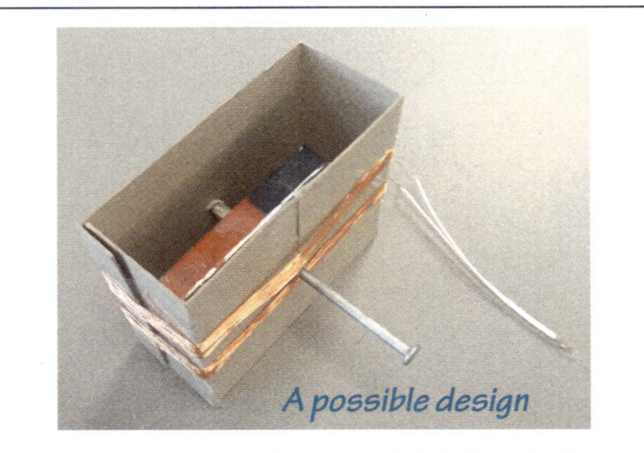

Student's design and description.

PS2.B PS3.A CE EM

©2020 BIOZONE International
ISBN: 978-1-927309-82-7
Photocopying Prohibited

- The relationship we have explored between magnetism and electricity is what allows the generator of a power plant to produce the electricity we are able to use.
- In generators, it does not matter whether it is the magnet that is turning or the coils of wire. It is the relative movement of the magnetic field to the coil that matters. However there will be slight differences in the design to accommodate the different moving parts.
- There are also different designs depending on whether alternating current (AC) or direct current (DC) is required.

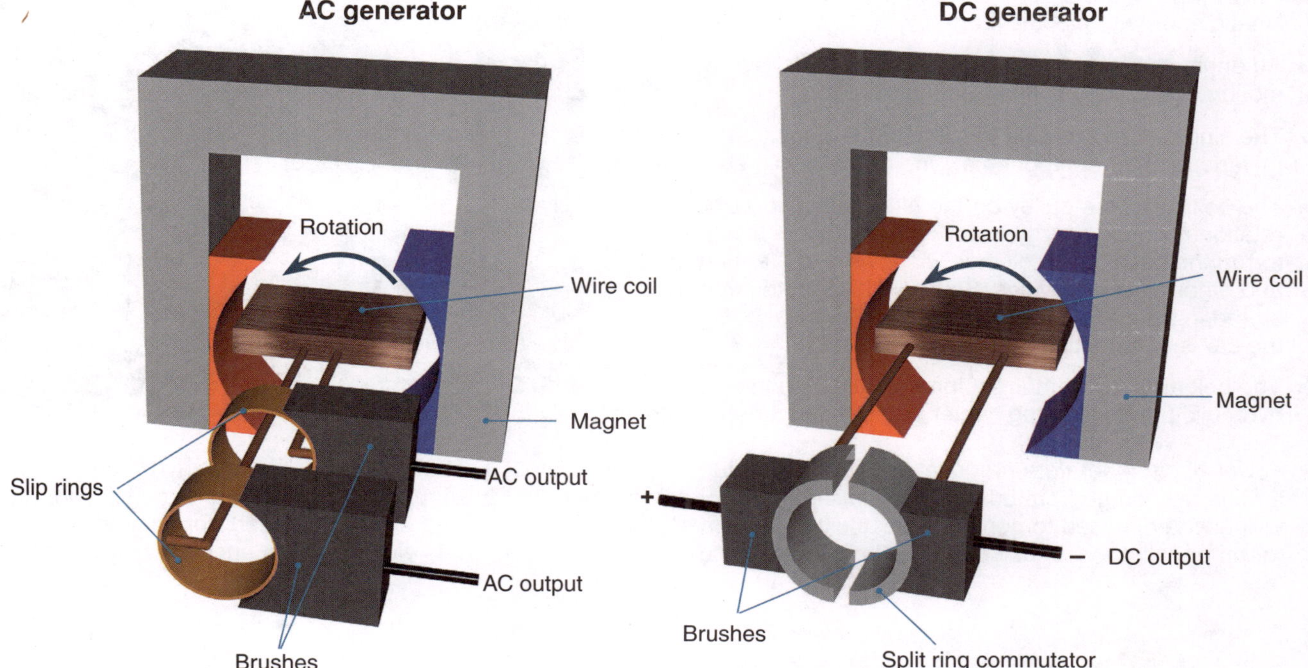

In an AC generator, the coil rotates through the magnetic field and current flows in the coil through to the slip rings and is passed to the brushes. Because one side of the coil first passes down through one side of the magnetic field and then up through the other side, the current alternates back and forth in the coil so each output will alternate from positive to negative.

In a DC generator, the coil is connected to a split ring called the commutator. The side of the coil moving down through the magnetic field is only ever in contact with the + brush, while the side moving up is only ever in contact with the – brush. This means a brush can only have either + or – output. The output rises and falls as the coil rotates through different parts of the magnetic field.

3. The diagram below shows an output wave for one of the generators above (AC or DC). Identify which type of generator produced the wave:

   AC generator.

4. On the axis below draw a graph (similar to that in Q3) for the type of generator (AC or DC) that you did not name in Q3.

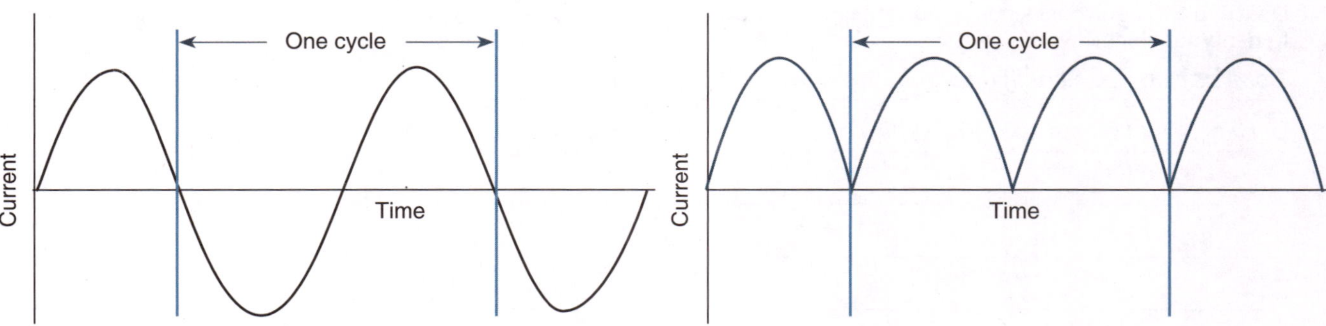

5. (a) What would happen to the output of a generator if the number of turns of wire in the coil was increased?

   Output would increase.

   (b) What are some other ways of increasing the generator's output? *Increase the rate at which the coil turns or increase the number of times the wire is wound around the coil.*

©2020 BIOZONE International
ISBN: 978-1-927309-82-7
Photocopying Prohibited

Electromagnetic induction

- The generators you have been learning about produces electricity through **electromagnetic induction**. This is the process of generating electric current with a magnetic field.
- Generators in cars (called alternators) and most power plants produce alternating current (AC).
- Regardless of the type of current, all generators change kinetic energy to electrical energy.
- An engine's crankshaft (connected to a pulley) converts the motion of the pistons into rotational motion.
- The single or multiple belts connect the engine's crankshaft to various other components. One of these is the alternator.
- The belt spins the pulley on the alternator and some of this rotational kinetic energy is converted into electricity used to run the car's lights, electric windows and steering, radio, and other electric devices. Some of this electric energy is stored in the car's battery to provide electrical energy when the car isn't running.
- The belt also runs other equipment including the pumps for hydraulic power steering, water pumps, and air conditioning.

- Power plant generators produce electricity using the kinetic energy of a rotating turbine. The energy to turn the turbine may come from burning fuel, falling water, or some other energy source. Below are diagrams showing how falling water is used to generate electricity in a hydroelectric power plant and steam produced from heating water (or other substances) using the heat produced in nuclear reaction generates electricity in a nuclear power plant.

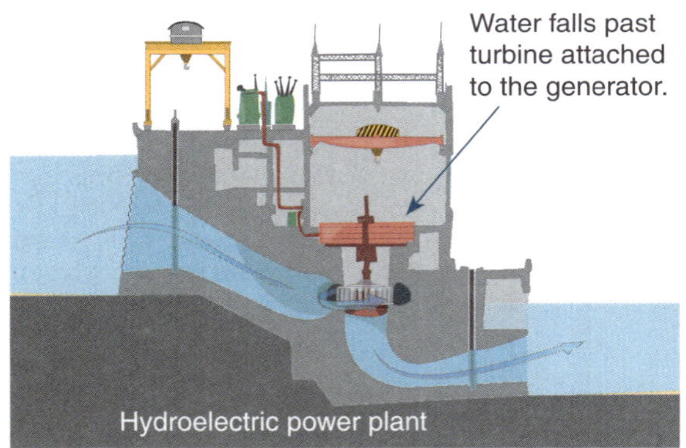

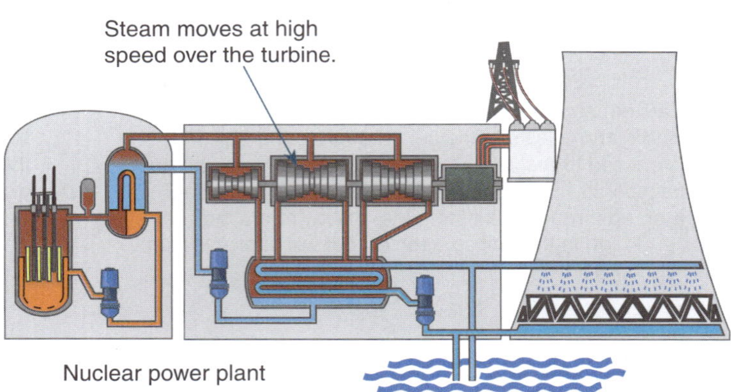

- Photovoltaic cells do not utilize generators to produce electricity. You will learn more about how these produce electricity later in this book.

6. Describe two additional kinds of power plants and what they use to provide the kinetic energy used to turn the turbines to generate electricity:

   Combustion (fossil fuel) power plants use heat from combustion to expand air molecules to do work on a turbine. Wind farms use moving air to turn rotors attached to generators.

7. The first battery was created in 1799 and batteries provided the main source of electricity before the development of electric generators in the mid to late 1800's. Explain the benefits of using changing magnetic fields to produce electricity rather than building many large batteries:

   Batteries would need to be continually replaced as the energy is used up and the sheer number of batteries that would be required to power modern cities would be prohibitive. Turbines only require the ability to rotate a magnet to induce electron flow.

©2020 **BIOZONE** International
ISBN: 978-1-927309-82-7
Photocopying Prohibited

Motors

- Generators convert kinetic energy into electrical energy. Electrical motors do the opposite. Motors take electrical energy and convert it into mechanical energy. *Note a motor converts various forms of energy into rotational energy. An engine converts thermal energy into rotational energy. An engine is therefore a specialized motor.*
- Just like generators, motors use magnets. Instead of using the magnet to produce electricity, it uses electricity to produce movement in the magnet. Let's create a motor and see how this works.

INVESTIGATION 6.11: Making a motor

See appendix for equipment list.

1. Start 4 cm from the end of a wire and wrap it about seven times around your finger to make a coiled loop about 1 cm in diameter. Without changing the shape, gently slide the coiled wire off your finger. Leave another 4 cm of wire on the other side and cut the wire.
2. Carefully wrap each of the two tails around the coils so that the coil is securely bound together and the tails extend perpendicular to the coil. The coiled wire should look similar to figure 1.
3. Use sandpaper to remove the insulation from the first 3 cm of one of the tails.
4. On the other tail, remove the first 3 cm of insulation from only the top of the wire. See figure 2.
5. Bend two paper clips to look like figure 3. Lightly sand the surfaces of the paper clips. This will ensure good contact with the battery.
6. Secure the large loop ends of each paper clip to the terminals of the D-cell battery. See figure 4.
7. Place the ceramic magnet to the side of the battery. See figure 5.
8. Place the coiled wire into the small hooks formed by the unattached ends of the paper clips. See figure 5.
9. Turn the coil slowly by hand, only touching the insulated part of the wire. Observe the magnetic attraction and repulsion between the coiled wire and the ceramic magnet.
10. Give the coil a little push to get to it started. Once spinning it should continue to spin on it own. It may take a little refining to the position or orientation of the magnet, depending on the magnet's strength, to get the coil spinning.
11. Once you've got the coil spinning consistently, place the motor on the edge of your lab bench. Bend the end of the wire with no insulation into a small, tight loop. Tie one end of a 0.3 m long piece of fine thread around this loop. Tie the other end to a small paper clip.
12. Give the coil a little push to get it start winding up the string. Use your finger as a guide as the added weight will throw the motor off balance. Once the motor picks up one paper clip, unwind it and try connecting more paper clips to the first. See how many paper clips your motor can lift.

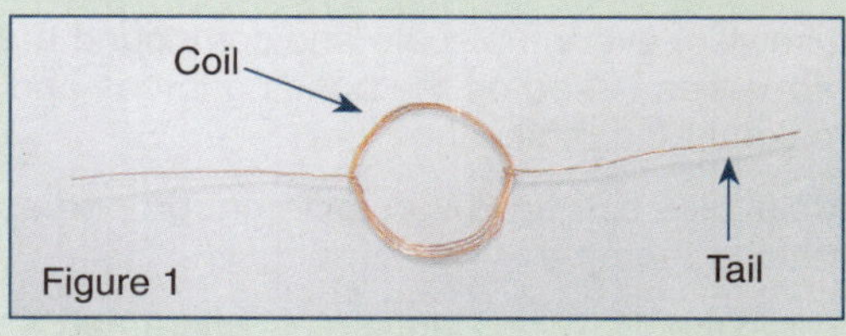

Figure 1

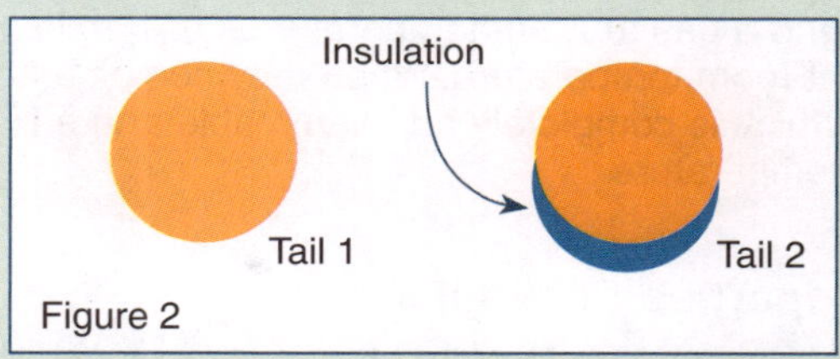

Figure 2

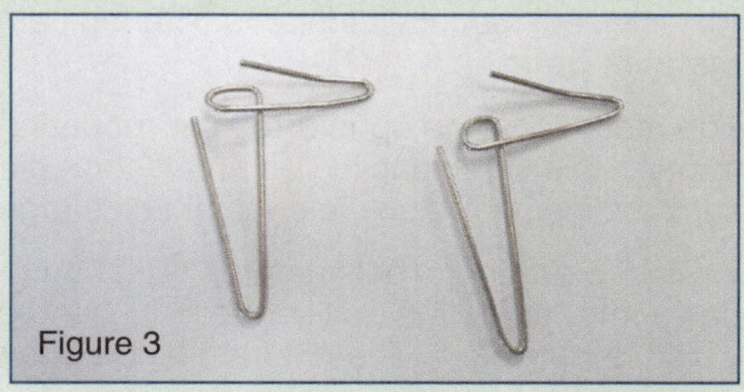
Figure 3

Figure 4

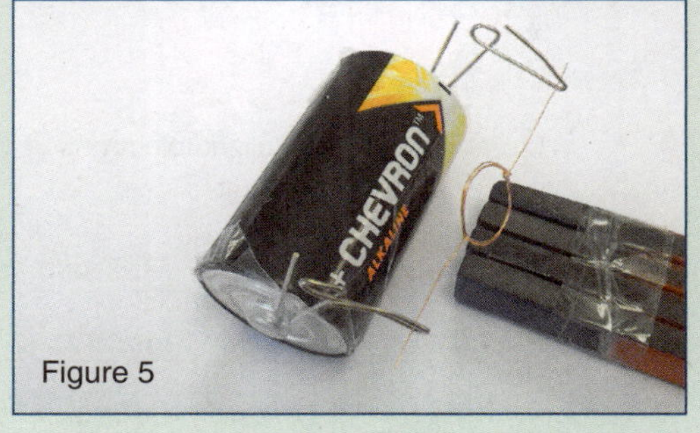

Figure 5

8. Explain why there was magnetic attraction and repulsion between the coiled wire and the ceramic magnet and how that was used to enable the motor to do work:

When the copper wire was placed on the two paper clips it completed the circuit and allowed a flow of electrons through it. The coil became an electromagnet. As the coil spun, the attractive side is pulled towards the ceramic magnet and the repulsive side is pushed away, causing the coil to continue spinning. This spinning (rotational mechanical energy) was able to do work.

©2020 **BIOZONE** International
ISBN: 978-1-927309-82-7
Photocopying Prohibited

89 Forces in Materials

Key Question: How do the atomic characteristics of materials affect the properties of the material?

Material properties

- Each element has distinct properties based on its structure: number of protons, electron configuration, etc.
- Likewise, every molecule and compound has its own distinct properties based on the bonding forces and atoms used to make up the molecule.
- Molecules can have very different properties to the atoms making them up.
- For example, hydrogen is highly flammable, extremely light, and a gas at room temperature. Oxygen is a reactive gas at room temperature. When they combine, they form water, which is completely non-flammable and a liquid at room temperature.

The space shuttle's main engines burned liquid hydrogen and oxygen, producing water vapor as the exhaust.

Properties of metals

- Metals can be thought of as positive ions (all of the atom except the valence electrons) in a "sea" of loose (delocalized) valence electrons. The mutual attraction of each metal atom to the others' valence electrons forms the **metallic bond**.
- The metal ions line up in a regular repeating pattern (a crystal lattice) and their loose valence electrons move through this crystal acting as an electron glue. Each of the ions is strongly attracted to all of the loose electrons surrounding it so the whole metal holds together as a crystal.
- While the attraction between the "electron glue" and the ions is strong, the attraction between the ions themselves has some "flexibility" (unlike the ionic bond between a sodium and chloride ion you learned about in Chapter 2, which is very rigid). The diagram below shows a general diagram of the atomic structure of a metal:

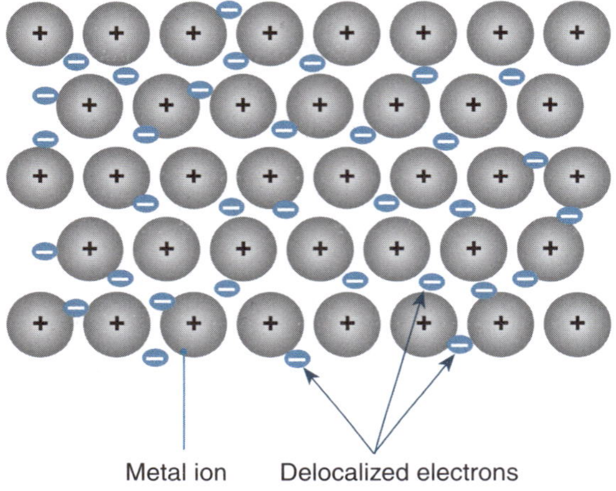

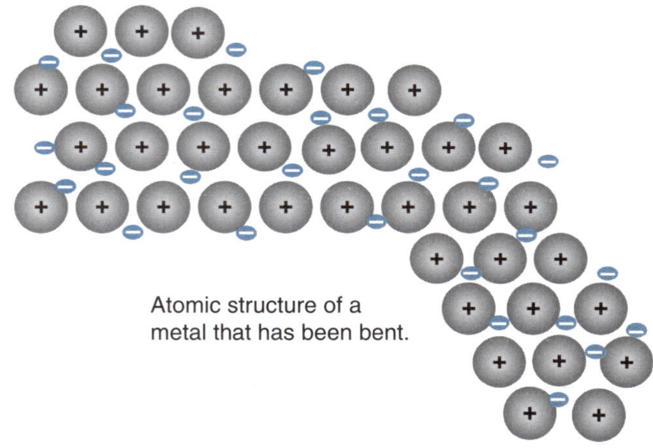

1. Explain these properties of metals:
   (a) All metals conduct electricity: *Metallic substances form metallic bonds where metal ions exist in a "sea" of valence electrons. These valence electrons are free to move around the substance. Electric current is able to flow freely through metals, utilizing the loosely held electrons in the metal.*

   (b) Metals are good conductors of heat: *The free electrons in the metal are able to collide with each other and the metal ions very frequently, transferring the kinetic energy (heat) through the material more easily.*

   (c) Metals are able to be shaped: *Metals are malleable because as the material is deformed, the free electrons flow with the material, allowing it to maintain its strength as its shape is changed.*

©2020 BIOZONE International
ISBN: 978-1-927309-82-7
Photocopying Prohibited

Metals and alloys

- Alloys are metals made from a mixture of different elemental metals and/or other elements to give the original metal different properties.
- Both physical and chemical properties change. The atomic structure and the electric fields change due to the presence of new elements.
- Carbon steel (or plain steel, the most common form of steel) is iron (Fe) with up to 2% addition of carbon (C). Plain steel is much stronger than iron, but is still ferromagnetic and rusts (the same as iron).
- But why is steel so much stronger than iron? What is the carbon doing to give steel its strength?

2. (a) How many valence electrons does carbon have? *4*

(b) How could it fill its valence shell? *A carbon atom could lose 4 electrons, gain 4 electrons, or share electrons with other neighboring atoms.*

(c) If carbon is added to iron, how could carbon get the electrons it needs to have a full valence shell?

Carbon could gain electrons from the delocalized electrons around the iron ions.

(d) What would this do to the charge of the carbon atom?

Gaining electrons would give carbon a negative charge.

(e) Iron is made of Fe ions surrounded by loose or delocalized electrons. The Fe ions are able to move because they are not attracted to any one Fe ion more than any other. What would happen to this ability to move if the Fe ions were strongly attracted to another kind of negative ion?

The electrons would not be able to move as freely through the substance. Some of the Fe ions would become locked with the negative ions (carbon) and limit the flow of the metal.

(f) Draw a diagram in the space below to show how Fe ions and carbon interact. Use your diagram to explain why adding carbon makes steel stronger than iron.

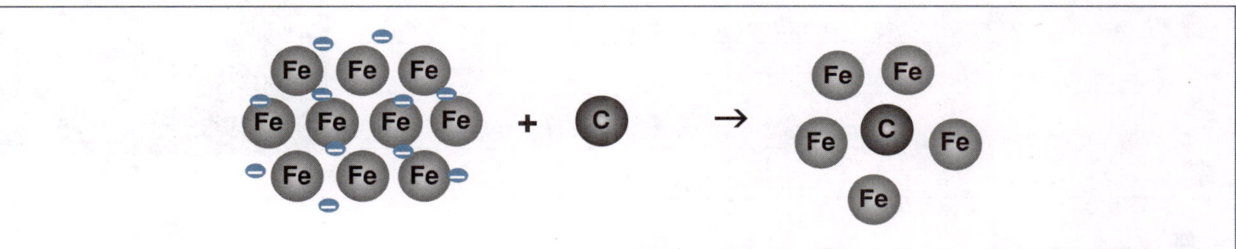

The positive Fe ions would have a strong attraction to the negative C ions. This would reduce the number of delocalized electrons in the material. The strong attraction would give the material more resistance to deformation allowing it to hold its shape under stress better.

3. (a) Based on the physical properties of iron and steel, explain why steel is a much more desirable material than iron for making structures:

Buildings support a lot of weight. Under stress, iron can bend, which would not be good for a building. Steel is more rigid and will support more weight without deforming, and is lighter than pure iron.

(b) Explain any difference you might observe if you were to hammer iron opposed to steel: *Iron would be easier to dent and you would see marks every time you hit it with a hammer. Steel would better resist deforming under the impacts from the hammer and you would have to hit a lot harder to get the same results.*

(c) By mass, steel is often only 1 – 2% carbon. Different types of steel use different quantities of carbon. Can you make any predictions about how the physical properties (malleability, ductility, etc) of steel might change as more carbon is added? Explain what you think is happening:

As more carbon is added to the steel, more of the delocalized electrons will be trapped by the carbon atoms. This will change the properties of the steel, making it less ductile and malleable the more carbon is added, while also making it more rigid.

©2020 **BIOZONE** International
ISBN: 978-1-927309-82-7
Photocopying Prohibited

Properties of ionic substances

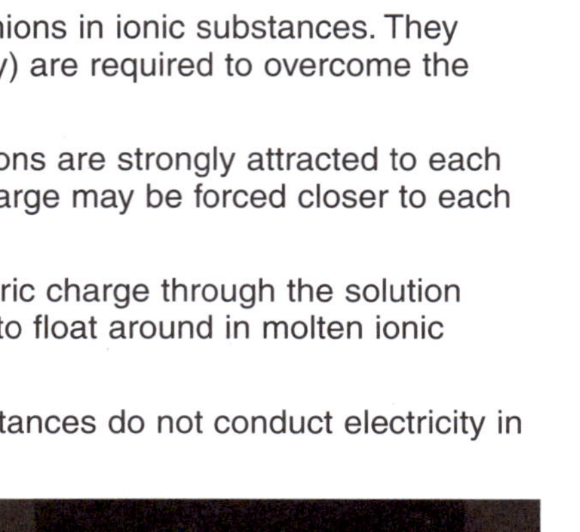

- Recall from Chapter 2 *Structure and Properties of Matter*, that ions formed by gaining an electron have a net negative charge (anions) and ions formed by losing an election have a net positive charge (cations).
- An ionic bond is formed when there is a large electronegative difference between the elements in the bond. The greater the difference, the stronger the attraction between the cation and the anion.
- The properties of ionic compounds relate to how strongly the cation and anion attract each other.
- Ionic substances form crystal lattices. The ionic crystal is a regular structure with cation and anion alternating with each other to form a three-dimensional structure. The shape of the structure is largely based on the smaller ions filling in the gaps between the larger ions.

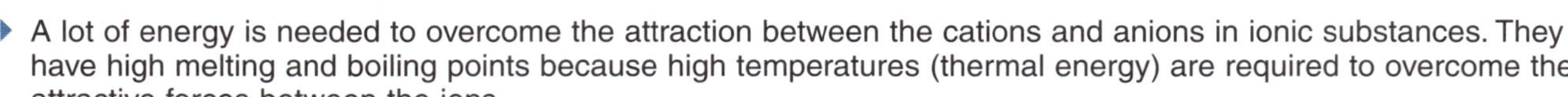

- A lot of energy is needed to overcome the attraction between the cations and anions in ionic substances. They have high melting and boiling points because high temperatures (thermal energy) are required to overcome the attractive forces between the ions.
- Ionic substances tend to be hard and brittle because the positive and negative ions are strongly attracted to each other and difficult to separate, however, when pressure is applied ions of like charge may be forced closer to each other causing electrostatic repulsion to split the crystal.
- When ionic substances are dissolved in water, the ions are free to conduct electric charge through the solution making ionic solutions good conductors of electricity. Because the ions are free to float around in molten ionic substances, molten salts are also good conductors of electricity.
- Because the ions in ionic substances are bound tightly to each other, ionic substances do not conduct electricity in solid form and are very good insulators.

4. The photo on the right shows a crystal of the ionic substance sodium chloride. It has the formula NaCl. Note the cubic structure of the crystal.

Hans-Joachim Engelhardt CC 4.0

(a) What do you think would happen if this crystal was hit very hard with a hammer? How does this compare to hitting a metal with a hammer?

It would shatter into many smaller cubic crystals. Metal deforms but stays together due to its metallic bonds.

(b) Why would this happen? *Striking the object shifts the ions This will lead to repulsive forces where like charges come in contact.*

5. (a) Ionic bonds are very strong, yet many ionic substances, including sodium chloride, dissolve in water. Recall that water is a polar molecule, but the effective charge on the opposite ends of a water molecule is small compared to the charge of a ion. The diagram below shows a sodium ion and a chloride ion. Draw water molecules onto the diagram to show how the water molecules would arrange themselves around each ion:

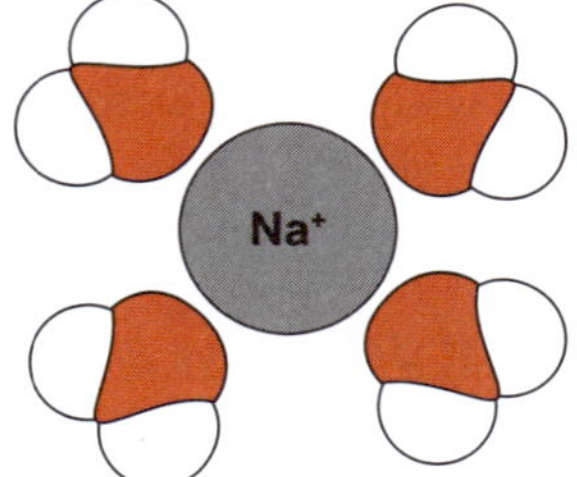

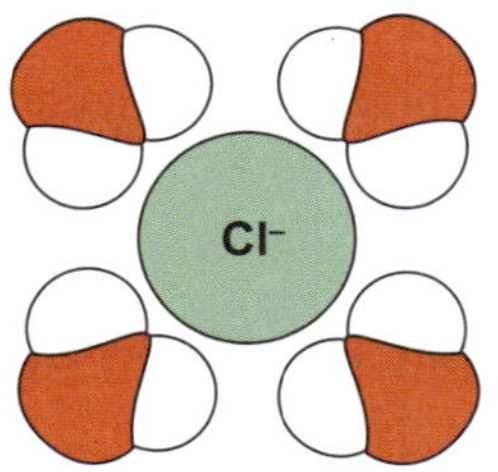

(b) Explain why sodium chloride dissolves in water: *The polar water molecules would arrange themselves so the negative side (oxygen) would face the Na. The combined charges of multiple water molecules would be stronger than that of a single ion.*

6. Solid sodium chloride is an electrical insulator, but molten sodium chloride and dissolved sodium chloride can conduct an electrical current. Explain these properties:

Current electricity is produced by the movement of charged particles (not necessarily just electrons). When in solid form, the ions in NaCl are locked in place in a crystal lattice. Ions (charged particles) are able to freely move in liquid form, or when dissolved in a liquid.

©2020 **BIOZONE** International
ISBN: 978-1-927309-82-7
Photocopying Prohibited

Properties of molecular substances

Discrete molecules

- Recall that molecules are formed when atoms share pairs of valence electrons creating covalent bonds.
- The covalent bonds in molecules form only between the atoms in the molecules. The forces holding the individual molecules together are much weaker.
- Discrete molecules are ones which have a fixed number of atoms per molecule e.g. NH_3 (ammonia) always has 4 atoms per molecule.
- The photos below show the substances propane (C_3H_8) and propanol (C_3H_7OH). Propane is a gas at 25°C and has to be kept in a cylinder with thick steel walls. Propanol is a liquid and can be kept in a lidded bottle.

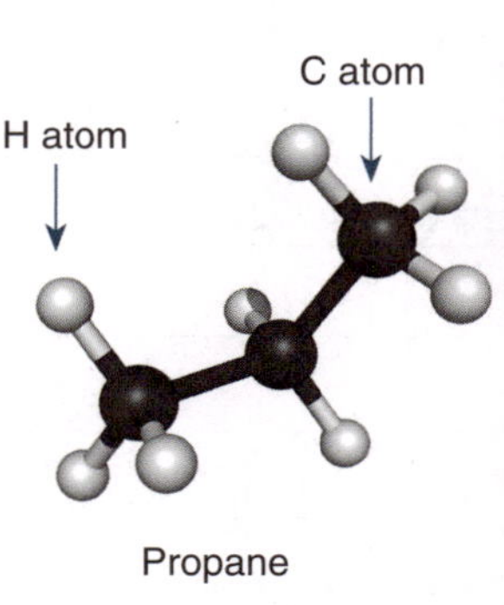

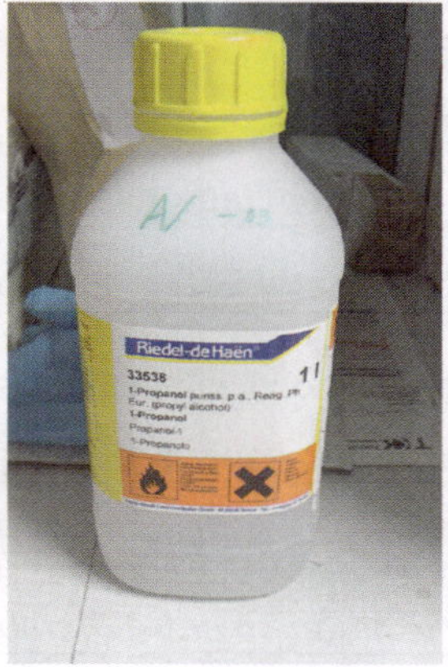

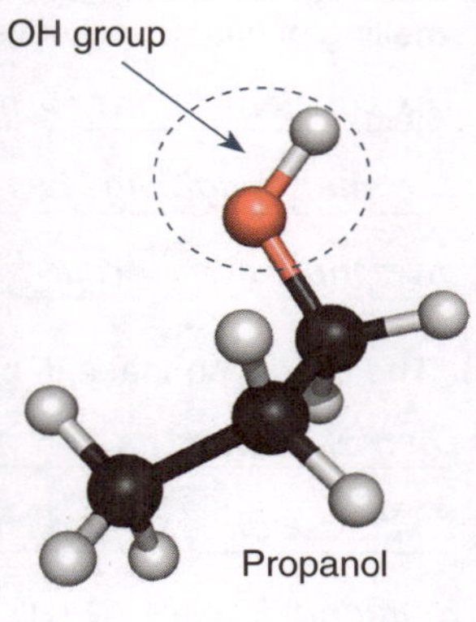

7. Study the diagrams showing the molecular structure of propane and propanol. Notice that they are virtually the same sized molecule, but that propanol has an OH group at the end. Explain why propane is a gas at 25°C whereas propanol is a liquid:

The OH group on one end of a propanol molecule makes it polar. Polar substances have more intermolecular forces holding them together giving them higher melting and boiling points. As a result, propanol will be a liquid at higher temperatures than propane.

- The graph below shows the melting and boiling points of a family of molecules called the alkanes. The simplest alkane is methane which has the formula CH_4. The alkanes have the general formula C_nH_{2n+2} when n is the number of carbon atoms. Thus the next alkane is ethane with the formula C_2H_6, followed by propane C_3H_8. In their simplest form the alkanes are long single chained molecules.

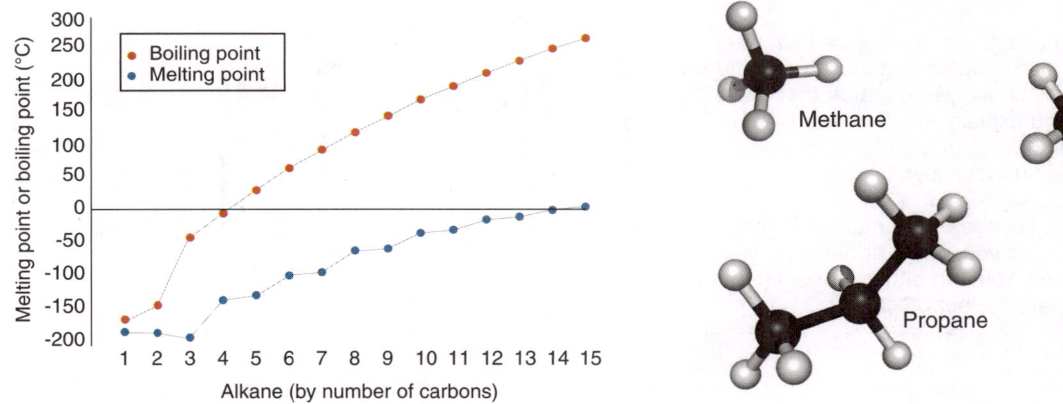

8. (a) What is the structural difference between each successive alkane from CH_4 to $C_{15}H_{32}$?

One carbon and two hydrogens are added as the molecule becomes longer.

(b) In general, the boiling point and melting point of alkanes increase as the molecules become bigger. Why might this happen?

As the molecule becomes larger, the number of intermolecular forces of attraction increase. Even though the molecule is nonpolar, the number of polar instances due to random electron movement increases.

©2020 **BIOZONE** International
ISBN: 978-1-927309-82-7
Photocopying Prohibited

Network molecules

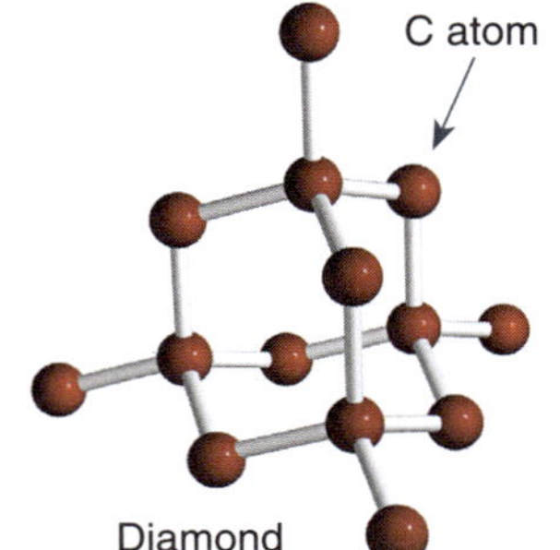

- Network molecules have no limit on the number of atoms in the molecule. For example silicon dioxide (quartz) has the formula SiO_2, but this simply shows the ratio of atoms in the molecule, not the number of atoms in the molecule.
- In SiO_2, every atom is covalently bonded to its neighbor and there can be any number of Si and O atoms bonded together. Other examples of network molecules include diamond and graphite.
- Diamond (right) is made of only carbon atoms covalently bonded together. It is the hardest natural substance known so diamonds (usually made synthetically) are embedded in diamond saws used for cutting materials such as concrete.

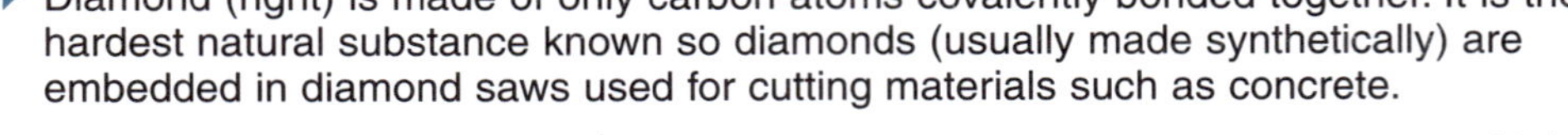

9. (a) Use your knowledge of bonding to explain why network molecules have very high melting points, usually over 1000°C:

The covalent bonds formed from shared electrons are very hard to break. In order to become a liquid, the energy holding the individual atoms has to be overcome unlike other substances where only the intermolecular forces between molecules have to be overcome.

(b) "The only thing that can cut a diamond, is another diamond." Explain:

Diamonds are the strongest naturally occurring material. This means specialized tools are needed to cut a diamond and these tools typically use diamond tips and diamond dust to cut and polish the diamond. Lasers can also be used to cut a diamond.

10. The strength of the electrostatic forces between atoms and molecules in a substance affects the properties of that substance. Using examples, explain how we use the specific properties of materials for specific tasks in our daily lives:

Students' answers will vary. Possible answers: copper in wires to conduct electricity, cloth oven mitts to protect hands from heat, wood to make furniture, because of rigidity of material etc.

11. Many different types of metals and metal alloys are used by humans. Some of these are: iron, aluminum, copper, zinc, chromium, tin, brass, steel, silver, and gold.

(a) Work in small groups of three or four. Each person in the group should pick three different metals or metal alloys, research their uses in industry, and explain why they are used in that way. Summarize your own findings about your three metals/metal alloys below.

(b) As a group, collaborate to make a digital presentation to your class about what you have learned. You can either present it to the class or share it electronically.

Student's own research. Typical student answer could include details such as:

Steel: An alloy of iron and carbon. Iron is abundant (as is carbon) and steel is easily produced and important in modern manufacturing and civil engineering, e.g., to make reinforced concrete. The addition of other elements can provide steel with other useful qualities. Nickel increases its durability and makes it more resistant to heat and acids;. Adding manganese makes it more durable. Adding tungsten helps it maintain hardness at high temperatures.

Copper: Copper's properties make it useful as a conductor of heat and electricity, as a building material, and as a constituent of various metal alloys, such as sterling silver used in jewellery. Its major applications are electrical wire (60%), roofing and plumbing (20%), and industrial machinery (15%). Copper is used mostly as a pure metal. When greater hardness is required, it is put into such alloys as brass and bronze.

Gold is relatively rare and resistant to most acids. It is the most malleable of all metals and has a lustrous appearance. Its rarity and beauty, as well as its malleable properties makes it suitable for its main purpose in jewellery and investments.

©2020 **BIOZONE** International
ISBN: 978-1-927309-82-7
Photocopying Prohibited

Choosing the correct material

- We don't we use salt to build houses or rubber to make plane wings? Clearly, certain materials are suited to certain tasks. Why?

INVESTIGATION 6.12: What is used where

See appendix for equipment list.

PS2-6

1. Carry out a survey around the school and sports fields. Look at different objects and what they are made from (e.g. chairs, bench seats, fences, etc). Are they made from wood, steel, aluminium, etc?
2. Record these in a table on a note pad. Staple your record to this page.

12. (a) From your survey, what materials were used to make chairs? *Student's answers may vary.*

    Students may find chairs made of plastic, metal, wood, or a combination of materials.

    (b) Can you think of any materials used to make chairs that you did not find at school? *Student's answers may vary.*

    (c) What properties did these materials share that made them useful for constructing a chair?

    Students should note rigid nature of the material.

13. From your survey in Investigation 6.12, select three objects made of different materials. For each one, describe the properties of the material that make it suitable for constructing the object:

    Student's answers will vary.

    Students may describe the rigid nature of the materials used to make desks and chairs in the classroom, the flexibility of material making balls in gym class, which allows them to be inflated.

14. New materials are being invented all the time for various purposes. An example of this would be polar fleece for light, yet warm clothing. Can you think of another material that was invented for a specific purpose? Do a little research if needed and describe the material's uses and why it was invented for that purpose:

    Student's answers will vary. Students may discuss semiconductors for use in electronics and solar cells and their ability to direct electrons in specific directions, the use of ceramics and kevlar in military body armor and how the hardness of the ceramics coupled with the strength of kevlar prevent penetration, or many space-age materials developed for space exploration.

15. For each of the following structures, identify a suitable material for its construction and how that material could be used:

    (a) Airplane wing: *Student's answers may vary. Students may realize aluminium is used because of its rigidity to resist deformation and metallic bonds to accommodate stress without breaking and is light.*

    (b) Cup: *Student's answers may vary.*

    Students may suggest plastic as it is a non-porous solid ideal for holding a liquid.

    (c) Knife blade: *Student's answers may vary.*

    Student may suggest metals because they can be sharpened easily.

    (d) Teething toy: *Food grade silicon, natural rubber etc. Material must be elastic and able to maintain integrity (hold together), and safe.*

©2020 **BIOZONE** International
ISBN: 978-1-927309-82-7
Photocopying Prohibited

90 Review Your Understanding

Key Question: How does a copper plate (a non-magnetic material) stop a strong magnet from hitting it and what forces are involved?

- Remember the magnet and copper plate demonstration at the beginning of the chapter? Have you figured out what is going on yet? Watch the video on the **BIOZONE Resource Hub** again and see if what you have learned in this chapter tells you anything more about what is going on.
- Copper is not a ferromagnetic material. Under normal conditions it does not react to a magnet nor can it become a permanent magnet.
- If you look at the electron configuration of copper (right), you will notice that all of copper's electrons are paired up except one. This single electron does produce a small magnetic field due to its spin, but it is not enough to produce a magnetic field on a macro level.
- However, copper did have a reaction to moving magnetic fields as you observed in this chapter. Copper is a good conductor of electricity and in the presence of a changing magnetic field a current of electricity is produced in the copper.
- In electromagnetism, a moving magnetic field can generate electric current. The reverse is also true: an electric current generates a magnetic field.
- When you created the motor, the copper wire was an electromagnet. It produced its own magnetic field. The electromagnet sometimes attracted and sometimes repelled the ceramic magnet on the battery. This is what caused the electromagnet to spin, allowing it to do work.

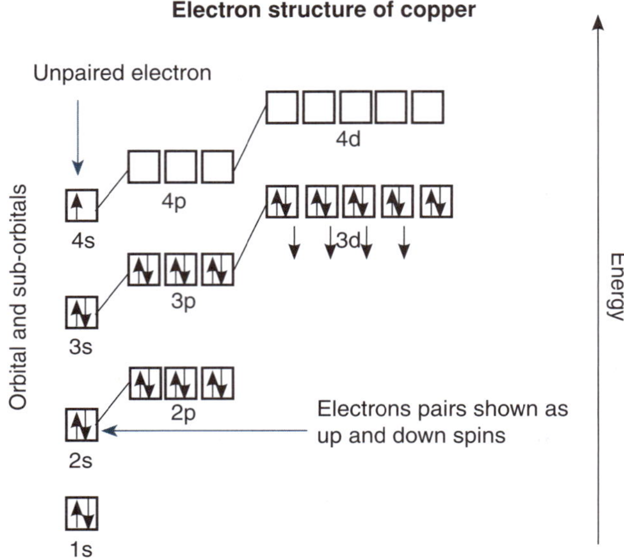

- Slowly place the magnet against the copper plate and no resistance is felt. However, swing the magnet at the copper plate and there is an observable force of resistance which prevents the magnet from hitting the copper.
- The movement of the magnet as it was swung towards the copper plate was enough to create a temporary flow of electrons on the surface of the copper. Remember that electron flow creates a magnetic field. Although temporary, the magnetic field produced by the movement of electrons in the surface of the copper repels the magnetic field of the magnet and prevents the magnet from hitting the copper plate.

1. Explain why there is no observable resistance between a magnet and a copper plate when the magnet moves slowly:

   Flow of electricity induced by movement of a magnetic field is dependent on the speed at which the magnetic field is moving. When the magnet is moving slowly, so is the electric current which means there is not enough field strength to affect the magnet.

2. What would you expect to happen if you increased the speed of the magnet as it approached the copper plate? Explain:

   Increasing the speed of the magnet increases the rate of change in the magnetic field and consequently the rate of electron flow. This increase in electron flow allows the EM field to affect the magnet and repel it.

3. How is the demonstration of the copper plate repelling the strong magnet related to the creation of electromagnets?

   The movement of the strong magnet towards the copper plate induces electron flow in the copper plate. Continuous motion in the magnetic field would yield an electromagnet from the copper for as long as the field is in a state of change.

4. Use the inverse square law to explain why the repulsive forces between the copper and magnet are not observable until they are very close together:

   $F_1 - kq_1q_2 \div r^2$. Halving the distance yields $F_2 = 4(kq_1q_2 \div r^2 = 4F$ and quartering the distance yields $F3 = 16F$. Closer together the field strength becomes very strong and noticeable.

©2020 BIOZONE International
ISBN: 978-1-927309-82-7
Photocopying Prohibited

91 Summing Up

PS2-4 **Forces at a distance**

1. Use electrostatics to explain the following:

(a) Why does water vapor in the air form clouds instead of just dispersing evenly throughout the atmosphere?

Electrostatic forces hold the particles in the clouds (water droplets and dust) together due to attraction between opposite charges.

(b) When oil (such as cooking oil) is added to water, it forms droplets that do not mix with the water. Explain why:

Water molecules are attracted to each other because they are polar. The positive pole of one water molecule will be electrostatically attracted to the negative pole of another water molecule. Water is not attracted to oil because oil is nonpolar and does not mix with water.

2. The diagram below shows the sizes of two ions from the ionic substances NaF and NaI. Use Coulomb's law to explain why NaF has a higher melting point than NaI:

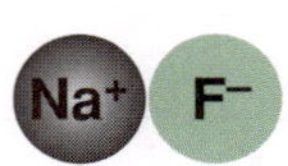

$$F = k\frac{q_1 q_2}{r^2}$$

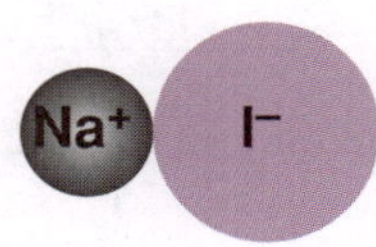

The charges between the ions in the ionic compounds have the same strength, but because of the size of the iodide ion, the distance between the Na and I is greater. This increased distance reduces the strength of the force of attraction allowing NaI to melt at lower temperatures.

3. Use Newton's law of gravity to compare the size of the gravitational force experienced by each pair of orbiting objects:

$$F = G\frac{M_1 M_2}{r^2}$$

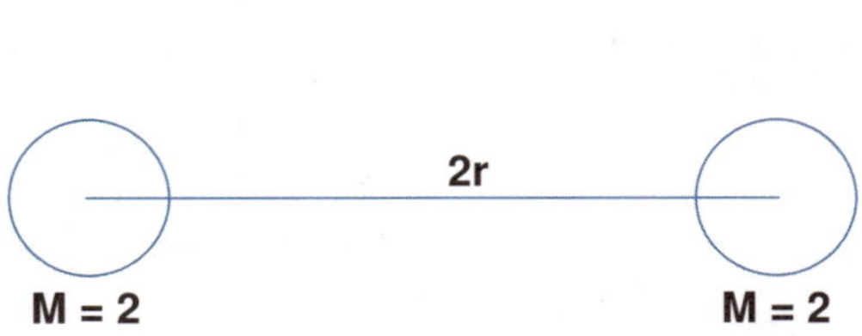

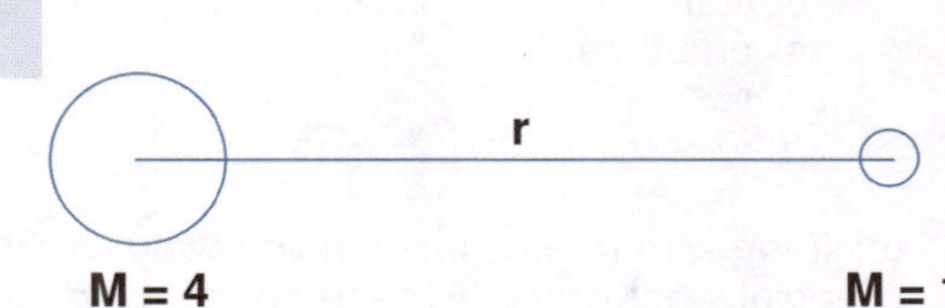

The gravitational force in the second diagram is 4x that of the 1st.

$F_1 = 4G \div 4r^2 = G/r^2$. $F_2 = 4G \div r^2 = 4G/r^2$. Substitute F_1 for G/r^2 and we get $F_2 = 4F_1$.

Or : Left diagram: F = G x 2x2 = 4G. 2 x r = 2. $2^2 = 4$. 4G/4 = 1G. F= 1G

Right diagram: F = G x 4 x 1 = 4G. 1x r = 1. $1^2 = 1$ 4G/1 = 4G F = 4G

4. The average distance from the Earth to the Sun is 1.5×10^{11} m. The mass of Earth is 5.97×10^{24} kg, and the mass of the Sun is 1.99×10^{30} kg. G = 6.673×10^{-11} Nm^2 /kg^2. Calculate the force acting between the Earth and Sun:

$F = (6.673 \times 10^{-11} \times 5.95 \times 10^{24} \times 1.99 \times 10^{20}) \div (1.5 \times 10^{11})^2 = 3.51 \times 10^{22}$ N

©2020 **BIOZONE** International
ISBN: 978-1-927309-82-7
Photocopying Prohibited

PS2-6 Inter and intra molecular forces

5. The table below lists some properties of three metals: iron (as steel), aluminum, and titanium, which are used for various purposes:

	Metal		
Property	**Iron (as carbon steel)**	**Aluminum (alloy 6061)**	**Titanium (grade 5)**
Strength (MPa)	841	300	950
Density (g/cm^3)	7.58	2.7	4.5
Resistance to corrosion	Medium-low	High	Very high
Price per tonne (approx)	$500	$2000	$30,000
Abundance in crust (ppm)	63,000	82,000	6,600
Ease of refinement from ore	Easy	Difficult	Difficult

Using the table above, explain why the metals are used in the following ways:

(a) Iron is commonly used in the construction industry to build large scale buildings (e.g. factory sheds or sky scrapers):

Iron is easy to find (abundant) and refine, relatively cheap (compared to other metals), and is strong.

(b) Aluminum and titanium are commonly used in the aerospace industry (e.g. building parts of planes):

Aluminum is much lighter than iron, which is important for something to fly. Titanium and aluminum are both highly resistant to corrosion, which is important due to the weather conditions planes can be subjected to.

Energy and forces

PS3-5 6. A wire is connected in a circuit to a galvanometer and placed into a magnetic field as shown in the diagram:

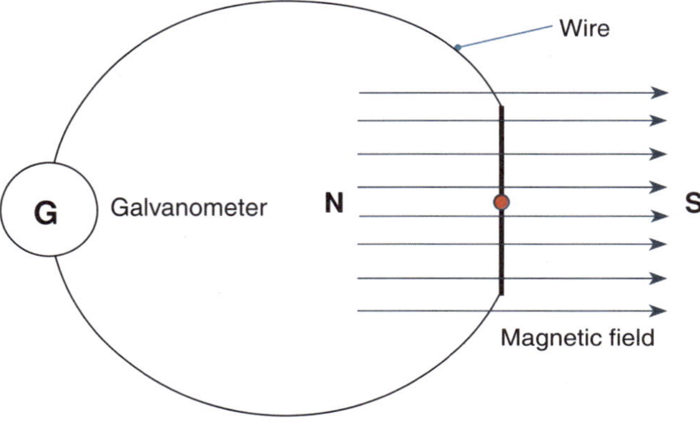

(a) If the wire is held stationary (not moving) will a current register in the galvanometer?

No

(b) If the magnetic field rotates one whole revolution about the red point what would be seen on the galvanometer?

A current will be detected.

(c) If the wire rotates one whole revolution about the red point what would be seen on the galvanometer?

A current will be detected.

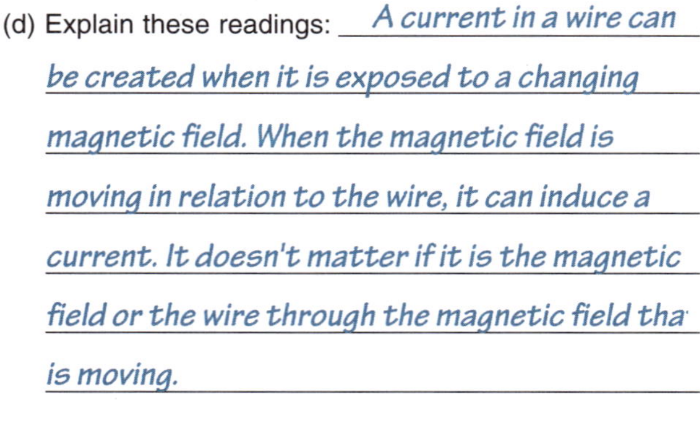

(d) Explain these readings: *A current in a wire can be created when it is exposed to a changing magnetic field. When the magnetic field is moving in relation to the wire, it can induce a current. It doesn't matter if it is the magnetic field or the wire through the magnetic field tha is moving.*

(e) What would happen to the strength of the current in the wire if the wire was coiled so a greater length of wire was in the rotating magnetic field?

The current would become stronger.

(f) What would happen to the strength of the current in the wire in the rotating magnetic field if the strength of the magnetic field was increased?

The current would become stronger.

(g) What would happen to the strength of the current in the wire in the rotating magnetic field if the magnetic field was rotated at a greater rate?

The current would become stronger.

©2020 BIOZONE International
ISBN: 978-1-927309-82-7
Photocopying Prohibited

Energy

Concepts and connections

Use arrows to make connections between related concepts in this section of the book

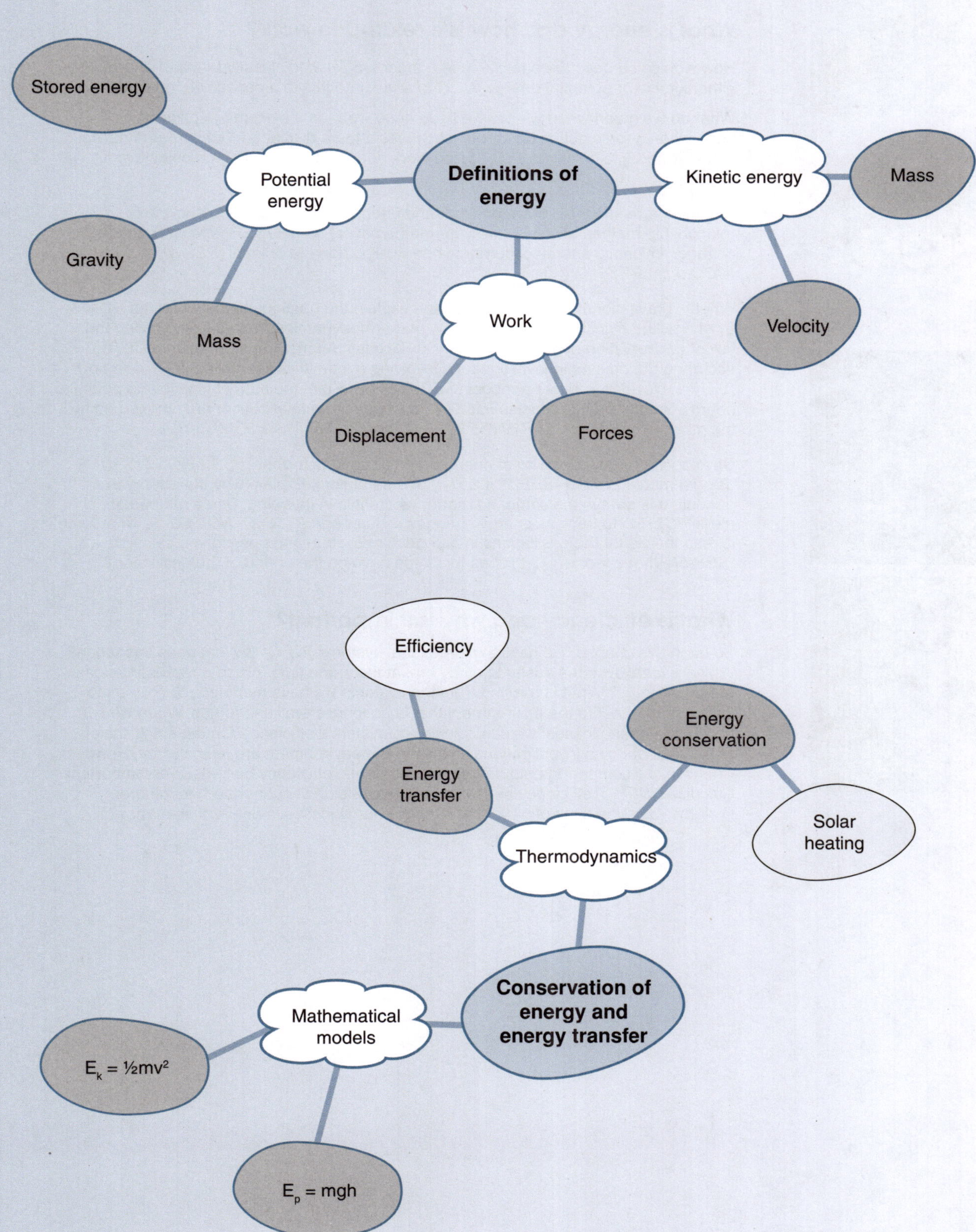

PS3.A
PS3.B

7: Definitions of Energy

Anchoring Phenomenon

The physics of amusement parks: How are the relationships between work, kinetic energy, and potential energy applied in amusement park rides? 92 101

What is energy and how is it related to work?

☐ 1 How would you describe energy? Use a model [SEP-2] to classify forms of energy as either kinetic or potential energy [CCC-5] and relate this to a capacity to do work [PS3.A]. 93

☐ 2 What do we mean when we say we have done work on a system? In physics, work is done when a force acting on an object causes it to be displaced. Use a mathematical model [SEP-2] to calculate [SEP-5] the work done by forces acting on objects in a system [PS3.A] [CCC-4]. 94

☐ 3 What is the relationship between work and energy and how can you show this relationship mathematically? Use your mathematical model [SEP-2] to calculate the changes in kinetic energy occurring when work is done to objects [PS3.A] [PS3.B] [CCC-5] [HS-PS3-1]. 95 101

☐ 4 What is gravitational potential energy and explain the relationship between the E_p and E_k of a falling object. Recall the first law of thermodynamics. Why is it also called the law of conservation of energy? Identify and explain energy transformations [CCC-5] occurring during simple investigations involving dynamic carts moving down a ramp [SEP-3]. Develop and use a model [SEP-2] to show that the sum of kinetic and potential energy of component particles [CCC-4] must equal the total bulk energy measured at the macroscopic level [SEP-5] [PS3.A] PS3.B] [CCC-5] [HS-PS3-2]. 96 97

☐ 5 Develop your understanding of energy transformations further [PS3.A] [CCC-5] by using a model system [SEP-2] [CCC-4] of a pendulum. Explain why the energy in a pendulum is always the same, no matter where it is in its swing. Use a mathematical model [SEP-2] to track its energy changes as it swings [CCC-5] [HS-PS3-2]. What force is responsible for the oscillations in a pendulum system and how do you calculate it? [SEP-5] What effect [CCC-2] does this force have on the period of the oscillation? 98

What is efficiency and why is it important?

☐ 6 So far in this chapter, we have explored ideal systems [CCC-4]. What does this mean? Why are ideal systems easier to study but, at the same time, not true representations of real systems? What is missing from the systems we have modeled [SEP-2] is the energy that is lost to the environment that is no longer available to use. When we can account for these losses, we can calculate a system's efficiency. On the small scales [CCC-3] of our own investigations, losses to the environment are insignificant. However, when we are comparing outputs from power plants, efficiency becomes very important. Calculate [SEP-5] efficiency of useful work (e.g. electricity generated) for different systems. Can you think of ways to minimize energy losses in different scenarios? [PS3.A] [PS3.B] [CCC-5] 99

Teacher's Notes

7. Definitions of Energy

Anchoring phenomenon

Amusement parks are great places to examine forces, motion, and energy. Rides like roller coasters are excellent examples of where potential energy (at the top of the hill) is converted to kinetic energy (at the bottom of the hill). Many important aspects of energy and efficiency are demonstrated in a roller coaster ride.

What is energy and how is it related to work?

1. Energy is a quantitative property that must be transferred from one object to another in order for work to be done or an object to be heated. Energy can be divided into potential or kinetic energy. Potential energy reflects the ability to do work or have motion and is a function of an object's position in a field (e.g. height from the ground in a gravitational field). Kinetic energy reflects the motion of an object. The greater the motion, the more energy the object has or is able to transfer. It is important to reiterate to students that energy cannot be created or destroyed, but it can be stored, transformed into another form, or transferred between objects.

2. Work is the product of force and displacement. At lower grades, students often use the simplified formula $W=Fd$. This accounts for the fact that students normally calculate work in line with the angle of the force (theta = 0), thus cos theta = 1 and so $W = Fd \times 1$ so $W=Fd$. At high school it is worth extending the model to include the angle at which the force is applied, $W = Fd \cos(\theta)$. This also helps extend student knowledge of forces on an incline (chapter 5).

3. It is important that students realize that because energy is a property of all objects above 0K, and that this energy can be transferred between objects, the equations that describe energy are all interchangeable. Thus $E_p = mgh$ and $W = Fd$ are different forms of essentially the same equation and variables can be substituted where appropriate. $W = Fd$ describes energy in any linear direction. $E_p = mgh$ describes energy in relation to distance above the ground where *m* and *g* combine to produce the force needed to lift an object. Both h and d are measured in meters and so $W = E_p = Fd = mgh$. Furthermore because E_k also describes an aspect of energy, it can be shown that $E_p = W = E_k$. Students should be able to move between these expressions and understand the aspect of energy each describes.

4. In a simple system of an object at a given height, the total energy is equal to $E_p + E_k$. Students should be able to use the formulae for E_k and E_p to calculate the work done when using a falling object to lift another, or to calculate the final velocity of a falling object from any height. It is also instructional to use these equations to show that the final velocity of a falling object (in a vacuum) is independent of its mass (e.g. what is the final velocity of a mass of 2 kg and a mass of 6 kg dropped from 10 m?).

5. Pendulums are a useful way of showing potential energy being converted into kinetic energy and back again. They are also useful for showing the forces involved in the work being done and to introduce oscillating motion and the period of waves, which students will encounter again in a later chapter. Students use their knowledge of energy and forces to describe the motion of a pendulum and demonstrate their knowledge by applying this to calculate the gravitational strength of an unknown planet.

What is efficiency and why is it important?

6. All systems lose energy to their surroundings, usually via heat transfer. This loss of energy describes why a system with no external energy input will eventually cease any motion (e.g. why a swinging pendulum stops swinging). The amount of energy lost from the system during a task is its efficiency. The lower the efficiency, the more energy is being lost from the system to the surroundings. Efficiency is an important aspect of today's world as the more efficient a device is, the less it affects the world's energy demand and thus indirectly the climate. Power plants need to be as efficient as possible to reduce fuel use and thus both running costs and effect on the climate. This is something they will explore in the following chapter.

ADVANCED LEARNING

Natural and man-made debris orbits Earth at speeds averaging 10 km/s

A 7 g object (center) shot at 7 km/s, the orbital velocity of the International Space Station, made this 15 cm crater in a solid block of aluminum.

Challenge question:

Use mathematics to explain why orbital debris, even when very small, is able to badly damage spacecraft and equipment.

Download this page from BIOZONE's Resource Hub

92 The Physics of Amusement Parks

Key Question: How are the relationships between work, kinetic energy, and potential energy applied in amusement park rides?

- Amusement parks are thrilling places to spend free time, but they are also huge physics classrooms. All of the rides are built with the laws of physics in mind and it is playing with these laws that makes these rides so much fun.
- Bumper cars exemplify Newton's three laws of motion that you learned about in Chapter 5, *Forces and Motion*.
- Rides such as the carousel or the "Gravitron" use the centripetal forces you learned about in Chapter 6, *Types of Interactions*.
- Free fall rides demonstrate the relationship between potential energy (specifically gravitational potential energy) and kinetic energy, which you will be exploring in this chapter.
- Roller coasters are perfect for demonstrating all of these laws, forces, and energy together. Well designed roller coasters are not powered by motors for the entire ride. In fact, most roller coasters are only pulled to the top of the first hill (the highest point in the ride). After that, the ride relies only on the potential energy gained due to the position at the top of this hill. Gravity and inertia are felt throughout the ride and influence how you experience the ride. When the ride cars travel up the hills, you feel heavier because your inertia wants you to stay behind. Then, when the car travels down the hills, you feel weightless because you are falling with the car. When loops and twists are built in the track, the track becomes the centripetal force that keeps the cars and passengers moving in circular motion.

1. When a roller coaster car is sitting at the station, it is at rest:

   (a) What is required to get the car moving? *A force. Something to pull it up the first hill.*

   (b) How does this affect the energy of the roller coaster car?

   The roller coaster will gain energy kinetic energy as it moves.

2. How do you think the potential energy of the roller coaster system is affected as it climbs the first hill? Explain your answer:

   The roller coaster will gain potential energy as it climbs the first hill. Energy is supplied by a motor (by doing work on pulling the roller coaster). As the roller coaster climbs higher it gains potential energy, as has further to fall and thus has a greater ability to do work (or gain speed).

3. What do you think happens to this energy as the roller coaster comes down the hill? *The potential energy stored in the roller coaster will be converted to movement (kinetic energy), sound and heat.*

4. The Claw, shown right, is like a huge swinging pendulum. Think of the energy at the end of the machine where the people sit. Describe what happens to this energy as the machine completes one swing back and forth:

   At the top of the swing, the potential energy is at its highest, kinetic energy is at its lowest. At the bottom of the swing the kinetic energy is at its highest, potential energy is at its lowest. Some students may realise that past 90° this does not hold strictly true.

©2020 **BIOZONE** International
ISBN: 978-1-927309-82-7
Photocopying Prohibited

93 Types of Energy

Key Question: What are the two categories used to describe energy and how are they related?

Energy

- Energy is one of the basic concepts in physics (with forces and motion), but it is one of the hardest to define. This is because energy is an abstract notion. In classical physics, there is no physical "essence" of energy and there is no such thing as "pure energy". Energy is always carried by an object in either the form of movement or position.
- You can think of energy as the capacity to cause movement. This is related to the definition of work which we will look at shortly and the reason why we define energy in physics as the capacity to do work.

Forms of energy

- Because energy is an abstract notion, it is found in many forms. Forms of energy include but are not limited to:

Mechanical energy	Energy due to an object's motion or position
Chemical energy	Stored energy released through chemical reactions
Electrical energy	Energy from the flow of electric charge
Thermal energy (heat)	The internal energy of an object due to the motion of its atoms and molecules
Nuclear energy	The energy stored in the nucleus of an atom
Radiant energy (light)	Energy that travels by waves or particles, particularly electromagnetic radiation.

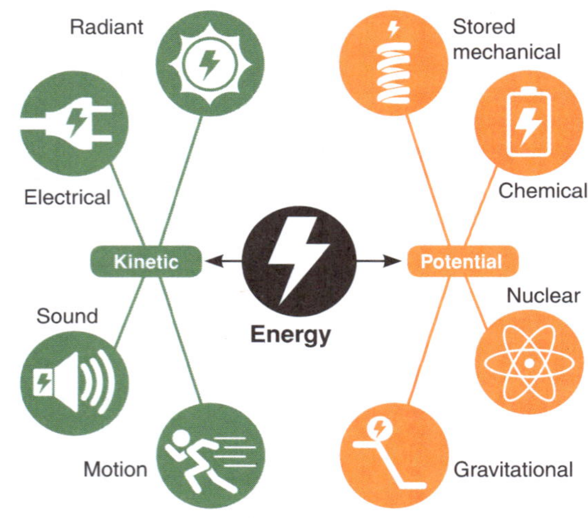

Kinetic vs potential energy

- We can divide all the many forms of energy into two broader types: **kinetic energy** and **potential energy**.
- Kinetic energy is energy due to the motion of an object. When a moving object comes in contact with another, it can transfer some of its energy to that object. A rolling bowling ball has energy, which it transfers to any pins it hits. The faster the ball is rolling, the more energy it has. The mass of the object also affects the energy it carries. A bowling ball has more energy than a softball at the same speed.
- Potential energy is energy an object has due to its position or arrangement. Recall ferromagnetic materials produced magnetic fields when its domains were aligned. The field produced by the specific arrangement of the domains is able to affect other objects. Objects in the field have a magnetic potential energy due to their position in the field. Two types of potential energy we will look at in more detail in this chapter are **gravitational potential energy** and **elastic potential energy**. These are both forms of potential energy due to position of the objects.

1. Write whether each image below shows potential or kinetic energy. Explain why you categorized it as such.

a Ball moving through air

b Single battery cells

c Rocks the at top of a cliff

d Flame from a candle

(a) *Kinetic energy. The ball is in motion and has energy due to the motion (the ball also has some potential energy due to its position above the ground).*

(b) *Potential energy. Battery cells store energy in chemical form for later use.*

(c) *Potential energy. The rocks posses potential energy due to their position above the ground.*

(d) *Kinetic energy. The visible flame is the result of radiant energy (light/ movement of photons). The heat results from radiant energy and the kinetic energy of the molecules.*

©2020 BIOZONE International
ISBN: 978-1-927309-82-7
Photocopying Prohibited

94 The Physics of Work

Key Question: When is work being done?

Work

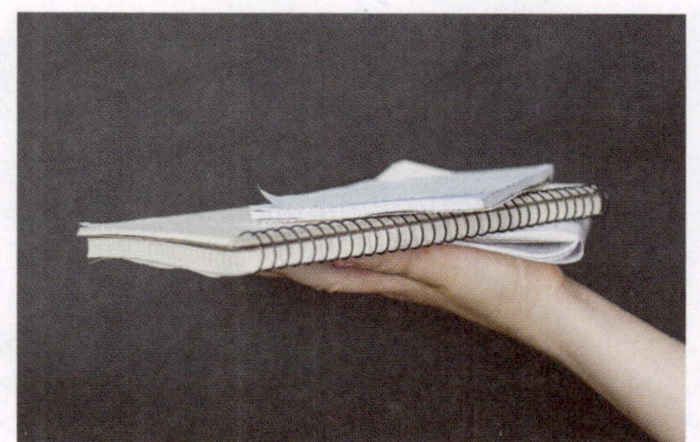

- Pick up your book and hold it steady out in front of you parallel to the ground for 1 minute. How do you feel? Did you apply a force to the book? Was work being done?
- Now, perform biceps curls for 1 minute while holding the book. How do you feel now? Was force applied to the book? Was work being done?
- In both situations, you applied an upward force on the book. When you held the book steady the net force on the book was zero as you balanced the downward pull of gravity to prevent the book from accelerating. When you moved the book during a biceps curl, the book changed velocity so there was a net force acting on the book.
- While you applied force in both situations, work was only done when you moved the book. This is because the force applied to the book caused it to move. In physics, work is done when a force acting on an object causes it to be displaced. Work is expressed as:

$$W = F d \cos \theta$$

- Where W is work, F is force, d is displacement, and θ is the angle between the force and the displacement vector.

- The equation is often simplified to simply $W = Fd$ because displacement is typically in the direction of the applied force. Whenever F and d are in the same direction, θ is 0° and the cos of 0° is 1. However, just like gravity on an incline in Chapter 5, sometimes only part of the force is applied in the direction of displacement.

Negative work

- As you learned in Chapter 5, a force can act on an object to slow it down. In such situations, the applied force doesn't cause the displacement but it does hinder it.
- When force is applied in the opposite direction of the displacement of the object, θ becomes 180°. The cos of 180° is -1. This gives a negative value for work, which means the force was applied in the opposite direction of displacement.
- Even though force and displacement are vectors and work can have a negative value, work is a scalar quantity. Because force and displacement are not necessarily in the same direction (cos θ), work does not have a direction.

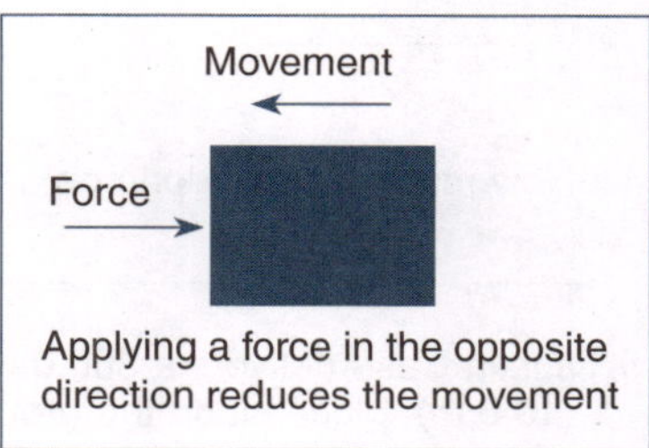

Applying a force in the opposite direction reduces the movement

Units of work

- Force is measured in Newtons (N) or kg m/s^2, displacement is measured in meters (m), and θ is a unitless measure, so work (W) is measured in Newton meters (Nm) or kg m^2/s^2.
- A Newton meter (Nm) (kg m^2/s^2) is the unit for a joule (J). Work is measured in joules.

1. What are the three key aspects of work in physics?
   Force, displacement, and cause (the force must cause displacement)

2. Is work dependent on the distance an object moves or its displacement? Explain your answer:
   Work is a scalar quantity. As such, work does not depend on the displacement but on distance.

3. In the following situations, explain whether or not work is being done:

   (a) A leashed dog strains at the limit of its leash to reach the mailman: *No work is being done. Even though the dog is applying a force to the leash there is no displacement in the direction of the force.*

   (b) A ball is dropped from waist height to the ground:
   Gravity is doing work on the ball causing it to move to the ground.

   (c) A Formula 1 race car accelerates at the start of a race: *A force is applied to the car causing it to accelerate forwards therefore work is being done.*

©2020 BIOZONE International
ISBN: 978-1-927309-82-7
Photocopying Prohibited

Calculating work done by forces

- $W = F\ d \cos\theta$
- It is important to note that the θ in work is different from the θ we used to solve for the component of g acting on an object on an incline in Chapter 5 (right). When calculating the component of g parallel to the incline, we used the angle of the incline as a frame of reference. However, when calculating work, we use the *angle between the force and the displacement vector*. This angle would be the angle between the component of g parallel to the incline and the vector for g, which equals 90° – θ. In both equations we are looking for the component of force acting on the object in the direction of the object's displacement.

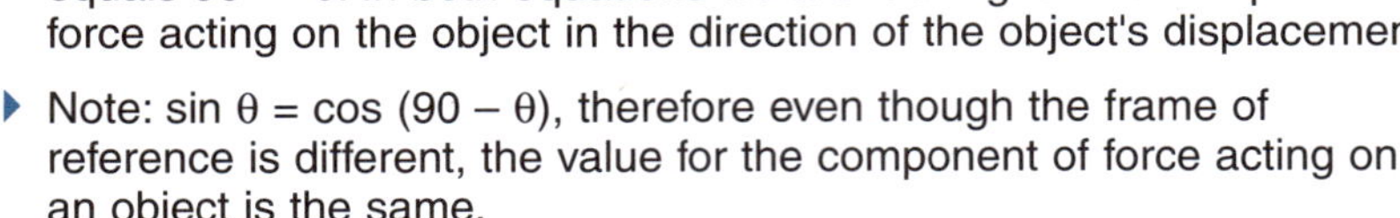

- Note: sin θ = cos (90 – θ), therefore even though the frame of reference is different, the value for the component of force acting on an object is the same.

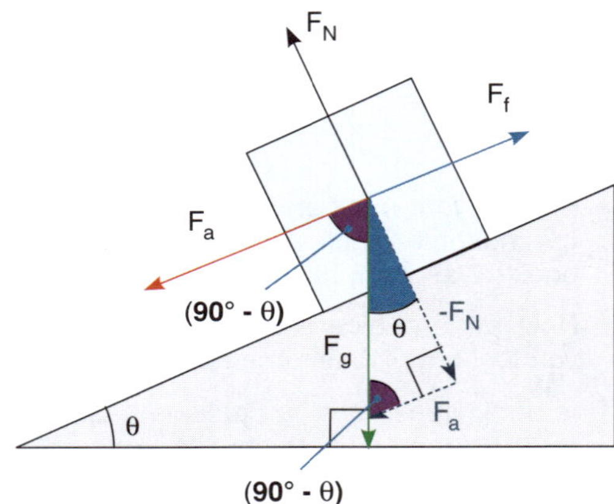

- The diagrams you will create to calculate work will look simpler than the force diagrams you started out using to calculate gravitational components in Chapter 5. You only need to identify the direction of motion, the direction the force was applied on the object and the angle (θ) between these vectors:

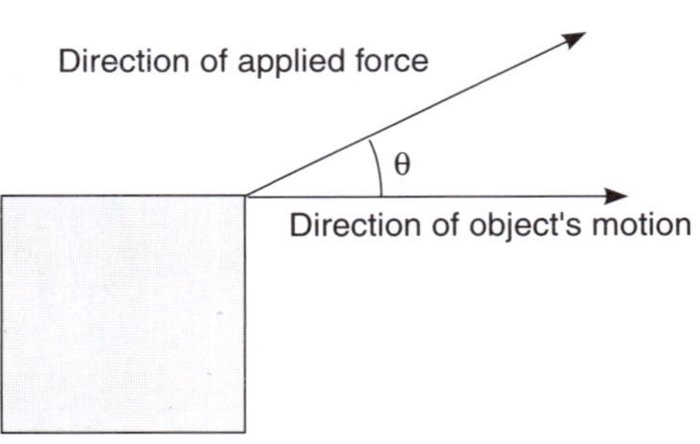

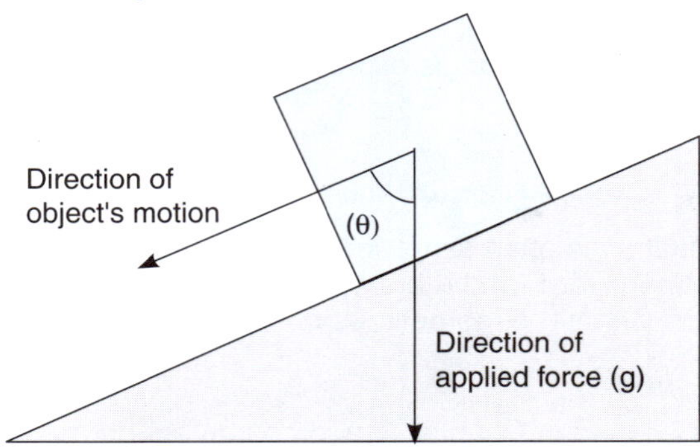

4. How much work is done on a 600 N rock you lift 1 m? *W = Fd = 600 N × 1 m = 600 J.*

5. Justin Case delivers a 300.0 N package to a fourth floor apartment. He carries the package up three flights of stairs (10.0 m), places it on the ground and pushes it with a horizontal force of 40.0 N at a constant speed of 0.5 m/s for 10.0 m to the apartment door. How much work does Justin on the package during this entire time?

   $W_1 = 300\ N \times 10\ m = 3000\ J$. $W_2 = 40\ N \times 10\ m = 400\ J$

   $W_{total} = W_1 + W_2 = 3400\ J$

6. Gina Rodriguez is doing push-ups with her soccer team. If Gina's mass is 45 kg and she raises her center of mass by 30 cm for every push-up, what is the minimum number of push-ups she needs to do to do at least 1300 J of work?

   $W = Fd = (45\ kg \times 9.8\ m/s^2) \times (X \times 0.3\ m) = 1300\ J$.

   Total distance = $1300 / (45\ kg \times 9.8\ m/s^2) = 2.95\ m$ X (total push ups) = $2.95 \div 0.3 = 9.8 = 10$ push ups.

7. You apply a 200 N force to a heavy crate through a rope which makes a 30° angle with the horizontal. If you pull the crate a distance of 10 m, how much work was done? Make a diagram in the space provided to illustrate the problem.

   $W = Fd\cos\theta$

   $= 200\ N \times 10\ m \times \cos 30$

   $= 1732\ J$

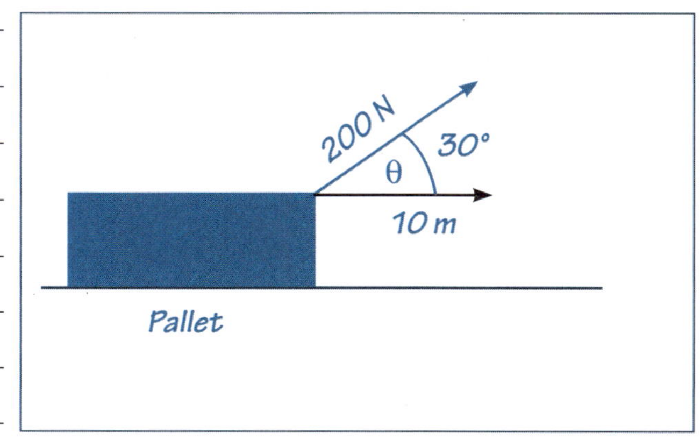

©2020 **BIOZONE** International
ISBN: 978-1-927309-82-7
Photocopying Prohibited

95 The Work - Kinetic Energy Theorem

Key Question: What is the relationship between work and energy?

Kinetic energy

- **Kinetic energy** is the energy due to motion. An object in motion has energy due to its movement. If a vehicle were to collide with a stationary object (right) then some of the original kinetic energy would become kinetic energy in the second object as well as heat and sound energy in the resulting crash.
- The amount of kinetic energy is dependent on the mass of the object and the speed of the object. The larger and faster the object is, the more energy it has.
- Kinetic energy is proportional to the square of the speed of the object and can be represented by the equation:

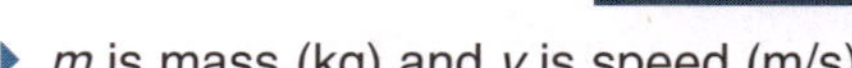

$$E_k = \frac{1}{2} m v^2$$

- m is mass (kg) and v is speed (m/s)
- Energy is a scalar quantity. It has no direction and is described by magnitude alone.

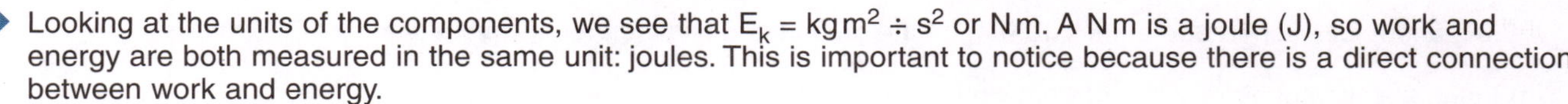

- Looking at the units of the components, we see that E_k = kg m^2 ÷ s^2 or N m. A N m is a joule (J), so work and energy are both measured in the same unit: joules. This is important to notice because there is a direct connection between work and energy.

1. A 1500 kg car has a speed of 2.0 m/s.

   (a) What is the kinetic energy of the car? $E_k = 0.5 \times 1500\text{ kg} \times (2.0\text{ m/s})^2 = 3000\text{ J}$

   (b) What would be the speed of a 5 kg bowling ball if it had the same kinetic energy as the car?

   $v = \sqrt{(2E_k \div m)}$ (OR $\sqrt{(E_k \div 0.5m)}$) $= \sqrt{(2 \times 3000\text{ J} \div 20\text{ kg})} = \sqrt{1200\text{ kg m}^2/\text{s}^2} = 34.6\text{ m/s}$

2. A boy pushes with a horizontal force of 400 N on a 20 kg box at rest on a level, slightly rough floor.

   (a) How much work has he done by the time the box traveled 10 m? $W = Fd = 400\text{ N} \times 10\text{ m} = 4000\text{ J}$

   (b) What happens to the work done by the boy? Work is converted to kinetic energy in the motion of the box.

   (c) What theoretical speed will the box have when it has traveled 10 m? $v = \sqrt{(2E_k \div m)}$

   $= \sqrt{(2 \times 4000\text{ J} \div 20\text{ kg})} = 20\text{ m/s}$

   (d) The box only reaches a speed of 15 m/s. Explain: Friction is a force that opposes motion. Friction did negative work on the box converting some work into thermal energy instead of mechanical.

3. A 1000 kg car travelling at 45 mph (20 m/s) accelerates to 65 mph (29 m/s).

   (a) What is the change in kinetic energy of the car? $\Delta E_K = E_{Kf} - E_{Ki} = 0.5\,mv_f^2 - 0.5mv_i^2$

   $= 0.5m\,(v_f^2 - v_i^2) = 0.5 \times 1000\text{ kg}\,((29\text{ m/s})^2 - (20\text{ m/s})^2) = 220{,}500\text{ J}$

   (b) The car then decelerates back to 45 mph. What is the change in kinetic energy?

   $\Delta E_K = 0.5 \times 1000\text{ kg}\ 0.5 \times 1000\text{ kg}\,((20\text{ m/s})^2 - (29\text{ m/s})^2) = -220{,}500\text{ J}$

   (c) Was work done on the car during this change in kinetic energy? Explain:

   Work was done by the engine on the car to increase its speed (kinetic energy) and then was done by the brakes on the car to decrease its speed.

©2020 **BIOZONE** International
ISBN: 978-1-927309-82-7
Photocopying Prohibited

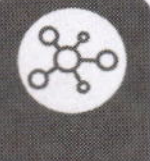

Work-energy relationship

- Recall W = F d cos θ. Whenever the object moves in the direction of the applied force, θ = 0. Since cos θ is a unitless value and cos 0 = 1, we usually simplify the equation to W = Fd.
- Also recall: F = *ma*. Substituting *m a* for F in our work formula gives us W = *m*a*d* (which is how work may make you feel).
- In Chapter 5, you derived a formula for displacement when time (Δt) was unknown: $\Delta x = (v_f^2 - v_i^2) \div 2a$.
  If we rearrange this equation to solve for a, we get: $a = (v_f^2 - v_i^2) \div 2\Delta x$. Substituting this value of a into our work formula and changing Δx to d Since Δx is distance travelled, we get $W = m\,d\,(v_f^2 - v_i^2) \div 2d$. Distance cancels out leaving us with $W = m\,(v_f^2 - v_i^2)\,½$.
- Using distribution, we get: $W = ½\,m v_f^2 - ½\,m v_i^2$
- $E_k = ½\,mv^2$, thus work is the change in kinetic energy of a system:

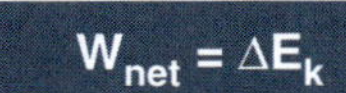

$W_{net} = \Delta E_k$

- Think of a car moving at a certain speed. It already has kinetic energy due to its mass and movement. If we were to speed the car up (accelerate) we would need to apply a force. Applying a force requires work to be done leading to a transfer of energy to the car. The car will now be moving at a new speed and have a new value for kinetic energy. The difference between the kinetic energy of the car before and after the force was applied is the work done on the car.
- This works whether energy is taken out of or added to the system. If a car were to slow down, the energy of the car would decrease. This will give a negative value for work simply because the work was done in the opposite direction to the car's motion.

4. A 7.0 g acorn falls from a tree branch 10.0 m off the ground and hits the ground with a speed of 10.5 m/s.

   (a) In the absence of air resistance, what would the speed of the acorn have been? $v_f = \sqrt{(2 \times a \times \Delta x)}$
   $= \sqrt{(2 \times 9.8\ m/s2 \times 10\ m)} = 14\ m/s$

   (b) Did air resistance do positive, negative, or no work on the acorn? Explain your answer: Air resistance did negative work on the acorn, opposing its motion to slow it down.

   (c) Prove your answer from (b) using W = Fdcosθ: $W_{air} = F_{air\ resist} \times d\cos\theta = F d\cos 180$. θ is 180° because displacement is in the opposite direction of the force of air resistance. Cos 180 = -1

   (d) How much work was done by the air resistance? $W_{net} = W_g + W_{air}$. $W_{air} = W_{net} - W_g$
   $= 0.5 \times 0.007\ kg \times (10 m/s)^2 - 0.5(0.007\ kg \times (14\ m/s)^2 = 0.386\ J - 0.686 = -0.3\ J$

   (e) What was the average force of the air resistance? W = Fd. F = w ÷ d = 0.3 J ÷ 10 m = -0.03 N

5. A 1000 kg car is moving at a constant speed of 30 m/s.

   (a) Determine the kinetic energy of the car: $E_K = 0.5\,mv^2 = = 0.5 \times 1000\ kg\ (30\ m/s)^2 = 450{,}000\ J$

   (b) What force is required to stop the car in 1000 m? W = Fd. F = W ÷ d = -450,000 ÷ 1000 m = -450 N

   (c) What force is required to stop the car in 100 m? F = w ÷ d = -450,000 J ÷ 100 m = -4500 N

   (d) Determine the work done in both (b) and (c): W_b = -450 N x 1000 m = -450,000 J
   W_c = -4500 N x 100 m = 450,000 J

6. A 20,000 kg truck is travelling at a constant speed of 65 miles per hour (approximately 30 m/s). In 100 m, the speed limit changes to 45 mph (approximately 20 m/s). The driver applies the brakes to slow the truck to this new speed.

   (a) Determine the kinetic energy of the truck before and after applying the brakes:
   E_{Ki} = 0.5 x 20,000 x (30 m/s)2 = 9,000,000 J
   E_{Kf} = 0.5 x 20,000 x (20 m/s)2 = 4,000,000 J

   (b) How much work is done by the brakes to slow the truck to the new speed? $W = \Delta E_K = E_{Kf} - E_{Ki}$
   = 4,000,000 J - 9,000,000 J = -5,000,000 J

©2020 **BIOZONE** International
ISBN: 978-1-927309-82-7
Photocopying Prohibited

96 Stored Energy

Key Question: In what ways can energy be stored and how can we calculate potential energy?

Potential energy

- **Potential energy** (E_p) is stored energy which has the *potential* to be used to do work.
- Chemical and nuclear energy are examples of potential energy. Chemical energy is stored in the bonds between atoms and can be released during a chemical reaction such as respiration where large molecules such as sugars are converted to smaller ones such as CO_2 and water. Nuclear energy is stored in the bonds between nucleons (protons and neutrons) in the nucleus of an atom. These bonds are very stable but when broken can release many times the amount of energy found in chemical bonds.
- Potential energy can also be seen in electromagnetism. A steel bar and a magnet are being held apart. If the bar is released, then the potential energy is converted to kinetic energy as the magnet and steel bar come together and becomes sound and heat energy when they join.
- In this chapter we will work with **gravitational potential energy** (E_g). You will be introduced to **elastic potential energy** (E_e) in a later chapter.

The energy released in a nuclear explosion is many magnitudes greater than a chemical explosion.

Gravitational potential energy

- The energy stored due to an object's position above the ground is called gravitational potential energy (E_g).
- Objects on top of a desk have greater gravitational potential energy than objects on the ground. If the desk were to disappear, the object would fall converting the potential energy into kinetic energy and then heat and sound energy.
- Gravitational potential energy is calculated using the equation: $E_g = mgh$ where *m* is the mass of the object, g is acceleration due to gravity and h is the height of the object.

1. (a) A 200 kg rock at the top of a cliff breaks loose and falls 55 m to the ground, where it shatters. How much work did the Earth do on the rock (remember g = 9.8 m/s^2)?

   $W = Fd = 200\ kg \times 9.8\ m/s^2 \times 55\ m = 107{,}800\ J$

   (b) Where did the energy for the work come from? *The rock possessed the potential energy due to its position.*

   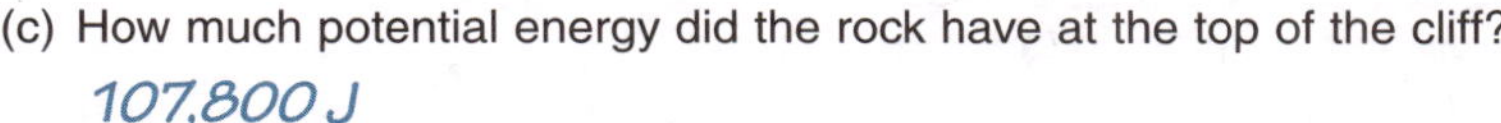

   (c) How much potential energy did the rock have at the top of the cliff?

   107,800 J

   (d) How much potential energy did the rock have at the bottom of the cliff? *0 J*

   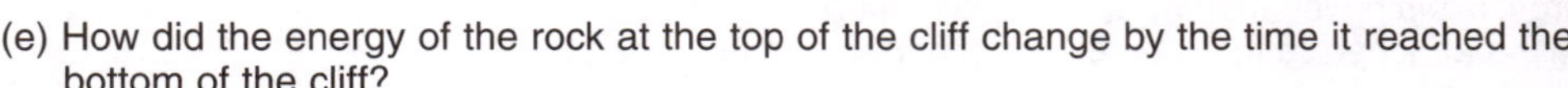

   (e) How did the energy of the rock at the top of the cliff change by the time it reached the bottom of the cliff?

   By the bottom of the cliff all of the potential energy had converted to kinetic energy

   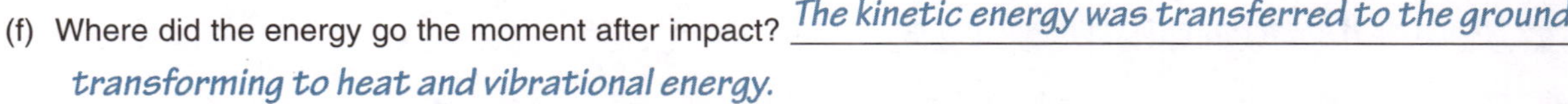

   (f) Where did the energy go the moment after impact? *The kinetic energy was transferred to the ground transforming to heat and vibrational energy.*

   (g) Calculate the speed of the rock at the bottom of the cliff: *$v_f = \sqrt{(2 \times a \times \Delta x)}$*

   $= \sqrt{(2 \times 9.8\ m/s^2 \times 55\ m)} = 32.8\ m/s$

   (h) Explain the relationship between the potential energy of the rock at the top of the cliff and its kinetic energy immediately before hitting the ground:

   At the top of the cliff the rock had all the potential energy. As the rock fell, gravity did work on the rock, converting the potential energy to kinetic energy. By the time the rock hit the ground, all the potential energy had converted to kinetic energy. The total energy of the rock remained the same throughout the fall so that at any point $E_{total} = E_P + E_K$

©2020 BIOZONE International
ISBN: 978-1-927309-82-7
Photocopying Prohibited

97 Conservation of Energy

Key Question: How does the law of conservation of energy explain changes in the energy that is available to do work?

PS3-2 Conservation of energy

- Can energy disappear? No. Energy lost in a system goes to the surrounding environment. A car (system) loses energy due to air resistance. This means air molecules (surrounding environment) gain energy and move faster. Energy is conserved because the amount of energy lost by the system is the same amount of energy gained by the surrounding environment.
- Energy cannot be created or destroyed but it can be changed from one form to another and transferred from one system to another. This is known as the law of conservation of energy (also the **first law of thermodynamics**).
- When work is done, the energy of the system changes. If work is done on a system, energy is added. If the system does work on the environment, energy goes out of the system. Suppose you were bowling and the ball slipped off your fingers and landed on your toes. The bowling ball had energy due to its position (gravitational potential energy), which converted to kinetic energy while falling. When the bowling ball lands on your toe, energy from the fall is transferred from the ball to your toe. The ball loses energy and your toe gains energy.
- In systems, energy is not perfectly conserved. Energy is always being lost to the environment, often due to frictional components. In many situations, we assume the energy in systems is conserved under specific situations, e. g. dense objects in free fall for short distances or balls rolling down smooth inclines.
- In such cases, we are able to equate work done by a system as equal to the energy used: $W = E_{initial} - E_{final}$
- Conversely, the work done on a system is equal to the energy added to the system: $W = E_{final} - E_{initial}$

Interconversion of mechanical energy

- If friction is negligible, then the total mechanical energy ($E_p + E_k$) of a system is constant. This means any change in E_p is equal and opposite to any change in E_k.
- In the diagram right, as the 20 kg mass falls, it loses E_g and gains E_k. However, some of the E_g the 20 kg mass loses also goes into a gain of E_g and E_k for the 10 kg mass.
- So: $E_{g(\text{lost by 20 kg})} = E_{k(\text{gained by 20 kg})} + E_{g(\text{gained by 10 kg})} + E_{k(\text{gained by 10 kg})}$
- Then: $E_{g(\text{lost by 20 kg})} - E_{g(\text{gained by 10 kg})} = E_{k(\text{gained by 20 kg})} + E_{k(\text{gained by 20 kg})}$
- So, in this case: $E_{p(\text{lost})} = E_{k(\text{gained})}$

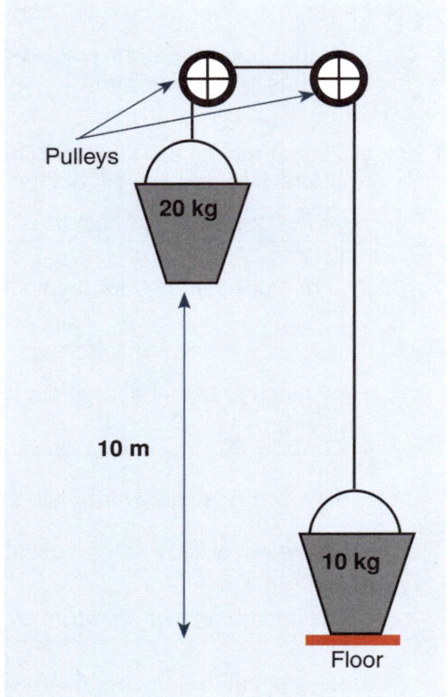

1. (a) Looking at the system on the side of the smaller mass, the smaller mass has no energy before the larger mass starts falling but gains potential energy. Where did the energy come from?

   From the work done by the larger mass to convert its E_k to E_g.

   (b) In terms of work and energy, explain why the larger mass falls more slowly when there is a smaller mass attached to the rope on the other side of the pulley than if it were to fall on its own:

   The larger mass does work on the smaller mass converting its kinetic energy to gravitational potential energy in the smaller mass. Less kinetic energy means the larger mass falls more slowly.

   (c) Calculate the gravitational potential energy of the 20 kg mass in the diagram: *$E_g = mgh$*

   $= 20\ kg \times 9.8\ m/s^2 \times 10\ m = 1960\ J$

   (d) Calculate the gravitational potential energy gained by the 10 kg mass when the 20 kg mass reaches the ground:

   $E_g = 10\ kg \times 9.8\ m/s^2 \times 10\ m = 980\ J$

   (e) Calculate the speed of the 20 kg mass the instant before it hits the ground:

   $W = E_i - E_f = 1960\ J - 980\ J = 980\ J$ $W = 0.5mv^2$ $v = \sqrt{(2W \div m)} = \sqrt{(2 \times 980 \div 20)} = 9.9\ m/s$

©2020 BIOZONE International
ISBN: 978-1-927309-82-7
Photocopying Prohibited

INVESTIGATION 7.1: Investigating work and energy

See appendix for equipment list.

1. Set up a ramp, measure the vertical distance to the top of the ramp, and record this height in Table 1.
2. Measure the mass of your dynamic cart and record in Table 1 below.
3. Calculate the change in potential energy in the cart if lifted to the height of the ramp. Record this value in Table 1.
4. Measure and mark a distance of 1 m from the end of your ramp.
5. Let your cart roll down the ramp and time how long it takes the cart to go from the **bottom of the ramp to the 1 m mark**. Run three trials and record the average time.
6. Calculate the average velocity and kinetic energy of your cart and record on the data table below.
7. The ratio of the kinetic energy gained to the work put in to lifting the cart (the potential energy gained) is our system efficiency. Use your calculations of kinetic energy and work (potential energy gained) to calculate efficiency and record it on Table 2.

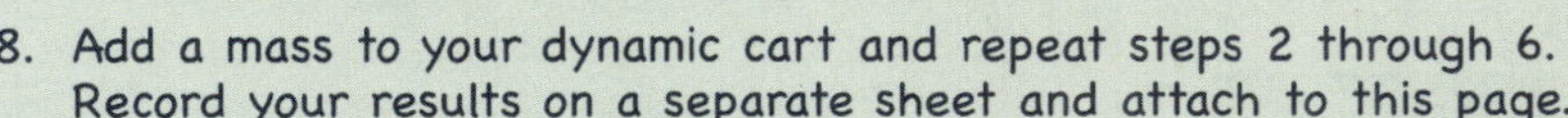

8. Add a mass to your dynamic cart and repeat steps 2 through 6. Record your results on a separate sheet and attach to this page.

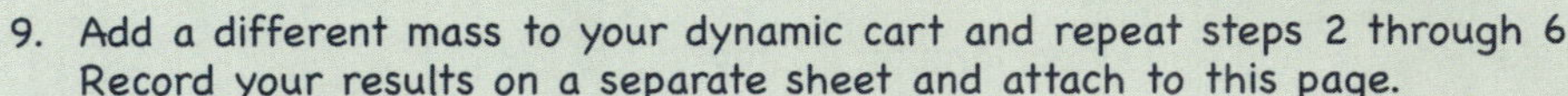

9. Add a different mass to your dynamic cart and repeat steps 2 through 6. Record your results on a separate sheet and attach to this page.

Table 1

Mass of cart	Height of ramp	Change in E_p
1 kg	0.2 m	1.96 J

Table 2

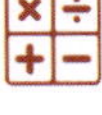

Mass of cart	Time to travel 1m	Average time	Average velocity	E_k	Efficiency
1 kg	0.5 s				
1 kg	0.5 s				
1kg	0.5 s	0.5 s	1.9 m/s	1.8	92%

2. Explain the relationship between work done and mass:
Work done is directly proportional to mass. The greater the mass the greater the work done.

3. Explain the energy transformations throughout the investigation (be sure to identify when work is being done):
Work was done to lift the cart up against gravity to the top of the ramp and add gravitational potential energy to the cart. When the cart was released at the top of the ramp, gravity did work on the cart to convert the gravitational potential energy to kinetic energy. At the bottom of the ramp friction does work on the cart to slow it down, converting kinetic energy into heat energy.

4. Explain why all the work done at the beginning of the investigation did not convert into kinetic energy (where did the missing energy go?):
Air resistance and friction did work on the cart throughout the investigation. As a result some energy was lost as thermal (heat) and vibrational (sound) energy.

©2020 **BIOZONE** International
ISBN: 978-1-927309-82-7
Photocopying Prohibited

98 Pendulums

Key Question: How is energy conserved in a pendulum?

PS3-2 The simple pendulum

- A simple pendulum consists of a mass (the bob) suspended from a fixed string that can move back and forth (oscillate) when displaced from its rest position.
- Pendulums are a useful tool for displaying the conversion of energy and the effect of forces on motion. Consider the diagram of the pendulum to the right:
- The potential energy in the bob at point A is at its maximum. As the bob moves to point B, it loses potential energy and gains kinetic energy. At point B, the kinetic energy of the bob is at is maximum and potential energy is at its minimum. As the bob moves to point C, kinetic energy is converted to potential energy.

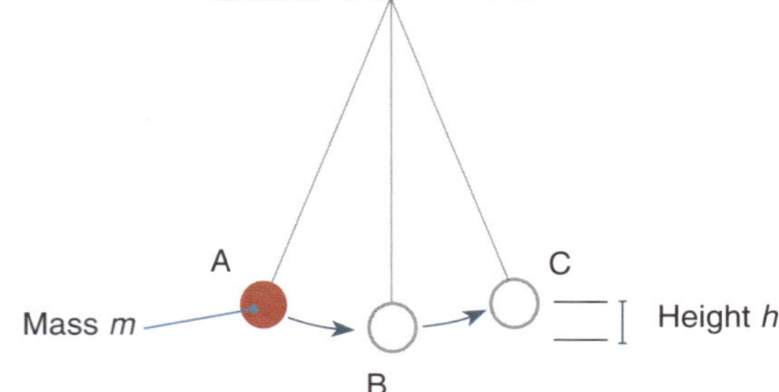

Energy in a simple pendulum

- The energy in a pendulum is always the same no matter where it is in its swing because as it loses potential energy it gains kinetic energy and vice versa. So we can express the energy in the system using the following equation:

$$E_{total} = E_p + E_k$$

- We know from earlier that potential energy = mgh and kinetic energy = ½mv². We can then substitute this into the equation:

$$E_{total} = mgh + \frac{1}{2}mv^2$$

- This equation makes it possible to track the energy changes of a pendulum as it swings. From there, other properties of the system can be followed as well. It is important to note that no system is truly isolated - there is always energy exchange with the environment. Friction is also a factor and causes energy to be lost from the system as heat. However, for small intervals of time, this equation for an ideal system approximates reality quite well.

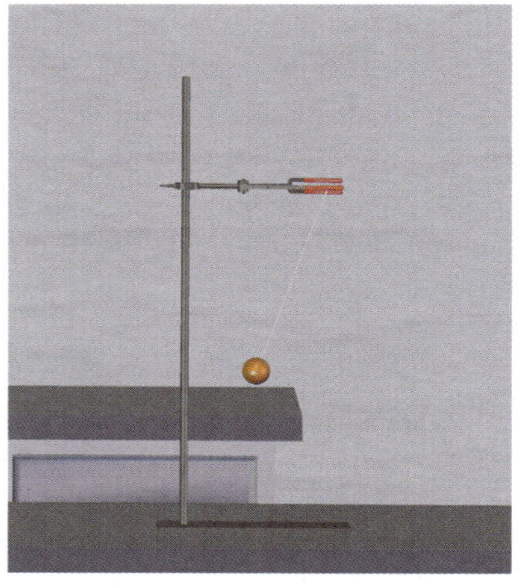

1. A pendulum with a bob mass of 0.25 kg is raised from its equilibrium position by a distance of 0.2 m. It swings with no noticeable friction. Complete the tables below to show the properties of the pendulum system as it swings:

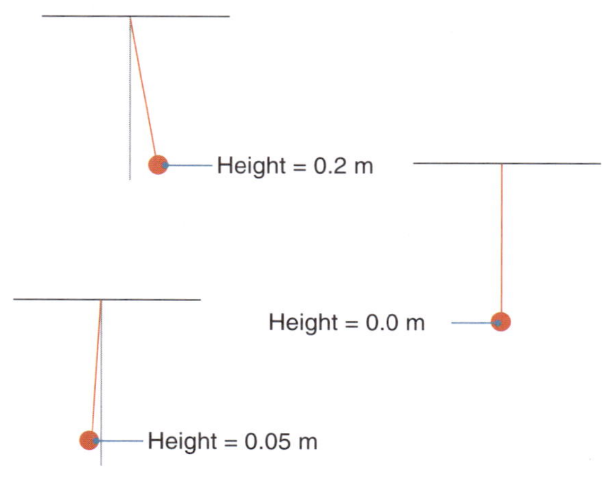

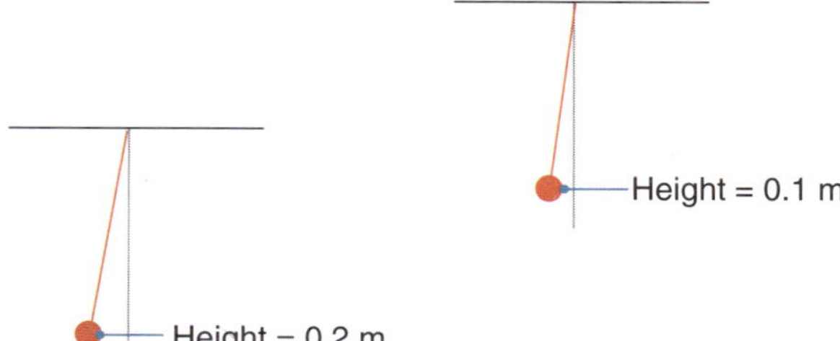

Height (m)	E_k (J)	E_p (J)	E_{total} (J)	Velocity of bob (m/s)
0.2	0	0.49	0.49	0

Height (m)	E_k (J)	E_p (J)	E_{total} (J)	Velocity of bob (m/s)
0.0	0.49	0	0.49	1.98

Height (m)	E_k (J)	E_p (J)	E_{total} (J)	Velocity of bob (m/s)
0.05	0.3675	0.1225	0.49	1.71

Height (m)	E_k (J)	E_p (J)	E_{total} (J)	Velocity of bob (m/s)
0.1	0.245	0.245	0.49	1.4

Height (m)	E_k (J)	E_p (J)	E_{total} (J)	Velocity of bob (m/s)
0.2	0	0.49	0.49	0

©2020 BIOZONE International
ISBN: 978-1-927309-82-7
Photocopying Prohibited

Forces in a simple pendulum

- All systems possess a state of lowest potential energy. This state is called **equilibrium**.
- To move a system out of equilibrium a force must be applied to the system to do work on it. Once work has been done on the system it has potential energy due to its new position outside of equilibrium.
- When a system is out of equilibrium, there is a force associated with returning the system to its rest state. This force is called the **restoring force**.
- In oscillating systems such as pendulums, the restoring force is the force responsible for the oscillations. Looking at the image to the right, the only force responsible for the oscillating motion of the pendulum is the *x*-component of the weight. Thus, the restoring force on the pendulum is:

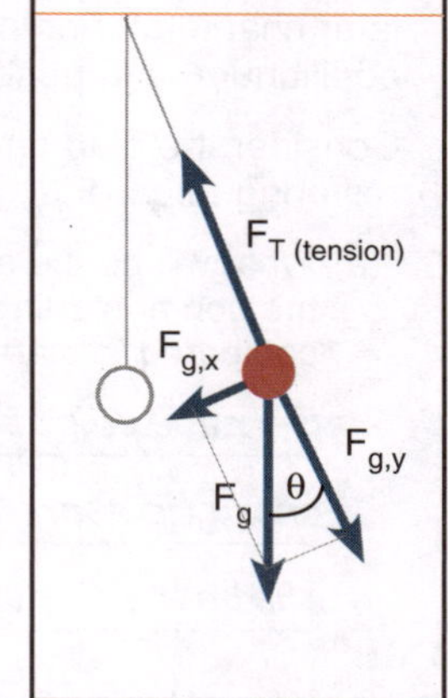

$$F_{g,x} = -mg \sin \theta$$

where θ is the angle between the bob at equilibrium and maximum displacement

Note: the negative sign indicates the downward acceleration due to gravity.

2. A magnetic field produces a force affecting magnetic objects in the field. The strength of this force is conditional on the position of the objects within the field. Is this magnetic force a restorative force? Explain your reasoning:

The force produced due to the position of an object in a magnetic field is a restorative force. This force is due to magnetic attraction/repulsion between the object and the source of the magnetic field. This force acts on the object to return the system to its equilibrium or lower energy state. As such it is a restorative force.

3. A gravitational field produces an attractive force on all objects with mass.

(a) Explain why gravitational force is a restoring force: *Gravity is an attractive property between objects with mass. Energy is required to separate masses. Without added energy, objects will return to their lowest energy states. In the case of gravity, this means falling towards the center of mass of a massive object. Thus gravitational force is restorative.*

(b) Explain how the restoring force inherent in gravity is related to the restoring force in a pendulum system: *Gravity is the restoring force in a pendulum system. Because of the tension force in the string only a portion of gravitational force restores the system to equilibrium. This portion is the y-component of gravity from the angle of tension in the pendulum system.*

(c) How would a pendulum system act differently in the absence of a gravitational field? *Without gravitational force, there would be no restorative force. As such, there would be no change in the movement of the pendulum. If a force were applied to the bob, it would continuously swing around the anchor point in a circular orbit.*

4. Consider a pendulum system where the maximum displacement of the bob gives θ = 30°.

(a) Calculate the restorative force of the system if the mass of the bob is 1 kg: *$F = -mg \sin\theta$.*
$= -1 \text{ kg} \times 9.8 \text{ m/s}^2 \sin 30 = -4.9 \text{ N}$

(b) Calculate the restorative force of the system if the bob is replaced with a 2 kg bob:
$-2 \text{ kg} \times 9.8 \text{ m/s}^2 \sin 30 = -9.8 \text{ N}$

(c) Compare the restorative force between the two masses in the system. How does mass affect the strength of the restorative force of a pendulum?
As the mass increases, the restoring force also increases.

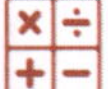

5. Calculate the force of gravity of an unknown planet where the known restorative force of a pendulum system with a 1 kg bob with a maximum angular displacement of 30° is 10 N:
$F = -mg \sin\theta$. $mg = F \div \sin 30 = 10 \div 0.5 = 20$. $g = 20 \div 1 \text{ kg} = 20 \text{ m/s}^2$

©2020 **BIOZONE** International
ISBN: 978-1-927309-82-7
Photocopying Prohibited

Period of a pendulum

- Oscillations in a pendulum follow properties of waves (we will explore waves in later chapters). For instance, they have amplitude (A), which is the height above equilibrium, and period (T), which is the time to complete a full back and forth swing.
- Consider the diagram right. The higher the pendulum bob is lifted, the greater the restoring force ($F_{g,x}$) and the potential energy in the bob will be.

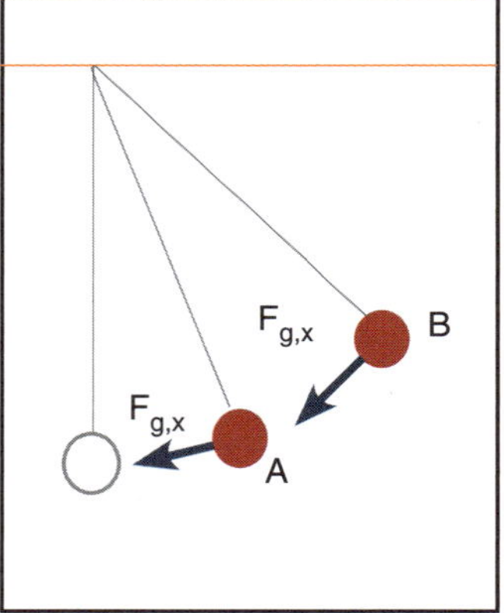

6. (a) What will be the effect of a greater restoring force or greater potential energy in the bob at starting height B on the velocity of the bob at the bottom of the swing compared to starting height A?

   Because point B is higher than A, a greater force is involved and more energy is stored. This will result in a greater velocity (speed) of the bob at the bottom of the swing.

   (b) How will this affect the time it takes the bob to complete a swing from height B and back compared to swinging from height A and back?

   The restoring force will have no effect on the time it takes for a swing back and forth (the period). The restoring force is less at A than at B. This means the bob will accelerate more slowly at A than at B and so reach a slower maximum speed, thus taking as long as B for a complete swing.

   (c) What is the effect of starting height on the time taken for a swing? *The time it takes to swing back and forth (the period) will not be affected by the height of the swing. Both A and B will have the same average speed. B will accelerate more quickly than A and so cover the greater distance in the same time as A covers the shorter distance.*

- Based on this, the equation for period we are left with is:

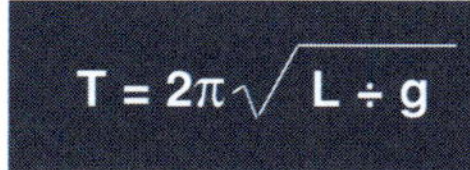

$$T = 2\pi\sqrt{L \div g}$$

- T is the period of the pendulum, L is the length of the pendulum, and *g* is the acceleration due to gravity.

7. Based on your answers from question 6, explain why the equation for the period of a pendulum does not include mass or displacement:

   These two variables do not affect period and so do not need to be included in the equation

8. A 1 kg mass is oscillating at a small angle on an unknown planet from a 0.2 m light string with a period of 0.6 s.

   (a) What is the acceleration due to gravity on this unknown planet?

   $T = 2\pi\sqrt{(L \div g)}$. $\sqrt{(L \div g)} = T \div 2\pi = 0.6\text{ s} \div 2\pi = 0.0955$. $L \div g = 0.0955^2 = 0.00912$

   $g = 0.2\text{ m} \div 0.00912 = 21.9\text{ m/s}^2$

   (b) What is the frequency of the pendulum? $f = 1 \div T = 1 \div 0.6\text{ s} = 1.67\text{ Hz}$

9. A 0.5 kg mass oscillates at a small angle from a light string with a period of 0.63 s. What is the length of the pendulum?

   $T = 2\pi\sqrt{(L \div g)}$ $0.63\text{ s} = 2\pi\sqrt{(L \div 9.8\text{ m/s}^2)}$. $L \div 9.8\text{ m/s}^2 = (0.63\text{ s} \div 2\pi)^2 = 0.01$.

   $L = 9.8\text{ m/s}^2 \times 0.01 = 0.098\text{ m} \sim 0.1\text{ m}$

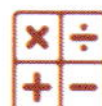

10. Determine the restorative force for a pendulum with a bob mass of 10 kg with an angular displacement of 15°:

   $F = -mg\sin\theta = -10\text{ kg} \times 9.8\text{ m/s}^2 \sin 15 = -25.4\text{ N}$

©2020 **BIOZONE** International
ISBN: 978-1-927309-82-7
Photocopying Prohibited

99 Efficiency

Key Question: Why is there never as much energy after transformation as before? Where did it go?

Work, heat, and efficiency

- Throughout this chapter we have been dealing with ideal systems - systems in which the energy output is equal to the energy input. In reality, there is no such thing as an ideal system.
- In most of the situations you have dealt with in this book, the scenarios presented are as close to ideal as possible, e.g. rolling balls down a smooth incline and high density objects free-falling over short distances. In such situations the energy lost by the system to the environment is insignificant.
- In every energy conversion, there is always energy lost to the environment that is no longer available to use. This lost energy is typically lost as thermal energy (heat).
- This is where the concept of efficiency comes in. Efficiency is a ratio of the work done (output) to energy input. The less energy lost to the system, the higher the efficiency.

Car engines lose energy via heat. Failure of the cooling system can cause catastrophic overheating of the engine.

Efficiency in power plants

- A consequence of the law of conservation of energy is that energy is conserved in a closed system. This means we cannot create or destroy energy, but it can change form.
- Power stations are good examples to demonstrate the concept of efficiency, but the same concept applies to all systems.
- Thermal power plants use the heat energy from the combustion of fuel to heat water and make steam. Nuclear power stations use heat from a nuclear fission reaction to heat water and make steam. The force of the steam on a turbine does work on a generator and produces electrical energy (electricity). Throughout this process some of the energy that originally came from the fuel is always lost to the surroundings, directly as heat energy, and also as a result of heat caused by frictional forces.
- The efficiency of a power plant can given as a percentage by the following calculation:

$$\text{Efficiency (\%)} = \frac{\text{useful work done (electrical energy generated)}}{\text{energy input from original source (fuel)}} \times 100$$

Using the diagram right: $\text{Efficiency} = \frac{6.0 \times 10^{11}}{1.5 \times 10^{12}} \times 100 = 40\%$

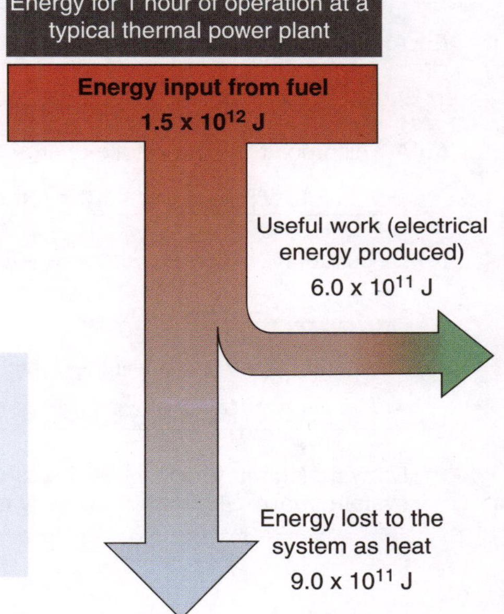

1. Assume that a power plant produces 470 kJ of electrical energy when a mole (16 g) of methane is burned. Burning one mole of methane produces 814 kJ of (total) energy.

   (a) What is the efficiency of the process? Efficiency = 470 kJ ÷ 814 kJ x 100 = 57.7%

   (b) How much heat is produced? 814 kJ - 470 kJ = 344 kJ.

2. Different parts of a power plant have different efficiencies of energy transformation. The boiler efficiency is about 90%, the turbine is about 75% efficient, and the electrical generator is about 95% efficient.

   (a) A fuel source in a power plant produces 5000 kJ of energy per kg. Use the information above to calculate the amount of work the fuel will produce:

   5,000 kJ x 0.9 x 0.75 x 0.95 = 3,206.25 kJ.

   (b) What is the overall efficiency of the power plant? 3,206.25 kJ ÷ 5,000 kJ x 100% = 64.125%.

3. Identify at least three ways in which energy is lost to the environment from a system: Directly as heat from combustion, as heat due to friction and air resistance, as vibrational energy such as sound heat, strain etc

©2020 **BIOZONE** International
ISBN: 978-1-927309-82-7
Photocopying Prohibited

EM SSM PS3.A

100 Review Your Understanding

Key Question: How are the relationships between work, kinetic energy, and potential energy utilized in amusement park rides?

- Amusement parks are based on the laws of physics. The concepts you have leaned in this chapter can help to explain the way in which many amusement park rides operate.
- Some roller coasters operate along a loop, moving in one direction at all times. Others move along a straight track, moving forwards then backwards (or the other way around). The image (right) shows one of these types of roller coasters:

Jeremy Thompson CC 2.0

1. The vehicle for the type of roller coaster shown on the right has a mass of 5000 kg and accelerates to 160 kilometers per hour on the level track before hurtling up the vertical track under its own momentum. It eventually comes to a halt before falling back down the track.

   (a) Describe the changes in energy as the vehicle moves along the track out of the station and back:

   Kinetic energy is added to the vehicle from electric motors. Kinetic energy is converted to gravitational potential energy as the vehicle rolls up the tower. As it falls back down, the Gp energy is converted back into kinetic energy. Kinetic energy is converted to heat when the brakes are applied.

   (b) At which place(s) along the track will the vehicle have no kinetic energy?

   At the top of the tower and at the starting point in the station.

   (c) A which point will the kinetic energy of the vehicle be at its maximum?

   At the bottom of the tower (on the flat)

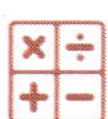

   (d) Calculate the maximum kinetic energy of the vehicle: $E_K = 0.5mv^2 = 0.5 \times 5000 \text{ kg} \times 44.4 \text{ (m/s)}^2$

   $= 4{,}938{,}172.84 \text{ J}$

   (e) How high up the tower will the vehicle go? $E_K = E_g = 4{,}938{,}172.84 = mgh = 5000 \text{ kg} \times 9.8 \text{ m/s}^2 \times h$

   $h = 4{,}938{,}172.84 \div 49{,}000 = 100.8 \text{ m}$

   (f) Draw a diagram to show the kinetic energy, potential energy transformations and total energy of the roller coaster system above. Show where energy may be lost from the system:

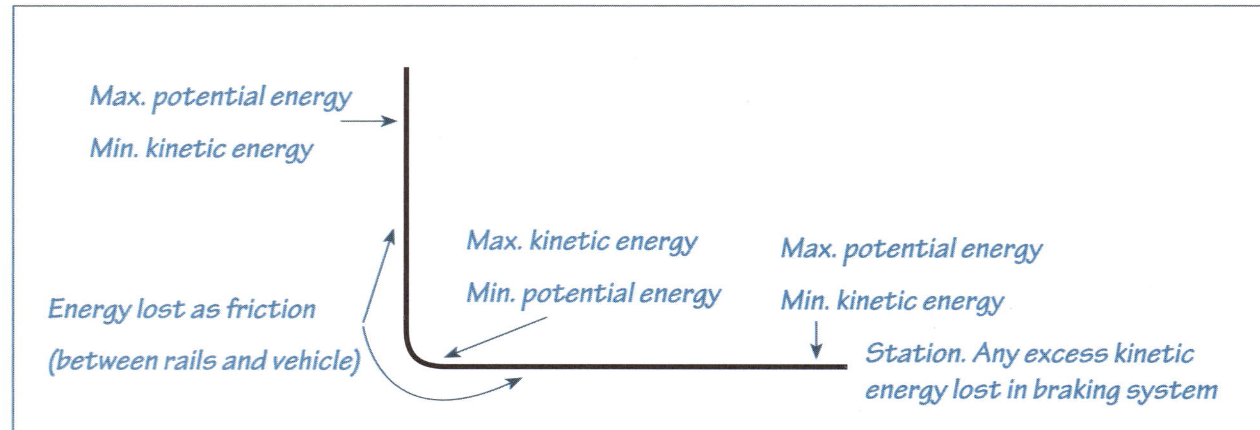

   (g) The roller coaster uses a magnetic braking system to slow the vehicle as it enters the station. Where the does the energy in the vehicle go as it slows down?

   The movement of the vehicle through the magnetic braking system produces EM forces that produce drag on the vehicle, which produces heat. Thus the kinetic energy of the vehicle is converted to heat.

©2020 **BIOZONE** International
ISBN: 978-1-927309-82-7
Photocopying Prohibited

101 Summing Up

Energy and energy conservation

1. A fully laden roller coaster cart train (4500 kg) pulls out of the station at 5 m/s. The base of the track is 3 m above the ground. To get to the top of the first lift hill 35 m above the ground, a force needs to be applied to the cart train over the 50 m length of the incline to the top. The speed of the cart train at the top of the hill is 8 m/s. Assume a near perfect system (losses due to friction and air resistance are negligible).

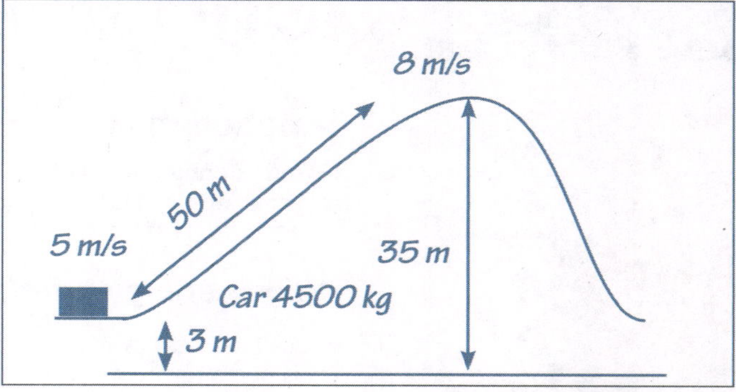

(a) Sketch a diagram of the problem in the space right (provide appropriate labels and values):

(b) What is the force required to pull the cart train to the top of the lift hill?

$E_{K(bottom)} + E_{P(bottom)} + W = E_{K(top)} + E_{P(top)} = 0.5mv_{(bottom)}^2 + mgh + Fd = 0.5\,mv_{(top)}^2 + mgh$

$= 0.5 \times 4500\text{ kg} \times (5\text{ m/s})^2 + 4500\text{ kg} \times 9.8\text{ m/s}^2 \times 3 + W = 0.5 \times 4500\text{ kg} \times (8\text{ m/s})^2$

$= 4500\text{ kg} \times 9.8\text{ m/s}^2 \times 35\text{ m} = 56{,}250\text{ J} + 132{,}300\text{ J} + W = 144{,}400\text{ J} + 1543{,}500$

$= 188{,}500 + W = 1{,}687{,}900.\ W = 1{,}687{,}900 - 188{,}500 = 1{,}499{,}500.$

$F = W/d = 1{,}499{,}500 \div 50 = 29{,}979\text{ N}$

(c) How long will it take for the cart train to reach the top of the lift hill? $v_f = v_i + at,\ t = (v_f - v_i) \div a v_f^2 = v_i^2 + 2ad,$

$a = (v_f^2 - v_i^2) \div 2d = (64\text{ m}^2/\text{s}^2 - 25\text{ m}^2/\text{s}^2) \div 100 = 0.39\text{ m/s}^2$

$t = (v_f - v_i) \div a = (8\text{ m/s} - 5\text{ m/s}) \div 0.39\text{ m/s}^2 = 7.7\text{ s}$

(d) The lift hill drops all the way down to base level (3 m above the ground). What is the theoretical speed of the cart train at the bottom of the hill?

$E_{K(top)} + E_{p(top)} = E_{K(bottom)} + E_{p(botom)},\ 0.5 \times 4500 \times (8\text{ m/s})^2 + 4500 \times 4500\text{ kg} \times 9.8\text{ m/s}^2 \times 35\text{ m}$

$= 0.5 \times 4500\text{ kg} \times v_{(bottom)}^2 + 4500 + 9.8\text{ m/s}^2 \times 3\text{m}$

$= 144{,}000\text{ J} + 1{,}543{,}500\text{ J} = 2250\text{ kg} \times v^2 + 132{,}300\text{ J} = 1{,}687{,}500 = 2250\text{ kg} \times v^2 + 132{,}300\text{ J}$

$2250\text{ kg} \times v^2 = 1{,}687{,}500 - 132{,}300\text{ J} = 1{,}555{,}200, \div 2250 = v^2,\ v = \sqrt{691.2} = 26.3\text{ m/s}$

(e) The actual speed of the cart train at the bottom of the lift hill is 24 m/s. How much energy is lost to heat due to friction and air resistance?

$W_{net} = \Delta E_K = 0.5\,mv_i^2 - 0.5\,mv_f^2 = 0.5\ 4500\text{ kg} \times (26.3\text{ m/s})^2 - 0.5\ 4500\text{ kg} \times (24\text{ m/s})^2$

$= 1{,}555{,}200\text{ J} - 1{,}296{,}000\text{ J} = 259{,}300\text{ J}$

(f) What is the efficiency of the roller coaster? $1{,}296{,}000\text{ J} \div 1{,}555{,}200 \times 100 = 83.3\%$

(g) Using the efficiency you calculated, what is the maximum height the next hill can be for the train to reach the top?

$E_{K(bottom)} \times 0.833 = E_{p(top\ hill2)}$

$= 0.5 \times 4500\text{ kg} \times (24\text{ m/s})^2 \times 0.833 = 1{,}080{,}000\text{ J} = mgh$

$1{,}080{,}000\text{ J} = 4500\text{ kg} \times 9.8\text{ m/s}^2 \times h$

$h = 1{,}080{,}000 \div 44{,}100 = 24.5\text{ m} + 3\text{ m} = 27.5\text{ m}$

(h) Explain why the second hill would need to be shorter than your calculation in (g) for the ride to continue:

For maximum height E_K will reduce to 0, meaning there is no motion at the top. The hill should be shorter so that some E_K will be left to roll the cart over the hill.

©2020 BIOZONE International
ISBN: 978-1-927309-82-7
Photocopying Prohibited

SSM PS3.A

PS3.A
PS3.B
PS3.D

8: Conservation of Energy and Energy Transfer

Anchoring Phenomenon

Winds of change: Where does the energy we use in our everyday lives come from and how will we be able to continue to provide that energy for a growing population? 102 109

How is energy transferred and conserved?

- ☐ 1 Recall the first law of thermodynamics from the previous chapter [CCC-5]. We saw that loss of energy from a system was work done on the system [PS3.B] [CCC-4] by the environment. Develop or use a model [SEP-2] to illustrate how energy is transformed and transferred [PS3.B] when electricity is generated and used [CCC-4] [HS-PS3-2]. How do we calculate [SEP-5] the energy [CCC-5] transferred into and out of a system [CCC-4] and explain [SEP-6] how work, energy and heat are related? What is the system and what is the environment and how do we tell? The Earth approximates a closed system [CCC-4], exchanging energy but (generally) not matter with the environment. How is the energy the Earth receives from the Sun used? Can you express these transformations as balanced chemical equations? [SEP-5] [PS3.B] [CCC-5] 103 110
- ☐ 2 Explain what is meant by entropy. The second law of thermodynamics states that the total entropy of an isolated system can never decrease over time [CCC-5]. Investigate [SEP-3] indirect and direct energy transfers [PS3.B] [PS3.D] in model systems [CCC-4] to provide evidence for the second law of thermodynamics. Use your findings to explain how the energy will flow between different components of a system. Repeat your investigations with different quantities of materials and use your findings to predict and explain [SEP-6] temperature changes and equilibrium conditions [HS-PS3-4]. 104
- ☐ 3 Use a computational model [SEP-2] [SEP-5] to investigate [SEP-3] heat transfer (energy flows) [PS3.A] [PS3.B] [CCC-3] [CCC-5]. Observe conduction in a model system [CCC-4] and simulate the effects of changing the properties of the system's components. 105

What is the relationship between work and power?

- ☐ 4 Recall from the previous chapter how we calculated work. Explain the relationship between work and power [PS3.B] and describe this relationship mathematically [SEP-5]. Investigate [SEP-3] this relationship by determining the power needed to use the energy in an oreo cookie when either walking or running. What do your findings tell you about the power of man-made systems [CCC-4] such as car engines? 106

How do energy conversion devices work?

- ☐ 5 A power plant is essentially a system [CCC-4] in which energy flows [CCC-5] out in the form of electricity. Explain how this is achieved in terms of energy transformations. Use your understanding of work and power to calculate [SEP-5] power output in systems when the efficiency of the system is known [PS3.B] [CCC-4] [HS-PS3-1]. 107 110
- ☐ 6 Recall from Chapter 6 that electrical generators are an essential component of nearly all types of electricity generation [PS3.B] [CCC-4] [CCC-5]. How do you think humans will meet future demands for electricity generation? What do you think needs to happen in order to meet future energy needs? 107
- ☐ 7 Think of the ways humans capture and convert energy. How important is it that these conversions are as efficient as possible? Use your understanding of energy conversion devices to design and build a Rube Goldberg machine [SEP-3] to demonstrate how energy is lost from a system [CCC-5] during energy conversions [PS3.B] [HS-PS3-3]. 108
- ☐ 8 Now use what you have learned to design and build your own energy conversion device to heat 4 L of water in 15 minutes [SEP-2]. The goal is to maximize both the temperature increase and efficiency. Create a design schematic [SEP-6] and submit it to your teacher for review. Evaluate your design and make any refinements as required. How efficient is your device before and after the refinements and how did you calculate this? [PS3.B] [PS3.D] [ETS1.A] [CCC-4] [CCC-5] [HS-PS3-3] [HS-ETS1-1] 108

8. Conservation of Energy and Energy Transfer

Anchoring phenomenon

Power plants of any kind are excellent examples of energy converting systems. Tracing the path of energy to the production of electricity is a good way of introducing the idea of energy conversion. Students can think about how they use energy in their everyday lives and how we might meet the increasing energy demands of the future.

How is energy transferred and conserved?

1. The total energy of an isolated system is constant. DeltaU, the internal energy of a system, equals Q, the energy supplied minus W, the work done *by* the system on its surroundings. The equation can be written in two ways. It is commonly seen as $\Delta U = Q - W$ in which the work is being done by the system (energy is moving out of the system). The IUPAC convention is $\Delta U = Q + W$ in which case work is being done *on* the system (energy is moving into the system). Both equations are correct as work can be positive or negative. Students should realize that a system must have a defined edge. We can think of the fuel/air mix in a cylinder as a system. When it explodes, it does work on a piston, moving it down the cylinder. We could then think of the engine in total as a system, where the moving components do work on the vehicle's transmission and wheels to make it move. Again we can think of the vehicle as a system. When moving, it has energy and so can do work on an object, such as pushing it along. It is important that students can identify what is doing work and what is having work done to it.

2. Entropy is often explained as the disorder of something or the inability of something to do work. The greater the entropy, the greater the disorder and the less the ability to do work. Entropy always increases over time. This is often described as "order tends to disorder". Entropy can be decreased in a particular system, but only by increasing it somewhere else. Thus separating objects into two piles increases order, but at the expense of using energy (e.g. fuel or food) and so increasing the disorder of that system. Students can investigate this by placing a beaker of cold water in a breaker of hot water. The system has order to start (different kinetic energy content or heat content). This order disappears over time as the temperatures become equal.

3. Energy2D is an excellent application for modeling energy and heat flow. A link to it can be found on **BIOZONE's Resource Hub**. It contains many in-built simulations but can also be used to build new models for investigating energy flow. The interface can take a little time to understand and unfortunately there is no online manual. In activity 105, we have outlined the steps involved in building a model using Energy2D. This provides students with a basic understanding of how to use the program so that they can build a model of their own. Note that Energy2D must be downloaded for use.

What is the relationship between work and power?

4. Power is an important extension of the concept of work by showing how fast the work is being done, i.e., how much energy (joules) are being used per second. The power of a device or machine is usually stated somewhere on it, e.g. a 100 watt light bulb or a 2000 watt kettle. Thus we know the amount of energy used per second and, if the total time the device is in use is known, we can also calculate the total energy used. Power is useful for comparing machines. A more powerful machine can use energy at a faster rate than a less powerful machine and so do a job (such as lifting or heating) more quickly.

How do energy conversion devices work?

5. All power plants, except photovoltaic cells, use the basic concept of using an energy source to turn a turbine connected to a generator to produce electricity. It is the difference in the energy source that defines the power plant. Fuel based plants, such as coal, oil, and nuclear plants, use the fuel to heat water to steam which turns the turbine. Hydroelectric power plants use falling water to turn the turbine. Wind turbines uses the force of the wind to turn the turbine. Different power plants have different efficiencies based on the number of steps the energy needs to go through to be converted to electricity and the efficiency of each step.

6. The increasing demand for electricity has seen the development of many very large electricity production schemes. This has culminated in huge developments such as the Three Gorges dam in China and the Grand Coulee Dam in Washington. However, these massive constructions can produce huge environmental problems, including loss of habitat and restriction of fish migration routes. In the last decade, there has been increasing investment in smaller personal electricity generation devices with the development of increasingly efficient small solar panels and wind turbines. Improving technology has also seen the development of more efficient devices, such as LED light bulbs, which use much less electricity than older incandescent bulbs. Combined, these innovations might help to meet the demand for electricity into the future and address the growing environmental problems associated with increased energy demand.

7. Rube Goldberg machines are designed to be intentionally over-complicated and inefficient. However, this makes them a fun and engaging way to study energy conversions. In general, the initial arrangement for a Rube Goldberg machine is irreversible, i.e., it can only be used once and does not return to its original state even with a continuous input of energy (e.g. a falling block can only fall a set distance and cannot then return to its original position). The machine must be reset each time via the input of a lot of energy, often at multiple stages. This energy is then released at each stage to do work by being tripped by the previous stage.

8. Students can now put their knowledge of energy transfer to work to build a device that will heat a continuous flow of water. They will need to investigate efficiency at each stage and overall, and decide the best way to heat the water. Different designs can be compared to help students understand the different result each of their devices produces.

ADVANCED LEARNING

★ Challenge question:

This 3D model depicts a reconstruction of an aeolipile (Hero's engine). Can you figure out how it works and what energy transformations are involved? Have a go at making one yourself.

Download this page from BIOZONE's Resource Hub

102 Winds of Change

Key Question: Where does the energy we use in our everyday lives comes from and how will we be able to continue to provide that energy for a growing population?

- Electricity is central to our lives and how we produce it is becoming increasingly important, especially in highly developed, industrialized societies with high power demands. As demand for electricity has increased, we have come to recognize the environmental damage done by large scale power plants, especially those that use fossil fuels.
- Technology now provides the means for people to meet their own needs for power generation more easily without relying on commercial suppliers. Many people find this independence appealing.

1. Think of the activities you do throughout the day. Which ones require electricity? Start from when you wake up in the morning and walk yourself through the day:

   Student answers will vary but some points may include:

   Turning on the lights in their room, the hot water for a shower, the refrigerator keeps food cool, using electricity to make breakfast (e.g. toaster or stove top), working on devices at school, the lights in the classroom, watching TV or going to the movies, cooking dinner or going to a restaurant, washing dishes in a dishwasher, washing a load of laundry, turning on the lights when it gets dark, charging mobiles phones.

2. Imagine your life without electricity. How would things be different? *Some answers could include:*

   Reduction in everyday technology (e.g. no phones, devices, televisions etc). Alternative lighting sources (e.g. candles and lamps), a change in cooking methods from microwaves and electric ovens and stove tops to open fires, woodfired ovens etc, heating the house with open fires instead of electric heating.

3. With the rapid growth of human population and our increasingly electricity-driven lives, there is a huge demand for more energy. How do you think we can prepare for the energy demands of the future?

   The focus should be on using renewable energy sources as the primary source and moving away from the use of non-renewable sources. Improving the efficiency of electricity generation from renewable sources is also important. Devices should be manufactured to be more efficient so they do not use as much electricity.

©2020 **BIOZONE** International
ISBN: 978-1-927309-82-7
Photocopying Prohibited

103 The First Law of Thermodynamics

Key Question: What does the law of conservation of energy say about society's increase in energy demands?

The law of conservation of energy

- In Chapter 7 (*Definitions of Energy*) we looked at the law of conservation of energy as a way to explain changes in systems due to work being done by or on a system. It was used to demonstrate the loss of energy in a system was actually work done on the system by the environment. For example: air resistance does work on a moving object resulting in a transfer of kinetic energy from the object to the molecules in the air it comes in contact with. The energy lost by the moving object is the energy gained by the air molecules, thus energy is conserved.
- At its most basic, the law of conservation of energy states: energy can not be created or destroyed, it can only be transformed from one form to another or transferred from one object to another. This law works well for the small systems we were dealing with. Kinetic energy transfer and the energy conversions in Chapter 7 resulted in heat loss at every energy transformation equal to the heat gained by the environment.
- In Chapter 4, we looked into nuclear processes and discovered that in a nuclear reaction some mass is converted into energy. As such, the law of conservation of energy can be rewritten to add an additional phrase "by normal chemical means". This meant that chemical reactions could not break the law of conservation of energy, but nuclear reactions could.
- In 1905, Einstein published a paper explaining the relationship between mass and energy. The paper described the famous equation $E = mc^2$ and the equivalence of mass and energy. This equation explains where the energy in the Sun and stars came from and showed that under specific circumstances mass could be converted into energy and vice versa.
- Looking at the universe, the law of conservation of energy deals with the total mass and energy available and states this total does not change. This means all the energy we will ever use already exists. It comes from somewhere and, when we are done with it, has to go somewhere.

Work being done by the air can be clearly be seen by the effect it has on an umbrella.

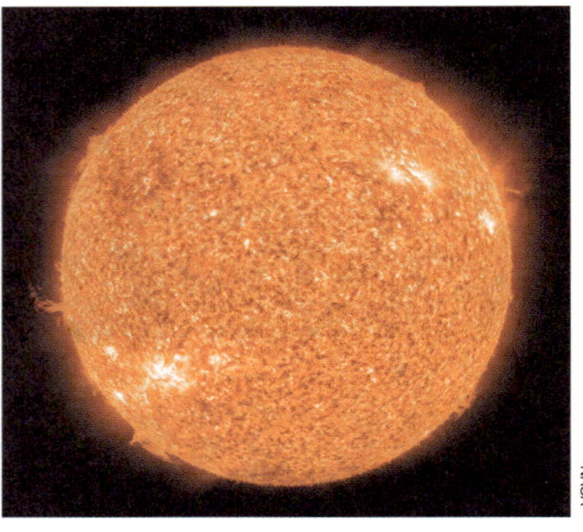

The Sun converts millions of tonnes of mass into energy every second.

1. Work in pairs to construct a model (diagram) in the space below to show how energy is transformed and transferred when electricity is generated by the movement of water and then used in an end product. Include as many steps as you can think of, and include where and how work is being done to the system and how and where energy is being lost from the system. Start with water in the ocean being heated by the Sun and finish with sound being produced by a speaker in a stereo plugged into a wall socket.

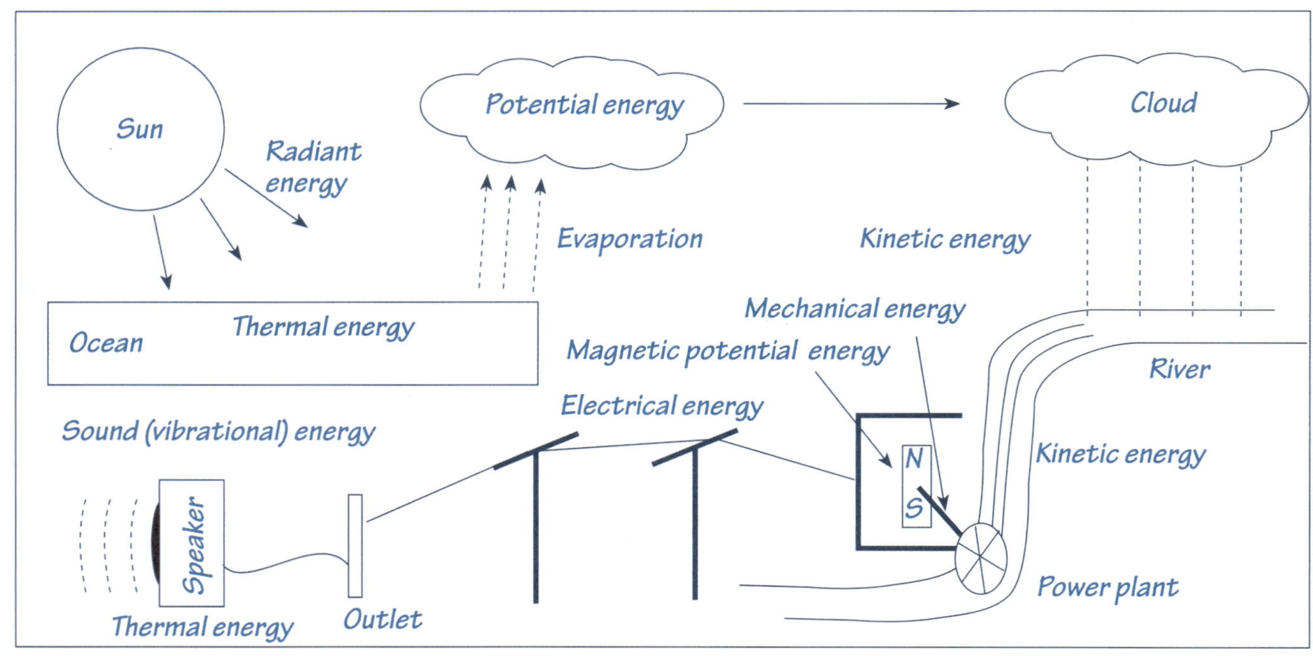

PS3.B PS3.D SSM EM

©2020 BIOZONE International
ISBN: 978-1-927309-82-7
Photocopying Prohibited

The first law of thermodynamics

- The first law of thermodynamics is a specific version of the law of conservation of energy that applies to systems where heat transfer and work are the means by which energy is transferred into and out of a system. It states that any change in the energy of a system (ΔU) is equal to the net transfer of heat from or to the system (Q) plus the net work done on or by the system (W). This is written:

$$\Delta U = Q - W$$

- It is important to be specific about the identified system to determine whether work is being done by the system or to the system and whether heat is going into or out of the system. Look at the example of the combustion engine shown right. Expanding gas due to heating from combustion does work on the piston causing it to move. The gas does work on the piston so the transfer of energy is from the gas to the piston. If we were looking at the gas as a system, the value for W is negative because work is being done **by** the gas, moving energy out of the system (in which case $\mathbf{\Delta U = Q - W}$). However, if we were looking at the piston as a system, the value for W is positive because work is being done **to** the piston ($\mathbf{\Delta U = Q + W}$).

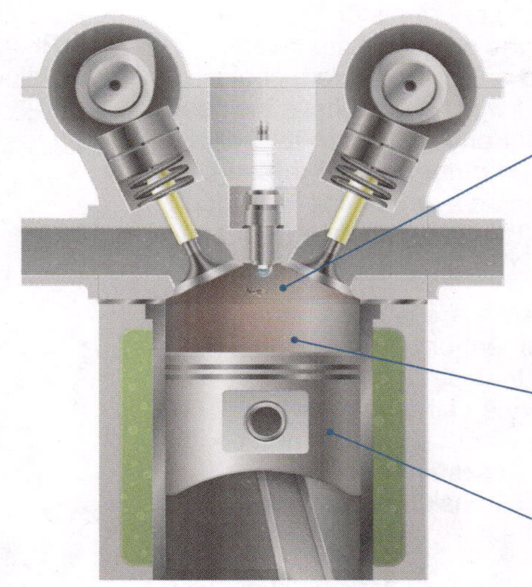

2. In a car, the combustion of gasoline heats up gas inside a fixed chamber with a tightly fitting movable piston.

   (a) What is the value of the work done by the gas if the work done on the piston is 200 J?

   200 J of work is done by the gas on the piston

   (b) Combustion adds 600 J of heat to the gas. 100 J of heat is transferred to the piston and 300 J of heat is transferred to the surrounding container. What is the net transfer of heat into the gas?

   $Q_{net} = Q_{gas} + Q_{piston} + Q_{environment} = 600\ J = Q + 100\ J + 300\ J,\ Q = 200\ J$

   (c) What is the change in the energy of the gas? 200 J

3. Explain the difference in the sign (negative or positive) of the value of work done by a system and the work done to a system.

   When a system does work energy is lost so there is negative value to work. When work is done on a system, energy is gained by that system so there is a positive value to work.

4. The diagram right shows an apparatus invented by James Joule to show how work, energy, and heat are all related. As it falls the mass pulls on a spindle attached to a paddle. The paddle is inside a sealed flask of water with a thermometer attached. The paddle spins as the mass falls.

   Explain how the device would show the relationship between work, energy, and heat:

   As the mass falls it does work on the system causing the paddles to rotate in the flask of water. Drag by the water does work on the paddle system, slowing it down. As work is done on the water some energy is lost as heat due to the drag (friction in a fluid). This heat is measured as a change in temperature in the thermometer.

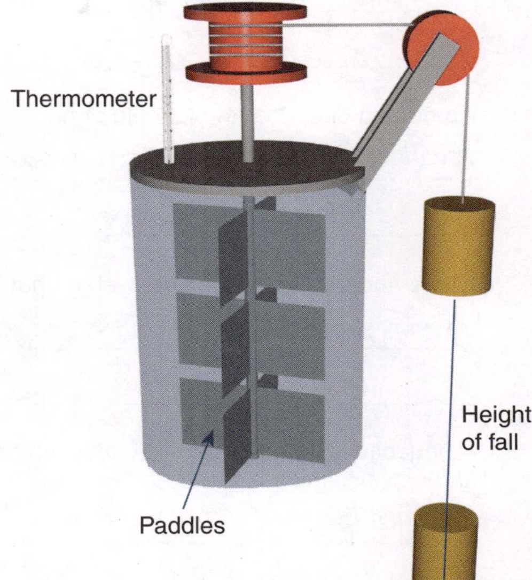

©2020 **BIOZONE** International
ISBN: 978-1-927309-82-7
Photocopying Prohibited

Systems and surroundings

- A system is an object or collection of objects we are interested in looking at. The definition is quite arbitrary as it depends entirely the limits set by the observer. A system could be very small like a molecule or large like Earth. On the previous page we looked at the interaction of heated gas and a piston. You were asked questions as if the gas and the piston were each their own system. However, the system could have been the engine itself which would have included both the gas in the chamber and the piston as well as the chamber, crankshaft and other parts of the engine. When looking at the engine as a whole, the individual components all contribute to the net change in energy of the system.
- When considering a system, there is often heat exchange with the environment or surroundings. The environment is everything that is not in the system. So, when considering the gas in the chamber as a system, the piston it does work on is part of the environment. However, if the system was the car engine, the piston would be considered part of the system.

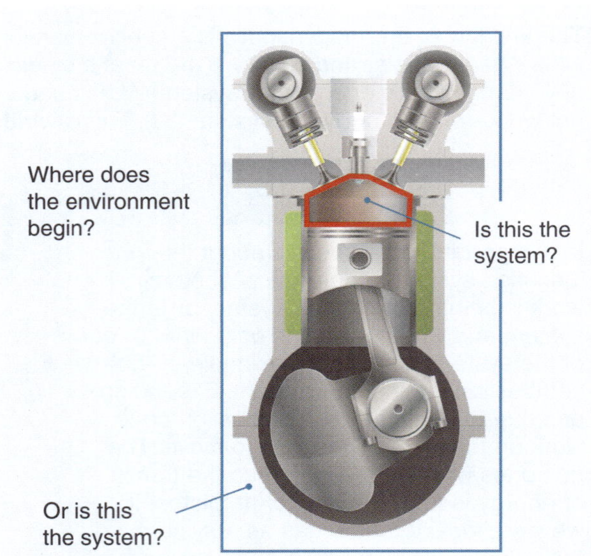

If the system is the gas then the piston is part of the environment. If the engine is the system then the environment is everything outside of the engine.

- There are three types of systems in thermodynamics:

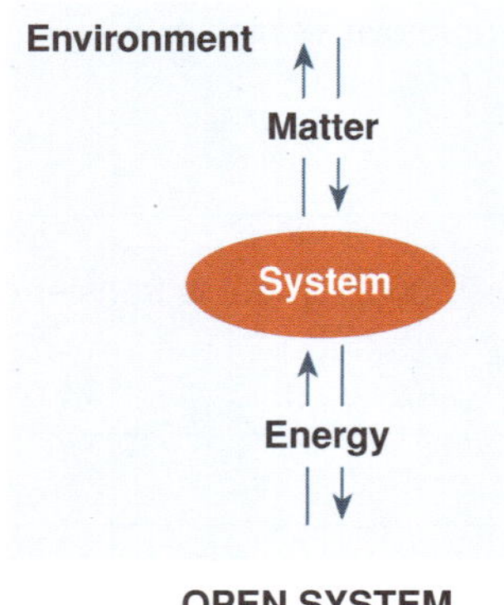

OPEN SYSTEM

Both energy and matter can be exchanged with its surroundings.

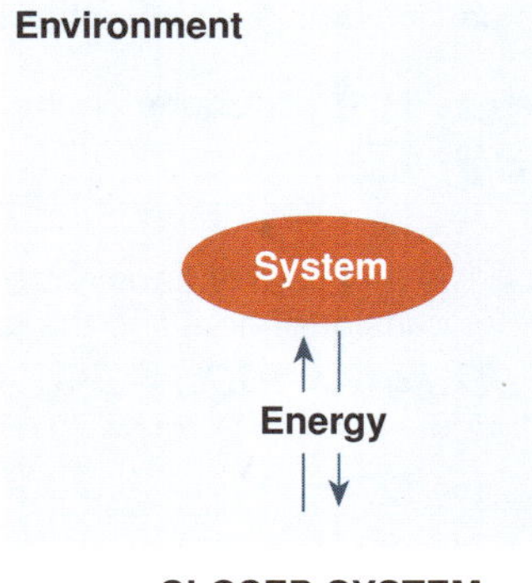

CLOSED SYSTEM

Only energy can be exchanged with its surroundings.

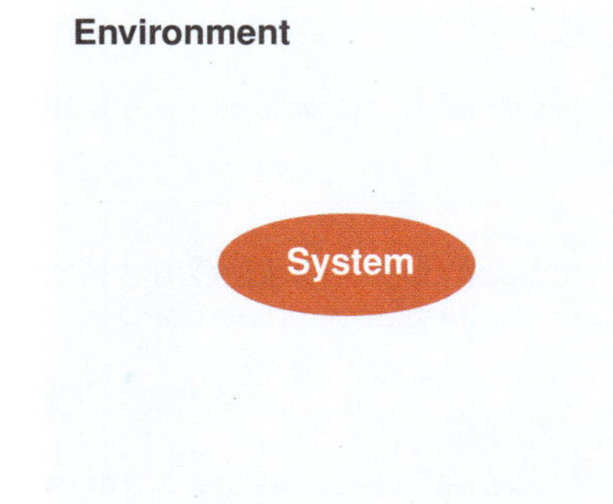

ISOLATED SYSTEM

Neither energy nor matter can be exchanged with its surroundings.

- Earth is an example of a closed system. For the most part, there is no exchange of matter with its surroundings (there is some exchange as hydrogen and helium gasses escape the atmosphere and meteorites enter). However, there is a continuous supply of added energy from the Sun.

5. Explain why the human body is an open system: *We constantly exchange matter (breathing, eating, excreting, etc) and energy (chemical energy in food, loss of heat to the environment).*

6. Name and describe an example of an open system that is not an organism: *A vehicle exchanges matter (fuel added, exhaust exists etc) and energy (intake chemical energy through fuel, loss of energy as heat, friction etc).*

7. Name and describe a closed system that is not a cosmic body: *An unopened bottle of water. Until opened there is no transfer of matter but it can still be heated or cooled, which are transfers of energy.*

8. A perfectly isolated system is difficult to find. There is possibly just one isolated system known to exist. Name this system and explain why it is isolated:
The universe. All matter and energy exchanges we know of occur within the system (universe). There is no evidence of exchanges with another system (universe)

©2020 **BIOZONE** International
ISBN: 978-1-927309-82-7
Photocopying Prohibited

Energy sources

- Most energy sources originate from the Sun (exceptions include nuclear, tidal, and geothermal energy).

Energy source	How the Sun's energy is utilized
Solar	Light energy excites electrons in solar panels resulting in an electric current.
Wood	Produced by plants through photosynthesis using sunlight.
Petroleum and natural gas	A fossil fuel formed by the fossilized remains of prehistoric organisms that settled in the oceans.
Coal	A fossil fuel formed by the fossilized remains of prehistoric plants buried in swamps.
Hydroelectric	Sunlight evaporates water from the oceans, which falls as rain and snow on top of mountains, and then returns to rivers. Water collected in dams stores potential energy.
Wind	Globally, winds are caused by differential heating of the Sun between the equator and poles.

- How is sunlight energy involved in the formation of these energy sources? Let's look at wood as an example. Photosynthesis traps light energy and stores it as chemical energy found in the chemical bonds of wood.
  - Unbalanced equation for producing wood using photosynthesis: $CO_2 + H_2O$ + energy (sunlight) $\rightarrow C_6H_{10}O_5$ (wood) $+ O_2$
  - Balanced equation for producing wood via photosynthesis: $6CO_2 + 5H_2O$ + energy $\rightarrow C_6H_{10}O_5 + 6O_2$
  - Burning the wood releases this stored energy in a reversal of the photosynthesis equation:

    Wood $+ O_2 \rightarrow CO_2 + H_2O$ + energy

Energy from the Sun is stored in the chemical bonds in the molecules that make up wood.

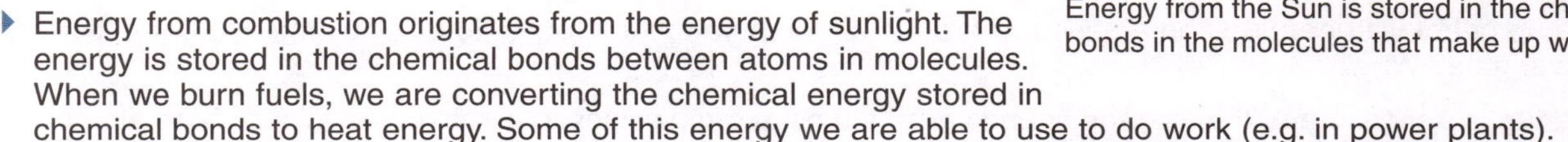

- Energy from combustion originates from the energy of sunlight. The energy is stored in the chemical bonds between atoms in molecules. When we burn fuels, we are converting the chemical energy stored in chemical bonds to heat energy. Some of this energy we are able to use to do work (e.g. in power plants).
- Thermodynamics deals with systems in which heat and work are the only ways that energy is transferred. This works well for energy. However, since the laws of conservation deal with both energy and matter, we do need to be aware of the matter involved in these energy transformations.
- Combustion releases stored energy we can use to do work, but the atoms in the fuel recombine with oxygen and to smaller produce molecules (CO_2 and H_2O). This results in a net change in the concentration of the reactants and products of the reaction *in the environment*. More combustion results in larger shifts in the Earth's atmospheric and geological equilibria which can have significant effects.

9. For each method of producing useful energy below, predict what unwanted by-products result during the conversion of energy sources into useful energy. Do not consider wastes produced in making the energy conversion device:

   (a) Combustion of fossil fuels: *CO_2, soot and CO (if combustion is incomplete)*

   (b) Hydroelectric power: *Waste heat (thermal pollution).*

   (c) Solar power: *None.*

   (d) Wind power: *Noise.*

10. Why is it important to consider the unwanted by-products of energy generation?

    Unwanted by-products have unintended consequences and can result in pollution of the environment. They need to be considered to ensure the benefit is worth the cost.

11. Why is it important to consider that Earth is a closed system when converting energy for human needs ?

    Because there is no exchange of matter with another system, the by-products (waste) produced by our energy generation methods remains with us. Waste can accumulate with negative consequences.

©2020 **BIOZONE** International
ISBN: 978-1-927309-82-7
Photocopying Prohibited

104 The Second Law of Thermodynamics

Key Question: What is entropy and how is it explained by heat flow?

Order and disorder

- The photo shows a log of wood burning in a campfire. The wood represents a concentration of energy. The energy is stored in the chemical bonds within the wood (e.g. between the bonds holding atoms together to form cellulose). By burning the wood, the energy is released as heat and becomes spread out (dispersed), heating the air molecules around it and causing them to move faster.
- This is an example of increasing **entropy**. Entropy is a measure of how much the energy is spread out in a system. This energy is not available to do work. Entropy in the universe never decreases. It will eventually increase to a point where all energy is evenly dispersed (maximum entropy). The ordered, concentrated energy in wood has low entropy. Burning it increases entropy as the energy in the wood becomes dispersed as heat.
- Entropy can be reduced in a closed system, but always by increasing entropy in the wider universe. The energy in the wood was stored as a product of photosynthesis, which is powered by the Sun. The Sun itself is releasing vast amounts of energy every second and thus rapidly increasing the entropy of the universe.

Energy flow

- Place a cup of hot water at 80°C on a bench where the air temperature is around 20°C. Does the cup get hotter or colder over time? Why?
- Temperature is a measure of the average speed of the atoms and molecules in a substance. When these atoms and molecules come in contact with each other, they transfer kinetic energy changing the speed of the individual particles. Slower molecules will speed up (heat up) and faster molecules will slow down (cool down).
- Heat is the primary means by which entropy increases in the universe. When objects and particles come into contact with each other there is an exchange of kinetic energy. This exchange between faster and slower moving particles results in a redistribution of heat such that the particles now have similar speeds (i.e. temperatures). This redistribution increases entropy as the energy is more evenly dispersed.
- The following investigations will help illustrate this concept.

1. Entropy never decreases unless it increases somewhere else by at least an equal amount. Name two everyday devices that reduce entropy somewhere while increasing it somewhere else:

   Air conditioning unit, refrigerator, freezer.

2. The second law of thermodynamics states that entropy always increases over time. You will sometimes hear it expressed as "order tends to disorder".

   As a class or group discuss what this means and why you think the entropy of the universe will always increase (to a maximum). Summarize your ideas below:

   The universe has a set amount of energy. To reorder something (decrease entropy) requires more energy than might be stored in the ordering energy. This dissipates some of the energy. Eventually this will lead to the even dispersal of all energy (pushing a ball up a hill uses more energy via friction etc than the ball stores at the top of the hill).

3. In terms of entropy, explain why it is important for life on Earth that Earth is not an isolated system:

   If the Earth was isolated, energy would spread out until it was evenly distributed. There would be no extreme differences in energy to exploit to do work. Life would stagnate and cease.

PS3.B PS3.D SSM EM

©2020 **BIOZONE** International
ISBN: 978-1-927309-82-7
Photocopying Prohibited

INVESTIGATION 8.1: Indirect energy transfer

See appendix for equipment list.

PS3-4

1. Fill a large beaker (~250 mL or similar vessel) with hot water around 50°C (measure the temperature to be precise).
2. Place a small beaker (~50 mL) filled with water at room temperature into the large beaker. Measure the temperature of the room temperature water to be precise.
3. Place the set up in an insulated box or wrap insulation around the larger beaker to reduce the energy loss from the system.
4. Measure and record the temperature of the water in the large and small beaker at two minute intervals and record the temperatures in the table provided.
5. Do this until the temperatures stabilize. Graph your result on the grid provided.

Large beaker
Small beaker
Water at 20°C
Water at 50°C

Time (minutes)	Temperature in large beaker (°C)	Temperature in small beaker (°C)
0 (initial)	48	24
2	46	27.5
4	43	30
6	42.5	32
8	40	33.3
10	38	34.5
12	37	35
14		
16		
18		
20		

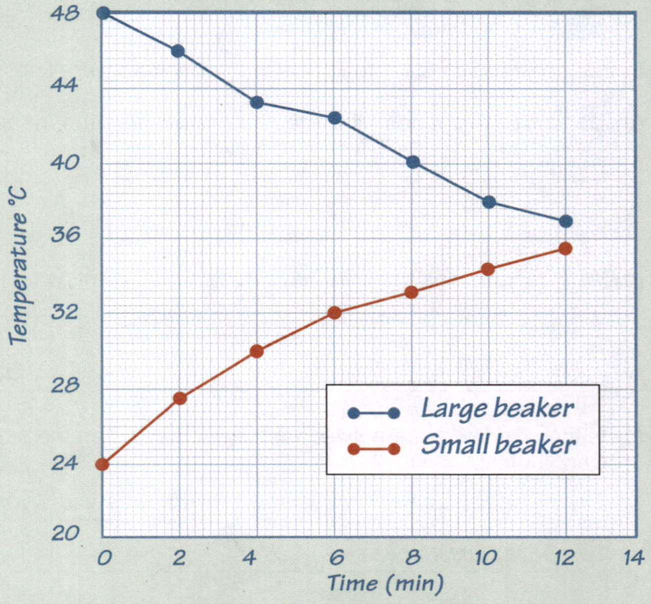

6. Design and carry out an investigation to measure the energy transfer between water and cooking oil. Attach a record of your design to this page and use the space below to record and plot your results.

Student's own design and results.
The design will be similar to that shown above, except the small beaker will contain cooking oil.
Typical results below and right

°C water	°C oil
48	26
46	32
45	35
43	38.5
41.5	39

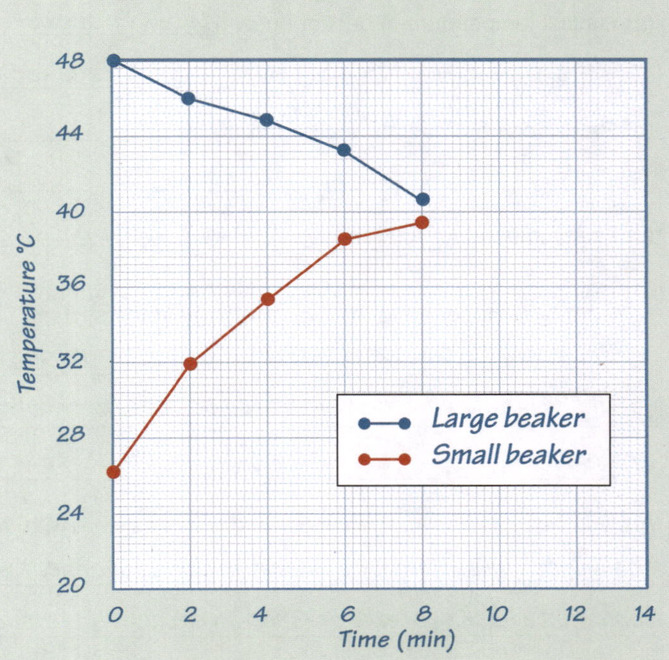

©2020 BIOZONE International
ISBN: 978-1-927309-82-7
Photocopying Prohibited

Mixing temperatures

- In the previous investigation, the liquids in the beakers were kept separated. This means that heat flowed between the liquids through the glass of the beaker instead of through direct contact. What would be different if instead of being separated the liquids were mixed together instead? Let's do another short investigation to see.

INVESTIGATION 8.2: Direct energy transfer

See appendix for equipment list.

1. Set up two 500 mL beakers, each containing 200 mL of water
2. In one beaker, set the water temperature to around 50°C (measure to be precise). In the second beaker, set the temperature to about 20°C (measure to be precise). Record the temperatures in 5(a) below.
3. Predict what will happen when the contents of the two beakers are mixed. Try to predict the precise temperature of the water after mixing based on your initial measurements. Write your prediction in 5(b).

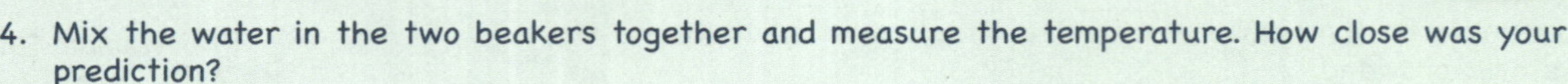

4. Mix the water in the two beakers together and measure the temperature. How close was your prediction?

4. In the Investigation 8.1, the liquids in the beakers were kept separated, but heat was exchanged between the systems.

(a) Describe the shape of the lines for the temperature in the large beaker and small beaker in the first investigation:

The lines converge in curves.

(b) Why do you think the lines have this shape? (Hint: what is happening to the energy in each beaker?)

Energy is being transferred from the hot water to the cold water.

(c) Is there a difference between the lines from when the small beaker contained water to when it contained cooking oil? Can you explain why?

The oil heats up faster because it absorbs energy more quickly and needs less energy per degree of temperature increase.

5. In Investigation 8.2, water of two different temperatures was mixed:

(a) i. Initial temperature in beaker one: *Student's answer*

ii. Initial temperature in beaker two: *Student's answer*

(b) Prediction of temperature after mixing:

Average of the two temperatures

(c) Final temperature after mixing: *Student's answer*

(d) How close was your prediction? In terms of the energy in the system, explain why the temperature changed and what has happened to the overall energy in the system:

The energy is the same before and after the mixing (minus some heat loss) The higher kinetic energy of the hot liquid was transferred to the lower energy particles of the cooler liquid. These reached an equilibrium.

6. Are you results for Investigation 8.1 and 8.2 consistent with the laws of thermodynamics? Explain:

Yes. In both investigations, energy in the form of heat was transferred from the warmer system to the cooler until both systems reached an equilibrium temperature.

©2020 **BIOZONE** International
ISBN: 978-1-927309-82-7
Photocopying Prohibited

105 Modeling Energy Flow

Key Question: How does modelling help us understand thermodynamics and energy flow?

Energy2D

- Computer simulations reproduce the behavior of a system using a computer. They provide a way to predict and model changes in a system by altering specific conditions within that system. Computer simulations allow us to examine complex systems and run many simulations in a short space of time without having to experiment with a physical system. The results of simulations are more accurate when the simulation closely models the real system.
- In this activity you will explore heat transfer using a simulation program called Energy2D. This is free, downloadable simulation software developed by Dr Charles Xie at the Concord Consortium. The software has many simulations (right) and a workspace where you can build your own thermal experiments.
- In this activity, you will use the program to model heat transfer by conduction.

Gifted and talented students may also like to explore further by modeling heat transfer by convection and radiation.

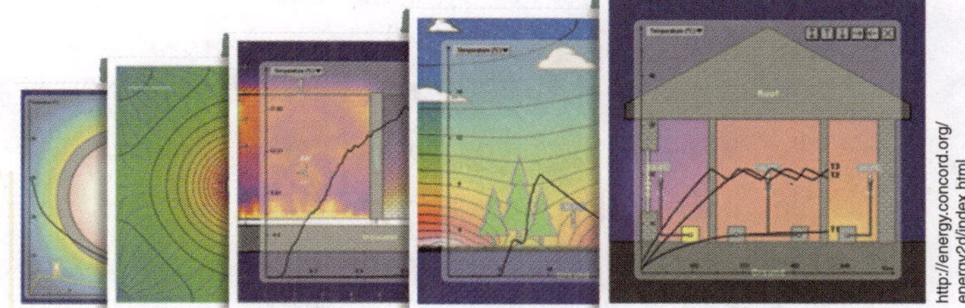

http://energy.concord.org/energy2d/index.html

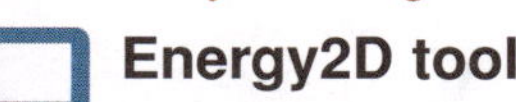

Energy2D tools

- Download Energy2D via the link on **BIOZONE's Resource Hub** or from http://energy.concord.org/energy2d/index.html
- When you open the program you will see the screen on the right. Become familiar with the tools (below).

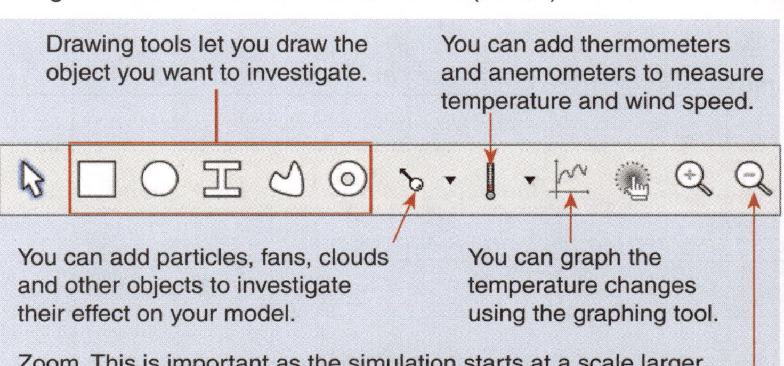

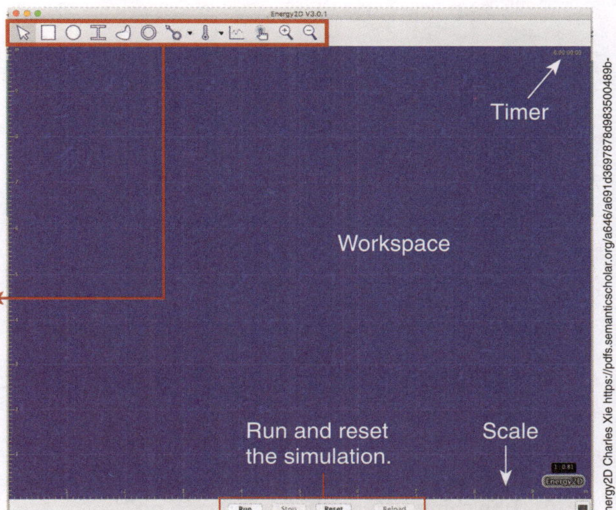

Energy2D Charles Xie https://pdfs.semanticscholar.org/a646/a691d3697878d983500489b-761477f06c2d0.pdf

Getting familiar with Energy2D

Follow the instructions below to run a simple simulation looking at heat and temperature and conduction. This will help you to become familiar with using Energy2D.

i. Click on the **Examples** tab at the top of the screen. Select **Heat and Temperature** then **Thermal Equilibrium Between Identical Objects** from the drop down menu (right). You will see the screen shown below right.

ii. Run the simulation for 10 seconds (push stop to end the simulation). You can stop and start the simulation at any time. For this example, the timer has been slowed down. You can do this by right-clicking on the workspace and selecting **Properties**. To set the timer to real-time, set the **Time Steplength** to 0.1. This allows you to easily observe the changes in both thermometers during the simulation period.

iii. You can view the simulation as a graph. After you have run your simulation click the graph icon to view the graph. If you would like to see the data graphed as the simulation runs, simply click on the graph icon *before* you click Run.

Now you are familiar with the simulation you will carry out some simple investigations on energy transfer.

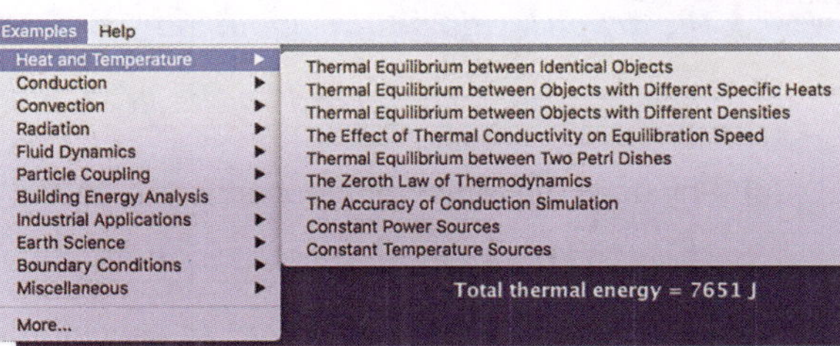

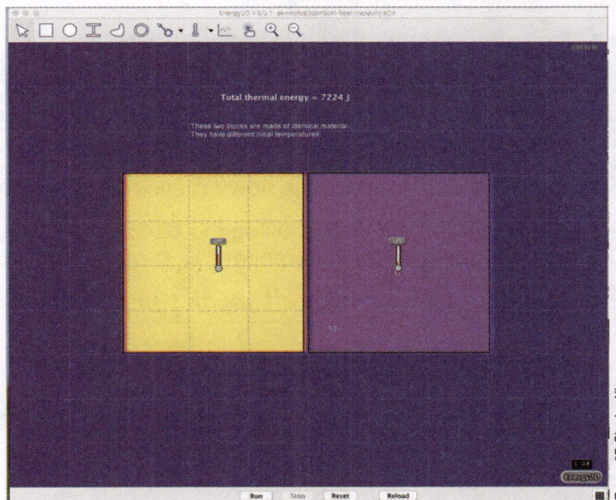

Energy2D Charles Xie

©2020 **BIOZONE** International
ISBN: 978-1-927309-82-7
Photocopying Prohibited

EM

SSM

PS3.B

PS3.A

Using Energy 2D to explore conduction

You will rebuild and modify the example simulation to understand how it works. You should get the same outcome as the example on the previous page.

INVESTIGATION 8.3: Conduction

1. Click on **File** then select **New**.
2. Reduce the scale by clicking on the + zoom tool 3 times.
3. Select the **rectangle** drawing tool and draw a box 2 squares wide by 3 squares high. **Copy** the box and **paste** it into the workspace right next to the original box (box 1). Do not leave any gaps between them.
4. **Right click** box 1 and select **Properties**. Click on the **Source** tab. Set the **Temperature** to 40. Under the **Thermal** tab, set **Thermal Conductivity** to 1, **Specific Heat** to 1000 and **Density** to 1.204. Click **OK.**
5. Repeat for box 2 but set the temperature to 10. Under the **Thermal** tab, set **Thermal Conductivity** to 1, **Specific Heat** to 1000 and **Density** to 1.204. Click **OK**.
6. Click **View** in the menu and select **See through**.
7. Click on the **thermometer icon** and then click on box 1 to add a thermometer. Repeat for box 2.
8. **Right click** on the workspace and select **Properties**. Deselect **Convective** under the **General** tab. This will set the simulation to conductive movement of energy. Set the **Time Steplength** to 0.02. Under the **Medium** tab set **Background Temperature** to 0, **Conductivity** to 0, **Specific Heat** to 1 and **Density** to 1.204. Click **OK**.
9. **Run** the simulation and produce a graph. Print your graphs and staple them to this page if you wish.
10. Save your work.

1. What happens to the temperature of the boxes over time? *They become even over time, reaching about 24 °C*

2. (a) Why is the conductivity of the workspace set to 0 in the simulation? *This stops energy being lost to the workspace and ensures all energy is kept within the two blocks being measured.*

   (b) Set the conductivity of the workspace to 4 and rerun the simulation. Produce a graph. How does this change the way the temperature changes over time?

   The graph levels out at about 11-12 °C The 40°C block loses most of its energy to the workspace and very little is transferred to the 10°C block.

   (c) Why does the temperature change like this? *Instead of the energy being transferred from one block to the other, the energy is transferred to the workspace (a greater surface area is open to the workspace, and the workspace is much colder (a thermometer placed in the workspace shows a change from 0 to 11°C)).*

- There are many other simulation examples that can be run.
- Explore the **Heat and Temperature**, and **Conductive** examples. You will find simulations of some of the investigations your have already carried out in the lab. For example, the **Thermal Equilibria Between Two Petri Dishes** is similar to the beaker inside another beaker experiment you carried out in Investigation 8.1.
- To see how the energy is flowing click on the **Heat flux** arrows in the **View** menu.

3. Carry out the simulation using the conditions you used during the Investigation 8.1. Do you get the same results? If not, can you explain the differences?

Results will be similar although temperature changes in the original experiment may be less due to energy being lost through the walls of the beaker. Unknowns such as the conductivity of the glass and air will also affect the results.

©2020 **BIOZONE** International
ISBN: 978-1-927309-82-7
Photocopying Prohibited

Changes in thermal energy

- You have seen that thermal energy flows from one object to another in the direction of highest thermal energy to lowest thermal energy (i.e. hot to cold). You have also measured the temperature change in substances as energy is transferred from one to the other.
- Thermal energy is given the symbol Q and is measured in joules (J).
- Recall the investigation in which a small beaker of cold water was placed into a large beaker of hot water (or the simulation on the previous page). This resulted in the water in the small beaker becoming hotter (gaining energy) and the water in the large beaker becoming cooler (losing energy). Assuming no energy was lost out of the system a simple equation can be written describing the transfer of thermal energy (heat) between the two beakers.

4. Write an equation showing the thermal energy transfer between two masses: *$\Delta Q1 = \Delta Q2$*

- Recall in Chapter 3 you calculated the energy changes when water was heated by burning alcohol. The equation you used was:

Energy produced = 4.2 x change in temperature (ΔT °C) x mL water

- This equation is specifically for calculating the energy in water. You can now produce a more general equation. Energy produced can be replaced by ΔQ, the change in thermal energy. The change in temperature remains the same (ΔT). The mass of the substance (m in grams) can be substituted for mL of water (1 mL of water = 1 g). 4.2 is the specific heat of water. The specific heat is the amount of heat energy needed to raise the temperature of 1 g of substance by 1°C. The symbol for specific heat is c. Its units are usually joules per gram per °C (J/g/°C).

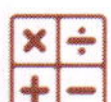

5. Using the information above write an equation to show how to calculate the change in energy in a substance:

$\Delta Q = m \Delta T c$

6. We can now explain the thermal transfer between two objects. Study the simple diagram below:

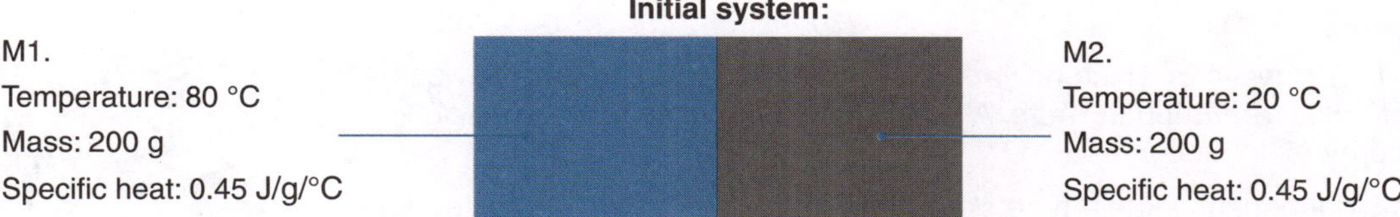

In the final system the temperatures of **M1** and **M2** were both **48°C**.

(a) What was the total thermal energy of the initial system? *80 × 200 × 0.45 + 20 × 200 × 0.45 = 9000 J*

(b) Calculate the thermal energy (heat) transferred between the two objects: *28 × 200 × 0.45 = 2520 J*

(c) How much energy was lost from the system? *48 x 200 x 0.45 + 48 x 200 x 0.45 = 8640 J*

9000 - 8640 = 360 J (or 4°C of "missing" temperature. 4 × 200 x 0.45 = 360 J)

7. (a) A 2.00 kg iron pot was placed on a fire and heated to 140°C. 1.00 kg of water at 20°C was added to the pot. Assume the pot loses no heat to the environment and negligible water is lost as steam.

Specific heat of water: 4.19 J/g/°C
Specific heat of iron: 0.452 J/g/°C

Predict the temperature of the system at thermal equilibrium:

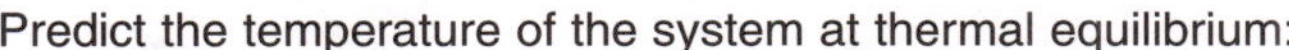

Student's answer

(b) The equation for determining the temperature at equilibrium can be derived from your equation in question 4. It is a weighted average of the temperatures (it takes into account mass and specific heat):

$$T_{final} = \frac{m_{iron}c_{iron}T_{iron} + m_{water}c_{water}T_{water}}{m_{iron}c_{iron} + m_{water}c_{water}}$$

Calculate the final temperature of the system: *(2000 x 0,452 × 140) + 1000 × 4.19 × 20) ÷ (2000 × 0.452) + (1000 × 4.19) = 210,560 ÷ 5094 = 41.3°C*

©2020 **BIOZONE** International
ISBN: 978-1-927309-82-7
Photocopying Prohibited

106 Work and Power

Key Question: What is power and how are energy, work, and power related to each other?

Work and Power

Recall from the previous chapter that work requires a force to cause displacement of an object. Notice that there is nothing about the amount of time the force acts to cause the displacement. Whether or not the work is done quickly or slowly the same amount of work is done.

- When we want to determine the rate at which work is done, we are looking at a new term called **power**. Power is the ratio between work done and the time it took to do that work. Power is measured in units of joules per second (J/s), also known as **watts** (W), and is calculated using the equation:

$$P = W / t$$

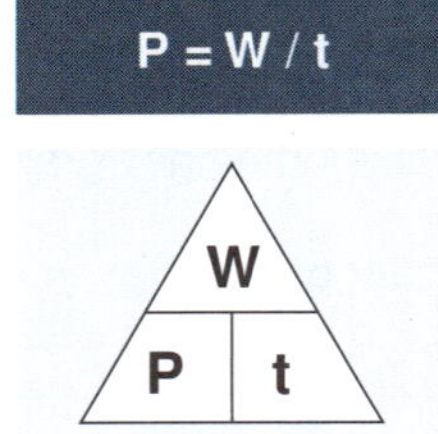

Where P = power (watts (W)), W = work (joules (J)), and t = time (seconds (s))

- Because work is an expression of force and displacement, the equation for power can be rewritten as force times displacement divided by time. Since displacement divided by time is velocity, power can also be expressed as:

$$P = F \times v$$

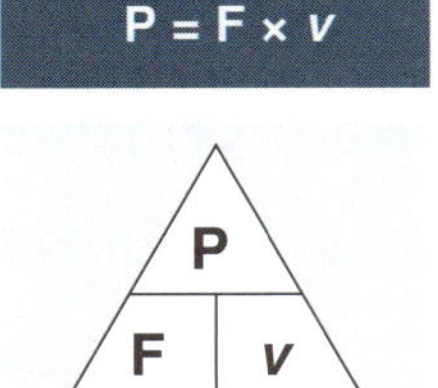

- This new equation reveals that powerful machines are both strong (large force) and fast (high velocity). The more powerful the machine, the greater the force that can be applied to cause a displacement in a short interval of time.

Machine power

- There are different ways in which power is described in devices. Some devices, such as light bulbs, are rated in watts. A 100 W light bulb uses 100 joules of electrical energy every second.
- A bulb does work because it is a type of **resistor**. A resistor is the opposite of a conductor. Where a conductor allows electricity to flow, a resistor works against the electric flow. The work done by the resistor on the electric current, slows the current down and releases energy in the form of heat and light.
- Many larger devices such as the engines of cars are measured in units of horsepower (hp). Horsepower is an imperial (non SI) unit of power equal to 550 foot-pounds per second. It still involves a force (pound), a displacement (foot), and an increment of time (second) and is equal to 746 watts.

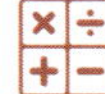

1. Dimetri and Lee are working out in the weight room before football practice. Dimetri bench presses 50 kg 10 times in 1 minute. Lee bench presses the same 50 kg 10 times in 30 seconds.

   (a) Which football player does the most work? Explain your answer:

   They both do the same amount of work as they are moving the same weight the same distance.

   (b) Which football player delivers the most power? Explain your answer:

   Lee delivers the power as he is able to do the same work in less time.

2. During a stairs workout, Chantelle and David run to the top of the stairs. Chantelle reaches the top in half the time it takes David. David has twice the mass of Chantelle.

   (a) Which person did the most work? Explain your answer:

   David did twice as much work because he went the same distance with twice the weight.

   (b) Which person delivered the most power? Explain your answer: *They both had the same power. David did twice the work as Chantelle but in twice the time, so he delivered the same power as Chantelle.*

3. An elevator with a maximum capacity of 800 kg takes 5 s to travel between the first and sixth floor, a distance of approximately 15 m. What is the power requirement of the elevator if it takes this trip with its maximum load?

   $P = W/t = mad/t = 800\ kg \times 9.8\ m/s^2 \times 15\ m \div 5\ s = 23{,}520\ W$

©2020 **BIOZONE** International
ISBN: 978-1-927309-82-7
Photocopying Prohibited

Energy to power

INVESTIGATION 8.4: The power of Oreos

See appendix for equipment list.

1. Find your mass in kg, determine your weight (W = mg), and record.
2. Go to the nearest stairwell or set of steps, measure the height of a single stair in meters, count the number of stairs to determine height of a flight of stairs, and record.
3. Determine the work done when ascending the stairs (only consider vertical displacement) and record.

Mass	Weight (F) (m × g)	Height of stair	# stairs	Height of flight of stairs (d)	Work (F × d)

4. Time yourself walking up the flight of stairs and record. Repeat while running up the stairs.
5. Calculate the power generated from walking and running and record.

Speed	Time (t)	Power (W ÷ t)
Walking		
Running		

6. From the nutritional information on the packages, determine the calories found in one original oreo and one double-stuf oreo, convert to joules (1 food calorie = 4,184 J) and record on the table below.

7. Determine the power needed to use the energy from the oreo when walking and running and record.

Oreo	Calories	Joules	Power needed to use energy (E ÷ t)	Distance needed to climb to use energy (d = Pt ÷ F)	Number of stairs
Original			walking:		
			running:		
Double-stuf			walking:		
			running:		

8. Calculate the distance needed to climb to use the energy, determine the number of stairs, and record.

4. Compare the number of stairs required to use the energy content in a single oreo when walking and running:

   (a) What does this comparison tell you about the amount of energy used and the distance over which work is done?

   No matter how fast the distance is covered, the energy consumed is the same, therefore the work done is the same.

   (b) What does this tell you about the difference increased power makes in relation to energy usage?

   Power is simply the rate at which energy is used up. Higher power means energy is consumed faster.

5. Compare the number of stairs required to use the energy content of an original oreo compared to a double-stuf one:

   (a) How many more steps do you need to do to use up the extra energy found in 1 double-stuf oreo?

   Student's answers will vary depending on student mass and height of the stairwell

   (b) Are the extra steps worth indulging in the double-stuf oreo? *Student's opinion*

6. Only in terms of energy consumption, explain the advantage of running to exercise instead of walking:

   Running allows you use excess energy in less time.

©2020 **BIOZONE** International
ISBN: 978-1-927309-82-7
Photocopying Prohibited

107 Energy and Power Plants

Key Question: How is energy harnessed to do work?

Mechanical energy

- In the process of doing work, the system doing the work exchanges energy with the system on which the work is done. Energy is lost by the system doing work and gained by the system on which the work is done. This acquired energy is known as mechanical energy. Mechanical energy can be defined as the ability to do work.
- Recall that mechanical energy is the energy due to an object's motion (kinetic) or position (potential). A system or object with mechanical energy is able to do work. Much of our ability to generate usable energy depends on converting various sources of energy into a form that we are able to transport and access easily.

Turbines

- A turbine is a machine used to extract energy from a flow of fluid (liquid or gas) and convert it into a form that can be used to do work. The fluid turns a wheel or rotor attached to a shaft, converting kinetic energy from the moving fluid into rotational mechanical energy on the shaft.

Water wheels

- The water wheel is an example of an early turbine. It provides mechanical work by harnessing the energy in falling water. Falling water does work by turning the wheel. The water falls more slowly than it would usually because it transfers kinetic energy to the water wheel.
- The kinetic energy lost by the falling water is the kinetic energy gained by the water wheel. The total energy of the system remains the same.
- Some of the first water wheels were used to grind grains. The kinetic energy transferred to the water wheel allowed it to do work on a series of gears which in turn worked to turn a millstone which rolled over the grains to crush them. The work done by each component of the system transferred the mechanical energy from the wheel to the millstone.

Windmills

- Another early example of a turbine is the windmill. Windmills extract energy from the flow of air (wind) and convert it into mechanical energy using blades instead of a wheel. The blades are angled to deflect wind in a specific direction causing it to do directional work on the blades and move the shaft. Windmills were typically used in areas where running water for a water wheel was not readily available.

1. Water wheels collect energy from falling water.

   (a) Where does the energy in falling water come from?

   Gravitational potential energy due to the water's position above the ground.

   (b) The water cycle describes the continuous movement of water on Earth. During the water cycle, water goes through different phases: liquid, solid, and gas. How does the water cycle factor into the energy captured by a water wheel?

   Energy from the Sun heats up water at ground level and the ocean. The excited water evaporates and rises in the atmosphere gaining potential energy due to gravity. Water wheels use the kinetic energy of falling water to capture energy.

   (c) Where does the energy driving the water cycle come from? *The Sun*

2. Windmills use the kinetic energy of wind to do work.

   (a) Explain where the energy causing the air particles to move comes from? *The Sun heats up the Earth. The Earth in turn heats up the air near the ground. The warm air rises and wind is the result of this moving air.*

   (b) Would windmills be more or less reliable than water wheels? Explain your answer:

   Windmills are less reliable than water wheels. Rivers continuously flow but there are periods when wind does not blow or does not blow hard enough to turn the windmill. Therefore, output can be quite variable.

PS3.B SSM

©2020 BIOZONE International
ISBN: 978-1-927309-82-7
Photocopying Prohibited

Modern turbines

- Instead of being used as part of mills, most turbines today are attached to generators to produce electricity. As the turbine turns, a strong magnet is rotated inside a coil of copper wires. The changing magnetic field induces electric current in the wires producing electricity (see Chapter 8, *Types of Interactions*).
- Hydroelectric power plants use falling water to turn a turbine in the same way waterwheels use falling water. However, in a hydroelectric power plant a dam provides water storage and determines the height from which the water will fall. The greater the mass of water passing through a turbine and the greater the height of water above it, the greater the energy that will be available (E = mgh).
- Hydrodams are normally situated in sections of a river where it is easier to build a high dam, which maximizes the height of water above the turbine (e.g. a deep narrow canyon). This can also be achieved by using the natural fall of a river where the dam is built well upstream and water is piped to a turbine well downstream.

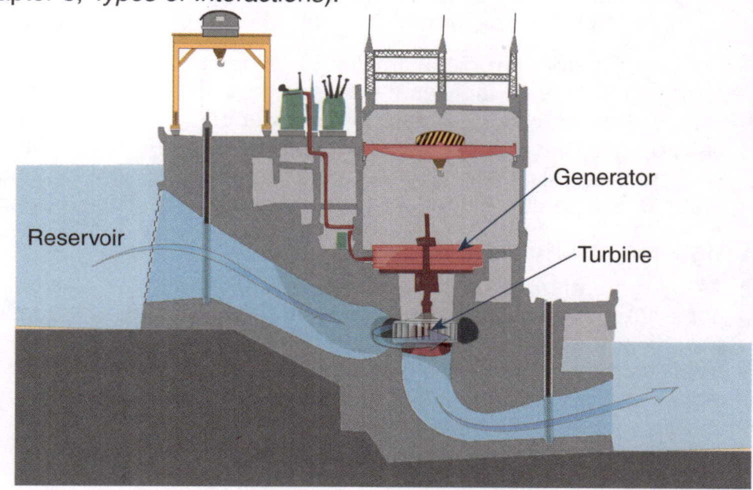

- Wind farms use moving air to turn a turbine in the same way windmills use moving air. However, wind farms can only be located where wind is near constant (e.g. beaches of mountain ridges).
- Consistency and speed of the wind are important factors in the location of wind farms. Because turbines do not generate electricity when the wind is not blowing many turbines are attached to a battery. Excess energy can be stored in the battery ready to be used when the wind is not blowing or not blowing hard enough to meet the energy consumption demands.

3. In terms of power, explain why the speed of the wind is important in the generation of electricity:
   Power is the rate at which work is done. Faster wind is able to do work on the blades faster, producing more electricity in a shorter time.

4. Describe the main factors in producing electricity in wind farms: *Speed of the wind, consistency of available wind, surface area of the turbine blades that the wind can act on.*

5. Why is it important that hydroelectric power plants are able to control the flow rate of water through the turbine?
   Control allows the power plant to meet energy consumption needs. Rate can be increased during the day when more people are awake and using power, or slowed at night.

6. A hydroelectric power plant needs to provide 9.8×10^9 J of electricity every hour to provide for the energy consumption needs of the community. The efficiency of the power plant is 90%. The water falls 40 m to the turbine.

   (a) What is the power output of the power plant in watts (W)? *$P = 9.8 \times 10^9 \text{ J} \div 3600 \text{ s} = 2700 \text{ W}$*

   (b) What is the actual power that must be supplied by the water to produce this power from the power station?
   $P \times 0.9 = 2700 \text{ W}$, $P = 3000 \text{ W}$

   (c) How much water (in kg) must pass through the power plant's turbine every second in order to provide this power?
   $P = W/t = mad/t$, $m = Pt/ad = 3000 \text{ W} \times 1 \text{ s} \div 9.8 \text{ m/s}^2 \times 40 \text{ m} = 7.65 \text{ kg}$

   (d) Explain what would happen if the power plant had insufficient power:
   Power outages, flickering electricity, or dimming lights etc.

©2020 **BIOZONE** International
ISBN: 978-1-927309-82-7
Photocopying Prohibited

Generators

- Generators are devices that convert energy. Typical generators convert mechanical energy into electricity. They are essential components of almost all power plants converting the rotational mechanical energy from a turbine into electrical energy through electromagnetic induction.
- Some generators are much smaller than their power plant counterparts, providing energy for individual buildings, houses, or equipment in the field where there is no access to the power grid. While some homes have back-up generators in the event of a power outage, in some buildings such as hospitals, back-up generators are a requirement for safety reasons. These generators are typically gasoline or diesel powered.

Small electricity generators

- As we depend more on our devices, we continue to seek better ways generate the necessary energy to keep them running. In the recent past, this was simply accomplished by exchanging energy depleted batteries for new ones. This practice is not only wasteful but the higher energy consumption of modern devices makes it difficult for standard batteries to keep up with the demand.
- Gasoline powered generators are sometimes used at camps to run devices, but they are loud and require fuel to keep running. Their size and fuel needs make them more difficult to move around as needed.
- Advances in solar cell technology have made it possible to create small solar panels, which can be transported easily and plug in to devices to run and charge (right). These can be attached to a battery pack to store energy for later use or for when the sun is not shining.
- Other energy generating devices include a small turbine that can be placed in a stream, providing enough energy to charge a phone.

7. Think about energy consumption (in terms of electricity) of developed countries.

(a) How do you expect the energy consumption of individuals will change as time goes by? Explain your answer:

Energy consumption per person will increase. The devices we use daily are becoming a larger part of our lives and use more energy as they become more and more advanced (computers, phones, etc).

(b) What do you think will happen to the world population over time and explain what you think this means in terms of global energy consumption:

The population of the Earth is continuing to increase. This will lead to greater global energy consumption as more people will be using more energy.

(c) As more countries become developed, explain how you expect development will change their energy consumption and subsequently global energy consumption:

Developed countries use more energy than undeveloped countries. As more countries become developed more energy will be consumed.

8. (a) Based on your answers from question 7, what do you expect future global energy consumption to look like:

Energy consumption will increase (at a growing rate)

(b) Do you expect that current methods for generating electricity will be sufficient to meet future consumption rates? Explain your answer:

No. Non renewable energy sources will run low are more is used and not enough energy is supplied by renewables (yet).

(c) What do you think needs to happen in order to meet future energy needs?

An increased reliance on renewable energy sources as well as an energy revolution with the development of other sources of energy (e.g. fusion reactors).

©2020 **BIOZONE** International
ISBN: 978-1-927309-82-7
Photocopying Prohibited

108 Energy Conversion Devices

Key Question: In what ways can energy be converted to accomplish a specific task?

Capturing energy

- Everywhere around us there is energy, some of it is stored (e.g. in food), and some it is being used (e.g. as rushing water). Being able to capture or convert this energy into a form we can use is one of the most important needs of modern civilization.
- Although we capture and convert a huge amount of energy into electrical energy, it is not the only way we use energy in the modern world. As the human population increases, so does the need for useful energy. For this reason, energy devices that require no electricity are becoming increasingly important.
- Solar ovens and solar water heaters are two of these devices. Using new technology and improved designs it is possible for these devices to cook food or heat water as easily as with an electric device.

Solar oven

Rube Goldberg machines

PS3-3

- The basic concept behind energy use is the transformation of energy into a useful form. In order to use as much of the original energy as possible, energy conversion devices should be as simple as possible. Since energy is lost as heat anytime there is a transfer, making fewer transfers increases the efficiency of the machine.
- Rube Goldberg machines take a different approach to typical energy conversion devices. Named after the cartoonist Rube Goldberg, these machines are complex contraptions that use chain reactions to perform simple tasks. They are so energy inefficient they are impractical for any real work, but they are a lot of fun to set up and watch operate.

INVESTIGATION 8.5: Building a Rube Goldberg machine — See appendix for equipment list.

1. Individually, in pairs, or in small groups, identify a simple task for your machine to achieve.
2. Working backwards, brainstorm a series of actions that will complete your task. Using internet sources would be a good way of researching ideas for the way your machine will work.
3. Draw a blueprint plan for what you expect to happen:

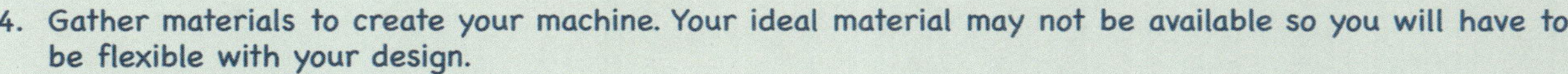

4. Gather materials to create your machine. Your ideal material may not be available so you will have to be flexible with your design.
5. Build your machine. Perform tests as you go to make sure each step of your machine works.
6. Demonstrate your machine to the rest of your class.

©2020 **BIOZONE** International
ISBN: 978-1-927309-82-7
Photocopying Prohibited

PS3.D

1. Watch the video on the **BIOZONE Resource Hub** in which an elaborate Rube Goldberg machine is used:

   (a) How many times was energy transferred from one object or system to another? *Well over 100 times*

   (b) The energy in the original starting collision appears to run the whole device, with numerous reactions, often with greater energy. Explain why this isn't actually the case:

   Each reaction is set up to begin with. Potential energy is already stored in the components due to their (unstable) positions. The action at the start of the device needs enough energy to trigger the first reaction. Potential energy is then transformed as the device proceeds through its task.

   (c) Why can this machine be used only once? *Once the reactions have finished, the PE is transformed and not returned to its original state. It must be reset by adding more energy to the machine.*

Engineering energy conversion devices

- Unlike Rube Goldberg machines, most devices are designed to be as efficient as possible - to get the most work out of the input energy. As we learned, the energy in the universe is limited. And most of that energy is in forms inaccessible to us. What's more, when we harvest energy we lose much of it in the form of unusable heat energy. This further limits the energy available for our use. When we consider the energy demands for society and the energy we have available, efficiency becomes very important.

- You will now have the opportunity to design your own energy conversion device by applying the basic principles of energy conversion. This design challenge can be done outside the classroom in groups or individually as your teacher decides. Your teacher will determine the amount of time you can take.

The goal of this design challenge is simple: **you must build a device that will produce the greatest temperature change in 4 liters of water in fifteen minutes**. Water must be able to flow through or around you device based on the conditions below.

Your device will be attached to a heating station set up by your teacher so that the testing conditions will be uniform for all the devices. A basic heating station will consist of two heat lamps on chemistry lab stands with clamps which can be easily adjusted, a submersible fountain pump with adjustable flow control, a container that will hold the 4 liters of water that will be heated and is able to submerge the fountain pump, a thermometer, and a stopwatch or timer. The actual layout of the heating station will depend on your teacher and available materials.

Make sure to note that your device will need to connect to the water pump - you will need appropriate hose connections to attach your device to the pump hoses.

You will need to:
- research and submit a design to your teacher
- build and test the design
- refine the design and make any changes necessary
- submit a plan showing any revisions of the design to your teacher
- present a report that describes how your device works, its power output and efficiency, and situations where it will be best used.
- provide a demonstration of the device so that it can be compared to other devices designed by your classmates.

Available materials may be limiting. Be sure you are aware of the materials you have available when designing your device. Keep note of any portion of your device where an unavailable material may have made a difference.

Note: **Efficiency = energy in / energy out**.

Energy out can be calculated from the temperature change in the water. 1 mL (1 g) of water requires 4.2 joules to raise its temperature by 1°C or $E = m \times 4.2 \times \Delta T$ (where m = mass in grams of water and ΔT = the change in temperature).

Energy in can be calculated using the power ratings of the heat lamps. Heat lamps are approximately 96% efficient, so the energy in will be the wattage of the heat lamps × 0.96 × time in seconds.

2. (a) In the space below, summarize your research notes on a possible device to build. You may use extra paper if needed:

   Student's research

©2020 BIOZONE International
ISBN: 978-1-927309-82-7
Photocopying Prohibited

(b) Your device will utilize a fountain pump to circulate water during testing. Can you think of any ways in which you could get water to circulate without using electricity?

Student's answer

3. (a) Draw a schematic of your device and add labels or notes explaining the various parts and what will be needed to build them. Submit this to your teacher for review:

Student's diagram

(b) Note down any changes you will need to make after your discussion with your teacher:

Student's answer

4. Build and test your device. Where is the energy coming from to power the device and how will you measure the energy input? Use the space below to record any measurements and refinements to your device:

Student's answer

©2020 **BIOZONE** International
ISBN: 978-1-927309-82-7
Photocopying Prohibited

5. (a) After refining your device, repeat your measurements of energy in and energy out. Use the space below. Has your device improved? By how much?

Student's answer

(b) Evaluate the efficiency of your device. Identify any components that increased or decreased the efficiency of the device. Explain how these components influenced the final results:

Student's answer

(c) Provide a final report to your teacher for marking, including a summary of your research and methods, your designs and calculations and evaluations of the device's efficiency and practicality. Demonstrate your device to the class.

6. After all the devices have been demonstrated, produce a short evaluation of each one. Which design proved to be the most efficient? Why?

Student's answer

©2020 BIOZONE International
ISBN: 978-1-927309-82-7
Photocopying Prohibited

109 Review Your Understanding

Key Question: Where does the energy we use in our everyday lives comes from and how will we be able to continue to provide that energy for a growing population?

- Fossil fuels are the primary source for global energy production. However, in addition to being nonrenewable sources of energy (when they are used up, they are gone), we are left with the combustion by-products which can and have caused additional problems.
- The population on Earth continues to grow and with it the demand for energy. So, how can we continue to provide energy to an increasingly energy dependent and growing population using depleting fuel sources? We simply cannot sustain our future energy needs solely on current practices.
- The demand for cleaner, more sustainable sources of energy is on the rise. Electricity generated from solar, hydroelectric, and geothermal plants and wind farms has increased over time and contribute a growing percentage of global energy production. These energy sources are renewable and do not directly contribute to increased pollution and emissions.

- As positive as renewable sources of energy are, there are limitations as to their use. Locations where usable energy can be extracted are limited (e.g. there are only so many locations where geothermal activity is near enough Earth's surface to access) and we cannot control how much energy is put in (we cannot make the wind blow faster or the Sun to shine brighter).
- Because of these limitations energy conversion devices require high efficiency in order to get the most usable energy out of the energy sources available. Experts say that current renewable technologies are not sufficient to meet today's energy demands. This can easily change as newer discoveries and technologies are developed.

1. Power plant efficiency is important because it allows for more usable energy to be generated for the energy put in.

   (a) The most efficient energy source we currently have is hydroelectric. Explain why there are not more hydroelectric power plants:

   Hydroelectric plants rely on a consistent supply of running water. This is not available in all locations so hydroelectric plants can only be used where there are rivers of sufficient capacity.

   (b) Explain why it is important for efficiency to improve in solar and wind farms: *Although the sun shines everywhere, there are still limited places where solar can be used (e.g. flat open land). Similarly, wind can only be used in places where specific conditions are met. Thus, these generation methods are currently limited. Increasing their efficiency would improve the power output from limited resources.*

2. Using the first law of thermodynamics, explain where the energy not converted to electricity goes:

   Energy not converted to electricity is converted to other forms, most of which is lost to the environment as heat and friction.

3. Explain how power plants increase the entropy in the universe: *Power plants take the high concentrations of energy and make use of them. Some of this energy is converted to (and lost) as unusable heat energy, which mixes with the environment and other systems. This increases entropy by decreasing order (separation of temperatures) and increasing disorder. The temperatures mix to create a new average temperature.*

©2020 **BIOZONE** International
ISBN: 978-1-927309-82-7
Photocopying Prohibited

110 Summing Up

Energy conservation and energy transfer

PS3-1

1. Combustion power plants are not very efficient with the most efficient plant approaching 60% efficiency. This means less than 60% of the energy produced through combustion is converted into usable energy.

(a) Explain this loss of usable energy in terms of the first law of thermodynamics:

Energy cannot be created or destroyed, only transformed from one form to another or transferred from one system to the other. Energy not converted to usable energy is converted to heat.

(b) Explain this loss of usable energy in terms of the second law of thermodynamics: *Some energy from a high energy system is lost as heat when transferring to another system. This lost energy moves the environment and system towards equilibrium, increasing entropy.*

2. A water wheel is connected to a generator that produces electricity.

(a) When the water is diverted past the wheel, a mass of 200 kg of water per second falls 5 m. What is the potential power of this diverted water?

$P = W/t = Fd/t = mad/t$, $200\ kg \times 9.8\ m/s^2 \times 5\ m \div 1\ s = 9800\ W$

(b) When the water is directed to flow over the water wheel so that the wheel starts turning, the generator produces 8 kW of power. What is the efficiency of the water wheel?

$9800\ W = 9.8\ kW$, efficiency = power out ÷ power in × 100% = $8\ kW \div 9.8\ kW \times 100\% = 81.6\%$

(c) What has happened to the "missing" energy?

The remaining energy is converted to other forms, such as heat.

(d) Write an algebraic expression to calculate the energy output from a given energy input for the waterwheel:

$E_{out} = E_{in} \times 0.816$

3. (a) LED light bulbs have generally replaced incandescent light bulbs due their greater efficiency. A 6 W LED light bulb produces about the same level of lighting as a 60 W incandescent light bulb.

Under controlled conditions it was found a 60 W incandescent bulb heated 1 L of water (specific heat of 4.2 J/g/°C) by 7.3°C in 10 minutes. What is the efficiency of the light bulb in terms of producing light?

$\Delta Q = 4.2 \times 1000 \times 7.3 = 30{,}660\ J$ absorbed by water. $60\ W = 60\ J/s$

Energy used by bulb $= 60\ J \times 600\ s$

$= 36{,}000\ J$. $30{,}660 \div 36{,}000 = 0.852$. $1 - 0.852 = 0.148$.

Only 14.8% of the electricity used by the bulb was converted to light.

(b) How does this compare to a modern LED light bulb which is above 95% efficient in terms of producing light?

14.8% is very low. Incandescent bulbs lose a lot of energy as heat, whereas as LED bulbs do not.

4. Two individuals with the same mass (60 kg) are using the stairwell to climb to the top floor of a skyscraper (150 m). One is walking and the other is jogging.

(a) Which individual consumed more energy during their climb?

Explain your answer: *Both used the same energy as they carried the same weight up the same distance.*

PS3.B SSM

©2020 BIOZONE International
ISBN: 978-1-927309-82-7
Photocopying Prohibited

(b) Which individual produced more power during their time climbing? Explain your answer:

The runner delivered more power as they were able to do more work (use more energy) in less time.

(c) Calculate the power in watts applied by each person if the jogger finished the climb in 6 minutes and the walker finished in 10 minutes:

$P_{jogger} = mad/t = 60\ kg \times 9.8\ m/s^2 \times 150\ m \div 360\ s = 245\ W$

$P_{walker} = mad/t = 60\ kg \times 9.8\ m/s^2 \times 150\ m \div 600\ s = 147\ W$

PS3-2 5. A fire piston (or fire syringe) is a device that has been used since about the 17th century to ignite tinder (e.g. cotton wool) for making fire. The tinder is placed inside the sealed syringe and the plunger is rapidly pressed down, usually by a hammer or strong thrust of the palm of a hand.

Explain why the tinder ignites when the plunger is pressed down rapidly in terms of power and the energy involved. Include a diagram to illustrate your explanation:

When the plunger is compressed, kinetic energy from pressing on the plunger is being transformed into potential energy in the position of the compressed air and thermal energy. Pressing the plunger down rapidly increases the power as the energy is transferred over a short time frame. Thermal energy does not have time to dissipate and can be hot enough to ignite the tinder in the piston.

Chocolateoak CC 3.0

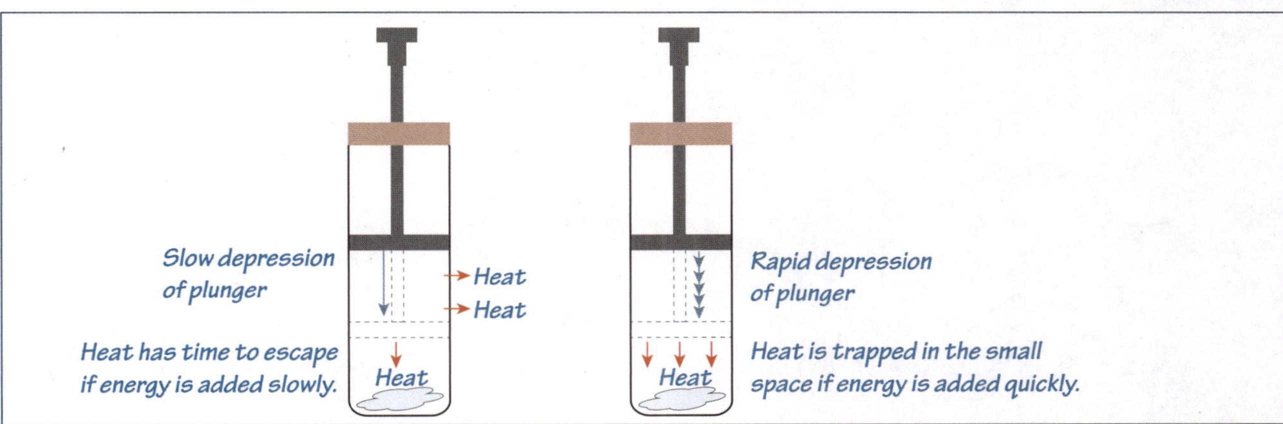

6. In a thermal power plant, steam is passed through a steam turbine. The steam is used to turn the turbine.

(a) What happens to the temperature of the steam as it passes through the steam turbine?

The temperature goes down.

(b) Why does this happen? *Steam molecules hit against the turbine and transfer heat (losing energy). The lost energy slows them down. This can be measured as a reduction in temperature.*

(c) The steam produced by the boiler is superheated to 540°C and the waste steam is rapidly cooled by the condenser to about 25°C. How does this help increase the power output of the power plant?

Gas expands at higher temperatures. Lowering the temperature of steam already used to turn the turbine reduces the pressure on that side allowing the high temperature steam on the other side to more easily do work on the turbine.

©2020 **BIOZONE** International
ISBN: 978-1-927309-82-7
Photocopying Prohibited

Waves and Their Application in Technologies and Information Transfer

Concepts and connections

Use arrows to make connections between related concepts in this section of the book

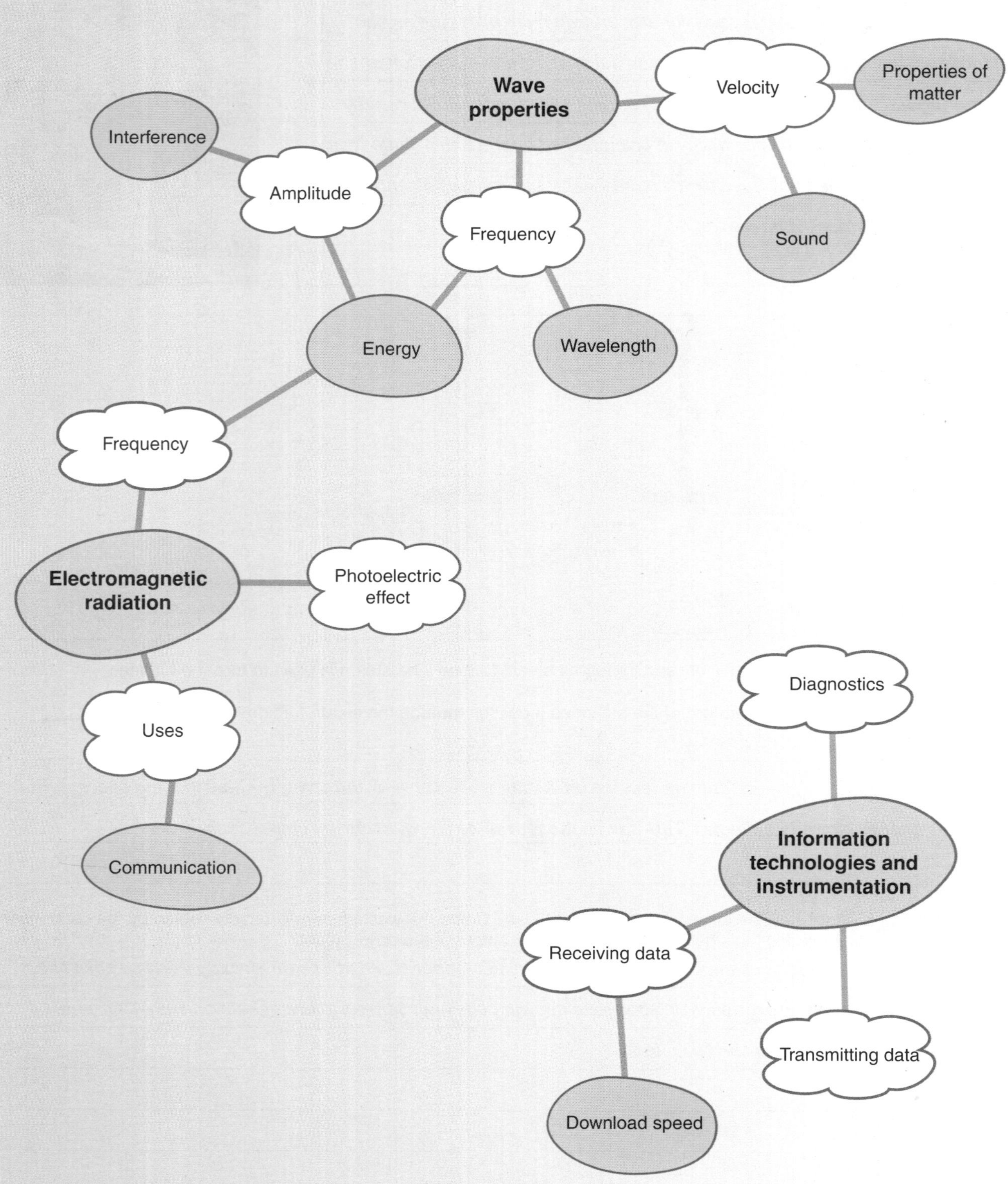

PS4.A

9: Wave Properties

Anchoring Phenomenon

Rogue waves: How might rogue waves in the ocean form? What principles are behind these phenomena? **111 119**

What are the properties of waves?

☐ 1 When you looked at energy in previous chapters, you learned that energy can be transformed from one form to another and transferred from one place to another. Think of examples of where energy is being transferred as waves. Which of your examples are mechanical waves and which are electromagnetic waves? Investigate the transfer of energy by mechanical waves in a slinky spring *[SEP-3]*. We can now identify the defining properties of waves and how these are related *[CCC-2]*. Use the findings of your investigations and mathematical representations *[SEP-5]* to describe the relationship between wave frequency, amplitude, and energy *[PS4.A]* *[HS-PS4-1]*. **112 120**

☐ 2 How do you think the medium would affect the speed at which energy is transferred? Use tuning forks or a frequency generator to investigate *[SEP-3]* the speed of sound in air. How do you think the speed of sound would change in a denser medium such as water or rock? How can you calculate the speed of sound by measuring wavelength and frequency? Develop an equation linking velocity, frequency, and wavelength *[CCC-2]* and use it calculate *[SEP-5]* unknown values for wave characteristics *[PS4.A]* *[HS-PS4-1]*. **113**

☐ 3 Volume is an important property of sound. How is it measured and what determines it? How is sound related to hearing and why can we not hear some of the sounds that other animals make? Recall your predictions for the speed of sound in different media. Now use the equation you developed to determine *[SEP-5]* how the speed of sound changes *[CCC-2]* in different media. Were your predictions correct? *[PS4.A]* *[HS-PS4-1]* **114 120**

☐ 4 We have seen how wave velocity is affected by the medium through which the wave travels. Now let's explore how waves are affected *[CCC-2]* by the movement of the object that produces them. This phenomenon is called the Doppler effect and it refers to the change in the frequency of sound, light, or other waves as the source and observer move towards (or away from) each other. The Doppler effect is noticed in everyday phenomena such as the change in pitch in a passing siren. Use mathematical thinking *[SEP-5]* to explain how the Doppler effect can be applied in everyday life (such as calculating the speed of vehicles) as well as in astronomy and medicine *[PS4.A]*. **115 120**

☐ 5 Investigate *[SEP-3]* interference, which is a property unique to waves. Describe what happens *[CCC-2]* when waves encounter each other and create a mathematical representation *[SEP-5]* of destructive and constructive interference *[PS4.A]* *[HS-PS4-1]*. **116 120**

☐ 6 A second property unique to waves is diffraction *[PS4.A]*. Investigate *[SEP-3]* how waves behave *[CCC-2]* when passing between barriers in a model system *[CCC-4]*. How is the wave affected when the gap is large relative to the wavelength of the wave? What about when the gap is similar to the wavelength? Use the findings of your investigation as evidence *[SEP-7]* for the properties of waves. Think of an everyday phenomenon where both diffraction and interference can be easily seen. Why do we consider these properties when designing structures such as breakwaters and marinas. **117 120**

How can we transmit information over wires and wirelessly?

☐ 7 Use your understanding of the properties and behavior of waves to describe how waves can encode information. Describe how technology makes use of this by (1) decoding wave interactions with media and (2) encoding signals using waves. Find out about these technologies and communicate *[SEP-8]* technical information about how the properties of waves and wave interactions with matter allow us to transmit and receive information *[PS4.A]* *[PS4.C]* *[CCC-2]* *[HS-PS4-5]*. **118**

☐ 8 Explain the difference between digital and analog technologies. Analyze and interpret data *[SEP-4]* about digital technologies and analog technologies with a similar role and assess the relative merits of each. Use what you have learned to evaluate questions *[SEP-1]* *[PS4.A]* about the advantages of using digital transmission and storage of information *[CCC-7]* *[HS-PS4-2]*. **118**

ADVANCED LEARNING

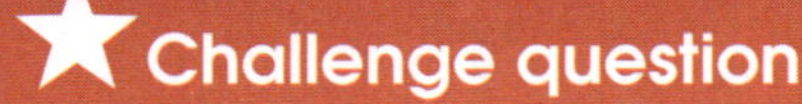

Challenge question:

How do the properties of waves explain the appearance of optical phenomena such as glories (top) and fog bows (bottom)?

Download this page from BIOZONE's Resource Hub

9. Wave Properties

Anchoring phenomenon

Rogue waves are an excellent example of a phenomenon with an unknown cause. Students may speculate that by adding waves together, a bigger wave can occur and this provides a starting point for wave properties. In reality, rogue waves do not follow the standard theory of wave formation at sea or the "linear model". Indeed it is possible rogue waves can be better explained by non linear quantum equations where instabilities occur in an otherwise stable or relatively flat surface.

What are the properties of waves?

1. Thunder and lightning and slinky springs are excellent vehicles for demonstrating wave properties. Thunder and lightning demonstrate the difference between waves traveling via different mediums and mechanisms. Slinky springs can visually illustrate both. Students should understand that although energy is moving along the wave, any part of the wave itself only moves back and forth from a central position. There are numerous on line visualizations for this. **BIOZONE's Resource Hub** has listed some particularly good examples.

2. The investigation for measuring the speed of sound has been designed so that students must find the relationship between the frequency and wavelength instead of simply calculating the speed of sound using a known equation. They do this by graphing frequency and wavelength and then using the equation from the graph to find the speed of sound. This is done to better show the investigative process rather than simply perform a series of steps to find an already known answer. Two methods are given. Although the experiment works when finding only one wavelength per frequency, space has been made available to find two wavelengths and so provide a way of confirming the first result.

3. Using the wave equation from the previous activity it is useful to calculate the speed of sound in various materials and then explain why the speed of sound is different in different materials. The speed of sound is affected by density and recoil ability of the material. Thus speed is higher in high density, high recoil material. Various animals have evolved to use sound in different ways. It is worth discussing with students why animals that use echolocation produce particularly high frequency calls and how the process works. Echolocation is the same as active sonar, using sounds made by the animal itself. The animal measures the time delay between its own sound emission and the echoes that return from the environment. Many echolocating species use their echolocation to forage and to navigate. The sound clicks are produced at a lower rate when searching for prey and a higher rate when prey is detected. Students could be challenged to ask why this is the case.

4. The Doppler effect is a commonly experienced phenomenon. Objects producing sound while moving cause the sound waves to bunch up in front of them, increasing the frequency and pitch of the sound. Behind the object, the waves spread out, decreasing the frequency. If the object is moving faster than the speed of sound, it catches up with the sound waves and one single wave front is produced. Where it touches the ground and sweeps over people, a single loud boom (the sonic boom) is heard. In aircraft, this is actually two booms, one produced by the front of the aircraft and one produced by the tail.

5. Interference is an important part of wave behavior. Students can develop a simple mathematical model of this behavior. Two positive or two negative wave amplitudes will add up to a larger amplitude (constructive) whereas a positive and a negative wave add up to a smaller amplitude (destructive). **BIOZONE's Resource Hub** contains interactive simulations where students can investigate the effect of constructive and destructive interference in different shaped waves.

6. Diffraction is the bending of a wave as it passes through a gap or past a barrier. When passing through a gap that is the same width or smaller than the wavelength there is a lot of bending in the diffracted wave. When the gap is wider than the wavelength, the diffraction is less and interference may occur on the edges of the diffraction pattern. There are many useful videos on the internet showing the effect of diffraction and interference and you can see examples of these on **BIOZONE's Resource Hub**.

How can we transmit information over wires and wirelessly?

7. Waves are how we transmit information. This includes talking (i.e. sound waves) and various electronic methods (e.g. flashing lights or using radio waves). Two important parts of this process are the encoding and decoding of information. Students should understand that in communications technology, encoding is the process of putting a sequence of characters into a specialized format for efficient transmission or storage. Decoding is the opposite process, i.e., converting an encoded format back into the original sequence of characters. Encoding and decoding are used in data communications, networking, and storage. Data can be stored as digital or analog. Students should be able to communicate how the properties of waves make it possible to encode and decode information, although this covered in more detail in the final chapter.

8. Show students a record, a CD, and a DVD as examples of how information is stored. Ask students if they think information is stored the same way on each device. They may think it is. Of course each format stores information differently. This can be used to introduce the difference between digital and analog. Students then have an opportunity to digitize their own analog wave (Activity 118). They see that the more sample points taken and the more precisely the points are recorded, the more precisely the reproduced wave is produced. To reconstruct a sine wave accurately we need to take at least two samples per cycle or wavelength, which means we must sample at a rate twice that of the frequency (i.e. two samples per wave). Human hearing is between 20 Hz and 20 kHz. Therefore, to capture the upper frequency of human hearing we must sample at a rate of at least 40 kHz. Generally sampling is done at around 44.1 kHz as this allows for various distortions, buffering, and filtering effects.

111 Rogue Waves

Key Question: How might rogue waves in the ocean form? What principles are behind these phenomena?

- Have you ever heard of the Draupner wave? It is famous as it provided direct evidence of a rogue wave at sea. The collection of the data (and since then data on many other rogue waves) has changed how ocean scientists understand wave generation at sea.
- In 1994, the Draupner E platform was built 160 km off the coast of Norway in the North Sea. As part of its construction it had instruments to measure wave height along with other meteorological and marine measurements.
- On the first of January 1995, a storm swept through the North Sea (as is common). Just after 3 pm, a massive wave of 25.6 m (from trough to crest), over twice the height of any other of the storm waves, smashed into the platform. Although the platform was only slightly damaged due to it standing over 20 m above mean sea level, damage to the underside of the platform confirmed the instrument's measurements.

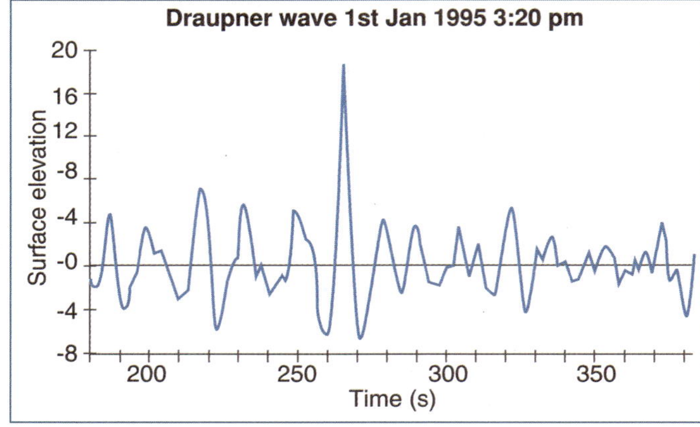

- Although the Draupner wave was not the first rogue wave ever recorded (that occurred in 1984) its height measured 6 m higher than models at the time estimated waves could reach. Soon after, in September of the same year, the cruise ship QE2 as hit by a 27 m high wave in the Atlantic. Clearly the wave models were not correct.

What are rogue waves?

- Rogue waves are any wave more than twice the height of the mean height of the largest third of the waves within the recording sequence. This means rogue waves don't have to be extremely high, rather they are an anomaly compared to other waves.
- Rogue waves tend to have very short wavelengths and as such are much steeper than other ocean waves. During storms, ships can usually ride over ocean waves, but rogue waves are so steep they are more likely to crash into and over a ship rather than sweep under it.
- What causes rogue waves is something of a mystery but they can be explained to some extent by some of the various wave properties you will study in this chapter. However, further study (e.g. from the internet) will provide other explanations of the appearance of rogue waves.

1. What is the biggest wave you have seen? *Student's answer. Students could use lake waves or ocean waves. If students are not in an area where large bodies of water are nearby, they could use internet examples.*

2. What was the weather like when this occurred? *Student's answer.*

3. How do waves on water bodies generally form? You may need to do some research. (Hint: why are waves on small lakes smaller than waves in the ocean?)
Wind is one of the major factors that affects wave size. Wind blowing over large distances can produce large forces on the surface of the water. Lakes have less open area than seas and so have smaller waves.

4. What might happen if one wave encounters or catches up with another wave?
The two waves will combine and become larger (taller).

5. How might this explain rogue waves? *If two or more waves meet, they could combine to form a very large wave.*

©2020 **BIOZONE** International
ISBN: 978-1-927309-82-7
Photocopying Prohibited

112 Properties of Waves

Key Question: How can we describe waves?

Thunder and lightning – sound waves versus light waves

- On March 5, 2019, residents in Southern California experienced a spectacular lightning event when the area from Santa Barbara to LA was hit by more than 2,200 lightning strikes. The unusual event was the result of the jet stream coming further south and pushing large amounts of moisture into the area.
- The sharp flash of lightning and the loud boom of thunder are phenomena you will be familiar with. When lightning strikes, we are able to see the flash of light immediately. It takes longer to hear the thunder because sound travels more slowly than light.
- Two types of wave form are created during a flash of lightning. The light wave emitted is a **transverse** wave. The sound wave produced by the sudden heating and expansion of the air is a **longitudinal** wave.

INVESTIGATION 9.1: Making waves

See appendix for equipment list.

1. This activity needs to be done in pairs.
2. On the floor, stretch a slinky spring between you and your partner until it is taut but be careful to not overstretch it.
3. Produce a **transverse wave** by flicking the slinky sideways one time to send a single wave to your partner. Observe what happens when the wave reaches your partner's end.
4. Now, flick the end of the slinky sideways repeatedly. Observe what happens as you vary the rate at which your hand moves.
5. Next, create a series of **longitudinal waves** by moving your hand towards and away from your partner. Observe this wave and how it travels.

1. (a) When you created the transverse wave, which direction did the wave move in comparison to the movement of your hand. Draw a diagram to help your description:

The wave travels perpendicular to the movement (displacement) of the slinky spring.

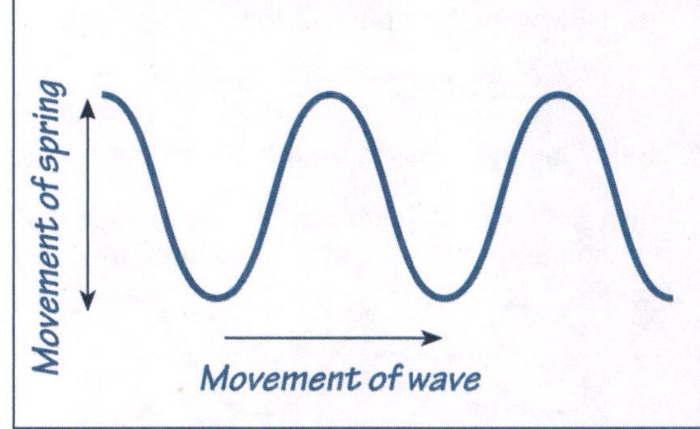

(b) When you created the longitudinal wave, which direction did the wave move in comparison to the movement of your hand. Draw a diagram to help your description:

The wave travels in the same direction as the movement (displacement) of the slinky spring.

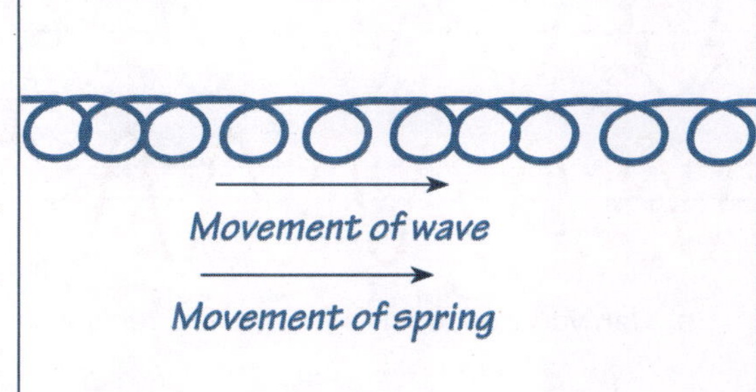

©2020 BIOZONE International
ISBN: 978-1-927309-82-7
Photocopying Prohibited

Properties of waves

- When we describe waves, we use four basic characteristics.
  - **Frequency** (f): the number of waves produced per second (measured in hertz, Hz)
  - **Wavelength** (λ): the distance between consecutive crests or consecutive troughs of a wave (in meters, m).
  - **Amplitude**: the displacement of a wave from the rest position. The amplitude is related to the amount of energy carried by a wave (more energy = greater amplitude).
  - **Period** (T), the time it takes for a single wavelength to pass a fixed point (1/f).
- Longitudinal and transverse waves are the two types of wave form. There are two types of wave that you will study. These are electromagnetic waves and mechanical waves.
  - **Mechanical waves** are waves that require a material medium to pass through. The energy is passed from atom to atom, molecule to molecule, through successive collisions of the particles in the material. As such, mechanical waves, e.g. sound, are unable to travel through a vacuum.
  - **Electromagnetic waves** do not require a material medium to pass through. They are able to traverse the matterless regions of space between cosmic bodies. Electromagnetic waves are affected by the presence of matter which can slow the waves down as the energy is absorbed, transmitted, reflected etc by the particles it comes in contact with. The speed of an electromagnetic wave changes as it changes mediums (the same as mechanical waves).

2. The diagram below shows a transverse wave:

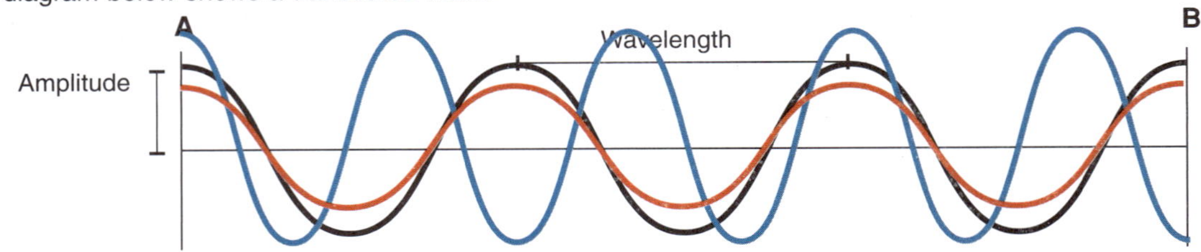

(a) Draw onto the diagram between points A and B a transverse wave with a shorter wavelength than shown:

(b) Would your wave have a higher or lower frequency than the wave shown? *Higher frequency.*

(c) In a different color, draw a wave with a smaller amplitude than the one shown:

3. The diagram below shows a longitudinal wave. It has the same wavelength as the one shown above:

(a) Mark on the diagram the three full wavelengths:

(b) How would you be able to tell what the amplitude of a longitudinal wave is? *The amplitude would be reflected in how close the compressed lines are in the crest (or inversely, by how far apart they are in the trough).*

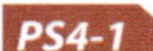

4. Study the three waves on the grid below.

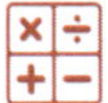

(a) Use the grid to measure the wavelengths of the waves in meters. The waves are the same type in the same medium so the frequency can be measured by counting the number of wavelengths shown. Fill in the table on the right:

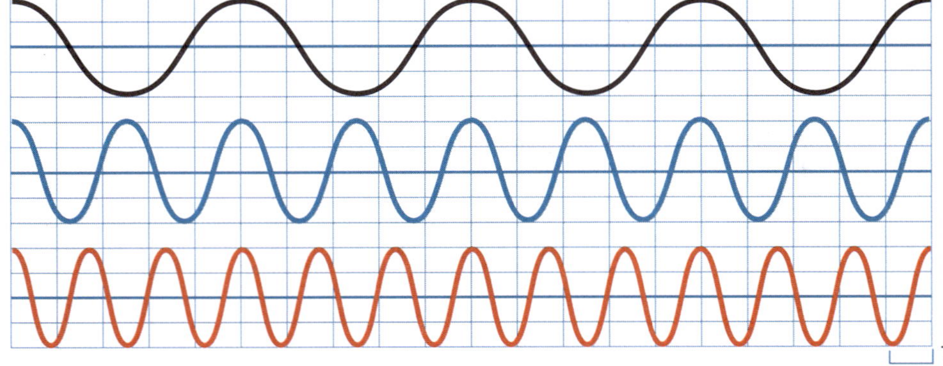

Wavelength (meters)	Frequency (Hz)
5.0	*4*
2.5	*8*
1.67	*12*

(b) Can you work out a mathematical relationship between the frequency and the wavelength?

Wavelength × frequency = a constant. m × 1/s = m/s. The constant must be velocity (m/s) so: $v = f\lambda$

©2020 BIOZONE International
ISBN: 978-1-927309-82-7
Photocopying Prohibited

What moves in a wave?

- Recall Investigation 9.1. You observed how a wave traveled along a slinky spring. But what actually traveled? Was it the spring or something else?

INVESTIGATION 9.2: More waves See appendix for equipment list.

1. This activity needs to be done in pairs.
2. This time, mark regular intervals on the spring using tape. On the floor, stretch the slinky spring between you and your partner until it is taut but be careful not to overstretch it.
3. Produce a **transverse wave** by flicking the slinky sideways one time to send a single wave to your partner. Observe how the marks on the spring move relative to the wave.
4. Next, create a series of **longitudinal waves** by moving your hand towards and away from your partner. Again, observe how the marks on the spring move relative to the wave.

- You will have noticed that the marks on the spring return to their stating point once the wave has passed. If this happens what is actually moving along the spring?
- A wave is actually a transfer of energy. In the spring example, energy delivered by your hand at one end of the spring is transmitted from one particle in the spring to the next until the energy reaches the end of the spring.
- Because the particles (atoms) in the spring are joined together by metallic bonds, the movement of one atom causes the movement of the one next to it and so on.
- As a wave passes through a gas, the particles collide, passing on the energy and returning to their original position.

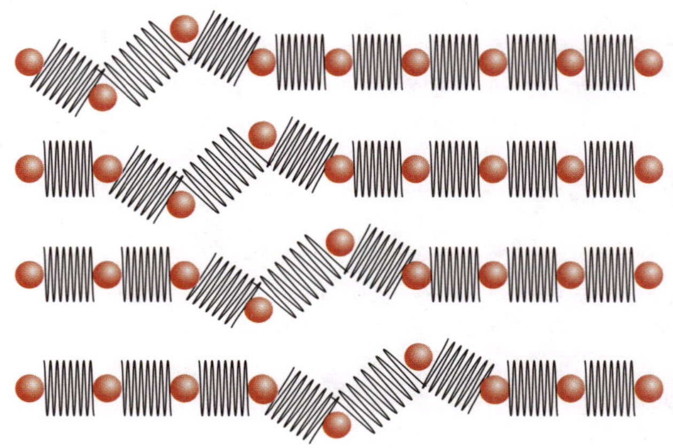

Movement of particles in transverse wave

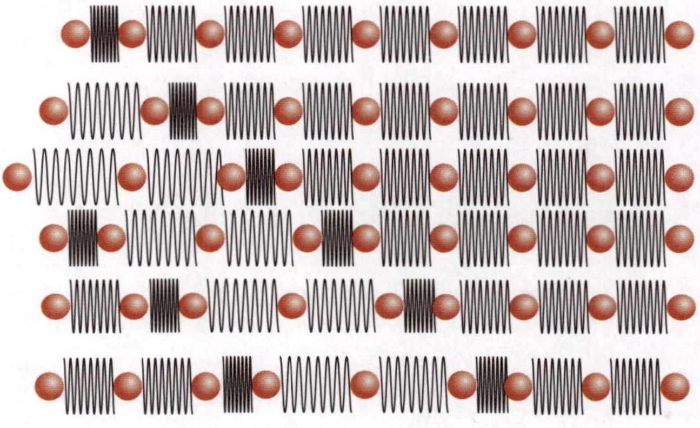

Movement of particles in longitudinal wave

Amplitude, frequency and energy

- In a mechanical wave, the energy, frequency, and amplitude are related. The greater the energy in the wave, the greater the movement of the particles from their resting position and so the greater the amplitude. Thus amplitude is a measure of wave energy.
- The energy in a wave is also related to frequency. The higher the frequency, the faster the particles oscillate and so the greater the amount of energy in the particles.
- Just as energy in a moving object is proportional to the square of the velocity, the energy in a mechanical wave is proportional to the square of the amplitude and the square of the frequency (this is not true for light waves).
- Consider the waves below.

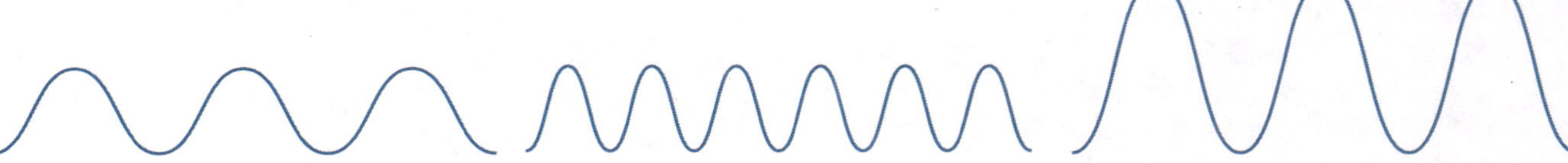

Amplitude = 1 frequency = 3. E~ $A^2 = 1$ and E~ $f^2 = 9$: 1 × 9 = 9

Amplitude = 1 frequency = 6. E~ $A^2 = 1$ and E~ $f^2 = 36$: 1 × 36 = 36

Amplitude = 2 frequency = 3. E~ $A^2 = 4$ and E~ $f^2 = 9$: 4 × 9 = 36

- Thus doubling the amplitude or the frequency quadruples the energy of the wave.

5. Which direction does energy move in a wave? *Energy moves in the same direction as the wave.*

6. How is energy transferred in a wave? *Energy is transferred from one particle to the next by either collisions or via the bonds holding particles together.*

©2020 **BIOZONE** International
ISBN: 978-1-927309-82-7
Photocopying Prohibited

113 The Speed of Sound

Key Question: How can we calculate the speed of sound by measuring its wavelength and frequency?

- Sound travels as a longitudinal wave. It is a pressure wave carried by the vibration of air molecules.
- Recall the example of lightning strikes. A lightning bolt can heat the air around it to 30,000°C. This extreme heating causes the air to expand explosively creating pressure on the air immediately around it. Due to electrostatic forces between the particles in the air, the area of high pressure pushes out on the air further away before the air molecules recoil back to their original position. This wave of pressure continues outward away from the origin, but as the affected area increases the further away from the source, the less intense and more distorted the sound gets (recall the inverse square law).
- Many mechanical waves are the result of variation of pressure in the material (air, water, rocks, etc.). These waves are called pressure waves. Examples of waves resulting in variations of pressure include sound waves and seismic waves.

Pressure waves from the expansion of gases from the guns of the battleship *USS Iowa* can be clearly seen flattening the surface of the water.

- In question 4 of the previous activity you should have noticed a relationship between the frequency of sound and the wavelength. When the wavelength decreases, the frequency increases.
- This relationship between frequency and wavelength can be used to work out the velocity of sound in air. We can vary frequency and record how wavelength changes. Graphing these variables will show us the relationship between them.

PS4-1

INVESTIGATION 9.3: The speed of sound in air See appendix for equipment list.

1. To start you will need tuning forks of various frequencies or a frequency generator with a speaker attached. You can download various cellphone apps that will do this and play the sound through the phone's earphones or speaker.
2. Set up the experiment as in the diagram. The PVC tube should be about 3-4 cm in diameter and about 30-60 cm in length. The cylinder of water also needs to be about the same depth (30-60 cm).
3. Record the frequency you are using (in hertz, Hz) (between 200 and 1000 Hz works).
4. Start with the PVC pipe fully down, with the top just out of the water.
5. Hold the frequency generator close to the opening of the pipe. While the sound is being generated, slowly move the pipe and the frequency generator up until you hear a loud humming in the pipe. This is called the first harmonic.
6. Clamp the pipe tight at this point and measure L (the distance from the top of the pipe to the water). Record this as L1 (in meters) in the table on the next page (L1 (m))
7. Now with the same frequency sounding continue to raise the pipe out of the water until a loud humming is heard again. This is called the second harmonic.
8. Clamp the pipe tight and measure L again.
9. Record this as L2 in the table on the next page (L2 (m)). Repeat steps 3-9 with at least four other frequencies.

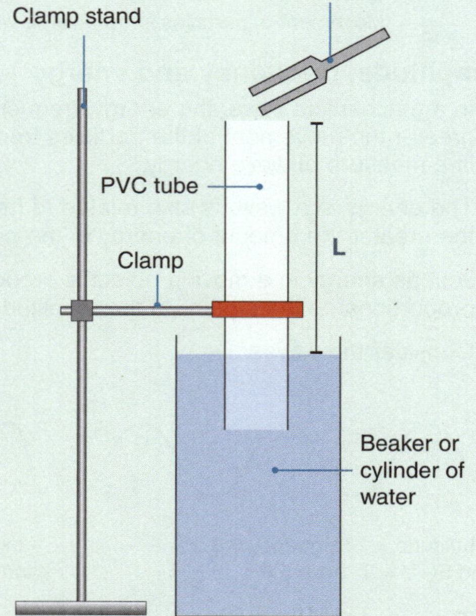

- An alternative set up for this experiment is provided on **BIOZONE's Resource Hub**

©2020 **BIOZONE** International
ISBN: 978-1-927309-82-7
Photocopying Prohibited

1. Record your results in Table 1 below:

Table 1

Frequency (Hz)	L1 (m)	L2 (m)
256	0.33	Student's
384	0.23	results
446	0.20	
512	0.16	
625	0.14	

2. (a) Using the equation $\lambda = 4L$ (where λ is the wavelength) calculate the wavelength of sound in the pipe using your value for L1. This formula works if only one value of L is obtained. Enter the wavelength into Table 2 below in the column labeled λ. Do this for all your frequencies.

(b) Now calculate $1/\lambda$ and enter this into Table 2. Because wavelength decreases as frequency increases a plot of frequency vs wavelength would produce a negative gradient. This can be reversed by plotting f vs $1/\lambda$ instead.

(c) Plot f vs $1/\lambda$ on the grid provided using the data in table 2.

(d) A second formula using both L1 and L2 (if L2 was obtained) can be used to calculate the wavelength: $\lambda = 2(L2 - L1)$. Calculate the wavelength and enter it into Table 3 in the column labeled λ. Do this for all your frequencies.

(e) Now calculate $1/\lambda$ for this second calculation of wavelength. It should match your calculations of wavelength using just L1, giving a way to check your measurements and calculations. Enter $1/\lambda$ into Table 3.

(f) Plot f vs $1/\lambda$ on the grid provided using the data in table 3 (use a different color and produce a key).

Table 2

Frequency (Hz)	λ	$1/\lambda$
256	1.32 m	0.76
384	0.92 m	1.08
446	0.80 m	1.25
512	0.64 m	1.56
625	0.56 m	1.78

Table 3

Frequency (Hz)	λ	$1/\lambda$
	Student's	
	results	

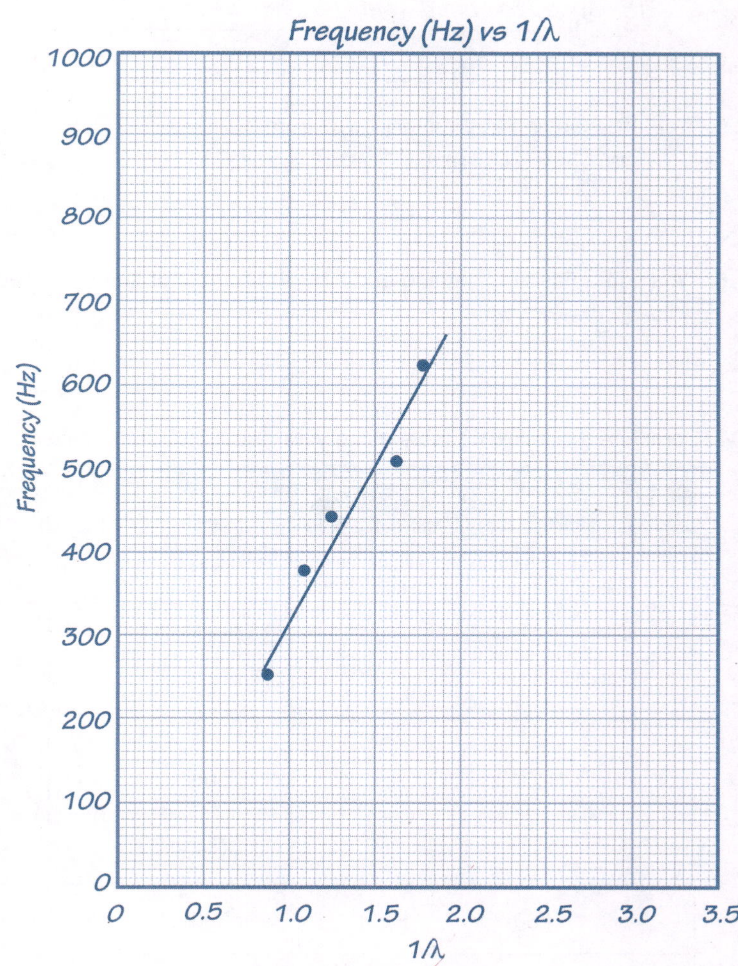

©2020 **BIOZONE** International
ISBN: 978-1-927309-82-7
Photocopying Prohibited

3. (a) Calculate the gradient of the graph (gradient = rise/run) for the line you produced (using either Table 2 or Table 3):

   170 ÷ 0.5 = 340 (using just L1) (table 2)

   (b) The equation for the gradient of your graph is: gradient = f ÷ 1/λ. Rewrite the equation as multiplication:

   gradient = f × λ.

   (c) Frequency is measured in cycles per second (1/s) and wavelength is measured in meters (m). What then is the gradient of the graph measured in and what does the gradient then represent? Show your working:

   g = f × λ = 1/s × m = m × 1/s = m ÷ s = m/s.

   Thus g = m/s or meters per second (velocity) The gradient is wave speed.

   (d) Based on the gradients of your lines, what is the speed of sound in air?

   340 m/s (using just L1)

   (e) Which line is closer to the speed of sound?

   Student answer. The lines should be the same.

   (f) Write the equation linking velocity (v), frequency (f) and wavelength (λ): *v = f × λ*

4. (a) Air particles are well spaced out and the restoring force (the force that pulls particles back into place) between the gas particles is not particularly high. What do you think would happen to the speed of sound in water?

   The speed of sound would be faster in water than in air.

   (b) Diamond is made of carbon atoms tightly bound together with covalent bonds. How would this affect the speed of sound passing through it? Explain your answer:

   The speed of sound would be even faster in a diamond than in water because of the restoring forces.

5. You are standing 200 meters away from a canyon wall. If you shout, how long will it take before you hear your echo?

   200 m × 2 = 400 m ÷ 340 m/s = 1.2 s

6. You see a flash of lightning and hear the rumble of thunder 5 seconds later. How far away did the lightning occur?

   5 × 340 = 1700 m

7. Complete the following table using your equation from question 3. (f):

Frequency (Hz)	Wavelength (m)	Velocity (m/s)	Period (s)
500	200	*100,000*	*2×10^{-3}*
3.75	120	450	*0.267*
1000	376	*376,000*	*1×10^{-3}*
8.4	305	2569	*0.12*
3521	*0.25*	897	*2.8×10^{-4}*
230	300	*69,000*	*4.3×10^{-3}*
2055	*0.22*	450	*4.9×10^{-4}*

©2020 **BIOZONE** International
ISBN: 978-1-927309-82-7
Photocopying Prohibited

114 Sound and Hearing

Key Question: How is sound perceived by animals? How is sound affected by different media?

- Sound is a longitudinal wave that vibrates objects it encounters. In humans, sound is heard when sound waves enter the ear. Bones in the middle ear transfer the vibrations to the inner ear, where they deform sensory cells and the deformation is converted into nerve impulses. The brain then interprets the nerve impulses as sound.
- As shown in the figure below, there is a wide range of frequencies for both sound production and reception amongst animals. The blue bars in the figure indicate sound production and the orange bars sound reception. Sharks have no sound producing organs as such and cannot make sounds through vocalization. There are reports of a few species that can 'bark' when threatened but this is the result of forceful expulsion of water.

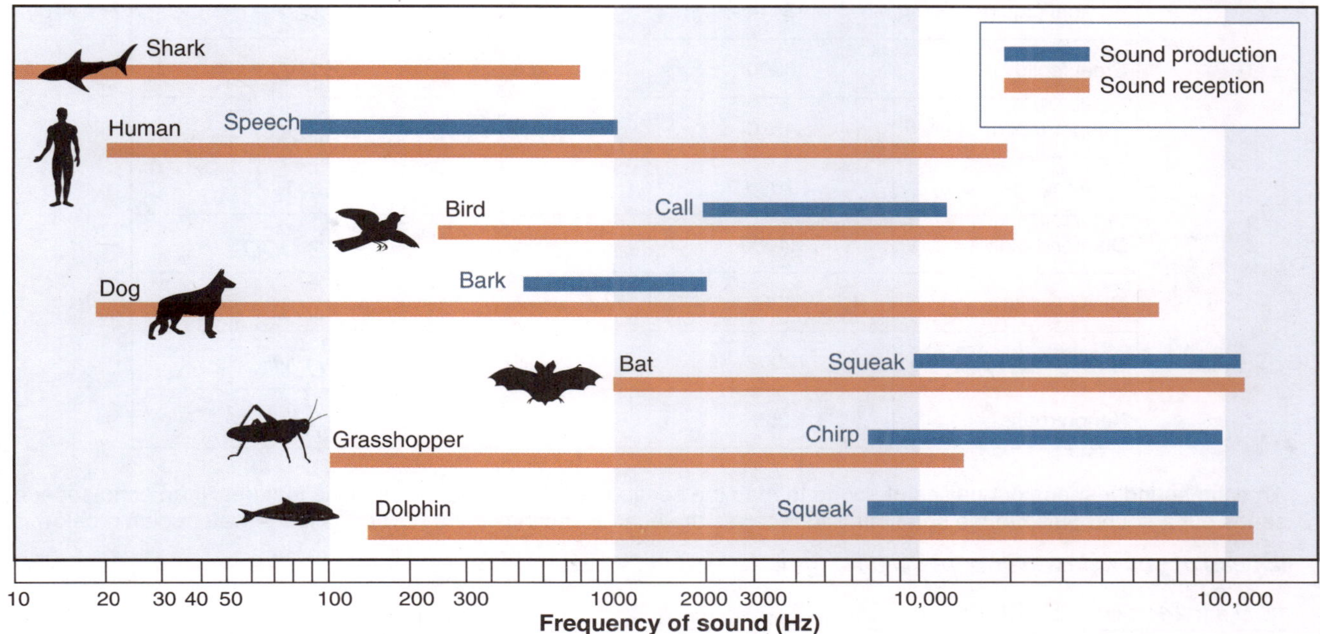

- Sound volume (how loud something is) is an important property of a sound wave.
- The greater the amount of energy put into the a sound wave the greater the vibrations of the air and thus the louder the sound.
- As the sound wave propagates outward from the source it loses energy (and thus volume) in accordance with the inverse square law.
- Sound intensity is measured in energy per second per square meter or watts/square meter (W/m^2). The unit decibel (dB) measures the change in intensity relative to a reference and is logarithmic. Thus a sound at 30 dB is 10 times more intense than at 20 dB and 100 times more intense than 10 dB.
- A difference in 10 dB is heard by humans as twice as loud.
- Exposure to sound levels above 85 dB can cause permanent hearing loss, while pain is experienced at around 120 dB.

Cause of sound	dB
Blue whale call	160
Space rocket launch	
Artillery fire	140
Small arms fire	
Jet aircraft at take-off	120
Rock concert	
Chainsaw	100
Lawn mower	
Alarm clock	80
Busy traffic	
Normal conversation	60
Moderate rainfall	
Quiet room	40
Isolated desert night	
Whisper	20
Normal breathing	
	0

©2020 BIOZONE International
ISBN: 978-1-927309-82-7
Photocopying Prohibited

CE
PS4.A

1. Sound travels at 343 m/s in air. Calculate the shortest and longest wavelength of sound heard by the following animals:

(a) Human: *Shortest = 17 m, Longest = 0.017 m*

(b) Bat: *Shortest = 0.343 m, Longest = 3.43×10^{-3} m*

(c) Dog: *Shortest = 17 m, Longest = 4.9×10^{-3} m*

2. Question 1 used the speed of sound in air. Air particles are quite far apart and loosely bound. The closer together and more tightly bound particles are, the faster a sound wave can travel through them. Use the data in the table below to calculate the speed of sound in the various media:

Medium	Frequency (Hz)	Wavelength (m)	Speed of sound (m/s)
Water (liquid)	2980	0.5	*1490*
Ethanol (liquid)	2414	0.5	*1207*
Steel (solid)	6300	0.5	*3150*
Copper (solid)	4520	0.5	*2260*
Polyethylene (solid)	1080	0.5	*540*
Diamond (solid)	24000	0.5	*12,000*
Glass (solid)	4760	0.5	*2375*
Helium (gas)	2002	0.5	*1001*
Xenon (gas)	356	0.5	*178*

3. An echo-sounder sends out pulses of sound to map the seafloor or other marine or aquatic features. If an echo-sounder sends out a sound pulse to the sea floor and receives the echo 5 seconds later, how deep is the water column below it?

Speed of sound in water = 1490 m/s. Time to bottom = 5/2 = 2.5 s.

2.5 s x 1490 m = 3725 m

4. (a) What properties of the medium the wave is traveling through might affect the speed at which a wave can travel?

Properties include density (mass per unit volume), temperature, state of matter, elasticity (the tendency of a material to maintain its shape and not deform when a force is applied to it).

(b) Choose two of the above properties and explain why you think they affect the speed of the wave:

Density: If a material is more dense because its molecules are larger, it will transmit sound slower because larger particles are harder to move.

Temperature: Particles are in higher states of motion at higher temperatures and this leads to faster wave speeds.

State of matter: Particles closer together can transmit waves faster. Waves travel faster in solids than in liquids and faster in both solids and liquids than in gases.

5. Calculate how many times more intense than a quiet room the sound of each the following has:

(a) Normal conversation: *100×*

(b) Artillery fire: *1×10^{10}*

6. Some animals can produce extremely loud noises even though we humans can't hear them. Explain how sounds can be loud yet inaudible to us:

If the frequency is outside our hearing range it can not be heard even if it is loud (although in some cases it could be detected simply due to the pressure wave it produces).

©2020 **BIOZONE** International
ISBN: 978-1-927309-82-7
Photocopying Prohibited

115 The Doppler Effect

Key Question: How are waves affected by the movement of the object producing them?

- The Doppler effect is something you have probably heard many times, especially on city streets, but also anywhere there are fast moving objects.
- Imagine a police car moving towards you at a constant speed with its siren on. The sound reaches you at a certain pitch. When the police car passes you the sound changes to a lower pitch. This is the Doppler effect in action.
- The Doppler effect is the change in the pitch of a sound in relation to an observer who is moving relative to the source of the sound.

Wave front diagrams

The Doppler effect is best represented by diagrams showing the waves as they are emitted or pass an observer.

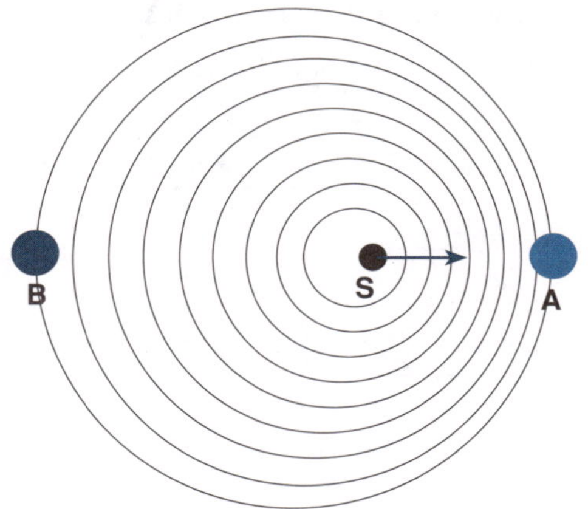

The source S is moving towards observer A. This bunches up the waves in front of the source and spreads them out behind. Observer A hears a higher pitch than observer B.

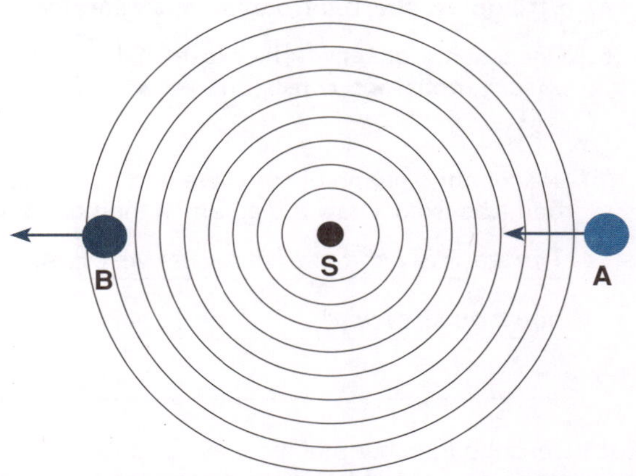

If the source is stationary, but the observers are moving the effect is the same. Observer A (moving towards the source hears a higher pitch than observer B (who is moving away).

Uses of the Doppler effect

The Doppler effect is used to detect speed. The difference between the frequency of light waves emitted by the radar and those reflected back by the car can be used to calculate the car's speed.

Red shift is a phenomenon seen in distant galaxies. Light from galaxies moving away from us is stretched into the red part of the visible spectrum. This is used to calculate the rate of expansion of the universe.

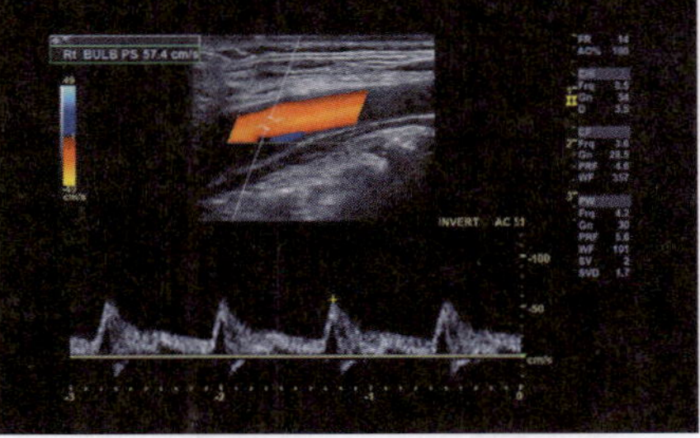

The Doppler effect can also be used to detect blood flow in the arteries. Ultrasonic sound waves are emitted and the change in frequency can be used to detect the direct of blood flow.

1. An observer is stationary while a siren sounding at 4000 Hz moves towards them at 20 m/s. A second observer moves at 20 m/s towards a stationary siren sounding at 4000 Hz. Explain why the observers will hear the same pitch.

   Because the observers are moving at the same speed relative to the siren they will both experience the same change in pitch (Doppler effect) and so hear the same pitch.

2. Astronomers observe a star they know to be a yellow star similar to the Sun to be changing in color in the repeating sequence blue - yellow- red- yellow. Describe the probable motion of this star:

   The star is probably moving in a circle (orbit), moving closer and then further away from us.

©2020 **BIOZONE** International
ISBN: 978-1-927309-82-7
Photocopying Prohibited

116 Wave Reflection and Interference

Key Question: How do waves interact with their surroundings and each other?

Reflection at boundaries

INVESTIGATION 9.4: Boundaries and wave reflection See appendix for equipment list.

1. This can be done with a slinky spring or a length of rope or hose. With a partner, lay the rope (at least 4 meters) on the ground, stretch it
2. Have your partner hold the rope at one end. At the other end of the rope, reach down and give it a quick horizontal flick perpendicular to its length. The flick need to be long enough for the wave to travel the length of the rope and reflect back to you (or a significant way along the rope).
3. Observe the wave created in the rope and its reflection. Repeat this 3-4 times to make sure the observations are consistent.
4. Straighten the rope again. This time leave the rope free at your partners end when you flick the rope.
5. Once again observe the wave in the rope and its reflection. Repeat this 3-4 times to make sure the observations are consistent.

1. (a) Describe the shape of the wave and its reflection when the end of the rope was fixed. Draw a diagram in the box to help your explanation:

The wave is reflected in the same phase (it appears upside down on the rope).

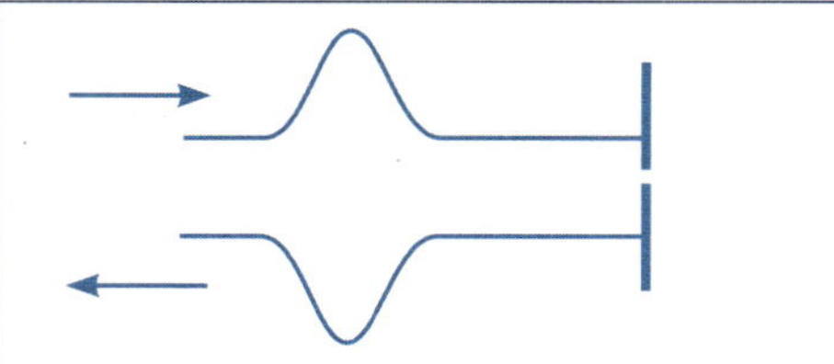

(b) Describe the shape of the wave and its reflection when the end of the rope was not fixed. Draw a diagram in the box to help your explanation:

The wave is reflected in the opposite phase (it is reflected on the same side of the rope as the original wave).

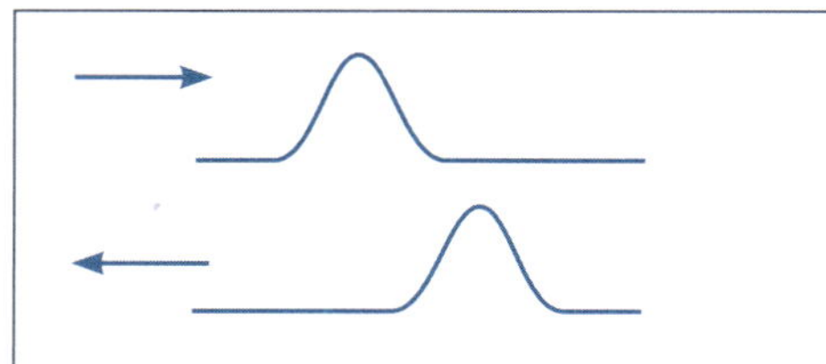

Wave interference

- The diagram below show two pairs of waves of waves. In each pair, the waves are traveling towards each other. Study them and think about what happens when the waves meet.

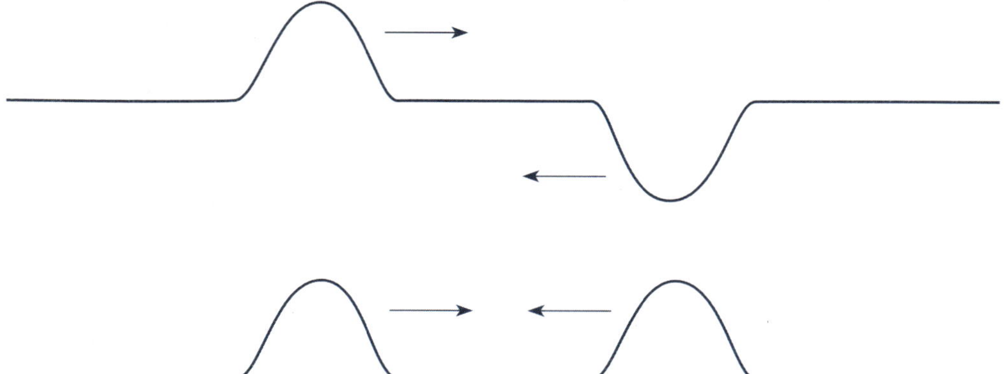

- We can predict what will happen if we add a y axis to the diagram and give each wave a height of 1 or -1 depending on if it is above of below the line (x axis).
- For the top wave 1 + -1 = 0. The waves will cancel out. This is called **destructive interference**.
- For the bottom wave 1 + 1 = 2. The wave amplitude will add together and in this case double the wave height. This is called **constructive interference**.

©2020 **BIOZONE** International
ISBN: 978-1-927309-82-7
Photocopying Prohibited

Waves can pass through each other

- The diagram on the previous page shows waves traveling towards each other. When they meet, they either cancel each other out or they add up to a larger amplitude. However, the waves do not stop there. Once they have met they pass through each other and continue on their way as if nothing happened.
- The diagram below shows how this can be demonstrated. Two speakers are placed far apart. Each one produces a pulse of sound with a slightly different frequency and wavelength but the same volume. The microphones pick up the sound as it passes.

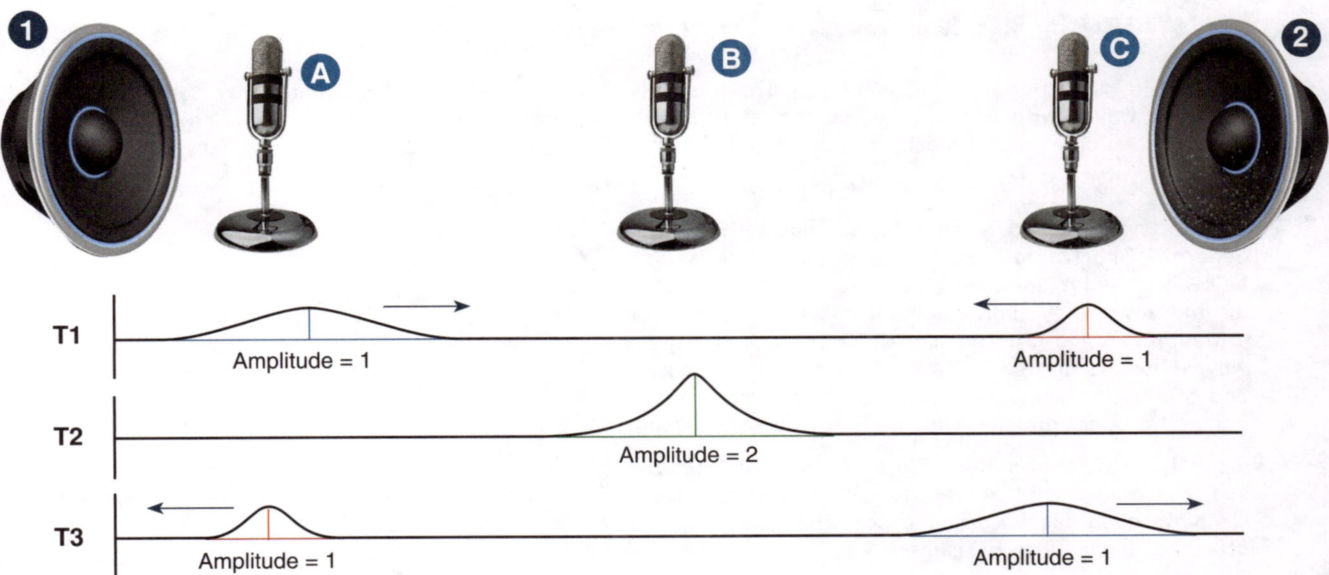

- At T1 (top graph) microphone A picks up the low frequency of speaker 1, while microphone C picks up the higher frequency of speaker 2. At T2 microphone B picks up the combined sound of both speakers. This produces a louder sound that is recorded by microphone B. At T3 microphone A picks up the higher frequency produced by speaker 2 and microphone C picks up the lower frequency produced by speaker 2.

2. The diagram below shows A wave as it approaches then reflects back the way it came:

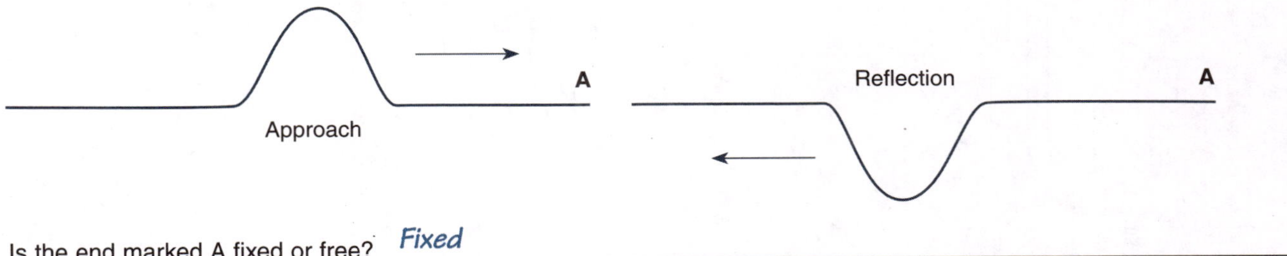

Is the end marked A fixed or free? *Fixed*

3. Complete the diagram bottom to show the end result of:

(a) Constructive interference between the two waves: (b) Destructive interference between the two waves:

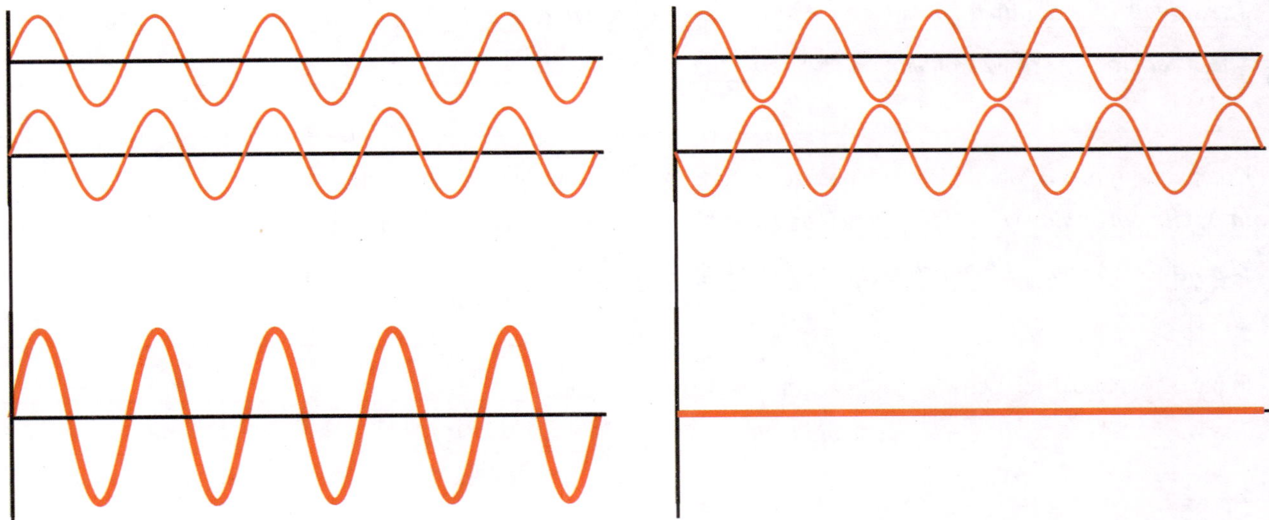

©2020 **BIOZONE** International
ISBN: 978-1-927309-82-7
Photocopying Prohibited

117 Diffraction and Interference

Key Question: What happens to waves when they pass through a narrow gap?

Investigating waves and gaps

INVESTIGATION 9.5: Two properties unique to waves — See appendix for equipment list.

1. For this investigation you will work in pairs or small groups. You will need a square tray, something to generate waves with (e.g. a ruler or thin stick slightly shorter than width of the tray), and two longer and three shorter wood blocks to act as barriers. You can use masses to hold the barriers down.
2. Add water to the tray so that the depth will be about 10 mm below the top of the barriers.
3. At one end of the tray, one student holds the "wave generator" horizontally so that its lower edge is just below the water surface. Move the wave generator up and down very slightly (about a couple of mm), at a steady rate, to create a series of regular straight waves flowing across the water towards the opposite end of the tray. Important: do not touch the tray or bump the bench as this will produce unwanted ripples.
4. Place the barriers in the middle of the tank as shown in the arrangements below (viewed from above) and complete each one by drawing how the waves are affected beyond the barriers.

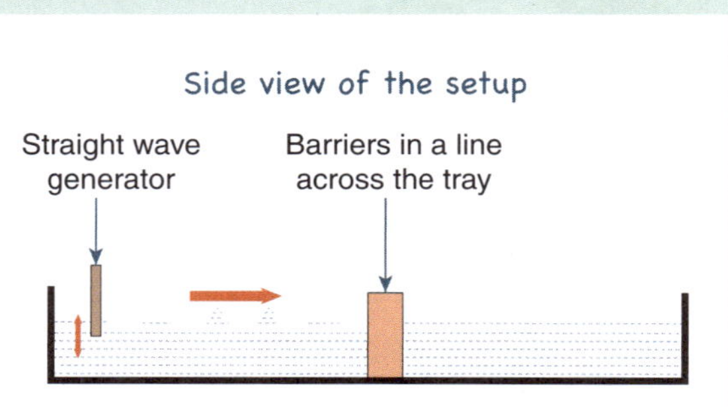

View of each arrangement from above

A

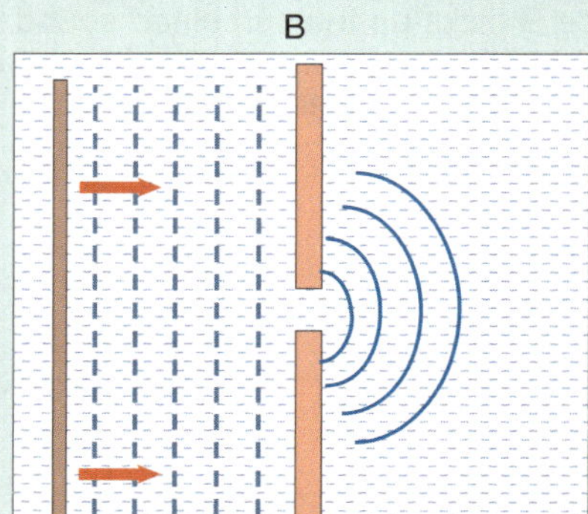

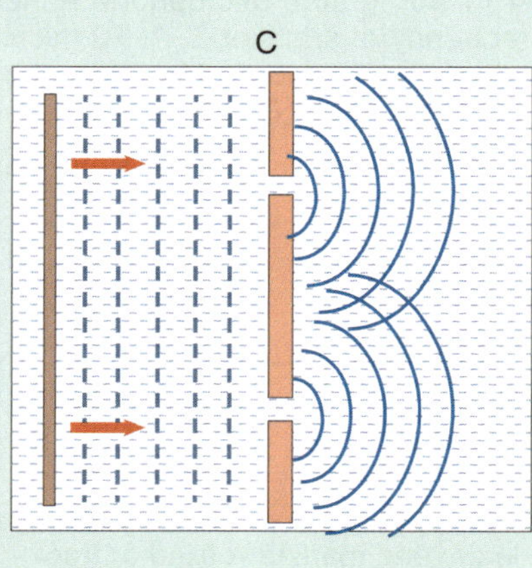

1. In A and B, what did you observe as waves passed through the gap (especially near the edges of the gap)?
The waves spread out from the gap. At the edges of the barriers, the wave started spreading out away from the direction of the wave through the gap.

2. Comparing A and B, what effect did the size of the gap have on what you observed?
In A, the spreading was more restricted and the waves just progress more or less straight forward. In B, the waves spreading from the edges of the barriers spread out more (as in the photograph opposite).

3. In part C, what did you observe happening at the two narrow gaps?
The waves start to spread out at each gap as for B.

©2020 **BIOZONE** International
ISBN: 978-1-927309-82-7
Photocopying Prohibited

4. In part C, what did you observe happening **beyond** the two narrow gaps?

As the waves spread out, they intersected, interfering with other's pattern. In places the waves appeared higher than each were on their own and at others there appeared to be no waves at all.

Wave diffraction

- **Diffraction** is the "bending" of waves as they pass through gaps or pass close to edges of obstacles. This can be seen on a large scale with sea waves around harbours(right) and ships. Diffraction is a property unique to waves (nothing else known to science is able to do this).
- If the gaps are large relative to the wavelength then the bending is only slight.
- If the gaps are similar in size to the wavelength (or smaller) then the bending is very pronounced.

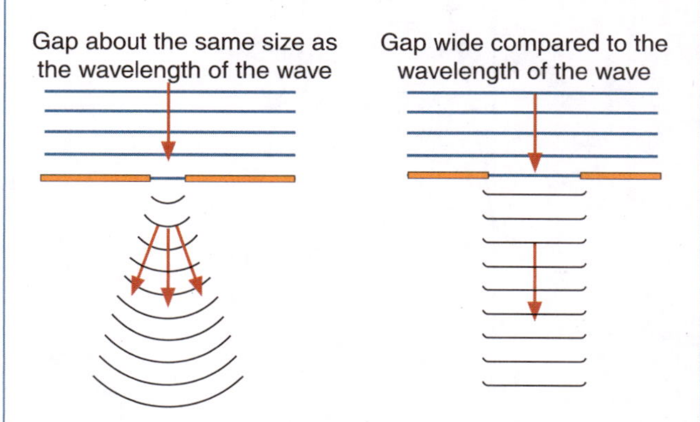

Diffraction and interference

- When waves pass through two gaps and both diffract they will eventually meet each other. Overlapping crests and troughs add together in some places and cancel each other out in other places. This interference is also a unique property of waves.
- Destructive interference occurs where a trough and a crest meet. There is no displacement. This is called a **node**.
- Constructive interference occurs where the crests overlap. The amplitude of the wave doubles. This is called an **antinode**.

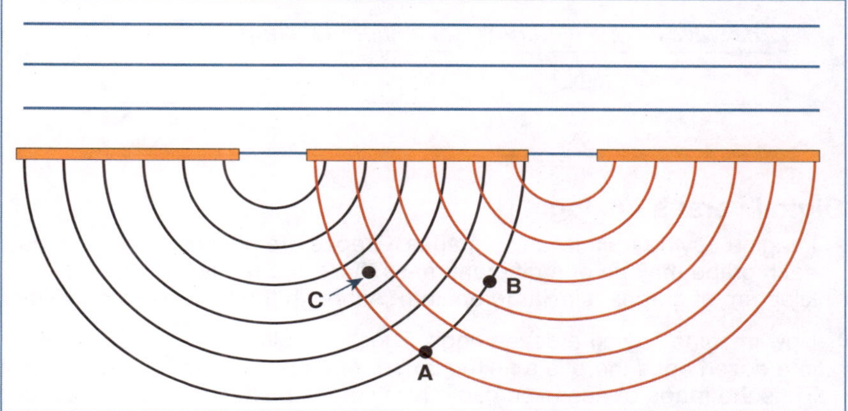

5. (a) In the diagram above right. What would you expect to see at point A? (assume lines represent wave crests):

Two crests come together so the height of the crest would twice the normal height

(b) What would you expect to see at point B?:

A crest and a trough come together so there would be no displacement (flat).

(c) What would you expect to see at point C?:

Two troughs come together so the trough would be twice the normal depth.

6. A boat club had a marina built with a breakwater like the one show in the photograph top right. They later decided a second entry was needed because of the volume of boat traffic (as in the diagram above right). Explain why this might not be a good idea:

A second gap would allow for interference to occur in the incoming waves. It is possible this could cause higher than expected waves and choppy waters in some areas of the marina and make mooring (tie-up) conditions difficult.

©2020 **BIOZONE** International
ISBN: 978-1-927309-82-7
Photocopying Prohibited

118 Digitizing Waves

Key Question: How can a sound wave be stored as data and be retrieved to reproduce the sound?

Graphing sound data

PS4-2 PS4-5

- Waves, whether they are transverse or longitudinal, can be displayed on a graph as a plot of the displacement of a particle from its original position relative to time.
- The data right shows an air particle's movement over time as a sound wave is played from a loud speaker. The units are arbitrary.
- Sound varies continuously, so a graph of the displacement of an air particle will display a complex wave. But this can still be measured and stored as data.
- The data to reproduce a certain sound can be stored as digital or analog data.

Time	Particle displacement
0	0
1	5
2	8
3	9
4	8
5	5
6	0
7	-5
8	-8
9	-9
10	-8
11	-5
12	0

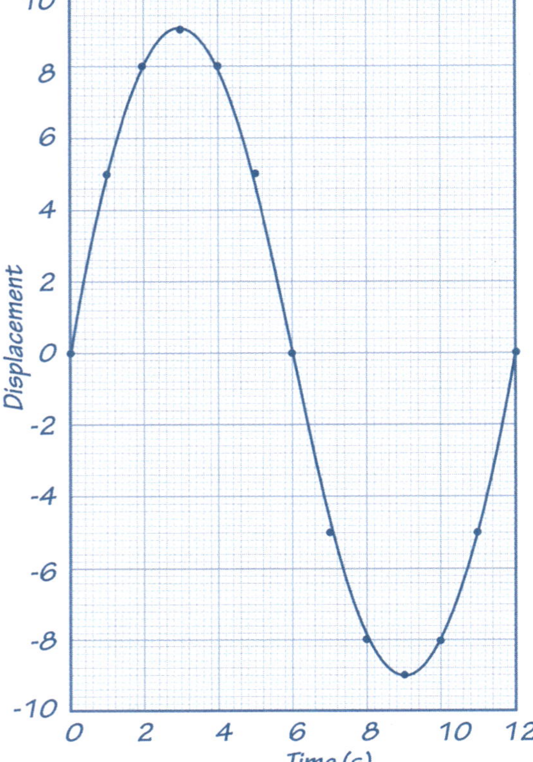

1. (a) Plot a line graph of the particle data right on the grid provided:

   (b) What can you say about the shape of the wave?

   The wave have a symmetrical sinusoidal shape.

Digital versus analog

- Imagine buying fruit, let's say grapes. Grapes are sold by mass, e.g. buying half a kilogram of grapes. Because each grape may be of a different mass there is a range (or scale) of the number of grapes that might be in half a kilogram of grapes (similar to analog). Although the range is not limitless, it is none-the-less very large.
- Now imagine buying a dozen eggs. Eggs are sold whole, never in halves or quarters and not usually by mass. Thus in a dozen eggs there is a finite number of eggs - 12. Additionally eggs are normally sold in groups of 6, 12, or 30. Thus the range of eggs you can buy is stepped. There are no in-between numbers of eggs (similar to digital).
- Another example of the difference between analog and digital is how time is displayed on analog and digital clocks.

Phrontis CC3.0

On an analog clock, the clock hands rotate continuously, around a continuous scale. As there are no steps in the time displayed, an analog clock can theoretically display any time at all (an infinite number of times) within the limits of our vision and the precision of the scale. In this case, the time displayed is a continuous or analog variable.

Digital clocks are programmed to change the time display once every 60 seconds (or once every second). This means the time displayed is "stepped". Any time between these steps cannot be read from the display. Thus, within each hour there are a finite number of times that can be displayed. In this case the time displayed is a digital variable.

©2020 **BIOZONE** International
ISBN: 978-1-927309-82-7
Photocopying Prohibited

Analog into digital

- A vinyl record is an analog music storage device. It has a continuous spiral groove stamped into it that rises and falls. A needle on the record player fits into the groove and vibrates as the groove passes under it. The vibrations are converted to an electric signal that is sent to the speakers.
- Much of the music we store today is digital. The original sound wave is stored as a collection of numbers which are snapshots of points on the original wave. Music on a CD or MP3 file is nothing more than a very long sequence of 0s and 1s (binary). The digital information is converted into an electric signal by the playing device (e.g. an MP3 player) and sent to the speakers.
- How do you reduce sound to a series of numbers? First the sound is received by a microphone, which converts the continuously varying sound wave to a continuously varying voltage. This varying voltage then goes to an electronic device called an ADC (analog to digital converter) which samples the voltage at very small, but regular, time intervals and produces a voltage that still varies, but in distinct steps (digitized). Each step is recorded as a unique binary number (encoded), depending on the size of its voltage.
- The rate at which the wave is sampled affects the final shape of the wave.

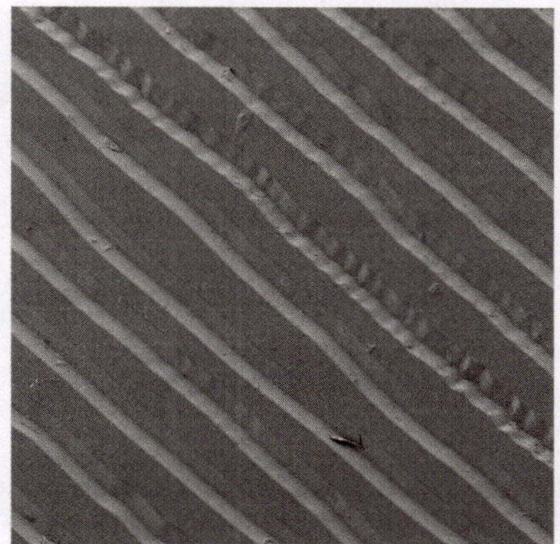

A scanning electron micrograph of the grooves in a vinyl record.

2. Study the wave pattern on the graph. It shows an analog wave form. To digitize this wave it needs to be sampled at discrete (non continuous) time periods. The rate of sampling will affect the shape of the digital reproduction of the wave:

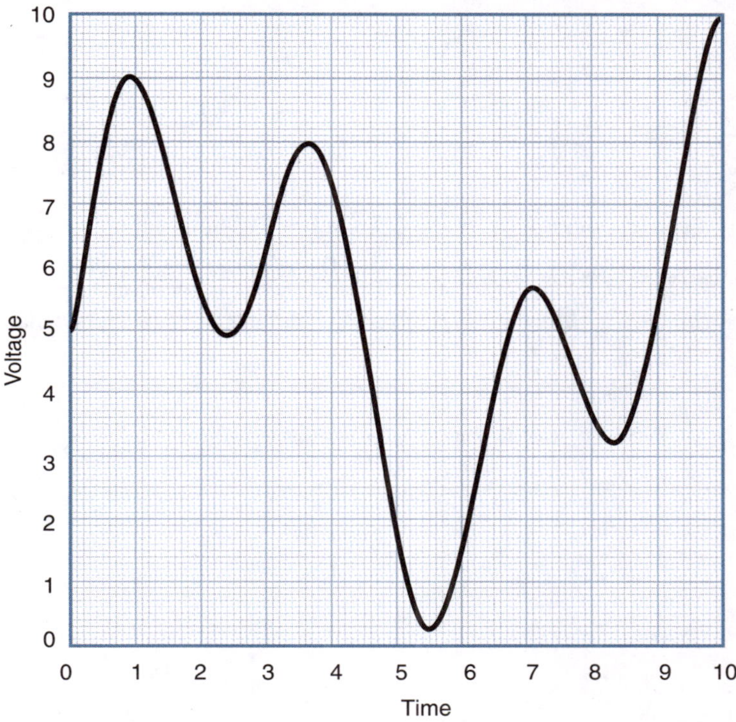

(a) Record the voltage to the nearest whole number for each of the times shown in Table 1 below:

Time	Voltage
0	5
1	9
2	6
3	6
4	7
5	2
6	2
7	6
8	4
9	5
10	10

(b) Now record the voltage to the nearest decimal place for each of the times shown in Table 2 below:

Time	Voltage	Time	Voltage
0.0	5.0	5.5	0.2
0.5	8.4	6.0	1.5
1.0	9.0	6.5	3.8
1.5	7.5	7.0	5.7
2.0	5.6	7.5	5.0
2.5	5.0	8.0	3.6
3.0	6.3	8.5	3.3
3.5	7.8	9.0	5.4
4.0	7.3	9.5	8.0
4.5	4.8	10	10.0
5.0	1.7		

©2020 **BIOZONE** International
ISBN: 978-1-927309-82-7
Photocopying Prohibited

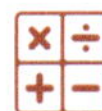

(c) Now plot the data from Tables 1 and 2 onto the grid below. Because this is a digital reproduction, time is discontinuous. The points are joined as steps rather than a continuous curved line:

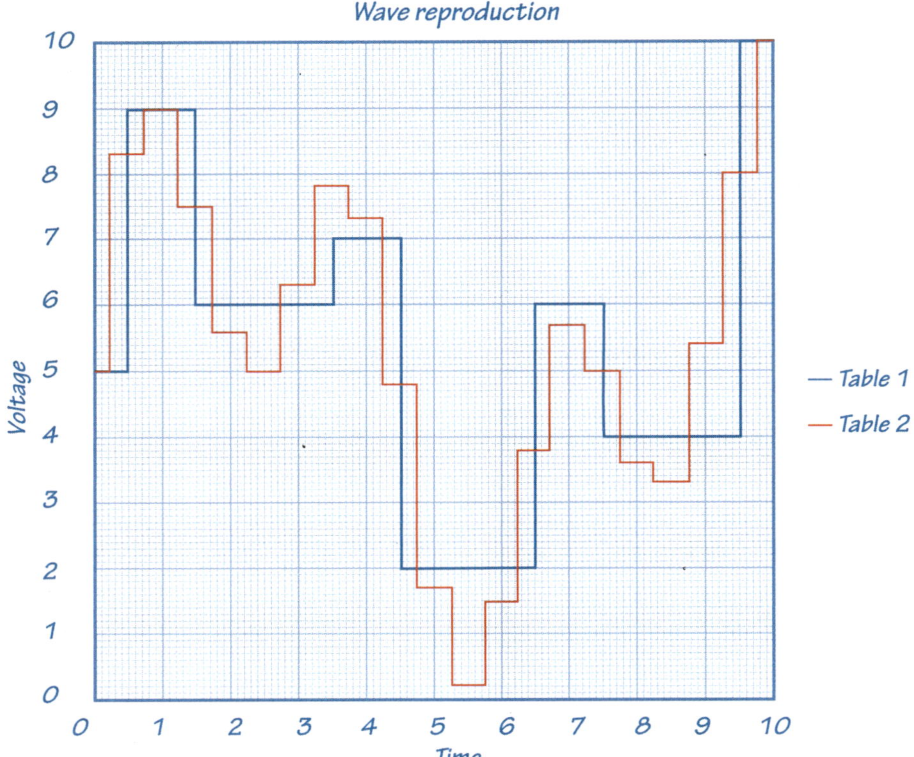

(d) Compare the two graphs you just made. Which one of them best resembles the original? *Table 2.*

(e) How could you efficiently reproduce the original graph? *The original graph will be better approximated with a larger data set of points. The more data points, the more accurate the translation and a smoother curve is produced.*

3. (a) Draw a wave on the graph below, similar to the one at the start of Q2.

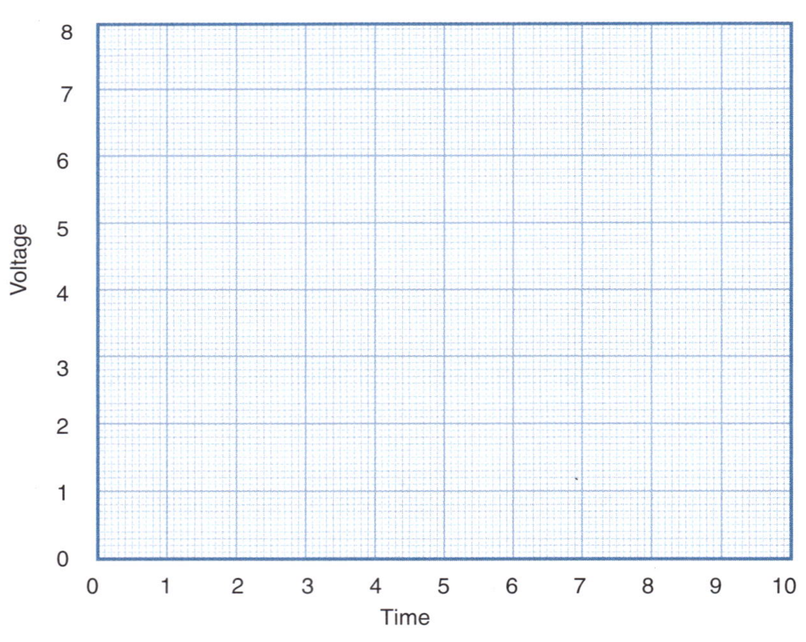

Student's graph

Time	Voltage	Time	Voltage
0.0		5.5	
0.5		6.0	
1.0		6.5	
1.5		7.0	
2.0		7.5	
2.5		8.0	
3.0		8.5	
3.5		9.0	
4.0		9.5	
4.5		10	
5.0			

(b) Complete the table for your wave, entering voltage values to the nearest decimal place. Read them out to your partner. As you read them out, your partner should plot the coordinates on the graph on the following page of their book and join the points with a curved line. *Student's answer*

(c) Once your partner has plotted your coordinates, swap over. Your partner reads out their coordinates for their wave while you plot them on the graph over the page.

©2020 BIOZONE International
ISBN: 978-1-927309-82-7
Photocopying Prohibited

Plot the coordinates for your partner's graph here: *Student's graph*

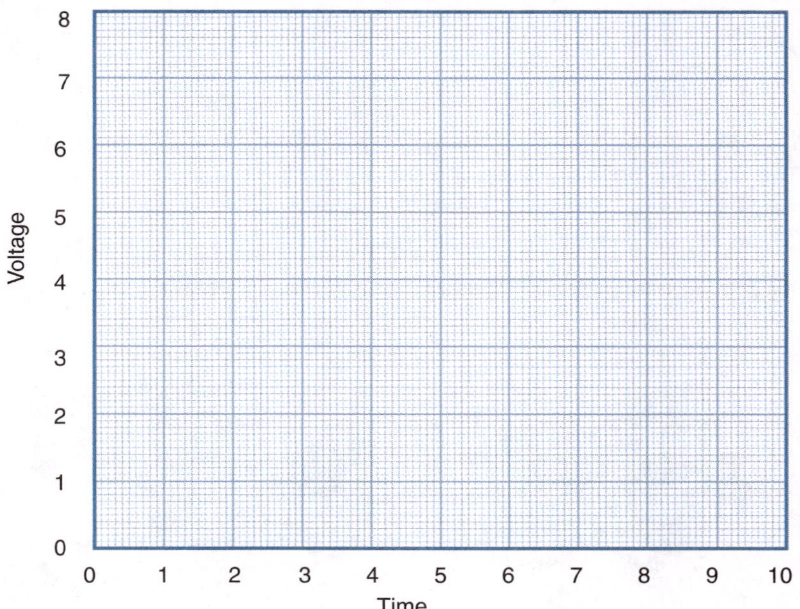

(d) Compare your original graph to the graph your partner created using the data points collected (your partner will also compare graphs). Describe any similarities and differences between the graphs:

Student's answer. Students should note places where curves between points do not match.

(e) How could your graphs be used to argue that analog recordings are more accurate than digital recordings?

Digital recordings take specific data points from the original data. They are essentially a collection of data points strung together and will always be missing the data between the points. Analog recordings are continuous therefore more accurate.

(f) If you repeated the process how could your version of your partner's graph be improved?

The graph would be improved with a more frequent sampling of data points.

(g) How could your graphs be used to argue that digital sampling could be just as accurate as an analog recording (and in some ways better)?

Just like movies are a series of pictures in fast enough succession we cannot perceive them, our ears cannot detect differences between recordings if the data points are frequent enough. Digital may be better because data points are easier to encode and transmit to devices.

4. Recording audio and visual information into digital information is common place today. Doing so has both advantages and disadvantages. There are many people who believe (with some valid arguments) that analog formats such as vinyl records produce a much "truer" sound than digital formats such as MP3 or even CD/DVDs.

(a) What are some advantages and disadvantages of storing data in digital formats?

Storing data in digital formats means relatively simple storage and retrieval of the information and the ability to convert between formats. Data can be transmitted more simply and securely over long distances. Storing data requires a lot of memory space (e.g. 3 min MP3 needs 32 MB of data).

(b) Why do some people argue analog formats produce a truer sound than digital formats?

Analog formats are essentially copies of the original wave shape. The sound produced is therefore exactly the same. A digital format misses some of the wave so the sound is slightly different.

©2020 **BIOZONE** International
ISBN: 978-1-927309-82-7
Photocopying Prohibited

119 Review Your Understanding

Key Question: How might rogue waves in the ocean form? What principles are behind these phenomena?

- At the beginning of this chapter we were introduced to the phenomenon of rogue waves. Waves at sea are generally created by wind. The stronger the wind blows and the greater the distance it can blow the larger the waves will be. This is why waves at sea are generally bigger than waves on lakes.
- Rogue waves however, are much bigger than this explanation would allow for, and tend to appear out of nowhere.
- Evidence has divided rogue waves into three groups. Single storm waves, broad walls of water, and what are called the three sisters.

A rogue wave may be a single wave peak that suddenly rises out of the sea and falls just as quickly.

There are reports of broad waves stretching across the sea. The QE2 hit one of these in 1995.

An apparently rarer rogue wave phenomenon is the appearance of the three sisters. Here one massive wave is preceded and then followed by a smaller but still significantly large wave.

- Just as there are rogue waves, there is the possibility of rogue holes. Just as a wave rises up out of the water, it is possible the opposite may occur, with a wave trough suddenly becoming much steeper and deeper than in the rest of the sea.

1. (a) How does a wave grow in amplitude? *Amplitudes can grow by: adding energy or shortening the wavelength.*

   (b) How would this occur on the sea or ocean? *Wind transfers energy to waves, increasing their height. When waves reach shore, wavelength decreases and amplitude increases. Underwater landslides can form waves.*

   (c) Waves at sea on a relatively calm day average around 2-3 m from crest to trough. Storm waves may reach 12 m from crest to trough. What causes storm waves to be so much bigger?
   Wind transfers energy to the waves. The stronger the wind and the longer it blows, the more energy is transferred to the waves and the higher the waves can get.

   (d) If a sea wave is measured at 6 m, what is its actual amplitude? *3 m*

2. (a) During this chapter you have studied the phenomenon of interference in waves. How might interference affect the height of a wave? Could it cause a rogue wave?
   Interference can cause wave amplitude to increase when two crests interfere. For a rogue wave to occur it would require perhaps three wave crests to interfere. Although this is unlikely it is not impossible in storms, especially if waves are traveling in different directions. There is some evidence to show two waves "colliding" can cause a sudden very high and sharp peak.

   (b) What problems can you see in using interference to explain rogue waves? *Interference alone cannot account for the height of a rogue wave unless an number of wave crests all converge on the same position. This is possible, but highly unlikely as the probability against it grows quickly with each wave added to the interference. Other factors must be involved (e.g. currents and non linear wave functions).*

©2020 **BIOZONE** International
ISBN: 978-1-927309-82-7
Photocopying Prohibited

120 Summing Up

Wave properties

PS4-1 1. (a) On the axis below draw three wavelengths of a transverse wave with a wavelength of 4 m and an amplitude of 2 m:

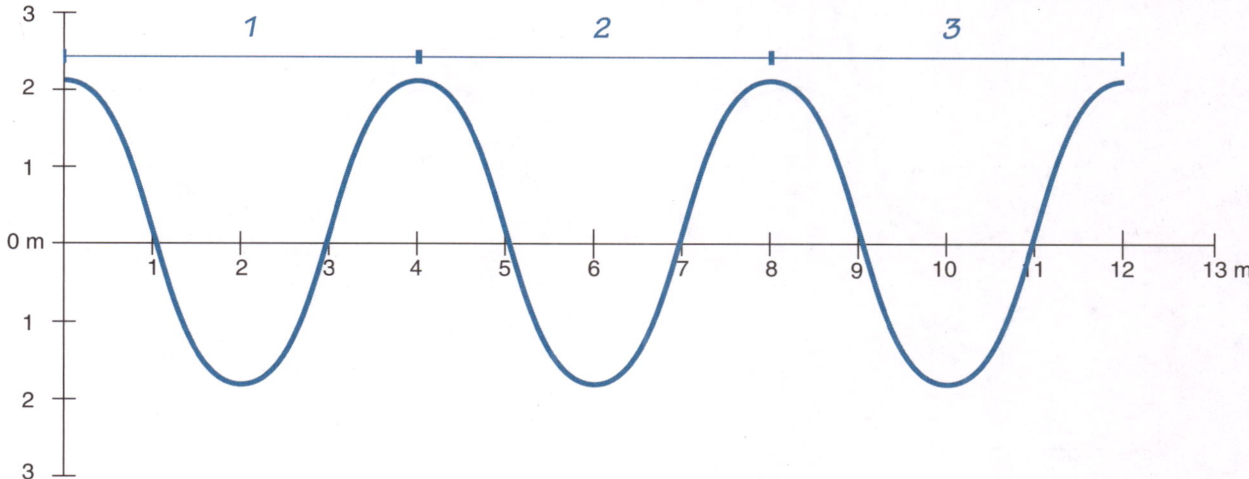

(b) For the wave above, the velocity of the wave was measured at 3 m/s. What is the frequency of the wave?

f = v ÷ λ = 3 m/s ÷ 4 m = 0.75 Hz.

(c) What is the period of the wave? *T = 1/f = 1 ÷ 0.75 Hz = 1.3 s.*

2. (a) Complete the table below:

Medium	Wavelength (m)	Frequency (Hz)	Speed (m/s)	Period (s)
1	1.7	2.0	*3.4*	*0.5 s*
2	0.9	*3.9*	3.5	*0.3 s*
3	*1.0*	2.1	2.2	*0.5 s*
4	0.6	3.6	*2.2*	*0.3 s*
5	*0.9*	2.2	2.0	*0.5 s*
6	1.8	*1.2*	2.2	*0.8*

(b) Based on the results above which media would have similar properties? Explain your answer: *Waves travel through different media at speeds specific to those media. Wave speed is based on the properties of the materials making up the medium. Media 3, 4, and 6 would have very similar properties as wave speed is the same through both. Media 1 and 2 would also be similar as their wave speeds are close.*

3. Describe the difference between a transverse and a longitudinal wave and give an example of each type:

In transverse waves, the particles in the material the wave is traveling through are displaced perpendicular to the direction the wave is traveling in, such as in ocean waves. The particles in longitudinal waves are displaced in the direction the wave is traveling in, such as in sound waves.

4. During observation of a test explosion, the explosion was seen 2.5 seconds before it was heard:

(a) Explain why the explosion was seen before it was heard: *Light waves travel faster than sound waves, so the light reaching our eyes occurs before the sound waves reach our ears.*

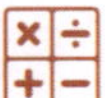

(b) If sound travels at 343 m/s how far away was the explosion? *d = v x t = 343 m/s x 2.5 s = 857.5 m.*

©2020 BIOZONE International
ISBN: 978-1-927309-82-7
Photocopying Prohibited

CE PS4.A

5. The diagram below shows two transverse waves with the same wavelength. The waves interfere with each other. Draw the resulting wave on to the diagram.

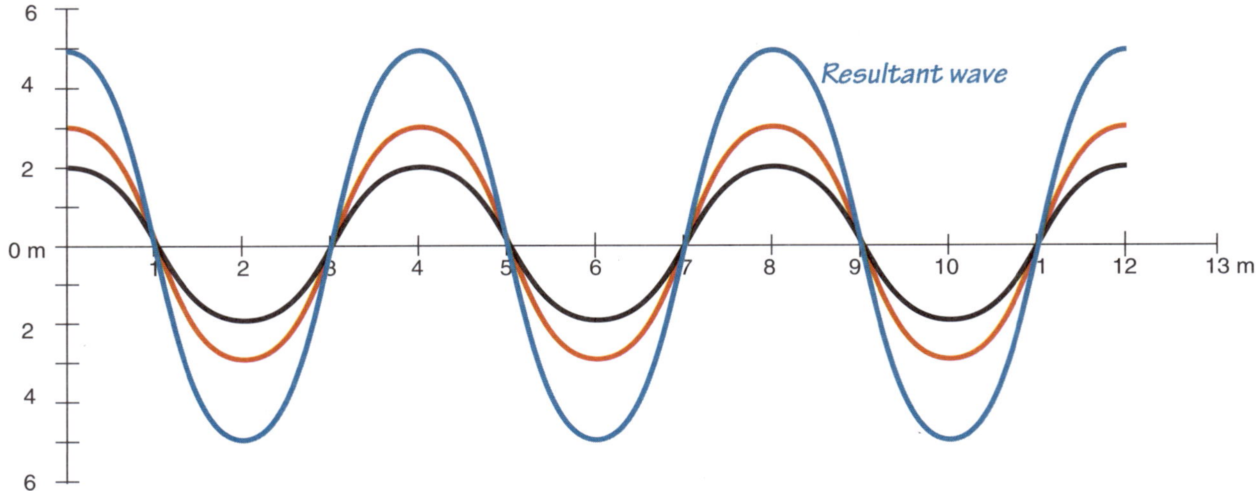

6. Study the diagram below:

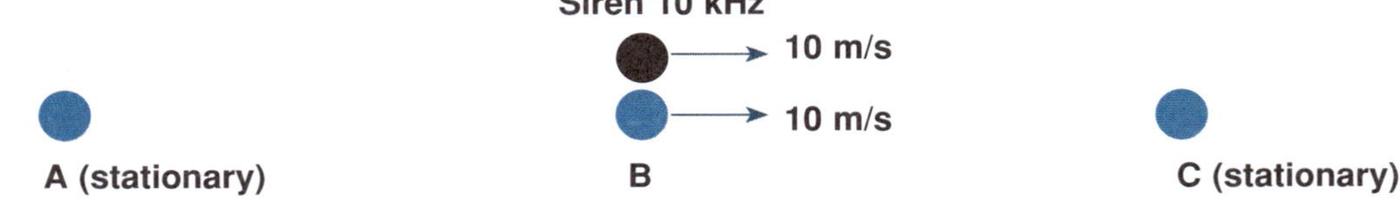

(a) Describe what observers A, B, and C, hear: *A will hear a lower pitch than the 10 kHz of the siren. B will hear the same pitch as the siren. C will hear a higher pitch than the siren.*

(b) Explain why this happens: *The sound waves have been stretched when they reach A due to the movement of the siren. At B, the sound waves be remain as they were produced because B is traveling with the siren. At C the waves will be compressed because the siren is traveling in the same direction as the waves.*

7. Study the diagram below:

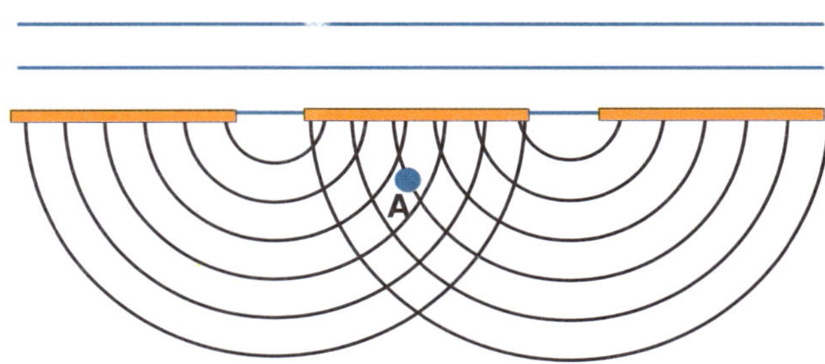

Describe what is occurring here and specifically what is occurring at point A: *The diagram shows diffraction and then interference is occurring. The waves bend as they move through the gaps. The resulting set of waves cross over each other and interfere. At point A a trough and a crest are interfering. This will cause no displacement at that point.*

©2020 **BIOZONE** International
ISBN: 978-1-927309-82-7
Photocopying Prohibited

PS4.B

10: Electromagnetic Radiation

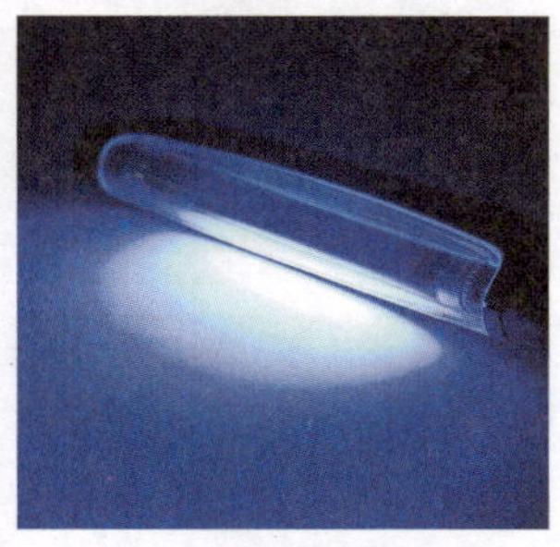

Anchoring Phenomenon

Photovoltaic cells: How does the behavior of light explain how photovoltaic cells work? 121 126

What is the electromagnetic spectrum?

☐ 1 What do understand by electromagnetic (EM) radiation? Explain how EM waves differ from the mechanical waves you explored in the previous chapter. In what ways are they similar? What evidence *[SEP-7]* do you have to support the idea that EM radiation can be described by a wave model? Relate the mathematical representations of amplitude and wavelength to electromagnetic waves by comparing light of different intensities *[CCC-5]* and colors. What is the relationship between changes in the wavelength of EM radiation and its color? *[PS4.B]* *[CCC-4]* Explain how the wavelength of EM radiation relates to how it is used both in the natural world and in technological applications. 122

☐ 2 Investigate *[SEP-3]* the wave properties of light using a laser pointer to perform a simple double-slit experiment. What was the effect of the slits on the pattern of light projected on to the screen? How does this model *[CCC-4]* provide further evidence *[SEP-7]* for the wave behavior of light? *[PS4.B]* *[HS-PS4-3]* 123

☐ 3 So far, we have described light as behaving like a wave, but some behaviors of light can only be explained if light is thought of as a particle (a photon). Many metals emit electrons when light shines on them. If light acted as a wave, electrons would be emitted from any frequency of light providing the wave has enough intensity (amplitude). However the ability of light to dislodge electrons depends on its frequency (each photon has an energy *[CCC-5]* proportional to its frequency). How does the photoelectric effect provide evidence for the model *[CCC-4]* of light as a particle? From your investigations, evaluate the claims for the duality of light (its behavior as both a wave and a particle) *[SEP-7]* *[PS4.B]* *[HS-PS4-3]*. 123 127

What are the health effects of EM radiation?

☐ 4 If you have ever been sunburned, you know that high energy (short wavelength) radiation harms biological tissues. However, what about longer-wavelength parts of the EM spectrum, such as microwave radiation? Apply your model of the particle nature of light from the photoelectric effect to evaluate published claims *[SEP-8]* about the effects *[CCC-2]* of different frequencies of EM radiation on health *[PS4.B]* *[HS-PS4-4]*. 124 125

You might like to consider the following:

- *Ionizing and non-ionizing radiation*: What is the difference between these types of electromagnetic radiation and how do the differences relate to the damage they can cause to living tissue?
- *Microwave ovens*: What is the frequency of microwaves and how far do they penetrate into material? How do they heat or thaw food? Why do some people think microwave cooking is dangerous?
- *Cellular phones*: Cellular phones also use microwaves and rely on producing and receiving EM radiation to work. Is there any basis to then claim that cell phone use causes brain cancer?
- *X-rays and gamma rays*: How are x-rays used in medical diagnostics and why are x-rays and gamma rays so damaging?

Ivanpah solar power plant near Las Vegas

★ Challenge question:

Explain why the efficiency of solar power generation varies with location around the globe. Where are the best places to build solar power plants and why?

Download this page from BIOZONE's Resource Hub

Teacher's Notes

10. Electromagnetic Radiation

Anchoring phenomenon

Photovoltaic cells have become extremely common in today's world. Student's may have seen them on roofs, garden ornaments, and even solar power street lighting. But students may find it difficult to explain how they work. They may know something about capturing light or using light to make electricity. Some may know something about photons and from there an introduction to what light actually is can begin.

What is the electromagnetic spectrum?

1. Electromagnetic waves are synchronized oscillations of electric and magnetic fields. They can be represented as two in-phase transverse waves at right angles to each other. Oscillating charges produce an electric field. At right angles to this, a magnetic field will be created by the oscillating electric field (in the same way that current flowing through a wire produces a magnetic field). The magnetic field then creates the next electric field (just as a changing magnetic field can create electricity) and so the wave self propagates. Thus EM waves do not need any media to keep the wave going and so can travel through a vacuum. Students may want to refer to electromagnetic waves as light waves, but this should be avoided. Each part of the EM spectrum is given its own name based on the wavelength or frequency, how it is produced, and the effect the wave has on other objects. Visible light only makes up a small part of this spectrum. We see colors because we have evolved sensory receptors that react to specific EM frequencies within a specific band of the spectrum. The particular sensory response is interpreted by the brain as a particular color. We have evolved to use this part of the spectrum because it is the most useful part to us. Ultraviolet, X-ray, and gamma radiation are not common enough on Earth to be useful, and radio and microwaves have wavelengths far too large to evolve a practical organ to receive them *"If we were to see radio waves our eyes would need to be as big as wagon wheels" – Carl Sagan – Cosmos.*

2. The double slit experiment, which students carry out themselves in Activity 123, was carried out by Thomas Young in 1801. It can be used to show that light forms diffraction patterns and is therefore a wave. Interestingly if the experiment is performed by sending photons one at a time through the double slit and mapping where the photons hit on the screen, the same pattern occurs over time. The same thing occurs with electrons and has even been carried out with molecules with up to 810 atoms. Indeed, as this experiment has been carried out with more and more sophisticated equipment, it has tended to confirm the wave-particle duality of light, rather than confirm Young's original conclusion that light is a wave.

3. The photoelectric effect is the classic piece of evidence that shows the particle nature of light. Electrons follow an all or nothing principle. Either the energy from the photon is absorbed and the electron is ejected from the material or the energy is re-emitted as a photon and the electron is not ejected. The development of semiconductors and specific ceramics and alloys that allow electrons to be ejected at relatively low frequencies (within the visible light spectrum) have allowed designers to use the photoelectric effect to develop photovoltaic cells.

What are the health effects of EM radiation?

4. The more complicated technology becomes, the less people understand how it works and so the less they understand about the pros and cons of its use or the risks involved. Social media tends to exacerbate this problem, often promoting unsubstantiated claims that are quickly circulated. Often explanations are misinterpreted. This is especially the case with statistics. Many science writers for newspapers and online blogs are guilty of using only the most extreme examples or alarming statistics. For example, one can never claim something is 100% safe (only safe within the limits of the test). This allows some to claim a device is in fact unsafe. Commonly reported claims include: "*cell phones will give people brain cancer*", "*microwave ovens will destroy the nutrients in your food*", and "*dental X-rays will give you cancer*". Students can research and investigate these claims and then evaluate them for themselves using their knowledge of EM waves and radiation.

121 Photovoltaic Cells

Key Question: How does the behavior of light explain how photovoltaic cells work?

- From generating electricity for homes and garden lights, to powering space probes millions of kilometers from Earth, photovoltaic cells (often simply called solar panels) have become an important part of our electricity generating network.
- The small solar panels on the garden lights below generate about 1 watt of power in full sunlight where as the Juno space probe generates about 14 kilowatts in Earth orbit (reducing to about 400 watts having reached Jupiter).
- But how is light converted to electricity? What properties of light allow this to happen?

Photovoltaic array

USAF

Solar garden 'rock' lights

Juno space probe

NASA

- Have you ever tried playing with a solar powered toy like the one shown on the right? A photovoltaic cell on the top captures sunlight and converts it to electricity to power a small electric motor. On a really sunny day these little toy cars can reach surprisingly high speeds. Such a toy is always on. To turn it off, you simply flip it on to its roof or cover the solar panel.

1. (a) What would happen to the speed of the solar car shown right if it moved into the shade?

   The car would slow down.

   (b) Would the solar car run faster or slower in winter compared to in summer? *Slower.*

   (c) Explain your answer: *The intensity of light from the Sun is less in winter. Less light would be available to convert to usable energy. The car would run slower with less available usable energy.*

2. Electricity is a flow of electrons. How does light from the sun hitting the solar panel achieve this? Discuss this as a group and summarize your ideas below:

   Student's answers.

3. The Juno space probe sent by NASA to Jupiter has three sets of solar panels that produced a large amount of electricity when it was near Earth, but this rapidly decreased as it went further into space. Explain why:

   The intensity of the light reaching the probe at Jupiter is much less that near Earth. Thus the solar panels are less effective at generating electricity.

©2020 **BIOZONE** International
ISBN: 978-1-927309-82-7
Photocopying Prohibited

122 The Electromagnetic Spectrum

Key Question: What are the properties of electromagnetic waves?

What is light?

- Light is a transverse wave, which is composed of electric and magnetic fields oscillating at 90° to each other and at 90° to the direction of travel. It is therefore described as electromagnetic (EM) radiation. In a vacuum a light wave travels at 299,792,000 meters per second (c).

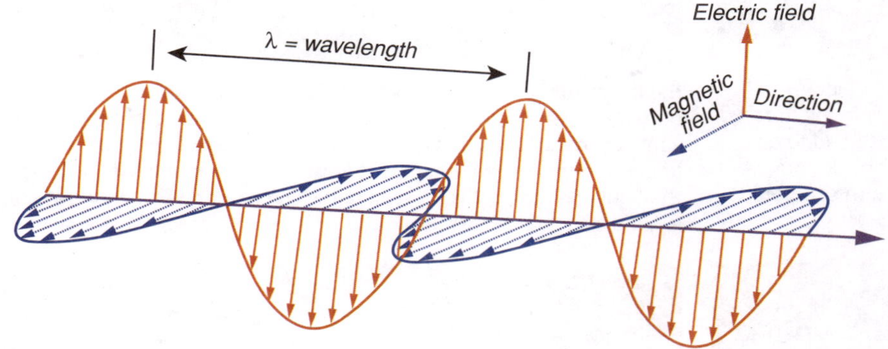

- Because EM radiation is a transverse wave it can be described by its amplitude (brightness or intensity), wavelength, frequency, and period, the same as a mechanical wave.
- Like other waves the energy (therefore intensity) of an EM wave is proportional to the square of its amplitude ($I \propto A^2$). If wave amplitude is doubled, it transports four times the energy (this is not true for frequency).

The EM spectrum

- Electromagnetic radiation exists on a spectrum where frequency (f) is inversely related to wavelength (λ) ($f = c/\lambda$).
- This spectrum is broadly grouped into categories based on its size. Size is measured by its wavelength. In order from shortest wavelength to longest, the electromagnetic spectrum includes:

  • gamma radiation • X-rays • ultraviolet light • visible light • infrared light • microwave radiation • radio waves

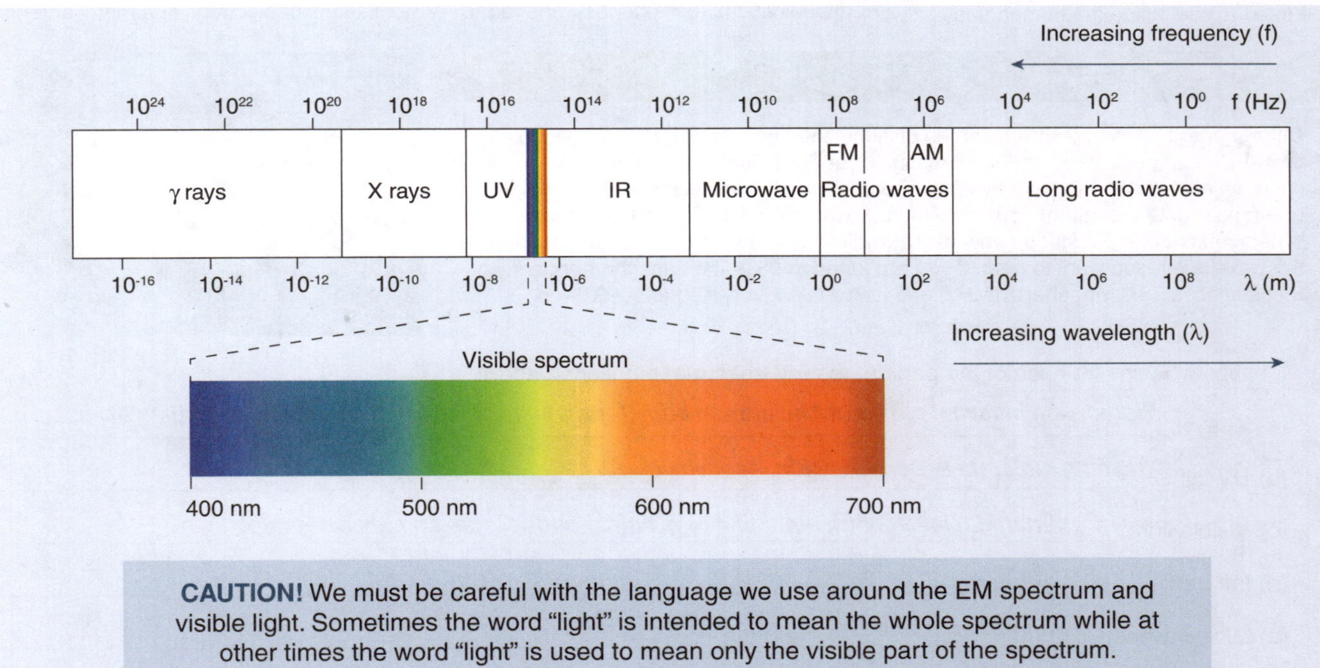

CAUTION! We must be careful with the language we use around the EM spectrum and visible light. Sometimes the word "light" is intended to mean the whole spectrum while at other times the word "light" is used to mean only the visible part of the spectrum.

1. Describe what happens to the frequency as wavelength on the EM spectrum changes:

   Frequency decreases as wavelength increases and increases as wavelength decreases.

©2020 **BIOZONE** International
ISBN: 978-1-927309-82-7
Photocopying Prohibited

EM SSM PS4.B

Some useful applications of the EM spectrum

Objects produce heat and at low temperatures this is mostly in the infrared wavelength. Thus it can be used to detect objects warmer than their background but in relatively cool situations (e.g. a person standing in a cool room).

Exposure to UV light can quickly cause burning of the skin and tissue damage. UV light is commonly used as a fluorescent tracer and to sterilize equipment or water. Sunlight contains UV light, much of which is filtered by the atmosphere.

Visible light is slightly different to different animals but generally has a wavelength between 380 nm and 750 nm. For humans this is violet to red. Many insects can see ultraviolet and others, e.g. pit vipers, can sense infrared.

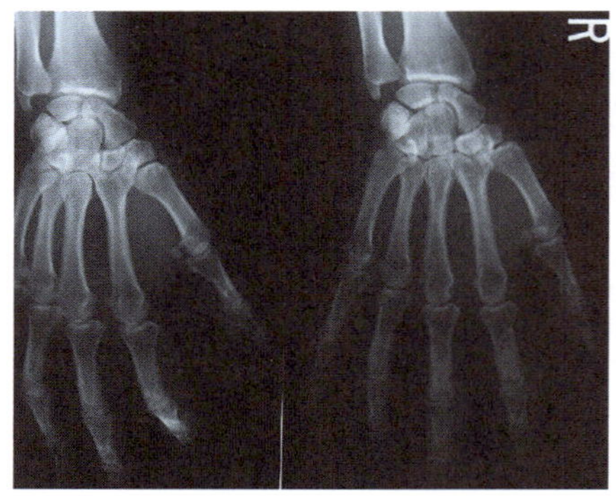

X-rays are highly energetic EM radiation. They can pass through soft materials such as muscle but are stopped by denser materials like bone. Thus they can be used to see inside the human body.

Microwave radiation has wavelengths of about a millimeter to a meter long. Microwaves are used for many purposes including cooking (microwave ovens), communications (cell phones), and radar.

Radio waves have wavelengths longer than a meter. AM, FM, and TV transmissions are obvious examples of the use of radio waves. Radio waves are generated by the oscillations of electrons in a metal antenna.

Different parts of the spectrum can be distinguished by referring to their frequency or wavelength depending on the situation. When we talk about light in terms of color, we often also relate it to the wavelength (e.g. blue has a wavelength of 450 nm and green has a wavelength of 550 nm). Radio waves are referred to by their frequency. When you change the station on your car radio, you are selecting the frequency of the wavelength you want to hear. If you are tuned to 98.3, it means the frequency of that station is 98.3 megahertz (MHz) and its wavelength will be about 3.04 m.

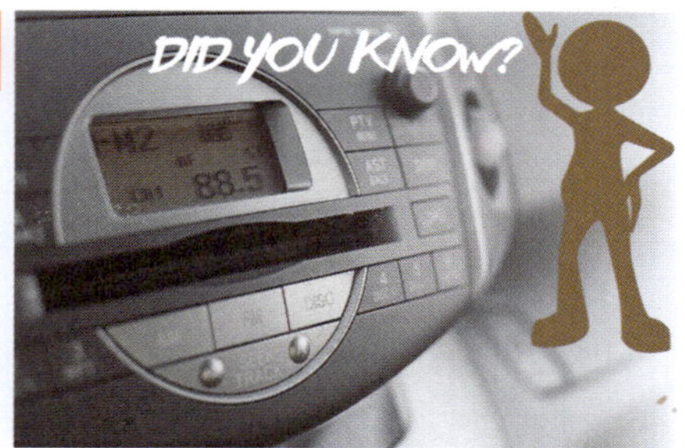

2. Identify two uses for each of the parts of the EM spectrum mentioned above:

(a) X-rays: *Visualizing internal parts of the human body (and other objects), determine molecular structure.*

(b) UV light: *Tracing molecules, sterilization, security*

(c) Visible light: *Communications (Wifi), signaling (e.g. light houses, ship to shore signals)*

(d) Infrared: *Night / heat vision cameras, heating / warming food, remote controllers*

(e) Microwaves: *Cooking food, communications, radar*

(f) Radio waves: *Communications, astronomy, navigation*

3. Pick one of the regions on the EM spectrum above and research how the waves are produced both by human technology and naturally.

Student's answer based on research.

©2020 **BIOZONE** International
ISBN: 978-1-927309-82-7
Photocopying Prohibited

The EM spectrum and the eye

- Visible light is only a small part of the EM spectrum. This means the information that we can access directly with our sense of sight is actually quite limited! The differences in color we can see are merely slight changes in the wavelengths within the visible spectrum.
- If visible light is a part of the EM spectrum, why can't our eyes detect other waves on this spectrum? Our eyes have evolved to respond to parts of the EM spectrum which are most useful to us. Of the wavelengths of light from the Sun that reach the Earth's surface, the intensity of visible light is the highest (see diagram below). Human eyes evolved to use the most abundant type of light available.

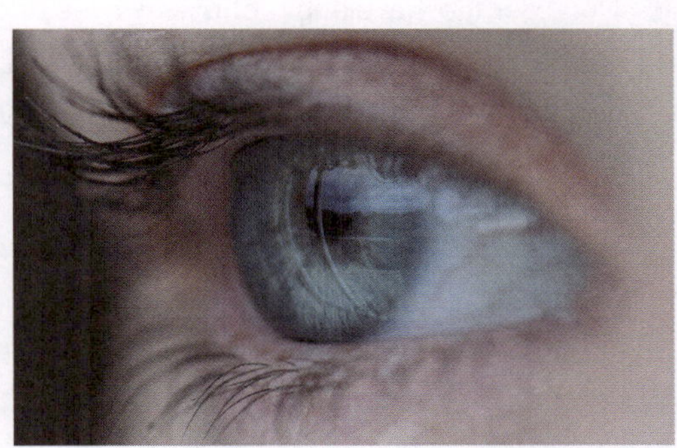

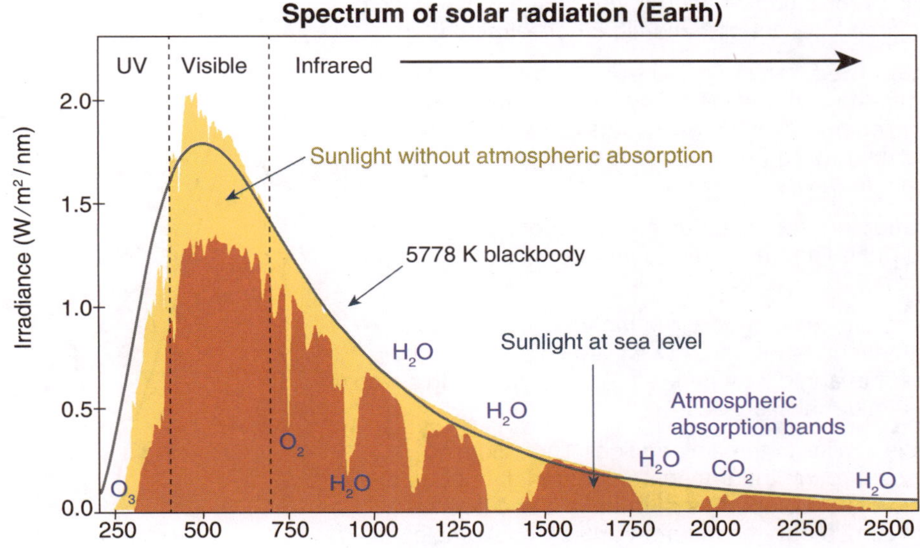

- Not all animals respond to the same wavelengths. If you look at the EM spectrum, you will see that visible light is sandwiched between UV light and infrared (IR) light (above). Some animals' senses allow them to detect these areas of the EM spectrum. For humans, it is unlikely that such an ability would provide a survival advantage.

The eyes of many insects, such as bees (above) are able to respond to (see) short wavelength UV light. This is useful to them because the plants they pollinate often have UV nectar guides to guide them to the nectar. *Mimulus* flower seen in visible light (left) and UV light (right) showing the dark nectar guide that is visible to bees but not to humans.

Pit vipers and pythons are able to detect infrared. This ability provides a survival advantage when hunting prey at night. The heat sensing pits can be seen below the nostril in this pit viper.

4. Suggest why eyes evolved to see the visible light part of the EM spectrum and not something like X-rays:

   Visible light is the most common form of EM radiation on Earth (see diagram above). Eyes have evolved to use the part of the spectrum most available.

5. Explain the advantages some animals have gained by extending their range of detection outside the visible spectrum:

   Animals that can access the infrared part of the EM spectrum have an adaptive advantage when hunting prey at night. Animals that can detect UV light are better able to find food because some sources, such as flowers, have nectar guides, which stand out in the UV spectrum.

©2020 **BIOZONE** International
ISBN: 978-1-927309-82-7
Photocopying Prohibited

6. Consider the wavelength of microwaves and radio waves. Why do AM radio stations need very tall (long) antennas whereas microwave antennas (e.g. in cell phones) are very short?

AM wavelengths are very long compared to microwaves. As a result, AM radio stations need a long antenna so that the electrons in it a free to oscillate efficiently. The shorter wavelengths of microwaves mean antennas can be shorter and so fit inside a cell phone.

What happens when light hits an object?

- When light comes in contact with a material substance any one of the following things (or a combination of them) can happen. The light can be absorbed, reflected, transmitted, or refracted.
- When light is **absorbed**, the energy is taken in by the material. The energy within the material increases, causing the particles to move faster. This energy eventually radiates as heat. An example of this would be heat rising from a pavement after the Sun has been shining on it for a while.
- When light is **reflected**, the material fails to absorb the energy from the light and the light rays are redirected. An example of this would be a mirror.
- When light waves are **transmitted**, the incoming light passes through the material unchanged. Light waves come out of the material at the same angle they entered. An example of this would be a transparent material like glass.
- **Refracted** light is similar to transmitted light. The incoming light again passes through the transparent material, but due to differences in the optical properties of materials and boundary angles, the light comes out at a different angle. An example would be when light from an object is seen after passing through a glass of water. The image is distorted.

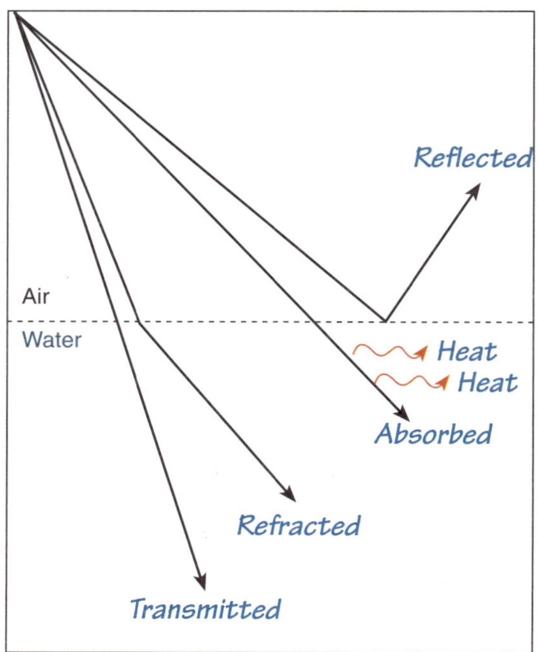

7. Use the descriptions above and in the box above draw arrows to show what is happening to incident light when it moves from one material to another (e.g. from air to water) and is absorbed, reflected, transmitted, and refracted.

8. Different wavelengths of light are absorbed and reflected depending on the material the light hits. How does this explain the color of different objects?

The wavelengths of light reflected off an object are the wavelengths seen. Objects appear to be the color of the wavelengths of light they do not absorb.

Light – wave or particle?

- Under specific circumstances, light acts as a wave and under others as a particle. This can be confusing when trying to determine exactly what light is.

9. Measurement of illumination at various distances from a point light source will confirm that light obeys an **inverse square law**.

   (a) Explain this relationship using what you know about particle behavior. Hint: think about spraying paint out of an aerosol can:

   The further away you are from the source of the particles, the more spread out the particles will be when they reach you.

   (b) Explain this relationship using what you know about wave behavior. Hint: Wave energy is related to amplitude which will attenuate as it passes through a material. But what about light moving outward through a vacuum?

   The intensity of a wave is reduced with distance. The energy is spread out over a greater area.

10. Based on your answers, do you think light acts as a wave or a particle? Explain: *Student's answer will vary. Particle = fewer photons hitting a specific area the further away it is. Wave = intensity of the wave becomes less with distance as would be expected as a wave spread out over a greater area.*

©2020 **BIOZONE** International
ISBN: 978-1-927309-82-7
Photocopying Prohibited

123 Duality of Light

Key Question: How can some of the stranger properties of EM radiation be explained?

EM radiation and waves

- Some of the properties of light can be explained if light acts as a wave. The experiment below replicates a famous but now common experiment to show the wave behavior of light.

INVESTIGATION 10.1: The double slit experiment

See appendix for equipment list.

1. Take a microscope slide and paint one side with graphite paint (or similar black paint).
2. When the paint is completely dry carefully use a craft knife and ruler to make a slit in the middle of the painted side.
3. Make a second slit parallel, and as close as possible, to the first slit – so now there is a "double slit".
4. Use plasticine or blu-tak to hold a laser pointer, and the slide, so that the laser beam is directed onto the slits and is then projected onto a wall 3-4 m beyond that.
5. Black out the room and turn the lights off.
6. Draw the pattern seen on the wall in the space provided below.

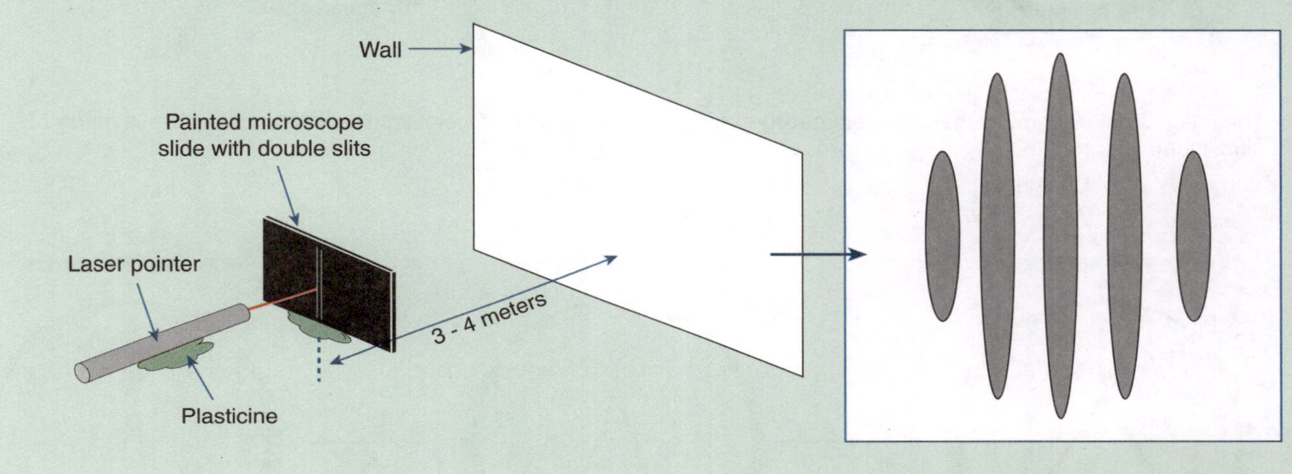

1. Describe what you saw when the laser beam passed through both holes:

   There will be a pattern of light and dark on the wall.

2. Based on what you know about waves, explain this pattern. Draw a diagram below to help you explanation:

   There is interference between two sets of (diffracted) waves. The bright coloured bands are where trough or crests constructively interfere. The dark or colorless bands are where a trough and a crest interfere.

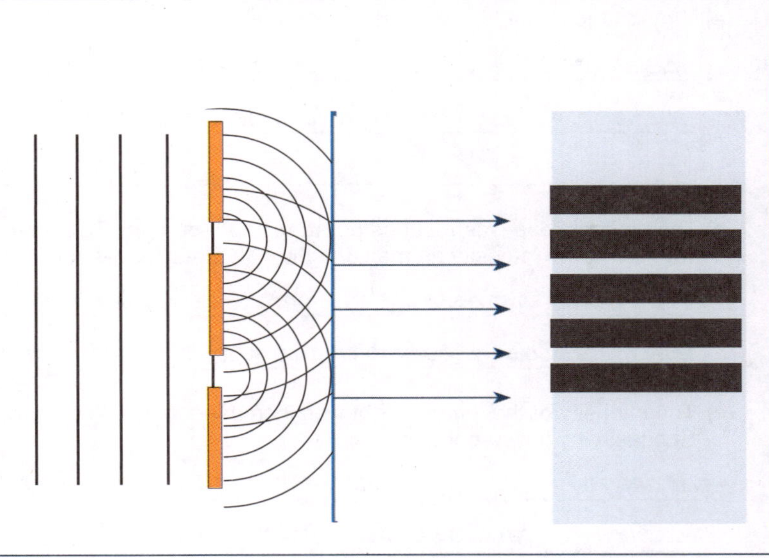

©2020 BIOZONE International
ISBN: 978-1-927309-82-7
Photocopying Prohibited

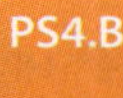

The effect of light on electrons

- In the previous activity it was stated that some wavelengths of light could eject electrons from their orbits around atoms. Here you will study this effect more closely.
- An electroscope is a device that can detect and show the effect of electrons. Simple ones consist of metal leaves hanging on a hook attached to a metal plate. When electrons are added to the metal plate they travel down to the metal leaves. The leaves repel each other and spread out (below).

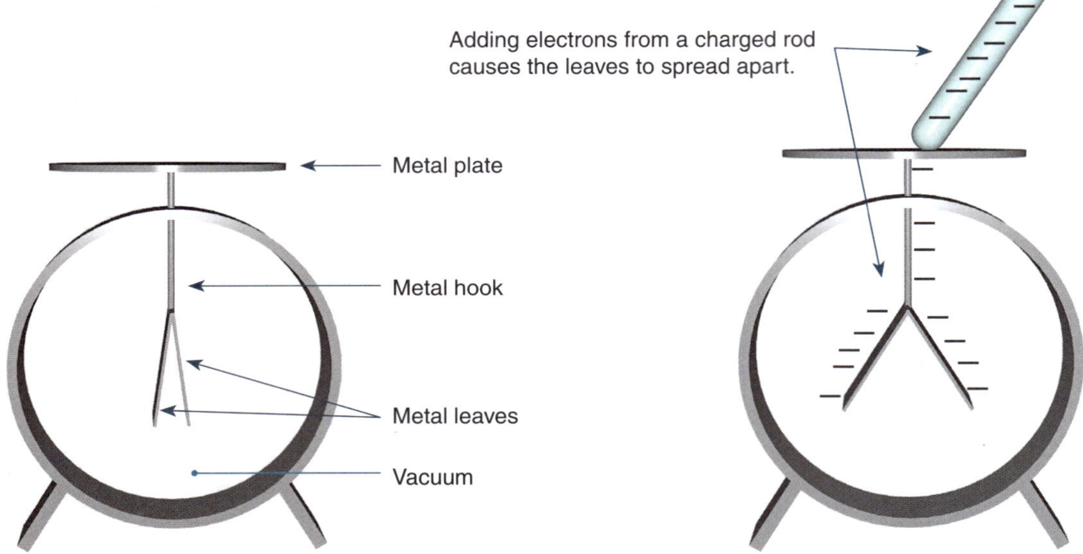

3. The images below show a charged electroscope being illuminated by ultraviolet light over time. The plate is made of aluminium.

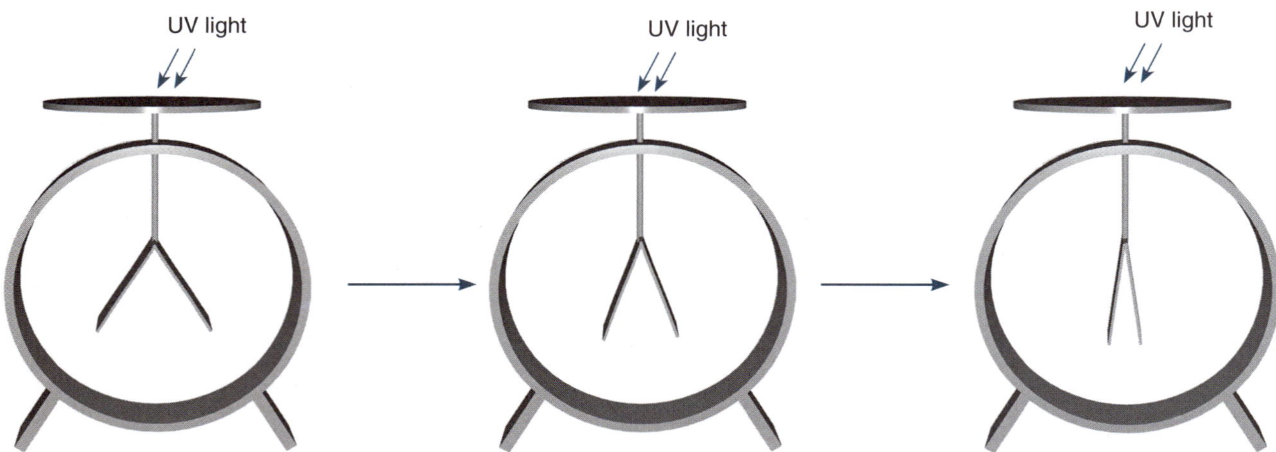

(a) Why do the metal leaves close back together over time? *The UV light causes electrons to be ejected from the metal plate. This leaves an electron hole to fill and creates a flow of electrons from the metal leaves to the metal plate. As the leaves lose electrons, the repulsive electrostatic forces lessen and they close.*

(b) The electroscope was reset as per the image top right. This time an infrared light was shone on the aluminum plate with the same intensity as the UV light. The metal leaves remained apart. Suggest why this happened:

The infrared light does not have enough energy per photon to knock electrons from the metal. Students may realize this by the fact that UV is ionizing radiation and infrared is not.

(c) The intensity of the infrared light was increased to three times that of the UV light. The metal leaves remained apart. Suggest why this happened:

It does not matter how many photons hit the metal plate (the intensity), what matters is the energy each photon carries (as related to its frequency).

©2020 **BIOZONE** International
ISBN: 978-1-927309-82-7
Photocopying Prohibited

The photoelectric effect

- So far we have described light as a wave, but some behaviors of light on materials and in laboratory tests can only be explained if light is thought of as a particle, called a **photon**.
- When light shines on a metal, the photons collide with electrons on the surface of the metal and knock them loose, releasing them. The greater the intensity (number of photons) of the light (in wave theory this would be a greater amplitude) the more electrons are knocked loose.
- The frequency of the light is also important. Short wavelength (high frequency) light, e.g. blue and UV light, carries more energy than long wavelength light, e.g. red and infrared (right). High energy photons are able to knock electrons loose whereas low energy photons cannot.

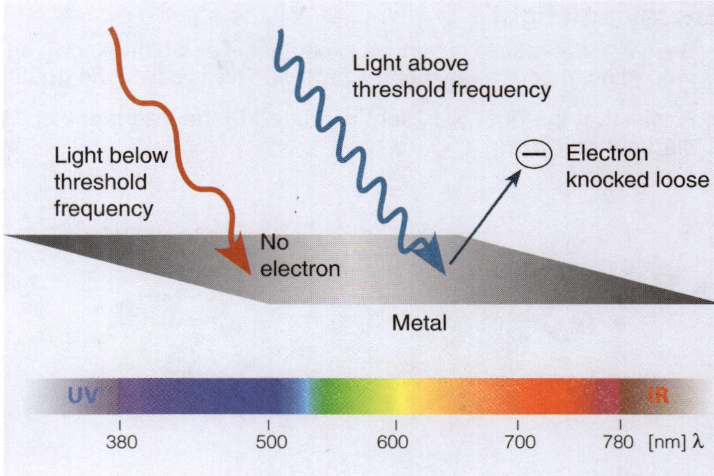

- Recall that in a mechanical wave, energy is related to the square of the frequency. This is not true for EM radiation.
- In EM radiation, energy is related to the frequency by the equation **E = hf**, where E is energy (in joules), f is the frequency of light, and h is Plank's constant ($6.62607015 \times 10^{-34}$ Js).
- Every metal has a different "threshold frequency". Below this frequency, photons of light cannot knock the electrons loose. Above this frequency, all photons can knock electrons loose.

4. Use the explanation above to explain why no matter how intense the infrared light from Question 3 on the previous page is shone on the aluminum plate, the metal leaves in the electroscope will remain spread apart:

   The frequency of the light determines whether or not electrons will be ejected from the metal. The intensity determines how many electrons will be ejected, if they will be ejected in the first place. Infrared light is not of a high enough frequency to knock electrons loose, so how intense it is does not matter.

5. What happens when light above the threshold frequency hits an electron?

   The electron will be ejected from is orbit

6. If light were a wave, what would happen to the energy of the electron as the wave continues to arrive and its energy continues to be absorbed?

   The wave would add energy to the electron in a continuous manner. Eventually the electron would gain enough energy to be ejected from its orbit, or climb to a higher orbital.

7. If an electron were to collect enough energy through this method, what should eventually happen? Explain:

   Gaining energy, the electron will move faster and slowly move up to a higher electron orbital, covering all the space between the lower and higher orbitals.

8. Explain how your last two answers, which assume light is a wave, conflict with photoelectric effects:

   From the photoelectric effect we see electrons only being ejected from an orbit above certain frequencies. This means that they cannot be receiving energy continuously from a wave of a lower frequency and slowly gaining energy as might be expected from Q6 and Q7

9. Explain what the photoelectric effect implies about the nature of light:

   From the photoelectric effect we see that in this instance light is not traveling or behaving as a wave, which would continuously add energy to an electron. Instead it must be traveling as a particle (called a photon). High frequency photons have enough energy to knock electrons loose, while low frequency photons do not. The existence of the photoelectric effect implies that light travels as a particle.

©2020 **BIOZONE** International
ISBN: 978-1-927309-82-7
Photocopying Prohibited

Energy and light

- Based on the results of various experiments and other known characteristics of light, we can see that light exhibits the properties of both a wave and a particle. This is called the **duality theory of light**.
- Recall that the energy in light is related to the frequency of light and the intensity (amplitude) as in the diagram below:

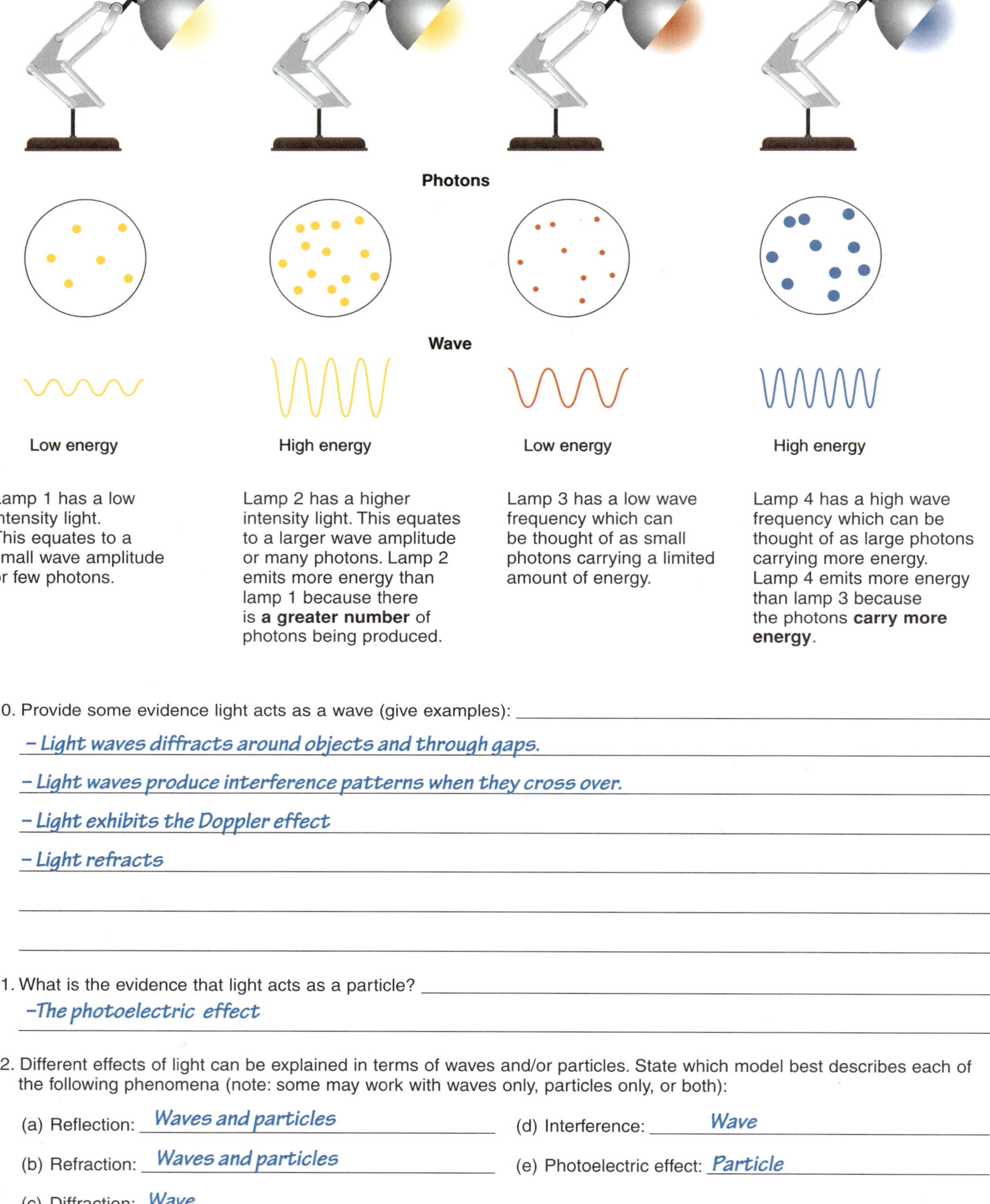

Lamp 1 has a low intensity light. This equates to a small wave amplitude or few photons.

Lamp 2 has a higher intensity light. This equates to a larger wave amplitude or many photons. Lamp 2 emits more energy than lamp 1 because there is **a greater number** of photons being produced.

Lamp 3 has a low wave frequency which can be thought of as small photons carrying a limited amount of energy.

Lamp 4 has a high wave frequency which can be thought of as large photons carrying more energy. Lamp 4 emits more energy than lamp 3 because the photons **carry more energy**.

10. Provide some evidence light acts as a wave (give examples):

– Light waves diffracts around objects and through gaps.

– Light waves produce interference patterns when they cross over.

– Light exhibits the Doppler effect

– Light refracts

11. What is the evidence that light acts as a particle?

–The photoelectric effect

12. Different effects of light can be explained in terms of waves and/or particles. State which model best describes each of the following phenomena (note: some may work with waves only, particles only, or both):

(a) Reflection: Waves and particles

(b) Refraction: Waves and particles

(c) Diffraction: Wave

(d) Interference: Wave

(e) Photoelectric effect: Particle

©2020 BIOZONE International
ISBN: 978-1-927309-82-7
Photocopying Prohibited

124 Evaluate a Claim

Key Question: How valid are the claims that microwaves from cellular towers and phones can cause human health problems?

Microwaves ovens cook food, so surely cell phone towers and cellular phones which use these wavelengths of the EM spectrum are dangerous to use, aren't they?

PS4-4

You may have heard various reports, stories, or rumors about cell phones causing cancer or some other health problem, or the dangers of microwaving food. But how likely are these dangers? Researching and evaluating the data will help you find the answer.

Ionizing and non-ionizing radiation

- An important aspect of EM radiation its ability ionize atoms. Recall that an ion is an atom that has lost or gained electrons, and so carries a positive or negative charge.
- The ability of EM radiation to ionize an atom is part of the **photoelectric effect** and is related to the frequency of light. The higher the frequency, the more likely it is that the EM radiation will ionize an atom.
- High frequency EM radiation includes UV light, X-rays, and gamma rays. These are able to knock electrons from atoms. If this occurs in body tissue free radicals (groups of highly reactive atoms) may be produced causing tissue damage.
- EM radiation with longer wavelengths and therefore lower frequencies are unable to ionize atoms. These wavelengths are therefore called non-ionizing radiation.

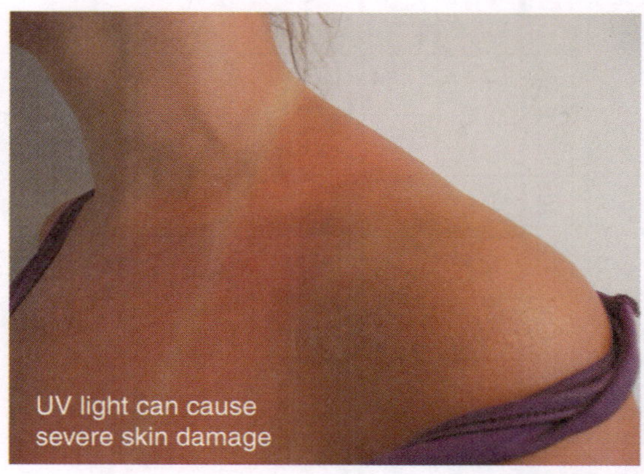

UV light can cause severe skin damage

How microwave ovens work

- About 95% of homes in America have a microwave oven. They are probably the most common appliance used for rapid heating or thawing of food. How is this done?
- The microwave oven produces microwaves using a device called a magnetron. The microwaves are directed into the food compartment of the microwave. Here, the waves cause water molecules to oscillate rapidly back and forth, which results in heating (heat is a measure of the vibrations of particles).
- Contrary to popular belief, microwave ovens do not heat food from the inside out. Microwaves usually only penetrate about a centimeter or so into the food. From there, heat is transferred into the middle of the food by conduction.

A magnetron

How cellular phones work

- Cellular phones produce microwaves in a similar way to a radio transmitter. Electrons oscillate in the cell phones' antenna, producing microwaves. These are sent in all directions (omnidirectional) and are received by the nearest cellular tower, which sends the signal to the appropriate transmitter tower.
- The transmitting tower broadcasts the microwave signal which the receiving cellular phone receives via its own antenna.
- Once the signal is established microwaves are transmitted back and forth between the two phones and the cellular towers at a specific frequency.
- In order for this to work cellular phones constantly send and receive signals to and from nearby cellular towers so that the network knows where the phone is.

©2020 **BIOZONE** International
ISBN: 978-1-927309-82-7
Photocopying Prohibited

1. The information on the previous page tells you a little about how EM radiation affects materials, and how microwaves and cell phones produce the EM radiation they use to carry out their tasks. Choose one of these topics to research: "***Using microwave ovens causes dangerous health effects***" or "***Cellular phone use causes brain cancer***"

   (a) Use the internet to locate claims about you chosen topic. Search using the quoted phrase or use key words.

   It may take a little time to sort through the various hits, but generally blogs and non-scientific websites will give you the "best" unscientific claims. Attach extra notes to this page if you wish.

   (b) In the space below summarize the claim and the evidence used to back it up:

   Student's answer

   (c) Comment on the evidence provided. Does it seem legitimate? Where did the evidence come from? Were the investigations carried out correctly? Were they repeatable? Are there any obvious flaws in the methods or reasoning?

   Student's answer. Students might note that many arguments are based on old data, data that is not significant (e.g. one experiment with small samples that was never repeated), data that is exaggerated, or sometimes even based on "urban legend" (often internet stories that cycle around where the original story or data can not be accurately traced).

   (d) Now search for websites that refute the claim or provide evidence against it. What evidence do they provide? Summarize it here:

   Student's answer (see above)

   (e) Based on your findings, evaluate the claim and decide if you think it is true. Justify your reasoning:

   Student's answer

©2020 **BIOZONE** International
ISBN: 978-1-927309-82-7
Photocopying Prohibited

125 High Energy Waves and Health

Key Question: What are the effects of X-rays and gamma rays on human health?

X-rays

- Have you ever had an X-ray? If you've fractured or broken a bone, or been to the dentist for a check up, it's likely you have. You may have noticed the operator of the X-ray machine (the radiographer) leaves the room or wears a lead lined apron when the X-ray is taken.
- Why is it safe for you to have one X-ray exposure but the radiographer leaves the room or is covered in lead? Does that seem right? The X-ray machine was pointed at you, not them.
- X-rays are produced by accelerating electrons in a magnetic field and have them collide with a metal target. The electron loses energy as an X-ray as it suddenly decelerates (right).
- One exposure to X-rays is relatively safe. It is about the same as 10 days worth of background radiation. However, having many exposures is not safe.
- Unlike many chemicals, which the body can remove or repair the effect of, the effect of radiation exposure is cumulative and builds up over time. Having multiple X-rays or CT scans can increase the chance of developing cancer.
- The X-ray operator leaves the room because although one X-ray may be safe, multiple X-ray exposures over time are not.

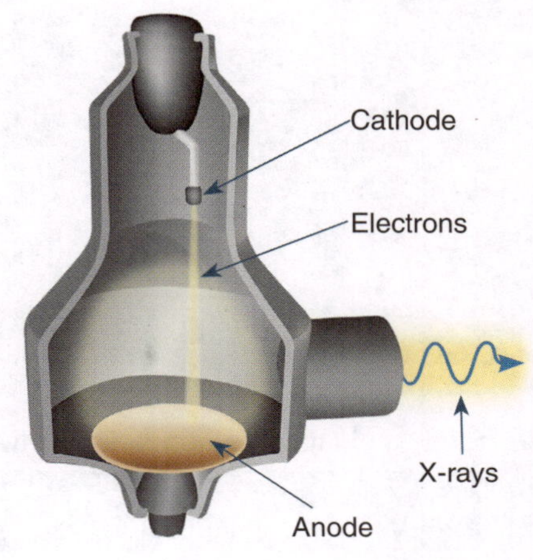
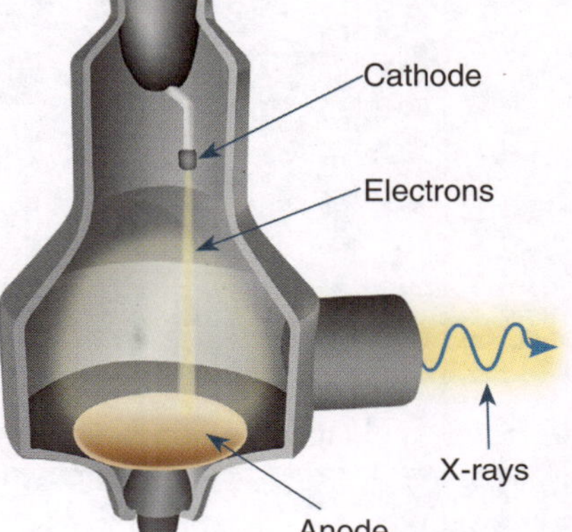

Gamma rays

- At 01:23:45 (am) on April 26 1986 the number 4 reactor at the Chernobyl nuclear power plant exploded. Radioactive material from the reactor's core was scattered around the outside of the reactor building and radioactive dust and gases were released into the atmosphere.
- In a monumental human effort, 500,000 people were engaged in the clean up of the disaster area, with a cost of US$68 billion dollars (in today's value). 68,000 people were evacuated and an exclusion zone with a radius of 30 km was set up around the power plant.
- Total related fatalities are unknown, but is thought to likely be in the tens of thousands. Direct deaths from radiation poisoning, however were quite low. The disaster is considered to be the worst nuclear disaster in history.
- But what made this such a disaster? What is it about the nuclear radiation that meant 68,000 people had to be permanently evacuated?

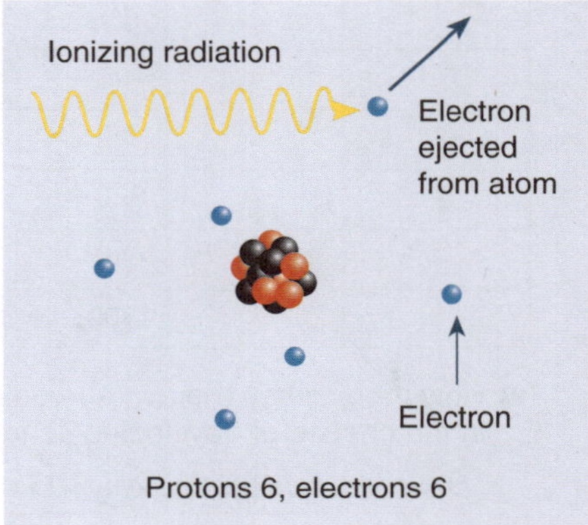

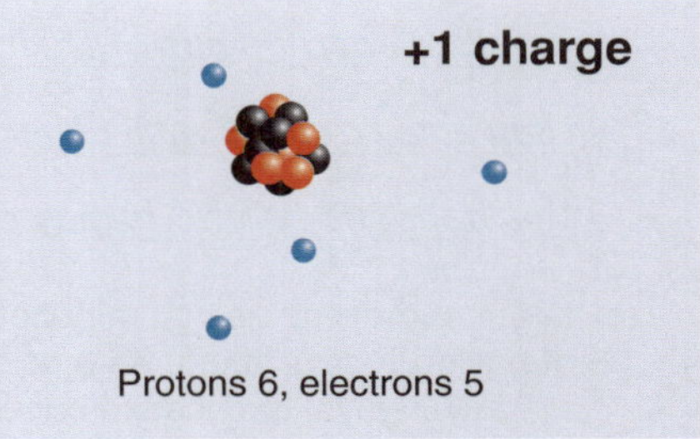

Biological damage

- Recall that radioactive material decays by alpha, beta, or gamma radiation. When alpha or beta decay occurs, alpha and beta particles are ejected from the nucleus. These may cause tissue damage, especially if alpha and beta emitters are ingested. After particle ejection, the remaining nucleus may be in a high energy state. It reduces this energy by emitting gamma radiation.
- Gamma radiation has much higher penetrating power than alpha or beta emissions. As a result, it can pass right through the body. As it travels through the body, the gamma ray may hit molecules of DNA, knocking electrons loose and producing ionization events. These damage the DNA (right) by affecting the bonds between atoms or producing highly reactive free radicals. Cellular repair mechanisms may repair this damage, but not always correctly. Thus mistakes in the DNA code may accumulate as more radiation is experienced and this can result in mutations and cancer.
- High exposure to gamma rays can damage so much cellular material it can be fatal within hours or days.

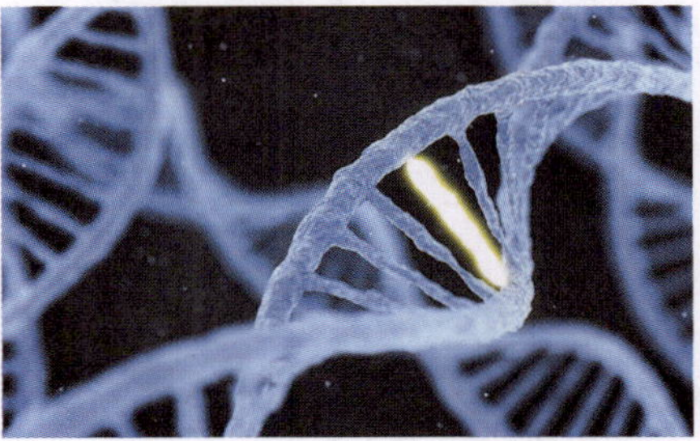

1. Why does the radiographer leave the room or wear lead-lined clothing when you have an X-ray image taken?

Accumulated multiple doses of X-rays can damage body tissues and possibly lead to cancer.

©2020 **BIOZONE** International
ISBN: 978-1-927309-82-7
Photocopying Prohibited

The Chernobyl disaster occurred, ironically, during a safety test. Steam to the turbine that produced the electricity was cut and external electricity to the pumps that circulated coolant water around the reactor core was switched off. This was to test if the steam turbine could provide the electricity for the pumps while it slowed down and before the emergency diesel generators spun up to full speed. Unfortunately the reactor had been put into an extremely unstable state before the test began due to the test being delayed and errors in the control of the reactor. When the pumps were switched off, the reactor output rapidly increased. An emergency shut down was initiated but this actually intensified the problem due to flaws in the reactor design. An initial steam explosion blew the lid off the reactor exposing the superheated interior to the air. This caused the larger explosion which destroyed most of the facility.

2. Study the data below on radioactivity and answer the questions that follow:

Sieverts	Annual dose of radiation relative to background	Biological result
0.006	1	Normal annual dose: no effect
0.01	2	Whole body CT scan (multiple X-ray exposures): no detectable effect
0.05	8	Total maximum annual dose for a worker in a radiation area. Slightly higher chance of some form of cancer (but statistically low) compared to no exposure.
0.1	16	Small risk of cancer but still very low.
0.4	64	Radiation sickness (if received as a single dose)
2	320	Acute radiation poisoning (if received as a single dose)
4	640	Mostly fatal within weeks
8	1300	Always fatal (and normally quickly)

(a) How many times the average daily background dose of radiation do you need to receive before there is an increase in the chance of developing cancer?

8 times, but more likely 16 times.

(b) How does ionizing radiation damage DNA and what is the result of this? *Ionizing radiation is radiation that knocks electrons out of their orbits and so produce ions (hence its name). The ions and free electrons are then able to react with other molecules in the cells. This may result in harmful reactions or molecules. If this happens to DNA it may not be repaired. This can affect cellular functions that may lead to cancer.*

(c) In the same way as you may hear claims about the safety of microwaves, you may also hear claims about the safety of having X-rays taken. As a scientist, reply to the statement from a concerned citizen "I don't want my dentist to give me any X-rays, they cause cancer":

The x-ray dosage received from dental X-rays is extremely low. Only the accumulated effect of multiple X-rays might cause cancer. You are more far likely to come to harm from a dental cavity revealed by an x-ray than the x-ray itself.

(d) You are probably aware that nuclear radiation is used by many comic book writers as a way of explaining why their ordinary person suddenly obtained super powers (want to get bitten by a radioactive spider, anyone?). Explain why this kind of scenario is extraordinarily unlikely:

Nuclear radiation (e.g alpha, beta or gamma rays) is ionizing radiation. It has a the ability to damage DNA which may cause (a) mutation(s). However this mutation will likely only lead to cancer. A total rearrangement of the DNA (i.e. enough to produce a new species) would be needed to gain super human strength or powers.

©2020 BIOZONE International
ISBN: 978-1-927309-82-7
Photocopying Prohibited

126 Review Your Understanding

Key Question: How does the behavior of light explain how photovoltaic cells work?

- At the start of this chapter you were introduced to photovoltaic cells. These capture light from the Sun and convert into electricity. You should now be able to answer at least part of the question of how this is done based on the properties of light.
- The functional part of a photovoltaic cell is two layers of silicon placed together. Each layer has a different combination of silicon and another element that produce a certain crystal structure. These structures are what produce and capture the electrons when the cell is exposed to sunlight.
- However, the key to how photovoltaic cells work is the behavior of EM radiation.

16 Huge photovoltaic panels power the ISS

1. Recall that light travels as both a wave and a particle.

   (a) What is the particle of light called? *A photon*

   (b) What happens when a light particle beyond the frequency threshold for a particular material hits that material?

   An electron is ejected from its orbital

   (c) Recall that electricity is a flow of electrons. How does your answer above explain the ability of solar panels to generate electricity?

   Electrons that are knocked loose by photons can be captured and fed into a circuit. By continually capturing electrons that have been knocked loose an flow of electrons in a circuit can be produced.

   (d) The diagram below shows the simplified structure of a solar cell. The materials that make up the solar cell are not important for this question. Use and add to the diagram, including notes, to explain how the cell generates electricity.

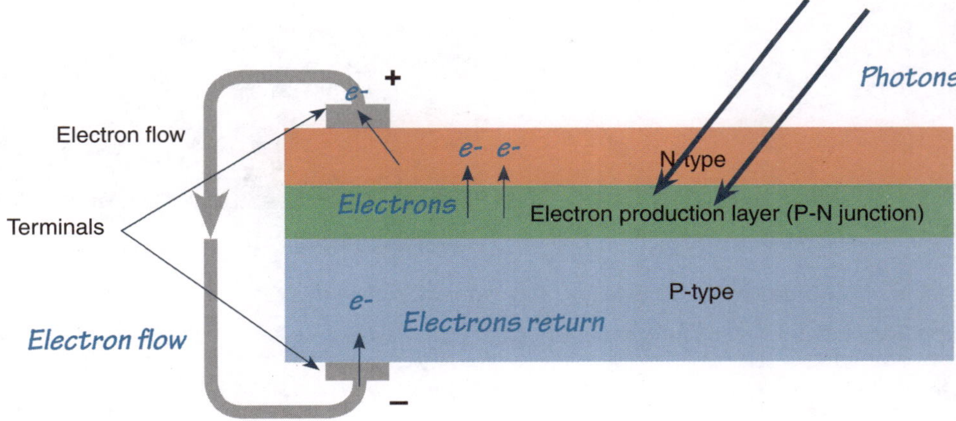

Photons of light hit the electron production layer. The electrons released are attracted to the N-type layer and the + terminal where they are "captured". They then flow around the circuit and return to the solar panel via the P-type layer, making a complete circuit.

©2020 **BIOZONE** International
ISBN: 978-1-927309-82-7
Photocopying Prohibited

127 Summing Up

Properties of electromagnetic waves

PS4-3 1. To our eyes, the vast majority of plant leaves appear green. Explain this phenomenon in terms of light waves:

When light waves come in contact with particles in a material, they can be absorbed or reflected (or refracted or transmitted in specific materials). Particles absorb specific wavelengths of light and reflect all others. Plant leaves appear green because they absorb all the wavelengths of visible light other than green, which they reflect. This green light is detected by our eyes when looking at the leaves.

2. Explain what is meant by the duality of light: *The duality of light refers to the evidence that light (electromagnetic radiation) behaves as both a wave and a particle (called a photon). Evidence that it acts as a wave includes it showing properties of diffraction and interference. Evidence that it acts as a particle includes the photoelectric effect.*

3. Light travels at 299,792,000 meters per second (c). For simplicity of calculations this is often rounded to 300,000,000 m/s (3×10^8 m/s). Calculate the following using the equation $f = c/\lambda$:

(a) A TV broadcast EM wave has a frequency of 890 MHz (megahertz). Calculate the wavelength of the wave:

$\lambda = v \div f$. λ = 300,000,000 m/s ÷ 890,000,000 Hz = 0.34 m

(b) An FM radio broadcast uses a carrier wavelength of 2.85 m. Calculate the frequency of the carrier wave:

$f = v \div \lambda$ f = 300,000,000 m/s ÷ 2.85 m = 105,263,157.9 Hz = 105 MHz

(c) Calculate the wavelength of UV light with a frequency of 3.6×10^{15} Hz:

$\lambda = v \div f$ λ = 300,000,000 m/s ÷ 3.6×10^{15} Hz = 8.33×10^{-8} m

4. The diagram below shows the effect of placing moist thermal printer paper in a microwave oven. Thermal paper changes color from white to black when exposed to heat. The paper what placed on a square sheet of card that fitted the shape of the microwave oven's interior and so did not rotate on the glass plate.

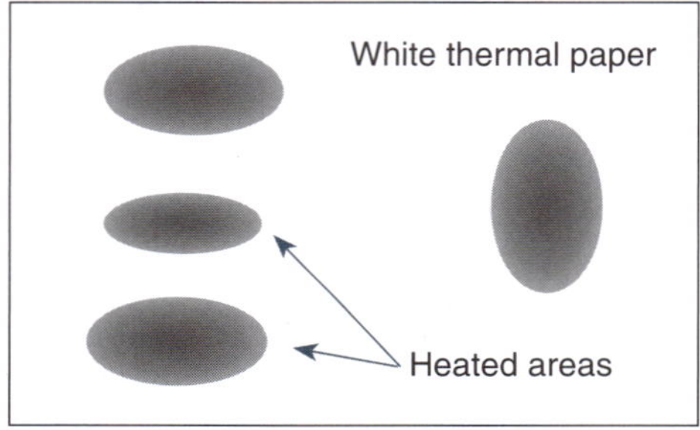

(a) Use you understanding of wave behavior, microwaves, and water to explain this result:

Microwaves cause water to vibrate. The vibrations can be measured as a rise in temperature (the water gets hotter). The pattern on the thermal paper can be explain by interference of the microwaves in the oven chamber. The darker areas are where waves constructively interfere. The lighter areas are areas where destructive interference has occurred and there is therefore no heating.

(b) How is this effect overcome in the design of a microwave oven? *The effect is overcome by having the oven plate rotate so that all parts of the object being heated pass through areas of heating.*

PS4.B SSM

©2020 BIOZONE International
ISBN: 978-1-927309-82-7
Photocopying Prohibited

5. Recall that photon energy is related to frequency. Dark rooms were used to process photographic film before the common use of digital cameras. These rooms were illuminated by dim red lights. Explain why:

The red light will not have enough energy to affect (bleach) the photographic film. The low light level (intensity) reduces the number of photons released and so reduces the effect of photons that may have a high enough frequency threshold. The red light still gives enough visible light to see by.

6. Explain why blue and violet colors fade more slowly in posters and photographs than red or yellow colors:

We see colors because pigments reflect certain wavelengths of light and absorb others. Blue and violet pigments reflect blue and violet wavelengths. Reds and yellow pigments absorb these wavelengths. These wavelengths have high frequencies and so have high energy photons that will break up the red and yellow pigments over time (via the photoelectric effect) causing them to fade.

7. A laser was shone at a screen after passing through a single then double slit as shown below:

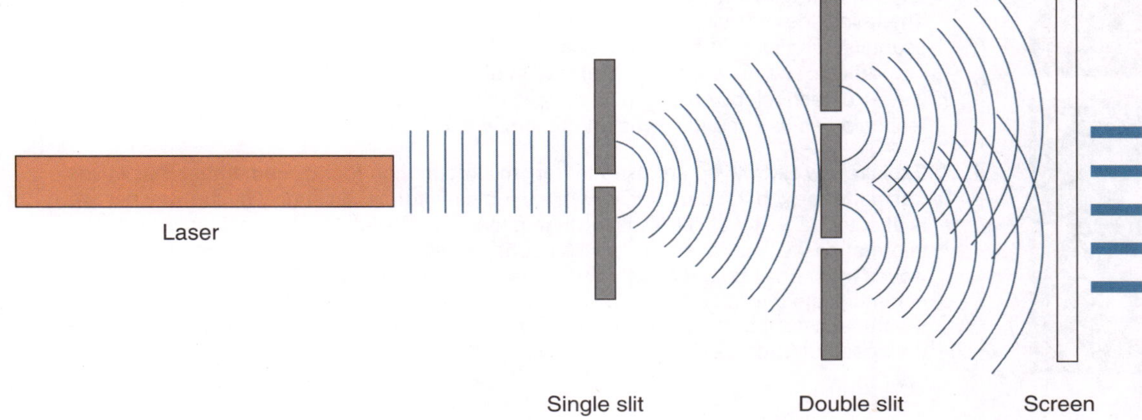

(a) What would you expect to see on the screen (annotating the diagram above may help your description)?

A pattern of alternating light and dark bands.

(b) The intensity of the laser was reduced until it was producing just a single photon at a time. The screen was replaced by particle detectors. Photons were sent through the first screen one at a time. The graph below left shows the number of photons detected as a function of distance along the screen after one second. The graph below right shows the number of photons detected as a function of distance along the screen after many seconds.

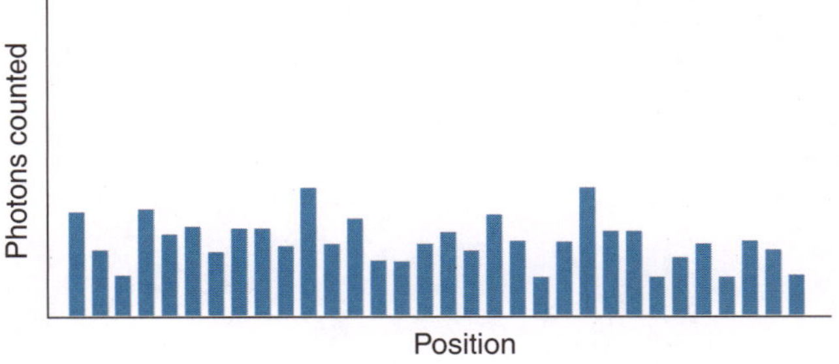

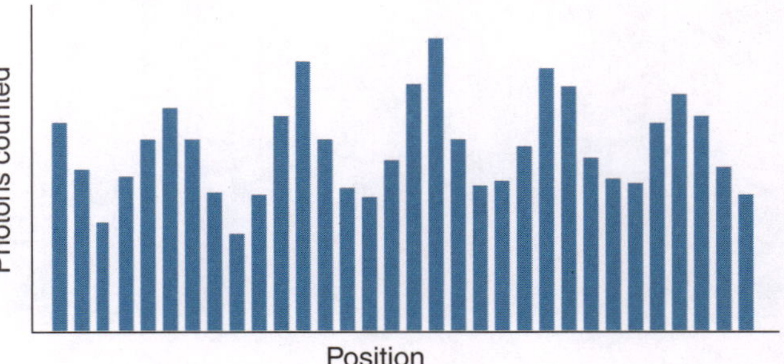

How does this data reconcile with your description in 7. (a)? *The graphs agree with the description in 7. (a), especially the graph on the right, taken as the number of photons counted increases. This shows more photons in the region where bright bands are seen and fewer photons in the region where dim or no bands are seen.*

©2020 **BIOZONE** International
ISBN: 978-1-927309-82-7
Photocopying Prohibited

PS4.C

11: Information Technologies and Instrumentation

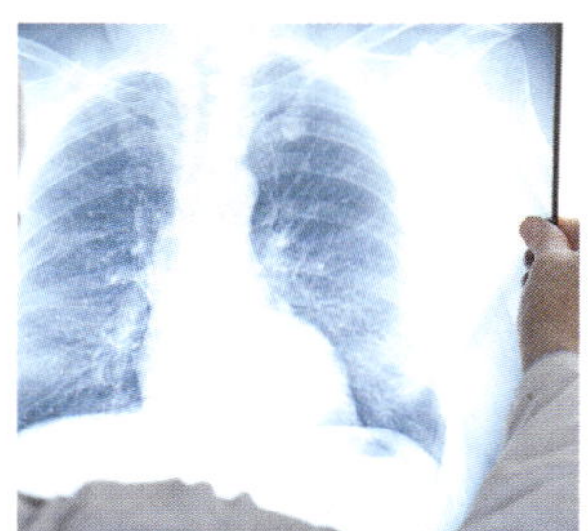

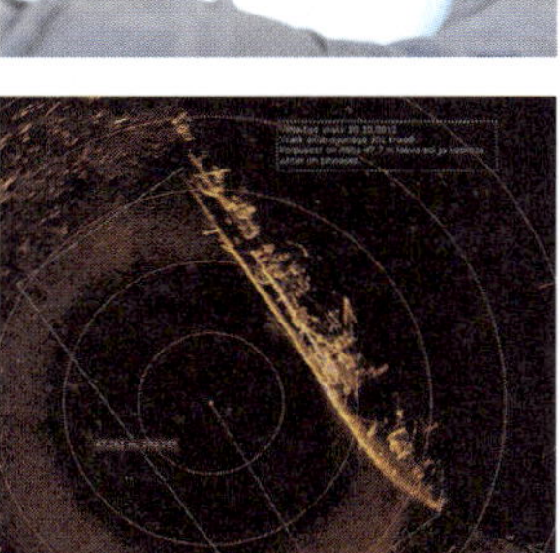

Anchoring Phenomenon

Data and EM waves: How can the properties of EM waves be exploited to transmit large amounts of data? 128 131

How does technology exploit the properties of waves?

- ☐ 1 In chapter 9 you explored the properties of waves and looked at how the information in waves can be stored and retrieved through digitization. In the previous chapter, you looked at the properties of electromagnetic radiation and how the wave-particle duality of light underlies the operation of solar cells and modern conveniences like microwave ovens and cellular phones. In this chapter, you will explore the technologies based on sound waves and EM radiation further and find out in more detail how they work. These technologies include medical imaging technology and communication devices. You will then communicate technical information about how they use the principles of wave behavior and wave interactions with matter to transmit and capture information and energy *[SEP-8] [PS4.C] [CCC-4] [HS-PS4-5]*. 132

- ☐ 2 Explain how technology uses the properties of waves to achieve diverse outcomes. These include medical diagnostics (X-rays and ultrasound imaging), radionuclide scanning, seafloor mapping with sonar, and fiber optic telecommunications. What properties of waves are being exploited in each of these technologies? How are these pieces of technology designed to achieve their desired effect? Explain why some technologies use sound waves and others use EM waves *[SEP-8] [PS4.C] [CCC-4]*. 129

- ☐ 3 Explain how EM waves are used to transmit information (data). Remember that waves transfer energy but not matter, so we can obtain information simply by the way a wave is affected *[CCC-4]* as it travels through a medium. Waves can also act as carriers for information we want to send. As with sound waves (chapter 9), the information in EM waves must be encoded and then decoded. Explain the principles of modulation and communicate this information *[SEP-8]*, distinguishing between AM and FM. What about the transmission of data using light waves (rather than radio waves)? Comment on how the speed of transmission can be increased and provide an example of this *[PS4.C]*. 130

11. Information Technologies and Instrumentation

Anchoring phenomenon

Data is information, and sending and receiving it are important aspects of today's world. As we develop more sophisticated technology, e.g. 5G technology, more data can be gathered and so more needs to be sent. But how is data sent? Why do some devices use radio waves and others microwaves? Why can submarines receive but not send ELF signals, and why do these signals only carry text, not voice? Students can discuss these questions. Some may have a better understanding of what acronyms such as 4G and 5G mean and why different parts of the spectrum are used for different means of communication, so a discussion of these terms before beginning the chapter could be beneficial.

How does technology exploit the properties of waves?

1. Technology can be used in many different ways to gather information and it has applications in many fields of science and medicine. The purpose of this chapter to extend the understanding developed in the previous chapters to explore the range of technologies based on sound waves and EM radiation. Some students will be familiar with a number of these technologies, while others may only be conversant with those they encounter on an everyday basis. The aim of this chapter is to develop a deeper technical understanding of these technologies.

2. Certain technology can be classed as "active". That is, the technology sends out certain waves and detects the resulting reflection or shadow depending on the relationship of the transmitter and receiver. Examples include sonar and CT scans. Other technology is "passive" and simply receives waves generated elsewhere. Examples include thermal cameras and some types of telescopes.

3. Finding ways to transmit more data, more quickly is a rapidly developing area of research.

   Modulation is an important aspect of information technology. Modulation can be defined as the process of impressing information from a modulating signal onto another signal called the carrier. The resulting signal is called the modulated signal. Students should recognize from studying waves earlier that it is wave properties that make this possible. They should be able to describe those properties and explain why it is possible to encode more information when shorter wavelengths and higher frequencies are used.

   High frequency waves are able to carry more data per second than low frequency waves because more wave fronts pass a receiver unit in a set amount of time. Similarly, in signals where a light pulse is 1 (or on) and no light is 0 (or off), higher light pulse rates or frequency of pulses will send data (as binary) faster than low pulse rates.

   More receivers receiving separated waves also increase the amount of data being sent. This might be thought of as "channels". Instead of sending data down one channel, data can be split up and sent down multiple channels and so the total volume of data is sent faster. More channels or waves can be sent at higher frequencies that at low frequencies. This is because the effect of the modulation is less at higher frequencies and so modulated waves do not cross into each other's frequency (and get mixed up). Thus a band of EM waves near the visible range can carry many more separate signals than a band of EM waves near the microwave range because of the much higher frequency of the waves in the visible range. As a result, fibre optics carry more data than 5G cellular networks, which carry more data than 4G and so on. However a limitation is that higher frequencies tend to be absorbed by obstacles such as walls and buildings.

ADVANCED LEARNING

★ Challenge question:

Why are the antennas used in communications technology sometimes long and sometimes short? Explain in terms of wavelength and frequency. How many different devices are being catered to in this image?

Image: Angus Thompson cc 4.0

Download this page from BIOZONE's Resource Hub

128 Data and EM Waves

Key Question: How can the properties of EM waves be exploited to transmit large amounts of data?

Cell phones

- Every few years, a new generation of cellular phone technology is produced that is advertised as being faster than the last. The latest generation, 5G, is claimed to be 10 times faster than the older 4G technology.
- But what does this mean? Faster actually refers to the rate of data transmission and download. A 5G phone will be able to download data 10 times faster than a 4G phone. How is this possible?
- You know from the previous chapter that cellular phones transmit and receive microwaves. But how is data sent or received by the phone? How is it different to radio?
- In this chapter you will look into the use of EM radiation in technology in order to answer these questions.

1. Microwaves travel at the same speed as radio waves. Give two reasons why cellular phones use microwaves instead of radio waves to transmit and receive data:

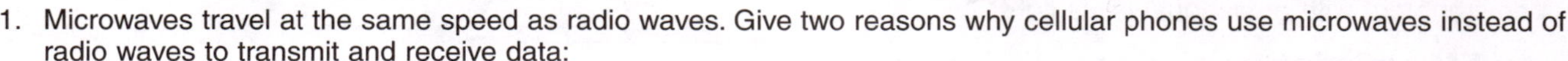

Student's answer. Reasons might include the length of the antenna needed to send signals (shorter in cell phones) or the speed of data transmission/download (faster in cell phones).

2. For a transmitter of a certain power output, radio waves travel a lot further than microwaves under real world Earth conditions. This is one reason cellular networks require many relay stations. Suggest a reason why microwaves don't travel as far as radio waves under Earth conditions:

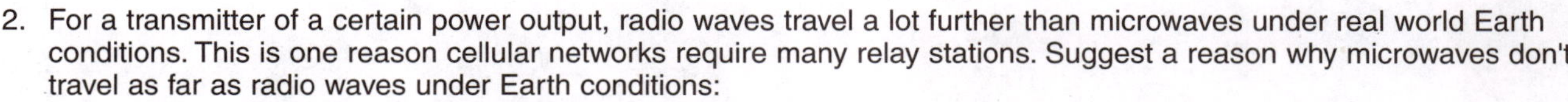

Because of the difference in wavelength and frequency between microwaves and radio waves they behave differently near the surface of the Earth. Microwaves tend to be absorbed by materials they hit (such as walls, hills, and clouds) so many relay stations are needed to maintain the signal. Radio waves pass through most objects and tend to bend around large objects better than microwaves so there are fewer areas of signal loss.

Submarines

- An important part of a military submarine's role is to evade detection. Part of doing this requires them to be submerged for long periods of time. This means they are out of communication range because radio waves don't penetrate very far into seawater. They could extend an aerial while still submerged to receive a radio message but this would allow other parties to locate them. Submarines therefore receive VLF (very low frequency) and ELF (extremely low frequency) radio (well below the commercial radio bands) which can penetrate seawater to a few hundred meters. However, VLF has very slow data download rates. To get around this, transmissions are very compact coded messages.

3. Why does VLF and ELF have such slow data download rates compared to microwaves?

Because the frequency is low (and the wavelength so long) data carried in the wave (e.g. at each wave crest) takes much longer to receive because the number of waves passing through the antenna is so low per unit of time (relative to waves with shorter wavelengths). Data therefore takes much longer to download using VLF and ELF. What's more, because of the low frequency, it is impossible to have many channels within the transmission band, so only one or two channels can be used to send data.

©2020 **BIOZONE** International
ISBN: 978-1-927309-82-7
Photocopying Prohibited

129 Waves and Technology

Key Question: How does technology use the properties of waves to achieve various outcomes?

- The ability to use electromagnetic waves has revolutionized the world. Our most important technologies are based on detecting and transmitting EM waves.
- Thanks to technology, we now have the ability to see into areas of the EM spectrum that we can't see naturally. We are able to build devices capable of detecting these other wavelengths and transforming them into information we can detect.
  - Radio receivers detect radio waves and transform them into sounds we can hear.
  - IR and UV glasses alter the wavelength of EM waves to allow us to see them.
  - Infrared radiation is emitted by all objects with a temperature above absolute zero so thermal imaging (right) makes it possible to see the environment with or without visible light.

Using EM radiation as a diagnostic tool

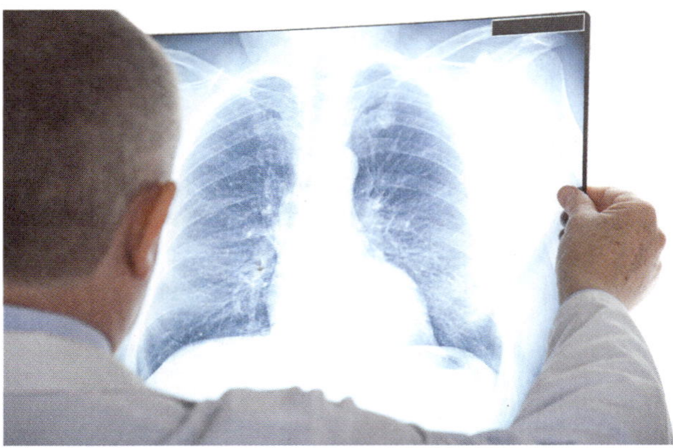

X-rays can pass through tissues and expose photographic film. The X-rays are absorbed by dense body tissues (e.g. bone) which appear as white areas, but they pass easily through less dense tissues (e.g. muscle), which appear dark. X-rays are used to identify fractures or abnormalities in bone. X-ray technology is also used in conjunction with computer imaging techniques (e.g. CT scans).

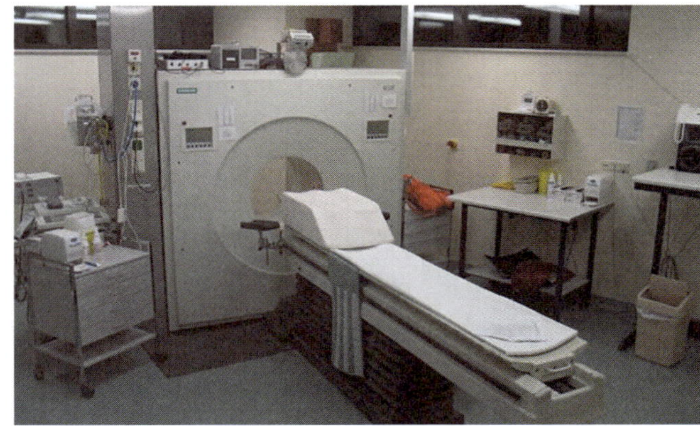

Radionuclide scanning involves introducing a radioactive substance (the radionuclide) into the body, where it is taken up in different amounts by different tissues (e.g. radioactive iodine is taken up preferentially by the thyroid gland). Gamma rays emitted by the tissues that take up the radionuclide are detected by a gamma camera or scintillator. Unlike X-rays which pass through the patient, the radiation comes from inside the patient.

1. Identify if the EM radiation is traveling through the object being studied or from the object being studied (it can be both).

   (a) The Sun: *Traveling from the Sun (mostly visible light with some ultraviolet and x-rays and some infrared)*

   (b) Using a infrared camera (as show above): *Infrared light emitted by the object being viewed*

   (c) Studying interstellar gas clouds: *Infrared light passing through the clouds*

   (d) Studying supernovae (stellar explosions): *From the supernova (including most of the EM spectrum)*

   (e) Producing a CT scan: *X-rays passing through the body from the x-ray emitter*

2. Explain how thermal imaging is different to X-ray imaging: *Thermal imaging uses light emitted by or reflected off objects. Thermal cameras detect the longer wavelengths of infrared light emitted by hot objects. X-ray imaging uses film or a camera that detects the x-rays passing through an object. An x-ray emitter, which is entirely separate from the camera or film must be used to produce the x-ray.*

PS4.C CE

©2020 BIOZONE International
ISBN: 978-1-927309-82-7
Photocopying Prohibited

Communications

- As you learned earlier in Chapter 10 (*Electromagnetic Radiation*), EM waves are used for communications, including TV, radio, and cellular phone networks.
- EM waves in the infrared spectrum are used for fiber optic telecommunications. This is because the loss of signal of the infrared light in the glass fiber is much less than for other EM waves. This means the signal can go further and fewer repeaters are needed. Multiple signals can be sent down the same fiber so large amounts of information can be sent very quickly. However, the speed of light in glass is only about 70% of the speed of light in air.

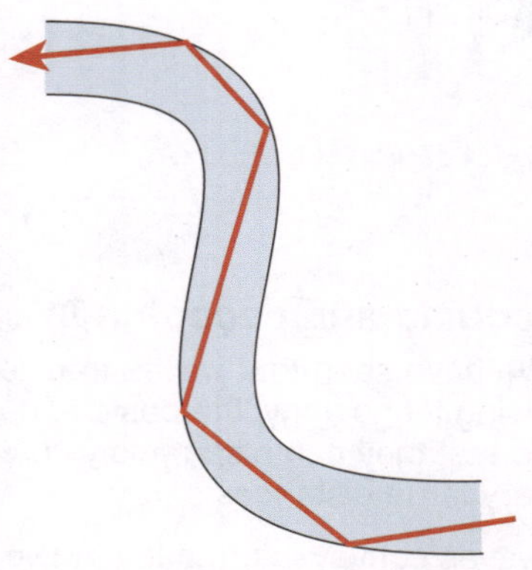

A light beam reflects inside the fiber (called total internal reflection) and so can follow the curve of the fiber.

Seeing with sound

- EM radiation can be used as a useful diagnostic tool based on how light travels through or is emitted from an object. However, sound can also be used as a diagnostic tool to see into the human body. It is also used in sonar to produce images of the seabed.

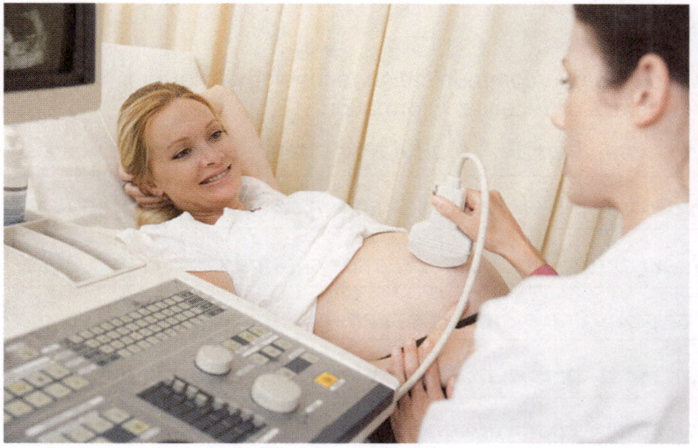

Ultrasound is a diagnostic tool for visualizing internal structures without surgery or X-rays. Ultrasound imaging is based on the fact that tissues of different densities reflect sound waves differently. Sound waves are directed towards a structure and the reflected sound waves are recorded. An image of the internal structures is analyzed by computer and displayed on a screen. Ultrasound is used to check the health of a fetus during pregnancy, or to diagnose disorders such as cirrhosis, cysts, blockages, and tumors.

Tuukritööde OÜ CC 4.0

Sonar is used to image objects in water. Sound is transmitted from an acoustic array at around 50 kHz and the echo is received by a receiver. Sonar is used for a variety of aquatic applications including military applications (e.g. submarine warfare), commercial fish finding, depth sounding, and scientific uses including mapping the seafloor and underwater archeology. The frequency used in sonar is important. Higher frequencies give greater image resolution.

3. The speed of communications is very important in financial markets. Being able to make a trade before someone else does could make you a lot of money. Explain why some remote traders now use point to point microwaves to communicate to financial centers rather than using fiber optic cables:

   Point to point transmission of data by microwaves is faster over a short distance than infrared light travelling in longer fiber optic cables at a slower speed despite the infrared having a shorter wavelength.

4. Why are sound pulses (sonar) more effective at viewing objects on the seafloor 2 km below the surface than EM waves?

   Light waves (of virtually any frequency) is absorbed quite quickly by water (within a few hundred meters). Sound travels further in water because it is a pressure wave, passed on by the displacement of water molecules. The density of water means it is travels very quickly with less loss of signal.

5. Why is using ultrasound to view a fetus a better option than X-ray? *Sound waves are safer (for a developing fetus to be exposed to) and simpler to produce and record allowing real-time viewing of a fetus. X-rays are potentially harmful to a fetus (although risk is small) and difficult to produce and record in real time due to the large, complex equipment needed to produce them and the film needed to record them.*

©2020 **BIOZONE** International
ISBN: 978-1-927309-82-7
Photocopying Prohibited

130 Transmitting Information Using Waves

Key Question: How can EM waves be used to transmit data and how does the frequency of the wave affect the data transmitted?

PS4-5 Encoding and decoding information

- We have seen that waves transfer energy but do not transfer matter. Using technology, the basic features of waves can be manipulated so that they can also rapidly convey vast amounts of information over very large distances.
- Waves convey information in two general ways:
  - Information about a medium can be obtained simply by the way a wave is affected as it travels through that medium, e.g. seismic waves, MRIs and X-rays.
  - Waves can be used as a "vehicle" for transmitting information that we want to send from one place to another, e.g. radio waves.
- As you saw earlier for sound waves, for waves to transmit and receive information, we must first be able to encode the information into waves and then decode the waves once they reach the receiver.

1. You may be aware of the various types of electromagnetic waves used in communication technology. In groups, discuss what you know about how each of the pieces of technology transmits or receives information:

   (a) AM radio: *A microphone translates sound to an electronic signal. A modulator changes a carrier wave's amplitude to include this signal. A receiver separates out the signal and translates the signal to sound.*

   (b) FM radio: *A microphone translates sound to an electronic signal. A modulator changes a carrier wave's frequency to include this signal. A receiver separates out the signal and translates the signal to sound.*

   (c) Cellular phone: *Phone converts sound, text, etc. to an electronic signal. This signal is transmitted via microwaves from the phone to the cell tower and on to the receiving phone.*

   (d) TV satellite: *Images and sounds are translated to electronic signals, which are added to a carrier wave. The carrier wave is sent to satellites which redirect it back to Earth where receivers translate the signals back to images and sounds to be enjoyed through the TV.*

   (e) Bluetooth emitter: *Transmitters produce radio waves and attach electronic sound signals to these waves. Receivers translate these signals back to sound.*

   (f) WIFI emitter: *Routers receive digital data and transmit the information on carrier waves so devices with receivers are able to translate the information.*

2. One of the simplest ways to encode information is something like Morse code where information is encoded as simple dots and dashes. Think of a short sentence. Write it out in Morse code below and see if a classmate is able to decode it:

   Student's answers will vary.

   __ ___ ._. ... . _._ ___ _.. . (Morse code).

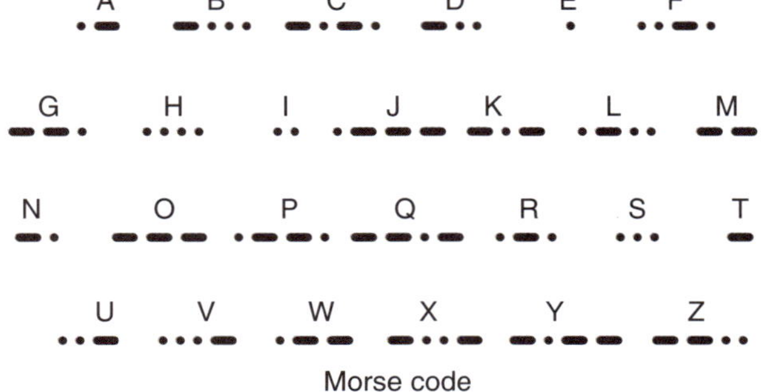

Morse code

PS4.C CE

©2020 BIOZONE International
ISBN: 978-1-927309-82-7
Photocopying Prohibited

Radio waves and modulation

- Recall that radio waves are part of the electromagnetic spectrum. They have longer wavelengths than visible light. But how do radio waves carry information?
- Each radio transmitting station broadcasts a radio wave, with a unique frequency, called the carrier wave. When there is an audio signal (human voice or music) to be sent out, that signal is used by the transmitting equipment to modify the carrier wave slightly. This is called **modulation**.
- You have heard of AM radio and FM radio, but what do these stand for and how do they work? What is the difference between them?

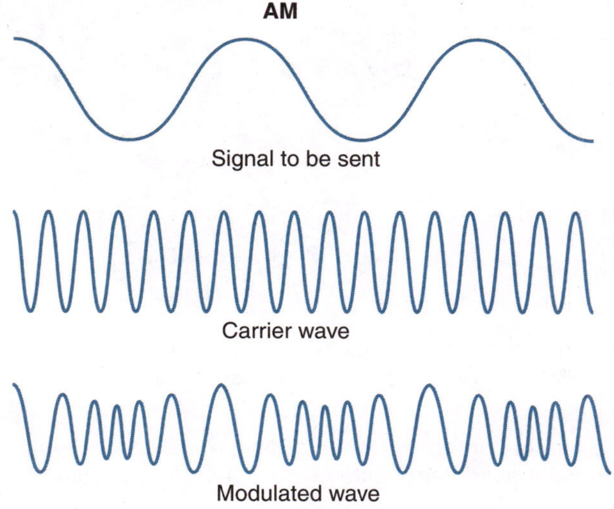

AM stands for **amplitude modulation**. The amplitude of the carrier wave is changed. AM wavelengths are much longer than FM wavelengths. This means they are less affected by structures such as buildings or mountains and hence have a greater range.

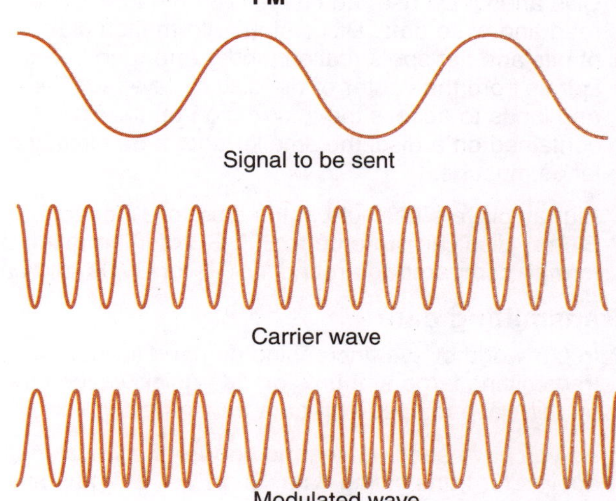

FM stands for **frequency modulation**. The frequency of the carrier wave is changed. More than one signal can be transported by one carrier wave (e.g. allows stereo sound) and it enables better reproduction of the original modulating wave.

3. (a) Describe how the crests and troughs of the wave being sent are represented on an AM carrier wave:

The crests of the signal wave are represented by higher amplitudes of the carrier wave in the modulated wave. Conversely, the troughs of the signal wave are represented by lower amplitudes of the carrier wave in the modulated wave.

(b) Describe how the crests and troughs of the wave being are represented on an FM carrier wave:

The crests and troughs of the signal wave are represented by higher and lower wave frequencies of the carrier wave in the modulated wave.

- In wireless (WIFI) communication, the router and network cards use more complicated forms of modulation. These types of modulation allow many different signals to be conveyed by each carrier wave and can enable multiple carrier waves to simultaneously use the same bandwidth.

4. (a) Explain why a router would need to engage multiple carrier waves to carry information:

The information running through a router is much greater and more complex than the simple sound and images associated with radios, TVs, etc. In order to transmit so much information in a reasonable time-frame, multiple carrier waves need to work simultaneously. Without enough carrier waves, information transmission slows down and you experience 'lag'.

(b) Why do radio stations only require a simple modulation to transmit information?

Radio stations are only transmitting sound waves. There is a lot less information needed to be transmitted so a single wave with simple modulation is all that is required.

©2020 **BIOZONE** International
ISBN: 978-1-927309-82-7
Photocopying Prohibited

Red and blue light

- As you saw earlier, radio waves can carry information, but visible light waves can also be made to carry information. This is the basis of CDs, DVDs, and Blu-ray discs. These all use lasers (intense and highly coherent beams of light) to read and write data.
- Visible light has shorter wavelengths (higher frequency) than radio waves. This means more information can be transmitted over the same time interval with these shorter waves.
- CDs and DVDs use red lasers (780 nm for CDs and 650 nm for DVDs) to read and write data. Discs store information digitally, using a sequence of pits and flat spots (called lands) arranged in a continuous line which spirals from the center of the disc to its edge. The laser reads the pits and lands to access the information on the disc. The more data that is contained on a disc, the smaller and more closely packed the pits and lands must be.
- Blu-ray players use light in the blue to UV spectrum (405 - 195 nm). A Blu-ray disc can store up to 300 GB of information depending on the format, compared to around 17 GB for DVDs and only 700 MB for a CD.

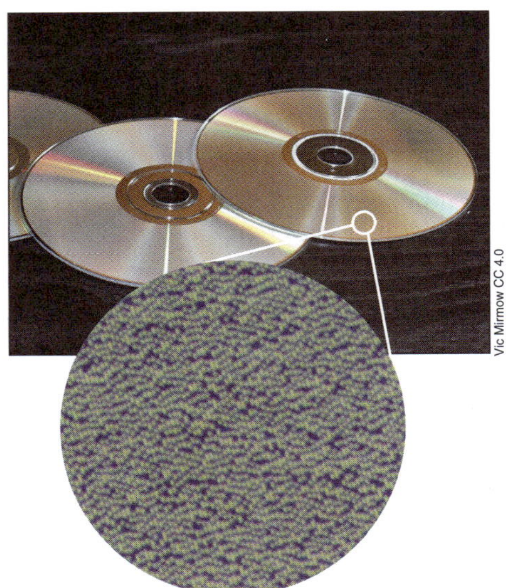
Vic Mirmow CC 4.0

Transmitting data

- In our world of ever increasing demand for data and processing speed, technology has had to solve the problem of transmitting large amounts of data quickly. Since the speed of light in constant, it is not a case of simply speeding up the wave to speed up data transmission.
- As you have seen earlier, the shorter the wavelength the more data can be stored on a disk. A similar concept applies for transmitting data. Short wavelengths and high frequencies transmit data faster than long wavelengths and low frequencies (below):

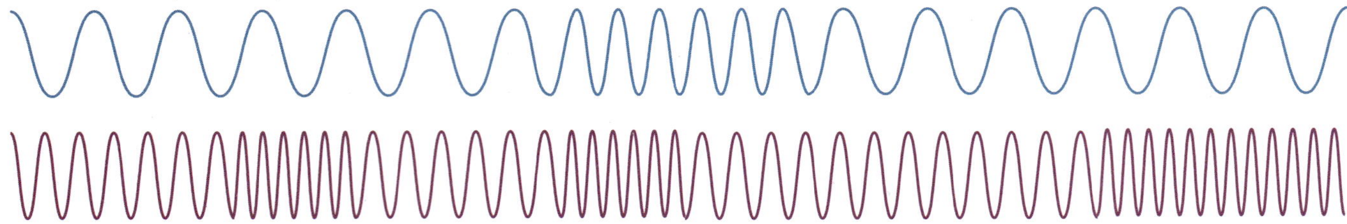

- In the diagram above data is sent as binary 1s and 0s. 1 is represented by a certain frequency and 0 by a shorter frequency (called frequency-shift-keying). In the top wave it can be seen that a signal of 101 has bent sent. Using the shorter wavelength in the bottom wave, the signal 10101100 has been sent in the same length of time.

- In addition, at higher frequencies more carrier waves can be fitted into a certain band of frequencies (e.g the Ka band is between 26 – 40 GHz). Imagine sending a 20 kHz signal along a 1 MHz carrier wave. This means the 1 MHz carrier wave changes frequency by up to ± 10 kHz or ± 1%. Now imagine using a 1 GHz carrier wave to send the same signal. In this case the frequency of the carrier wave changes by just ± 0.001%. Thus many more signals can be sent without crossing into each other's frequency range.
- Data transmission (and download) speed can be increased by breaking up one long signal (e.g. data and voice encoded together) into separate shorter signals (e.g. multiple data channels and a voice channel) that can be sent all at once and then reassembled at the receiver. This is how broadband works.

5. Suggest why more information can be stored, read, or transmitted using higher rather than lower frequencies:

Higher frequencies allow more information (i.e 1s and 0s) to be squeezed into a smaller space (either in a wave or as physical markings on a disc). It also means that more channels (carrier waves) can be transmitted in the same frequency band, increasing data transmission and reception speeds.

6. Are there any dangers associated with using shorter and shorter wavelengths for data storage and transmission?

As wavelengths get shorter, the energy carried by the waves increases. In the UV wavelength range, we start approaching ionizing wavelengths. Ionizing radiation could be a health hazard to humans and might even be damaging to the data being stored.

7. Explain why transmitters that send and receive microwaves sometimes suffer interference from "rain fade" (signal loss):

Water molecules absorb certain frequencies of microwaves. Thus when microwaves pass through heavy rain or thunder storms they can be absorbed and so the signal is lost.

©2020 BIOZONE International
ISBN: 978-1-927309-82-7
Photocopying Prohibited

131 Review Your Understanding

Key Question: How can the properties of EM waves be exploited to transmit large amounts of data?

Downloading data

- A computer using optical fiber based broadband might have a download rate of 100 Mbps (megabits per second). 5G networks have download speeds of about 20 Mbps and 4G networks have download speeds of about 1-2 Mbps.
- All these data transmission methods are based on EM radiation. Fiber optics use infrared light, with wavelengths of around 850 nanometers. 5G networks use millimeter wavelengths whereas 4G microwaves with wavelengths of slightly shorter than a meter.

1. Light travels at the same speed for any kind of wavelength. In fact, light in fiber optics travels only 70% of its speed in air. Explain the difference in the download rate between fiber optics, 5G and, 4G:

Nanometers are a million times shorter than millimeters. As a result, infrared waves can have many times more channels or carrier frequencies to carry data . Higher frequencies also transmit more data over the same amount of time than lower frequencies. Despite the fact that infrared light travels only 70% of the speed of light in optical fibers, the much greater amount of data that can be transmitted and downloaded makes up for this. 4G networks use even longer microwaves wavelengths than 5G. This means there are fewer available channels and 4G is slower again than 5G networks at transmitting and receiving data. However, 4G networks need fewer repeaters because they are less likely to be affected by buildings and other obstacles. Infrared can only be used in line of sight transmission outside of cables (e.g. T.V. remotes only work when pointing more or less directly at the television and can't work through walls).

2. Submarines receive transmissions by VLF and ELF. These have very slow download rates and are limited in what can be sent. Typically the data is only coded as text and usually limited to 400 to 700 words per minute. Additionally the antennas used to transmit the frequencies are huge, covering many hectares (which is why submarines can only receive and not transmit).

Using your knowledge of EM waves explain the information above.

VLF and ELF are very long wavelength of radio waves, measuring in the tens to hundreds of kilometers long. This means that have a very low frequency. Not only does this mean the modulations in the carrier wave arrive very slowly (but virtue of being spaced far apart), but there is no space in a channel for transmitting multiple carrier waves or even voice (ELF is a 3 - 30 kHz band, voice is 20 kHz, so transmitting a voice message on one channel would take up almost the entire band).

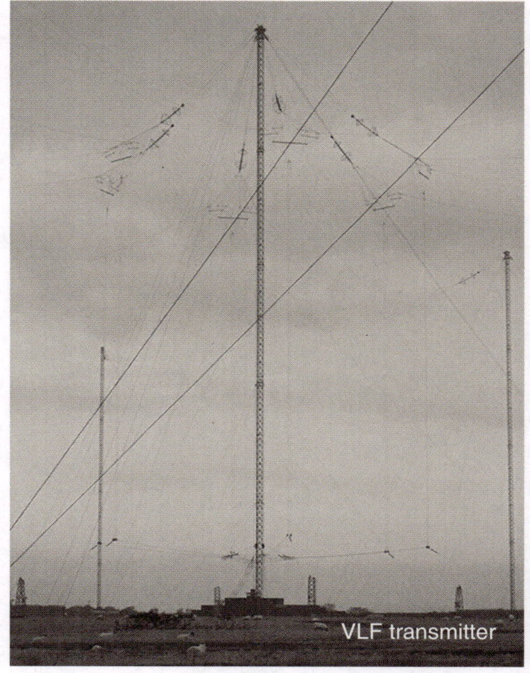
VLF transmitter

James T M Towill

©2020 BIOZONE International
ISBN: 978-1-927309-82-7
Photocopying Prohibited

132 Summing Up

Information technology

PS4-5

1. Select one of the following topics to research. Use the space below for planning purposes:

   (a) Form a group of four. Each person should select a different technology based on electromagnetic waves (e.g. medical imaging such as CT scans, or communication technology such as cell phones). Each person should research their topic and report back to the group on how their technology was developed, how it works, and any future possible developments or technologies based on the original technology (e.g. 5G in cell phones).

   (b) Alternatively a group of four should choose one type of technology based on electromagnetic waves and research how the technology was developed, how it works, and any future possible developments or technologies based on the original technology. This should then be presented to the class as a poster, PowerPoint, or other kind of presentation.

Student's own research based on their topic of choice.

PS4.C CE

©2020 **BIOZONE** International
ISBN: 978-1-927309-82-7
Photocopying Prohibited

Appendix: Glossary

acceleration: The change in velocity of an object divided by the time taken to make the change. A vector quantity with both magnitude and direction.

activation energy: The minimum energy needed by reactants for a reaction to occur.

atomic number (Z): The number of protons in the nucleus of an atom. It is also equal to the number of electrons in a neutral atom.

Avogadro's number: The number of objects in 1 mole of those objects. It is equal to 6.02×10^{23}.

concentration: The number of moles of a substance (solute) in solution per liter. Given as moles per liter (mol/L).

delta (Δ): Symbol meaning "the change in" or "the difference between". For example: Δv, meaning the change in velocity or the difference between initial velocity and final velocity.

diffraction the bending or warping of a wave. A wave will be diffracted if the wavelength is equal to or greater than the size of the object in its path.

element: A substance that cannot be separated into simpler components by chemical means. A substance composed only of atoms with the same atomic number.

empirical formula: Chemical formula that shows the simplest ratio of elements in the substance.

enthalpy (H): A thermodynamic quantity that is equal to the heat content of the system plus the pressure times volume. In a chemical reaction, the change in enthalpy (ΔH) is a measure of the energy (in joules) released into ($-\Delta H$) or absorbed from ($+\Delta H$) the environment.

fissile: A subset of fissionable nuclei that are able to sustain a fission chain reaction. For example uranium-235 is fissile.

fissionable: Nuclei that are able to undergo fission after capturing a neutron. For example uranium-238 is fissionable but not fissile.

fission: The splitting or break up of a large nucleus into smaller nuclei.

force: A push or a pull. A force will accelerate an object in the direction of the push or pull.

fusion: The combining of two smaller nuclei to produce a larger nucleus.

hydrated: An ion or molecule to which water is attached.

hydrogen bond: A intermolecular bond formed between hydrogen from a molecule X-H (in which X is strongly electronegative) and a strongly electronegative atom from a separate molecule.

indicator: a solution that can be used to identify the presence of an acid or base.

interference: the interaction of two waves to form a resultant wave of greater, lower, or the same amplitude.

ion: An atom or group of atoms that carry a positive or negative charge. (e.g. Li^+ or OH^-).

isotope: One of two or more atoms of an element that have different atomic mass but the same atomic number. For example uranium- 235 and uranium have the same atomic number (92) but different atomic masses due to a difference in neutrons.

joule: (J) The SI unit of energy. 1 joule is equal to 1 newton acting through a distance of one meter.

Law of conservation of mass: The mass of an object or objects never changes, no matter how the parts are rearranged.

Law of definite proportions: A chemical compound will always have the same proportions of each element by weight.

Le Châtelier's principle: When a stress is placed upon a system in a state of dynamic equilibrium the system will respond in a way that minimizes the stress.

mass: The amount or quantity of matter in a system. Measured in kilograms (kg). The derived unit is grams (g).

mass number (A): The number of neutron and protons together in the nucleus of an atom. Also referred to as atomic mass.

molar mass: The mass (in grams) per mole of atoms of an element. The mass per mole of molecules of a compound. The mass per mole of formula units of an ionic compound. Measured in grams per mole (g/mol).

mole: (mol). The unit for the amount of a substance. 1 mole is equal to the number of atoms in exactly 12 grams of carbon-12.

molecular formula: The chemical formula showing the exact numbers of atoms in the molecule., e.g. $C_6H_{12}O_6$.

newton: (N) The unit of force. 1 newton will accelerate a 1 kg mass by 1 meter per second per second ($1/ms^2$)

nuclear decay: The break up of a nucleus by the release of high energy particles. This includes alpha, beta, and gamma decay as well as spontaneous fission.

nucleus: The central part of the atom containing most of the mass. It is composed of protons and neutrons.

radioactive: Emitting ionizing radiation (caused by the break up of unstable atoms).

relative atomic mass: A dimensionless quantity defined as the ratio of the average mass of all the atoms of an element to 1/12 the mass of a carbon-12 atom (or 1 atomic mass unit).

relative molecular mass: a dimensionless quantity defined as the ratio of the average mass of all the atoms of a molecule to 1/12 the mass of a carbon-12 atom (or 1 atomic mass unit).

pH: the negative log of the hydrogen ion concentration of a solution. It is used to measure the acidity of a solution.

quark: elementary particle that comes in one of three colors and six flavors. Three quarks combine to form a hadron. The type of hadron (proton or neutron) depends on the combination of quark flavours. All three colors of quark must be present.

refraction: the alteration of wave velocity as a wave passes from one medium to another.

solution: A mixture containing a solute (a dissolved substance) in a solvent (a substance that dissolves other substances).

strong force: The force responsible for holding nucleons and quarks together.

valence shell: The outer most layer of electrons around an atom.

vector: A quantity with both magnitude and direction.

watt: (W) The SI unit power. 1 watt equals 1 joule per second.

weak force (or weak interaction): The force that governs beta decay. A neutrino exchanges a W- boson with a neutron, turning the neutrino into an electron and the neutron into a proton.

weight: The supportive force needed to maintain an object's rest state relative to the support. The mass of an object multiplied by the acceleration due to gravity.

Appendix: Equipment list

The equipment list provides the material and equipment needed per student, pair, or group

2: Structure and Properties of Matter

INVESTIGATION 2.1:
Repulsion theory
4 x balloons
Marker pens

INVESTIGATION 2.2:
Polarity
Per student/pair:

50 mL burette
Clamp
Distilled water
Cyclohexane
100 mL beaker
Glass rod
Silk or polyester fabric or towel

INVESTIGATION 2.3:
Properties of matter
Per group:

11 x watch glasses
11 x 50 mL beakers
Conductivity meter
Stirring rod
Test tube
2 x deflagrating spoon
Bunsen burner
Small heat resistant pad or glass plate
Glass jar

1 g sodium chloride (NaCl),
1 g magnesium (Mg)
1 g magnesium sulfate ($MgSO_4$)
1 g sulfur (S)
1 g copper (Cu
1 g copper sulfate ($CuSO_4$)
Distilled water
Ice
1 mL cyclohexane (C_6H_{12})
1 g quartz or glass (SiO_2)

3: Chemical Reactions

INVESTIGATION 3.1:
Molar mass
Per group:

6 x petri dishes or
6 x filter paper (any will do)
Balance

Sodium chloride (NaCl)
Sulfur (S)
Copper chloride ($CuCl_2.2H_2O$)
Calcium carbonate ($CaCO_3$)
Carbon (C)
Glucose ($C_6H_{12}O_6$)

INVESTIGATION 3.2:
Find the formula
Per group:

Crucible and lid
Balance
Tripod
Clay triangle
Bunsen burner
Tongs

Magnesium ribbon (10 cm)

INVESTIGATION 3.3:
Hydrated copper sulfate
Per group:

Crucible and lid
Balance
Tripod
Clay triangle
Bunsen burner
Tongs

6 g hydrated copper sulfate

INVESTIGATION 3.4:
The cycling of copper ions through a series of reactions
Per group:

Filter paper (any will do)
Balance
2 x test tubes
Rubber cork
Delivery tube
Clamp stand and clamp
Glass beaker (big enough to fit the test tube)
Bunsen burner
100 mL beaker
Glass stirring rod
Steel wool or sandpaper
Large iron nail
Buchner funnel
Drying oven (optional)

2 g copper II carbonate ($CuCO_3$)
Limewater
80 mL 1 mol/L sulfuric acid solution

INVESTIGATION 3.5:
Making a standard
Per group:

Small beaker
250 mL volumetric flask
Glass funnel
Dropper

1.3 g anhydrous Na_2CO_3
Distilled water

INVESTIGATION 3.6:
Standardizing HCl
100 mL beaker
25 mL pipette
250 mL volumetric flask
50 mL burette
Clamp stand and clamps
4 x 100 mL conical flasks

50 mL 1 mol/L HCl solution
Na_2CO_3 solution (prepared in investigation 3.5)
Methyl orange indicator
Distilled water

INVESTIGATION 3.7:
Standardizing NaOH
Per group:

100 mL beaker
250 mL volumetric flask
25 mL pipette
Balance
50 mL burette
20 mL pipette
Clamp stand and clamps
4 x 100 mL conical flask

50 mL 1 mol/L NaOH solution or 1 g NaOH
80 ml standardized HCl solution
Phenolphthalein indicator

INVESTIGATION 3.8:
Analyzing vinegar
Per group:

50 mL burette
Clamp stand and clamps
25 mL pipette
100 mL volumetric flask
4 x 100 mL conical flask

50 mL standardized NaOH solution
25 mL vinegar (any will do)
Phenolphthalein indicator

INVESTIGATION 3.9:
Reaction rates
Per group:

4 50 mL beakers
250 mL beaker
1 mol/L HCl
2 mol/L HCl
0.5 g Calcium carbonate ($CaCO_3$)

INVESTIGATION 3.10:
Endothermic / exothermic reactions
Per pair/group:

3 x 25 mL beakers
Watch glass
Thermometer

Sodium hydroxide (solid)
Ammonium chloride (solid)
1 M HCl solution
Ammonium nitrate (solid)

INVESTIGATION 3.11:
Investigating Enthalpy
Per group:

Styrofoam cup
100 mL measuring cylinder
Thermometer
100 mL 0.1 mol/L copper sulfate solution
Zinc powder

INVESTIGATION 3.12:
Energy from alcohols
Per group:

Balance
Alcohol burner
Soda can (empty)
Thermometer
Ring stand
Wire gauze or clamp stand
Ring clamp and clamp
Stopwatch

Methanol (CH_3OH)
Ethanol (CH_3CH_2OH)
Propan-1-ol ($CH_3CH_2CH_2OH$)
Distilled water

INVESTIGATION 3.13:
Reversibility
Per student/pair:

2 x test tubes
Red litmus paper
Blue litmus paper

0.1 M HCl
0.1 M NaOH

INVESTIGATION 3.14:
Equilibrium
Per group:

2 x 1 L minimum beaker or basin
100 mL beaker
50 mL beaker

INVESTIGATION 3.15:
pH scale
Per pair:

Watch glasses (or spotting tile)

0.1 mol/L hydrochloric acid (HCl)
0.1 mol/L nitric acid (HNO_3)
0.1 mol/L ethanoic acid (CH_3COOH)
0.1 mol/L citric acid ($C_6H_8O_7$)
0.1 mol/L sodium hydroxide (NaOH)
0.1 mol/L ammonia solution (NH_3)
0.1 mol/L sodium carbonate (Na_2CO_3)
0.1 mol/L sodium hydrogen carbonate ($NaHCO_3$)

Distilled water
pH paper

4: Nuclear Processes

INVESTIGATION 4.1:
Modeling the strong nuclear force
Per group/class

An old click pen
2 x small neodymium magnets with central holes

NOTE: The diameter of the spring needs to be greater than the diameter of the hole in the magnets otherwise 2 washers (outer diameter bigger than hole in the magnets and inner diameter bigger than the tube but smaller than the spring) may be needed.

5: Forces and Motion

INVESTIGATION 5.1:
Distance, displacement, and velocity
Per class

Measuring tape (at least 10 m)
Flag or bright fabric as starter signal
12 stop watches (watches/cell phone)

INVESTIGATION 5.2:
Marble momentum
Per student/pair/group

Marble and ball bearing of similar size but different mass
Ramp

Tape measure
Carpeted area for run

INVESTIGATION 5.3:
Investigating impulse
Dual range force sensor + logging program (e.g. Logger Pro)
Mass to act as "jumper"
Rubber bands or bungee cord
Clamp stand

INVESTIGATION 5.4:
Building a lander
Per student/pair/group

1 egg
1 plastic egg (for testing)
Tape
5 medium sized rubber bands
1 small garbage or plastic bag
10 paper clips
String
20 plastic or paper straws
Equipment can be modified with equivalent replacements

6: Types of Interactions

INVESTIGATION 6.1:
Balloon electrostatics
Per student/pair

Two balloons
Fabric (wool or synthetic)
Nylon thread or fishing line

INVESTIGATION 6.2:
Threading the needle
Per student

Sewing needle
Nylon thread and cotton thread of same gauge

INVESTIGATION 6.3:
Charges and magnets
Per student/pair

Magnet
2 x plastic rulers
Wool
Small bottle lid

INVESTIGATION 6.4:
Magnetic fields
Per student/pair

Two bar magnets
Zip lock sandwich bag
Iron filings (or powder)
Index card (or ruled card)

INVESTIGATION 6.5:
Strength of a magnetic field
Per student/pair

Sheets of copy paper
Magnet
Metal paper clip

INVESTIGATION 6.6:
Making a magnet
Per student/pair

Bar magnet
Iron object, e.g. nail or needle
Pen to mark nail center
Compass or iron objects to test

INVESTIGATION 6.7
Electricity affects a compass part 1
Per student/pair/group

1 x 1.5 volt AA cell
Insulated wire
Electrical tape
Cardboard
1 x compass

INVESTIGATION 6.8
Electricity affects a compass part 2
Per student/pair/group

1 x 1.5 volt AA cell
Insulated wire
Electrical tape
Cardboard
1 x compass
Clear tape

INVESTIGATION 6.9
Electricity affects a compass part 3
Per student/pair/group

1 x 1.5 volt AA cell
Insulated wire
Electrical tape
Cardboard
1 x compass

INVESTIGATION 6.10
Electricity from magnets
Per student/pair/group

Insulated wire
1 x bar magnet
1 x cardboard tube
1 x galvanometer

INVESTIGATION 6.11
Making a motor
Per student/pair/group

D size battery
30 cm thin copper wire
Sandpaper
2 paper clips
Tape
Strong bar magnet
Sewing thread

INVESTIGATION 6.12:
What is used where
No special equipment required

7: Definitions of Energy

INVESTIGATION 7.1:
Investigating work and energy
Per student/pair/group

Dynamic trolley
Tape measure
Smooth bench top
Spring meter

8: Conservation of Energy and Energy Transfer

INVESTIGATION 8.1:
Indirect energy transfer
Per student/pair/group

Large beaker (~250 mL)
Small beaker (~50 mL)
Thermometer
Insulated box or insulation
50 mL cooking oil

INVESTIGATION 8.2
Direct energy transfer
2 x 500 mL beakers
Thermometer

INVESTIGATION 8.3
Conduction
Computers and online access to Energy 2D software
http://energy.concord.org/energy2d/index.html

INVESTIGATION 8.4
The power of Oreos
Stairwell or set of steps
Stopwatch
Packaging from Oreos and Double Stuffed Oreos.

INVESTIGATION 8.5
Building a Rube Goldberg machine
No special equipment needs to be supplied. Students gather equipment themselves. Teachers might like to supply items a such as tape or rubber bands.

9: Wave Properties

INVESTIGATION 9.1
Making waves
Per pair

1 x Slinky spring

INVESTIGATION 9.2
More waves
Per pair

1 x Slinky spring

INVESTIGATION 9.3
The speed of sound in air
Per group/class

Set of tuning forks or frequency generator with speaker
1 x clamp stand and clamp
Water
1 x large beaker
1 x PVC pipe (30-60 cm)

INVESTIGATION 9.4
More waves
Per pair

1 x Slinky spring

INVESTIGATION 9.5
Investigating two properties unique to waves
Per pair/group

1 x square tray
1 x ruler or thin stick
Blocks of wood (3 shorter, 3 longer)
Water

10: Electromagnetic Radiation

INVESTIGATION 10.1
The double slit experiment
Per student/group/class

1 x microscope slide painted black
1 x craft knife
1 x ruler
Plasticine or blu-tak
1 x laser pointer

Image credits

We acknowledge the generosity of those who have provided photographs or diagrams or developed freeware for this edition including: • Charles Xie for Energy2D • Felix Hicks for original drawings of the ice skaters and astronauts, and cation and anion diagrams • Biomimetics and Dexterous Manipulation Lab, Stanford University for the image of the person climbing glass. Electrolysis of water photo from Science Photo Library

We also acknowledge the photographers that have made their images available through **Pixabay** (pixabay.com) or **Wikimedia Commons** under Creative Commons Licences 2.0, 2.5, 3.0, or 4.0:
• James Hedberg • Graphtecinst • Jan Fjaldowski • Angus Thompson • Hans-Joachim Engelhardt • Géry Parent • Eurico Zimbres • Wilco Oelen • Tavoromann • Cyberchemist • Cjp24 • Toby Hudson • Benutzer:the-viewer • Jeff Kubina • Capaccio • Maximilien Brice • gamsiz • Z22 • Jeremy Thompson • Chocolateoak • Tbraunstein • Tuukritööde OÜ • Vic Mirmow • James T M Towill • Plantsurfer • Mark Mannetti • _indica Garcinia indica

Contributors identified by coded credits are: **ISS**: International Space Station, **KP**: Kent Pryor, **NASA**: National Aeronautics and Space Administration, **NASA/ESA**: National Aeronautics and Space Administration / European Space Agency **NASA/JPL**: NASA Jet Propulsion Laboratory, **NHTSA**: National Highway Traffic Safety Administration, **NOAA**: The National Oceanic and Atmospheric Administration, **USDE**: United States Department of Energy, **USGS**: United States Geological Survey, **USAF**: United States Air Force

Royalty free images, purchased by BIOZONE International Ltd, are used throughout this workbook and have been obtained from the following sources:
• Adobe Stock , stock.adobe.com • iStock images • Corel Corporation from their Professional Photos CD-ROM collection • PhotoDisc®, Inc. USA, www.photodisc.com • 3D images created using Bryce, Poser, and Pymol

Index

Long Answers

Science Practices:

Activity 1. The Nature of Science

2. Each student's answer will vary, but it may contain the following points and ideas:

Historically, the development of scientific ideas was not only subject to change based on new scientific evidence but was also reflective of the social and religious thoughts dominant at the time. Often this meant there was conflict between science and society and scientific progress was delayed as a result. For example, in the 1500s the Church supported a geocentric model to explain the rotation of planets relative to the Sun. This meant that they thought the Sun and planets orbited the Earth. When **Nicolaus Copernicus** proposed the heliocentric model where the Earth and planets orbited the Sun. The Church vigorously argued against this. Around a 100 years later **Galileo** published his observations of the moons of Jupiter. He used this as evidence that the idea that everything orbited the Earth was wrong because here were objects orbiting something else. The Church again opposed these ideas and he spent the rest of his life under house arrest. This power to shut down ideas not in line with the thought of the day prevented scientific ideas developing as quickly as they might have if there had not be the fear of retaliation.

Fortunately as the body of scientific evidence grew, the acceptance of the heliocentric model became widely accepted. This is largely due to the overwhelming body of observations and evidence collected by scientists. While early models were based on observations, ideas and experience, they were not often tested under controlled conditions. For example, **Aristotle's** idea that light objects fall more slowly than heavy objects was accepted until at the 15th century because it appeared to match everyday observation. No controlled testing was carried out. This idea was not disputed until **Galileo** proved scientifically that all objects fall at the same rate no matter their mass. He did this by slowing down the motion of falling objects using an inclined plane and balls made of different materials and masses. He then accurately measured the time it took them to roll a set distance with a water clock. His conclusion was also explained using mathematics. This allowed predictions to be tested experimentally. With solid evidence, scientifically generated ideas began to gain popular support.

Students should recognize that scientists build models to explain phenomena, but that the scientific model is not static, it is dynamic. This means that when new evidence is discovered, the model is changed to incorporate the new information. For this reason science is generally a collaborative endeavor, the work of earlier scientists is often the basis for those who follow. For example, even though Galileo's observations on orbiting bodies was not well received, **Joahnnes Kepler** added to the body of knowledge collected by **Tycho Brahe** and the idea was accepted. Another example is how **Isaac Newton** altered the explanation of orbiting planets described by Aristotle. Aristotle believed that the different motions of planets was explained by different causes, whereas Newton proved that the same causes (gravity) explained the different motions. However not all orbits could be predicted using Newton's law. For example, the orbit of Mercury did not match the predicted orbit. However the new ideas of scientists such as **Albert Einstein** were instrumental in redefining the model. Einstein's theory of relativity was revolutionary and has stood up to every scientific test, including gravitational waves.

Activity 44. Verifying a Claim

8. Suggested method as follows:
   - Pipette 25 mL of the orange juice into a 250 mL volumetric flask and dilute with distilled water to the mark.
   - Pipette 10 mL of the dilute orange juice into a conical flask and add two drops of phenolphthalein.
   - Rinse a burette with 0.1 mol/L NaOH and then titrate the NaOH with the orange juice until the indicator just turns (and remains) pink.
   - Repeat until the volumes of NaOH used agree to within 0.2 mL.
   - Calculate the moles of NaOH used. This is equal to the acid content (assuming 1:1 reaction) of the conical flask.
   - Calculate the concentration of dilute acid (orange juice) using $c = n \div V$. Then calculate the concentration of acid in the undiluted orange juice.

Activity 59. Radioactive decay

5. Summary of experiments for Rutherford and Curie are described below. The information for Becquerel is provided on the page.

Marie Curie

Marie Curie continued with the work of uranium started by Becquerel. Using an electrometer she discovered that uranium rays caused the air around a sample to conduct electricity and the activity of the uranium compounds depended only on the quantity of uranium present. Other factors, such as whether the uranium was solid or pulverized, pure or in a compound, wet or dry, or exposed to light or heat did not matter. More importantly, she proposed that radioactivity was a property of the atom itself (not something external). Her research revealed that two uranium ores, pitchblende and chalcolite, were much more radioactive than pure uranium itself. Working with her husband Pierre, and through a series of rigorous chemical separations, they isolated two new elements radium and polonium.

Ernest Rutherford

Through the work of earlier scientists, Rutherford already knew that uranium emitted ionizing radiation. Rutherford hypothesized that the rays emitting from uranium would be composed of different types of rays. He developed experiments to help show what the characteristics of the rays were. Between some uranium and an electroscope detector he placed a single thickness of aluminum foil and measured the intensity of the radiation by measuring the time required to discharge the electroscope. These experiments found different types of rays (positively charged alpha and negatively charged beta rays). By adding additional layers of foil to block the radiation he could study the properties of the different rays (e.g. alpha rays could be blocked with a few sheets of foil). The experiment was repeated using different metal foils and different radioactive sources. He also confirmed that Becquerel's rays were electromagnetic gamma rays.

Activity 70. Newton's Third Law

2. In the second image there are now two important sets of reaction pairs operating. The Earth still pulls on the SLS and the SLS still pulls on the Earth. However the boosters are now pushing on the SLS, and the SLS presses back on the boosters (this is simplified, but essentially the net outcome of other force pairs). The forces acting between the SLS and boosters (resulting in thrust) are greater than the forces acting between the SLS and the Earth. As a result the SLS accelerates into the air. The free body diagram would show the thrust of the boosters as being greater than the weight of the SLS.